THERMAL
MEASUREMENTS
and
INVERSE
TECHNIQUES

HEAT TRANSFER
A Series of Reference Books and Textbooks

Editor

Afshin J. Ghajar

Regents Professor
School of Mechanical and Aerospace Engineering
Oklahoma State University

Engineering Heat Transfer: Third Edition, *William S. Janna*

Conjugate Problems in Convective Heat Transfer, *Abram S. Dorfman*

Thermal Measurements and Inverse Techniques, *Helcio R.B. Orlande; Olivier Fudym; Denis Maillet; Renato M. Cotta*

Upcoming titles include:

Introduction to Thermal and Fluid Engineering, *Allan D. Kraus, James R. Welty, and Abdul Aziz*

THERMAL MEASUREMENTS and INVERSE TECHNIQUES

Edited by
Helcio R.B. Orlande
Olivier Fudym
Denis Maillet
Renato M. Cotta

CRC Press
Taylor & Francis Group
Boca Raton London New York

CRC Press is an imprint of the
Taylor & Francis Group, an **informa** business

CRC Press
Taylor & Francis Group
6000 Broken Sound Parkway NW, Suite 300
Boca Raton, FL 33487-2742

First issued in paperback 2017

© 2011 by Taylor and Francis Group, LLC
CRC Press is an imprint of Taylor & Francis Group, an Informa business

No claim to original U.S. Government works

ISBN-13: 978-1-4398-4555-4 (hbk)
ISBN-13: 978-1-138-11386-2 (pbk)

Library of Congress Cataloging-in-Publication Data

Thermal measurements and inverse techniques / edited by Helcio R.B. Orlande ... [et al.].
    p. cm. -- (Heat transfer)
    Summary: "The use of inverse problems constitutes a new research paradigm in which groups of theoretical, computational, and experimental researchers synergistically interact to better understand the physical phenomena being studied. This book presents a comprehensive research-based survey of modern inverse techniques and their applications to engineering heat transfer. Written by top-notch contributors, the text provides a concise, single-source reference on the mathematical modeling, measurements, and solution of inverse problems in heat transfer. It covers topics such as multiscale modeling, thermophysical properties, heat flux, and temperature measurements"-- Provided by publisher.
    Includes bibliographical references and index.
    ISBN 978-1-4398-4555-4 (hardback)
    1. Heat--Transmission. 2. Heat equation. 3. Inverse problems (Differential equations) 4. Heat engineering. I. Orlande, Helcio R. B., 1965- II. Title. III. Series.

TJ260.T4938 2011
621.402'201515357--dc22

        2010045340

Visit the Taylor & Francis Web site at
http://www.taylorandfrancis.com

and the CRC Press Web site at
http://www.crcpress.com

Contents

Part III   Applications

Preface

The design and operation of modern technological systems and the proper comprehension of their interaction with nature (e.g., in pollution control and global warming issues) require the permanent processing of a large amount of measured data. Nowadays, progress in the mathematical modeling of complex industrial or environmental systems, associated with the continuous increase in memory and calculation power of computers, has made numerical simulations of almost any physical phenomena possible. These facts bring about the need for an appropriate tool that rigorously bridges the gap between the information stemming from measurements and that corresponding to theoretical predictions, aiming at the better understanding of physical problems, including real-time applications. Inverse analysis is such a tool.

Heat transfer permanently takes part in our daily life. Examples can be found in natural phenomena, such as the solar heating of Earth, meteorology or thermoregulation of biological activity, as well as in a wide range of man-made applications, such as the conversion of energy in heat engines, thermal control of chemical reactors, air conditioning, cooling of electronic equipment, development of micro- and nano-technologies with the associated thermal challenges, etc. Recent advances in both thermal instrumentation and heat transfer modeling permit the combination of efficient experimental procedures and of indirect measurements within the research paradigm of inverse problems. In this paradigm, the groups of theoretical, computational, and experimental researchers synergistically interact during the course of the work in order to better understand the physical phenomena under study. Although initially associated with the estimation of boundary heat fluxes by using temperature measurements taken inside a heated body, inverse analyses are nowadays encountered in single- and multi-mode heat transfer problems dealing with multiscale phenomena. Applications range from the estimation of constant heat transfer parameters to the mapping of spatially and timely varying functions, such as heat sources, fluxes, and thermophysical properties.

In heat transfer, the classical inverse problem of estimating a boundary heat flux with temperature measurements taken inside a heat-conducting medium has many practical applications. For example, the heat load of the surface of a space vehicle reentering the atmosphere can be estimated through inverse analysis by using temperature measurements taken within the thermal protection shield. If a technique that sequentially estimates such boundary heat flux is used, inverse analysis may allow for online trajectory corrections in order to reduce the heat load. Therefore, overheating of the structure of the spacecraft can be avoided, reducing the risk of fatal accidents. Moreover, modern engineering strongly relies on newly developed materials, such as composites, and inverse analysis can be used for the characterization of the unknown properties of such nonhomogeneous materials. The use of nonintrusive measurement techniques with high spatial resolutions and high measurement frequencies, such as temperature measurements taken with infrared cameras, allows the characterization of nonhomogeneous materials even at small scales, including crack or defect detection. The latest research in heat transfer follows a trend toward small scales, at micro- and nano-levels. This requires that physical phenomena be taken into consideration, which may be negligible and, hence, not accounted for at macroscales. By the same token, modern techniques now permit nonintrusive measurements to be taken at small space and time scales, thus allowing the observation of such complex physical phenomena.

All subjects required for the understanding and solution of the physical situations described above are available in this book, including the modeling of heat transfer problems, even at micro- and nano-scales, modern measurement techniques, and the solution of inverse problems by using classical and novel approaches. This book is aimed at engineers, senior undergraduate students, graduate students, researchers both in academia and industry, in the broad field of heat transfer. It is assumed, however, that the reader has basic knowledge on heat transfer, such as that contained in an undergraduate heat transfer course.

This book is intended to be a one-source reference for those involved with different aspects of heat transfer, including the modeling of physical problems, the measurement of primary heat transfer variables, and the estimation of quantities appearing in the formulation (indirect measurements) through the solution of inverse problems. Keeping this main objective in mind, the book was divided into three parts, namely: Part I—Modeling and Measurements in Heat Transfer, Part II—Inverse Heat Transfer Problems, and Part III—Applications. Parts I and II provide a concise theoretical background along with examples on modeling, measurements, and solutions of inverse problems in heat transfer. Part III deals with applications of the knowledge built up in Parts I and II to several practical test cases. Each chapter contains its own lists of variables and references. Hence, depending on the reader's background and interest, they can be read independently.

This book results from the Advanced Schools METTI (Thermal Measurements and Inverse Techniques) held in 1995, 1999, 2005, and 2009. Started under the auspices of SFT—French Heat Transfer Society, the last METTI School was co-organized with ABCM—Brazilian Society of Mechanical Engineering and Sciences, and held in Angra dos Reis (state of Rio de Janeiro) as one of the activities of the Year of France in Brazil. However, the book was intended to be self-consistent and didactic, not being at all the single collection of lectures previously given during the METTI schools.

We would like to thank all the contributors for their diligent work that made this book possible. We are indebted to Professor Afshin J. Ghajar, the Heat Transfer series editor for CRC Press/Taylor & Francis, for his encouragement and support to pursue this book project. We also appreciate the valuable recommendation by Professor Sadik Kakac, who carefully reviewed our book proposal. The cooperation of the staff at CRC Press/Taylor & Francis is greatly appreciated, especially that from Jonathan W. Plant, the senior editor for mechanical, aerospace, nuclear, and energy engineering, and from our project coordinator, Amber Donley. Finally, we would like to express our deepest gratitude for the financial support provided for the publication of this book by CAPES, an agency of the Brazilian government for the fostering of science and graduate studies.

For MATLAB® and Simulink® product information, please contact:

The MathWorks, Inc.
3 Apple Hill Drive
Natick, MA, 01760-2098 USA
Tel: 508-647-7000
Fax: 508-647-7001
E-mail: info@mathworks.com
Web: www.mathworks.com

Contributors

Stéphane Andre
Energy and Theoretic and Applied Energy
   Laboratory (LEMTA)
University of Nancy and CNRS
Vandœuvre-lès-Nancy, France

Liliane Basso Barichello
Institute of Mathematics
Federal University of Rio Grande do Sul
Porto Alegre, Brazil

Elena Palomo del Barrio
TREFLE Laboratory
Ecole Nationale Supérieure des Arts et
   Métiers
University of Bordeaux
Talence, France

Jean-Christophe Batsale
TREFLE Laboratory
Ecole Nationale Supérieure des Arts et
   Métiers
University of Bordeaux
Talence, France

Jean-Luc Battaglia
TREFLE Laboratory
Ecole Nationale Supérieure des Arts et
   Métiers
University of Bordeaux
Talence, France

Valério L. Borges
School of Mechanical Engineering
Federal University of Uberlândia
Uberlândia, Brazil

Marcelo J. Colaço
Department of Mechanical Engineering
Federal University of Rio de Janeiro
Rio de Janeiro, Brazil

Renato M. Cotta
Department of Mechanical Engineering
Federal University of Rio de Janeiro
Rio de Janeiro, Brazil

Manuel Ernani Cruz
Department of Mechanical Engineering
Federal University of Rio de Janeiro
Rio de Janeiro, Brazil

Morgan Dal
Materials Engineering Laboratory of
   Brittany (LIMATB)
University of Southern Brittany
Lorient, France

Jean-Luc Dauvergne
TREFLE Laboratory
Ecole Nationale Supérieure des Arts et
   Métiers
Talence, France

George S. Dulikravich
Department of Mechanical and Materials
   Engineering
Florida International University
Miami, Florida

Ana P. Fernandes
School of Mechanical Engineering
Federal University of Uberlândia
Uberlândia, Brazil

Olivier Fudym
Research in Albi on Particulate Solids,
   Energy and the Environment
University of Toulouse
Albi, France

Eric Gavignet
Department of Energy & Engineering of
    Multiphysic Systems
FEMTO-ST
University of Franche-Comté
Belfort, France

Manuel Girault
Department of Fluid, Thermal and
    Combustion Sciences
PPRIME Institute
Chasseneuil-du-Poitou, France

Gilmar Guimarães
School of Mechanical Engineering
Federal University of Uberlândia
Uberlândia, Brazil

Saulo Güths
Department of Mechanical Engineering
Federal University of Santa Catarina
Florianópolis, Brazil

Yvon Jarny
Department of Thermal and Energy
    Sciences
Polytechnic School of the University of
    Nantes
Nantes, France

Jean-Claude Krapez
Department of Theoretical and Applied
    Optics
ONERA—The French Aerospace Lab
Salon de Provence, France

François Lanzetta
Department of Energy & Engineering of
    Multiphysic Systems
FEMTO-ST
University of Franche-Comté
Belfort, France

Denis Maillet
Energy and Theoretic and Applied Energy
    Laboratory
National Center for Scientific Research
University of Nancy
Vandœuvre-lès-Nancy, France

Philippe Le Masson
Materials Engineering Laboratory of
    Brittany (LIMATB)
University of Southern Brittany
Lorient, France

Carlos Frederico Matt
Department of Equipment and Installations
Electric Power Research Center
Rio de Janeiro, Brazil

Thomas Metzger
Thermal Process Engineering
Otto-von-Guericke University
Magdeburg, Germany

Luís Mauro Moura
Thermal System Laboratory
Pontifical University Catholic of Paraná
Curitiba, Brazil

Carolina P. Naveira-Cotta
Department of Mechanical Engineering
Federal University of Rio de Janeiro
Rio de Janeiro, Brazil

Christophe Le Niliot
University Institute of Industrial Thermal
    Systems (IUSTI)
Technopole de Château Gombert
Marseille, France

Helcio R. Barreto Orlande
Department of Mechanical Engineering
Federal University of Rio de Janeiro
Rio de Janeiro, Brazil

Marina Silva Paez
Department of Statistical Methods
Federal University of Rio de Janeiro
Rio de Janeiro, Brazil

Daniel Petit
Department of Fluid, Thermal and
    Combustion Sciences
PPRIME Institute
Chasseneuil-du-Poitou, France

Benjamin Remy
Energy and Theoretic and Applied Energy
    Laboratory (LEMTA)
University of Nancy and CNRS
Vandœuvre-lès-Nancy, France

Fabrice Rigollet
University Institute of Industrial Thermal
    Systems (IUSTI)
Technopole de Château Gombert
Marseille, France

Paulo Seleghim, Jr.
Mechanical Engineering Department
University of São Paulo
São Carlos, Brazil

Priscila F.B. Sousa
School of Mechanical Engineering
Federal University of Uberlândia
Uberlândia, Brazil

Haroldo F. de Campos Velho
Laboratory of Computing and Applied
    Mathematics
National Institute for Space Research
São José dos Campos, Brazil

Etienne Videcoq
Department of Fluid, Thermal and
    Combustion Sciences
PPRIME Institute
Chasseneuil-du-Poitou, France

Part I

Modeling and Measurements in Heat Transfer

1

Modeling in Heat Transfer

Jean-Luc Battaglia and Denis Maillet

CONTENTS

1.1 Introduction

Modeling constitutes a very general activity in engineering. A system can be considered as modeled if its behavior or its response to a given excitation can be predicted. So prediction is one of the natural characteristics of modeling.

In the second section of this chapter, the basics on heat transfer physics are presented. The existence of temperature and more specifically of temperature gradient must be discussed carefully when time and length scales become very small. This is the case for new applications in the field of inverse heat conduction problems. This point is known for a long time at very low temperature. It becomes also particularly true at the nanoscale when temperature is greater than the Debye temperature (above this temperature, the quantum effects are generally neglected). Classical Fourier's law, at the basis of standard heat transfer models, is no longer valid, and either a new model or a definition of the reliable time range for the pertinent use of Fourier's law is thus required. In the third section of this chapter, the concept of homogenization for heterogeneous materials through macroscopic homogenized models is presented. This topic is also studied in Chapter 2. An illustration of such a problem is represented in Figure 1.1.

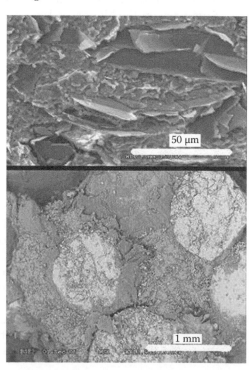

FIGURE 1.1
Phase change material for energy storage (double porosity carbon graphite/salt porous media with phase change material [PCM]). Scanning electron microscopy (SEM) imaging illustrates different heterogeneity levels according to the observation scale and shows that a specific model is required for each.

Another important feature of a model, which is only a theoretical representation of the physical reality in the case of a material system, is its structure (we do not deal here with information systems). In heat transfer, the choice is quite large, and the model structure should be selected according to the objective of the model. The model-builder can have in mind an optimal design problem, a parameter estimation problem using measurements, a control problem to define the best excitation shape for a given desired output, or a model reduction problem, just to quote a few applications.

The choice of the structure of a model in heat transfer depends on many things:

- State variable and observed quantities
  In a heat diffusion problem, temperature is the quantity that constitutes the state variable, in the thermodynamics sense. In order to calculate temperature and heat flux at any time t and at any point P, one has to know the initial temperature field (at time $t = 0$) at the local scale, as well as the history of the different thermal disturbances between times 0 and t. So, one has to define what is a *local* point P and a local scale. For instance, if heat transfer is intended to be studied at the very small scale in a metal (smaller than the grain size), Fourier's law, relating heat flux to temperature gradient, may no longer be valid. In such a case, two temperatures (respectively for the electron gas and the lattice) are required to describe heat transfer at this scale (see Section 1.2.3). Such a *detailed* state model will be necessary if *observations* or predictions are looked for at the nanoscale or at the picosecond timescale. The upper thresholds of both scales depend on the considered material. A similar effect appears in a heterogeneous medium composed of two homogeneous materials (grains made of one material embedded in a matrix made of the other material, for example): instead of using temperature at the local scale (grain or matrix), some averaging, that is a space filtering, will be used at the macroscopic scale (see Sections 1.3 and 1.4 and Chapter 2).

- State definition
  The continuous state equations have then to be defined for the modeling problem at stake: it can be a partial differential equation, the heat equation (state = temperature), or an integro-differential equation, the radiative transfer equation (state = radiative intensity), or both coupled equations. Their solution, that is constituted by both temperature and intensity fields in the third case, should be calculated everywhere and any time past the initial time (see Section 1.5.2).

- Quantities of the direct problem
  We focus on the diffusion heat equation in a medium composed of one or several homogeneous materials, with its associated initial, boundary, and interface equations. Its solution, the state variable, here the continuous temperature field $T(P, t)$, has first to be found, and the desired observed quantities, that is, the (theoretical) output of the model at a given point P, $y_{mo}(t) = T(P, t)$, have to be calculated next (see Section 1.5.1). Here the quantities that are required for solving the *direct problem* are the structural parameters of the system (conductivities, volumetric heat capacities, heat exchange coefficients, emissivities of walls, . . .), the thermal excitation, and the initial temperature field $T(P, t = 0)$. Let us note here that it is possible to make a *physical reduction* of a model based on the three-dimensional (3D) transient heat equation to get simpler models of lower dimensionality. The thermal fin (1D) or the bulk temperature (0D) types (see Section 1.6.2) constitute such reduced models. This type of reduction may also reduce the number of parameters defining the excitations.

- Numerical/analytical model
  There are many ways for solving the heat equation and finding a *state model* for the observations: *analytical solutions* provide the temperature field explicitly as a function of the structural parameters of the system, the excitation, and the initial state. They can be constructed if the heat equation in each material and the associated conditions are all linear and the corresponding geometry simple. The other class of state models relies on the *discrete formulation* of the heat equation: one can quote the nodal, boundary element, finite elements, and finite volume methods, for example. State models rely on an internal representation of the system: the temperature field has to be found first and the observations are calculated next. External representations that short circuit the state variable and link directly the observation to the excitation(s), for example, through a time or space transfer function, in the linear case, constitute another class of models (see Section 1.5.2.1).

- Parameterization for inverse problem solution
  Parameterization of the data of the direct problem constitutes another characteristic of the structure of a model: structural parameters, thermal excitations, and the initial temperature field are, in the very general case, functions of different explanatory variables: space, time, and temperature. The conversion of functions into vectors of finite dimensions does not involve any problem in the *direct problem* (calculation of the observations, the model output, as a function of the input). It is no more the case when the *inverse problem* is considered. This point will be discussed in Section 1.5.2.2. The interested reader can also consult Chapter 14, where reduction of experimental data is studied. One of the objectives of *mathematical reduction* methods is to construct a *reduced model* that will have a reduced number of structural parameters, starting from a *detailed reference model* (see Chapter 13 for details on model reduction), while *physical reduction* also changes the definitions of both output and excitations (see Section 1.6.2).

1.2 Pertinent Definition of a Direct Model for Inversion of Measurements

1.2.1 Heat Conduction at the Macroscopic Level

Heat transfer by diffusion takes place in solids and motionless fluids and was mathematically described for the first time by Joseph Fourier (1828) in his "Mémoire sur la théorie analytique de la chaleur" (Treatise on the analytical theory of heat). Fourier's relation is phenomenological, that is, derived experimentally. It relates the heat flux density (a vector) to the temperature gradient inside the material under the form of the following linear relationship:

$$\vec{\varphi} = -k\vec{\nabla}T \tag{1.1}$$

where operator $\vec{\nabla}T = (\partial T/\partial x, \partial T/\partial y, \partial T/\partial z)$ denotes the temperature gradient. Consequently, the heat flow rate $d\phi$ traversing an elementary surface of area dS, centered at this location with an orientation defined by a unit length outward pointing vector $\vec{n}$, is

$$d\phi = \varphi_n\, dS \quad \text{with } \varphi_n = \vec{\varphi} \cdot \vec{n} = -k\vec{\nabla}T \cdot \vec{n} \tag{1.2}$$

where

the direction of $\vec{n}$ is arbitrary (two choices are possible)

φ_n is the normal flux (a scalar, sometimes called normal flux density) expressed in W m^{-2}

In order to recover the heat flux ϕ (W) going through a finite surface (not necessary planar) of area S, Equation 1.2 has to be integrated over its whole area. In the particular case of a one-dimensional heat transfer through a planar surface of area S, normal to the x-direction (a cross section), the heat flux is

$$\phi = -kS\frac{\partial T}{\partial x} \tag{1.3}$$

Finally, k is defined as the thermal conductivity of the material. It can be viewed as an intrinsic thermal property of the material. However, it is expressed from more fundamental quantities such as the mean free path of heat carriers (phonons, electrons, and fluid particles), the velocity group as well as fundamental constants (the reduced Planck constant $\hbar$ and the Boltzmann constant k_B).

In many cases encountered in nature or in man-made objects, thermal conductivity is no longer isotropic but orthotropic, or more generally anisotropic. In the orthotropic case (for composite materials, for example, and in the principal axes of the tensor), Fourier's law becomes

$$\vec{\varphi} = -k_x\frac{\partial T}{\partial x}\vec{x} - k_y\frac{\partial T}{\partial y}\vec{y} - k_z\frac{\partial T}{\partial y}\vec{z} \tag{1.4}$$

The three components of the heat flux are expressed according to the three corresponding values for the thermal conductivity in each direction. In case of an anisotropic medium, the symmetrical thermal conductivity tensor can be introduced:

$$\overline{\overline{k}} = \begin{bmatrix} k_{xx} & k_{xy} & k_{xz} \\ & k_{yy} & k_{yz} \\ \text{sym} & & k_{zz} \end{bmatrix} \tag{1.5}$$

Thermal conductivity of materials can vary significantly with temperature. In a general manner, materials act as superconductors at very low temperature (in the 1–10 K temperature range) whereas the thermal conductivity decreases as the temperature increases. The thermal conductivity varies slightly when temperature is greater than the Debye temperature up to the phase change. In the molten state, the thermal conductivity does not change significantly, but in such a configuration, heat transport by convection becomes as important as conduction.

Thermal diffusivity is defined as the ratio of the thermal conductivity and the specific heat per unit volume:

$$a = \frac{k}{\rho c_p} \tag{1.6}$$

It is thus possible to estimate the diffusion time $t_{diff} = L^2/a$ when heat diffuses in the direction defined by its characteristic length L as reported in Table 1.1.

TABLE 1.1

Characteristic Diffusion Times (Thermal Diffusivity
Is $a = 10^{-6}$ m^2 s^{-1})

L	Sphere (Radius 6400 km)	0.3 m	1 cm	100 nm	1 nm
t_{diff}	10^{12} years	10^5 s	100 s	10^{-8} s	10^{-12} s $= 1$ ps

Fourier's law becomes unappropriate to simulate heat transfer by conduction at very short times of the order of the picosecond that are related to the nanoscale (according to Table 1.1). If one considers the response to a localized heat pulse on the material, Fourier's law shows that the temperature field is modified instantaneously at every point of space since the pulse start. However, at a later time t, temperature cannot have been modified beyond a distance equal to the quantity: $c\,t$, otherwise the effect of the pulse would have propagated faster than the speed of light c. The relationship relating heat flux and temperature gradient must therefore be modified. It has been done by Caetano who introduced a form involving a relaxation time τ:

$$\tau \frac{\partial \vec{\varphi}}{\partial t} + \vec{\varphi} = -k \vec{\nabla} T \tag{1.7}$$

This relaxation time τ depends on the nature of the heat carriers (phonons, electrons, or fluid particles) and more generally on the collision processes between them.

Equivalently, we may compare a characteristic length scale for evolution of the system with the other intrinsic property: the mean free path of the heat carriers. If the latter is much greater than the characteristic length of the medium, the local Fourier law is no longer valid.

1.2.2 An Experimental Observation

Before presenting theoretical developments, it would be interesting to start with an experimental result obtained using the femtoseconds (1 fs $= 10^{-15}$ s) time domain thermoreflectance (TDTR) technique. This experiment consists in applying a very short pulse (some tenths of femtoseconds) at the front face of a material and to measure the transient temperature response on the heated area (see Figure 1.2). The pulse laser is called the pump. A probe laser beam is focused on the heated area, and a photodiode allows measuring the reflected beam intensity from the surface. Since the intensity of the reflected beam is known to vary linearly with temperature (for small pump intensity), the measured signal is proportional to the variation of the time-dependent surface temperature. An ad hoc postprocessing of the output signal allows building a normalized impulse response for the sample.

This experiment is known as the front face method (the thermal disturbance and the temperature measurement are realized at the same location). In a sense, the TDTR can be viewed as an extension of the classical "flash" method for very short times. In the experimental configuration described in Figure 1.2, the TDTR technique is used for characterizing a very thin layer (100 nm thick) of a semiconducting alloy: Ge$_2$Sb$_2$Te$_5$ (commonly denoted GST) whose thermal effusivity is $b_{GST} = \sqrt{k_{GST}(\rho c_p)_{GST}}$. A thermal

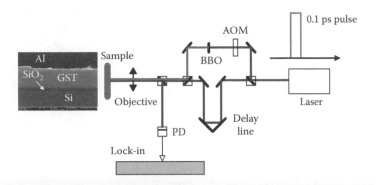

FIGURE 1.2
Radiation of pump is doubled by a β-BaB₂O₄ (BBO) nonlinear optical crystal. The probe pulse is delayed according to the pump pulse up to 7 ns with a temporal precision of a few tens of femtoseconds by means of a variable optical path. The pump beam, whose optical path length remains constant during the experiment, is modulated at a given frequency of 0.3 MHz by an acousto-optic modulator (AOM). In order to increase the signal over noise ratio, a lock-in amplifier synchronized with the modulation frequency is used. Probe and pump beams have a Gaussian profile. The experimental setup is described in Battaglia et al. (2007). An example of sample is represented on the SEM image; the Al layer is used as a thermal transducer to absorb the incident radiation of the pump.

transducer is an aluminum film (denoted Al), of thickness e_{Al} and specific heat per unit volume $(\rho c_p)_{Al}$, deposited on the GST layer in order to increase the signal–noise ratio during the TDTR experiment. For the duration of the experiment (a tenth of nanoseconds), the GST layer is viewed as a semi-infinite medium. Using the classical heat diffusion model, based on Fourier's law, an analytical expression is obtained for the average (with respect to the spatial distribution of temperature on the heated area) normalized impulse response as follows:

$$\overline{TDTR} = \exp\left(\frac{t}{t_c}\right) \mathrm{erfc}\left(\sqrt{\frac{t}{t_c}}\right) \quad \text{with } t_c = \left(\frac{e_{Al}(\rho c_p)_{Al}}{b_{GST}}\right)^2 \tag{1.8}$$

Experimental measurements are reported in Figure 1.3, as well as the simulation obtained from the analytical solution (1.8).

It clearly appears that the measured impulse response fits very well with the simulated semi-infinite behavior when time becomes higher than $t_c = 0.3$ ns. This result comes from the fact that thermal equilibrium, also called thermalization, between the electrons gas and the lattice in the aluminum film must be taken into account in the model for short times just after the pulse. This effect can be modeled through a specific model: the two-temperature model (see Section 1.2.3.4). This time is defined as the thermalization time of the heat carriers: electrons and phonons. It can be viewed as the relaxation time that has been introduced in Equation 1.6. However, as it will be shown in Section 1.2.3.4, the relaxation time τ is lower than time t_c, estimated from Figure 1.3, since the thermal resistance at the Al–GST interface was not taken into account in this equation.

This observation leads us to take care of the direct model formulation that will be used to solve an inverse problem. It should be adapted to the timescale concerned within the experiment. Indeed, in the example presented above, one can only estimate the thermal effusivity of the layer for time t such as $t > t_c$. This last point has a significance since the

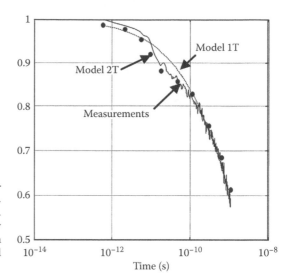

FIGURE 1.3

Impulse response obtained using the TDTR experiment on a GST layer capped with an Al transducer. Plain line is the measurement from 1 ps up to 2 ns. The dotted line is the simulation using the Fourier law (Model 1T), and the plain circles are obtained from the simulation of the two-temperature model (Model 2T, described later in this text).

concepts of thermal conductivity and even of temperature do not make sense anymore at the very small scales. Finally, it is also clear that different thermal parameters, in terms of their physical meaning, will be introduced according to the direct model formulation.

1.2.3 How Can Heat Transfer Be Modeled at the Nanoscale?

1.2.3.1 Discussion

We have highlighted above the intimate link between temperature gradient and mean free path of the carriers in solids: phonons and electrons. In particular, if the characteristic dimension of the material is smaller than the mean free path Λ of these carriers, only a thermal conductance K can be used for relating heat flux to the temperature difference ΔT at the material surface as $\varphi = K\Delta T$. In other words, expressing the thermal conductance as the classical ratio k/e when $e \leq \Lambda$ does not make any sense (see Figure 1.4).

Nevertheless, current challenges for miniaturization force engineers to implement materials in structures whose dimensions lie between several nanometers and a few hundreds of nanometers (see Figure 1.5). Study of the heat transfer in these structures requires using

FIGURE 1.4

Thermal characterization using the front face experiment. (a) If the sample thickness e is less than the mean free path Λ of the heat carriers (electrons/phonons) the experiment allows identification of the thermal resistance (or conductance) of the layer only. (b) In the opposite case, the method allows identification of the thermal effusivity of the layer.

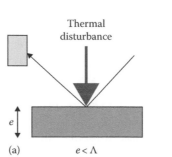

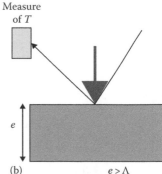

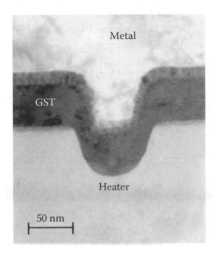

FIGURE 1.5
Nanoelectronics: a nonvolatile memory cell based on phase change chalcogenide alloy (GST stands for germanium–antimony–tellurium). The characteristic dimension of the cell is 50 nm.

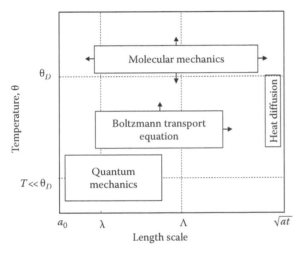

FIGURE 1.6
At dimensions comparable to the phonon wavelength λ and temperatures much smaller than the Debye temperature θ_D, heat transfer rests essentially on quantum mechanics. For larger dimensions and room temperatures, the BTE and the classical MD are well adapted for modeling heat transfer inside the studied structure. For even larger dimensions, Fourier's law can be efficiently implemented with a denoting thermal diffusivity.

specific tools that will be developed now. According to the scale, four types of methods will be used for constructing a heat transfer model, as described schematically in Figure 1.6.

We will first present the transport of heat through molecular dynamics (MD). We will then skip to Boltzmann transport equations (BTE) and present the two-temperature model further on. These models allow taking local thermal nonequilibrium into account. This nonequilibrium occurs between the thermal states of the electron gas and the crystal lattice, for metals and for semiconductors and only of the lattice for insulators. We will come finally to the model of heat diffusion designed by Fourier nearly 200 years ago. We will pinpoint, for each type of approach, the possibilities of measurement inversion. In other words, we will seek to define what are the physical parameters accessible to measurement and what are the thermal properties inherent to each.

1.2.3.2 Molecular Dynamics

MD aims at calculating the position, speed, and acceleration of ions or molecules that make up the material according to the classical Newtonians' mechanics equations, that is, the fundamental principle of dynamics (FPD). For a detailed description of the method, see the book of Volz (2007) as well as that of Frenkel and Berend (1996). MD also leads to reliable results when quantum effects are predominant by using ab initio calculation starting from the Schrödinger relation. These quantum effects appear at low temperature and more precisely below Debye temperature. One will be able thus to use the FPD in MD only for $T > \Theta_D$. Another criterion to validate the use of FPD consists in calculating the ratio λ/a_0 where λ is the average wavelength of the ion (or molecule) vibration and a_0 is the interatomic distance. The relationship between wavelength λ, particle mass m, and temperature T is

$$\lambda = \frac{\hbar}{\sqrt{2\pi m k_B T}} \tag{1.9}$$

In this relation, $\hbar$ is Planck's constant and k_B is Boltzmann's constant. We note $\vec{r}_i$, $\vec{v}_i = \mathrm{d}\vec{r}_i/\mathrm{d}t$, and $\vec{a}_i = \mathrm{d}\vec{v}_i/\mathrm{d}t$ the position, speed, and acceleration of particle i, respectively. The total energy of particle i is the sum of its kinetic and potential energy:

$$E_i = E_{c_i} + E_{p_i} \tag{1.10}$$

The potential energy is itself the sum of an external potential field (such as an electromagnetic field) and of an internal field (caused by mutual interactions of the particles).

The force that is exerted on each particle thus derives from the potential energy:

$$\vec{F}_i = -\vec{\nabla} E_{p_i}(\vec{r}_i) \tag{1.11}$$

FPD applied to one particle is then

$$\vec{F}_i = m\vec{a}_i \tag{1.12}$$

Solving this vector-relationship (three scalar equations in three dimensions) for each particle (see Figure 1.7) leads to the position and then to the velocity of each particle. Calculation of its kinetic energy derives from knowledge of its speed:

$$E_{c_i} = \frac{m_i v_i^2}{2} \tag{1.13}$$

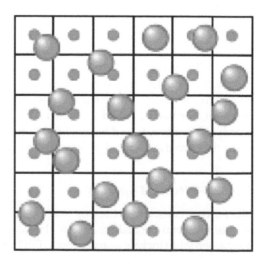

FIGURE 1.7
Classical configuration used for particle motion simulation using the MD. Periodic boundary conditions on the cell are generally assumed.

The kinetic theory provides temperature as follows:

$$T = \frac{2}{3k_B} E_{c_i} \tag{1.14}$$

Temperature is not subscripted by i deliberately because the notion of temperature relies on a large number of particles. Even if the mass of the particle is not present explicitly in the expression of temperature above, that is not true any more when various elements make up the material. In this case, one of the masses is taken as a reference and a mass correction is made for the other elements.

The theoretical difficulty in MD stems from the calculation of the potential of interaction between the particles.

MD can be implemented at the very low scale in order to calculate thermal conductivity of solids using non-homogeneous non-equilibrium molecular dynamics (NEMD) (see Figure 1.8). This is certainly the simplest technique (compared to the Green–Kubo calculation at equilibrium) to understand and implement, for it is analogous to the well-known guarded hot plate experiment. The idea is to simulate steady-state one-dimensional heat transfer in a system by inserting a hot and a cold source and then calculating the flux exchanged between the sources as well as the temperature gradient. The most widely used approach consists in adapting the velocity field of the atoms belonging to the heat sources in such a way as to impose the thermal power exchanged between the hot and cold sources. This method requires a large computation time: the number of particles that must be retained in this simulation is large since temperature is a statistical quantity.

Moreover, since thermal conductivity calculation requires defining a thermal gradient, the number of required particles increases dramatically in order to get a precise enough corresponding derivative. Moreover, this simulation always leads to the value of the thermal resistance R_{th} (the inverse of thermal conductance K defined in Section 1.2.3.1) of the material inserted between the hot and cold plates. This quantity is certainly as

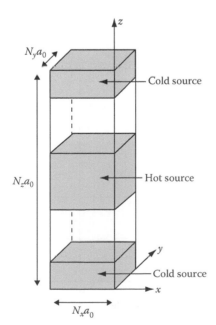

FIGURE 1.8
Nonequilibrium MD simulation for thermal conductivity simulation.

interesting as thermal conductivity in practical configurations encountered in engineering. As we said previously, if one wants to relate thermal resistance R_{th} to thermal conductivity k from the classical relationship $R_{th} = L/k$, then the dimension of the simulation box must be chosen as $L \gg \Lambda$, where Λ is the mean free path of the phonons.

We introduce now some basic ideas concerning the statistical nature of temperature since it is not always clear at very small scales. Using statistical mechanics arguments, temperature in a perfect gas can be defined for each particle of the gas. For liquids or solids, another definition, based on the interactions between the particles, must be given. Thus, the true question is the lowest size down to which the average energy of the phonons can be calculated. The answer is related to the value of the mean free path introduced in the preceding paragraph that is the distance separating two successive collisions of a phonon. If two areas in space have different temperatures, then they have also a different distribution of phonons. We know that this distribution can be modified only through the process of collisions. Anharmonic processes (processes where the assumption of small oscillations of particles around their equilibrium state is no longer valid) are responsible on thermal conductivity itself. The low frequency phonons have a large mean free path and correspond to low temperatures. In the so-called Casimir limit, for low temperatures, the mean free path size is about the same as the dimension of the material system. For high temperatures, on the contrary, phonons have a high frequency and mean free paths become much smaller. An illustration is given in Figure 1.9, where it is clearly demonstrated that the thermal conductivity and thus the temperature gradient take sense only when the number of particles involved in the MD simulation is high enough.

MD can be efficiently used as the direct model in an inverse procedure. Since inversion calls upon the model several times, it seems that it will take huge computational times. In order to answer the question about the parameters than can be estimated, it clearly appears that the unknown parameters in the model relate to the potential functions between each particle. Thus, one can imagine measuring the thermal conductance of a thin layer and then using this result as the minimizing function. To our knowledge, no work has ever been published on such a topic.

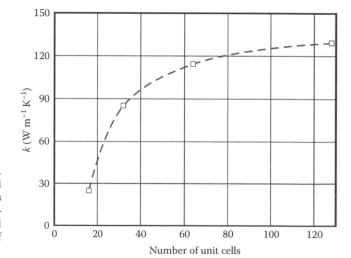

FIGURE 1.9
Result of NEMD simulation for silicon. Thermal conductivity is calculated according to the number of unit cells (crystal cell). As expected, thermal conductivity tends asymptotically toward the experimental value as the number of unit cell becomes high enough.

1.2.3.3 Boltzmann Transport Equation

The phonon BTE describes the rate of change of a statistical distribution function for phonons. The fundamental assumption in deriving the phonon BTE is that a distribution function, $N_q(\vec{r}, t)$, exists. It describes the average occupation of phonon mode q (this mode is associated with frequency ω_q and with wave vector $\vec{k}_q$ that are related through the dispersion curve for the studied material) in the neighborhood of a location $\vec{r}$ at time t. This equation relies on the assumption that phonon position and momentum can simultaneously be known with an arbitrary precision. However, in quantum mechanics, these quantities correspond to noncommuting operators and hence obey the uncertainty principle. The BTE is formally written as follows (see Volz [2007]):

$$\frac{\partial N_q(\vec{r}, t)}{\partial t} + \vec{v}_q \cdot \vec{\nabla} N_q(\vec{r}, t) = \left.\frac{\partial N_q(\vec{r}, t)}{\partial t}\right]_c \tag{1.15}$$

where $\vec{v}_q$ is the group velocity associated to phonon of wave vector $\vec{k}_q$. The term on the right-hand side is the rate of change due to collisions. Solution of the phonon BTE requires evaluation of the collision term, which constitutes the challenging problem here. The relaxation time approximation, associated with mode q, is widely used to model it. Under this approximation, the BTE is rewritten using the average distribution function $\overline{N}$ as follows:

$$\frac{\partial N_q(\vec{r}, t)}{\partial t} + \vec{v}_q \cdot \vec{\nabla} N_q(\vec{r}, t) = -\frac{N_q(\vec{r}, t) - \overline{N}}{\tau_q} \tag{1.16}$$

A key conceptual problem in using the relaxation time approximation is the requirement for a thermodynamic temperature that governs the scattering rate. Since phonons are not in an equilibrium distribution, there is no temperature to strictly speak of. The usual practice in such nonequilibrium problems is to define an ad hoc equivalent temperature based on the local energy.

The BTE can be efficiently used in order to compute the thermal conductivity of solids. Indeed, it is demonstrated that thermal conductivity can be related to thermal capacity c_v (J kg^{-1}) as follows:

$$k = \int_0^{q_{max}} v_q^2 c_v(q) \tau_q \, dq \tag{1.17}$$

Specific heat can also be expressed analytically in terms of frequency mode ω_q and of temperature T as follows:

$$c_v(\omega_q) = \frac{3\hbar^2}{2\pi^2 k_B T^2 v_q} \int_0^{\omega_{max}} \frac{e^{\hbar\omega_q/k_B T}}{\left(e^{\hbar\omega_q/k_B T} - 1\right)^2} \omega_q^2 q^2 \, d\omega_q \tag{1.18}$$

The frequency mode is related to the wave vector through the dispersion curves of the material. However, we must insist on the fact that this definition of thermal conductivity rests on the fact that the use of Fourier's law is allowed. In other words, time t must verify $t \gg t_c$, and characteristic dimension L of the medium must be such as $L \gg \Lambda$, in order to define the temperature gradient inside the medium. These conditions are less restrictive for the definition of the specific heat since it only involves temperature and not its gradient.

The question now is as follows: does the BTE can be considered as the direct model in an inverse procedure and for identifying what? The answer is clearly yes since, as viewed previously, there is an analytical model for both specific heat and thermal conductivity. This model could be implemented in order to estimate the mean relaxation time of the phonons inside the material, which is generally unknown. Again, to our knowledge, such a work has not been made or published yet.

1.2.3.4 The Two-Temperature Model

We conclude this first part with the two-temperature model that constitutes a very good transition to homogenization methods at the macroscopic scale that will be described further on. The two-temperature model describes the time-dependent electron and lattice temperatures, T_e and T_l, respectively, in a metal or in a semiconductor during the thermalization process as follows:

$$c_e(T_e)\frac{\partial T_e}{\partial t} = \vec{\nabla} \cdot (k_e(T_e, T_l)\vec{\nabla}T_e) - G(T_e - T_l) + q_{vol} \tag{1.19}$$

$$c_l\frac{\partial T_l}{\partial t} = G(T_e - T_l) \tag{1.20}$$

In these equations, c_e and c_l are the electronic and lattice specific heat per unit volume, k_e is the electronic thermal conductivity that can be assimilated to bulk thermal conductivity for metals, and q_{vol} is the volumetric heat source in the lattice. These two nonlinear equations are coupled through the electron–phonon coupling constant G that can be explicitly defined starting from the BTE for both electrons and phonons. A detailed explanation of this model foundation can be found in the paper of Anisimov et al. (1974, 1975).

Regarding the TDTR reference experiment (see Section 1.2.2), Equation 1.19 means that, after the pulse, hot electrons will move inside the medium while losing their energy to the lattice. Let us insist on the fact that this model has a physical meaning only during the thermalization process, since Equation 1.20 shows that the lattice temperature remains constant as soon as $T_e = T_l$ or, in other words, when the thermalization process between electrons and lattice ends.

It must be also emphasized that this model involves a temperature gradient in the electron gas whereas thermal conduction in the lattice is neglected with respect to heat exchange between electrons and the lattice. It means that the characteristic length of the medium is such as $L \gg \Lambda_e$, where Λ_e is the mean free path of the electrons. Indeed, we saw previously that the mean free path for electrons is larger than for phonons. However, the constraint on time is just related to the relaxation time for electrons, which is of the order of some tenth of femtoseconds. In other words, the simulation time range for the two-temperature model can be (and should be) shorter than the relaxation time for the phonons.

When implementing the model in relation with the thermoreflectance experiment, the heat source q_{vol} is a function of the heated area (laser beam radius) of the optical penetration depth of the beam inside the material (related to the extinction coefficient) and of the intensity of the source.

We used the finite element method in order to simulate the two-temperature model starting from parameters given in the literature for aluminum. This simulation remains

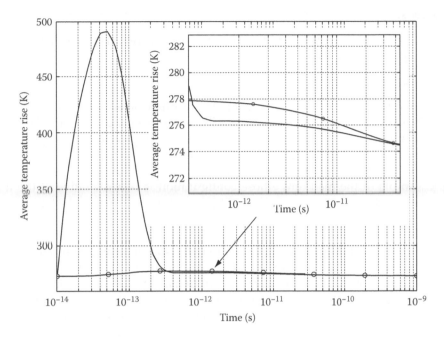

FIGURE 1.10

Two-temperature model simulation for the aluminum sample using the finite element method. Line with circles represents the lattice temperature; plain line is the electron gas temperature.

coherent with the definition of temperature for the electron gas and the lattice since the sample thickness has been chosen larger than the mean free path of the electrons in aluminum, which is approximately 10 nm (in other words, the minimum distance between two nodes of the mesh should be larger than this critical length). The resulting time-dependent temperatures of the lattice and of the electron gas are reported in Figure 1.10. The electron gas temperature increases very quickly and reaches its maximum at 50 fs. Temperature of the lattice begins to increase at 20 fs and reaches the electrons gas temperature at $t_c = 200$ fs. The calculation shows the undercooling of the electrons relative to the lattice at the surface. This undercooling comes from the high value of the coupling factor for aluminum. It is also observed for gold or copper whose coupling factors are smaller, but it is less pronounced than for aluminum. Figure 1.10 shows that complete thermalization between electron gas and lattice is reached at times between 25 and 30 ps. It demonstrates what was said in Section 1.2.2, that is, the relaxation time is lower than time t_c that has been estimated through the TDTR experiment.

The use of the two-temperature model as a model to invert has been made by Orlande et al. (1995) in order to estimate the coupling factor G for several kinds of metals. In fact, analytical expressions for this parameter are generally inaccurate: knowledge of the dispersion curve for the studied material is required. It is then interesting to estimate it directly from measurements similar to those given by the TDTR.

Our reference experiment shows that at the thermalization end, the TDTR measured response is only sensitive to the lattice cooling, which means that use of the classical one-temperature model becomes appropriate in order to describe heat diffusion inside the medium. Let us note that the two-temperature model degenerates naturally toward the one-temperature model when $t > t_c$.

1.3 Heat Diffusion Model for Heterogeneous Materials: The Volume Averaging Approach

1.3.1 Model at Local Scale

Thermal properties of heterogeneous materials are often determined experimentally by assuming the sample behaves macroscopically like a homogeneous medium. Therefore, the reliability of measurements depends heavily on the validity of the "homogeneous medium" assumption (see also Chapter 2 on the same subject). This is particularly true for measurements based on transient heat conduction. Let us consider now an elementary volume (a sample of the medium) whose configuration is representative of the material. Such a representative elementary volume (REV) is shown in Figure 1.11 for a medium constituted of two phases σ and β.

The shape of the REV is arbitrary but its size is not: if the REV is a sphere of diameter $D = 2r_0$, this diameter should be much smaller than the size of the whole system L: $D/L \ll 1$; this sphere constitutes a sample of the material and its diameter must be larger than the scale representative of the distribution of the two phases in space (an averaged distance l_β separating the "grains" of the discontinuous phase σ embedded in the continuous phase β in Figure 1.11, for example): $D/l_\beta \gg 1$. If the local structure of the material within this REV does not change too much when this sphere is moved anywhere in the whole medium, this medium can be homogenized.

One assumes here that Fourier's law is applicable for both phases at any point whose location is determined by its position vector $\vec{r}$ and for each time t. Thermal conductivities are denoted k_σ and k_β, and specific heat per unit volume is denoted $(\rho c_p)_\sigma$ and $(\rho c_p)_\beta$, for phases σ and β, respectively.

The heat transfer model at the local scale is as follows:

$$(\rho c_p)_\sigma \frac{\partial T_\sigma(\vec{r}, t)}{\partial t} = \vec{\nabla} \cdot (k_\sigma \vec{\nabla} T_\sigma(\vec{r}, t)) \quad \text{for } \vec{r} \text{ in the } \sigma\text{-phase} \tag{1.21}$$

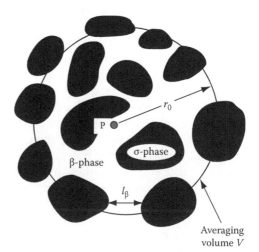

FIGURE 1.11
REV of a two-phase heterogeneous medium.

for the σ-phase, and

$$(\rho c_p)_\beta \frac{\partial T_\beta(\vec{r}, t)}{\partial t} = \vec{\nabla} \cdot (k_\beta \vec{\nabla} T_\beta(\vec{r}, t)) \quad \text{for } \vec{r} \text{ in the β-phase} \tag{1.22}$$

for the β-phase. Heat transfer between the two phases appears at the boundary condition at the σ–β interface.

Two homogenized models that transform this two-phase model into one single homogeneous (equivalent) phase can be now introduced. This homogenized medium may exist or not.

1.3.2 The One-Temperature Model

A volume averaging operator, noted $\langle \rangle$, can be defined here for any space field f at a point $\vec{r}$ located at the center $\vec{r}$ of the REV as follows:

$$\langle f \rangle(\vec{r}, t) = \frac{1}{V(\vec{r}, D)} \int_{V(\vec{r}, D)} f(\vec{r}', t)\, dV(\vec{r}') \tag{1.23}$$

where
$V(\vec{r}, D)$, $\pi D^3/6$ here, is the volume of an REV centered at point $\vec{r}$
$dV(\vec{r}')$ is a microscopic volume centered at any point $\vec{r}'$ located inside the REV

Thus, an averaged "enthalpic" temperature T_H can always be defined:

$$T_H(\vec{r}, t) - \frac{1}{\langle \rho c_p \rangle V(\vec{r}, D)} \int_{V(P, D)} \rho c(\vec{r}') T(\vec{r}', t)\, dV(\vec{r}') = \frac{1}{\rho c_t} \langle H \rangle (\vec{r}, t) \tag{1.24}$$

where $H(\vec{r}, t)$ is the local enthalpy by unit volume: $H(\vec{r}, t) = \rho c_t(\vec{r}) T_H(\vec{r}, t)$, the total volumic heat ρc_t being defined by

$$\rho c_t(\vec{r}) = \langle \rho c_p \rangle(\vec{r}) = \varepsilon_\sigma (\rho c_p)_\sigma + \varepsilon_\beta (\rho c_p)_\beta \tag{1.25}$$

Here ε_σ and ε_β are the local volume fractions of the σ and β phases ($\varepsilon_\sigma + \varepsilon_\beta = 1$). These volume fractions are derived from the characteristic functions χ_α of each phase α (for $\alpha = \sigma$ or β), where $\chi_\alpha(\vec{r}) = 1$ if $\vec{r}$ belongs to the fluid phase and $\chi_\alpha(\vec{r}) = 0$ otherwise

$$\varepsilon_\beta(\vec{r}) = \langle \chi_\beta \rangle; \quad \varepsilon_\sigma(\vec{r}) = \langle \chi_\sigma \rangle = 1 - \varepsilon_\beta(\vec{r}) \tag{1.26}$$

One can notice that if the medium can be homogenized, its specific heat per unit volume ρc_t defined above should not depend on location $\vec{r}$.

The one-temperature model requires the definition of a thermal conductivity tensor $\overline{\overline{k}}$ whose coefficients can be considered as conductivities depending on the nature, thermophysical properties and geometry of the distribution of phases σ and β. A diffusion energy equation for the space and time variations of the averaged temperature can be written in the case of a homogenized medium (Moyne et al., 2000):

$$\rho c_t \frac{\partial T_H}{\partial t} = \nabla \cdot (k \nabla T_H) + q_{vol} \tag{1.27}$$

where
q_{vol} is a volumetric source term
k is an effective (or equivalent) conductivity of the material that is supposed to be locally isotropic here (otherwise k has to be replaced by $\overline{\overline{k}}$)

This model can be extended to take fluid flow into account (see Testu et al. [2007]).

1.3.3 The Two-Temperature Model

At this stage, we introduce now the notion of intrinsic phase average, noted $\langle \rangle^\alpha$ here, for any time–space field $f(\vec{r}, t)$ defined in the α-phase:

$$\langle f_\alpha \rangle^\alpha (\vec{r}, t) = \frac{1}{V_\alpha(\vec{r}, D)} \int\limits_{V_\alpha(\vec{r}, D)} f(\vec{r}', t) \, dV(\vec{r}') \quad \text{for } \alpha = \sigma \text{ or } \beta \tag{1.28}$$

where $V_\alpha(\vec{r}, D) \subset V(\vec{r}, D)$ designates the volume occupied by the α-phase ($\alpha = \sigma$ or β) in the REV shown in Figure 1.11. Subscript α of f_α indicates that integration is made for $\vec{r}'$ belonging to the $V_\alpha(\vec{r}, D)$ volume, while superscript α, in $\langle \rangle^\alpha$, is related to division by volume $V_\alpha(\vec{r}, D)$ in the right-hand member of this equation: $\langle . \rangle = \varepsilon_\alpha \langle . \rangle^\alpha$.

One can therefore introduce two different average temperatures $\langle T_\alpha \rangle^\alpha$ at the same point $\vec{r}$. These two temperatures are related to the previous averaged "enthalpic" temperature T_H through the definition of the average enthalpy:

$$\langle H \rangle = \rho c_t(\vec{r}) T_H = (\rho c_p)_\sigma \langle T_\sigma \rangle^\sigma + (\rho c_p)_\beta \langle T_\beta \rangle^\beta \tag{1.29}$$

In the case of local thermal equilibrium the temperatures both $\langle T_\alpha \rangle^\alpha$ are equal, which implies that they are also both equal to the average enthalpic temperature, because of the previous equation and of the definition of ρc_t: $\langle T_\sigma \rangle^\sigma = \langle T_\beta \rangle^\beta = T_H$. In the opposite case, the enthalpic temperature still exists but its observation is somewhat involved because a perfect temperature detector would provide a temperature that will be either close to $\langle T_\alpha \rangle^\alpha$ or to $\langle T_\beta \rangle^\beta$, depending on the quality of its coupling with either of each phase. In any case, the sensor temperature would be close to T_H, because, by definition, this temperature lies in between these two temperatures.

The macroscopic description of heat transfer in heterogeneous media by a single energy equation does not imply the assumption of local thermal equilibrium between the two phases. However, in order to get such an equilibrium, as described by Carbonell and Whitaker (1984), some criteria must be verified.

We use now the following notation: D denotes the characteristic dimension of the REV of volume V, $a_v = A_{\sigma-\beta}/V$ is its specific area, that is the ratio of the area of the interface $A_{\sigma-\beta}/V$ between the two phases by its volume, V_σ is the volume of the σ-phase, $\varepsilon = V_\sigma/V$ is its volume fraction, and L denotes the characteristic dimension of the heterogeneous medium.

Then time t must verify (see Carbonell and Whitaker [1984])

$$\frac{\varepsilon(\rho C_p)_\sigma D^2}{t}\left(\frac{1}{k_\sigma}+\frac{1}{k_\beta}\right)\ll 1 \quad \text{and} \quad \frac{(1-\varepsilon)(\rho C_p)_\beta D^2}{t}\left(\frac{1}{k_\sigma}+\frac{1}{k_\beta}\right)\ll 1 \tag{1.30}$$

And the characteristic dimension L of the REV must verify

$$\frac{\varepsilon k_\sigma D}{a_v L^2}\left(\frac{1}{k_\sigma}+\frac{1}{k_\beta}\right)\ll 1 \quad \text{and} \quad \frac{(1-\varepsilon)k_\beta D}{a_v L^2}\left(\frac{1}{k_\sigma}+\frac{1}{k_\beta}\right)\ll 1 \tag{1.31}$$

Obviously, another factor that can affect the assumption of local thermal equilibrium is the location of the considered point with respect to the heat source: equilibrium cannot occur in the vicinity of this source. Such a situation is met, for example, for front face heat pulse excitation of a multilayer slab made of layers of different thermophysical properties.

For situations in which local thermal equilibrium is not valid, models have been proposed based on the concept of two macroscopic continua. Intrinsic average temperatures for the σ-phase and the β-phase are denoted by $\langle T_\sigma(\vec{r},t)\rangle^\sigma$ and $\langle T_\beta(\vec{r},t)\rangle^\beta$, respectively (see Equation 1.28).

The pore-scale temperature deviation in the σ-phase is defined by

$$T_\sigma(\vec{r},t)=\langle T_\sigma(\vec{r},t)\rangle^\sigma+\tilde{T}_\sigma(\vec{r},t) \tag{1.32}$$

One can introduce this decomposition into the pore-scale equation for the σ-phase and then form the volume average in order to obtain the macroscopic equation. After extensive use of the averaging theorem, the following energy equation emerges for the σ-phase:

$$\varepsilon(\rho c_p)_\sigma\frac{\partial\langle T_\sigma\rangle^\sigma}{\partial t}=\nabla\cdot\left(\mathbf{K}_{\sigma\beta}\cdot\nabla\langle T_\beta\rangle^\beta+\mathbf{K}_{\sigma\sigma}\cdot\nabla\langle T_\sigma\rangle^\sigma\right)-a_v h\left(\langle T_\sigma\rangle^\sigma-\langle T_\beta\rangle^\beta\right) \tag{1.33}$$

Equivalently, the same procedure for the β-phase leads to

$$(1-\varepsilon)(\rho c_p)_\beta\frac{\partial\langle T_\beta\rangle^\beta}{\partial t}=\nabla\cdot\left(\mathbf{K}_{\beta\beta}\cdot\nabla\langle T_\beta\rangle^\beta+\mathbf{K}_{\beta\sigma}\cdot\nabla\langle T_\sigma\rangle^\sigma\right)-a_v h\left(\langle T_\beta\rangle^\beta-\langle T_\sigma\rangle^\sigma\right) \tag{1.34}$$

The macroscopic conductivity tensors $\mathbf{K}_{\beta\beta}, \mathbf{K}_{\beta\sigma}, \mathbf{K}_{\sigma\beta}, \mathbf{K}_{\sigma\sigma}$ and the volumetric exchange coefficient $a_v h$ are given by the solution of three closure problems that have to be solved over unit cells representative of the medium characteristics (see the paper of Quintard et al. [1997]).

Let us note that the previous one- or two-temperature models have been derived using the volume-averaging technique. The same kind of results can be set using the homogenization technique, where two different independent coordinate systems can be defined, one at the local scale and the other one at the mesoscopic scale. The interested reader can refer to Auriault and Ene (1994) for an example of practical application of this type of technique.

1.3.4 Application to a Stratified Medium

Here, we are interested by the macroscopic thermal behavior of a stratified medium subjected to a Dirichlet boundary condition, the flux being parallel to the strata (see Figure 1.12).

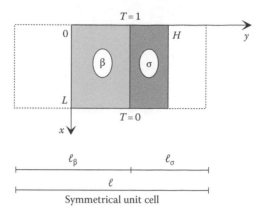

FIGURE 1.12
Stratified medium unit cell.

This choice is due to the fact that in this particular geometry, reference analytical exact solutions exist for the macroscopic effective properties, and only four effective parameters are independent and have to be identified.

Since the stratified medium is orthotropic and the main tensor axis coincides with the direction normal to the layers, the two-equation model is reduced to

$$\frac{\partial \langle T_\sigma \rangle^\sigma}{\partial t} = \frac{K_{\sigma\sigma}}{\varepsilon_\sigma (\rho c_p)_\sigma L^2} \frac{\partial^2}{\partial x^{*2}} \langle T_\sigma \rangle^\sigma - \frac{a_v h}{\varepsilon_\sigma (\rho c_p)_\sigma} \left(\langle T_\sigma \rangle^\sigma - \langle T_\beta \rangle^\beta \right) \tag{1.35}$$

and

$$\frac{\partial \langle T_\beta \rangle^\beta}{\partial t} = \frac{K_{\beta\beta}}{\varepsilon_\beta (\rho c_p)_\beta L^2} \frac{\partial^2}{\partial x^{*2}} \langle T_\beta \rangle^\beta - \frac{a_v h}{\varepsilon_\beta (\rho c_p)_\beta} \left(\langle T_\beta \rangle^\beta - \langle T_\sigma \rangle^\sigma \right) \tag{1.36}$$

where x^* is the dimensionless space variable x/L. Then, in this configuration, the four independent parameters to be identified are defined by

$$A_\beta = \frac{K_{\beta\beta}}{\varepsilon_\beta (\rho c_p)_\beta L^2}; \quad H_\beta = \frac{a_v h}{\varepsilon_\beta (\rho c_p)_\beta L^2}; \quad A_\beta = \frac{K_{\sigma\sigma}}{\varepsilon_\sigma (\rho c_p)_\sigma}; \quad H_\sigma = \frac{a_v h}{\varepsilon_\sigma (\rho c_p)_\sigma} \tag{1.37}$$

This study has been the subject of a paper of Gobbé et al. (1998).

1.4 Summary on the Notion of Temperature at Nanoscales and on Homogenization Techniques for Heat Transfer Description

We have seen above that in a solid material, temperature can be considered as a potential that "explains" transfer of energy and, at scales large enough, transfer of heat. At the nanoscale, its definition requires the presence of a high enough number of particles of each phase (ions in a lattice, electrons) because of its statistical nature. Once this condition is fulfilled, the studied medium can be considered as continuous, which means that any potential field or physical quantity can be assigned to any space point in the geometrical 3D Euclidian domain. Two different temperatures can be defined then, one for each phase, at the same location. These can degenerate to one single temperature, if the two phases locally present at the same point reach equilibrium, depending on the time–space scales considered.

The same type of approach can be adopted at larger space scales, when solid materials composed of two phases are considered. At these larger scales, let us say above 10 nm, (1) the material is considered as continuous and (2) Fourier's law becomes valid at any point in space. Both previous conditions are not equivalent, since the second condition requires validity of the first one.

The use of an REV allows "filtering" the locally heterogeneous material, which leads to the definition of either one single "average enthalpic temperature" or two "intrinsic average temperatures," verifying one or two coupled heat equations.

If the structure of the REV is not modified by its translation in space, the material can be considered as homogenous. Modification by rotation leads to anisotropic properties, but this notion does not derive from the spatial distribution of the two phases only. Under this condition of invariance by translation, the REV averaged thermophysical properties of the material become constant that is uniform in space. These properties are

- Its volume fractions ε_σ and ε_β defining its total volumetric heat ρc_t, and its effective thermal conductivity k (or a thermal conductivity tensor $\bar{\bar{k}}$ in the more general anisotropic case), for the one-temperature model (see Equation 1.29).
- The macroscopic thermal conductivity tensors $\mathbf{K}_{\beta\beta}, \mathbf{K}_{\beta\sigma}, \mathbf{K}_{\sigma\beta}, \mathbf{K}_{\sigma\sigma}$ and volumetric exchange coefficient, $a_v h$, for the two-temperature model (see Equations 1.35 and 1.36).

Homogenization techniques are presented in Chapter 2.

1.5 Physical System, Model, Direct and Inverse Problems

We will consider now on, in the presentation of inverse problems in heat transfer and in the remaining part of this chapter, the generic case of heat diffusion in an isotropic or anisotropic material that verifies the one-temperature model heat equation (based on Fourier's law), but its (continuous) material thermophysical properties (conductivity tensor $\bar{\bar{k}}$ and total volumic heat denoted ρc now) may vary in space (nonhomogeneous case) and possibly with temperature (thermodependent properties of the material).

1.5.1 Objective of a Model

The model-builder has a given objective: he tries to represent the real physical system by a model M that will be used to simulate its behavior. This model requires the knowledge of a given number of structural parameters that are put inside a parameter vector $\boldsymbol{\beta}$. Its objective is to get identical responses of both system $y(t)$ and model $y_{mo}(t, \boldsymbol{\beta}, u)$, under the excitation by an identical time-varying stimulus $u(t)$ (see Figure 1.13).

If the control science terminology is used, this stimulus is called « input » and the response « output ». These two terms have no geometrical meaning here.

In heat transfer, the stimulus is produced either by a source, that is, for example, a surface thermal power (absorption of a radiative incident flux by a solid wall, for example)

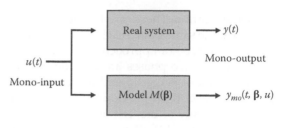

FIGURE 1.13
Real system and its representation by a model.

or by an internal power (Joule effect produced by an electrical current, heat of reaction of a chemical reaction,...). It can also be an imposed temperature difference (temperature difference between the inside and outside air environments on both sides of a solid wall, for example).

Let us note that if steady-state regime is considered, both stimulation u and responses y and y_{mo} do not vary with time.

1.5.2 State Model, Direct Problem, Internal and External Representations, Parameterizing

1.5.2.1 Example 1: Mono Input/Mono Output Case

Figure 1.14 shows a semi-infinite medium in the x-direction, whose front face ($x=0$) is stimulated by a heat flux u (W m^{-2}) at initial time $t=0$. The initial temperature distribution $T_0(x)$ may be nonuniform. A temperature sensor is embedded at a depth x_s inside the medium and delivers a signal y. So, starting at initial time, a transient 1D temperature field $T(x, t)$ develops inside the medium.

This temperature field, also called "state" of the system, is the solution of the heat equation, a partial derivative equation here, as well as of its associated boundary and initial conditions.

These equations are called *state equations* of this thermal system.

Different structural parameters appear in these equations: the medium heat conductivity k (W m^{-1} K^{-1}) and its thermal diffusivity $a=k/\rho c$ (m^2 s^{-1}), where ρ and c are its density (kg m^{-3}) and its specific heat (J kg^{-1} K^{-1}), respectively. The theoretical signal of the sensor y_{mo} (response of the model), caused by the medium stimulation u, is given by the *output equation*.

$$y_{mo}(t) = T(x_s, t) \tag{1.38}$$

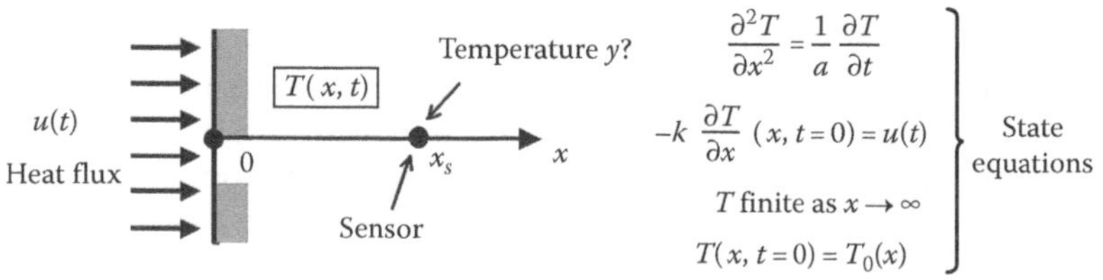

FIGURE 1.14
Model for the response of a temperature sensor embedded in a semi-infinite medium. The interrogation mark (?) designates what is looked for.

The state equations give an **internal representation** of the **direct problem** that allows the calculation of the system response everywhere, for a known excitation, while the sensor response is given by the output equation.

The state equations can be solved analytically here, and calculation of the output can be directly implemented, because the system is *causal*, *linear*, and *invariant in time* (see Ozisik [1980]):

$$y_{mo}(t) = \int_0^\infty G(x_s, x, t)T_0(x)\,dx + \int_0^t Z(t-\tau)u(\tau)\,d\tau = y_{mo\ relax}(t) + y_{mo\ forced}(t) \qquad (1.39)$$

with

$$G(x_s, x, t) = \frac{1}{2\sqrt{\pi a t}}\left[\exp\left(-\frac{(x_s - x)^2}{4at}\right) + \exp\left(-\frac{(x_s + x)^2}{4at}\right)\right] \qquad (1.40)$$

$$Z(t) = \frac{1}{b\sqrt{\pi t}}\exp\left(\frac{-x_s^2}{4at}\right) \qquad (1.41)$$

where

$G(x_s, x, t)$ is Green's function associated to relaxation, at location x_s, of the initial temperature field $T_{0(x)}$

$Z(t)$ is the transfer function of the system, while $b = (k\rho c)^{1/2}$ is the thermal effusivity of the medium

Equation 1.39 indicates that two effects overlap: the first term corresponds to relaxation of the initial temperature field (free solution that vanishes for long times) while its second term, a convolution product, corresponds to the response ("forced" solution) to the heat flux excitation. Transfer function Z that links a temperature response to an excitation power is called a time impedance, the same way as in AC electrical circuits. This function, once convoluted with the flux excitation u, yields the forced component of the temperature signal of the model. This can be expressed by a simple product of the corresponding Laplace transforms:

$$\bar{y}_{mo\ forced}(p) = \bar{Z}(p)\bar{u}(p) \quad \text{with } \bar{f}(p) = \int_0^\infty f(t)\exp(-pt)\,dt \qquad (1.42)$$

If initial temperature T_0 is uniform in the medium, the first term in $y_{mo}(t)$ in Equation 1.39 becomes equal to T_0.

This last equation constitutes an *external representation* of the direct problem. It makes calculation of the state $T(x, t)$ of the modeled system needless.

The (theoretical) output of the model depends on three parameters: the two thermophysical properties of the medium's material, a and b, and a parameter that relates to the sensor, that is, its location x_s. These three parameters can be gathered in a specific parameter vector $\boldsymbol{\beta} = [a \ b \ x_s]^T$. This parameter vector $\boldsymbol{\beta}$ contains the structural parameters is of the problem: it does not change when input $u(t)$ and/or initial state $T_0(x)$ changes.

1.5.2.1.1 Important Point on Notation
Let us precise the notation that will be adopted now on

- A scalar or a scalar function depending continuously on an other scalar or vector variable (time t or temperature T, for example) will be noted in lower or upper case italic characters (k, or $T(t, x)$, for example).
- A column vector ($\boldsymbol{\beta}$, or $\boldsymbol{u}$, or $\boldsymbol{U}$ [see Equation 1.46] further down) or a column vector function will be noted in bold lower or upper case italic characters.
- A matrix or a matrix function will be noted in bold upper case characters (matrix A or matrix function E [see Equation 1.47] further down, except if this matrix function is a standard explicit function, such as the exponential of a matrix, noted **exp**(.) here).

The previous structural parameters $\boldsymbol{\beta}$, input u, and initial state T_0 can be assembled in a unique **list** (not a column vector made of scalar quantities here) of explanatory quantities $x = \{\boldsymbol{\beta}, u(t), T_0(x)\}$, gathering all the data necessary for the calculation of output y_{mo}.
Result of this modeling is sketched in Figure 1.15.

1.5.2.2 Parameterizing a Function

In the previous list x of explanatory quantities, one can find scalar parameters (diffusivity, lengths, ...) corresponding to structural parameters, as well as a time function $u(t)$, here a heat flux. Other functions can appear such as a nonuniform initial state $T_0(P)$ or a nonuniform structural parameter $\beta(P)$ or a parameter depending on temperature $\beta(T)$.
We suppose here that such a function is a time-depending input $x = u(t)$. In order to be able to deal with this kind of function, in the simulation (direct) problem and also in the inverse problem (finding u from measured y's, where this aspect becomes of prime importance), this function has to be parameterized by its projection on a selected basis of n chosen functions $f_j(t)$:

$$u_{param}(t) = \sum_{j=1}^{n} u_j f_j(t) \tag{1.43}$$

The new function u_{param}, replaced now by a vector $\mathbf{u} = [u_1 \ u_2 \ \cdots \ u_n]^T$ of finite size n, is an approximation of the original u function that can consequently be considered as a vector with an infinite number of components. This approximation, that we will call **parameterization** now on, generates an a priori error that depends both on the chosen basis as well as on its size.

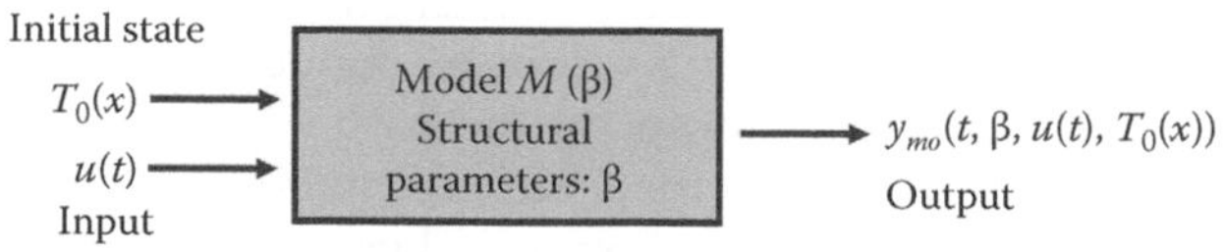

FIGURE 1.15
Input–output model for a thermal system.

Initial state
$T_0(x) \longrightarrow$ Model M (β) Structural parameters: β $\longrightarrow y_{mo}(t, \beta, u(t), T_0(x))$
$u(t) \longrightarrow$ Output
Input

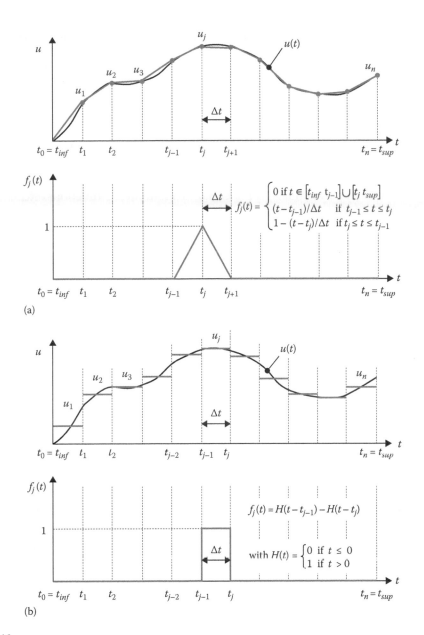

FIGURE 1.16
Two examples of function parameterization in a local basis: (a) parameterizing with a hat function basis and (b) parameterizing with a door function basis.

Figure 1.16 shows two possible choices, using a constant time step $\Delta t = t_j - t_{j-1}$:

- In case (a) the u_j components are the discrete values of the original function on the time grid and « hat » functions are selected as basis functions (see Figure 1.16a).
- In case (b) these components are averaged values of this function over one time step and « door » functions are selected for this basis (see Figure 1.16b).

The choice for the basis is not unique and strongly depends on the problem at stake.

So hat function *parameterization* of case (a) corresponds to linear interpolation using a table of discrete values; this parameterization choice is appropriate if a temperature dependency has to be modeled, for thermal conductivity $\lambda(T)$, for example. In that case, time t has to be replaced by temperature T in the basis functions that become $f_j(T)$.

In case (b), a piecewise constant function basis has been chosen. It suits deconvolution inverse problems, such as a time-varying source estimation using an experimental temperature response.

In both cases, each u_j component requires, for its calculation, knowledge of function $u(t)$ within the neighborhood of time t_j only. The use of such *local* bases is convenient because they directly derive from the time–space gridding. It is also possible to use projections on *nonlocal* bases such as polynomials, exponentials, trigonometric functions, etc.

The choice for a type of parameterization is very large. Constraints can be a priori set for the functions of the basis: they can present various properties such as monotony, regularity (continuous function with continuous first and second derivatives), and positivity, or they can be assigned fixed values on part of their time domain $[t_{inf}\ t_{sup}]$. One can also think of B-splines bases, wavelets bases. . . .

Remark

The use of orthogonal function bases is possible: they correspond to functions $f_j(t)$ such as

$$\int_{t_{inf}}^{t_{sup}} f_j(t)f_k(t)\,\mathrm{d}t = N_j\delta_{jk} \tag{1.44}$$

where

δ_{jk} is Kronecker symbol ($\delta_{jk} = 0$ if $k \neq j$ and $\delta_{jk} = 1$ otherwise)

N_j is the square of the norm of function f_j

This kind of orthogonal projection, as well as its implementation, is deeply discussed in Chapter 14.

Door functions shown in Figure 1.16b are orthogonal, but it is not the case for hat functions shown in Figure 1.16a.

It is very interesting to choose the eigenfunctions of the heat equation (found using the method of separation of variables, see Ozisik [1980] for these f_j functions). In that case, the components of the corresponding u vector become integral transforms, that is, the different harmonics, of the original function (see the book *Thermal Quadrupoles*, by Maillet et al. [2000]). This method is related to singular value decomposition (see Press et al. [1992]).

1.5.2.3 State-Space Representation for the Heat Equation

The one-temperature heat equation can be written for a thermal diffusion problem in an anisotropic medium as the following partial differential equation:

$$\mathrm{div}\left(\overline{\overline{k}}\ \mathbf{grad}\ T\right) + q_{vol} = \rho c\frac{\partial T}{\partial t} + \text{Boundary, interface and initial conditions} \tag{1.45}$$

Here, q_{vol} designates the volumic heat sources (W m^{-3}) but other surface sources may be present in the boundary or interface conditions. $\overline{\overline{k}}$ designates the conductivity tensor here.

This partial differential equation system is of the evolution type and can be considered as a dynamical system. So, its solution, the temperature field $T(P, t)$, that is, continuous in time, constitutes the state of the system, which can be noted here $T_P(t)$, that is, for a given time t, a vector in an infinite dimension space.

This system that corresponds to a *distributed parameter system* can be discretized in space, using N nodes, the discretized state becoming a vector $T(t)$ in a N dimension space. The resulting state equation of this system takes the form of a *lumped parameter system* that corresponds to a system of first ordinary differential equations:

$$\frac{dT}{dt} = E(t, T, U) \quad \text{with } T(t = t_0) = T_0 \tag{1.46}$$

where vector $U(t) = [u_1(t), u_2(t), \ldots, u_p(t)]^T$ corresponds to a local parameterization in space, but not in time, of the volumetric distributed source $q_{vol}(P, t)$ and of the other sources possibly present in the boundary or interface conditions. The number of different parameterized sources is called p here.

Let us note that this equation is written here in the very general case of a fully nonlinear system where temperature is the only state variable: conductivity or volumetric heat may depend on temperature, or the associated interface/boundary conditions may not be linear (radiative surface heat losses, for example). In that case, matrix E depends on temperature $T(t)$ in a nonlinear way. In a similar way, stimulation vector U may also be temperature dependent. In that case, each of the p components u_j of U is an implicit function of time, since it depends on the present and past states of the system, that is, on T on the $[0\ t_0]$ interval.

We assume to be in the linear case (linear heat equation system and linear source) now on

$$E(t, T, U) = AT + BU \quad \text{with } A \text{ and } B \text{: constant matrices} \tag{1.47}$$

The different vectors and matrices present in the linear form of the state equation (1.47) are thus defined in Figure 1.17.

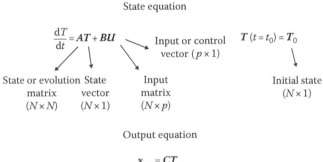

State equation

$$\frac{dT}{dt} = AT + BU$$

Input or control vector $(p \times 1)$

$T(t = t_0) = T_0$

State or evolution matrix $(N \times N)$

State vector $(N \times 1)$

Input matrix $(N \times p)$

Initial state $(N \times 1)$

Output equation

$$y_{mo} = CT$$

State vector $(N \times 1)$

Output vector $(q \times 1)$

Output matrix $(q \times N)$

FIGURE 1.17
State and output equations for a linear dynamical thermal system.

An analytical solution for the state vector $T(t)$ of this *state-space representation* of a linear system can be found formally using the exponential function of a matrix:

$$T(t) = \exp(A(t - t_0))T_0 + \int_{t_0}^{t} \exp(A(t - \tau))BU(\tau)\,\mathrm{d}\tau \tag{1.48}$$

In practice, and in the case of implementation of an inverse technique, all the N components of the state vector (temperatures at the different nodes of the model here) do not present the same interest: only a subset of it, composed of a selected number $q(q \le N)$ of its components, constitutes the model output. They can correspond to observations provided by q sensors, for example. These outputs are numbered and called $y_{mo,i}$, and they are put in an *output vector* y_{mo}:

$$y_{mo} = \begin{bmatrix} y_{mo,1} & \cdots & y_{mo,i} & \cdots & y_{mo,q} \end{bmatrix}^T \tag{1.49}$$

Output vector y_{mo} is linked to state vector T through an output matrix (or observation matrix) C, of $q \times N$ dimensions: the coefficients of this observation matrix are either 0 or 1's, according to the observed nodes:

$$y_{mo} = CT \tag{1.50}$$

This equation is also called the *output equation*.

The response of the system, which is the observed output, can be calculated thanks to Equations 1.48 and 1.50 as

$$y_{mo}(t) = C \exp(A(t - t_0))T_0 + C \int_{t_0}^{t} \exp(A(t - \tau))BU(\tau)\,\mathrm{d}\tau \tag{1.51}$$

One notices, in a very similar way as in the previous example (1.39), that this response is the sum of a term corresponding to relaxation of initial state T_0, which is the free regime, and a convolution product term corresponding to response to stimulation $U(t)$, the forced regime.

The meaning of the notion of state appears clearly here: knowledge of the state of the system at a given time $T(t_0)$ as well as the history of the different sources for the $[t_0\ t]$ time interval allows calculating the current state $T(t)$ of the material system. So, at a given time, the thermal state contains the whole past of the system.

Remark 1.1

Equation 1.45 can easily be generalized to the case of heat transport in a pure fluid:

$$\mathrm{div}(\overline{\overline{k}}\ \mathbf{grad}\ T) - \rho c_f v \cdot \mathbf{grad}\ T + q_{vol} = \rho c \frac{\partial T}{\partial t}$$

$$+ \text{boundary, interface, and initial conditions} \tag{1.52}$$

where the advection term based on the volumetric heat of the fluid $\rho c_f = \rho c$ and on the fluid velocity v (solution of the Navier–Stokes and continuity equations) has been added and where, in this case, $\overline{\overline{k}}$ reduces simply to the thermal conductivity k of the fluid.

In the case of heat dispersion in a porous medium, this velocity has to be replaced by a local Darcy velocity, temperature T becomes an average "enthalpic" temperature at the local scale (for the one-temperature model), while $\overline{\overline{k}}$ becomes the thermal dispersion tensor, whose coefficients depend on this local Darcy velocity. In this case, ρc, the volumetric heat in the transient storage term, differs now from ρc_f. This total volumetric heat ρc results from a mixing law and represents the total volumetric heat of both fluid and solid phases, using the local volume fractions as weights (see Testu et al. [2007]).

Remark 1.2

State of a thermal system is not always composed of the sole temperature T. Two different examples of a composite state are given next.

If a physical or chemical transformation occurs inside the modeled material, a polymerization of a thermoset resin, for example, heat source is produced by the heat of reaction and usually depends on the degree of advancement of the reaction, through a kinetic law. This degree of advancement constitutes the second state variable. In that case, the state equations are composed of the heat diffusion Equation 1.45 completed with a coupled mass balance equation for each of the species present in the reacting system.

Another example can be given for coupled conduction and radiation heat transfer in semitransparent media. The radiative intensity is the second state variable, and the radiative transfer equation (an integro-differential of equation) will be associated with the heat diffusion equation in order to constitute the new state equations.

Remark 1.3

When a steady-state T_{ss} corresponding to an input vector U_{ss} exists, Equation 1.46 allows its calculation: it is written with $dT/dt = 0$, which yields in the fully linear case (see Equations 1.46 and 1.47):

$$T_{ss} = -A^{-1}BU_{ss} \Rightarrow y_{mo,ss} = -CA^{-1}BU_{ss} \tag{1.53}$$

1.5.2.4 Model Terminology and Structure

All the equations and necessary conditions for calculating the output of the model constitute the structure of the model, which can be written as a functional relationship, for a single output variable:

$$y_{mo} = \eta(t, x) \tag{1.54a}$$

or

$$y_{mo} = \eta(t, \mathbf{x}) \tag{1.54b}$$

where x is either a *list* (1.54a) of explanatory quantities, including functions, $x = \{\boldsymbol{\beta}, u(P, t), T_0(P)\}$ or its vector version $\mathbf{x} = [\boldsymbol{\beta}\, U\, T_0]^T$ (1.54b), built with functions *parameterized* in space *and* time (or in temperature, for nonlinear problem with thermal dependency of either input u or structural parameters β_j s). For output variables, one deals with an output vector (not a scalar y_{mo} anymore), which requires the use of a vector function $\boldsymbol{\eta}(.)$ whose arguments are time t and either the x list or its vector version x:

$$y_{mo} = \boldsymbol{\eta}(t, x) \tag{1.55a}$$

or

$$y_{mo} = \boldsymbol{\eta}(t, x) \tag{1.55b}$$

A wider meaning can be given for vector U in this last definition of parameter vector x: this vector can represent, in a nonlinear case, a temperature-dependent stimulus $u(T)$ that has been parameterized. Let us note that a temperature-dependent thermophysical property $\beta_j(T)$, once parameterized, gives rise to constant coefficients of parameter vector $\boldsymbol{\beta}$. Coefficients of vector $\boldsymbol{\beta}$ can also stem from a space-dependent property $\beta_j(P)$ that has been parameterized in the case of a heterogeneous medium.

The "direct problem" consists in finding model output $y_{mo}(t; x)$ at a given time t in the $[t_0, t_{final}]$ interval, for known data $x = \{\boldsymbol{\beta}, u(P, t), T_0(P)\}$. Solution of this problem can allow further numerical simulations of the output behavior.

A model relies on a given *structure*, that is, a functional relationship, noted η above, between the output variable (or explained or dependent variable) y_{mo} (an observed temperature here) and the independent variable (time t for transient problems) and a parameter vector x, whose components are the parameterized explanatory quantities. It is important to remind that aside the previous structure, parameters x of the model should be defined accordingly (see Figure 1.18). They can either have a physical meaning if a state modeling is performed or simply a mathematical meaning without clear physical interpretation if an identified modelization is implemented.

One can notice that a model, in case of a single output, can provide not only a scalar output y_{mo} depending continuously on time t but also a vector output y_{mo}. This output column vector y_{mo} is associated with the same output variable, a local temperature, for example, sampled at different times $t_1, t_2, \ldots, t_m$, or can result from a sampling of the explanatory variable that can be a space coordinate for a steady-state problem. It can

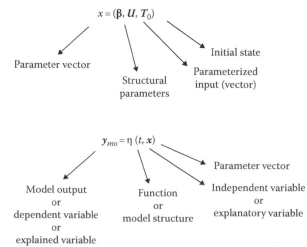

FIGURE 1.18
Parameter vector and structure of a model.

also gather in a single column vector, of length qm, several output temperatures observed at different points P_i ($i = 1$ to q), sampled for m different times t_k.

Let us note here that a general introduction to inverse problems is proposed in Chapter 7, and general methods and skills for their solution are discussed in Part II.

1.5.3 Direct and Inverse Problems

1.5.3.1 Direct Problem

We have seen above that when the studied problem allows it, the usual approach of the thermal science scientist consists in constructing a **knowledge-based model**, such as Equation 1.45, in order to be able to simulate the behavior of the physical system.

This leads to a numerical or analytical solution of a partial differential equation in the case of a heat diffusion problem (or an integro-differential system of equations for radiation heat transfer in semitransparent media, temperature, and radiative intensity being the state variables) that represents the corresponding transfer of heat. The solution of these equations also requires the knowledge of the conditions at the boundaries (Dirichlet, Neumann, Fourier, etc.) or at the internal interfaces (for a medium composed of different materials) as well as the initial condition in the system.

If an internal representation is adopted, several quantities of different nature have to be introduced in the state (1.45) and output equations $y_{mo}(t) = T(P_i, t)$ of the model, written for a single temperature sensor located at point P_i. If the output is observed at q such points for m times that constitute a time vector $t = [t_1 \ t_2 \ \cdots \ t_m]^T$, it becomes an output vector $y_{mo}(t; x)$ that depends also on parameter vector x, where this vector is composed of

- The raw $u(P, t)$ or parameterized $U(t)$ excitation
- Vector β_{struct} of structural parameters, a and b in Example 1 or coefficients of matrices A and B in the linear state equations (1.46) and (1.47)
- Vector β_{pos} describing the position of the observation, x_s in Example 1 and coefficients of matrix C in output equation (1.50)
- The initial temperature field $T_0(P)$ or its parameterized version T_0

Input variables $u(P, t)$ are controlled by the user: they are either power sources or imposed temperature differences, inside or outside the system, that make temperature and output depart from a zero value in case of zero initial temperature $T_0(P)$.

Structural parameters β_{struct} characterize the system. They can be

- Geometrical quantities (shape and dimensions of the system)
- Thermophysical properties: conductivities, volumetric heat capacities, heat transfer coefficients, emissivities, contact or interface resistances, etc.

The relationship between output variables, generally a subset of the state, and state variables, the temperature field, makes the previous position parameter vector β_{pos} appear in this output equation.

A functional scheme corresponding to linear state and output heat equations is shown in the lower line of Figure 1.19.

This corresponds to the usual process of a model user: for a known initial state $T(t_0)$, a known excitation $U(t)$, and known structural parameters, the heat equation and the output

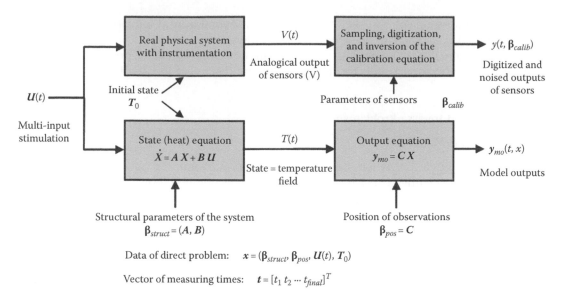

FIGURE 1.19
Linear model and material system with temperature measurement.

equations are solved sequentially to calculate the theoretical response y_{mo} of the sensors. This output corresponds to a possible real temperature measurement at the same locations (upper line in Figure 1.19). The **direct problem** can thus be solved.

1.5.3.2 Inverse Problem Approach

The preceding analysis shows that any variation in the data represented inside the x vector (including structural and position parameters β_{struct} and β_{pos}) will produce a variation of the y_{mo} output.

Conversely, any variation of this output y_{mo} is necessarily caused by variation of some data inside x.

The inverse approach is based on this principle. When knowledge of part of the variables that are necessary to solve the direct problem is lacking, data vector x of this problem can be split into two vectors the following way:

$$x = \begin{bmatrix} x_r \\ x_c \end{bmatrix} \tag{1.56}$$

where
  x_r now represents the (column) vector gathering the unknown part of the data that are
    sought (*researched*)
  x_c is its *complementary* part that contains *known* data

In that case, solving the direct problem constitutes an impossible task. Any process aimed at finding x_r requires some *additional information*.

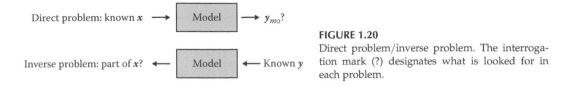

FIGURE 1.20
Direct problem/inverse problem. The interrogation mark (?) designates what is looked for in each problem.

Problems whose objective is to find a value for x, starting from additional information, are called *inverse problems*.

Any inverse problem consists in making the model work in the « backwards » way: if outputs y as well as model structure η are known, part x_r of x will be sought, its complementary part being known (see Figure 1.20).

A general introduction to inverse problems is proposed in Chapter 7, and general methods and skills are discussed further in Sections 7.2 and 7.3 of the same chapter.

1.5.3.3 Inverse Problems in Heat Transfer

1.5.3.3.1 Different Types of Inverse Problems in Heat Transfer
The nature of additional information necessary for solving the inverse problem allows bringing out three main types of problems:

1. *Inverse measurement problems*, where this information stems from output signal y of sensors.
2. *Control problems*, where the previous measurements are replaced by desired values of either the state $T(P, t)$ or output variables y: data or y are the targets. In this class of problem, the sought quantity is generally the stimulus $u(P, t)$ or the initial state $T_0(P)$, but it can also be a structural parameter (a velocity or a flow rate in a forced convection cooling problem, for example). In this class of problems, it is not always possible to reach the targets, for physical or mathematical reasons, and it may be necessary to specify a certain number of constraints on the sought solution.
3. *System identification problems*, that is, model construction for simulating the behavior of a system (see Chapters 13 and 14). These can be classified into two categories:
   a. *Model reduction*: y is the output of a detailed model $\eta_{det}(t; x_{det})$ completely known, and the structural parameters (part of x_{red}) of a reduced model $\eta_{red}(t; x_{red})$ of given structure η_{red} are sought, both models sharing either identical or close stimulations $u(P, t)$ and initial state $T_0(P)$ that are parts of x_{det} and x_{red}. This can be written as follows:

$$\eta_{det}(t; x_{det}) \approx \eta_{red}(t; x_{red}) \quad \text{where } x_{det} = [\boldsymbol{\beta}_{det} \quad \boldsymbol{U}_{det} \quad \boldsymbol{T}_{0\,det}]^T$$
$$\text{and} \quad x_{red} = [\boldsymbol{\beta}_{red} \quad \boldsymbol{U}_{red} \quad \boldsymbol{T}_{0\,red}]^T \tag{1.57}$$

with, for mathematical reduction:

$$u_{red}(P, t) = u_{det}(P, t) \Rightarrow \boldsymbol{U}_{red} = \boldsymbol{U}_{det}$$
$$T_{0\,red}(P, t) = T_{0\,det}(P, t) \Rightarrow \boldsymbol{T}_{0\,det} = \boldsymbol{T}_{0\,red} \tag{1.58}$$

or, for physical reduction:

$$u_{red}(P, t) \approx u_{det}(P, t) \Rightarrow U_{red} = f_U(U_{det})$$
$$T_{0\,red}(P, t) \approx T_{0\,det}(P, t) \Rightarrow T_{0\,red} = f_{T0}(T_{0\,det}) \tag{1.59}$$

In both cases, *mathematical* or *physical model reduction*, the structural parameters of the reduced model depend on the corresponding parameters of the detailed model:

$$\boldsymbol{\beta}_{red} = f_\alpha(\boldsymbol{\beta}_{det}) \quad \text{for } \alpha = u \text{ or } T_0 \tag{1.60}$$

but this relationship, function f_α, is explicit for physical reduction (see Section 1.6), while it is not generally the case for mathematical reduction.

b. *Experimental model identification*: y, U, and T_0 are measured, or supposed to be known, and the structural parameters (part of x) of a model $\eta(t; x)$ of given structure η are known, U and T_0 being their complementary part in x.

Let us note that system identification leads to models that can be of the *white box* type, which means models based on first principles, for example, a model for a physical process from the Newton's equations. The previous state-space model (1.46), based on a heat balance and on Fourier's law defining heat flux, belongs to this category. The nature of the parameters in this class of models is perfectly known, which explains why they are used for thermophysical property estimation. Conversely, an identified model on an experimental basis, without a priori information on its structure, is also called a *black box* model: parameters of such a model have only a mathematical, but not physical, meaning. Such black box models may, for example, derive from neural network modeling. In between, one can find *gray box* or *semi-physical* models: the model, that is, the structure/parameter couple, is chosen according to a certain physical insight on what is happening inside the system, and these parameters are estimated on an experimental basis.

1.5.3.3.2 Inverse Measurement Problems in Heat Transfer

We will now focus on *inverse measurement problems* where model structure (the equations) η is known and where measurements $y(t)$ are available on the time interval $[t_0, t_{final}]$.

According to the nature of the explanatory variables x_r that are sought, solution finding for inverse problems may differ. One can distinguish in particular

1. Inverse problems of *structural parameters estimation*: $x_r \equiv \boldsymbol{\beta}_r$
   System identification problems, of the black or gray box type, belong to this category: structural parameters (part of x) of an ad hoc $\eta(t; x)$ model are sought through experimental characterization. *Thermophysical property estimation* belongs to the white box category: *intrinsic* parameters, that is, parameters that can be used for completely different simulation/experimental configurations are sought through experimental characterization. In both types of problems, several experiments on the same setup, for the same sample, can be repeated in order to estimate the same unknown parameter(s).

2. Inverse *input* problems: $x_r \equiv u(P, t)$
   In heat transfer, this type of problem consists in finding the locations and values of the sources. Such a source, or excitation, is either a volumetric, surface, line, or

point heat source or simply a temperature difference imposed inside or at the boundaries of the system. It differs from the previous problem because the solutions sought are specific to each experiment made.

3. Inverse *initial state* problem: $x_r \equiv T_0(\mathrm{P})$
   This problem is very close to the inverse input problem, since each sought solution is relative to a single given experiment.

4. Inverse *shape reconstruction* problems
   In the previous types of inverse problems, boundaries of the domain are usually fixed and known. In certain cases (problems with change of phase, in welding or in solidification applications, for example) shape of the domain (its boundary) or location of an interface between sub-domains (a change of phase moving front, for example) has to be taken into account in the variables defining the direct problem. In the corresponding inverse problem, the shape of this boundary has to be first parameterized In order to reconstruct it through inversion.

5. Inverse problems of *optimal design/control*
   A usual process aimed at reducing estimation errors, in a characterization process of type (1), consists in coupling it to an optimal conception/control problem for the characterization experiment. This optimization allows the design and the sizing of the experimental setup as well as the procedure for the trials that will bring additional information necessary for this characterization. This approach can provide a methodology for a pertinent choice of inputs, locations of measurement points, time observation windows, etc. The choice of these design quantities can be made in order to maximize a criterion based on the sensitivity of the output observations to the parameters that are sought. In heat transfer, characterization problems (that are structural parameter estimation or system identification problems) are usually nonlinear, which means that optimization of any design has to be implemented on the prior assumption that the sought parameters are known, with an iterative approach, once a first estimation has been found. This means that *nominal* values of these parameters are necessary for such a design.

Remark

The use of any sensor that very often delivers an electrical output quantity (a tension V, for example) requires the construction of a relationship between the quantity one wants to measure, temperature T here, and this instrument output.

It is therefore necessary to find, on the basis of the physical principle the sensor and the whole instrumental chain rely on, a model structure $V_{mo}(T; \beta_{calib})$ where temperature is now the explanatory variable and where vector β_{calib} gathers all the parameters required for calculating the theoretical output temperature signal (thermoelectric power and cold junction temperature, in the case of a thermocouple sensor). Construction of the V_{mo} model and estimation of parameters present in β_{calib} starting from simultaneous measurements of both V and T (using a reference temperature sensor) constitute a *calibration problem*, that is, by nature, a parameter estimation problem, that is a type (1) inverse problem (see section above) that has to be dealt with this way.

1.5.3.4 Measurement and Noise

In *inverse measurements problems*, the additional information is brought by the measured output that differs from the model output y_{mo}.

The difference $\varepsilon(t)$ between a sensor measurement y and the output of an ideal sensor y^* giving the true temperature at the sensor location can be introduced:

$$y(t) = y^*(t) + \varepsilon(t) \tag{1.61}$$

The sensor giving y^* is ideal for two reasons: (1) its presence does not affect the local temperature of the medium (*non-intrusive* detector) and (2) it provides the *true value* of its *own* temperature.

Equation 1.61 defines the measurement noise $\varepsilon(t)$ that can be considered as a random variable caused by the imperfect character of both instrumentation and of digitization of the signal. This noise is present, but its deterministic value can not be reached in practice.

This equation also shows that the measured signal is a random variable whose variance is the same as noise ε.

The assumption of a pertinent, that is, non-biased, model is made in practice:

$$y^*(t) = y_{mo}(t, x^*) \tag{1.62}$$

where x^* is the true value of the explanatory variables.

Verifying this assumption of consistency between model and measurements is crucial. Corresponding tools exist (study of the residuals).

Remark

Form (1.61) should be defined for discrete values $y_i = y(t_i)$, $\varepsilon_i = \varepsilon(t_i)$, and $y_i^* = y^*(t_i)$ corresponding to the sampling times t_i of the measured signal, of the exact temperature, and of the noise, respectively.

1.6 Choice of a Model

1.6.1 Objectives, Structure, Consistency, Complexity, and Parsimony

Before constructing a model, the model-builder has to be clear about the way his model will be used, that is, about the objective of such a modeling. The objectives depend on the application and can belong to one of the following categories that can be listed in a non-limitative way:

- Estimation of thermophysical properties
- Heat source/flux estimation
- Initial temperature field estimation
- Defect detection and nondestructive testing
- Simulation of the system behavior for better design or future state forecasting
- Model reduction for faster computation or use for heat source/flux estimation
- Conception of a model for closed-loop (feedback) control

So the type of model will not be the same for each application, because the required model precision will differ: defect detection in a composite slab using infrared thermography (Maillet et al., 1993) does not require a model with the same temperature resolution as in thermophysical property estimation, such as the flash method for liquid diffusivity estimation (Rémy and Degiovanni, 2005).

The accuracy of a model is determined by its consistency with the physical situation modelized, that is, its ability to simulate closely the behavior of the studied system. *Internal representation*, with the use of state-space models, should be generally favored, because it provides a mathematical structure linked to the physics of the modeled problem « for free ». In addition, this type of representation allows highlighting the intrinsic parameters of the system, that is, its thermophysical properties or thermal resistances and impedances.

The purpose of the model that is used for inverting measurements is not to reproduce or to mimic the whole temperature field: it should only provide an output that can be compared to the sensor output signal at the location where this one is embedded. Structure, that is, scalar or vector function η used above, is what defines a model. Its complexity should be adapted to the uncertainties associated with any description of a physical system: the use of a model that is too much simplified (simple structure with a low number of structural parameters, such as a *lumped parameter* model see Section 1.6.2) can introduce a systematic error, a bias, in its output variables, that could depart too much from model predictions and from the experimental observations to be used the inverse way. Conversely, the choice of a too-*detailed* model, with a high number of parameters

- Tends to make implementation of the inversion algorithm involved or to make it numerically impossible or very difficult.
- May lead to unstable solutions for the inverse problem, because of noise amplification (in case of inversion of measurements): the inverse problem becomes ill-posed.

This dilemma pleads in favor of the purpose of *parsimonious* models for inverse use, that is, models that provide a good balance between antagonist criteria of the use of a minimum number of parameters on the one hand and maximum agreement with reality (fidelity to measurements) on the other hand.

Up-to-date capacities of numerical simulation tools as well as structure of the *optimization* and *regularization* algorithms allow solving inverse problems with more and more complex models, using *mathematical model reduction* techniques. These allow a very significant reduction of the size of the state vector (temperature at different nodes of the numerical grid here). So reduction of a model, followed by its implementation in an inverse procedure, can bring an efficient approach for the most difficult cases, such as 3D heat transfer with change of phase or advecto-diffusive transfers within flowing fluids, for example (Girault et al., 2008). We will now focus on a different type of reduction technique, *physical model reduction*.

1.6.2 Example 2: Physical Model Reduction

In order to show that a thermal model can be reduced on a physical basis and that many models of different complexity and resolution are available to simulate the same heat

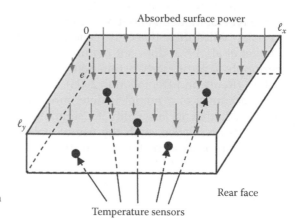

FIGURE 1.21
Model for temperature response of a slab heated on
one of its faces.

transfer situation (nonuniqueness of a model), we will consider heat transfer in a slab, whose characteristics are defined as follows:

- Homogeneous rectangular slab, thickness e, lengths ℓ_x and ℓ_y in its plane
- Thermal diffusivity and conductivity a and k, respectively, volumetric heat $\rho c = k/a$

This slab is stimulated by a surface power (absorption of solar radiation, for example) on its front face, and temperature is measured at q points by sensors either embedded in the material or located on the front or rear face of the slab (see Figure 1.21). The slab is supposed to be insulated on its four (lateral) sides and exchanges heat with the surrounding environment T_∞ only on its rear face through a uniform heat transfer coefficient h that represents its losses (convection and linearized radiative losses). Its initial temperature T_0, at time $t = 0$, when heating starts, is supposed to be uniform.

A model allowing to find the temperature response $y_{mo,i}(t)$ of sensor number i ($i = 1$ to q) at time t is sought.

1.6.2.1 3D Model

Heat source $u(x, y, t)$ (W m^{-2}) is supposed to be nonuniform at the front face. Evolution with time of the temperature field can be described by a three-dimension transient model (see Figure 1.22a):

$$\frac{\partial^2 T}{\partial x^2} + \frac{\partial^2 T}{\partial y^2} + \frac{\partial^2 T}{\partial z^2} = \frac{1}{a}\frac{\partial T}{\partial t} \tag{1.63}$$

$$T = T_0 \quad \text{for } t = 0 \tag{1.64}$$

$$\frac{\partial T}{\partial x} = 0 \quad \text{at } x = 0, \ell_x; \quad \frac{\partial T}{\partial y} = 0 \quad \text{at } y = 0, \ell_y \tag{1.65}$$

$$-k\frac{\partial T}{\partial z} = u(x, y, t) \quad \text{at } z = 0; \quad -k\frac{\partial T}{\partial z} = h(T - T_\infty) \quad \text{at } z = e \tag{1.66}$$

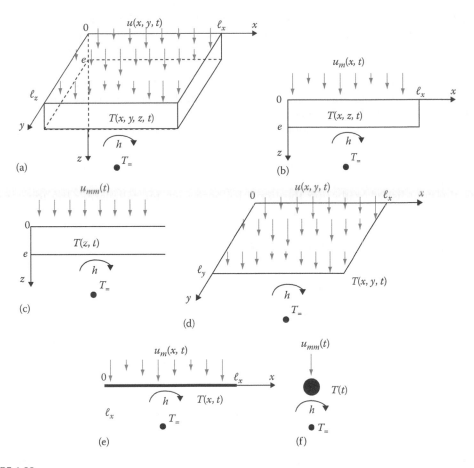

FIGURE 1.22
"Physical" model reduction. (a) 3D model, (b) 2D model, (c) 1D model, (d) 2D fin model, (e) 1D fin model, (f) 0D lumped model: « small » body.

This system of eight equations constitutes model M_a that will be called "detailed model", whose solution, noted $T - T_a$ here, determines the response of each sensor:

$$y_{mo,i} = \eta_i(t,x) = T_a(x_i, y_i, z_i, t; u(x,y,t), T_0, T_\infty, h, \ell_x, \ell_y, e, \lambda, a) \qquad (1.67)$$

In this equation, u, T_0, and T_∞ are input quantities of the model, independent from the structure of the material system (if they are all equal to zero, temperature stays to a zero level everywhere in the slab), while the other quantities are the structural parameters β, either linked to geometry (ℓ_x, ℓ_y, e), or to the thermophysical properties (k, a) of the slab material and to its coupling with the outside environment (h), or linked to the location of the sensors (x_i, y_i, z_i, for $i = 1$ to q).

List $x = \{\beta, u, T_0, T_\infty\}$ can be introduced now. It gathers *structural parameters* β, *inputs* u and T_∞, and initial state T_0 of this dynamical system composed of $(3q + 9)$ quantities.

1.6.2.1.1 Dimensionless 3D Model
The number of quantities present in Equations 1.63 to 1.66 can be reduced if they are written in a dimensionless form: dimensionless temperature $T^* = (T - T_\infty)/\Delta T$ appears,

with $\Delta T = T_0 - T_\infty$, and it is the same for dimensionless time, Fourier number $t^* = t/\tau_{diff}$, and dimensionless heat transfer coefficient, Biot number $H = he/k$. In a similar way, dimensionless observation locations $x_i^* = x_i/e$, $y_i^* = y_i/e$, $z_i^* = z_i/e$ and dimensions $\ell_x^* = \ell_x/e$ and $\ell_y^* = \ell_y/e$ are introduced.

Here, $\tau_{diff} = e^2/a$ is the characteristic time, related to the duration of thermal diffusion in the thickness of the slab. The resistance of the slab in the thickness direction, related to a unit area, $R = e/k$, can be introduced.

This new model M_a^* that corresponds to the same response of the sensors becomes

$$y_{mo,i} = \eta^*(t, x^*) = \Delta T \cdot T^*\left(x_i^*, y_i^*, z_i^*, t/\tau_{diff}, R, u(x,y,t)/\Delta T, H, \ell_x^*, \ell_y^*\right) + T_\infty \qquad (1.68)$$

where the new list x^*, gathering the variables necessary for calculating the temperature response at a given time t, comprises one less parameters than the original x list (1.67):

$$x^* = \{\boldsymbol{\beta}^*, u, \Delta T, T_\infty\} \quad \text{with } \boldsymbol{\beta}^* = ((x_i^*, y_i^*, z_i^*) \quad \text{for } i = 1 \text{ to } q), \tau_{diff}, R, H, \ell_x^*, \ell_y^*) \qquad (1.69)$$

1.6.2.2 2D Model in X- and Z-Directions

Model M_a can be simplified: if one knows that stimulus u does not vary much in direction y, or if the sensor whose response has to be simulated is not a point sensor but integrates the temperature signal in this direction, a y-direction average temperature field T_b can be rebuilt, with the definition of a new model M_b (see Figure 1.22b):

$$T_b(x, z, t) = \frac{1}{\ell_y} \int_0^{\ell_y} T_a(x, y, z, t) \, dy \qquad (1.70)$$

This 2D temperature field is produced by a source that varies in one single space direction, instead of two previously. This new source $u_m(x, t)$ does not depend on y and, as temperature, is the mean, in this direction, of the previous stimulus:

$$u_m(x, t) = \frac{1}{\ell_y} \int_0^{\ell_y} u(x, y, t) \, dy \qquad (1.71)$$

This mean temperature field verifies the following equations:

$$\frac{\partial^2 T}{\partial x^2} + \frac{\partial^2 T}{\partial z^2} = \frac{1}{a} \frac{\partial T}{\partial t} \qquad (1.72)$$

$$T = T_0 \quad \text{at } t = 0, \quad \frac{\partial T}{\partial x} = 0 \quad \text{in } x = 0, \ell_x \qquad (1.73)$$

$$-k\frac{\partial T}{\partial z} = u_m(x, t) \quad \text{at } z = 0; \quad -k\frac{\partial T}{\partial z} = h(T - T_\infty) \quad \text{at } z = e \qquad (1.74)$$

Once put in a dimensionless form, this M_b model comprises $(2q + 7)$ independent variables:

$$x = \{\boldsymbol{\beta}, u_m, \Delta T, T_\infty\} \quad \text{with } \boldsymbol{\beta} = \left((x_i^*, z_i^* \quad \text{for } i = 1 \text{ to } q), \tau_{diff}, R, H, \ell_x^*\right) \qquad (1.75)$$

Let us note now that in order for this model to show really no bias for sensor i, this detector should not be a point sensor, but a line sensor.

This is possible if the rear face $(z_i^* = 1)$ temperature field is measured by infrared thermography. In that case, output of model M_b at location (x_i, y_i) is

$$y_{mod,i}(t_k) = T_b(x_i, z_i = e, t_k) \tag{1.76}$$

Its experimental counterpart can be scrutinized: one notes now $T_k^{exp}(x^m, y^j)$ the temperature signal at time t_k, for pixel (x^m, y^j) of the infrared frame, where (m, j) designates a pixel located in the mth line and jth column.

The output (y-averaged) temperatures of the model have to be compared with the corresponding experimental response $y_i(t_k)$ of the ith detector: this can be obtained through simple addition:

$$y_i(t_k) = \frac{1}{n_i} \sum_{j=1}^{n_i} T_k^{exp}(x^m, y^j = y_i) \tag{1.77}$$

where n_i is the number of pixels in the ith column (constant x^m). The reader should not be confused by the present notation in Equation 1.77: $y_i(t_k)$ is the experimental temperature signal of the ith detector, while y_i is its coordinate, in the y-direction.

If the average temperature in the y-direction is really measured by a line sensor, there will be no model error in the estimation of $u_m(x, t)$. However, the information on the variation of u in the y-direction is lost by this reduced modeling, which means that the description of u will be made with no resolution in this direction: people in charge of this estimation would have therefore to reduce also their initial objective, that is, estimation of $u_m(x, t)$ instead of $u(x, y, t)$.

1.6.2.3 1D Model in Z-Direction

Such an averaging can be pursued if one considers now the averaged value of the source over the whole front face area. The same type of averaging is made for the temperature field. This leads to model M_c, shown in Figure 1.22c:

$$u_{mm}(t) = \frac{1}{\ell_x} \int_0^{\ell_x} u_m(x, t) \, dx \tag{1.78}$$

$$T_c(z, t) = \frac{1}{\ell_x} \int_0^{\ell_x} T_b(x, z, t) \, dx \tag{1.79}$$

$$\frac{\partial^2 T}{\partial z^2} = \frac{1}{a} \frac{\partial T}{\partial t} \tag{1.80}$$

$$T = T_0 \quad \text{for } t = 0 \tag{1.81}$$

$$-k \frac{\partial T}{\partial z} = u_{mm}(t) \quad \text{at } z = 0; \quad -k \frac{\partial T}{\partial z} = h(T - T_\infty) \quad \text{at } z = e \tag{1.82}$$

Once model M_c is put in a dimensionless form, only $(q+6)$ independent variables remain in the x list:

$$x = \{\boldsymbol{\beta}, u_{mm}, \Delta T, T_\infty\} \quad \text{with } \boldsymbol{\beta} = \left((z_i^*, \quad \text{for } i = 1 \text{ to } q), \tau_{diff}, R, H\right) \tag{1.83}$$

This reduction in the number of variables is made at the expense of the space resolution for u that is completely lost here since it is replaced by its space average u_{mm}.

1.6.2.4 2D Fin Model in X- and Y-Directions

If the Biot number $H = he/k$ is much lower than unity, temperature variations in the z-direction, corresponding to the slab thickness, can be considered as negligible and, consequently, heat transfer in the slab becomes two-dimensional (2D). The resulting 2D temperature field stems from an integration, with respect to z, of the 3D temperature field (see Figure 1.22d):

$$T_d(x, z, t) = \frac{1}{e} \int_0^e T_a(x, y, z, t) \, dz \tag{1.84}$$

This reduced model M_d corresponds to a 2D fin whose temperature verifies the following equations:

$$\frac{\partial^2 T}{\partial x^2} + \frac{\partial^2 T}{\partial y^2} - \frac{h(T - T_\infty)}{ke} + \frac{u(x, y, t)}{ke} = \frac{1}{a} \frac{\partial T}{\partial t} \tag{1.85}$$

$$T = T_0 \quad \text{at } t = 0 \tag{1.86}$$

$$\frac{\partial T}{\partial x} = 0 \quad \text{in } x = 0, \ell_x; \quad \frac{\partial T}{\partial y} = 0 \quad \text{in } y = 0, \ell_y \tag{1.87}$$

List x is now composed of $(2q+8)$ independent variables:

$$x = \{\boldsymbol{\beta}, u, \Delta T, T_\infty\} \quad \text{with } \boldsymbol{\beta} = \left((x_i^*, y_i^* \text{ for } i = 1 \text{ to } q), \tau_{diff}, R, H, \ell_x^*, \ell_y^*\right) \tag{1.88}$$

This relatively high number of variables allows however to keep the initial spatial resolution of stimulus u.

1.6.2.5 1D Fin Model in X-Direction

The 2D-reduced model M_b can be used now to construct a 1D fin model, noted M_e, with the same condition on the Biot number H, through an integration in the z-direction (the same model M_e can be obtained through integration of model M_d in y-direction [see Figure 1.22e]):

$$T_e(x, t) = \frac{1}{e} \int_0^e T_b(x, z, t) \, dz \tag{1.89}$$

$$\frac{\partial^2 T}{\partial x^2} - \frac{h(T - T_\infty)}{ke} + \frac{u(x, y, t)}{ke} = \frac{1}{a}\frac{\partial T}{\partial t} \tag{1.90}$$

$$T = T_0 \quad \text{at } t = 0 \tag{1.91}$$

$$\frac{\partial T}{\partial x} = 0 \quad \text{in } x = 0, \ell_x \tag{1.92}$$

List x of the independent variables of the model is composed of $(q + 7)$ quantities:

$$x = \{\boldsymbol{\beta}, u_m, \Delta T, T_\infty\} \quad \text{with } \boldsymbol{\beta} = ((z_i^* \quad \text{for } i = 1 \text{ to } q), \tau_{diff}, R, H, \ell_x^*) \tag{1.93}$$

1.6.2.6 0D Lumped Model

If the source is nearly uniform in space, with a low Biot number in direction z, or if the sensor provides the volume-averaged temperature of the slab, one obtains a 0D M_f model, also called lumped model or « small body » model. It corresponds to integration of model M_e in x-direction (see Figure 1.22f):

$$T_f(t) = \frac{1}{\ell_x} \int_0^e T_e(x, t)\, dx \tag{1.94}$$

This temperature field is produced by a point source whose intensity $u_{mm}(t)$ varies with time, with

$$u_{mm}(t) = \frac{1}{\ell_x} \int_0^{\ell_x} u_m(x, t)\, dx \tag{1.95}$$

The heat equation becomes

$$\rho ce \frac{dT}{dt} + h(T - T_\infty) = u_{mm}(t) \tag{1.96}$$

The x list of this model is now composed of only five independent variables, including a convective resistance (based on a unit area) $G = 1/h$ and a time constant $\tau = \rho ce/h = \tau_{diff}/H$:

$$x = \{\boldsymbol{\beta}, u_{mm}, \Delta T, T_\infty\} \quad \text{with } \boldsymbol{\beta} = (\tau, G) \quad \text{and} \quad \Delta T = T_0 - T_\infty \tag{1.97}$$

An analytical solution can easily be found:

$$T = T_\infty + \Delta T \exp\left(\frac{-t}{\tau}\right) + \frac{G}{\tau} \int_0^t u_{mm}(t') \exp\left(-\frac{t - t'}{\tau}\right) dt' \tag{1.98}$$

This model is a limit model, only valid if the Biot number, based on the largest of the three dimensions ℓ_x, ℓ_y or e, is much lower than unity. If not, it is a biased model, but its output T_f can always be compared to the average temperature of the q sensors. This averaged experimental temperature brings an interesting information on the time variation of the average absorbed power density on the front face, $u_{mm}(t)$.

1.6.2.7 1D Local Model

A last model, noted M_g here, can be used. It is a 1D « local » temperature defined by

$$y_{mo,i} = T_g(x_i, y_i, t) = T_c(z_i, t; u(x_i, y_i, t), \Delta T, T_\infty, \boldsymbol{\beta}) \tag{1.99}$$

with

$$\boldsymbol{\beta}_i = (z_i^*, \tau_{diff\,i}, R_i, H_i) \tag{1.100}$$

It corresponds to the previous 1D model M_c, applied locally for each sensor. Its response depends on the sole excitation $u(x_i, y_i, t)$ that prevails on the front face at the same (x, y) location (see Figure 1.23).

This allows considering a 3D problem as a set of independent 1D problems, each individual problem being associated to a specific sensor. Structural parameters belonging to vector $\boldsymbol{\beta}_i$ differ for each sensor. This vector is composed of a diffusion characteristic time $\tau_{diff\,i}$, a resistance R_i, and a Biot number H_i that have all local values corresponding to location of sensor i. These structural parameters are related to local thickness e_i, local heat transfer coefficient h_i, and local conductivity k_i and diffusivity a_i.

For the whole set of sensors, this model is composed of $(q + 6)$ independent variables if these sensors are embedded at the same depth in the slab and if the thermophysical parameters, h, and the slab thickness do not vary in the x–y plane.

This model is valid only if heat transfer is negligible in the directions of this same plane, that is, if the slab is made of a composite material that is homogenized but anisotropic: the principal directions of conductivity tensor $\bar{\bar{k}}$ should be those of the slab, with principal

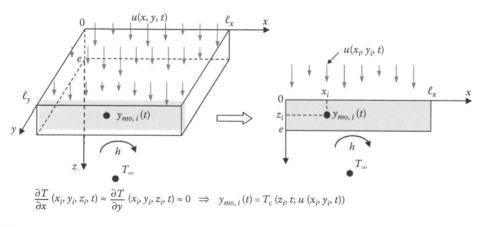

$$\frac{\partial T}{\partial x}(x_i, y_i, z_i, t) \approx \frac{\partial T}{\partial y}(x_i, y_i, z_i, t) \approx 0 \quad \Rightarrow \quad y_{mo,i}(t) = T_c(z_i, t; u(x_i, y_i, t))$$

FIGURE 1.23
1D local model M_g.

components $k_x = k_y = 0$, $k_z = k$. However, it is possible to use it with a reasonable bias for sensors facing front face locations where stimulus u does not vary much (low gradient in the plane of this face) and for low thickness and thermophysical local variations. This model is also very interesting in nondestructive testing of composite slabs by infrared thermography (see Benítez et al. [2008]).

Remarks

- The six reduced models M_b to M_g are all derived from the detailed model M_a and have lower order dimensions than this original 3D model. They are also characterized by a lower number of structural parameters (see Chapters 13 and 14 for more details concerning the model reduction).

- Structural parameters of the slab and of the sensors either disappear or are transferred from one model to a more reduced one along this progressive physical reduction process. So, passing from model M_e to model M_f makes parameter ℓ_x^*, R, and x_i^* disappear while parameters H and τ_{diff} merge into a single parameter $\tau = \tau_{diff}/H$. This reduction of the parameters number is an irreversible one, which means that it is not possible to rebuild values of H and τ_{diff} starting from the knowledge of τ only.

- One can also note that during this reduction process, relationships between former and new parameters are linear if the logarithms of these parameters are considered: $\ln(\tau) = \ln(\tau_{diff}) - \ln(H)$. This gives an interesting relationship between reduced sensitivities (see the corresponding course in this series).

- In parallel with the reduction in the number of parameters, a reduction of the space dimension necessary for reproducing the sensor behavior appears: from an initial $u(x, y, t)$ stimulus for models M_a and M_d, one gets a $u_m(x, t)$ stimulus for models M_b and M_e to finally $u_{mm}(t)$ for models M_c and M_f and $u_i(t) = u(x_i, y_i, t)$ for model M_g.

- All these models rely on specific physical assumptions, and none of them corresponds to the absolute reality, even model M_a: this one neglects convecto-radiative losses on the front face and on the four sides of the slab, coefficient h is supposed to be uniform in the rear face plane, and the same is true for the initial temperature inside the slab.

This example shows that the user has to make his or her own choice for the model, since several representations are generally possible. Accordingly, a more reduced model conveys less information about the spatial distribution of the heat source. However, this inconvenience in direct modeling can become an asset when inversion to reconstruct the source takes place.

1.6.3 Linear Input–Output Systems and Heat Sources

This section is devoted to the definition of what can be considered as a thermal power stimulus u. It can be later used for the purpose of estimating u, in an inversion procedure.

It has been shown above, for two geometries, semi-infinite medium (Example 1), and plane wall (Example 2), that the system-forced response u, to a surface heat flux stimulus, can be written, for any point P inside the medium, as a convolution product in time (see Equations 1.39 and 1.96), with a degenerate lumped body model in the second case.

In a very similar way, a continuous stimulus, that is, a power volumetric density, $u(P, t)$, once discretized in space (or, more generally parameterized, using any basis of functions in 3D) as an input vector $U(t)$, yields a forced response, in any point of the system, that corresponds to a convolution product in time, if the heat equation as well as its associated conditions are linear, with coefficients that can vary in space (nonhomogeneous system), but not in time (time invariant system). Let us notice that we consider only, in this section, linear heat sources, that is, sources that do not depend on state, here temperature, in the system.

This very general result can be applied to such a system in the specific case of a stimulus u whose time dependency can be separated from its space dependency:

$$u(P, t) = f(t)g(P) \tag{1.101}$$

We assume here that the source intensity (W m^{-3}) is associated to its time component $f(t)$, while its distribution in space $g(P)$ is its characteristic function (no unit): its value is 1, if point P belongs to the source and zero otherwise.

If the model is linear (in terms of the input/output relationship) and if its coefficients do not vary with time, model response $y_{mo,i}(t)$ at time t, in any point P_i in the system, can be written as a convolution product (Ozisik, 1980), for a zero initial temperature:

$$y_{mo,i}(t) = \int_0^t \int_V Z(t - \tau, P_i; g(P), \boldsymbol{\beta}) f(\tau) \, dV(P) \, d\tau \tag{1.102}$$

In this equation, Z is a transfer function (impedance or space Green's function) that depends on location of the observed point P_i, on the model structural quantities, as well as on the space distribution $g(P)$ of the source, P being any point inside the system. The convolution product is implemented between this impedance and the intensity $f(t)$ of the source.

If stimulus $u(P, t)$ cannot be separated into a product of space and time distributions, this means that several different sources coexist in the system. Each of them can be "separated" and is noted $u_k(P, t) = f_k(t)g_k(P)$, where k is the number of the individual source. One can think, for example, of two heating electrical resistances, embedded in a solid, and that are not turned on at the same time. So, a superposition of solutions of the previous form (1.102) can be implemented to get the global response in point P_i:

$$u(P, t) = \sum_k f_k(t)g_k(P) \tag{1.103}$$

$$y_{mo,i}(t) = \sum_k \int_0^t \int_V Z_k(t - \tau, P_i; g_k(P), \boldsymbol{\beta}) f_k(\tau) \, d\tau \, dV(P) \tag{1.104}$$

Forms (1.102) and (1.104) remain valid in the quite general case where thermophysical properties of the constitutive materials, as well as the heat transfer coefficients and interface resistances used in the model, vary in space (system composed of heterogeneous materials).

However, if these parameters vary with time, the heat equation and its associated conditions may be still linear, but convolution products or transfer functions cannot be used for calculating the sensor responses anymore.

Nomenclature

a	thermal diffusivity ($m^2\,s^{-1}$)
$\vec{a}$	acceleration ($m\,s^{-2}$)
A	state matrix
b	thermal effusivity ($J\,s^{1/2}\,m^{-2}\,K^{-1}$)
B	input matrix
c, c_p	specific heat ($J\,kg^{-1}\,K^{-1}$)
C	output matrix
D	diameter (m)
e	thickness (m)
E	energy (J)
$E(\dots)$	vector function
$\mathbf{exp}(.)$	exponential of a matrix
$f(.)$	function (for time variable)
$\vec{F}$	force (N)
$g(.)$	function (for space variable)
G	coupling factor for the two-temperature model ($W\,m^{-3}\,K^{-1}$) or convective thermal resistance for a unit area of 0D lumped model ($m^2\,K\,W^{-1}$)
$G(.)$	Green's function
$\mathbf{grad}(.)$	gradient vector
h	heat transfer coefficient ($W\,m^{-2}\,K^{-1}$)
$\hbar$	Planck's constant (J s)
H	Biot number, or enthalpy by unit volume ($J\,m^{-3}$)
k	thermal conductivity ($W\,m^{-1}\,K^{-1}$)
k_B	Boltzmann's constant ($J\,K^{-1}$)
$\vec{k}$	wave vector (m^{-1})
$\bar{\bar{k}}$	conductivity tensor, or thermal dispersion tensor ($W\,m^{-1}\,K^{-1}$)
K	thermal conductance for a unit area ($W\,m^{-2}\,K^{-1}$)
$\mathbf{K}$	macroscopic conductivity tensor of the two-temperature model ($W\,m^{-1}\,K^{-1}$)
ℓ, L	length (m)
m	number of data samples, or mass (kg)
M	model
n	number of parameterized input function components
N	size of the state vector
N_q	distribution function for mode q
N_j	square of the function f_j norm
p	dim(B), or Laplace parameter (s^{-1})
$P = (x, y, z)$	point coordinates
q	number of measurement points
q_{vol}	distributed volumic heat source ($W\,m^{-3}$)
$\vec{r}$	position vector
R	thermal resistance per unit area ($m^2\,K\,W^{-1}$)
t	time (s)
t_c	characteristic time (s)
t'	dumb integration variable (s)
T	temperature (K)
T_0	initial temperature (K)

T_∞	temperature of the fluid environment (K)
$T(t)$	column-vector of the discretized state (temperature)
$T_P(t)$	state (temperature) of the system, continuous version
$u(.)$	single input function (W m^{-2} or W m^{-3})
$\pmb{u}_{param}$	single input parameterized vector
$\pmb{U}(t)$	inputs column-vector (dim p)
$\vec{v}, \pmb{v}$	velocity vector (m s^{-1})
x	list of data of direct problem
x	data list for the direct problem
y	measured signal (output of a single sensor)
y_{mo}	theoretical signal, output of a model
$\pmb{y}$	experimental output column-vector (dim m)
$\pmb{y}_{mo}$	simulated column-vector (dim m)
Z	thermal impedance (m^2 K J^{-1})

Greek Variables

$\pmb{\beta}$	parameter vector
χ	characteristic function
ΔT	temperature difference (K)
ε	volume fraction or porosity, or measurement noise
ε_i	noise at time t_i (K)
$\eta(.)$	function, output model structure
$\pmb{\eta}(.)$	multiple-output model structure
λ	wavelength (m)
Λ	mean free path (m)
∇	nabla operator (gradient)
ϕ	heat flux (W)
$\vec{\varphi}$	heat flux density (W m^{-2})
ρ	mass density (kg m^{-3})
τ	time constant or relaxation time (s)
τ_{diff}	characteristic diffusion time (s)
Θ_D	Debye's temperature (K)

Subscripts

c	complementary (known)
$calib$	calibration
det	relative to a detailed model
e	electron
H	enthalpic
l	lattice
m	space average
mm	double space average
mo	model
n	normal
0	initial
$param$	parameterized
pos	position

P	relative to point P
q	mode number
r	researched
red	relative to a reduced model
s	sensor
ss	steady state
struct	structural
t	total
th	thermal
x	direction *x*
y	direction *y*
z	direction *z*
β	β-phase
λ	relative to wavelength λ
σ	σ-phase

Superscripts

_	time Laplace transform
=	tensor
*	exact value, or dimensionless quantity of a dimensionless model
T	transposed matrix

References

Anisimov, S.I., B.L. Kapeliovich, and T.L. Perel'man. 1974. *Zh. Eksp. Teor. Fiz.* **66**, 776.

Anisimov, S.I., B.L. Kapeliovich, and T.L. Perel'man. 1975. *Sov. Phys. JETP* **39**, 375.

Auriault, J.L. and H.I. Ene. 1994. Macroscopic modelling of heat transfer in composites with interfacial thermal barrier. *Int. J. Heat Mass Transf.* **37**(18), 2885–2892.

Battaglia, J.L., A. Kusiak, C. Rossignol, and N. Chigarev. 2007. Thermal diffusivity and effusivity of thin layers using the time-domain thermoreflectance. *Phys. Rev. B* **76**, 184110.

Benítez, H.D., C. Ibarra-Castanedo, A. Bendada, X. Maldague, H. Loaiza, and E. Caicedo. 2008. Definition of a new thermal contrast and pulse correction for defect quantification in pulsed thermography. *Infrared Phys. Technol.* **51**(3), 160–167.

Carbonell, R.G. and S. Whitaker. 1984. Heat and mass transfer in porous media. In: *Fundamentals of Transport Phenomenon Porous Media*, eds. J. Bear and M.Y. Corapcioglu, pp. 121–198. Dordrecht, the Netherlands: Martinus Nijhof.

Fourier, J. 1828. Mémoire sur la théorie analytique de la chaleur. http://www.academie-sciences. fr/membres/in_memoriam/Fourier/Fourier_pdf/Mem1828_p581_622.pdf

Frenkel, D. and S. Berend. 1996. *Understanding Molecular Simulation*. San Diego, CA: Academic Press.

Girault, M., D. Maillet, F. Bonthoux, B. Galland, P. Martin, R. Braconnier, and J.-R. Fontaine. 2008. Estimation of time-varying pollutant emission rates in a ventilated enclosure: Inversion of a reduced model obtained by experimental application of the modal identification method. *Inverse Probl.* **24**(February), 01 5021, 22.

Gobbé C., J.-L. Battaglia, and M. Quintard. 1998. Advanced concepts and techniques in thermal modelling, In: *Proceedings of Eurotherm Seminar No. 53*, Elsevier, (ed.), pp. 75–82, Mons, Belgium.

Maillet, D., S. André, J.C. Batsale, A. Degiovanni, and C. Moyne. 2000. *Thermal Quadrupoles—Solving the Heat Equation through Integral Transforms*. Chichester, U.K.: Wiley.

Maillet, D., A.S. Houlbert, S. Didierjean, A.S. Lamine, and A. Degiovanni. 1993. Nondestructive thermal evaluation of delaminations inside a laminate—Part I: Identification using the measurement of a thermal contrast and Part II: The experimental Laplace transforms method. *Compos. Sci. Technol.* **47**(2), 137–154, 155–172.

Moyne, C., S. Didierjean, H.P. Amaral Souto, and O.T. Da Silveira. 2000. Thermal dispersion in porous media: One equation model. *Int. J. Heat Mass Transf.*, **43**, 3853–3867.

Orlande, H.R.B., M.N. Ozisik, and D.Y. Tzou. 1995. Inverse analysis for estimating the electron-phonon coupling factor in thin metal films. *J. Appl. Phys.* **78**(3), 1843–1849.

Ozisik, M.N. 1980. *Heat Conduction*. Chichester, U.K.: Wiley.

Press, W.H., B.P. Flannery, S.A. Teulkosky, and W.T. Vetterling. 1992. *Numerical Recipes—The Art of Scientific Computing*. New York: Cambridge University Press.

Quintard, M., M. Kaviany, and S. Whitaker. 1997. Two-medium treatment of heat transfer in porous media: Numerical results for effective properties. *Adv. Water Resour.* **20**, 11–94.

Rémy, B. and A. Degiovanni. 2005. Parameters estimation and measurement of thermophysical properties of liquids. *Int. J. Heat Mass Transf.* **48**(19–20), 4103–4120.

Testu, A., S. Didierjean, D. Maillet, C. Moyne, T. Metzger, and T. Niass. 2007. Thermal dispersion coefficients for water or air flow through a bed of glass beads. *Int. J. Heat Mass Transf.* **50**(7–8), 1469–1484.

Volz, S. (Ed.). 2007. Microscale and nanoscale heat transfer. *Top. Appl. Phys.* **107**, 333–359.

2

Multiscale Modeling Approach to Predict Thermophysical Properties of Heterogeneous Media

Manuel Ernani Cruz and Carlos Frederico Matt

CONTENTS

2.1 Introduction

The purpose of this chapter is twofold: first we present, in a didactic form, the main ideas underlying the method of homogenization (also called homogenization theory) and, second, we use the method as a tool to develop a multiscale modeling approach, able to analyze a wide spectrum of transport phenomena in random heterogeneous media (media whose microstructures may be described appropriately by non-trivial joint probability density functions [JPDFs]). The approach is also based on variational calculus and the finite element method and leads to the prediction of macroscopic effective properties of heterogeneous media. Here, the multiscale approach is exposed in the context of the heat conduction problem in composite materials, whose components are all thermally conducting. An expression for the tensorial effective thermal conductivity of such materials is derived, and some properties of the effective conductivity are shown.

In this chapter, we present in detail the continuous formulations of the heat conduction problems, which are part of the multiscale approach. On the other hand, we only summarize the main steps for numerical solution of these problems via the finite element method. Sample numerical results for the effective thermal conductivity of the 2D square array of circular cylindrical fibers and of the 3D simple cubic array of spheres are presented up to maximum packing. The reader is referred to the works by Cruz and Patera (1995), Cruz et al. (1995), Cruz (1997, 1998), Machado and Cruz (1999), Matt (1999, 2003), Rocha (1999), Machado (2000), Rocha and Cruz (2001), Matt and Cruz (2001, 2002a, 2002b, 2004, 2006, 2008), and Pereira et al. (2006) for more details of the numerical solutions and for the presentations and analyses of numerical results for the effective thermal conductivities of 2D and 3D, ordered and random composites. Various computational techniques developed to address the heat conduction problem in composite materials are reviewed by Pereira et al. (2006), Matt and Cruz (2006, 2008), and Cruz (2001).

It should be remarked that there are several other approaches to analyze transport phenomena in heterogeneous and multiphase systems. Phenomenological effective medium approaches (see Torquato 2002) do not tackle the underlying physics at the microstructural level, such that they attempt to establish the macroscopic properties by proposing ad hoc assumptions. Another much employed technique is volume averaging, as discussed in Chapter 1 and in the monograph by Whitaker (1999). The main objective of volume averaging is to formulate the spatially smoothed governing equations that are valid everywhere in the heterogeneous medium of interest. The development of closure problems is then necessary to permit the prediction of the medium's effective transport properties, which relate macroscopic fluxes to intensity gradients. Regarding both volume averaging and homogenization approaches, it appears that much more research effort has been devoted to formulating several different classes of transport problems in heterogeneous media than to computing the associated macroscopic properties. Therefore, a comparative analysis of effective property results arising from these alternative methods is beyond the scope of the present work.

The outline of this chapter is as follows. In Section 2.2, the method of homogenization is introduced didactically. We first offer a formal definition and then illustrate with physical examples the mathematical problems involved in the definition. Next, we give a brief overview of the analytical techniques that may be employed in the homogenization procedure. Finally, we apply the method to a general elliptic model problem in strong form.

In Section 2.3, we apply the method of homogenization to the heat conduction problem of interest, adopting a variational approach and exploiting the analysis of Section 2.2. Although some of our results are also shown, in a different form, in Auriault (1983), we not only present a more detailed derivation here, but also the variational treatment makes the final expressions directly suitable for subsequent numerical treatment using the finite element method. In Sections 2.4 through 2.7, we describe the multiscale modeling approach, which decomposes the original multiscale problem into the macroscale, mesoscale, and microscale (sub)problems. In Section 2.8, we briefly discuss the numerical treatment of the pertinent problems, and in Section 2.9 we present some representative results stemming from solutions to the mesoscale and microscale problems. Finally, in Section 2.10, we state the conclusions.

2.2 Method of Homogenization

The method of homogenization can be applied to analyze a variety of periodic heterogeneous systems—those composed of several macroscopic phases and/or dissimilar constituents and characterized by a repetitive elementary structure. A comprehensive treatment of the subject is given in Bensoussan et al. (1978), and a survey of applications of homogenization theory to a wide spectrum of problems can be found in Babuška (1975). The method has been applied to study neutron and radiative transport (Larsen 1975, Bensoussan et al. 1979), to tackle the problem of dynamic fluid–structure interactions in large rod bundles (Schumann 1981) and to develop a procedure for shape optimization of structures (Bendsøe and Kikuchi 1988). In Mei and Auriault (1989), the method is the essence of the formulation of the creeping flow problem through periodic porous media with several spatial scales, and in Mei and Auriault (1991) the approach is extended to include the effect of weak inertia. Kamiński and Kleiber (2000) have also employed homogenization to investigate the behavior of random elastic composites with stochastic interface defects.

In the heat transfer (or rather conduction) context, the objective is to determine the effective thermal conductivity of an equivalent homogeneous medium, which will thermally behave, in a macroscopic sense, as the original heterogeneous medium (Milton 2002). Auriault (1983) and Auriault and Ene (1994) have used homogenization to determine the effective conductivity of certain types of laminated composites. More recently, homogenization theory has been applied in Cruz (1998) to derive an expression for the effective conductivity of particulate composites whose continuous (the matrix) and dispersed (the particles) components are thermally conducting. The dependence of the thermal conductivity of composite materials on temperature has been considered in Chung et al. (2001) by applying homogenization.

2.2.1 Definition

In short, the method of homogenization employs volume averaging (see Chapter 1) to yield a mathematically rigorous mixture-type model for a heterogeneous medium with periodic microstructure and separated length scales. A formal definition may be offered by first introducing three types of boundary value problems (BVPs).

1. BVP-1

$$A^{\varepsilon} u_{\varepsilon} = f \quad \text{in } \Omega, \tag{2.1}$$

u_{ε} subject to boundary conditions on $\partial\Omega$. $\tag{2.2}$

The domain Ω is an open bounded set of $\mathcal{R}^n$, $\partial\Omega$ is the bounding surface of Ω in $\mathcal{R}^{n-1}$, A^{ε} is a general partial differential operator with periodically varying and *continuous* coefficients, $f: \Omega \rightarrow \mathcal{R}^m$, $m \leq n$, is the source term, and u_{ε} is subject to Dirichlet, Neumann, and/or mixed boundary conditions in (2.2).

The characteristic length scales of the domain Ω and of the periods of the coefficients are, respectively, L and λ; the positive parameter ε is the ratio of such scales, and it is assumed here that the scales are well separated, that is,

$$\varepsilon \equiv \frac{\lambda}{L} \ll 1, \tag{2.3}$$

implying statistical homogeneity. BVP-1 is said to have rapidly oscillating coefficients.

2. BVP-2

$$A^{\mathrm{H}} u_{\mathrm{H}} = f \quad \text{in } \Omega, \tag{2.4}$$

u_{H} subject to boundary conditions on $\partial\Omega$. $\tag{2.5}$

The partial differential operator A^{H} has *constant* coefficients, that is, A^{H} is a homogeneous operator; thus, this BVP is said to be homogenized.

3. BVP-3

$$A^C u_C = f^C \quad \text{in } \Omega^C, \tag{2.6}$$

u_C subject to boundary conditions on $\partial\Omega^C$. $\tag{2.7}$

The domain Ω^C, an open bounded set of $\mathcal{R}^n$, is a periodic cell of characteristic size λ, that is, with dimensions proportional to λ in all n coordinate directions. The partial differential operator A^C may have constant or variable coefficients within Ω^C, and u_C and f^C are λ-periodic functions (functions that admit period $C_j\lambda$, $C_j = O$ (1) $\in \mathcal{R}$, in the direction x_j, $j = 1, \ldots, n$). This BVP is called a cell problem.

We are now in a position to offer a formal definition of the method of homogenization: the method is a rigorous mathematical technique whereby one can replace, in the limit $\varepsilon \rightarrow 0$, a BVP with rapidly varying coefficients (type BVP-1) with a homogenized problem (type BVP-2), whose coefficients must be determined through the solution of a cell problem (type BVP-3). Although all three problems are, in general, hard to solve analytically, the method of homogenization has the distinct advantage that problems of the types BVP-2 and BVP-3 are much easier to solve numerically than those of the type BVP-1, since the latter not only require $O(1/\varepsilon^n)$ more degrees of freedom but are also much stiffer.

From the point of view of physics, problem BVP-1 may describe heat transfer, creeping flow, or a neutron transport process in a heterogeneous medium of typical macroscale L with a spatially periodic microstructure of period λ. Problems BVP-2 and BVP-3 may describe the same aforementioned phenomena, respectively, in a homogeneous, effective

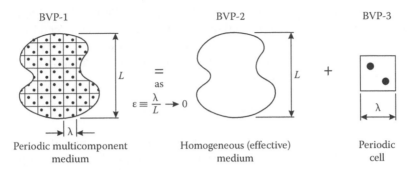

FIGURE 2.1
Diagrammatic representation of the method of homogenization.

medium (in general anisotropic) of typical macroscale L and in a periodic cell of size λ. Note that the coefficients of A^H in BVP-2 correspond, by definition, to the effective macroscopic properties of the original heterogeneous medium considered in BVP-1. Because the determination of such coefficients demands that a solution be found to a cell problem defined in the periodic microstructure of the medium, of size $\lambda \ll L$, it is said that the method of homogenization allows one to describe macroscopically the behavior of a heterogeneous medium through the analysis of the behavior of its underlying microscopic structure; Figure 2.1 illustrates this process.

The lack of a unique precise definition of effective property of a heterogeneous medium led to many reports in the past with discrepant results (Babuška 1975). The method of homogenization not only provides a consistent way of computing effective properties for heterogeneous materials with periodic microstructures, but it also relates global quantities (e.g., bulk heat flow) defined for the original medium to those computed for the equivalent homogeneous medium. It should be pointed out that, typically, real random heterogeneous media possess no period λ, in which case homogenization theory does not *directly* apply. The concept of the correlation length (Cruz and Patera 1995, Cruz 2005), developed in Section 2.6.4, may be used to bridge the transition *periodic → random*, provided such length is small compared to the macroscale L.

2.2.2 Additional Considerations

As previously discussed in Chapter 1, the elaboration of a mathematical model to describe a given physical phenomenon is relative to the desired scale of observation, and is a typical product of scientific investigation. Frequently, the model leads to a problem of the type BVP-1, particularly when one is dealing with heterogeneous systems; homogenization theory can thus be employed to solve such model. In order to replace the operator A^ε of BVP-1 with the operator A^H of BVP-2, several mathematical techniques can be used, based on (see, e.g., Bensoussan et al. 1978) the following:

1. Asymptotic expansions using multiple scales, the fast scale proportional to λ, and the slow scale proportional to L
2. Energy estimates
3. Probabilistic arguments
4. Spectral decomposition of A^ε

The method of asymptotic expansions is attractive when dealing with problems of the type BVP-1 because of the presence of a natural *separation of scales*, as evidenced by Equation 2.3; note that such clear separation of scales is not present in turbulence. The procedure is then to look for the solution $u_\varepsilon = u_\varepsilon(\mathbf{x})$, $\mathbf{x} \in \mathcal{R}^n$, of BVP-1 in the form of an asymptotic expansion in terms of the small positive parameter ε:

$$u_\varepsilon = u_0 + \varepsilon u_1 + \varepsilon^2 u_2 + \cdots, \tag{2.8}$$

where the functions u_j, $j = 0,1,\ldots$, are now the new unknowns, having all the same order of magnitude. Next, by inserting (2.8) into (2.1) and (2.2) and collecting equal powers of ε, a problem of the type BVP-2 is obtained for u_0, with boundary conditions dependent on those prescribed for the original problem. The main result of the method, shown by Bensoussan et al. (1978), is that u_ε converges weakly to u_0 as $\varepsilon \to 0$ (weak convergence means convergence of suitable averages). The explicit analytical construction of the homogeneous operator A^H is crucial for the actual solution of the problem and involves solving a λ-periodic cell problem (type BVP-3), which yields the correct constant coefficients of A^H. In general, the homogenized and cell problems have to be solved numerically. In the following section, we apply the asymptotic expansion technique to a typical elliptic model problem.

2.2.3 Application to a Model Problem

Let us apply the method of homogenization to the following model problem in strong form: in BVP-1, let

$$A^\varepsilon = -\frac{\partial}{\partial x_i}\left(a_{ij}(\mathbf{y})\frac{\partial}{\partial x_j}\right) + a_0(\mathbf{y}), \tag{2.9}$$

where $\mathbf{x} \in \mathcal{R}^3$, $\mathbf{y} \equiv \mathbf{x}/\varepsilon$, and $a_{ij}(\mathbf{y})$, $i,j = 1,2,3$, and $a_0(\mathbf{y})$ are continuous λ-triply periodic functions; we remark that the summation convention is adopted throughout this chapter. Formally, a function is said to be λ-triply periodic if it admits periods proportional to λ in all three coordinate directions. The second-order elliptic operator A^ε in (2.9) models many physical phenomena (e.g., heat or electrical conduction) in composite materials with periodic microstructure. We are now interested in determining the behavior of the solution u_ε of BVP-1, with A^ε given in (2.9), as $\varepsilon \to 0$.

The presence of the two disparate scales L and λ in BVP-1, and the λ-periodicity of A^ε motivate the application of the method of asymptotic expansions using multiple scales (Bensoussan et al. 1978, Mei and Auriault 1989), whereby we look for the solution $u_\varepsilon(\mathbf{x})$ in the form

$$u_\varepsilon(\mathbf{x}) = u_0\left(\mathbf{x},\frac{\mathbf{x}}{\varepsilon}\right) + \varepsilon u_1\left(\mathbf{x},\frac{\mathbf{x}}{\varepsilon}\right) + \varepsilon^2 u_2\left(\mathbf{x},\frac{\mathbf{x}}{\varepsilon}\right) + \cdots, \tag{2.10}$$

where $u_j(\mathbf{x},\mathbf{y})$, $\mathbf{y} \equiv \mathbf{x}/\varepsilon$, $j = 0,1,2,\ldots$, are λ-triply periodic in $\mathbf{y}$. The "fast" variable $\mathbf{y}$ scales (magnifies) the period λ to L and is introduced here to separate the periodic and nonperiodic parts of u_ε, which vary, respectively, rapidly over λ and slowly over L. The new BVPs for the unknown functions u_j are determined by first inserting (2.10) into (2.1), with A^ε given in (2.9), and then by collecting the terms with equal powers of ε. Note that care is necessary with the operator $\partial/\partial x_j$: when operating on a function $G = \hat{G}(\mathbf{x}) = \check{G}(\mathbf{x},\mathbf{y})$, we must first treat $\mathbf{x}$ and $\mathbf{y}$ as independent variables, and subsequently replace $\mathbf{y}$ with $\mathbf{x}/\varepsilon$ to obtain

$$\frac{\partial}{\partial x_j}(G) = \frac{\partial \hat{G}}{\partial x_j} = \frac{\partial \check{G}}{\partial x_j} + \frac{1}{\varepsilon}\frac{\partial \check{G}}{\partial y_j}. \tag{2.11}$$

If, furthermore, G can be expanded as $G = G_0 + \varepsilon G_1 + \varepsilon^2 G_2 + O(\varepsilon^3)$, then from (2.11)

$$\frac{\partial}{\partial x_j}(G) = \frac{\partial \check{G}_0}{\partial x_j} + \frac{1}{\varepsilon}\frac{\partial \check{G}_0}{\partial y_j} + \varepsilon\frac{\partial \check{G}_1}{\partial x_j} + \frac{\partial \check{G}_1}{\partial y_j} + \varepsilon^2\frac{\partial \check{G}_2}{\partial x_j} + \varepsilon\frac{\partial \check{G}_2}{\partial y_j} + O(\varepsilon^2). \tag{2.12}$$

Inserting Equation 2.9 into 2.1, and using Equation 2.12, one obtains

$$A^{\varepsilon}u_0 = (\varepsilon^{-2}A_2 + \varepsilon^{-1}A_1 + \varepsilon^0 A_0)u_{\varepsilon} = f, \tag{2.13}$$

where

$$A_2 = -\frac{\partial}{\partial y_i}\left(a_{ij}(\mathbf{y})\frac{\partial}{\partial y_j}\right), \tag{2.14}$$

$$A_1 = -\frac{\partial}{\partial y_i}\left(a_{ij}(\mathbf{y})\frac{\partial}{\partial x_j}\right) - \frac{\partial}{\partial x_i}\left(a_{ij}(\mathbf{y})\frac{\partial}{\partial y_j}\right), \tag{2.15}$$

$$A_0 = -\frac{\partial}{\partial x_i}\left(a_{ij}(\mathbf{y})\frac{\partial}{\partial x_j}\right) + a_0. \tag{2.16}$$

Now inserting Equation 2.10 into 2.13, and collecting the powers ε^{-2}, ε^{-1}, and ε^0, the following equations involving A_0, A_1, A_2 and u_0, u_1, u_2 result:

$$A_2 u_0 = 0, \tag{2.17}$$

$$A_2 u_1 + A_1 u_0 = 0, \tag{2.18}$$

$$A_2 u_2 + A_1 u_1 + A_0 u_0 = f. \tag{2.19}$$

Before proceeding further, we state a result to be used in the development to follow. The *solvability condition* (i.e., uniqueness up to an additive constant) for the problem

$$\begin{cases} A_2\phi = F \text{ in } Y, \\ \phi \text{ periodic in } Y, \end{cases} \tag{2.20}$$

where A_2 is given in (2.14) and Y is a region in $\mathcal{R}^3$, is (see Bensoussan et al. 1978)

$$\int_Y F(\mathbf{y})\, d\mathbf{y} = 0. \tag{2.21}$$

To arrive at (2.21), we integrate (2.20) over Y, apply the first form of Green's theorem (Hildebrand 1976), and then use the periodicity of ϕ.

Noting that the operator A_2 involves $\mathbf{y}$ only, and considering the solvability condition (2.21), we conclude that Equation 2.17 implies that u_0 is a function of $\mathbf{x}$ only, that is,

$$u_0 = u_0(\mathbf{x}). \tag{2.22}$$

Inserting Equations 2.15 and 2.22 into 2.18, we obtain

$$A_2 u_1 = \left(\frac{\partial}{\partial y_i} a_{ij}(\mathbf{y}) \right) \frac{\partial u_0(\mathbf{x})}{\partial x_j}; \tag{2.23}$$

the separation of the variables $\mathbf{x}$ and $\mathbf{y}$ on the right-hand side (RHS) of (2.23) allows one to represent u_1 in the following simple form: if $\chi^j = \chi^j(\mathbf{y})$ is defined as the λ-triply periodic solution (up to an additive constant) of

$$A_2 \chi^j = -\frac{\partial}{\partial y_i} a_{ij}(\mathbf{y}), \tag{2.24}$$

then the general solution of (2.23) is given by

$$u_1(\mathbf{x}, \mathbf{y}) = -\chi^j(\mathbf{y}) \frac{\partial u_0}{\partial x_j} + \tilde{u}_1(\mathbf{x}). \tag{2.25}$$

The problem for u_1 then reduces to finding $\chi^j(\mathbf{y})$; since A_2 involves $\mathbf{y}$ only and both $a_{ij}(\mathbf{y})$ and $\chi^j(\mathbf{y})$ are λ-triply periodic, Equation 2.24 (with proper boundary conditions) constitutes the *cell problem* BVP-3.

From the condition (2.21), it is easily seen that one can solve (2.19) for u_2, treating $\mathbf{x}$ as a parameter, if

$$\int_Y (A_1 u_1 + A_0 u_0)\, d\mathbf{y} = \int_Y f\, d\mathbf{y} \tag{2.26}$$

(note that, here, Y has dimensions proportional to λ in all coordinate directions); using (2.15), (2.16), and (2.25) and the fact that $f = f(\mathbf{x})$, (2.26) becomes

$$\int_Y \left\{ -\frac{\partial}{\partial y_i} \left(a_{ij}(\mathbf{y}) \frac{\partial u_1(\mathbf{x}, \mathbf{y})}{\partial x_j} \right) - \frac{\partial}{\partial x_i} \left(a_{ik}(\mathbf{y}) \frac{\partial}{\partial y_k} \left(-\chi^j(\mathbf{y}) \frac{\partial u_0}{\partial x_j} + \tilde{u}_1(\mathbf{x}) \right) \right) \right.$$
$$\left. -\frac{\partial}{\partial x_i} \left(a_{ij}(\mathbf{y}) \frac{\partial u_0}{\partial x_j} \right) + a_0 u_0 \right\} d\mathbf{y} = f \int_Y d\mathbf{y}, \tag{2.27}$$

or, since $\mathbf{x}$ is a parameter,

$$-\frac{1}{|Y|} \left\{ \int_Y \left(a_{ij} - a_{ik} \frac{\partial \chi^j}{\partial y_k} \right) d\mathbf{y} \right\} \frac{\partial^2 u_0}{\partial x_i \partial x_j} + \frac{1}{|Y|} \left\{ \int_Y a_0(\mathbf{y}) d\mathbf{y} \right\} u_0 = f, \tag{2.28}$$

where $|Y|$ is the measure of the entire region Y,

$$|Y| \equiv \int_Y d\mathbf{y}. \tag{2.29}$$

Clearly, the coefficients

$$C_{\text{eff}_{ij}} \equiv -\frac{1}{|Y|} \int_Y \left(a_{ij} - a_{ik} \frac{\partial \chi^j}{\partial y_k} \right) d\mathbf{y}, \tag{2.30}$$

and

$$C_0 \equiv \frac{1}{|Y|} \int_Y a_0(\mathbf{y}) d\mathbf{y} \tag{2.31}$$

are constants ($\mathbf{y}$ is integrated out); therefore, Equation 2.28 (with proper boundary conditions) constitutes the *homogenized problem* BVP-2. We can thus write the homogenized operator A^H explicitly as

$$A^H = C_{\text{eff}_{ij}} \frac{\partial^2}{\partial x_i \partial x_j} + C_0; \tag{2.32}$$

defining, in general, the average

$$m(\phi) \equiv \frac{1}{|Y|} \int_Y \phi(\mathbf{y}) d\mathbf{y}, \tag{2.33}$$

then

$$C_{\text{eff}_{ij}} = -m(a_{ij}) + m \left(u_{ik} \frac{\partial \chi^j}{\partial y_k} \right) \tag{2.34}$$

and

$$C_0 = m(a_0). \tag{2.35}$$

Mathematically, $C_{\text{eff}_{ij}}$ and C_0 are the *effective coefficients* of the operator A^ε; physically, they are the *effective bulk* properties of the heterogeneous medium, associated with the physical process for which BVP-1 is the appropriate model.

It is worthwhile to conclude this section by stating the following results, which are proved by Babuška (1975) and Bensoussan et al. (1978).

1. *Symmetry.* If A^ε is symmetric ($a_{ij} = a_{ji}$), then A^H is also symmetric.
2. *Ellipticity.* For our model problem, the operator A^H, which does not depend on Ω, is elliptic.

3. *Anisotropy.* If A^ε is diagonal ($a_{ij} = 0$ for $i \neq j$), A^H is not necessarily diagonal.

4. *Error.* The method of asymptotic expansions is justified for both Dirichlet and Neumann boundary conditions on $\partial\Omega$; the end result for the error estimate is

$$\|u_\varepsilon - u_0\|_{L^\infty(\Omega)} \leq C\varepsilon,$$

where C depends on Ω, f, and a_{ij}, but not on ε.

5. *Variational formulation.* Using "energy" ("weak") arguments, it can be proved that the solution u_ε of BVP-1 converges weakly to u_0 as $\varepsilon \to 0$, u_0 being the solution to the problem: find $u_0 \in V(\Omega)$ such that

$$a^H(u_0, v) = (f, v) \quad \forall v \in V(\Omega),$$

where $V(\Omega)$ is an appropriate function space $\left(H_0^1(\Omega) \subseteq V \subseteq H^1(\Omega)\right)$, $a_0 = 0$, and the bilinear form and inner product are defined, respectively, as

$$a^H(u_0, v) \equiv \int_\Omega -C_{\text{eff}_{ij}} \frac{\partial u_0}{\partial x_j} \frac{\partial v}{\partial x_i} d\mathbf{x},$$

$$(f, v) \equiv \int_\Omega fv \, d\mathbf{x}.$$

To obtain the expression for a^H, multiply (2.28) by $v \in V$ and integrate by parts over Ω.

6. *Fluxes.* The fluxes associated with u_ε and u_0,

$$q_{\varepsilon_i} = a_{ij} \frac{\partial u_\varepsilon}{\partial x_j} \quad \text{and} \quad q_{0_i} = C_{\text{eff}_{ij}} \frac{\partial u_0}{\partial x_j},$$

are not close, since the partial derivatives $\partial u_\varepsilon/\partial x_i$ do not converge strongly to $\partial u_0/\partial x_i$; however,

$$\left\| q_{\varepsilon_i} - q_{0_j} \frac{\partial \chi^i}{\partial x_j} \right\|_{L^\infty(\Omega)} \leq C\varepsilon.$$

The interpretation of q_{0_i} is that it represents average fluxes as $\varepsilon \to 0$.

2.3 Homogenization Applied to Heat Conduction in Composites

Several techniques are available to address transport phenomena problems in heterogeneous media (e.g., Beran 1968, Kohn and Milton 1989, Torquato 2002). In this section, the method of homogenization (Bensoussan et al. 1978) is applied to the problem of heat conduction in a composite medium.

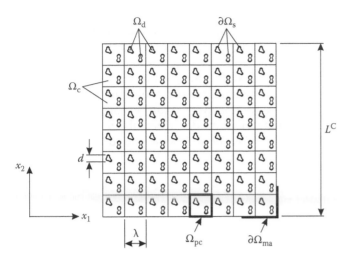

FIGURE 2.2
Periodic composite medium in two dimensions.

2.3.1 Description of the Multiscale Problem

We consider a composite material, periodic (Figure 2.2) or not (Figure 2.3), whose continuous and distributed components are, respectively, a matrix of thermal conductivity k_c and randomly and homogeneously distributed inclusions (fibers or particles) of thermal conductivity k_d; for the sake of simplicity, the conductivities k_c and k_d are taken constant. Both components are assumed to be solid, homogeneous, and isotropic and have perfect thermal contact (for defective thermal contact, see Auriault and Ene 1994, Rocha and Cruz 2001, Matt and Cruz 2008). We define the conductivity ratio α, $0 \leq \alpha < \infty$, as

$$\alpha \equiv \frac{k_d}{k_c}. \tag{2.36}$$

The space coordinates are $(x_1, x_2, x_3) = \mathbf{x} \in \mathcal{R}^3$, and the geometric regions occupied by the continuous and dispersed components are, respectively, Ω_c and Ω_d. Physically, the composite extends throughout a characteristic length L^C; temperature gradients $\Delta T/L$ are imposed over the large scale L, called the *macroscale*, which is $O(L^C)$. The macroscale region is indicated by $\Omega_{ma} = \Omega_c \cup \Omega_d$. In general, the volume fraction of inclusions is specified as a concentration function, $c(\mathbf{x}): \Omega_{ma} \to [0, 1]$, varying significantly on the macroscale only. The smallest scale present is the characteristic size of the inclusions, d, called the *microscale*.

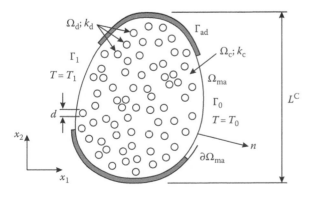

FIGURE 2.3
Random unidirectional fibrous composite in 2D.

The representative volume element (RVE) of the composite, which may be a periodic cell (Figure 2.2), contains many inclusions and is denoted by Ω_{pc}; the size of the RVE, λ, is the intermediate length scale, called the *mesoscale* (Section 2.6). The external boundaries of the composite, where boundary conditions are specified, are denoted as $\partial\Omega_{ma}$. In Figure 2.3, $\partial\Omega_{ma} = \Gamma_0 \cup \Gamma_1 \cup \Gamma_{ad}$, where Γ_0, Γ_1, and Γ_{ad} are subregions of the boundary on which, respectively, uniform temperatures T_0 and T_1, and adiabatic conditions are imposed.

The multiscale heat conduction problem in the medium described above, under steady-state conditions, can be mathematically expressed by the following equations:

$$-\frac{\partial}{\partial x_j}\left(k_c\frac{\partial T^c}{\partial x_j}\right) = \dot{g}_c \quad \text{in } \Omega_c, \tag{2.37}$$

$$-\frac{\partial}{\partial x_j}\left(k_d\frac{\partial T^d}{\partial x_j}\right) = \dot{g}_d \quad \text{in } \Omega_d, \tag{2.38}$$

$$T^c = T^d \quad \text{on } \partial\Omega_s, \tag{2.39}$$

$$-k_c\frac{\partial T^c}{\partial x_j}n_j = -k_d\frac{\partial T^d}{\partial x_j}n_j \quad \text{on } \partial\Omega_s, \tag{2.40}$$

$$T^c \text{ and } T^d \text{ subject to boundary conditions on } \partial\Omega_{ma}. \tag{2.41}$$

Here T^c, $\dot{g}_c$ and T^d, $\dot{g}_d$ are, respectively, the temperature field and the volumetric rate of heat generation in the continuous and distributed components; $\partial\Omega_s$ is the union of all the interfaces between the matrix and the inclusions; and $\mathbf{n}$ is the unit vector locally normal to $\partial\Omega_s$ and pointing into Ω_d. Note that the external temperature gradients are imposed through the boundary conditions in (2.41). Problems (2.37) through (2.41) can also be written as

$$A^\varepsilon T_\varepsilon = -\frac{\partial}{\partial x_j}\left(k\frac{\partial T_\varepsilon}{\partial x_j}\right) = \dot{g} \quad \text{in } \Omega_{ma}, \tag{2.42}$$

$$[T_\varepsilon]_{\partial\Omega_s} = 0, \tag{2.43}$$

$$\left[-k\frac{\partial T_\varepsilon}{\partial x_j}\right]_{\partial\Omega_s} n_j = 0, \tag{2.44}$$

$$T_\varepsilon \text{ subject to boundary conditions on } \partial\Omega_{ma}, \tag{2.45}$$

where, respectively,

$$k, T_\varepsilon, \dot{g} = \begin{cases} k_c, T^c, \dot{g}_c & \text{in } \Omega_c \subset \Omega_{ma} \\ k_d, T^d, \dot{g}_d & \text{in } \Omega_d \subset \Omega_{ma} \end{cases} \tag{2.46}$$

and the notation $[\phi]_{\partial\Omega_s}$ is used to indicate the discontinuity (or jump) of the function ϕ at $\partial\Omega_s$. We note that, for this problem, the oscillating coefficient $a_{ij} = k\delta_{ij}$ (δ_{ij} is the Kronecker delta) of the operator A^ε in Equation 2.42, in contrast to the coefficient in Equation 2.9 of Section 2.2.3, is isotropic and discontinuous, assuming different (constant) values in the two components; also, here $a_0 = 0$.

2.3.2 Variational Formulation

A variational formulation (Cruz and Patera 1995, Rocha and Cruz 2001, Alzina et al. 2006) of problem (2.42) through (2.45) is advantageous for carrying out the homogenization procedure, because first, by transposing derivatives, we bypass the difficulty introduced by the discontinuity of the coefficient $a_{ij} = k\delta_{ij}$, which prevents us from using Equations 2.24 and 2.25 directly; second, the flux boundary condition at the inclusion surfaces, Equation 2.44, is automatically taken care of.

We consider the function space $X(\Omega_{ma}) = \{w \in H_0^1(\Omega_{ma}) | w_{|\Omega_c \subset \Omega_{ma}} = w^c, \ w_{|\Omega_d \subset \Omega_{ma}} = w^d,$ $[w]_{\partial\Omega_s} = 0\}$, where $H_0^1(\Omega_{ma})$ is the space of all functions which vanish on the *portions* of $\partial\Omega_{ma}$ where Dirichlet boundary conditions apply, and for which both the function and derivative are square integrable over Ω_{ma} (Adams 1975). Multiplying (2.42) by $v \in X(\Omega_{ma})$, we obtain

$$-v\frac{\partial}{\partial x_j}\left(k\frac{\partial T_\varepsilon}{\partial x_j}\right) = v\dot{g} \quad \forall v \in X(\Omega_{ma}). \tag{2.47}$$

Integrating (2.47) over Ω_{ma}, it follows that

$$\int_{\Omega_{ma}} -v\frac{\partial}{\partial x_j}\left(k\frac{\partial T_\varepsilon}{\partial x_j}\right)d\mathbf{x} = \int_{\Omega_{ma}} v\dot{g}\,d\mathbf{x} \quad \forall v \in X(\Omega_{ma}). \tag{2.48}$$

Next, applying the first form of Green's theorem (Hildebrand 1976) to (2.48), and considering the continuity condition (2.44) and the definition of the space X, we derive

$$\int_{\Omega_{ma}} k\frac{\partial T_\varepsilon}{\partial x_j}\frac{\partial v}{\partial x_j}\,d\mathbf{x} - \int_{\partial\Omega_{ma}} vk\frac{\partial T_\varepsilon}{\partial x_j}n_j\,ds = \int_{\Omega_{ma}} v\dot{g}\,d\mathbf{x} \quad \forall v \in X(\Omega_{ma}). \tag{2.49}$$

Note that, due to the space X, only the portions of $\partial\Omega_{ma}$ subject to Neumann boundary conditions contribute to the integral on $\partial\Omega_{ma}$. To facilitate the presentation, such *known* contributions are henceforth considered to be summed to the inhomogeneities on the RHS of (2.49), and we can thus omit the integral on $\partial\Omega_{ma}$.

2.3.3 Asymptotic Expansion

We now introduce the multiple-scale asymptotic expansions

$$T_\varepsilon(\mathbf{x}) = T_0(\mathbf{x}, \mathbf{y}) + \varepsilon T_1(\mathbf{x}, \mathbf{y}) + \varepsilon^2 T_2(\mathbf{x}, \mathbf{y}) + O(\varepsilon^3), \tag{2.50}$$

$$v(\mathbf{x}) = v_0(\mathbf{x}, \mathbf{y}) + \varepsilon v_1(\mathbf{x}, \mathbf{y}) + \varepsilon^2 v_2(\mathbf{x}, \mathbf{y}) + O(\varepsilon^3), \tag{2.51}$$

where, as before, $\varepsilon = \lambda/L$ and $\mathbf{y} = \mathbf{x}/\varepsilon$. Combining Equations 2.49 through 2.51, and 2.12 yields, to order ε,

$$\int_{\Omega_{ma}} k\left(\frac{\partial T_0}{\partial x_j} + \frac{1}{\varepsilon}\frac{\partial T_0}{\partial y_j} + \varepsilon\frac{\partial T_1}{\partial x_j} + \frac{\partial T_1}{\partial y_j} + \varepsilon\frac{\partial T_2}{\partial y_j}\right)\left(\frac{\partial v_0}{\partial x_j} + \frac{1}{\varepsilon}\frac{\partial v_0}{\partial y_j} + \varepsilon\frac{\partial v_1}{\partial x_j} + \frac{\partial v_1}{\partial y_j} + \varepsilon\frac{\partial v_2}{\partial y_j}\right)d\mathbf{x}$$

$$= \int_{\Omega_{ma}} (v_0 + \varepsilon v_1)\dot{g}\,d\mathbf{x} \ \forall v_0, v_1 \in X(\Omega_{ma}). \tag{2.52}$$

Inserting (2.50) into (2.43) and (2.44), we obtain

$$[T_0 + \varepsilon T_1 + \varepsilon^2 T_2]_{\partial \Omega_s} = 0, \tag{2.53}$$

$$\left[-k \frac{\partial}{\partial x_j}(T_0 + \varepsilon T_1 + \varepsilon^2 T_2)\right]_{\partial \Omega_s} n_j = 0; \tag{2.54}$$

applying (2.11) to (2.54), it follows that, to order ε,

$$\left[-k\left(\frac{\partial T_0}{\partial x_j} + \frac{1}{\varepsilon}\frac{\partial T_0}{\partial y_j} + \varepsilon\frac{\partial T_1}{\partial x_j} + \frac{\partial T_1}{\partial y_j} + \varepsilon\frac{\partial T_2}{\partial y_j}\right)\right]_{\partial \Omega_s} n_j = 0. \tag{2.55}$$

The next step is simply to identify terms in (2.52), (2.53), and (2.55) which have equal powers of ε. The analysis presented below applies, provided the ratio $k_d/\dot{g}_d$, $\dot{g}_d \neq 0$, is of the same order of magnitude as the ratio $k_c/\dot{g}_c$, $\dot{g}_c \neq 0$; equivalently, the analysis is valid if $\alpha \equiv k_d/k_c = O(\dot{g}_d/\dot{g}_c)$.

Collecting terms of order $1/\varepsilon$ in (2.52), of order 1 in (2.53), and of order $1/\varepsilon$ in (2.55), we obtain

$$\int_{\Omega_{ma}} k\left(\frac{\partial T_0}{\partial x_j}\frac{\partial v_0}{\partial y_j} + \frac{\partial T_0}{\partial y_j}\frac{\partial v_0}{\partial x_j} + \frac{\partial T_0}{\partial y_j}\frac{\partial v_1}{\partial y_j} + \frac{\partial T_1}{\partial y_j}\frac{\partial v_0}{\partial y_j}\right)dx = 0 \quad \forall v_0, v_1 \in X(\Omega_{ma}), \tag{2.56}$$

$$[T_0]_{\partial \Omega_s} = 0, \tag{2.57}$$

$$\left[-k\frac{\partial T_0}{\partial y_j}\right]_{\partial \Omega_s} n_j = 0. \tag{2.58}$$

Choosing $v_0 = 0 \in X(\Omega_{ma})$ in (2.56), we derive

$$\int_{\Omega_{ma}} k\left(\frac{\partial T_0}{\partial y_j}\frac{\partial v_1}{\partial y_j}\right)dx = 0 \quad \forall v_1 \in X(\Omega_{ma}). \tag{2.59}$$

From (2.59) and considering the nontrivial case with $k \neq 0$, we conclude that

$$\frac{\partial T_0}{\partial y_j} = 0 \tag{2.60}$$

and thus, from (2.57) (Auriault 1983),

$$T_0 = T_0^c = T_0^d = T_0(\mathbf{x}). \tag{2.61}$$

Equation 2.61 can be motivated physically: the behavior of the function $T_0(\mathbf{x})$ is dictated by the external boundary conditions on $\partial \Omega_{ma}$, so that on the macroscale, such behavior is the same in both components.

Collecting terms of order 1 in (2.52), of order ε in (2.53), and of order 1 in (2.55), and using $\partial T_0/\partial y_j = \partial v_0/\partial y_j = 0$, we obtain

$$\int_{\Omega_{ma}} k\left(\frac{\partial T_0}{\partial x_j} + \frac{\partial T_1}{\partial y_j}\right)\left(\frac{\partial v_0}{\partial x_j} + \frac{\partial v_1}{\partial y_j}\right) dx = \int_{\Omega_{ma}} v_0 \dot{g} \, dx \quad \forall v_0, v_1 \in X(\Omega_{ma}), \tag{2.62}$$

$$[T_1]_{\partial\Omega_s} = 0, \tag{2.63}$$

$$\left[-k\left(\frac{\partial T_0}{\partial x_j} + \frac{\partial T_1}{\partial y_j}\right)\right]_{\partial\Omega_s} n_j = 0. \tag{2.64}$$

We can break Equation 2.62 into two equations: we choose, first, $v_1 = 0$ and, second, $v_0 = 0$ to obtain

$$\int_{\Omega_{ma}} k\left(\frac{\partial T_0}{\partial x_j} + \frac{\partial T_1}{\partial y_j}\right)\left(\frac{\partial v_0}{\partial x_j}\right) dx = \int_{\Omega_{ma}} v_0 \dot{g} \, dx \quad \forall v_0 \in X(\Omega_{ma}), \tag{2.65}$$

$$\int_{\Omega_{ma}} k\left(\frac{\partial T_0}{\partial x_j} + \frac{\partial T_1}{\partial y_j}\right)\left(\frac{\partial v_1}{\partial y_j}\right) dx = 0 \quad \forall v_1 \in X(\Omega_{ma}). \tag{2.66}$$

As we show in Section 2.4, Equations 2.65 and 2.66 will lead to the homogenized and cell problems.

2.4 Multiscale Modeling Approach

For the multiscale composite material described in Section 2.3.1, it is apparent that finding a solution to the heat conduction BVP in $\Omega_c \cup \Omega_d$ is an enormous task: analytically, because of the geometrical complexity and numerically, because of the excessive number of degrees of freedom required to resolve both the microscale and the macroscale. Fortunately, however, it is not usually necessary to solve the problem down to local detail: in engineering practice, one is typically interested in the macroscopic behavior and determination of bulk quantities. For a particular set of positions of the inclusions, an effective (macroscopic) property of the medium associated with a given transport phenomenon can be viewed as the ratio of the volume-averaged (bulk) flux through the medium and the volume-averaged externally imposed gradient of the corresponding potential (Milton 2002).

In this section, we present the multiscale modeling approach to predict effective properties and statistical correlation lengths of heterogeneous media. The approach is a first-principle analytical–numerical methodology based upon the following (see Cruz and Patera 1995, Cruz et al. 1995, Machado and Cruz 1999, Matt and Cruz 2002a, Cruz 2005): (1) a variational, homogenization-based hierarchical decomposition procedure which recasts the original multiscale problem as a sequence of three scale-decoupled (sub) problems; (2) a variation-bound nip-element technique by means of which microscale

START
__Level 4:__ for each dispersed-phase volume fraction c_i, $i = 1, \ldots, m_c$
   __Level 3:__ for each periodic cell size λ_j, $j = 1, \ldots, m_\lambda$
      __Level 2: Outer Monte-Carlo Loop__
         draw samples, $\{y\}_N$, from dispersed-phase JPDF
         construct periodic cell domain for configuration $\{y\}_N$
         __Level 1: Inner Finite-Element Solution Kernel__
            loop: for each realization $\{y\}_N$
               construct mesh
               effect discretization
               solve periodic cell problem
               compute effective property $k_e' = K_e(c_i, \lambda_j, \{y\}_N)$
            endloop
         __End of Level 1__
         perform statistical analysis (estimate $<K_e>(c_i, \lambda_j)$ and uncertainties)
      __End of Level 2__
      determine correlation length, $\lambda^C(c_i)$
      estimate $k_e(c_i) \equiv <K_e>(c_i, \infty) \approx <K_e>(c_i, \lambda^C(c_i))$
   __End of Level 3__
   construct functional relation $k_e(c)$
__End of Level 4__
END

FIGURE 2.4
Scheme of the four-level numerical algorithm to solve the mesoscale problem.

models are incorporated into the mesoscale problem; and (3) numerical solution of the resulting mesoscale problem by nested Monte Carlo and finite element methods (see Figure 2.4). In the *macroscale problem* (Section 2.5) for heat conduction in composites, the effective thermal conductivity of the homogenized medium is supplied (*input*) to the energy equation in order to calculate the bulk heat flow rate of interest. In the *mesoscale problem* (Section 2.6), the effective conductivity, as well as the statistical correlation length, is determined (*output*) by solving an appropriate sequence of many-inclusion periodic-cell problems generated in the Monte Carlo loop (Figure 2.4). The finite element procedure to solve the mesoscale problem suffers from the severe geometric stiffness that arises when treating the distorted domains associated with the presence of very close inclusions. The boundaries of very close inclusions form nip regions that may be hard, or even impossible, to mesh, rendering numerical solutions either prohibitively expensive, due to excessive degrees of freedom and ill conditioning, or hopeless. In the *microscale problem* (Section 2.7), the nearfield behavior of clustered inclusions is modeled, alleviating the difficulties caused by geometric stiffness. The microscale problem is treated by a variational-bound nip-element technique (Cruz et al. 1995, Machado and Cruz 1999, Machado 2000, Matt and Cruz 2002a): an inner–outer decomposition of the geometrically stiff problem is effected, by means of which analytical approximations in inner nip regions—the microscale models—are folded into a modified outer problem defined over a geometrically more homogeneous domain. As a result, by virtue of the variational nature of the problem, rigorous upper and lower bounds for the configuration effective property may be designed. This technique is rigorously applicable to problems for which the effective property of interest is the extremum of a quadratic, symmetric, positive-(semi) definite functional.

2.5 Macroscale Problem

The key realization needed to derive the macroscale problem and, also, the mesoscale cell problem is to assume, motivated by solvability (see Section 2.2.3, Equations 2.24 and 2.25), that we can separate the functional dependence of T_1 on the variables $\mathbf{x}$ and $\mathbf{y}$:

$$T_1(\mathbf{x}, \mathbf{y}) = -\chi_p(\mathbf{y}) \frac{L}{\Delta T} \frac{\partial T_0(\mathbf{x})}{\partial x_p} = 0, \tag{2.67}$$

$$\chi_p(\mathbf{y}) = \begin{cases} \chi_p^c(\mathbf{y}) & \text{in } \Omega_c \subset \Omega_{\text{ma}}, \\ \chi_p^d(\mathbf{y}) & \text{in } \Omega_d \subset \Omega_{\text{ma}}, \end{cases} \tag{2.68}$$

where the unknown function χ_p, $p = 1, 2, 3$, is a λ-triply periodic solution (to (2.66)) corresponding to a temperature gradient $\Delta T/L$ imposed in the x_p direction (summation over p is implied). Note that, since χ_p is a temperature, the inverse of the external temperature-gradient scaling factor, $L/\Delta T$, is necessary on the RHS of (2.67) to preserve dimensionality.

In order to derive the macroscale, or homogenized, problem for the function $T_0(\mathbf{x}) = T_0^c(\mathbf{x}) = T_0^d(\mathbf{x})$, we first insert (2.67) into (2.65) to obtain

$$\int_{\Omega_{\text{ma}}} k\left(\frac{\partial T_0}{\partial x_j} - \frac{L}{\Delta T} \frac{\partial \chi_p}{\partial y_j} \frac{\partial T_0}{\partial x_p}\right) \frac{\partial v_0}{\partial x_j} d\mathbf{x} = \int_{\Omega_{\text{ma}}} v_0 \dot{g} \, d\mathbf{x} \quad \forall v_0 \in X(\Omega_{\text{ma}}), \tag{2.69}$$

where χ_p, $p = 1, 2, 3$, are now taken as *known* functions. Equation 2.69 can be shortened to

$$\int_{\Omega_{\text{ma}}} k\left(\delta_{jp} - \frac{L}{\Delta T} \frac{\partial \chi_p}{\partial y_j}\right) \frac{\partial T_0}{\partial x_p} \frac{\partial v_0}{\partial x_j} d\mathbf{x} = \int_{\Omega_{\text{ma}}} v_0 \dot{g} \, d\mathbf{x} \quad \forall v_0 \in X(\Omega_{\text{ma}}). \tag{2.70}$$

Because our heterogeneous medium is (assumed) periodic, we can use the *periodicity property* (see Keller 1980, Auriault 1983, Bendsøe and Kikuchi 1988, Rocha and Cruz 2001), which for our purposes can be expressed as

$$\lim_{\varepsilon \to 0} \int_{\Omega_{\text{ma}}} g(\mathbf{x}, \mathbf{y}) d\mathbf{x} = \int_{\Omega_{\text{ma}}} \left(\frac{1}{|\Omega_{\text{pc}}|} \int_{\Omega_{\text{pc}}} g(\mathbf{x}, \mathbf{y}) d\mathbf{y}\right) d\mathbf{x}, \tag{2.71}$$

where $|\Omega_{\text{pc}}| \equiv \int_{\Omega_{\text{pc}}} d\mathbf{y}$ is the total volume measure of a many-inclusion periodic-cell Ω_{pc}. Property (2.71) expresses the fact that as $\varepsilon \to 0$, integration of a quantity over $\Omega_{\text{ma}} = \Omega_c \cup \Omega_d$ can be performed by just capturing the average of the quantity over a representative periodic cell, since the latter becomes essentially a point relative to Ω_{ma}.

Applying the periodicity property (2.71) to Equation 2.70 yields the weak-form *homogenized*, or *macroscale*, problem

$$\int_{\Omega_{\text{ma}}} \left\{\frac{1}{|\Omega_{\text{pc}}|} \int_{\Omega_{\text{pc}}} k\left(\delta_{jp} - \frac{L}{\Delta T} \frac{\partial \chi_p}{\partial y_j}\right) d\mathbf{y}\right\} \frac{\partial T_0}{\partial x_p} \frac{\partial v_0}{\partial x_j} d\mathbf{x} = \int_{\Omega_{\text{ma}}} \left\{\frac{1}{|\Omega_{\text{pc}}|} \int_{\Omega_{\text{pc}}} v_0 \dot{g} \, d\mathbf{y}\right\} d\mathbf{x} \quad \forall v_0 \in X(\Omega_{\text{ma}}),$$

$$\tag{2.72}$$

which can also be written as

$$
\int_{\Omega_{\mathrm{ma}}} \frac{1}{|\Omega_{\mathrm{pc}}|} \left\{ \int_{\Omega_{\mathrm{pc,c}}} k_{\mathrm{c}} \left(\delta_{jp} - \frac{L}{\Delta T} \frac{\partial \chi_p^{\mathrm{c}}}{\partial y_j} \right) dy + \int_{\Omega_{\mathrm{pc,d}}} k_{\mathrm{d}} \left(\delta_{jp} - \frac{L}{\Delta T} \frac{\partial \chi_p^{\mathrm{d}}}{\partial y_j} \right) dy \right\} \frac{\partial T_0}{\partial x_p} \frac{\partial v_0}{\partial x_j} d\mathbf{x}
$$

$$
= \int_{\Omega_{\mathrm{ma}}} \frac{1}{|\Omega_{\mathrm{pc}}|} v_0 \left(\int_{\Omega_{\mathrm{pc,c}}} \dot{g}_{\mathrm{c}} \, dy + \int_{\Omega_{\mathrm{pc,d}}} \dot{g}_{\mathrm{d}} \, dy \right) d\mathbf{x} \quad \forall v_0 \in X(\Omega_{\mathrm{ma}}). \tag{2.73}
$$

In (2.73), $\Omega_{\mathrm{pc,c}}$ and $\Omega_{\mathrm{pc,d}}$ are, respectively, the portions of Ω_{pc} in the continuous and dispersed components. The macroscale region Ω_{ma} (see BVP-2 in Figure 2.1) is much less geometrically complex than the individual regions Ω_{c} and Ω_{d} of the components, such that it is easy to conclude that the homogenized problem will require only a small fraction of the number of degrees of freedom demanded by the original problem. The macroscale problem can then be routinely solved with state-of-the-art commercial software packages.

2.5.1 Nondimensional Homogenized Problem

In practice, a numerical treatment of the macroscale problem is based on the nondimensional form of Equation 2.73. Choosing, arbitrarily, the inclusion size d as the characteristic length, and $T^C = \Delta T(d/L)$ as the characteristic temperature, we define $T_0^* \equiv T_0/T^C$, $\chi^* \equiv \chi/T^C$, $v^* \equiv v/T^C$, $v_0^* \equiv v_0/T^C$, $x^* \equiv x/L$, $\mathbf{y}^* \equiv \mathbf{y}/d$, and $\dot{g}^* \equiv \dot{g}\, L^2/k_{\mathrm{c}}\, T^C$. Using these definitions to normalize Equation 2.73, and considering k_{c} and k_{d} to be constants, we obtain

$$
\int_{\Omega_{\mathrm{ma}}} \left\{ \int_{\Omega_{\mathrm{pc,c}}} \left(\delta_{jp} - \frac{\partial \chi_p^{*\mathrm{c}}}{\partial y_j^*} \right) dy^* + \int_{\Omega_{\mathrm{pc,d}}} \alpha \left(\delta_{jp} - \frac{\partial \chi_p^{*\mathrm{d}}}{\partial y_j^*} \right) dy^* \right\} \frac{\partial T_0^*}{\partial x_p^*} \frac{\partial v_0^*}{\partial x_j^*} d\mathbf{x}^*
$$

$$
= \int_{\Omega_{\mathrm{ma}}} v_0^* \left(\int_{\Omega_{\mathrm{pc,c}}} \dot{g}_{\mathrm{c}}^* \, dy^* + \int_{\Omega_{\mathrm{pc,d}}} \dot{g}_{\mathrm{d}}^* \, dy^* \right) d\mathbf{x}^* \quad \forall v_0^* \in X(\Omega_{\mathrm{ma}}). \tag{2.74}
$$

The reader will note that problem (2.74) is analogous to (the nondimensional version of) problem (2.49), that is, the weak form for problems (2.42) through (2.45), provided it is identified that the effective thermal conductivity, introduced in Section 2.6.2, plays the role of k.

2.6 Mesoscale Problem

The assumed statistical homogeneity of the analyzed (ordered or random) heterogeneous medium leads to a natural assumption: there exists an intermediate length scale, called the *mesoscale* λ, $d < \lambda < L$, which represents the size of the region around one inclusion, inside which most of the interactions of the inclusion with neighboring inclusions occur. This region is treated in the mesoscale problem of the multiscale approach. As illustrated in Figure 2.4, the mesoscale problem encompasses four nested loops (Cruz and Patera 1995,

Cruz 2005). The solution to this problem yields the appropriate function $\underline{\mathbf{k}}_e(c(\mathbf{x}))$ to the macroscale problem. The microscale analysis, described in Section 2.7, must be incorporated into the mesoscale equations to circumvent the difficulty associated with the geometrical stiffness arising from the presence of close inclusions.

2.6.1 Level 1—The Cell Problem

In level 1, the microstructure of the composite medium must be prescribed: a volume fraction, or concentration, of inclusions, c, is assumed; next, a particular configuration (i.e., realization) of the composite is considered, by introducing a representative periodic cell of specified edge length λ, which contains many inclusions, whose geometrical centers are located at $\{\mathbf{y}\}_N = \mathbf{y}_1, \ldots, \mathbf{y}_N$; N is the number of inclusions, for example, $N = 4c\lambda^2/\pi d^2$ for circular cylinders. An illustration of a 2D periodic cell with many inclusions (long, circular, cylindrical fibers) is shown in Figure 2.5. Given the model for the medium's microstructure, we now need to derive the appropriate BVP for the transport phenomenon being investigated.

To derive the periodic-cell problem, we first insert (2.67) into (2.66) to arrive at

$$\int_{\Omega_{ma}} k \left\{ \frac{\partial T_0}{\partial x_j} + \frac{\partial}{\partial y_j} \left(-\chi_p(\mathbf{y}) \frac{L}{\Delta T} \frac{\partial T_0}{\partial x_p} \right) \right\} \frac{\partial v_1}{\partial y_j} d\mathbf{x} = 0 \quad \forall v_1 \in X(\Omega_{ma}). \tag{2.75}$$

Equation 2.75 can be rewritten as

$$\int_{\Omega_{ma}} k \left(\delta_{jp} - \frac{L}{\Delta T} \frac{\partial \chi_p}{\partial y_j} \right) \frac{\partial T_0}{\partial x_p} \frac{\partial v_1}{\partial y_j} d\mathbf{x} = 0 \quad \forall v_1 \in X(\Omega_{ma}), \tag{2.76}$$

where δ_{ij} is the Kronecker delta. Because the heterogeneous medium is considered periodic, we can, again, apply the periodicity property (2.71) (Keller 1980, Auriault 1983, Bendsøe and Kikuchi 1988, Rocha and Cruz 2001) to Equation 2.76 to yield

$$\int_{\Omega_{ma}} \left\{ \frac{1}{|\Omega_{pc}|} \int_{\Omega_{pc}} k \left(\delta_{jp} - \frac{L}{\Delta T} \frac{\partial \chi_p}{\partial y_j} \right) \frac{\partial v_1}{\partial y_j} d\mathbf{y} \right\} \frac{\partial T_0}{\partial x_p} d\mathbf{x} = 0 \quad \forall v_1 \in X(\Omega_{ma}). \tag{2.77}$$

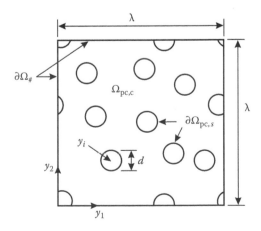

FIGURE 2.5
Realization of a 2D periodic cell with many inclusions.

Equation 2.77 implies that the inner integral must be zero for any λ-triply periodic test function v:

$$\int_{\Omega_{\mathrm{pc}}} k \left(\delta_{jp} - \frac{L}{\Delta T} \frac{\partial \chi_p}{\partial y_j} \right) \frac{\partial v}{\partial y_j} \, d\mathbf{y} = 0 \quad \forall v \in Y(\Omega_{\mathrm{pc}}), \tag{2.78}$$

or

$$\int_{\Omega_{\mathrm{pc}}} k \frac{L}{\Delta T} \frac{\partial \chi_p}{\partial y_j} \frac{\partial v}{\partial y_j} \, d\mathbf{y} = \int_{\Omega_{\mathrm{pc}}} k \frac{\partial v}{\partial y_p} \, d\mathbf{y} \quad \forall v \in Y(\Omega_{\mathrm{pc}}), \tag{2.79}$$

where $Y(\Omega_{\mathrm{pc}}) = \left\{ w \in H^1_{\#}(\Omega_{\mathrm{pc}}) | w_{|\Omega_{\mathrm{pc,c}} \subset \Omega_{\mathrm{pc}}} = w^{\mathrm{c}}, w_{|\Omega_{\mathrm{pc,d}} \subset \Omega_{\mathrm{pc}}} = w^{\mathrm{d}}, [w]_{\partial \Omega_{\mathrm{pc,s}}} = 0 \right\}$, $H^1_{\#}(\Omega_{\mathrm{pc}})$ is the space of all λ-triply periodic functions (subscript #) in Ω_{pc} for which both the function and derivative are square integrable over Ω_{pc} ($H^1(\Omega_{\mathrm{pc}})$), and $\partial \Omega_{\mathrm{pc,s}}$ is the portion of $\partial \Omega_{\mathrm{s}}$ in the cell. Equation 2.79 is the appropriate mesoscale *cell problem*, which is clearly solvable: setting $v = 0 \in Y(\Omega_{\mathrm{pc}})$, both sides of (2.79) vanish. Note that the left-hand side of (2.79) is the standard (negative) Laplacian operator, and the RHS, although slightly nonstandard, is easily computed for a chosen test function. The cell-problem boundary conditions imposed by (2.79) and the space $Y(\Omega_{\mathrm{pc}})$ are λ-triple periodicity for χ_p; from (2.61), (2.63), and (2.67),

$$[\chi_p]_{\partial \Omega_{\mathrm{pc,s}}} = 0; \tag{2.80}$$

and, from (2.64) and (2.67), the following flux condition is naturally enforced:

$$\left[-k \left(\delta_{jp} - \frac{L}{\Delta T} \frac{\partial \chi_p}{\partial y_j} \right) \frac{\partial T_0}{\partial x_p} \right]_{\partial \Omega_{\mathrm{pc,s}}} n_j = 0. \tag{2.81}$$

We can rewrite the cell problem in the following way:

$$\int_{\Omega_{\mathrm{pc,c}}} k_{\mathrm{c}} \frac{L}{\Delta T} \frac{\partial \chi_p^{\mathrm{c}}}{\partial y_j} \frac{\partial v^{\mathrm{c}}}{\partial y_j} d\mathbf{y} + \int_{\Omega_{\mathrm{pc,d}}} k_{\mathrm{d}} \frac{L}{\Delta T} \frac{\partial \chi_p^{\mathrm{d}}}{\partial y_j} \frac{\partial v^{\mathrm{d}}}{\partial y_j} d\mathbf{y} = \int_{\Omega_{\mathrm{pc,c}}} k_{\mathrm{c}} \frac{\partial v^{\mathrm{c}}}{\partial y_p} d\mathbf{y} + \int_{\Omega_{\mathrm{pc,d}}} k_{\mathrm{d}} \frac{\partial v^{\mathrm{d}}}{\partial y_p} d\mathbf{y} \quad \forall v \in Y(\Omega_{\mathrm{pc}}). \tag{2.82}$$

The imposed boundary conditions, besides the λ-triple periodicity of χ_p, become

$$\chi_p^{\mathrm{c}} = \chi_p^{\mathrm{d}} \quad \text{on } \partial \Omega_{\mathrm{pc,s}} \tag{2.83}$$

and from (2.61),

$$-k_{\mathrm{c}} \left(\delta_{jp} - \frac{L}{\Delta T} \frac{\partial \chi_p^{\mathrm{c}}}{\partial y_j} \right) n_j = -k_{\mathrm{d}} \left(\delta_{jp} - \frac{L}{\Delta T} \frac{\partial \chi_p^{\mathrm{d}}}{\partial y_j} \right) n_j \quad \text{on } \partial \Omega_{\mathrm{pc,s}}. \tag{2.84}$$

In (2.82) through (2.84), the functions χ_p^c and χ_p^d are determined up to a (common) constant; thus we further require for uniqueness that

$$\int_{\Omega_{pc,c}} \chi_p^c \, d\mathbf{y} + \int_{\Omega_{pc,d}} \chi_p^d \, d\mathbf{y} = 0. \tag{2.85}$$

2.6.1.1 Nondimensional Cell Problem

In practice, a numerical treatment of the cell problem is based on the nondimensional form of Equations 2.82 and 2.85. Using the same nondimensional variables as before (Section 2.5), and considering k_c and k_d to be constants, we obtain

$$\int_{\Omega_{pc,c}} \frac{\partial \chi_p^{*c}}{\partial y_j^*} \frac{\partial v^{*c}}{\partial y_j^*} dy^* + \int_{\Omega_{pc,d}} \alpha \frac{\partial \chi_p^{*d}}{\partial y_j^*} \frac{\partial v^{*d}}{\partial y_j^*} dy^*$$

$$= \int_{\Omega_{pc,c}} \frac{\partial v^{*c}}{\partial y_p^*} dy^* + \int_{\Omega_{pc,d}} \alpha \frac{\partial v^{*d}}{\partial y_p^*} dy^* \quad \forall v^* \in Y(\Omega_{pc}) \tag{2.86}$$

and

$$\int_{\Omega_{pc,c}} \chi_p^{*c} \, dy^* + \int_{\Omega_{pc,d}} \chi_p^{*d} \, dy^* = 0. \tag{2.87}$$

2.6.2 The Configuration Effective Conductivity

For the particular cell configuration of level 1, by simply inspecting Equation 2.72, we easily recognize the tensorial *effective thermal conductivity* to be

$$k'_{e_{pq}} = \frac{1}{|\Omega_{pc}|} \int_{\Omega_{pc}} k \left(\delta_{pq} - \frac{L}{\Delta T} \frac{\partial \chi_q}{\partial y_p} \right) d\mathbf{y}; \tag{2.88}$$

alternatively, $k'_{e_{pq}}$ can be written as

$$k'_{e_{pq}} = \frac{1}{|\Omega_{pc}|} \left\{ \int_{\Omega_{pc,c}} k_c \left(\delta_{pq} - \frac{L}{\Delta T} \frac{\partial \chi_q^c}{\partial y_p} \right) d\mathbf{y} + \int_{\Omega_{pc,d}} k_d \left(\delta_{pq} - \frac{L}{\Delta T} \frac{\partial \chi_q^d}{\partial y_p} \right) d\mathbf{y} \right\}. \tag{2.89}$$

The prime in Equations 2.88 and 2.89 is used to designate the *configuration* effective conductivity, which corresponds to the particular realization of the representative cell of the heterogeneous medium. Note the presence of the factor $1/|\Omega_{pc}|$ in Equation 2.89 multiplying the integrals over $\Omega_{pc,c}$ and $\Omega_{pc,d}$, irrespective of whether k_c or k_d is zero. A physical motivation for the division by the *total* periodic cell measure, $|\Omega_{pc}|$, is that the homogenized medium occupies the total extension of the cell; therefore, division by $|\Omega_{pc}|$ yields the correct average over the periodic cell (Cruz and Patera 1995).

2.6.2.1 Nondimensional Effective Conductivity

The nondimensional effective conductivity is given by

$$
k'^*_{e_{pq}} = \frac{1}{|\Omega_{pc}|^*} \left\{ \int_{\Omega_{pc,c}} \left(\delta_{pq} - \frac{\partial \chi_q^{*c}}{\partial y_p^*} \right) dy^* + \int_{\Omega_{pc,d}} \alpha \left(\delta_{pq} - \frac{\partial \chi_q^{*d}}{\partial y_p^*} \right) dy^* \right\},
\tag{2.90}
$$

where $k'^*_{e_{pq}} \equiv k'_{e_{pq}}/k_c \; |\Omega_{pc}|^* \equiv |\Omega_{pc}|/d^3$.

2.6.2.2 Properties of $k'_{e_{pq}}$

We can now show some properties of $k'_{e_{pq}}$:

1. *Symmetry.* From Equation 2.82, taking $v \in Y(\Omega_{pc})$ such that $v^c = \chi_q^c$ and $v^d = \chi_q^d$, we obtain

$$
\int_{\Omega_{pc,c}} k_c \frac{\partial \chi_q^c}{\partial y_p} \, dy + \int_{\Omega_{pc,d}} k_d \frac{\partial \chi_q^d}{\partial y_p} \, dy = \int_{\Omega_{pc,c}} k_c \frac{L}{\Delta T} \frac{\partial \chi_p^c}{\partial y_j} \frac{\partial \chi_q^c}{\partial y_j} dy + \int_{\Omega_{pc,d}} k_d \frac{L}{\Delta T} \frac{\partial \chi_p^d}{\partial y_j} \frac{\partial \chi_q^d}{\partial y_j} dy;
\tag{2.91}
$$

therefore, switching p and q in (2.91), we conclude that

$$
\int_{\Omega_{pc,c}} k_c \frac{\partial \chi_q^c}{\partial y_p} \, dy + \int_{\Omega_{pc,d}} k_d \frac{\partial \chi_q^d}{\partial y_p} \, dy = \int_{\Omega_{pc,c}} k_c \frac{\partial \chi_p^c}{\partial y_q} \, dy + \int_{\Omega_{pc,d}} k_d \frac{\partial \chi_p^d}{\partial y_q} \, dy
\tag{2.92}
$$

or, in short,

$$
\int_{\Omega_{pc}} k \frac{L}{\Delta T} \frac{\partial \chi_p}{\partial y_j} \frac{\partial \chi_q}{\partial y_j} \, dy = \int_{\Omega_{pc}} k \frac{\partial \chi_q}{\partial y_p} \, dy = \int_{\Omega_{pc}} k \frac{\partial \chi_p}{\partial y_q} \, dy.
\tag{2.93}
$$

From (2.89) and (2.92), and the fact that $\delta_{ij} = \delta_{ji}$, we conclude that $k'_{e_{pq}}$ is *symmetric.*

2. *An equivalent expression.* We now show that $k'_{e_{pq}}$ can alternatively be written as

$$
k'_{e_{pq}} = \frac{1}{|\Omega_{pc}|} \int_{\Omega_{pc}} k \left\{ \frac{\partial}{\partial y_j} \left(y_p - \frac{L}{\Delta T} \chi_p \right) \frac{\partial}{\partial y_j} \left(y_q - \frac{L}{\Delta T} \chi_q \right) \right\} dy.
\tag{2.94}
$$

Expanding the RHS of (2.94), we obtain

$$
\text{RHS} = \frac{1}{|\Omega_{pc}|} \int_{\Omega_{pc}} k \left(\delta_{jp}\delta_{jq} - \frac{L}{\Delta T} \frac{\partial \chi_p}{\partial y_j} \delta_{jq} - \frac{L}{\Delta T} \frac{\partial \chi_q}{\partial y_j} \delta_{jp} + \frac{L^2}{\Delta T^2} \frac{\partial \chi_p}{\partial y_j} \frac{\partial \chi_q}{\partial y_j} \right) dy;
\tag{2.95}
$$

simplifying the terms with the Kronecker delta, we get

$$
\text{RHS} = \frac{1}{|\Omega_{pc}|} \int_{\Omega_{pc}} k \left(\delta_{pq} - \frac{L}{\Delta T} \frac{\partial \chi_p}{\partial y_q} - \frac{L}{\Delta T} \frac{\partial \chi_q}{\partial y_p} + \frac{L^2}{\Delta T^2} \frac{\partial \chi_p}{\partial y_j} \frac{\partial \chi_q}{\partial y_j} \right) dy.
\tag{2.96}
$$

Using (2.93), (2.96) simplifies to

$$\text{RHS} = \frac{1}{|\Omega_{pc}|} \int\limits_{\Omega_{pc}} k\left(\delta_{pq} - \frac{L}{\Delta T}\frac{\partial \chi_q}{\partial y_p}\right) d\mathbf{y};$$

(2.97)

from (2.88) and (2.97), we deduce that (2.94) is true.

3. *Positive definiteness.* From Equation 2.94, it follows that for any vector function $\psi_p \in \mathcal{R}^3$,

$$\psi_p k'_{e_{pq}} \psi_q = \frac{1}{|\Omega_{pc}|} \int\limits_{\Omega_{pc}} k\left\{\frac{\partial}{\partial y_j}\psi_p\left(y_p - \frac{L}{\Delta T}\chi_p\right)\frac{\partial}{\partial y_j}\psi_q\left(y_q - \frac{L}{\Delta T}\chi_q\right)\right\} d\mathbf{y},$$

(2.98)

which is essentially the square of the modulus of the gradient of $\psi_p(y_p - \chi_p)$ integrated over Ω_{pc}; hence, $\psi_p k'_{e_{pq}} \psi_q \geq 0$. We conclude that $k'_{e_{pq}}$ is *positive definite*, which guarantees the well posedness and uniqueness for the homogenized problem (2.72) (Lax–Milgram lemma).

2.6.2.3 Extremizing Property

We now show an extremizing property of $k'_{e_{pq}}$. Defining the functional I_Ω^p,

$$I_\Omega^p(v) = \int\limits_\Omega k\frac{L}{\Delta T}\frac{\partial v}{\partial y_p} d\mathbf{y}.$$

(2.99)

Equation 2.88 can be rewritten as

$$k'_{e_{pq}} = \frac{1}{|\Omega_{pc}|}\left\{\int\limits_{\Omega_{pc}} k\delta_{pq} \, d\mathbf{y} - I_{\Omega_{pc}}^p(\chi_q)\right\}.$$

(2.100)

We also introduce the functional J_Ω^p,

$$J_\Omega^p(v) = \int\limits_\Omega k\frac{L^2}{\Delta T^2}\frac{\partial v}{\partial y_j}\frac{\partial v}{\partial y_j} d\mathbf{y} - 2I_\Omega^p(v),$$

(2.101)

the first term of which is a positive-definite bilinear form. From (2.99) and (2.101), we have that

$$J_{\Omega_{pc}}^p(\chi_p) = \int\limits_{\Omega_{pc}} k\frac{L^2}{\Delta T^2}\frac{\partial \chi_p}{\partial y_j}\frac{\partial \chi_p}{\partial y_j} d\mathbf{y} - 2\int\limits_{\Omega_{pc}} k\frac{L}{\Delta T}\frac{\partial \chi_p}{\partial y_p} d\mathbf{y};$$

(2.102)

but from (2.93) we know that

$$\int\limits_{\Omega_{pc}} k\frac{L^2}{\Delta T^2}\frac{\partial \chi_p}{\partial y_j}\frac{\partial \chi_p}{\partial y_j} d\mathbf{y} = \int\limits_{\Omega_{pc}} k\frac{L}{\Delta T}\frac{\partial \chi_p}{\partial y_p} d\mathbf{y}.$$

(2.103)

Thus, from (2.102) and (2.103), we conclude that

$$J^p_{\Omega_{pc}}(\chi_p) = - \int_{\Omega_{pc}} k \frac{L}{\Delta T} \frac{\partial \chi_p}{\partial y_p} \, d\mathbf{y},$$

(2.104)

or

$$I^p_{\Omega_{pc}}(\chi_p) = -J^p_{\Omega_{pc}}(\chi_p) = \int_{\Omega_{pc}} k \frac{L^2}{\Delta T^2} \frac{\partial \chi_p}{\partial y_j} \frac{\partial \chi_p}{\partial y_j} \, d\mathbf{y}.$$

(2.105)

Now note that the solution χ_p in (2.79) can also be written as (Bendsøe and Kikuchi 1988, Cruz and Patera 1995, Matt and Cruz 2002a)

$$\chi_p = \arg \min_{v \in Y(\Omega_{pc})} J^p_{\Omega_{pc}}(v).$$

(2.106)

Clearly, the weak form for χ_p presented in (2.79) derives from the first variation of the functional $J^p_{\Omega_{pc}}(v)$:

$$\delta J^p_{\Omega_{pc}}(v) = 2 \int_{\Omega_{pc}} k \frac{L^2}{\Delta T^2} \frac{\partial v}{\partial y_j} \frac{\partial w}{\partial y_j} \, d\mathbf{y} - 2 \int_{\Omega_{pc}} k \frac{L}{\Delta T} \frac{\partial w}{\partial y_p} \, d\mathbf{y},$$

(2.107)

where $w = \delta v \in Y(\Omega_{pc})$ is the variation of v; the function that minimizes $J^p_{\Omega_{pc}}(v)$, denoted as χ_p, must be such that $\delta J^p_{\Omega_{pc}}(\chi_p) = 0$, and, therefore, dividing (2.107) through by $2(L/\Delta T)$, we obtain

$$\int_{\Omega_{pc}} k \frac{L}{\Delta T} \frac{\partial \chi_p}{\partial y_j} \frac{\partial w}{\partial y_j} \, d\mathbf{y} - \int_{\Omega_{pc}} k \frac{\partial w}{\partial y_p} d\mathbf{y} = 0 \quad \forall w \in Y(\Omega_{pc}),$$

(2.108)

which is the same as Equation 2.79.

From (2.100) and (2.105), it follows that

$$\begin{aligned}
k'_{e_{pp}} &= \frac{1}{|\Omega_{pc}|} \left\{ \int_{\Omega_{pc}} k \, d\mathbf{y} - I^p_{\Omega_{pc}}(\chi_p) \right\} \\
&= \frac{1}{|\Omega_{pc}|} \left\{ \int_{\Omega_{pc}} k \, d\mathbf{y} + J^p_{\Omega_{pc}}(\chi_p) \right\} \\
&= \frac{1}{|\Omega_{pc}|} \left\{ \int_{\Omega_{pc}} k \, d\mathbf{y} - \int_{\Omega_{pc}} k \frac{L^2}{\Delta T^2} \frac{\partial \chi_p}{\partial y_j} \frac{\partial \chi_p}{\partial y_j} d\mathbf{y} \right\};
\end{aligned}$$

(2.109)

thus, from (2.106) and (2.109), we derive

$$k'_{e_{pp}} = \frac{1}{|\Omega_{pc}|} \left\{ \int_{\Omega_{pc}} k \, d\mathbf{y} + \min_{v \in Y(\Omega_{pc})} J^p_{\Omega_{pc}}(v) \right\}. \tag{2.110}$$

Finally, defining the bilinear form

$$a^{pq}_{\Omega_{pc}}(v, w) = \frac{1}{|\Omega_{pc}|} \int_{\Omega_{pc}} k \left\{ \frac{\partial}{\partial y_j} \left(y_p - \frac{L}{\Delta T} v \right) \frac{\partial}{\partial y_j} \left(y_q - \frac{L}{\Delta T} w \right) \right\} d\mathbf{y}, \tag{2.111}$$

we can rewrite $k'_{e_{pq}}$ as given in (2.94) in the equivalent form

$$k'_{e_{pq}} = a^{pq}_{\Omega_{pc}}(\chi_p, \chi_q). \tag{2.112}$$

Since, using (2.103) and (2.106),

$$
\begin{aligned}
a^{pp}_{\Omega_{pc}}(\chi_p, \chi_p) &= \frac{1}{|\Omega_{pc}|} \int_{\Omega_{pc}} k \left\{ \frac{\partial}{\partial y_j} \left(y_p - \frac{L}{\Delta T} \chi_p \right) \frac{\partial}{\partial y_j} \left(y_p - \frac{L}{\Delta T} \chi_p \right) \right\} d\mathbf{y} \\
&= \frac{1}{|\Omega_{pc}|} \int_{\Omega_{pc}} k \left\{ 1 - \frac{L}{\Delta T} \frac{\partial \chi_p}{\partial y_p} - \frac{L}{\Delta T} \frac{\partial \chi_p}{\partial y_p} + \frac{L^2}{\Delta T^2} \frac{\partial \chi_p}{\partial y_j} \frac{\partial \chi_p}{\partial y_j} \right\} d\mathbf{y} \\
&= \frac{1}{|\Omega_{pc}|} \left\{ \int_{\Omega_{pc}} k \, d\mathbf{y} + J^p_{\Omega_{pc}}(\chi_p) \right\} \\
&= \frac{1}{|\Omega_{pc}|} \left\{ \int_{\Omega_{pc}} k \, d\mathbf{y} + \min_{v \in Y(\Omega_{pc})} J^p_{\Omega_{pc}}(v) \right\}.
\end{aligned} \tag{2.113}
$$

It finally follows that

$$k'_{e_{pp}} = \min_{v \in Y(\Omega_{pc})} a^{pp}_{\Omega_{pc}}(v, v) = a^{pp}_{\Omega_{pc}}(\chi_p, \chi_p). \tag{2.114}$$

The extremizing property (here, minimum) (2.114) can be extended to the off-diagonal terms of the effective conductivity tensor in a form similar to the inequalities derived in Nir et al. (1975) for the components of the shearing tensor. Property (2.114) is crucial for the development of the microscale models (see Section 2.7).

2.6.3 Level 2—The Sample of Cell Configurations

In level 1, as described in Section 2.6.1, the volume fraction of inclusions, c, and the periodic cell size, λ, are prescribed quantities; furthermore, the configuration of the N inclusions in the cell, expressed by their positions $\{\mathbf{y}\}_N$, is assumed to be given. The configuration effective conductivity of the cell is then determined as $k'_{e_{pq}}(c, \lambda, \{\mathbf{y}\}_N)$, Equation 2.89.

In level 2, we consider the spatial distribution of the dispersed phase (i.e., the inclusions) in the periodic cell (Cruz and Patera 1995). The positions of the geometric centers of the inclusions in the cell are now treated as continuous random variables, $\{\mathbf{Y}\}_N = \mathbf{Y}_1, \ldots, \mathbf{Y}_N \in \Omega_{pc}$, and the spatial distribution of the centers is specified by means of a JPDF, $P_N(\{\mathbf{y}\}_N)$. Naturally, any results obtained for the effective property based on an assumed JPDF for the microstructure will be practically relevant, only if the random medium under study is well characterized with respect to the geometry and distribution of inclusions, or if the behavior of the medium is (known to be) rather insensitive to the geometry and distribution of the inclusions.

For a real random medium, there appears to be no unique approach to determine the JPDF which (best) characterizes its microstructure. The JPDF can, in some cases, be known a priori, for example, when the heterogeneous medium is manufactured via a well-known controlled process. In general, it is very difficult to ascertain an assumed JPDF experimentally a posteriori.

A much utilized JPDF is the one corresponding to the random sequential addition process (Torquato 2002), illustrated in Figure 2.6. This JPDF is defined and constructed recursively from conditional JPDFs. Each conditional JPDF is uniform over the available region in the cell, in a manner to impose the condition that any two inclusions must not overlap. The statistical properties of this JPDF approximate well-defined limits as λ, and therefore N, tend to infinity. The JPDF corresponding to the random sequential addition process is isotropic and homogeneous, and is equivalent to the equilibrium distribution of the hard disk fluid up to third-order moments. Still, one may conjecture that this JPDF is similar to the one associated with the (physically intuitive) hypothesis that all possible configurations in which there is no overlap of inclusions are equally likely to occur.

The effective conductivity of the composite medium, in this level, is thus a random variable, expressed by $K_{e_{pq}} = k'_{e_{pq}}(c, \lambda, \{\mathbf{y}\}_N)$. The objective in level 2 is, therefore, to determine the average effective conductivity of the composite over the *ensemble* of possible configurations, $\langle K_{e_{pq}} \rangle (c, \lambda)$,

$$\langle K_{e_{pq}} \rangle (c, \lambda) = \int_{[\Omega_{pc}]^N} k'_{e_{pq}}(c, \lambda, \{\mathbf{y}\}_N) P_N(\{\mathbf{y}\}_N) \, d\mathbf{y}_1 \cdots d\mathbf{y}_N. \tag{2.115}$$

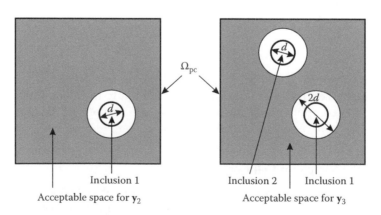

FIGURE 2.6
Illustration of the random sequential addition process.

As shown in Figure 2.4, in order to solve the multidimensional integral in (2.115), Monte Carlo methods can be used (see, e.g., the studies by Ghaddar 1995 and Lisboa 2000 on flow through fibrous porous media), which require repetition of the level-1 procedure for many realizations of the medium.

2.6.4 Level 3—The Size of the Cell

In level 3, we progressively increase, for each concentration value c, the size λ (and, thus, N) of the periodic cell, with the objective to determine the correlation length of the composite, $\lambda^C(c)$ (Cruz and Patera 1995). The concept of the correlation length is based on the regularity assumption, according to which the limit $\lim_{\lambda \to \infty} \langle K_{e_{pq}} \rangle (c, \lambda) \equiv k_{e_{pq}}(c)$ exists. The correlation length is defined as that value of the edge length of the periodic cell for which two conditions are satisfied:

1. For $\lambda > \lambda^C(c)$, the value of $\langle K_{e_{pq}} \rangle (c, \lambda)$ does not change appreciably, such that the quantity

$$\frac{|\langle K_{e_{pq}} \rangle (c, \lambda > \lambda^C(c)) - k_{e_{pq}}(c)|}{k_{e_{pq}}(c)} \tag{2.116}$$

   is smaller than a small prescribed tolerance, ε_1.

2. The standard deviation of $\langle K_{e_{pq}} \rangle (c, \lambda > \lambda^C(c))$ is smaller than a small prescribed fraction, ε_2, of $k_{e_{pq}}(c)$.

The first condition guarantees that, as the edge length of the periodic cell increases beyond λ^C (therefore incorporating more inclusions), the average $\langle K_{e_{pq}} \rangle$ does not change appreciably; the condition on the standard deviation of $K_{e_{pq}}$ guarantees that, for $\lambda > \lambda^C$, a particular realization of the medium will have an effective conductivity sufficiently close to the mean $\langle K_{e_{pq}} \rangle$. The correlation length is, thus, a key quantity of the multiscale modeling approach and establishes the connection between the behaviors of periodic and random media. We observe, furthermore, that an important practical application of the correlation length λ^C is that it indicates whether a given heterogeneous body is large enough to apply $k_{e_{pq}}(c) \approx \langle K_{e_{pq}} \rangle (c, \lambda^C)$ to compute global (engineering) quantities.

2.6.5 Level 4—The Volume Fraction of Inclusions in the Cell

Finally, in level 4, we determine, by repeating the evaluation procedure of level 3 for different values of c, the functional dependencies $\lambda^C(c)$ and $\langle K_{e_{pq}} \rangle (c, \lambda^C(c))$ for $0 < c < c_{max}$, where c_{max} is an appropriate maximum packing for the particular inclusion geometry and distribution under consideration.

Given the four-level mesoscale procedure described in this section, the problem of heat conduction in a composite material—or, in general, the transport phenomenon problem in a heterogeneous medium—for which the concentration distribution, $c(\mathbf{x})$, varies appreciably only on the macroscale, L, and for which the macroscale is large compared to $\max_{x \in \Omega_{ma}} \lambda^C(c(x))$, is basically solved: with high quantifiable probability, the macroscale result for the bulk flux will accurately predict the one for the original problem for any particular realization of the random composite medium.

2.7 Microscale Problem

In Cruz et al. (1995), Machado and Cruz (1999), Machado (2000), and Matt and Cruz (2002a), the microscale component of the macro-meso-microscale approach for the heat conduction problem in composite materials is described and discussed in detail. Similarly, in Ghaddar (1995), Cruz et al. (1995), and Lisboa (2000), the macro-meso-microscale approach for the fluid flow problem in fibrous porous media is presented. Here, we formulate isotropic microscale models to avoid the nip regions between close inclusions which hamper mesh generation. In this section, the nip-region models are presented and applied to the 3D *isotropic* heat conduction problem in particulate media (i.e., the inclusions are particles) with thermally conducting phases; for a particular configuration containing nips, the models lead to lower and upper bounds for the corresponding effective conductivity $k'_e \equiv k'_{e_{pp}}$. The bounds rely on the minimization property, Equation 2.114, of k'_e. The variational forms of the microscale-prepared mesoscale problems associated with the lower and upper bounds resemble Equation 2.106, and are respectively defined over the modified ("less stiff") domains $\mathcal{L}$ and $\mathcal{U}$, as shown in the following.

2.7.1 Nips Geometries

When dealing with random media, as the concentration increases, it is more likely that one particle in a cell will get very close to other particles in the same cell or in neighboring cells; in ordered media, the number of neighbors of one particle in the cell is fixed, and regular clusters of very close particles will be formed when the concentration is high enough. Here, we postulate that a pair of close unitary diameter particles forms a nip region when the center-to-center (nondimensional) separation distance $1 + \gamma$ is less than $1 + \gamma_c$, where γ_c is a "small" prescribed parameter. A nip region between two close spherical particles, as illustrated in Figure 2.7, is delimited by a circular cylindrical surface of radius β and with the axis parallel to the line joining the particles' centers, and two spherical end caps on the particle surfaces. Figure 2.8 shows the geometries of vertical y_1-y_2 cuts of the nip regions for the lower (Figure 2.8a) and upper (Figure 2.8b) bounds, for which we, respectively, define that $\mathcal{D}_{LB,n}$ and $\mathcal{D}_{UB,n}$ are the domains associated with nip region n, $n = 1, \ldots, N$, N is the number of nips in the cell (note that $N = 3$ for the simple cubic array); $\mathcal{L} = \Omega_{pc} \backslash \cup_{n=1}^{N} \mathcal{D}_{LB,n}$ and $\mathcal{U} = \Omega_{pc} \backslash \cup_{n=1}^{N} \mathcal{D}_{UB,n}$ are the associated modified cell domains; and $\partial\mathcal{D}_{LB,n}$ and $\partial\mathcal{D}_{UB,n}$ are the boundary surfaces of nips $\mathcal{D}_{LB,n}$ and $\mathcal{D}_{UB,n}$.

In the next two sections, we employ isotropic microscale models to construct rigorous lower and upper bounds for the effective conductivity, $k_{LB} \leq k'_e \leq k_{UB}$, based only on

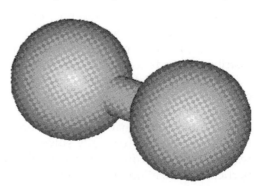

FIGURE 2.7
One 3D nip region between two proximal spherical particles; a circular cylindrical surface and two spherical end caps delimit the nip.

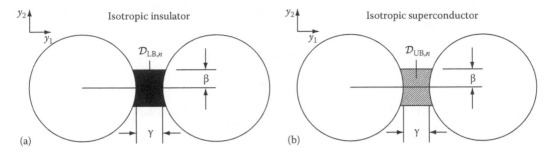

FIGURE 2.8
Geometries of vertical cuts across the y_1–y_2 plane of (a) one lower-bound nip region filled with an isotropic insulator and (b) one upper-bound nip region filled with an isotropic superconductor.

solutions defined over $\mathcal{L}$ and $\mathcal{U}$, respectively: we avoid the hard- or impossible-to-mesh nip regions, while maintaining strict control over the resulting error.

2.7.2 Lower Bound

A lower bound for k'_e, k_{LB}, can be obtained by simply assuming that the material in the nip regions $\mathcal{D}_{LB,n}$, $n = 1, \ldots, N$, is an *isotropic insulator*, Figure 2.8a; thus, since the total available volume for heat flow is decreased, we physically expect k_{LB} to be a lower bound. Because the thermal conductivity is zero inside the nips, the inner problems in $\cup_{n=1}^{N} \mathcal{D}_{LB,n}$ are irrelevant. The lower bound k_{LB} will depend on the temperature field χ_{LB} inside the modified cell domain $\mathcal{L} = \mathcal{L}_c \cup \Omega_{pc,d}$, $\mathcal{L}_c = \Omega_{pc,c} \backslash \cup_{n=1}^{N} \mathcal{D}_{LB,n}$, which is given by the variational form

$$\chi_{LB} = \arg \min_{w \in X_{\#,LB(\mathcal{L})}} J_{\mathcal{L}}(w), \tag{2.117}$$

where $X_{\#,LB}(\mathcal{L}) = \left\{ w \in H^1_{\#}(\mathcal{L}) | w_{|\mathcal{L}_c} = w^c, w_{|\Omega_{pc,d}} = w^d, \int_{\mathcal{L}_c} w^c \, d\mathbf{y} + \int_{\Omega_{pc,d}} w^d \, d\mathbf{y} = 0 \right\}$. From Equation 2.101, we write $J_{\mathcal{L}}(w)$ as

$$J_{\mathcal{L}}(w) = \int_{\mathcal{L}_c} \frac{\partial w^c}{\partial y_j} \frac{\partial w^c}{\partial y_j} \, d\mathbf{y} + \int_{\Omega_{pc,d}} \alpha \frac{\partial w^d}{\partial y_j} \frac{\partial w^d}{\partial y_j} \, d\mathbf{y} - 2 \left(\int_{\mathcal{L}_c} \frac{\partial w^c}{\partial y_1} \, d\mathbf{y} + \int_{\Omega_{pc,d}} \alpha \frac{\partial w^d}{\partial y_1} \, d\mathbf{y} \right). \tag{2.118}$$

Therefore, from the first variation of $J_{\mathcal{L}}(w)$, we derive the weak form for the field of the microscale-prepared mesoscale lower-bound problem: Find $\chi_{LB} \in X_{\#,LB}(\mathcal{L})$ such that $\forall v \in X_{\#,LB}(\mathcal{L})$:

$$\int_{\mathcal{L}_c} \frac{\partial \chi_{LB}^c}{\partial y_j} \frac{\partial v^c}{\partial y_j} \, d\mathbf{y} + \int_{\Omega_{pc,d}} \alpha \frac{\partial \chi_{LB}^d}{\partial y_j} \frac{\partial v^d}{\partial y_j} \, d\mathbf{y} = \int_{\mathcal{L}_c} \frac{\partial v^c}{\partial y_1} \, d\mathbf{y} + \int_{\Omega_{pc,d}} \alpha \frac{\partial v^d}{\partial y_1} \, d\mathbf{y}. \tag{2.119}$$

The main difference between problem (2.119) and the original problem (2.82) is that $\mathcal{L}_c$ in the former substitutes $\Omega_{pc,c}$ in the latter. Equation 2.119 naturally enforces the appropriate Neumann boundary conditions on χ_{LB} at the three curved surfaces of $\partial \mathcal{D}_{LB,n}$ of each insulating nip region $n, n = 1, \ldots, N$, such that the global (macroscopic) heat flux is zero at these surfaces.

In view of the results in the previous section for the effective conductivity k'_e, we now define, based on the solution χ_{LB} of the modified problem (2.119) in $\mathcal{L}$, the quantity k_{LB} as

$$k_{LB} \equiv a_{\mathcal{L}}(\chi_{LB}), \tag{2.120}$$

which is shown below to be a lower bound for k'_e. From Equations 2.117 through 2.120 and (2.111), it follows that (Cruz et al. 1995, Machado 2000)

$$k_{LB} = \min_{w \in X_{\#, LB}(\mathcal{L})} a_{\mathcal{L}}(w). \tag{2.121}$$

Also, from Equations 2.119 through 2.120 and (2.111), we can rewrite k_{LB} as

$$k_{LB} = (1 - \tilde{c}) + \alpha c - \frac{1}{|\Omega_{pc}|} \left(\int_{\mathcal{L}_c} \frac{\partial \chi_{LB}^c}{\partial y_1} \, d\mathbf{y} + \int_{\Omega_{pc,d}} \alpha \frac{\partial \chi_{LB}^d}{\partial y_1} \, d\mathbf{y} \right), \tag{2.122}$$

where $\tilde{c}$ is an "effective concentration" given by $\tilde{c} = 1 - \left(1/|\Omega_{pc}| \right) \int_{\mathcal{L}_c} d\mathbf{y}$.

Finally, we now prove mathematically the physically expected bounding property of k_{LB}, by using domain embedding arguments:

$$k_{LB} = a_{\mathcal{L}}(\chi_{LB}) = \min_{w \in X_{\#, LB(\mathcal{L})}} a_{\mathcal{L}}(w)$$

$$\leq a_{\mathcal{L}}(\chi|_{\mathcal{L}} + s) = a_{\mathcal{L}}(\chi|_{\mathcal{L}}) \tag{2.123}$$

$$\leq a_{\Omega_{pc}}(\chi) = k'_e. \tag{2.124}$$

In (2.123), $\chi|_{\mathcal{L}}$ is the solution to the original mesoscale problem (2.82) restricted to $\mathcal{L}$, and $s \in \mathcal{R}$ is the required shift such that $\int_{\mathcal{L}} (\chi|_{\mathcal{L}} + s) d\mathbf{y} = 0$. The inequality (2.123) follows from the fact that $(\chi|_{\mathcal{L}} + s) \in X_{\#, LB}(\mathcal{L})$; the inequality (2.124) follows from the positive (semi) definiteness of the quadratic form defined in Equation 2.111, which leads to a positive contribution over $\Omega_{pc} \backslash \mathcal{L}$.

2.7.3 Upper Bound

An upper bound for k'_e, k_{UB}, can be obtained by simply assuming that the material in the nip regions $\mathcal{D}_{UB,n}$, $n = 1, \ldots, N$, is an *isotropic superconductor*, Figure 2.8b; thus, since the total volumetric capacity for heat flow is increased, we physically expect k_{UB} to be an upper bound. Because the thermal conductivity is infinite inside the nip regions, the inner problems in $\cup_{n=1}^{N} \mathcal{D}_{UB,n}$ have trivial solutions: the nips are isothermal, so that the temperature field χ_{UB} over the cell domain Ω_{pc} is constant inside each superconducting nip. The upper bound k_{UB} will depend on χ_{UB}, whose variational form is

$$\chi_{UB} = \arg \min_{w \in W_{\#, UB}(\Omega_{pc})} J_{\Omega_{pc}}(w), \tag{2.125}$$

where $W_{\#, UB}(\Omega_{pc}) = \left\{ w \in H_{\#}^1(\Omega_{pc}) | w_{|\mathcal{U}_c} = w^c, w_{|\Omega_{pc,d}} = w^d, w_{|\mathcal{D}_{UB,n}} = C_n, n = 1, \ldots, N, \int_{\Omega_{pc}} w \, d\mathbf{y} = 0 \right\}$, the constants $C_n \in \mathcal{R}$ are part of the solution, and $\mathcal{U}$ is the modified cell domain, $\mathcal{U} = \mathcal{U}_c \cup \Omega_{pc,d}$, $\mathcal{U}_c = \Omega_{pc,c} \setminus \cup_{n=1}^{N} \mathcal{D}_{UB,n}$; it is important to note that $W_{\#,UB}(\Omega_{pc}) \subset Y(\Omega_{pc})$ (function space restriction). We can express the functional $J_{\Omega_{pc}}(w)$ as

$$J_{\Omega_{pc}}(w) = J_{\mathcal{U}}(w_{|\mathcal{U}}) + \sum_{n=1}^{N} J_{\mathcal{D}_{UB,n}}(w_{|\mathcal{D}_{UB,n}}), \tag{2.126}$$

where $w_{|\mathcal{U}}$ and $w_{|\mathcal{D}_{UB,n}}$ are the restrictions of $w(\mathbf{y})$ to $\mathcal{U}$ and $\mathcal{D}_{UB,n}$, respectively.

We can now break the problem (2.125) into N inner (microscale) problems defined over the nip regions,

$$\chi_{UB,in}\{\mathbf{y}; \bar{C}_n\} = \arg \min_{w \in W_{UB}(\mathcal{D}_{UB,n})} J_{\mathcal{D}_{UB,n}}(w), \quad n = 1, \ldots, N, \tag{2.127}$$

and one outer problem defined over $\mathcal{U}$,

$$\chi_{UB,out} = \arg \min_{w \in \bar{W}_{\#,UB}(\mathcal{U})} \left(J_{\mathcal{U}}(w) + \sum_{n=1}^{N} J_{\mathcal{D}_{UB,n}}(\chi_{UB,in}\{\mathbf{y}; w_{|\partial \mathcal{D}_{UB,n}}\}) \right), \tag{2.128}$$

where $W_{UB}(\mathcal{D}_{UB,n})$ is the rather trivial set of all functions $w(\mathbf{y}) \in H^1(\mathcal{D}_{UB,n})$ for which $w = \bar{C}_n$, $\bar{C}_n \in \mathcal{R}$ given (inner nip regions are isothermal); $\bar{W}_{\#,UB}(\mathcal{U}) = \{w \in H_{\#}^1(\mathcal{U}) | w_{|\mathcal{U}_c} = w^c,$ $w_{|\Omega_{pc,d}} = w^d, w_{|\partial \mathcal{D}_{UB,n}} = C_n, n = 1, \ldots, N, \int_{\mathcal{U}} w \, d\mathbf{y} = 0\}$, $C_n \in \mathcal{R}$ part of the (outer) solution; and

$$\chi_{UB,out} = \chi_{UB}|_{\mathcal{U}} + s', \quad \chi_{UB,in}\{\mathbf{y}; \chi_{UB,out|\partial \mathcal{D}_{UB,n}}\} = \chi_{UB}|_{\mathcal{D}_{UB,n}} + s', \quad n = 1, \ldots, N, \tag{2.129}$$

$s' \in \mathcal{R}$ is a constant shift such that $\int_{\Omega_{pc}} \chi_{UB} \, d\mathbf{y} = 0$ and $\int_{\mathcal{U}} \chi_{UB,out} \, d\mathbf{y} = 0$ may be obtained.

The inner problems have trivial solutions, since by assumption $\chi_{UB,in}\{\mathbf{y}; \bar{C}_n\} = \bar{C}_n$, $n = 1, \ldots, N$. The outer problem thus becomes

$$\chi_{UB,out} = \arg \min_{w \in \bar{W}_{\#,UB}(\mathcal{U})} J_{\mathcal{U}}(w), \tag{2.130}$$

since $J_{\mathcal{D}_{UB,n}}(\chi_{UB,in}\{\mathbf{y}; \bar{C}_n\}) = 0$, $n = 1, \ldots, N$. Taking the first variation of $J_{\mathcal{U}}(w)$, we obtain the weak form for the field of the microscale-prepared mesoscale upper-bound problem: Find $\chi_{UB,out} \in \bar{W}_{\#,UB}(\mathcal{U})$ such that $\forall v \in \bar{W}_{\#,UB}(\mathcal{U})$:

$$\int_{\mathcal{U}_c} \frac{\partial \chi_{UB,out}^c}{\partial y_j} \frac{\partial v^c}{\partial y_j} \, d\mathbf{y} + \int_{\Omega_{pc,d}} \alpha \frac{\partial \chi_{UB,out}^d}{\partial y_j} \frac{\partial v^d}{\partial y_j} \, d\mathbf{y} = \int_{\mathcal{U}_c} \frac{\partial v^c}{\partial y_1} \, d\mathbf{y} + \int_{\Omega_{pc,d}} \alpha \frac{\partial v^d}{\partial y_1} \, d\mathbf{y}. \tag{2.131}$$

Problem (2.131) differs from the original problem (2.82) in that $\mathcal{U}_c$ and $\bar{W}_{\#,UB}(\mathcal{U})$ in the former, respectively, substitute $\Omega_{pc,c}$ and $Y(\Omega_{pc})$ in the latter.

In view of the previous results for the effective conductivity k_e and lower bound k_{LB}, we now write, based on the solution $\chi_{UB,out}$ of the modified problem (2.131) in $\mathcal{U}$, the quantity k_{UB} as

$$k_{UB} \equiv a_{\Omega_{pc}}(\chi_{UB}) = \min_{w \in W_{\#,UB}(\Omega_{pc})} a_{\Omega_{pc}}(w), \tag{2.132}$$

which is shown below to be an upper bound for k'_e. From Equations 2.131, 2.132, and 2.111, and the fact that $J_{\mathcal{D}_{\text{UB},n}}(\chi_{\text{UB,in}}\{\mathbf{y}; \bar{C}_n\} = 0, n = 1, \ldots, N, k_{\text{UB}})$ can be rewritten as (see algebraic details in Machado 2000)

$$k_{\text{UB}} = 1 + (\alpha - 1)c - \frac{1}{|\Omega_{\text{pc}}|}\left(\int_{\mathcal{U}_c} \frac{\partial \chi^c_{\text{UB,out}}}{\partial y_1}\, d\mathbf{y} + \int_{\Omega_{\text{pc,d}}} \alpha \frac{\partial \chi^d_{\text{UB,out}}}{\partial y_1}\, d\mathbf{y}\right). \tag{2.133}$$

Finally, we now prove mathematically the physically expected bounding property of k_{UB}, by using function space restriction arguments:

$$k_{\text{UB}} = a_{\Omega_{\text{pc}}}(\chi_{\text{UB}}) = \min_{w \in W_{\#,\text{UB}}(\Omega_{\text{pc}})} a_{\Omega_{\text{pc}}}(w)$$

$$\geq \min_{w \in Y(\Omega_{\text{pc}})} a_{\Omega_{\text{pc}}}(w) = a_{\Omega_{\text{pc}}}(\chi) = k'_e, \tag{2.134}$$

where χ is the solution to the original cell problem (2.82). The inequality (2.134) follows from the fact that $W_{\#,\text{UB}}(\Omega_{\text{pc}}) \subset Y(\Omega_{\text{pc}})$.

2.7.4 Application of the Bounds

In Figure 2.9, we show an illustrative periodic cell with 10 fibers in which a medium-grained triangular finite element mesh has been generated and which has been prepared for the lower- and upper-bound microscale models (Machado 2000). In Figure 2.10, we show a 3D cell for the simple cubic array, containing one sphere in which a tetrahedral finite element mesh has been generated and which has been prepared with three nips for the microscale models (Matt and Cruz 2002a). As previously remarked, the cell must be prepared for the microscale models in two situations: when the cell possesses nip regions that prevent the generation of a mesh, or when a mesh can be generated, but the geometrical stiffness is so high as to prevent that a numerical solution be found. Therefore, for a level-1 periodic cell configuration, or realization, such as those in Figures 2.9 and 2.10, the

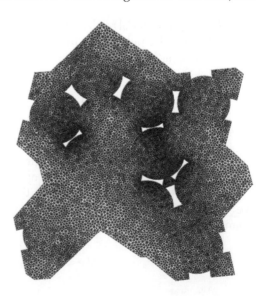

FIGURE 2.9
Microscale-prepared medium-grained triangular finite element mesh for a 2D periodic cell with 10 fibers.

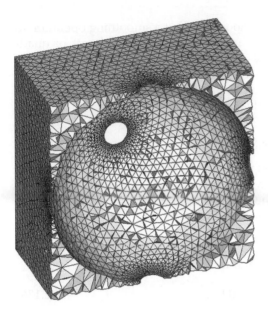

FIGURE 2.10
Microscale-prepared tetrahedral finite element mesh for a 3D periodic cubic cell with one spherical particle.

configuration effective conductivity k'_e cannot be directly computed; in this case, the configuration effective conductivity must be substituted by its *estimate*, $k'_{e,est}$, given by

$$k'_{e,est} = \frac{k_{LB} + k_{UB}}{2}. \tag{2.135}$$

The absolute error incurred with the substitution of k'_e by $k'_{e,est}$ is equal to half of the difference $(k_{UB} - k_{LB})$, and it can be made to be of the same order of magnitude as the discretization error (Machado 2000, Matt and Cruz 2002a). The corresponding relative error is equal to the absolute error divided by $k'_{e,est}$, multiplied by 100%.

2.8 Numerical Solution

Numerical solution of problems (2.82), (2.119), and (2.131) requires three steps: geometry and mesh generation (Section 2.8.1), finite element discretization, and solution of the resultant linear system of algebraic equations (Section 2.8.2).

2.8.1 Geometry and Mesh Generation

Finite element discretization requires that the physical domain of interest be meshed, that is, subdivided into a collection of nonoverlapping conforming subdomains called the elements. Thus, geometry and mesh generation are needed for the domains representing the microstructures of the class of composite materials under study. An automatic or semiautomatic geometry and mesh generation procedure must be developed, preferably based on third-party accredited software; for example, in Cruz and Patera (1995) and Machado (2000), the program MSHPTG developed at INRIA (Hecht and Saltel 1990) is used, while in Matt and Cruz (2002a) and Matt (2003), the program NETGEN developed in Austria (Schöberl 1997, 2001) is employed. NETGEN can perform boolean operations with many different primitive solids and can generate 2D and 3D unstructured meshes using the advancing front algorithm. The user may also choose to effect a smoothing operation to optimize the shape of the finite elements. For planar and surface meshes, linear or quadratic triangles can be chosen; for volume meshes, linear or quadratic tetrahedra can be chosen. The reader is referred to Machado (2000) and Matt (2003) for detailed descriptions, as well as several illustrative figures, of the domain and mesh generation procedures developed to study heat conduction in 2D and 3D composites, respectively.

2.8.2 Finite Element Discretization and Iterative Solution

In this section, we discretize the heat conduction problems formulated previously, but for the *isotropic* case. Therefore, the associated effective conductivities k'_e, k_{LB}, and k_{UB} are scalar quantities, and χ_p is simply χ. We first present the discretization procedure for the problem in the standard cell domain, Ω_{pc}, and then describe the procedural differences for the problems in the modified cell domains, $\mathcal{L}$ and $\mathcal{U}$.

The field variable of interest in the periodic cell Ω_{pc} is the temperature $\chi(\mathbf{y}) \in Y(\Omega_{pc})$, given by Equation 2.82, rewritten here in the general form

$$a(v, \chi) = \ell(v) \quad \forall v \in Y(\Omega_{pc}), \tag{2.136}$$

where $a(v, w) \equiv \int_{\Omega_{pc}} f(\mathbf{y})(\partial v/\partial y_j)(\partial w/\partial y_j)\, d\mathbf{y}$ is the symmetric bilinear form, where $f(\mathbf{y}) = 1$ if $\mathbf{y}$ belongs to the continuous-phase portion of the domain and $f(\mathbf{y}) = \alpha$ if $\mathbf{y}$ belongs to the particle-phase portion of the domain; $\ell(v) \equiv \int_{\Omega_{pc}} f(\mathbf{y})(\partial v/\partial y_1)\, d\mathbf{y}$ is the linear functional on the RHS of Equation 2.82.

An accurate representation of the geometry is increasingly necessary as the conductivity ratio α increases. It is thus appropriate to effect quadratic isoparametric discretization (Bathe 1982, Hughes 2000, Reddy and Gartling 2001), for which the Galerkin approximation to (2.136) can be written as

$$a(v, \chi_h) = \ell(v) \quad \forall v \in Y_h(\Omega_{pc,h}), \tag{2.137}$$

where

χ_h is the discrete approximation to χ
$Y_h(\Omega_{pc,h}) = \{w|_{t_k} \in P_2(t_k)\} \cap H^1_\#(\Omega_{pc,h})$, where $P_2(t_k)$ is the space of all polynomials of degree 2 defined on the kth element t_k
numerical domain $\Omega_{pc,h}$ is the quadratic representation of Ω_{pc}

All the quadratic-element midside nodes, generated by the mesh generator, which belong to edges whose extremities lie on a curved surface in Ω_{pc} are thus *moved* to the curved surface by changing their (y_1, y_2, y_3) coordinates appropriately.

Expressing the space coordinates y_j, $j = 1, 2, 3$; χ_h and v in (2.137) in terms of the usual nodal second-order Lagrangian interpolants (or shape functions), and performing all the quadratures numerically using Gauss integration (Bathe 1982), the discrete linear system of equations is obtained:

$$\underline{A}\underline{\chi}_h = \underline{F}, \tag{2.138}$$

where

$\underline{A}$ is the global system matrix corresponding to the discrete (negative) Laplacian operator
$\underline{\chi}_h$ and $\underline{F}$ are, respectively, the global vector of unknown nodal values of the scalar field $\underline{\chi}_h$ and the global vector of nodal values of the inhomogeneity $\ell(v)$

The uniqueness condition, given in continuous form in the definition of the space $Y(\Omega_{pc})$, is discretely imposed by requiring that χ_h have zero algebraic average.

The discrete equation for the numerical equivalent of the effective conductivity, $k'_{e,h}$, nondimensionalized with respect to k_c, is obtained by substituting χ_h for χ and $\Omega_{pc,h}$ for Ω_{pc} in Equation 2.90,

$$k'_{e,h} = \frac{1}{(\lambda/d)^3} \left\{ \int_{\Omega_{pc,h}} f(\mathbf{y}) \left(1 - \frac{\partial \chi_h}{\partial y_1} \right) d\mathbf{y} \right\}. \tag{2.139}$$

As verified in Matt and Cruz (2002a) and Matt (2003), $k'_{e,h}$ is optimally approximated by the finite element method, in that cubic convergence of $k'_{e,h}$ is obtained when quadratic isoparametric elements are used.

Solution of the discrete problem (2.138) can be carried out iteratively, using the well-known conjugate gradient algorithm (Golub and Van Loan 1989), with or without preconditioning. The global system matrix $\underline{A}$ is not formed; instead, the memory-efficient technique of elemental evaluation of the operator (Fischer and Patera 1994, Cruz and Patera 1995, Machado 2000) is used. The iteration proceeds until a criterion for the incomplete-iteration error, based on the Euclidean norm of the residual, is satisfied; the stopping criterion should be such that the incomplete-iteration error is made much smaller than the discretization error.

Finally, we now describe the differences of the numerical solutions of problems (2.119) and (2.131) with respect to that of problem (2.82). For problem (2.119) in the modified cell domain $\mathcal{L}$, equations similar to (2.137), (2.138), and (2.139) are obtained: in Equation 2.136, we substitute χ_{LB}, $\mathcal{L}$, and $X_{\#,LB}(\mathcal{L})$ for χ, Ω_{pc}, and $Y(\Omega_{pc})$, respectively, and follow the same discretization and iterative solution procedures indicated above. Note that the appropriate Neumann boundary conditions on χ_{LB} at the curved surfaces of $\partial\mathcal{D}_{LB,n}$ of each insulating nip region n, $n = 1, \ldots, N$, are naturally enforced. All the midside nodes that belong to element edges whose extremities lie on the curved surfaces of $\partial\mathcal{D}_{LB,n}$, $n = 1, \ldots, N$, in $\mathcal{L}$ are moved to the curved surfaces. For problem (2.131) in the modified cell domain $\mathcal{U}$, equations similar to (2.137), (2.138), and (2.139) are also obtained: in Equation 2.136, we substitute $\chi_{UB,out}$, $\mathcal{U}$, and $\bar{W}_{\#,UB}(\mathcal{U})$ for χ, Ω_{pc}, and $Y(\Omega_{pc})$, respectively, and slightly modify the discretization procedure to impose the appropriate boundary conditions on the nips surfaces. Constant temperature conditions are enforced by making all the finite element global nodes on the boundaries of each nip region $\mathcal{D}_{UB,n}$ to correspond to the *same* temperature degree of freedom C_n, $n = 1, \ldots, N$ (Cruz et al. 1995, Machado and Cruz 1999, Machado 2000). All the midside nodes that belong to element edges whose extremities lie on the curved surfaces of $\partial\mathcal{D}_{UB,n}$, $n = 1, \ldots, N$, in $\mathcal{U}$ are moved to the curved surfaces. The iterative solution procedure is the same as that for χ_h.

2.9 Sample Results

In this section, for completeness of the chapter, we present some sample numerical results for the effective conductivity (1) of the 2D square array of circular cylindrical fibers (Cruz 1997, Machado and Cruz 1999) in Table 2.1 and (2) of the 3D simple cubic array of spheres (Matt and Cruz 2002) in Table 2.2. Both sets of results are made nondimensional with

TABLE 2.1

Numerical Results, $k_{e,h}$, for the Transverse Effective Conductivity of the Square Array of Circular Cylindrical Fibers, as a Function of the Dispersed Phase Volume Fraction, c, and Conductivity Ratio, α

	$k_{e,h}$		
c	$\alpha = 2$	$\alpha = 10$	$\alpha = 50$
0.10	1.069	1.178	1.213
0.20	1.143	1.391	1.476
0.30	1.222	1.652	1.813
0.40	1.308	1.980	2.263
0.50	1.401	2.415	2.915
0.60	1.503	3.037	3.990
0.70	1.615	4.063	6.342
0.75	1.677	4.946	9.546
0.77	1.702	5.469	12.75
0.78	1.715	5.805	16.32
$\pi/4$	1.714 ($\pm$0.53%)	5.9 ($\pm$4.1%)	18 ($\pm$26%)

Source: Cruz, M.E., Two-dimensional simulation of heat conduction in ordered composites with a thermally-conducting dispersed phase, *Proceedings of the 14th Brazilian Congress of Mechanical Engineering (COBEM)*, Paper COB288, December 8–12, Bauru, Sao Paulo, Brazil, 1997; Machado, L.B. and Cruz, M.E., Bounds for the effective conductivity of unidirectional composites based on isotropic microscale models, *Proceedings of the 15th Brazilian Congress of Mechanical Engineering (COBEM)*, Paper AACEDD, November 22–26, Sao Paulo, Brazil, 1999.

TABLE 2.2

Numerical Results, $k_{e,h}$, for the Effective Conductivity of the Simple Cubic Array of Spheres, as a Function of the Dispersed Phase Volume Fraction, c, and Conductivity Ratio, α

	$k_{e,h}$			
c	$\alpha = 2$	$\alpha = 5$	$\alpha = 10$	$\alpha = 50$
0.05	1.0380	1.0883	1.1169	1.1484
0.10	1.0769	1.1819	1.2434	1.3123
0.15	1.1169	1.2817	1.3814	1.4954
0.20	1.1580	1.3883	1.5324	1.7018
0.25	1.2003	1.5035	1.6998	1.9399
0.30	1.2438	1.6278	1.889	2.220
0.35	1.2887	1.7649	2.106	2.568
0.40	1.3351	1.9173	2.364	3.016
0.45	1.3836	2.094	2.692	3.674
0.50	1.434	2.304	3.147	4.920
0.51	1.444	2.353	3.269	5.411
$\pi/6$	1.458	2.420 ($\pm$0.06%)	3.465 ($\pm$0.3%)	6.9 ($\pm$6.9%)

Source: Matt, C.F. and Cruz, M.E., Effective conductivity of longitudinally-aligned composites with cylindrically orthotropic short fibers, *Proceedings of the 12th International Heat Transfer Conference (IHTC)*, Vol. 3, pp. 21–26, August 18–23, Grenoble, France, 2002a.

respect to the matrix thermal conductivity and are given as functions of the dispersed phase volume fraction c and phase conductivity ratio α. The volume fraction c increases all the way up to the corresponding maximum packing values, such that the techniques for the conductivity lower and upper bounds, described in Section 2.7 and illustrated in Figures 2.7 through 2.10, have been applied. To obtain the results in Tables 2.1 and 2.2, respectively, linear triangles and isoparametric quadratic tetrahedra have been used. The results in the tables are shown with the proper number of significant digits and have been validated in Cruz (1997) and Machado and Cruz (1999) for the 2D case, and in Matt and Cruz (2002a) for the 3D case. For the maximum packing volume fractions in Tables 2.1 and 2.2, the relative errors incurred with the substitution of the configuration conductivity with the estimated conductivity are also indicated.

For numerical effective conductivity results for other 2D and 3D, ordered and random geometries, the reader is referred to the works by Cruz and Patera (1995), Cruz et al. (1995), Cruz (1997, 1998), Machado and Cruz (1999), Matt (1999, 2003), Rocha (1999), Machado (2000), Rocha and Cruz (2001), Matt and Cruz (2001, 2002a, 2002b, 2004, 2006, 2008), and Pereira et al. (2006).

2.10 Conclusions

In this chapter, we have presented the continuous formulations of the problems that are part of the multiscale modeling approach, a technique applicable to the analysis of transport phenomena in random heterogeneous media. The approach consists in the variational hierarchical decoupling of the length scales of the original multiscale problem, such that the macroscale, mesoscale, and microscale (sub)problems are derived. In the macroscale problem, for which the effective property is input data, one seeks to compute global quantities. In the mesoscale problem, in four levels, the random nature of the medium is considered, and one seeks to calculate not only the effective property of interest but also the statistical correlation length. Finally, in the microscale problem, local effects are modeled, mitigating the difficulty associated with the geometrical stiffness present in some realizations of the periodic cells with many inclusions.

Acknowledgments

M.E.C. would like to thank the Brazilian Council for Development of Science and Technology (CNPq) for Grant PQ-302725/2009-1. The authors would also like to thank Dr. Joachim Schöberl, from Johannes Kepler Universität Linz, Austria, for the free academic license of NETGEN 4.4, and Eng. Toseli Matos for his assistance with the figures.

Nomenclature

a	bilinear form
A	operator
c	dispersed phase volume fraction

d	microscale
f, F, G	generic functions
I, J	functionals
k	thermal conductivity
ℓ	linear functional
L	macroscale
P	space of polynomials
q	flux
$\Re$	set of real numbers
T	temperature
u	solution function
W, X, Y	function spaces
x, y	space coordinates

Greek Variables

α	conductivity ratio
β	radius of 3D nip region
χ	mesoscale temperature
ε	ratio of fast scale to slow scale
λ	mesoscale
Ω	domain

Superscripts

$'$	quantity pertaining to a cell configuration
$*$	nondimensional quantity
c	continuous phase
C	pertaining to correlation length
d	dispersed phase

Subscripts

c	continuous phase
d	dispersed phase
e	effective
in	pertaining to inner problem
LB	lower bound
out	pertaining to outer problem
pc	periodic cell
UB	upper bound

References

Adams, R. 1975. *Sobolev Spaces*. New York: Academic Press, Inc.

Alzina, A., E. Toussaint, A. Béakou, and B. Skoczen. 2006. Multiscale modelling of thermal conductivity in composite materials for cryogenic structures. *Compos. Struct.* 74: 175–185.

Auriault, J.-L. 1983. Effective macroscopic description for heat conduction in periodic composites. *Int. J. Heat Mass Transfer* 26: 861–869.

Auriault, J.-L. and H. I. Ene. 1994. Macroscopic modelling of heat transfer in composites with interfacial thermal barrier. *Int. J. Heat Mass Transfer* 37: 2885–2892.

Babuška, I. 1975. Homogenization and its application. Mathematical and computational problems. Technical Note BN-821, University of Maryland, Maryland.

Bathe, K. J. 1982. *Finite Element Procedures in Engineering Analysis*. Englewood Cliffs, NJ: Prentice-Hall, Inc.

Bendsøe, M. P. and N. Kikuchi. 1988. Generating optimal topologies in structural design using a homogenization method. *Comput. Methods Appl. Mech. Eng.* 71: 197–224.

Bensoussan, A., J.-L. Lions, and G. C. Papanicolaou. 1978. *Asymptotic Analysis for Periodic Structures*. Amsterdam, the Netherlands: North-Holland Publishing Co.

Bensoussan, A., J.-L. Lions, and G. C. Papanicolaou. 1979. Boundary layers and homogenization of transport processes. *Publ. RIMS, Kyoto Univ.* 15: 53–157.

Beran, M. J. 1968. *Statistical Continuum Theories*. New York: John Wiley & Sons, Inc.

Chung, P. W., K. K. Tamma, and R. R. Namburu. 2001. Homogenization of temperature-dependent thermal conductivity in composite materials. *J. Thermophys. Heat Transfer* 15: 10–17.

Cruz, M. E. 1997. Two-dimensional simulation of heat conduction in ordered composites with a thermally-conducting dispersed phase. *Proceedings of the 14th Brazilian Congress of Mechanical Engineering (COBEM)*, Paper COB288, December 8–12, Bauru, Sao Paulo, Brazil.

Cruz, M. E. 1998. Computation of the effective conductivity of three-dimensional ordered composites with a thermally-conducting dispersed phase. *Proceedings of the 11th International Heat Transfer Conference (IHTC)*, Vol. 7, pp. 9–14, August 23–28, Taylor and Francis, Inc., Levittown, PA.

Cruz, M. E. 2001. Computational approaches for heat conduction in composite materials. In *Computational Methods and Experimental Measurements X*, eds. Y. V. Esteve, G. M. Carlomagno, and C. A. Brebbia, pp. 657–668. Southampton, U.K.: WIT Press.

Cruz, M. E. 2005. Introduction to stochastic homogenization (in Portuguese). In *First Meeting on Multiscale Computational Modeling*, eds. M. A. Murad, F. Pereira, H. A. Souto, M. Cruz, and G. Braga, Chap. 5. Petrópolis, Brazil: Gráfica LNCC.

Cruz, M. E., C. K. Ghaddar, and A. T. Patera. 1995. A variational-bound nip-element method for geometrically stiff problems; application to thermal composites and porous media. *Proc. R. Soc. Lond. A Math. Phys. Sci.* 449: 93–122.

Cruz, M. E. and A. T. Patera. 1995. A parallel Monte-Carlo finite-element procedure for the analysis of multicomponent random media. *Int. J. Numer. Meth. Eng.* 38: 1087–1121.

Fischer, P. F. and A. T. Patera. 1994. Parallel simulation of viscous incompressible flows. *Ann. Rev. Fluid Mech.* 26: 483–527.

Ghaddar, C. K. 1995. On the permeability of unidirectional fibrous media: A parallel computational approach. *Phys. Fluids* 7: 2563–2586.

Golub, G. H. and C. F. Van Loan. 1989. *Matrix Computations*, 2nd edn. Baltimore, MD: The Johns Hopkins University Press.

Hecht, F. and E. Saltel. 1990. Emc^2: Editeur de maillages et de contours bidimensionnels. Manuel d'Utilisation, Rapport Technique No. 118, INRIA, France.

Hildebrand, F. B. 1976. *Advanced Calculus for Applications*, 2nd edn. Englewood Cliffs, NJ: Prentice-Hall, Inc.

Hughes, T. J. R. 2000. *The Finite Element Method: Linear Static and Dynamic Finite Element Analysis*. New York: Dover Publications, Inc.

Kamiński, M. and M. Kleiber. 2000. Numerical homogenization of N-component composites including stochastic interface defects. *Int. J. Numer. Meth. Eng.* 47:1001–1027.

Keller, J. B. 1980. Darcy's law for flow in porous media and the two-space method. In *Nonlinear Partial Differential Equations in Engineering and Applied Science: Proceedings of a Conference Sponsored by ONR Held at University of Rhode Island*, eds. R. L. Sternberg, A. J. Kalinowski, and J. S. Papadakis, pp. 429–443. New York: Marcel Dekker, Inc.

Kohn, R. V. and G. W. Milton. 1989. *Random Media and Composites*. Philadelphia, PA: SIAM.

Larsen, E. W. 1975. Neutron transport and diffusion in inhomogeneous media. I. *J. Math. Phys.* 16: 1421–1427.

Lisboa, E. F. A. 2000. A multi-scale approach for the calculation of the longitudinal permeability of random fibrous porous media. MSc dissertation (in Portuguese), UFRJ-PEM/COPPE, Rio de Janeiro, Brazil.

Machado, L. B. 2000. Determination of the effective thermal conductivity of unidirectional fibrous random composites. MSc dissertation (in Portuguese), UFRJ-PEM/COPPE, Rio de Janeiro, Brazil.

Machado, L. B. and M. E. Cruz. 1999. Bounds for the effective conductivity of unidirectional composites based on isotropic microscale models. *Proceedings of the 15th Brazilian Congress of Mechanical Engineering (COBEM)*, Paper AACEDD, November 22–26, Sao Paulo, Brazil.

Matt, C. F. 1999. Heat conduction in three-dimensional ordered composites with spherical or cylindrical particles. MSc dissertation (in Portuguese), UFRJ-PEM/COPPE, Rio de Janeiro, Brazil.

Matt, C. F. 2003. Effective thermal conductivity of composite materials with three-dimensional microstructures and interfacial thermal resistance. DSc thesis (in Portuguese), UFRJ-PEM/COPPE, Rio de Janeiro, Brazil.

Matt, C. F. and M. E. Cruz. 2001. Calculation of the effective conductivity of ordered short-fiber composites. *35th AIAA Thermophysics Conference, 2001 Summer Co-Located Conferences*, Paper AIAA 2001–2968, June 11–14, Anaheim, CA.

Matt, C. F. and M. E. Cruz. 2002a. Application of a multiscale finite-element approach to calculate the effective conductivity of particulate media. *Comput. Appl. Math.* 21: 429–460.

Matt, C. F. and M. E. Cruz. 2002b. Effective conductivity of longitudinally-aligned composites with cylindrically orthotropic short fibers. *Proceedings of the 12th International Heat Transfer Conference (IHTC)*, Vol. 3, pp. 21–26, August 18–23, Grenoble, France.

Matt, C. F. and M. E. Cruz. 2004. Calculation of the effective conductivity of disordered particulate composites with interfacial resistance. *37th AIAA Thermophysics Conference*, Paper AIAA 2004-2458, June 28–July 1, Portland, OR.

Matt, C. F. and M. E. Cruz. 2006. Enhancement of the thermal conductivity of composites reinforced with anisotropic short fibers. *J. Enhanc. Heat Transfer* 13: 17–38.

Matt, C. F. and M. E. Cruz. 2008. Effective thermal conductivity of composite materials with 3-D microstructures and interfacial thermal resistance. *Numer. Heat Transfer, Part A: Appl.* 53: 577–604.

Mei, C. C. and J.-L. Auriault. 1989. Mechanics of heterogeneous porous media with several spatial scales. *Proc. R. Soc. Lond. A* 426: 391–423.

Mei, C. C. and J.-L. Auriault. 1991. The effect of weak inertia on flow through a porous medium. *J. Fluid Mech.* 222: 647–663.

Milton, G. W. 2002. *The Theory of Composites*. Cambridge, U.K.: Cambridge University Press.

Nir, A., H. F. Weinberger, and A. Acrivos. 1975. Variational inequalities for a body in a viscous shearing flow. *J. Fluid Mech.* 68: 739–755.

Pereira, A. C., C. F. Matt, and M. E. Cruz. 2006. Numerical prediction of the effective thermal conductivity of fibrous composite materials. *9th AIAA/ASME Joint Thermophysics and Heat Transfer Conference*, Paper AIAA2006-3429, June 5–8, San Francisco, CA.

Reddy, J. N. and D. K. Gartling. 2001. *The Finite Element Method in Heat Transfer and Fluid Dynamics*, 2nd edn. Boca Raton, FL: CRC Press LLC.

Rocha, R. P. A. 1999. Heat conduction in unidirectional fibrous composites with interfacial thermal resistance. MSc dissertation (in Portuguese), UFRJ-PEM/COPPE, Rio de Janeiro, Brazil.

Rocha, R. P. A. and M. E. Cruz. 2001. Computation of the effective conductivity of unidirectional fibrous composites with an interfacial thermal resistance. *Numer. Heat Transfer, Part A: Appl.* 39: 179–203.

Schöberl, J. 1997. *NETGEN—An Advancing Front 2D/3D-Mesh Generator Based on Abstract Rules*. Linz, Austria: Institute of Mathematics, Johannes Kepler Universität Linz.

Schöberl, J. 2001. *NETGEN—4.0, Numerical and Symbolic Scientific Computing*. Linz, Austria: Johannes Kepler Universität Linz.

Schumann, U. 1981. Homogenized equations of motion for rod bundles in fluid with periodic structure. *Ingenieur-Archiv*. 50: 203–216.

Torquato, S. 2002. *Random Heterogeneous Materials, Microstructure and Macroscopic Properties*. New York: Springer-Verlag.

Whitaker, S. 1999. *The Method of Volume Averaging*. Dordrecht, the Netherlands: Kluwer Academic Publishers.

3

Temperature Measurements: Thermoelectricity and Microthermocouples

François Lanzetta and Eric Gavignet

CONTENTS

3.1 Introduction

What is temperature exactly? How to measure it? These are two simple questions but the answers are complex. A brief history of the temperature measurement gives us the keys to understand how the intuitive concept of temperature becomes a scientific reality [1]. Temperature "measures" hot and cold and the word is Latin in origin: *temperare*—to mix. It was mostly used when liquids are mixed that cannot afterward be separated, like wine and water. The "*-tur*" of the present tense indicates that some liquid is being mixed with another one. For Hippocrates of Cos, the Greek physician, proper mixing represented an imbalance of the bodily fluids blood, phlegm, and black and yellow bile that was supposed to lead to disease that made the body unusually hot or cold or dry or moist. Klaudios Galenos, another Greek physician, took up the idea and elaborated on it. He assumed an influence of the climate on the mix of body fluids that would then determine the character, or temperament, of a person.

Until about 260 years ago, temperature measurement was very subjective. Intuitively, people have known about temperature for a long time: fire is hot and snow is cold, and the first temperature measurement was mainly indicated to confirm the presence or absence of fever. In the time of Hippocrates, only the hand was used to detect the hot or cold of the human body, although fever and chills were known as signs of morbid medical processes. In the Middle Ages, the four humors were assigned the qualities of hot, cold, dry, and moist, and thus fever again acquired importance. Galileo in 1592 devised a crude temperature-measuring instrument, but it had no scale and therefore no numerical readings; further, it was affected by atmospheric pressure. A large step forward was achieved by Sanctorio Sanctorius who invented a mouth thermometer [2]. He described his inventions in 1625. He produced several designs, but all were cumbersome and required a long time to measure the oral temperature. To this day, the time to get an accurate, stable reading remains difficult.

By the early eighteenth century, as many as 35 different temperature scales had been devised. In 1714, Daniel Gabriel Fahrenheit invented both the mercury and the alcohol thermometer. Fahrenheit's mercury thermometer consists of a capillary tube that after being filled with mercury is heated to expand the mercury and expel the air from the tube. The tube is then sealed, leaving the mercury free to expand and contract with temperature changes. Although the mercury thermometer is not as sensitive as the air thermometer, by being sealed, it is not affected by the atmospheric pressure. Mercury freezes at −39°C, so it cannot be used to measure temperature below this point. Alcohol, on the other hand, freezes at −113°C, allowing much lower temperatures to be measured.

At the time, thermometers were calibrated between the freezing point of salted water and the human body temperature. Fahrenheit subdivided this range into 96 points, giving his thermometers more resolution and a temperature scale very close to today's Fahrenheit scale. Later in the eighteenth century, Anders Celsius realized that it would be advantageous to use more common calibration references and to divide the scale into 100 increments instead of 96. He chose to use 100° as the freezing point and 0° as the boiling point of water. This scale was later reversed and the centigrade scale was born.

Lord Kelvin, or William Thomson, postulated the existence of an absolute zero and established the absolute scale of temperature. Sir William Hershel discovered that when sunlight was spread into a color swath using a prism, he could detect an increase in temperature when moving a blackened thermometer across the spectrum of colors.

Hershel found that the heating effect increased toward and beyond the red in the region we now call "infrared." He measured radiation effects from fires, candles, and stoves and deduced the similarity of light and radiant heat. However, it was not until well into the following century that this knowledge was exploited to measure temperature.

In 1821, Thomas Johann Seebeck discovered that a current could be produced by unequally heating two junctions of two dissimilar metals, the thermocouple effect. Seebeck assigned constants to each type of metal and used these constants to compute total amount of current flowing. Also in 1821, Sir Humphrey Davy discovered that all metals have a positive temperature coefficient of resistance and that platinum could be used as an excellent resistance temperature detector (RTD). These two discoveries marked the beginning of serious electrical sensors. The late nineteenth century saw the introduction of bimetallic temperature sensor. These thermometers contain no liquid but operate on the principle of unequal expansion between two metals.

The twentieth century has seen the discovery of semiconductor devices, such as the thermistor, the integrated circuit sensor, a range of noncontact sensors, and also the fiber-optic temperature sensors. Also, Lord Kelvin was finally rewarded for his early work in temperature measurement. The increments of the Kelvin scale were changed from degrees to kelvins. The twentieth century also saw the refinement of the temperature scale. Temperatures can now be measured to within about 0.001°C over a wide range, although it is not a simple task. The most recent change occurred with the updating of the International Temperature Scale in 1990 [3] to the International Temperature Scale of 1990 (ITS-90).

The twenty-first century will see the new definition of the kelvin. The international measurement community, through the International Committee for Weights and Measures, is considering updating the International System of Units (SI). This update, which will probably occur in 2011, will redefine the kilogram, the ampere, and the kelvin in terms of fundamental physical constants. The kelvin, instead of being defined by the triple point of water as it is currently, will be defined by assigning an exact numerical value to Boltzmann's constant. The change would generalize the definition, making it independent of any material substance, measurement technique, and temperature range, to ensure the long-term stability of the unit. This new definition will allow the accuracy of temperature measurements to gradually improve without the limitations associated with the manufacture and use of triple point of water cells. For some temperature ranges at least, true thermodynamic methods are expected to eventually replace the International Temperature Scale as the primary standard of temperature. The unit of thermodynamic temperature, also referred to as Kelvin temperature or absolute temperature, is kelvin (K) [4]. It is defined in terms of the interval between the absolute zero and triple point of pure water, 273.16 K. Kelvin is the fraction 1/273.16 of that temperature. In addition to the thermodynamic temperature, the Celsius (°C) temperature is defined as equal to the thermodynamic temperature minus 273.15, and the magnitude of 1°C is numerically equal to 1 K.

The numerous measurement techniques can be classified into three different categories depending on the nature of contact between the sensor and the external system (gaseous, liquid, or solid) [5]. The first one is invasive: the sensor is in direct contact with the medium (i.e., sensor in liquid). The second one is semi-invasive: the medium is treated in some manner to enable remote observation (i.e., temperature sensitive paint). The third category concerns noninvasive method: the medium is observed remotely (i.e., pyrometry, infrared thermography). Besides, the usual temperature sensors present a lot of limitations: the different thermal processes allowing to obtain the equilibrium of sensors are based on relatively slow phenomena, inducing a low time resolution; the association of the three basic modes of heat transfer in these processes induces difficulties for the theoretical modeling. Then, most often, a temperature sensor only gives its own temperature, which can be quite different from the fluid one. Of course, the sensor volume appears always as a main parameter: it governs the intrusive character of the measurement and its spatial resolution. Being omnipresent in the heat transfer equations, its effect on the thermal inertia and thus on the time resolution is preponderant too. This is why a reduction of the volume of the sensors is a very interesting way.

Among the numerous families of sensors, thermoelectric junctions present a well-known disadvantage: their limited sensitivity, due to the low level of the thermal electromotive force (EMF), imposes efficient electronic devices in order to amplify the signal and a lot of precautions in order to maintain a good signal-to-noise ratio. But they present also a lot of advantages, particularly a large temperature range and a good linearity. Above all, they make very good competitors in the race for the volume reducing, and the term "micro-thermocouple" is now usual in the scientific literature.

3.2 Measurement of Thermocouple Voltage

3.2.1 Thermoelectric Effects

The thermocouple is the most widely used electrical sensor in thermometry, and it appears to be the simplest of electrical transducers. Thermocouples are inexpensive, small in size, rugged, and remarkably accurate when used with an understanding of their peculiarities. Accurate temperature measurements are typically important in many scientific fields for the control, the performance, and the operation of many engineering processes. A simple thermocouple is a device that converts thermal energy to electric energy. Its operation is based upon the findings of Seebeck [6]. When two different metals A and B form a closed electric circuit and their junctions are kept at different temperatures T_1 and T_2 (Figure 3.1), a small electric current appears.

The electromotive force, EMF, produced under these conditions is called the Seebeck EMF. The amount of electric energy produced is used to measure temperature. The EMF depends on materials used in the couple and the temperature difference $T_1 - T_2$. Seebeck effect is actually the combined result of two other phenomena, Peltier effect [7] and Thomson effect [8]. Peltier discovered that temperature gradients along conductors in a circuit generate an EMF. Thomson observed the existence of an electromotive force due to the contact of two dissimilar metals. Thomson effect is normally much smaller in magnitude than the Peltier effect and can be minimized and disregarded with proper thermocouple design.

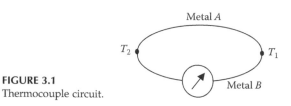

FIGURE 3.1
Thermocouple circuit.

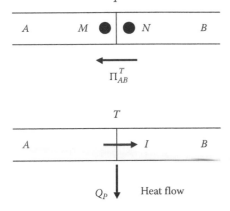

FIGURE 3.2
Peltier effect without current flow.

FIGURE 3.3
Peltier effect with current flow.

3.2.1.1 Peltier Effect

A Peltier EMF $V_M - V_N$ is created at the junction of two different materials (wire or film) A and B, at the same temperature T, depending on the material and the temperature T (Figure 3.2):

$$V_M - V_N = \Pi_{AB}^T \tag{3.1}$$

where Π_{AB} is the Peltier coefficient at temperature T.

When a current I flows through a thermocouple junction (Figure 3.3), heat Q_P is either absorbed or dissipated depending on the direction of current. This effect is independent of Joule heating.

$$dQ_P = (V_M - V_N) \cdot I \cdot dt = \Pi_{AB}^T \cdot I \cdot dt \tag{3.2}$$

where Q_P is the heat quantity exchanged with the external environment to maintain the junction at the constant temperature T.

The phenomenon is reversible, depending on the direction of the current flow and

$$\Pi_{AB}^T = -\Pi_{BA}^T \tag{3.3}$$

3.2.1.2 Volta's Law

In an isothermal circuit composed by different materials, the sum of the Peltier EMFs is null (Figure 3.4) and

$$\Pi_{AB} + \Pi_{BC} + \Pi_{CD} + \Pi_{DA} = 0 \tag{3.4}$$

3.2.1.3 Thomson Effect

Thomson EMFs correspond to the tension $e_A(T_1, T_2)$ between two points M and N of the same conductor, submitted to a temperature gradient, depending only on the nature of the conductor (Figure 3.5):

$$e_A(T_1, T_2) = \int_{T_1}^{T_2} \tau_A dT \tag{3.5}$$

where τ_A is the Thomson coefficient of the material A.

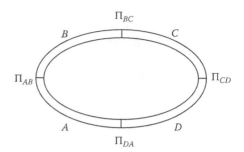

FIGURE 3.4
Volta's law with four materials.

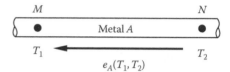

FIGURE 3.5
Thomson effect without current flow.

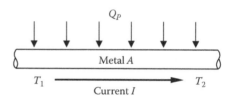

FIGURE 3.6
Thomson effect with current flow.

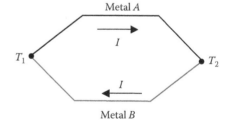

FIGURE 3.7
Seebeck effect.

When a current I flows through a conductor within a thermal gradient $(T_1 - T_2)$, heat Q_T is either absorbed or dissipated (Figure 3.6):

$$dQ_T = e_A(T_1, T_2)I\,dt = \int_{T_1}^{T_2} \tau_A dT\,I\,dt \tag{3.6}$$

3.2.1.4 Seebeck Effect

When a circuit is formed by a junction of two different metals A and B and the junctions are held at two different temperatures T_1 and T_2, a current I flows in the circuit caused by the difference in temperature between the two junctions (Figure 3.7).

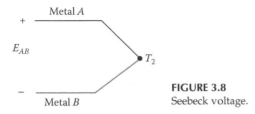

FIGURE 3.8
Seebeck voltage.

The sum of the different Peltier and Thomson EMFs for the circuit corresponds to the Seebeck EMF:

$$E_{AB}(T_2, T_1) = \Pi_{AB}^{T_1} + \Pi_{BA}^{T_2} + \int_{T_1}^{T_2} \tau_B \, dT + \int_{T_2}^{T_1} \tau_A \, dT$$

$$E_{AB}(T_2, T_1) = \Pi_{AB}^{T_1} - \Pi_{AB}^{T_2} + \int_{T_2}^{T_1} (\tau_A - \tau_B) \, dT$$

(3.7)

Then, the Seebeck EMF becomes

$$E_{AB}(T_1, T_2) = \sigma_{AB}(T_1 - T_2) \tag{3.8}$$

where σ_{AB} is the Seebeck coefficient for the A and B metals of the couple (μV °C^{-1} or μV K^{-1}). This coefficient corresponds to a constant of proportionality between the Seebeck voltage and the temperature difference.

If the circuit is open at the center (Figure 3.8), the net open voltage is a function of the junction temperature and the composition of the two metals.

The thermoelectric power, or sensitivity, of a thermocouple is given by (Table 3.1)

$$\sigma_{AB} = \frac{dE_{AB}}{dT} \tag{3.9}$$

Thermocouples are made by the association of dissimilar materials producing the biggest possible Seebeck. In industrial processes, the common thermocouples are presented in Table 3.2.

3.2.2 Practical Measurement of Thermocouple Voltage

3.2.2.1 Measurement of Junction Voltage

We consider a thermocouple composed by two dissimilar materials A and B (Figure 3.9) [9,10]. The hot junction is submitted to a medium whose temperature needs to be measured (T_2). The voltage E_{AB} is measured by a voltmeter placed at the cold end of the thermocouple at temperature T_1. The Seebeck EMF is

$$E_{AB} = \int_{T_1}^{T_2} \sigma_A \, dT + \int_{T_2}^{T_1} \sigma_B \, dT = (\sigma_A - \sigma_B)(T_2 - T_1) \tag{3.10}$$

$$E_{AB} = \sigma_{AB}(T_2 - T_1) \quad \text{where } \sigma_{AB} = \sigma_A - \sigma_B \tag{3.11}$$

TABLE 3.1

Seebeck Coefficients of Various Thermocouple Materials Relative to Platinum at 0°C

Material	Seebeck Coefficient (μV °C^{-1})	Material	Seebeck Coefficient (μV °C^{-1})
Bismuth	−72	Silver	6.5
Constantan	−35	Copper	6.5
Alumel	−17.3	Gold	6.5
Nickel	−15	Tungsten	7.5
Potassium	−9	Cadmium	7.5
Sodium	−2	Iron	18.5
Platinum	0	Chromel	21.7
Mercury	0.6	Nichrome	25
Carbon	3	Antimony	47
Aluminum	3.5	Germanium	300
Lead	4	Silicium	440
Tantalum	4.5	Tellurium	500
Rhodium	6	Selenium	900

Source: Rathakrishnan, E., *Instrumentation, Measurements and Experiments in Fluids*, CRC Press, Boca Raton, FL, 2007.

Platinum = Platinum is the reference material for calculating the Seebeck coefficient of all other materials, because its value is 0 μV °C^{-1}.

3.2.2.2 Intermediate Metal

A third isothermal metal (i.e., metal C) is inserted between the two metals A and B (Figure 3.10). Then, the Seebeck EMF E_{AB} becomes

$$E_{AB} = \int_{T_1}^{T_2} \sigma_A \, dT + \int_{T_2}^{T_1} \sigma_B \, dT \tag{3.12}$$

$$E_{AB} = \sigma_{AB}(T_2 - T_1)$$

The output voltage E_{AB} is not influenced by the additional isothermal metal C.

3.2.2.3 Temperature Gradient along a Metal Element

Heating or cooling is provided along a metal element (Figure 3.11). This phenomenon creates a temperature gradient ($T_3 - T_4$). The Seebeck voltage E_{AB} is

$$E_{AB} = \int_{T_1}^{T_2} \sigma_A \, dT + \int_{T_2}^{T_3} \sigma_B \, dT + \int_{T_3}^{T_4} \sigma_B \, dT + \int_{T_4}^{T_1} \sigma_B \, dT \tag{3.13}$$

$$E_{AB} = \sigma_A(T_2 - T_1) - \sigma_B(T_2 - T_1) \tag{3.14}$$

$$E_{AB} = \sigma_{AB}(T_2 - T_1) \tag{3.15}$$

The temperature gradient along the element of metal B does not affect the output voltage.

TABLE 3.2
Thermocouple Types

Type	Metal A (+)	Metal B (−)	Temperature Range (°C)	Seebeck Coefficient α (μV/°C) at T°C	Standard Error (%)	Minimal Error (%)	Comments
B	Platinum–30% rhodium	Platinum–6% platinum	0 to 1820	5.96 μV at 600°C	0.5	0.25	Idem R type (glass industry)
E	Nickel–10% chromium	Copper–nickel alloy (constantan)	−270 to 1000	58.67 μV at 0°C	1.7–0.5	1–0.4	Interesting sensitivity
J	Iron	Copper–nickel alloy (constantan)	−210 to 1200	50.38 μV at 0°C	2.2–0.75	1.1–0.4	For atmosphere reduced (plastic industry)
K	Nickel–chromium alloy (chromel)	Nickel–aluminum alloy (alumel)	−270 to 1372	39.45 μV at 0°C	2.2–0.75	1.1–0.2	The most widely used because of its wide temperature range, supports an oxidizing atmosphere
N	Nickel–chromium–silicium alloy (Nicrosil)	Nickel–silicium alloy (nisil)	−270 to 1300	25.93 μV at 0°C	2.2–0.75	1.1–0.4	New combination very stable
R	Platinum–13% rhodium	Platinum	−50 to 1768	11.36 μV at 600°C	1.5–0.25	0.6–0.1	High temperature applications, resists oxidation
S	Platinum–10% rhodium	Platinum	−50 to 1768	10.21 μV at 600°C	1.5–0.25	0.6–0.1	Idem R type
T	Copper	Copper–nickel alloy (constantan)	−270 to 400	38.75 μV at 0°C	1–0.75	0.5–0.4	Cryogenic applications
W	Tungsten	Tungsten–26% rhenium	+20 to +2300				Sensitive to oxidizing atmospheres, linear response and good performance in high temperature
W3	Tungsten–3% rhenium	Tungsten–25% rhenium	+20 to +2000				Idem W type
W5	Tungsten–5% rhenium	Tungsten–26% rhenium	+20 to +2300				Idem W type

Source: Devin, E., *Techniques de l'Ingénieur,* tome R2594:1–26, 1997.

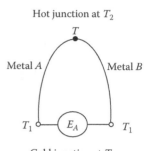

FIGURE 3.9
Basic thermocouple measurement voltage.

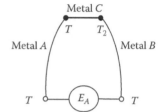

FIGURE 3.10
Isothermal intermediate material.

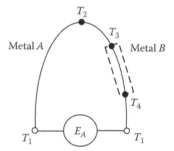

FIGURE 3.11
Temperature gradient along a metal element.

3.2.2.4 External Reference Junction

The circuit contains two thermocouples in differential mode (Figure 3.12). This circuit maintains one thermocouple at an external reference temperature $T_{ref} = 0°C$. The output EMF voltage E_{AB} is

$$E_{AB} = \int_{T_1}^{T_{ref}} \sigma_B \, dT + \int_{T_{ref}}^{T_2} \sigma_A \, dT + \int_{T_2}^{T_1} \sigma_B \, dT \tag{3.16}$$

$$E_{AB} = \sigma_{AB}(T_2 - T_{ref}) \tag{3.17}$$

If $T_{ref} = 0°C$, then the output voltage becomes $E_{AB} = \sigma_{AB} T_2$. This method is used to determine the Seebeck coefficient for any thermocouple. For example, if one metal A is Platinum (for which Seebeck effect is null), this method provides Seebeck coefficient for the metal B and allows the construction of tables [4].

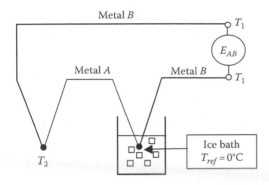

FIGURE 3.12
Reference junction kept at $T_{ref} = 0°C$.

3.2.2.5 Thermocouple Extension/Compensation Wire

The wire connecting the thermocouple to the instrument may be made of the same material as the thermocouple with the same physical characteristics. But, this is not always the most cost-effective method. So, a less expensive substitute material is selected. Thermocouples can be connected to an instrument by the following:

- Thermocouple wires: The same material as that used to manufacture the thermo-couple. This solution can be expensive.
- Extension wires: Wires with chemical composition and EMF characteristics similar to the thermocouple materials over a limited temperature range.
- Compensating wires (Figure 3.13): Alloys that have EMF characteristics similar to the thermocouple alloy, less expensive than thermocouples. Then, the output voltage is

$$E_{AB} = \int_{T_1}^{T_3} \sigma_{A'} \, dT + \int_{T_3}^{T_2} \sigma_A \, dT + \int_{T_2}^{T_3} \sigma_B \, dT + \int_{T_3}^{T_1} \sigma_{B'} \, dT \tag{3.18}$$

with $\sigma_A \approx \sigma_{A'}$ and $\sigma_B \approx \sigma_{B'}$

$$E_{AB} \approx \int_{T_1}^{T_2} \sigma_A \, dT + \int_{T_2}^{T_1} \sigma_B \, dT \tag{3.19}$$

$$E_{AB} \approx \sigma_{AB}(T_2 - T_1) \tag{3.20}$$

The correction appears at the cold junction at temperature T_1.

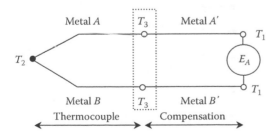

FIGURE 3.13
Compensation wires.

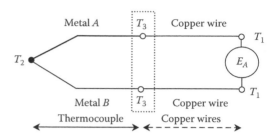

FIGURE 3.14
Connection with copper wires.

3.2.2.6 Connection with Copper Wires

The thermocouple is connected to the instrument with two copper (Cu) wires (Figure 3.14). In this case, the output voltage is

$$E_{AB} = \int_{T_1}^{T_3} \sigma_{Cu}\, dT + \int_{T_3}^{T_2} \sigma_A\, dT + \int_{T_2}^{T_3} \sigma_B\, dT + \int_{T_3}^{T_1} \sigma_{Cu}\, dT \tag{3.21}$$

$$E_{AB} = \int_{T_3}^{T_2} \sigma_A\, dT + \int_{T_2}^{T_3} \sigma_B\, dT \tag{3.22}$$

$$E_{AB} = \sigma_{AB}(T_2 - T_3) \tag{3.23}$$

The correction of the cold junction is at temperature T_3. Then, if $T_3 = T_1$, the output voltage is equal to

$$E_{AB} = \sigma_{AB}(T_2 - T_1) \tag{3.24}$$

3.2.2.7 Thermopile Connection

Figure 3.15 presents the circuit with n thermocouples $A - B$ ($n = 4$ in this example) placed in series arrangement. The objective is to multiply the EMF of one thermocouple by the number of thermocouples. The output voltage E_{AB} for four thermocouples is equal to

$$E_{AB} = \int_{T_1}^{T_3} \sigma_{Cu}\, dT + 4\int_{T_3}^{T_2} \sigma_A\, dT + 4\int_{T_2}^{T_3} \sigma_B\, dT + \int_{T_3}^{T_1} \sigma_{Cu}\, dT \tag{3.25}$$

$$E_{AB} = 4\int_{T_3}^{T_2} \sigma_A\, dT + 4\int_{T_2}^{T_3} \sigma_B\, dT \tag{3.26}$$

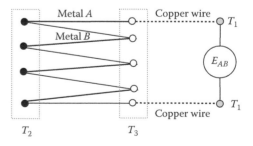

FIGURE 3.15
Thermopile connection (serial connection of thermocouples).

Finally,

$$E_{AB} = n\sigma_{AB}(T_2 - T_3) \quad \text{and} \quad \text{for } n = 4 \; E_{AB} = 4\sigma_{AB}(T_2 - T_3) \tag{3.27}$$

If $T_3 = T_1$, then,

$$E_{AB} = n\sigma_{AB}(T_2 - T_1) \quad \text{and} \quad \text{for } n = 4 \; E_{AB} = 4\sigma_{AB}(T_2 - T_1) \tag{3.28}$$

3.2.2.8 Parallel Thermocouple Arrangement

The thermocouples are placed in a parallel arrangement (Figure 3.16). They are connected to a common cold junction. This method needs equal electrical resistance R_i for each i thermocouple. A thermocouple is associated to an electric generator with voltage E and an electric resistance R. The parallel arrangement presents an equivalent electrical resistance R_{equ} (Figure 3.17):

$$\frac{1}{R_{equ}} = \frac{1}{R_1} + \frac{1}{R_2} + \cdots + \frac{1}{R_n} \quad \text{and} \quad \text{if } R_1 = R_2 = \cdots = R_n = R \text{ then } R_{equ} = \frac{R}{n} \tag{3.29}$$

The measured temperature T_{mes} becomes

$$E_{equ} = \frac{\sum_1^n (E_i/R_i)}{\sum_1^n (1/R_i)} = \frac{(1/R)\sum_1^n E_i}{(n/R)} = \frac{1}{n} \sum_1^n E_i \Rightarrow T_{mes} = \frac{1}{n} \sum_1^n T_i \tag{3.30}$$

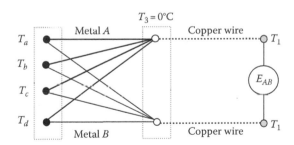

FIGURE 3.16
Parallel connections of thermocouples.

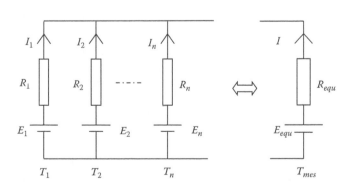

FIGURE 3.17
Analog circuit for the parallel. Connection of thermocouple.

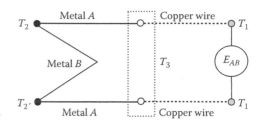

FIGURE 3.18
Differential thermocouple.

In this example, $n = 4$ for the four temperatures T_a, T_b, T_c, and T_b:

$$T_{mes} = \frac{1}{4}(T_a + T_b + T_c + T_d) \tag{3.31}$$

This complicated method can be used to measure a mean temperature.

3.2.2.9 Differential Thermocouple

In this method, two thermocouples are mounted in a series arrangement (Figure 3.18). The differential thermocouple is composed of two similar wires A joined to a single dissimilar wire B with the two measuring junctions normally at different temperatures. The resulting EMF is the difference between the two junctions, commonly referred to as the differential temperature. So that

$$E_{AB} = \int_{T_1}^{T_3} \sigma_{Cu}\, dT + \int_{T_3}^{T_{2'}} \sigma_A\, dT + \int_{T_{2'}}^{T_2} \sigma_B\, dT + \int_{T_2}^{T_3} \sigma_A\, dT + \int_{T_3}^{T_1} \sigma_{Cu}\, dT \tag{3.32}$$

$$E_{AB} = \sigma_A(T_{2'} - T_3) + \sigma_A(T_3 - T_2) + \sigma_B(T_2 - T_{2'}) \tag{3.33}$$

$$E_{AB} = \sigma_{AB}(T_{2'} - T_2) \tag{3.34}$$

3.3 Wire Microthermocouple Measurements

A microthermocouple has two major interests: small size and good response to transient phenomena. The small size is the most important geometric parameter for systems with very small sizes like microsystems. For the last 10 years, the intense development of micro electromechanical systems (MEMS) has prompted the growth of modern hydrodynamics, thermodynamics, and heat transfer with applications in areas as various as aerospace, mechanical engineering, biology, chemical analysis, and optics. The development of the first transistor by Shockley in 1948 [11], Bardeen and Brattein [12] has opened the way of electronic miniaturization called microelectronics, and very high density of components has been achieved ever since. Today, a MEMS includes a variety of devices, structures, and systems that contains both electrical and mechanical components with characteristic sizes ranging from nanometers to millimeters [13]. In this chapter, we limit our discussion to temperature measurement with microthermocouple wires in systems for which the miniaturization of components introduces quantitative downscaling effects [14,15]. Then, it becomes very difficult to measure physical quantities like temperature, heat flux, pressure drop, fluid flows, and velocities in microchannels. These measurements are a real challenge for designing new mechanical and physical systems.

A good response to transient phenomena is the second major interest for a thermocouple to accurately monitor the time–temperature history. However, when a thermocouple is placed in a gas flow, or on the external surface of a material or embedded, and when its temperature suddenly changes, it indicates a temperature different from that of the true value at any time before the thermodynamic equilibrium has been reached.

This chapter presents experimental and theoretical results for the dynamic calibration of microthermocouples ranging in wire size from 0.5 to 50 μm applied to fluid temperature measurements.

3.3.1 Typical Microthermocouple Designs

Different methods are used to design a thermocouple probe. It consists of a sensing element assembly, a protecting tube, and terminations. Two dissimilar wires are joined at one end to form the measuring junction, which can be a bare thermocouple element twisted and welded or butt welded. The protecting tube protects the sensing element assembly from the external atmosphere by a non-ceramic insulation, a hard fired ceramic insulator, or a sheeted compact ceramic insulator [16–19].

The thermocouple probe consists of two wires inserted in a ceramic double bore tube with length and external diameter depending on the experimentation. The wires are cut with a razor blade to produce a flat edge perpendicular to the axis. To realize the junction, the thermocouple wires are connected to a bank of condensers (Figure 3.19).

The two extremities are approached together in the same time, and the beaded junctions are made by a sparking method. The energy release produced by the voltage–capacitance pair is sufficient to weld together the wires. One advantage of this technique is that the resulting junction diameter is not significantly greater than the wires diameter (Figure 3.20). Aside from the low heat capacity effect, another consequence is that the cross-sectional area of the wire itself can be used to calculate time constants. A drop of glue is deposited at the tube extremity and pushed down around both wires to minimize the probe fragility (Figure 3.21).

Materials used for thermocouples are numerous and classified in terms of thermoelectric polarity. A thermocouple associates a positive wire and a negative wire. A positive material

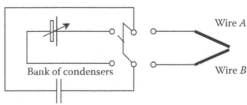

FIGURE 3.19
Thermocouple spark welding device (FEMTO-ST Belfort).

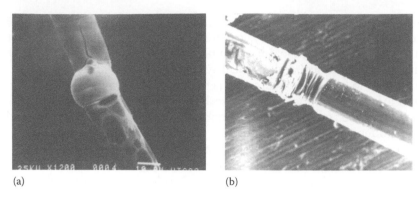

(a)   (b)

FIGURE 3.20
K type thermocouples. (a) Diameter = 12.7 μm. (b) Diameter = 53 μm.

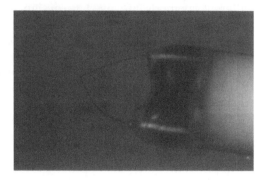

FIGURE 3.21
Thermocouple probe (diameter of the wires = 1.27 μm).

has an EMF that increases with temperature along its length, and a negative material has an EMF that decreases with temperature along its length. This chapter deals only with micro-thermocouples designed with two kinds of couples, chromel–alumel couples and platinum–rhodium couples. The chromel–alumel thermocouple, called K type thermocouple, with a positive chromel wire and a negative alumel wire, is recommended for use in clean oxidizing atmospheres. It is the thermocouple that is most widely used in industrial applications. The operating range for this alloy is 1260°C for the largest wire sizes. Smaller wires should operate at lower temperatures correspondingly. The K type thermocouples exhibit a number of instabilities and inaccuracies at higher temperatures, changing their EMF versus temperature characteristics.

The commercial wire diameters are 7.6, 12.7, 25.4, and 50 μm. The Seebeck coefficient is 40 μV °C^{-1} in the linear region at 20°C. The temperature–EMF data have been extracted from NIST Monograph 175 [19] (Figure 3.22 and Table 3.2). The platinum–rhodium thermocouple, called S type thermocouple, is a noble-metal thermocouple in common use. The S type thermocouples show a positive wire of 90% platinum and 10% rhodium used with a negative wire of pure platinum. Both metals have a high resistance to oxidation and corrosion. However, hydrogen, carbon, and many metal vapors can contaminate a platinum–rhodium thermocouple. The recommended operating range for the platinum–rhodium alloys is 1540°C. The commercial wire diameters are 0.5, 1.27, 5.4, 25.4, and 50 μm. The Seebeck coefficient is 6 μV °C^{-1} in the linear region at 20°C [20] (Figure 3.22).

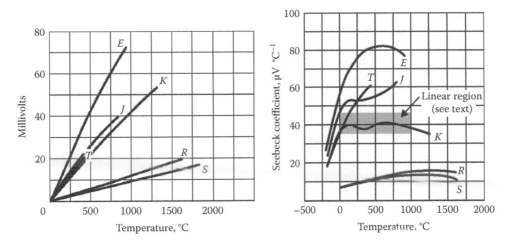

FIGURE 3.22
EMF and Seebeck coefficients versus temperature for different thermocouples (ITS-90).

3.3.2 Dynamic Temperature Measurements in Fluids

3.3.2.1 Introduction

Transient phenomena appear in many industrial processes, and many researchers and engineers have been paying attention to the measurement of temperature fluctuations in turbulent-reacting flows, compressible flows, boiling, cryogenic apparatus, fire environments, under the condition of simultaneous periodical variations of velocity, flow density, viscosity, and thermal conduction in gas [21–28].

There has been considerable progress in recent years in transient thermometry techniques. Some of these techniques are applicable for solid material characterization while others are suitable only for fluid thermometry. This chapter deals only with temperature thermocouple measurements in fluids (gases and liquids). Many concepts involved in the temperature measurements in fluids are common to both types and they are discussed here. The techniques for temperature measurement in a fluid consist in inserting a thermocouple, allowing it to come to thermal equilibrium, and measuring the generated electrical signal. When a thermocouple is submitted to a rapid temperature change, it will take some time to respond. If the sensor response time is slow in comparison with the rate of change of the measured temperature, then the thermocouple will not be able to faithfully represent the dynamic response of the temperature fluctuations. Then, the problem is to measure the true temperature of the fluid because a thermocouple gives its own temperature only. The temperature differences between the fluid and the sensor are also influenced by thermal transport processes taking place between the fluid to be measured, the temperature sensor, the environment, and the location of the thermocouple. Consequently, the measured temperature values must be corrected. Whereas in steady conditions only the contributions of the conductive, convective, and radiative heat exchanges with the external medium occur, unsteady behavior introduces another parameter, which becomes predominant: the junction thermal lag that is strongly related to its heat capacity and thermal conductivity. The corrections generally decrease with the thermocouple diameters, and both temporal and spatial resolutions are improved. However, while spatial resolution is fairly directly connected with the thermocouple dimensions, the temporal resolution does not only depend on the dimensions and the thermocouple physical characteristics, but also on the rather complex heat balance of the whole thermocouple. To obtain the dynamic characteristics of any temperature probe, we analyze its response to an excitation step from which the corresponding first time constant τ can be defined as

$$\tau = \frac{\rho c V}{hA} \tag{3.35}$$

where
  τ is the time constant
  ρ is the density
  c is the specific heat
  V is the volume of the thermocouple
  A is the area of the fluid film surrounding the thermocouple while h is the heat transfer coefficient

The goal of this work consists in calculating or measuring time constants of thermocouples and comparing their behavior according to different dynamical external heating like convective, radiative, and pseudo-conductive excitations.

3.3.2.2 Theory

An accurate calibration method is an essential element of any quantitative thermometry technique, and the goal of any measurement is to correctly evaluate the difference between the "true" temperature and the sensor temperature. Figure 3.23 shows the energy balance performed at the butt-welded junction of a thermocouple for a junction element dx resulting from the thermal balance between the rate of heat stored by the junction $d\dot{Q}_{th}$ and heat transfer caused by the following:

- Convection in the boundary layer around the thermocouple $d\dot{Q}_{cv}$
- Conduction along the wires $d\dot{Q}_{cd}$
- Radiation between the wires and the external medium $d\dot{Q}_{rad}$
- Contribution of another source of heat power (a laser source in our work) $d\dot{Q}_{ext}$

During a transient period, because of its thermal capacity, the thermocouple temperature will lag behind any gas temperature variation. This leads to an error from which a thermocouple time constant can be defined. The general heat balance for a junction of length dx is expressed as

$$d\dot{Q}_{th} = d\dot{Q}_{cv} + d\dot{Q}_{cd} + d\dot{Q}_{rad} + d\dot{Q}_{ext} \tag{3.36}$$

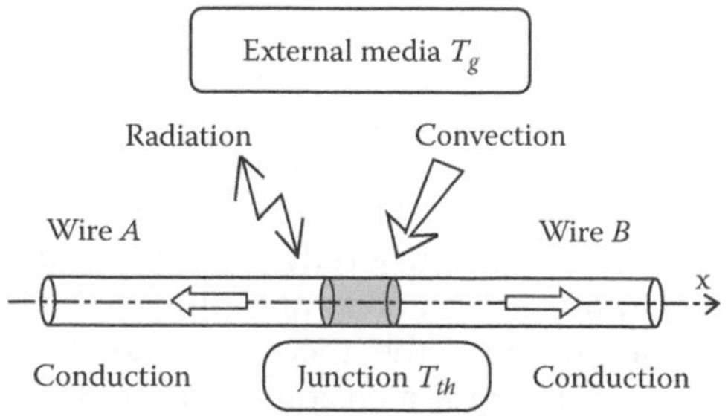

FIGURE 3.23
Heat balance of the probe.

The thermoelectric junction stores the heat by unit time $d\dot{Q}_{th}$:

$$d\dot{Q}_{th} = \rho_{th}c_{th}\frac{\pi d^2}{4}\frac{\partial T_{th}}{\partial t}dx \qquad (3.37)$$

where ρ_{th}, c_{th}, and T_{th} are the density, the specific heat, and the temperature of the junction, respectively.

The junction is approximated by a cylinder whose diameter equals the wire diameter d. This does not exactly fit reality but remains currently used in numerical calculations [29–35]. Moreover, if the wires are uniformly curved, the observation near the junction confirms the previous assumption (Figures 3.20 and 3.21). Newton's law of cooling is

$$d\dot{Q}_{cv} = \pi \cdot dx \cdot Nu \cdot \lambda_g(T_g - T_{th}) \qquad (3.38)$$

where λ_g and T_g are the thermal conductivity and the static temperature of the gas, respectively. The difficulty is to obtain an accurate relation between the Nusselt number Nu and the flow characteristics around the junction assumed as a cylinder [31,36–39].

Indeed, such a thermocouple is surrounded by both a thermal and aerodynamic gradient that acts as a thermal resistance that is estimated from empiric approaches. A purely convective heat transfer coefficient h is generally deduced from correlations about the Nusselt number that is generally expressed as a combination of other dimensionless numbers, such as Eckert, Reynolds, Prandtl, or Grashof numbers. Table 3.3 presents a list of dimensionless numbers relevant to heat transfer. However, if many cases have been

TABLE 3.3

Selected Dimensionless Groups of Heat Transfer

Group	Definition	Interpretation
Eckert number	$E_c = \dfrac{V^2}{c_p(T - T_\infty)}$	Kinetic energy of the flow relative to the boundary layer enthalpy difference
Grashof number	$Gr_x = \dfrac{g\beta(T - T_\infty)L}{\nu^2}$	Ratio of buoyancy to viscous forces
Knudsen number	$Kn = \dfrac{\ell}{L}$	Ratio between the mean free path of a gas molecule and the macroscopic dimension
Mach number	$Ma = \dfrac{V}{c}$	Ratio between the speed of matter relative to the local speed of sound
Nusselt number	$Nu = \dfrac{hL}{\lambda}$	Dimensionless temperature gradient at the surface, it represents the ratio between the heat transfer by convection and the transfer by conduction alone
Prandtl number	$Pr = \dfrac{\nu}{a}$	Ratio of the momentum and thermal diffusivities
Reynolds number	$Re = \dfrac{VL}{\nu}$	Ratio of the inertia and viscous forces

V is the fluid velocity, T and T_∞ are the surface and fluid temperatures, c_p is the specific heat capacity, g is the gravitational acceleration, β is the fluid dilatation coefficient, L is the representative macroscopic dimension (i.e., diameter, local or mean length), c is the speed of sound, ℓ is the mean free path of a molecule, ν is the kinematic viscosity, h is the convection coefficient, λ is the thermal conductivity, and a is the thermal diffusivity.

TABLE 3.4

Heat Transfer Laws

Author	Temperature for λ, ρ, and μ	Correlation	Reynold's Number Domain
Andrews	T_f	$Nu = 0.34 + 0.65\,Re^{0.45}$	$0.015 < Re < 0.20$
Bradley and Mathews	T_f	$Nu = 0.435\,Pr^{0.25} + 0.53\,Pr^{0.33}\,Re^{0.52}$	$0.006 < Re < 0.05$ $0.7 < Pr < 1$
Churchill and Brier	T_f	$Nu = 0.535\,Re^{0.50}(T_f/T_{th})^{0.12}$	$300 < Re < 2{,}300$
Collis and Williams	T_{film}	$Nu = (0.24 + 0.56\,Re^{0.45})(T_{film}/T_{gaz})^{0.17}$	$0.02 < Re < 44$
Collis and Williams	T_{film}	$Nu = (0.48\,Re^{0.45})(T_{film}/T_{gaz})^{0.17}$	$44 < Re < 140$
Davies and Fisher	T_f	$Nu = (2.6/\gamma\pi)Re^{0.33}$	$0.01 < Re < 50$
Eckert and Soehngen	/	$Nu = 0.43 + 0.48\,Re^{0.5}$	$1 < Re < 4{,}000$
Glawe and Johnson	T_f	$Nu = 0.428\,Re^{0.50}$	$400 < Re < 3{,}000$
King	T_{film}	$Nu = 0.318 + 0.69\,Re^{0.5}$	$0.55 < Re < 55$
Kramers	T_{film}	$Nu = 0.42\,Pr^{0.2} + 0.57\,Pr^{0.33}\,Re^{0.5}$	$0.01 < Re < 10{,}000$ $0.7 < Pr < 1{,}000$
McAdams	T_{film} and T_f for ρ	$Nu = 0.32 + 0.43\,Re^{0.52}$	$40 < Re < 4{,}000$
Olivari and Carbonaro	T_{film}	$Nu = 0.34 + 0.65\,Re^{0.45}$	$0.015 < Re < 20$ $L/d > 40$
Parnas	T_f	$Nu = 0.823\,Re^{0.5}(T_{th}/T_f)^{0.085}$	$10 < Re < 60$
Richardson	/	$Nu = 0.3737 + 0.37\,Re^{0.5} + 0.056\,Re^{0.66}$	$1 < Re < 10^5$
Scadron and Warshawski	T_f	$Nu = 0.431\,Re^{0.50}$	$250 < Re < 3{,}000$
Van den Hegge Zijnen	T_{film}	$Nu = 0.38\,Pr^{0.2} + (0.56\,Re^{0.5} + 0.01\,Re)Pr^{0.33}$	$0.01 < Re < 10^4$

These laws describe the heat transfer from a cylinder of infinite length. The film temperature T_{film} is defined as the mean value between the fluid temperature T_f and the thermocouple temperature T_{th} [30–32,34–39,43–47].

investigated, the example of thin cylinders cooling process is still an open question. Table 3.4 gives a list of the main Nusselt correlations in this particular case.

Conduction heat transfer $d\dot{Q}_{cd}$ that occurs along the wires to the thermocouple supports has the following general expression:

$$d\dot{Q}_{cd} = \lambda_{th}\frac{\pi d^2}{4}\frac{\partial^2 T_{th}}{\partial x^2}dx \tag{3.39}$$

However, different studies and experiments have shown that conduction dissipation effects along cylindrical wires can be neglected when the aspect ratio between the length and the diameter is large enough [17,40–46]. Indeed, practical cases of anemometry and thermometry have led to fix such a condition:

$$\frac{L}{d} > 100 \tag{3.40}$$

Hence, the temperature gradient can be considered null in the axial direction of the thermocouple wire. The thermocouple is placed in an enclosure at temperature T_w.

The enclosure dimensions are assumed to be large with respect to the probe dimensions. Then, the influence of the radiative heat transfer can be expressed by the simplified form:

$$d\dot{Q}_{rad} = -\sigma\varepsilon(T_{th})\left(T_{th}^4 - T_w^4\right)dS_{ray} \tag{3.41}$$

where
σ is the Stefan–Boltzmann constant
$\varepsilon(T_{th})$ is the emissivity of the wire at the temperature T_{th}

The exchange surface of the radiative heat transfer $dS_{rad} = \pi d\,dx$ nearly equals the surface exposed to the convective heat flux. This supposes that the radiative heat transfer between the sensor and the walls is greater than between the gas and the sensor. Here, the assumption is that the gas is transparent; however, it is not satisfied in several practical applications like temperature measurements in flames.

In Section 3.3.2.3.2, we consider a radiative calibration so that the thermocouple junction is submitted to an external heat contribution $d\dot{Q}_{ext}$ produced by a laser beam [41].

$$d\dot{Q}_{ext} = \sqrt{\frac{2}{\pi}}\frac{(1-\overline{R})}{a}P_L\,\mathrm{erf}\left[\frac{d}{a\sqrt{2}}\right]\exp\left[-2\frac{x^2}{a^2}\right]dx \tag{3.42}$$

where
P_L is the laser beam power
$\overline{R}$ is the mean reflection coefficient of the thermocouple junction surface
d is the diameter of the junction
a is the laser beam radius (this value corresponds to the diameter for which one has 99% of the power of the laser beam)

The total heat balance of the thermocouple may be written as follows [18]:

$$\rho_{th}c_{th}\frac{\pi d^2}{4}\frac{\partial T_{th}}{\partial t} = Nu\,\lambda_g(T_g - T_{th}) + \lambda_{th}\frac{\pi d^2}{4}\frac{\partial^2 T_{th}}{\partial x^2}$$

$$-\sigma\varepsilon(T_{th})\left(T_{th}^4 - T_w^4\right)\pi d + \sqrt{\frac{2}{\pi}}\frac{(1-\overline{R})}{a}P_L\,\mathrm{erf}\left[\frac{d}{a\sqrt{2}}\right]\exp\left[-2\frac{x^2}{a^2}\right] \tag{3.43}$$

The expression of the gas temperature T_g is deduced from Equation 3.43:

$$T_g = T_{th} + \tau_{cv}\left[\begin{array}{c}\dfrac{\partial T_{th}}{\partial t} - \dfrac{\lambda_{th}}{\rho_{th}c_{th}}\dfrac{\partial^2 T_{th}}{\partial x^2} + \dfrac{4\sigma\varepsilon(T_{th})}{\rho_{th}c_{th}d}\left(T_{th}^4 - T_w^4\right) \\[2ex] -\dfrac{4}{\rho_{th}c_{th}d^2}\sqrt{\dfrac{2}{\pi}}\dfrac{(1-\overline{R})}{a}P_L\,\mathrm{erf}\left[\dfrac{d}{a\sqrt{2}}\right]\exp\left[-2\dfrac{x^2}{a^2}\right]\end{array}\right] \tag{3.44}$$

Equation 3.44 represents a general expression of the thermocouple dynamic behavior, including each of the heat transfer modes. In this expression, the time constant τ_{cv} of the thermocouple junction is defined by

$$\tau_{cv} = \frac{\rho_{th}c_{th}d^2}{4Nu\,\lambda_g} = \frac{\rho_{th}c_{th}d}{4h} \tag{3.45}$$

If the radiation, the conduction, and the external heat supply are neglected, the gas temperature simplifies to

$$T_g = T_{th} + \tau_{cv} \frac{\partial T_{th}}{\partial t} \tag{3.46}$$

The time-response of a temperature sensor is then characterized by a simple first order equation. This is a common but erroneous way. For a step change in temperature, Equation 3.46 reduces to

$$\frac{T_g - T_{th}}{T_g - T_i} = \exp\left[-\frac{t}{\tau_{cv}}\right] \tag{3.47}$$

where T_i is the initial temperature.

Conventionally, the time constant τ_{cv} is defined as the duration required for the sensor to exhibit a 63% ($= 1 - e^{-1}$) change from an external temperature step, in the case of a single-order equation. Actually, the fact that different kinds of heat transfers are involved should lead to a global time constant in which the different phenomena contributions are included [30,43]. As a consequence, the ability of a thermocouple to follow any modification of its thermal equilibrium is resulting from a multi-ordered time response where the most accessible experimental parameter remains the global time constant. The multi-ordered temperature response of a thermocouple can be represented by the general relation:

$$\frac{T_g - T_{th}}{T_g - T_i} = K_1 \exp\left[-\frac{t}{\tau_1}\right] - K_2 \exp\left[-\frac{t}{\tau_2}\right] - \cdots - K_n \exp\left[-\frac{t}{\tau_n}\right] \tag{3.48}$$

where
   T_i is the initial temperature
   T_g is the fluid temperature

The values of the constants $K_1, K_2, \ldots, K_n$ as well as the time constants $\tau_1, \tau_2, \ldots, \tau_n$ depend on the heat flow pattern between the thermocouple and the surrounding fluid

If experiments have shown that most configurations involve nearly first-order behaviors, the measured time constant does not allow isolating each of the different contribution modes.

Therefore, the remaining problem of experiments is to relate this global time constant to the different implied heat transfer modes. Then, our contribution in this section will be to show the influence of the heat transfer condition on the measured time constant value through three different methods of dynamic calibration.

Classical testing of thermocouples often involves plunging them into a water or oil bath and for providing some information only about the response of the thermocouple under those particular conditions. It does not provide information about the sensor response under process operating conditions where the sensor is used. In order to improve thermo-couple transient measurements, a better understanding of the dynamic characteristics of the sensor capability is necessary.

3.3.2.3 Dynamic Calibration with Single-Wire Thermocouple Technique

The calibration methods consist of a series of heating and cooling histories performed by submitting the thermocouple to different excitation modes. Then, the resulting exponential rise and decay times of the thermocouple signals allow estimating the time constant τ. The thermocouple signal is amplified with a low-noise amplifier having a -3 dB bandwidth of 25 kHz (Gain $= 1000$). The output voltage is finally recorded by a digital oscilloscope.

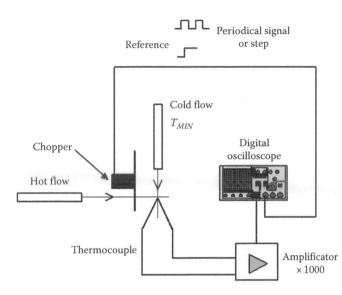

FIGURE 3.24
Experimental device for the convective calibration.

3.3.2.3.1 Convective Calibration

Figure 3.24 illustrates the convective experimental device. The thermocouple junction is exposed continuously to a constant cold airstream at constant temperature T_{MIN}. A second hot airflow excites periodically the thermocouple and creates a temperature fluctuation of frequency f [18,47]. The response of a thermocouple submitted to successive steps of heating or cooling is close to a classical exponential first-order response from which the time constant can be determined (Figure 3.25). It can be deduced from the measurement of four temperatures, T_{MAX}, T_{MIN}, $T_{th\,max}$, and $T_{th\,min}$:

T_{MAX}: the maximal temperature when the thermocouple is submitted to a constant hot flow

T_{MIN}: the minimal thermocouple temperature in a constant cold flow

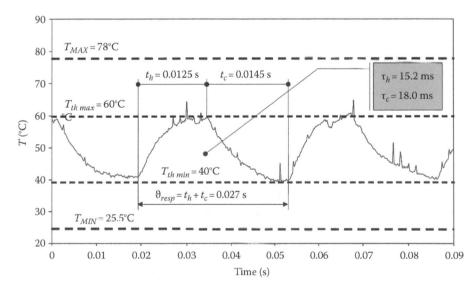

FIGURE 3.25
Typical exponential responses: temperature histories for a 12.7 μm K type thermocouple.

$T_{th\,max}$: the maximal temperature when the thermocouple is submitted to the periodic hot flow

$T_{th\,min}$: the minimal temperature when the thermocouple is submitted to the periodic hot flow

For the heating period t_h, we define the temperature differences δ_{1h} and δ_{2h}:

$$\delta_{1h} = T_{MAX} - T_{th\,min} \tag{3.49}$$

and

$$\delta_{2h} = T_{MAX} - T_{th\,max} \tag{3.50}$$

For the cooling period t_c, the temperature differences δ_{1c} and δ_{2c} by

$$\delta_{1c} = T_{th\,max} - T_{MIN} \tag{3.51}$$

and

$$\delta_{2c} = T_{th\,min} - T_{MIN} \tag{3.52}$$

Then, the two convective time constants are defined while the thermocouple is heating (τ_h) and cooling (τ_c). If we consider a first-order response of the sensor, we obtain the expressions as follows:

$$\tau_h = \frac{t_h}{\ln(\delta_{1h}/\delta_{2h})} \tag{3.53}$$

and

$$\tau_c = \frac{t_c}{\ln(\delta_{1c}/\delta_{2c})} \tag{3.54}$$

Then the period of the thermocouple response is as follows:

$$\vartheta_{resp} = t_c + t_h \tag{3.55}$$

Figure 3.25 presents temperature histories for a 12.7 μm K type thermocouple. The excitation frequency is 37 Hz. The velocities of both hot and cold air are 13 m s^{-1} at the outlet of the airflow tubes. In any case, the measured time constants are longer during the heating phase than during the cooling one. This phenomenon corresponds to a greater magnitude of the convection coefficient (h). Table 3.5 presents convective time constants for the different thermocouple diameters, resulting from heating periods only and for two airflow velocities (13 and 23 m s^{-1}) and for a 5–72 Hz explored frequency bandwidth. One can notice that time constants decrease when increasing the flow velocity because of a larger surface over volume ratio exposed to the flow. Finally, even if the repeatability is good, such a calibration method remains however quite difficult to perform because the fragility of the sensor increases when the wires' dimension decreases and the fluid flow increases.

TABLE 3.5

Convective Time Constant τ_{cv} (ms) and Bandwidth Δf (Hz) versus Junction Diameters

Junction Diameter		Air Velocity: 13 m s^{-1}		Air Velocity: 23 m s^{-1}	
	d (μm)	τ_{cv} (ms)	Δf (Hz)	τ_{cv} (ms)	Δf (Hz)
S	0.5	—	—	—	—
	1.27	—	—	—	—
	5	2.9	55	2.2	72
K	12.7	15.2	10.5	8.5	18.7
	25	20	8	17	9.4
	250	32	5	25	6.4

The thermocouple mechanical resistance is not sufficient for the flows with 13 and 23 m s^{-1} air velocities.

3.3.2.3.2 Radiative Calibration

This calibration method is based on a radiative excitation produced by a continuous argon laser [18,48,49]. A set of two spherical lenses allows locating the beam waist on the junction and an optical chopper generates a periodic modulation of the continuous laser beam. In order to avoid parasitic turbulences around the junction, the sensor is placed in a transparent enclosure (Figure 3.26). The signal obtained is close to a first-order response, which gives immediately the sensors dynamic performances. Time constant decreases as diameter and heat transfer (the laser power) increase (Figures 3.27 and 3.28). This is consistent with the effect of an increasing value of the power density or a decreasing of the beam radius that both act on the power to heated mass ratio. Table 3.6 presents the radiative time constant for all the thermocouple junction diameters, and the explored frequency bandwidth ranges from 5 to 2274 Hz.

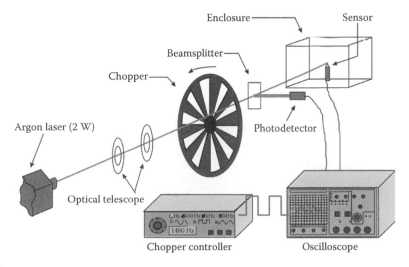

FIGURE 3.26
Radiative excitation device.

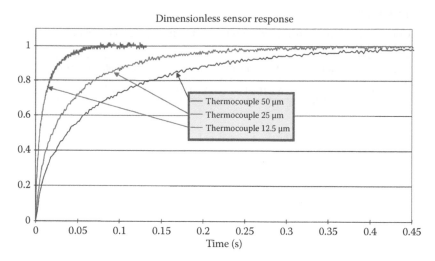

FIGURE 3.27
Calibration of K thermocouples with a laser beam step.

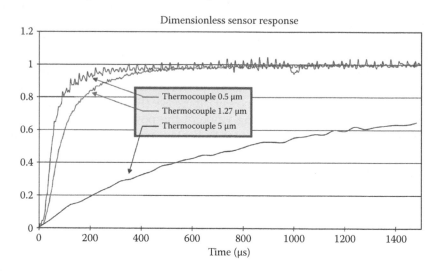

FIGURE 3.28
Calibration of S thermocouples with a laser beam step.

TABLE 3.6

Radiative Time Constant τ_{rad} (ms) and Bandwidth Δf (Hz) versus Junction Diameters

Junction Diameter d (μm)	Radiative Time Constant τ_{rad} (ms)	Bandwidth Δf (Hz)
S 0.5	0.07	2274
1.27	0.18	884
5	1.3	123
K 12.7	8.5	19
25	34	5
50	64.5	2.5

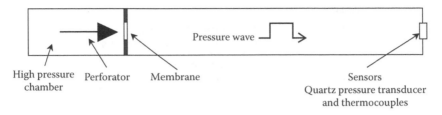

FIGURE 3.29
Radiative excitation device.

3.3.2.3.3 *Shock Tube Calibration*

The experimental device is based on a shock tube. This process is generally used for pressure sensors calibration. Experiments were performed in a tube with an overall length of 20 m and a diameter of 0.5 m as illustrated in Figure 3.29. The shock tube and high pressure chamber (filled with dry air) are isolated from each other with cellophane membranes. A quartz pressure transducer and the thermocouples are mounted flush with the end wall of the shock tube. Nevertheless, in this experiment, the pressure step, which propagates in the tube and reflects off the end wall of the shock tube, produces a suitable temperature step used to test the thermocouple [41] (Figure 3.30). The quartz pressure transducer signal is used to identify the transient pressure step and to compare with the response of the different thermocouples. The thermal exchange between the junction and the gas is not radiative. In fact, both conduction and convection take place. Table 3.7 presents the pseudo-convective time constant for thermocouple whose time responses are compatible with the excitation duration produced by the shock tube. This one being limited to about 6 ms, this technique is not available for the larger thermocouples (K types). The explored frequency bandwidth ranges from 100 to 758 Hz.

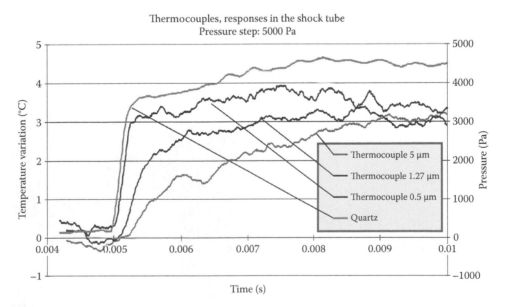

FIGURE 3.30
Calibration of S thermocouples with a pressure step.

TABLE 3.7

Pseudo-Convective Time Constant τ_{pc} (ms)
and Bandwidth Δf (Hz) versus Junction Diameters

	Junction Diameter d (μm)	Pseudo-Convective Time Constant τ_{pc} (ms)	Bandwidth Δf (Hz)
S	0.5	0.21	758
	1.27	0.45	354
	5	1.50	106
K	12.7	—	—
	25	—	—
	50	—	—

The thermocouple is not sensitive to the temperature variation
and the signal is totally integrated during the pressure step.

3.3.3 Measurements in Fluids with Multi-Wire Thermocouple Technique

Measuring fluid temperatures with a single thermocouple requires knowledge of the fluid
velocity and fluid properties to determine the global heat transfer coefficient of the wire,
integrating convection, conduction and radiation, and its natural frequency. In such a case,
the thermal inertia of the thermocouples acts as first-order low-pass filters attenuating the
high frequency fluctuations. Another technique consists of temperature measurements
with a probe using two [19,27,50–54] or three thermocouples [55,56] of same nature but
different in diameter located close together at the measurement point (Figure 3.31). This
method was first used to characterize fluctuating gas flows for combusting flows. A two or
multithermocouple probe allows simultaneously the estimation of thermocouple time
constants and the compensation of thermocouple response.

3.3.3.1 Basic Analysis of a Two-Thermocouple Probe

The basic analysis neglects the effects of the radiative and conductive environment in
which the thermocouple junction is located and the effects of catalytic reaction on the
thermocouple wire surface [19,50].

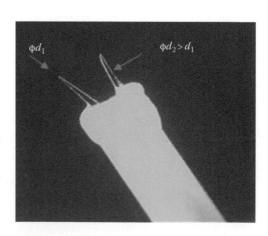

FIGURE 3.31
Two-thermocouple probe with K (chromel–alumel) type
wires (FEMTO-ST Belfort). (From Lanzetta, F. et al., Two-
microthermocouple probe for temperature and velocity
measurements in an oscillating flow in a heat exchanger
of Stirling machine, *ASME ATI Conference, Energy: Production, Distribution and Conservation*, Milan, Italy, May 14–17,
2006, pp. 633–642.)

The instantaneous heat balance written for each thermocouple permits to estimate the gas temperatures T_{g1} and T_{g2} function of the measured temperatures T_1 and T_2 and the time constants τ_1 and τ_2, respectively:

$$T_{g1} = T_1 + \tau_1 \frac{dT_1}{dt} \tag{3.56}$$

and

$$T_{g2} = T_2 + \tau_2 \frac{dT_2}{dt} \tag{3.57}$$

The two thermocouples are assumed to be exposed to identical flow conditions. Since the velocity V is the same for both thermocouple junctions, the time constants can be written from Equation 3.45 as follows:

$$\tau_1 = K d_1^{2-m} V^{-m} \tag{3.58}$$

and

$$\tau_2 = K d_2^{2-m} V^{-m} \tag{3.59}$$

where
  K is a constant
  d_1 and d_2 are the thermocouple junction diameters
  m is an exponent function of the ratio between the Nusselt and Reynolds numbers and is generally assumed to lie in the range

$$0.3 \leq m \leq 0.7 \tag{3.60}$$

From Equation 3.58, a constant α can be defined as the ratio of the time constants:

$$\alpha = \frac{\tau_1}{\tau_2} = \left(\frac{d_1}{d_2}\right)^{2-m} \tag{3.61}$$

3.3.3.1.1 Frequency Domain Reconstruction
The Fast Fourier Transform (FFT) of Equations 3.56 and 3.57 is used for reconstruction of signals from a two-thermocouple measuring rig [52,53,55,56] to obtain the following frequency domain equation, assuming $T_g = T_{g1} = T_{g2}$:

$$\overline{T}_g = \frac{\overline{T}_1 \overline{T}_2 (\alpha - 1)}{\alpha \overline{T}_1 - \overline{T}_2} \tag{3.62}$$

where $\overline{T}$ denotes the FFT.
  Then, the time–domain representation is found by taking the inverse FFT:

$$T_g = FFT^{-1}(\overline{T}_g) \tag{3.63}$$

This method is simple, but it presents the inconvenient to be dependent on singularities and noise.

3.3.3.1.2 Time–Domain Reconstruction

In order to reconstruct the gas temperature, another technique consists in data smoothing followed by parameter estimation in the time domain [19,27,51,54,55,58]. Data smoothing is implemented through filtering the data with a low-pass filter or through a moving window averaging.

The next step consists in estimating the time constants τ_1 and τ_2, assuming $T_g = T_{g1} = T_{g2}$, by minimizing the time-average difference between the two reconstructed temperatures T_{g1}, T_{g2} and then $(T_{g1} - T_{g2})^2$.

For a given data window, an ordinary linear least square estimation based on a one parameter (τ_2) model y_{mo} for the difference of the two observable quantities $T_i = T_g + \tau_i(dT_i/dt)$ (for $i = 1, 2$), see Equations 3.56, 3.57, and 3.61, can be constructed. It is based on the following least square sum, see Chapter 7:

$$S_{OLS}(\tau_2) = \sum_{i=1}^{N} (y_i - y_{mo}(t; \tau_2))^2 \tag{3.64}$$

with

$$y_{mo}(t; \tau_2) = -\tau_2 \left(\alpha \frac{dT_1}{dt} - \frac{dT_2}{dt} \right)$$

and with their experimental counterparts: $y_i = T_1^i - T_2^i$ with

$$T_1^i = T_1^{exp}(t_i); \quad T_2^i = T_2^{exp}(t_i); \quad \frac{dT_1^i}{dt} = \frac{T_1^i - T_1^{i-1}}{\Delta t}; \quad \frac{dT_2^i}{dt} = \frac{T_2^i - T_2^{i-1}}{\Delta t}$$

The sensitivity coefficient to τ_2 can be calculated with experimental temperatures:

$$X = \frac{\partial y_{mo}}{\partial \tau_2} = \alpha \frac{dT_1}{dt} - \frac{dT_2}{dt} \Rightarrow X_i \approx \alpha \frac{dT_1^i}{dt} - \frac{dT_2^i}{dt} \tag{3.65}$$

Estimation of τ_2 is calculated with the linear estimator, see Chapter 7, where X and y are column vectors constructed with the different values in time of both X and y.

$$\hat{\tau}_2 = (X^T X)^{-1} X^T y = \frac{\sum_{i=1}^{N} \left[(T_2^i - T_1^i) \left(\alpha \frac{dT_1^i}{dt} - \frac{dT_2^i}{dt} \right) \right]}{\sum_{i=1}^{N} \left(\alpha \frac{dT_1^i}{dt} - \frac{dT_2^i}{dt} \right)} \tag{3.66}$$

$$\hat{\tau}_1 = \alpha \, \hat{\tau}_2 \tag{3.67}$$

Finally, after determining the two time constants τ_1 and τ_2, the temperatures T_{g1} and T_{g2} are calculated from Equations 3.56 and 3.57.

3.3.3.2 Fluid Velocity Measurement with a Two-Thermocouple Probe

The goal of this technique is to measure the temperature and the velocity of the periodic flow inside an engine (Stirling machine in our case) with different two-microthermocouple

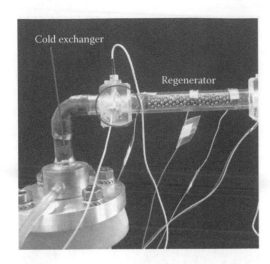

FIGURE 3.32
Experimental setup.

probes [57]. Once the time constants τ_1 and τ_2 are determined, it is possible to extract the value of the fluid velocity V from Equations 3.58 and 3.59, and

$$V = \left(\frac{\tau_2}{Kd_2^{2-m}} \right)^{-1/m} = \left(\frac{\tau_1}{Kd_1^{2-m}} \right)^{-1/m} \tag{3.68}$$

The experimental apparatus consists of a rigid circular tube with a diameter of 10 mm placed before and after the regenerator of a Stirling machine (Figure 3.32). A sinusoidal flow is generated by the way of a compression cylinder, a piston, and a crankshaft with adjustable stroke lengths. The crankshaft is driven by a dc electric motor with variable speed. In the present experiments, oscillating frequencies vary from 1 to 10 Hz. With this maximal frequency and the diameter of the tube, the mean flow can be considered laminar. The sinusoidal pulsating flow in a rigid circular tube may be described by the frequency of pulsation, the mean-flow velocity and temperature, and the magnitude of the harmonic velocity and temperature. The reconstructed signal from the two-wire microthermocouple probe allows the determination of the temperature of the fluid. The frequency response of the fine-wire thermocouples can be described as a first-order lag system in the presence of convective heat transfer only without radiation and negligible conduction along the wires. For this last point, each thermocouple presents the ratio wire length on diameter greater than 200 (Figure 3.33), and thus, the conduction can be effectively neglected.

Figures 3.34 through 3.36 give the temperature measurements with the three different probes composed by the wires with 7.6, 12.7, 25.4, and 50 μm diameters. After having smoothed the curves, we calculate the mean time constant on a cycle for each thermocouple with Equation 3.67.

The results are given in Table 3.8. They show that the smallest sensors answer with a very small response time. Table 3.9 gives the values of flow velocities calculated by Equation 3.68. The values are relatively close. They show a difference of 4% between the extreme values.

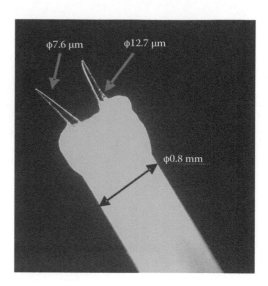

FIGURE 3.33
Two-microthermocouple probe (FEMTO-ST Belfort).

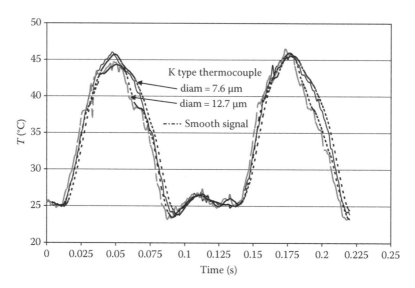

FIGURE 3.34
Measurements with a 7.6/12.7 μm probe (frequency = 8 Hz).

3.3.3.3 More Realistic Analysis of a Two-Thermocouple Probe

The basic analysis neglects radiation and catalytic effects. We consider a fluid at high temperature and an ambient medium at T_{amb} temperature [19,27,51]. Heat is exchanged between the wires and the fluid only by convection and radiation. In these conditions, the heat balance, Equations 3.56 and 3.57, becomes

$$T_{g1} = T_1 + \tau_1 \frac{dT_1}{dt} + \frac{\sigma \varepsilon}{h_1} \left(T_1^4 - T_{amb}^4 \right) \tag{3.69}$$

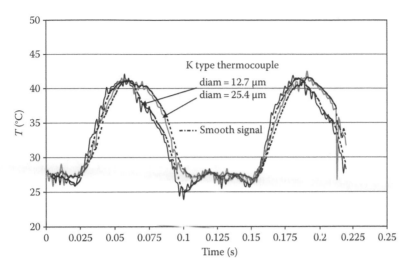

FIGURE 3.35
Measurements with a 12.7/25.4 μm probe (frequency = 8 Hz).

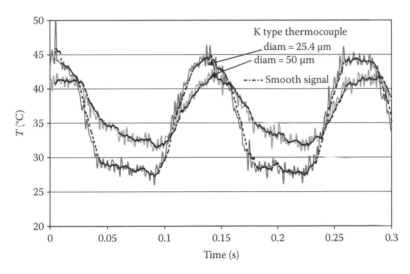

FIGURE 3.36
Measurements with a 25.4/50 μm probe (frequency = 8 Hz).

TABLE 3.8

Experimental Time Constants

d (μm)	τ (ms)
7.6	5.2
12.7	12
25.4	34
50	91

TABLE 3.9

Velocity Calculated by Equation 3.68

$\dfrac{d_2}{d_1}$	$\alpha = \dfrac{\tau_1}{\tau_2}$	$\alpha = \left(\dfrac{d_1}{d_2}\right)^{2-m}$	V (m s^{-1})
7.6/12.7	2.30	2.20	0.86
12.7/25.4	2.83	2.90	0.83
25.4/50	2.68	2.83	0.86

Thermocouple: K type (chromel–alumel): $\rho = 8600$ kg m^{-3}; $c = 480$ J kg^{-1} K^{-1}; fluid = air (35°C): $\lambda = 0.0268$ W m^{-1} K^{-1}; $\rho = 1.146$ kg m^{-3}; $\mu = 1.78 \times 10^{-5}$ Pa s.

and

$$T_{g2} = T_2 + \tau_2 \frac{dT_2}{dt} + \frac{\sigma \varepsilon}{h_2}\left(T_2^4 - T_{amb}^4\right) \tag{3.70}$$

where
  σ is the Stefan–Boltzmann constant
  ε is the emissivity of the thermocouple
  T_{amb} is the ambient temperature with the assumption $T_{amb} < T_{1,2}$

Then, the set of two Equation 3.68 becomes

$$T_{g1} = T_1 + \tau_1 \left(\frac{dT_1}{dt} + \frac{\beta}{d_1} T_1^4\right) \tag{3.71}$$

and

$$T_{g2} = T_2 + \tau_2 \left(\frac{dT_2}{dt} + \frac{\beta}{d_2} T_2^4\right) \tag{3.72}$$

where $\beta = (4\tau\varepsilon/\rho c)$ is an independent constant, taking account of the fluid flow conditions.

The two time constants are calculated as previously in the time domain, and the new corresponding expressions are as follows:

$$\tau_1 = \frac{\sum_{i=1}^{N}\left(R_2^i\right)^2 \sum_{i=1}^{N}\left[R_1^i\left(T_2^i - T_1^i\right)\right] - \sum_{i=1}^{N}\left[R_1^i R_2^i\right] \sum_{i=1}^{N}\left[R_2^i\left(T_2^i - T_1^i\right)\right]}{\sum_{i=1}^{N}\left(R_1^i\right)^2 \sum_{i=1}^{N}\left(R_2^i\right)^2 - \left[\sum_{i=1}^{N}\left(R_1^i R_2^i\right)\right]^2} \tag{3.73}$$

$$\tau_2 = \frac{\sum_{i=1}^{N}\left[R_1^i R_2^i\right] \sum_{i=1}^{N}\left[R_1^i\left(T_2^i - T_1^i\right)\right] - \sum_{i=1}^{N}\left(R_1^i\right)^2 \sum_{i=1}^{N}\left[R_2^i\left(T_2^i - T_1^i\right)\right]}{\sum_{i=1}^{N}\left(R_1^i\right)^2 \sum_{i=1}^{N}\left(R_2^i\right)^2 - \left[\sum_{i=1}^{N}\left(R_1^i R_2^i\right)\right]^2} \tag{3.74}$$

with

$$R_1^i = \frac{dT_1^i}{dt} + \frac{\beta}{d_1}\left(T_1^i\right)^4 \tag{3.75}$$

and

$$R_2^i = \frac{dT_2^i}{dt} + \frac{\beta}{d_2}\left(T_2^i\right)^4 \tag{3.76}$$

Finally, after determining the two time constants τ_1 and τ_2, the temperatures T_{g1} and T_{g2} are calculated from Equations 3.71 and 3.72.

3.3.4 Fluid Velocities and Static Pressure Measurements with Thermocouples

This paragraph presents the development of a microthermocouple sensor for velocity/temperature measurements or pressure/temperature measurements [59–62]. The transient thermocouple sensor consists of two type K microthermocouples used in an electronic oscillator. One is placed in the system and the other takes place in a reference volume. During a half period, the two microthermocouples are heated to a suitable temperature above ambient. During the next half period, when the supply of current is interrupted, the microthermocouple placed in the system measures the flow (or pressure) while the other compensates for the ambient temperature changes. Cooling caused by experimental conditions under variable flow (or pressure) results in a change in the oscillator frequency. The sensor is developed in order to measure flows (or pressures) and temperatures in microsystems like small channels (width < 500 μm), microtubes (diameter < 50 μm), and small structures (volume <100 μm^3).

The working principle of the anemometer consists in heating two identical K type thermocouples (Figure 3.37) during a predetermined time t_h by means of an electrical current step. One of the thermocouples, Th_{sens}, is put inside the fluid in which the measurement of the velocity (or the pressure) is required, while the other one, Th_{ref}, is disposed in a closed volume to avoid the external disturbances. The probe Th_{ref} is used for the generation of a reference signal (Figure 3.38). At the end of the heating state, the thermocouple Th_{sens} attains a temperature higher than that of the fluid temperature T_f. The time t_r necessary to the sensor during the cooling phase to reach the value T_f depends directly on the fluid velocity V (or the pressure P). The general form of the signal detected from the

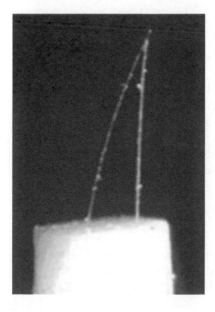

FIGURE 3.37
Microthermocouple (K type, diameter 25.4 μm).

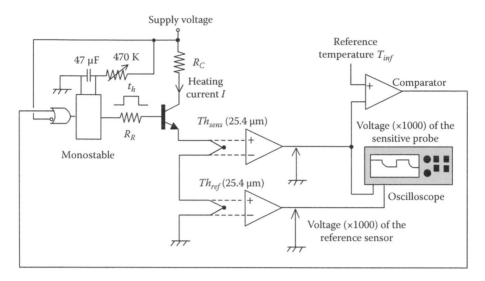

FIGURE 3.38
Block diagram of the sensor.

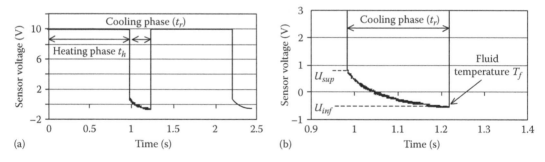

FIGURE 3.39
Sensor response in ambient conditions and without flow ($I = 32$ mA, $t_h = 1$ s). (a) Heating and cooling phases. (b) Zoom of the cooling phase.

thermocouple Th_{sens} placed in the flow is shown in Figure 3.39a in which the two periods are clearly distinguishable.

The first, whose issued voltage from the thermocouple is constant, corresponds to the heating time t_h. The amplifier is then saturated resulting in a voltage nearly equal to the supply one. The second period corresponds to the relaxation state of the sensor. Its duration t_r is not constant, and it depends directly upon the velocity V (or the pressure) of the fluid. The action of the fluid is then traceable only during the cooling phase, and the frequency f of the obtained signal is directly proportional to the fluid velocity V.

3.3.4.1 Velocity and Temperature Measurements

3.3.4.1.1 Theory

The theoretical analysis is conducted through the dynamic behavior of the resistive wire subjected to a periodic heating and cooling by forced convection. The two thermocouple wires are modeled by a single wire whose average thermo-physics characteristics are considered as constants. Thermocouple wires are assembled for forming a wire of length $2L$ and diameter d placed in the fluid. The wire is heated by Joule effect by intensity of current I and dissipates the heat produced into the surrounding fluid of temperature T_f by convection. The air fluid characteristics are supposed constant. The one-dimensional thermal balance interprets the equality between the accumulated power by the cylindrical volume of the wire and the sum of the following modes of heat transfer: the convection through the boundary layer surrounding the wire, the conduction along the wires towards the support of the thermocouple, the radiation between the probe and the surrounding wall of the conduit, and the internal heating by Joule effect:

$$\frac{\partial T}{\partial t} = a\frac{\partial^2 T}{\partial x^2} - \frac{4(h_{cv} + h_{rad})}{\rho c_p d}(T - T_f) + A[1 + \alpha_{th}(T - T_f)]I^2 \tag{3.77}$$

with α_{th} the coefficient of thermal dependency of the resistivity of the wire material

$$A = \left(\frac{4}{\pi d^2}\right)^2 \frac{\rho_0}{\rho c_p} \tag{3.78}$$

with ρ_0 the electrical resistivity at reference temperature that corresponds to the fluid temperature T_f.

Equation 3.36 interprets the thermal balance between a wire of circular section heated by Joule effect and the surrounding fluid and wall. This equation brings in two thermo-physics parameters representing the quality of the thermal exchange between the wire surface and the fluid: the convection coefficients h_{cv} and the linear radiative heat transfer coefficient h_{rad}.

The convective time constant of the wire is defined by the classical expression

$$\tau_{cv} = \frac{\rho c_p d}{4h_{cv}} = \frac{\rho c_p d^2}{4\lambda_f Nu} \tag{3.79}$$

with

$$Nu = \frac{h_{cv}d}{\lambda_f} = C_1 + C_2 Re^n \tag{3.80}$$

the Nusselt number of the flow where the Reynolds number Re is based on the wire diameter d and the local velocity V. We consider the relation of Olivari and Carbonaro (Table 3.4), valid for a Reynolds number $0.015 < Re < 20$ with

$$C_1 = 0.34, \quad C_2 = 0.65 \quad \text{and} \quad n = 0.45 \tag{3.81}$$

The radiative time constant of wire is also defined by the expression:

$$\tau_{rad} = \frac{\rho c_p d}{4h_{rad}} = \frac{\rho c_p d}{16\varepsilon\sigma T_f^3} \tag{3.82}$$

with $h_{rad} = 4\varepsilon\sigma T_f^3$ the linear radiative heat transfer coefficient (the heating of the wire is very small, its surface temperature remains near to that of the fluid flow).

Moreover, the time constant corresponding to the heating of the wire next to the heat accumulation by the Joule effect is defined as follows:

$$\tau_{el} = \frac{1}{A\alpha I^2} \tag{3.83}$$

Finally, we call τ_g the global time constant of the sensor that can be defined by the relation:

$$\frac{1}{\tau_g} = \frac{1}{\tau_{cv}} + \frac{1}{\tau_{rad}} - \frac{1}{\tau_{el}} \tag{3.84}$$

Then, Equation 3.77 can be modified by introducing the global time constant τ_g:

$$\frac{\partial T}{\partial t} = a \frac{\partial^2 T}{\partial x^2} - \frac{(T - T_f)}{\tau_g} + A I^2 \tag{3.85}$$

The conduction effect along the wire can be neglected with respect to the convection for the aspect ratio $L/d > 200$ sufficiently large. Taking into account the simplifying hypotheses, the thermal balance is finally written in the form of a differential equation of first degree whose solution gives the temperature profile of the wire during the heating and relaxation states of the wire:

$$\frac{dT}{dt} + \frac{(T - T_f)}{\tau_g} - A I^2 = 0 \tag{3.86}$$

3.3.4.1.2 Heating Phase

The solution of the differential equation (3.86) describes the temperature rising of the wire during the time:

$$T(t) = T_f + A I^2 \tau_g + (T_{inf} - T_f - A I^2 \tau_g) e^{(-t/\tau_g)} \tag{3.87}$$

with the initial condition:

$$T(t = 0) = T_{inf} \tag{3.88}$$

where T_{inf} is the reference temperature corresponding to the threshold temperature for heating the wire. The temperature T_{sup} can be expressed at the end of the heating period t_h. $T(t = t_h) = T_{sup}$ and Equation 3.88 written as (3.89)

$$T_{sup} = T_f + A I^2 \tau_g + (T_{inf} - T_f - A I^2 \tau_g) e^{(-t_h/\tau_g)} \tag{3.89}$$

If, under convective conditions, duration t_h of heating is much larger than characteristic time τ_g ($\tau_g \ll t_h$), the sensor can reach thermodynamic equilibrium with the fluid, at temperature level T_{sup} (Figure 3.38):

$$T_{sup} \approx T_f + A I^2 \tau_g \tag{3.90}$$

hence, for a heating with respect to the average temperature of the flowing fluid,

$$T_{sup} - T_f \approx AI^2 \tau_g \tag{3.91}$$

3.3.4.1.3 Relaxation Phase

The relaxation phenomenon, of duration t_r (Figure 3.38), is due to the cooling of the thermocouple wire caused by the fluid flowing at the velocity V. In this way, the duration of dynamic cooling produces two informations relative to the local velocity of the flow and the temperature of the thermocouple when its thermal equilibrium is reached with the fluid. The wire is cooled by forced convection from the new temperature T_{sup}, obtained at the end of heating period t_h. It tends towards the corresponding temperature T_{inf}, either to the threshold temperature where the voltage is fixed by the electronic system or to the regime of thermal equilibrium if the inertia of the wire is sufficient. The differential equation that governs the relaxation phenomenon is obtained from the balance equation (3.86) for which the heating current is set to zero.

So, for $I = 0$:

$$\frac{dT}{dt} + \frac{(T - T_f)}{\tau_g^*} = 0 \tag{3.92}$$

with the initial condition:

$$T(t = t_h) = T_{sup} \tag{3.93}$$

and now the global time constant of the sensor becomes

$$\frac{1}{\tau_g^*} = \frac{1}{\tau_{cv}} + \frac{1}{\tau_{rad}} \tag{3.94}$$

Considering Equations 3.93 and 3.94, the solution of the wire temperature has the following expression:

$$T(t) = T_f + (T_{sup} - T_f)e^{\left(-(t-t_h)/\tau_g^*\right)} \tag{3.95}$$

From Equation 3.95, the temperature T_{inf} corresponds to the threshold temperature attained by the sensor at the end of relaxation time t_r:

$$T(t = t_r) \equiv T_{inf} = T_f + (T_{sup} - T_f)e^{\left(-(t_r-t_h)/\tau_g^*\right)} \tag{3.96}$$

Or on heating

$$T_{inf} - T_f = (T_{sup} - T_f)e^{\left(-(t_r-t_h)/\tau_g^*\right)} \tag{3.97}$$

From Equation 3.97, we can express the relaxation time t_r at the end when the temperature tends towards the value T_{inf}. Then, it occurs $T(t = t_r) = T_{inf}$ and, finally,

$$t_r = t_h - \tau_g^* \ln\left(\frac{T_{inf} - T_f}{T_{sup} - T_f}\right) \tag{3.98}$$

3.3.4.1.4 Oscillation Frequency

The sensor is heated by an electric current during a constant time t_h. Next, it is cooled by convection during a variable time t_r, which depends upon the nature of the flow and the fluid properties. The temperature then attains the threshold value T_{inf} fixed by the electronic system. A new heating is thus achieved. By this way, the sensor is subjected to an oscillation whose frequency is proportional to the cooling and thus to the velocity of the flow as well. The oscillation frequency f of the sensor can be related by putting the heating time as a parameter fixed by the electronic system. So we get

$$f = \frac{1}{t_h + t_r} = \frac{1}{2t_h - \tau_g^* \ln\left(\dfrac{T_{inf} - T_f}{T_{sup} - T_f}\right)} \tag{3.99}$$

Finally, the oscillation frequency of the sensor can be expressed as a function of the following parameters: flow velocity V and wire diameter d (taking place in the expressions of the time constant τ_g^* and τ_g), heating duration t_h, heating current I, and relaxation heating $T_{inf} - T_f$.

$$f = \left[2t_h - \tau_g^* \ln\left(\frac{T_{inf} - T_f}{AI^2\tau_g + \left(T_{inf} - T_f - AI^2\tau_g\right)e^{(-t_h/\tau_g)}}\right)\right]^{-1} \tag{3.100}$$

3.3.4.1.5 Applications

In the case of gas velocity measurement, the sensor calibration system is presented in Figure 3.40. The calibrated airflow is at ambient temperature, and the flow rate may be regulated with a *Brooks* calibration mass flowmeter. This flow is realized in a long cylindrical tube of length 20 cm in order to obtain a sufficient length of flow establishment, and of circular section of diameter 2 mm. The thermocouple Th_{sens} is then installed in the center of the section. The *Brooks* flowmeter being used as a reference, it allows fixing a flow rate in the tube and therefore measuring the signal frequency for different values of flow rate. For these operating conditions, the velocity range obtained is 0–3.5 m s^{-1}, corresponding to a laminar flow regime (Figure 3.41). The measured frequency is then directly proportional to the maximum velocity since the thermocouple is introduced at the center of the tube.

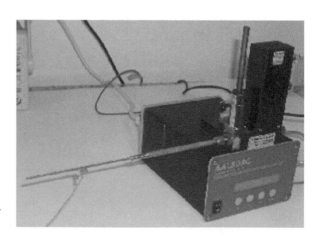

FIGURE 3.40
Microflow in a circular tube (internal diameter 2 mm).

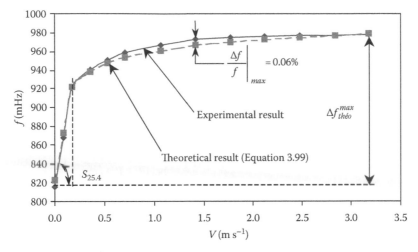

FIGURE 3.41
Flow measurement in a circular tube.

3.3.4.2 Pressure and Temperature Measurements

3.3.4.2.1 Theory
The sensor is used for the measurements of pressure from 10^{-1} to 10^5 Pa range. In this pressure range, the pressure dependence of the conductivity can be split up into three regimes: molecular, viscous slip, and viscous. In such type of sensor, heat from the heating element is dissipated to the ambient through thermal conduction of physical parts of the sensor, radiation, convection, and thermal conduction of surrounding gas. Heat dissipation by convection and thermal conduction of gas has an essential effect on the sensor characteristics versus pressure [63–65].

3.3.4.2.2 Heating Phase
The heat balance is the same than in the previous paragraph. We obtain a differential equation where the convection effects are estimated by a general Morgan correlation (Table 3.10). The initial condition is $T(0) = T_{inf}$. The heating time t_c is a given data.

$$\frac{dT}{dt} + (T - T_f)\left(K(P)(T - T_f)^m + \frac{1}{\tau_{rad}} - \frac{1}{\tau_{él}}\right) - AI^2 = 0 \tag{3.101}$$

TABLE 3.10

Morgan Relations

$Nu = C(Gr\,Pr)^m$		
C	m	Gr Pr Range
0.675	0.058	$10^{-10} < Gr\,Pr < 10^{-2}$
1.020	0.148	$10^{-2} < Gr\,Pr < 10^2$
0.850	0.188	$10^2 < Gr\,Pr < 10^4$
0.480	0.250	$10^4 < Gr\,Pr < 10^7$
0.125	0.333	$10^7 < Gr\,Pr < 10^{12}$

Physical characteristics at $T_{film} = (T_f + T)/2$.

where

$$A = \left(\frac{4}{\pi d^2}\right)^2 \frac{\rho_0}{\rho c_p}; \quad \tau_{rad} = \frac{\rho c_p d}{16\varepsilon\sigma T_f^3}; \quad \tau_{\hat{e}l} = \frac{1}{A\alpha I^2}$$

$$Gr = \frac{g \cdot \beta \cdot (T - T_f) \cdot L^3}{v_f^2}; \quad Pr = \frac{\mu_f \cdot c_{pf}}{\lambda_f} \tag{3.102}$$

$$\frac{1}{\tau_{cv}} = K(P)(T - T_f)^m \quad \text{with } K(P) = \frac{4\lambda_f C}{\rho c_p d^2}\left(\frac{g c_{pf} d^3}{T_f^2 \lambda_f v_f r}P\right)^m$$

3.3.4.2.3 Relaxation Phase

The relaxation phase corresponds to the cooling of the wire by natural convection varying with the mass of gas in the system, that is, the pressure. This dynamical phase gives two informations: one corresponds to the gas pressure and the other to the gas temperature at steady state. The wire is cooled from the temperature T_{sup}, obtained at the end of the heating period by natural convection, and drops to the temperature T_{inf}, corresponding to the thermal equilibrium of the wire. The differential equation is obtained with a heat balance without electrical current ($I = 0$):

$$\frac{dT}{dt} + (T - T_f)\left(K(P)(T - T_f)^m + \frac{1}{\tau_{rad}}\right) = 0 \tag{3.103}$$

The relaxation phase is represented by a nonlinear differential equation, the solution of which gives the wire temperature along the time.

3.3.4.2.4 Oscillation Frequency

The oscillation frequency f of the sensor can be related by putting the heating time as a parameter fixed by the electronic system:

$$f = \frac{1}{t_h + t_r} \tag{3.104}$$

3.3.4.2.5 Applications

Figure 3.42 shows the schematic diagram of the calibration system for pressure measurement application. It consists of the microthermocouple Th_{sens} introduced in a measurement chamber, a vacuum pump, and two valves to control the vacuum. The measurement chamber is a copper cylinder of interior volume equal to 90 cm^3, and the interior pressure is regulated by two valves. The first valve O_1 allows connecting directly the pump to the measure chamber and so as to establish the vacuum in this volume. The second valve O_2 is connected to the ambient pressure and is used to adjust the working pressure in the cylinder. At the beginning of the experiment, O_2 is closed and O_1 open until reaching with the pump a primary vacuum. Then O_1 isolates the pump to the chamber and the valve O_2 is used to increase the pressure in the chamber. The measure range obtained with this device corresponds to 10^{-1} to 10^5 Pa. A Pirani vacuum gauge (Thermovac TM 20) introduced close to the measurement volume gives the value of the pressure and is used as reference. The characterization of the sensor

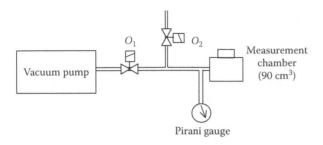

Pirani gauge

FIGURE 3.42
Frequency versus pressure in a test volume.

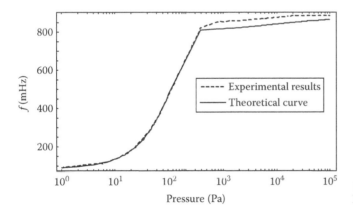

FIGURE 3.43
Frequency versus vacuum.

consists in measuring the signal frequency as a function of the static pressure. Figure 3.43 gives the experimental and theoretical results of the oscillatory frequency versus pressure realized in the test section presented above.

3.4 Conclusion

The thermocouple is one of the most widely used devices for temperature measurement. It presents advantages: inexpensive, rugged, simply constructed, fast in the response to changes in temperature (microthermocouples), and capable of being used to directly measure temperatures from −200°C up to 2600°C. But, disadvantages exist too: temperature measurement with a thermocouple requires in fact independent measurements of two temperatures, the junction at the hot junction and the junction where wires meet the instrumentation copper wires (cold junction). To avoid error, the cold junction temperature is in general compensated in the electronic instruments by measuring the temperature at the terminal block using with a semiconductor, thermistor, or RTD. Thermocouple operation is relatively complex with potential sources of error. The materials of thermocouple wires are not inert, and the thermoelectric voltage developed along the length of the thermocouple wire may be influenced by corrosion, etc. The relationship between the process temperature and the thermocouple signal (millivolt) is not linear. The calibration of the thermocouple should be carried out while it is in use by comparing it to a nearby comparison thermocouple.

The size reduction of thermal sensors has been significant during the last 20 years. So, the reducing of the time and spatial resolutions and the increasing of the nonintrusive character of measurements have opened the way to a real improvement of performances and to various new applications. Wire microthermocouples, well adapted to fluid investigations, are champions in the field of low inertia measurements: micronic and even submicronic junctions are operative today. A diameter of 1.3 μm is almost usual in advanced research laboratories and a diameter of 0.5 μm possible, but both are subject to specific cautions because of their weakness. The present limit (0.5 μm) is essentially due to the commercial unavailability of smaller thermoelectrical wires.

Nomenclature

a	thermal diffusivity ($m^2\ s^{-1}$)
A	area (m^2)
c	heat capacity ($J\ kg^{-1}\ K^{-1}$)
C	constant
d	diameter (m)
E	tension (V)
Ec	Eckert number
EMF	thermocouple electromotive force (V)
f	frequency (Hz)
Gr	Grashof number
h	convection coefficient ($W\ m^{-2}\ K^{-1}$)
I	electric intensity (A)
K	constant
Kn	Knudsen number
ℓ	mean free path (m)
L	length (m)
m	constant
Ma	Mach number
Nu	Nusselt number
Pr	Prandtl number
Q	heat quantity (J)
$\dot{Q}$	rate of heat (W)
R	electric resistance (Ω)

Re	Reynolds number
S	least square sum (Equation 3.64)
t	time (s)
T	absolute temperature (K)
$\overline{T}$	fast Fourier transform of the temperature
V	volume (m^3)
X	sensitivity coefficient
X	column vector of the sensitivity coefficients
y	difference of observable quantities (Equation 3.64)
y	column vector of the difference of observable temperatures

Greek Symbols

α	ratio of time constant
δ	temperature difference (K)
ε	emissivity
λ	thermal conductivity (W m^{-1} K^{-1})
Π	Peltier voltage (V)
ρ	density (kg m^{-3})
σ	Seebeck coefficient (V K^{-1})
τ	time constant (s)
ν	kinematic viscosity (m^2 s^{-1})
ϑ	period (s)

Subscripts

cd	conduction
cv	convection
el	electric
equ	equivalent
ext	external
f	fluid
$film$	film
g	gas
h	heating
inf	inferior
L	laser
mea	measured
mod	model
OLS	ordinary least square
P	Peltier effect
r	relaxation
rad	radiation
ref	reference
sup	superior
S	Seebeck effect
th	thermocouple
T	Thomson effect
w	wall
0	temperature reference

References

1. Müller, I. 2007. *A History of Thermodynamics, The Doctrine of Energy and Entropy*, Springer, New York.
2. Santorio, S., *Commentaria in Primam Fen Primam Libri Canonis Avicenna*. 1625. Cited and illustrated by Lyons, A.S. and Petrucelli, R.J. 1987. *Medicine, An Illustrated History*, Abrams, New York.
3. Preston-Thomas, H. 1990. The international scale of temperature, *Metrologia* 27(1):3–10.
4. Mangum, B.W., Furukawa, G.T., Kreider, K.G., Meyer, C.W., Ripple, D.C., Strouse, G.F., Tew, W.L., Moldover, M.R., Carol Johnson, B., Yoon, H.W., Gibson, C.E., and Saunders, R.D. 2001. The kelvin and temperature measurement, *J. Res. Natl. Inst. Stand. Technol.* 106(1):105–149.
5. Childs, P.R.N., Greenwood, J.R., and Long, C.A. 2000. Review of temperature measurement, *Rev. Sci. Instrum.* 71(8):2959–2978.
6. Seebeck, T.J. 1823. Magnetische polarisation der metalle und erze durch temperatur-differenz, *Abh. K. Akad. Wiss*, Berlin, 265:289–346.
7. Peltier, J.C.A. 1834. *Ann. Chim. Phys.* 56:371–470.
8. Thomson, W. 1848. On an absolute thermometric scale founded on Carnot's theory of the motive power of heat, and calculated from Regnault's observations, *Philos. Mag.* 33:313–317.
9. Rathakrishnan, E. 2007. *Instrumentation, Measurements and Experiments in Fluids*, CRC Press, Boca Raton, FL.
10. Devin, E. 1997. Couples thermoélectriques, données numériques d'emploi, *Techniques de l'Ingénieur*, tome R2594:1–26.
11. Shockley, W., Bardeen, J., and Brattein, W.H. 1948. Electronic theory of the transistor, *Science* 108:678–679.
12. Bardeen, J. and Brattein, W.H. 1948. The transistor: A semi-conductor triod, *Phys. Rev.* 74 (2):230–231.
13. Löfdahl, L. and Gad-el-Hak, M. 1999. MEMS applications in turbulence and flow control, *Prog. Aero. Sci.* 35:101–203.
14. Anduze, M. 2000. Etude expérimentale et numérique de microécoulements liquides dans les microsystèmes fluidiques, PhD dissertation, INSA Toulouse, France.
15. Oosterbroek, E. 1999. Modeling, design and realization of microfluidic components, PhD dissertation, University of Twente, Enschede, the Netherlands.
16. Secco, R.A. and Tucker, R.F. 1992. Thermocouple butt-welding device, *Rev. Sci. Instrum.* 63 (11):5485–5486.
17. Forney, L.J., Meeks, E.L., Ma, J., and Fralick, G.C. 1993. Measurement of frequency response in short thermocouple wires, *Rev. Sci. Instrum.* 64(5):1280–1286.
18. Lanzetta, F. 1997. Etude des transferts de chaleur instationnaires au sein d'une machine frigorifique de Stirling, PhD dissertation, University of Franche-Comté, France.
19. Tagawa, W. and Ohta, Y. 1997. Two thermocouple probe for fluctuating temperature measurement in combustion—Rational estimation of mean and fluctuating time constants, *Combust. Flame* 109:549–560.
20. Burns, G.W., Scroger, M.G., Strouse, G.F., Croarkin, M.C., and Guthrie, W.F. 1993. Temperature-electromotive force reference functions and tables for the letter-designated thermocouple types based on the ITS-90, NIST Monographe 175, National Institut of Standards and Technology.
21. Yule, A.J., Taylor, D.S., and Chigier, N.A. 1978. On-line digital compensation and processing of thermocouples signals for temperature measurements in turbulent flames, *AIAA 16th Aerospace Sciences Meeting*, Huntsville, Alabama, January 16–18, pp. 78–80.
22. Lenz, W. and Günther, R. 1980. Measurement of fluctuating temperature in a free-jet diffusion flame, *Combust. Flame* 37:63–70.
23. Lockwood, F.C. and Moneib, H.A. 1980. Fluctuating temperature measurements in a heated round free jet, *Combust. Sci. Technol.* 22:63–81.

24. Voisin, P., Thiery, L., and Brom, G. 1999. Exploration of the atmospheric lower layer thermal turbulences by means of microthermocouples, *Eur. Phys. J. Appl. Phys.* 7(2):177–187.
25. Pitts, W.M., Braun, E.B., Peacock, R.D., Mitler, H.E. et al. 1998. Temperature uncertainties for bare-bead and aspirated thermocouple measurements in fire environments, *Proceedings of the 14th Meeting of the United States Japan Conference on Development of Natural Resources (UJNR) Panel on Fire Research and Safety*, 8 May–3 June, Tsukuba, Japan.
26. Blevins, L.G. and Pitts, W.M. 1999. Modeling of bare and aspirated thermocouples in compartment fires, *Fire Safety J.* 33(4):239–259.
27. Santoni, P-A., Marcelli, T., and Leoni, E. 2002. Measurement of fluctuating temperatures in a continuous flame spreading across a fuel bed using a double thermocouple probe, *Combust. Flame* 131(1–2):47–58.
28. Rakopoulos, C.D., Rakopoulos, D.C., Mavropoulos, G.C., and Giakoumis, E.G. 2004. Experimental and theoretical study of the short term response temperature transients in the cylinder walls of a diesel engine at various operating conditions, *Appl. Therm. Eng.* 24(5–6):679–702.
29. Bardon, J.P., Raynaud, M., and Scudeller, Y. 1995. Mesures par contact des températures de surface, *Rev. Gén. Therm.* 34(HS95):15–35.
30. Paranthoen, L. and Lecordier, J.C. 1996. Mesures de température dans les écoulements turbulents, *Rev. Gén. Therm.* 35:283–308.
31. Olivari, D. and Carbonaro, M. 1994. Hot wire measurements. *Measurements Techniques in Fluid Dynamics. An Introduction*, Annual Lecture Series, Vol. 1, Von Karman Institute for Fluid Dynamics, Belgium, pp. 183–218.
32. Million, F., Parenthoën, P., and Trinite, M. 1978. Influence des échanges thermiques entre le capteur et ses supports sur la mesure des fluctuations de température dans un écoulement turbulent, *Int. J. Heat Mass Transfer* 21:1–6.
33. Bradley, D. and Mathews, K. 1968. Measurement of high gas temperature with fine wire thermocouple, *J. Mech. Eng. Sci.* 10(4):299–305.
34. Collis, D.C. and Williams, M.J. 1959. Two dimensional convection from heated wires at low Reynolds numbers, *J. Fluid Mech.* 6:357–384.
35. Knudsen, J.G. and Katz, D.L. 1958. *Fluid Dynamics and Heat Transfer*, McGraw-Hill Book Co., New York.
36. Van der Hegg Zijnen, B.G. 1956. Modified correlation formulae for the heat transfer by natural and by forced convection from horizontal cylinders, *Appl. Sci. Res. A* 6:129–140.
37. Mac Adams, W.H. 1956. *Heat Transmission*, McGraw-Hill Book Co., New York.
38. Eckert, E.R. and Soehngen, E. 1952. Distribution of heat transfer coefficients around circular cylinders, Reynolds numbers from 20 to 500, *Trans. ASME, J. Heat Transfer* 74:343–347.
39. Scadron, M.D. and Warshawski, I. 1952. Experimental determination of time constants and Nusselt numbers for bare-wire thermocouples in high velocity air streams and analytic approximation of conduction and radiation errors, NACA Technical Note 2599.
40. Tarnopolski, M. and Seginer, I. 1999. Leaf temperature error from heat conduction along the wires, *Agric. For. Meteorol.* 93(3):185–190.
41. Bailly, Y. 1998. Analyse expérimentale des champs acoustiques par méthodes optiques et microcapteurs de température et de pression, PhD dissertation, University of Franche-Comté, France.
42. Fralick, G.C. and Forney, L.J. 1993. Frequency response of a supported thermocouple wire: effects of axial conduction, *Rev. Sci. Instrum.* 64(11):3236–3244.
43. Sbaibi, H. 1987. Modélisation et étude expérimentale de capteurs thermiques, PhD dissertation, University of Rouen, France.
44. Singh, B.S. and Dybbs, A. 1976. Error in temperature measurements due to conduction along the sensor leads, *J. Heat Transfer* 491:491–495.
45. Kramers, H. 1946. Heat transfer from spheres to flowing media, *Physica* 12:61–80.
46. King, L.V. 1914. On the convection of heat from small cylinders in a stream of fluid, *Philos. Trans. R. Soc. Lond., Ser. A* 214(14):373–432.

47. Hilaire, C., Filtopoulos, E., and Trinite, M. 1991. Mesure de température dans les flammes turbulentes. Développement du traitement numérique du signal d'un couple thermoélectrique, *Rev. Gén. Therm.* 354/355:367–374.

48. Castellini, P. and Rossi, G.L. 1996. Dynamic characterization of temperature sensors by laser axcitation, *Rev. Sci. Instrum.* 67(7):2595–2601.

49. Hostache, G., Prenel, J.P., and Porcar, R. 1986. Couples thermoélectriques à définition spatio-temporelle fine. Réalisation. Réponse impulsionnelle de microjonctions cylindriques, *Rev. Gén. Therm.* 299:539–543.

50. Cambray, P., Vachon, M., Masciaszek, T., and Bellet, J.V. 1985. *Proceedings of the 23rd ASME National Heat Transfer Conference*, 4–7 August, Denver, Colorado.

51. Cambray, P. 1986. Measuring thermocouple time constants: A new method, *Combust. Sci. Technol.* 45:221–224.

52. Forney, L.J. and Fralick, G.C. 1994. Two wire thermocouple: Frequency response in constant flow, *Rev. Sci. Instrum.* 65(10):3252–3255.

53. O'Reilly, P.G., Kee, R.J., Fleck, R., and McEntee, P.T. 2001. Two-wire thermocouples: A non linear state estimation approach to temperature reconstruction, *Rev. Sci. Instrum.* 72(8):3449–3457.

54. Marcelli, T., Santoni, P.A., Leoni, E., and Simeoni, A. 2002. *Forest Fire Research & Wildland Fire Safety*, Viegas (ed.), Millpress, Rotterdam, the Netherlands.

55. Forney, L.J. and Fralick, G.C. 1995. Multiwire thermocouples in reversing flow, *Rev. Sci. Instrum.* 66(10):5050–5054.

56. Forney, L.J. and Fralick, G.C. 1995. Three wire thermocouple: Frequency response in constant flow, *Rev. Sci. Instrum.* 66(5):3331–3336.

57. Lanzetta, F., Boucher, J., and Gavignet, E. 2006. Two-microthermocouple probe for temperature and velocity measurements in an oscillating flow in a heat exchanger of Stirling machine, *ASME ATI Conference, Energy: Production, Distribution and Conservation*, Milan, Italy, May 14–17, 2006, pp. 633–642.

58. Warshawsky, I. 1995. On-line dynamic gas pyrometry using two-thermocouple probe, *Rev. Sci. Instrum.* 66(3):2619–2624.

59. Gavignet, E., Lanzetta, F., and Nika, P. 2003. Thermocouple flow sensor with a-c heating for simultaneous temperature and gas flow measurements, *ITBM-RBM* 24(2):98–100.

60. Khan, M., Lanzetta, F., Gavignet, E., and Nika, P. 2004. A transiently heated microthermocouple anemometer, *International Conference on Thermal Engineering Theory and Applications*, Berut, Liban, May 31–June 4, 2004, pp. 1–10.

61. Lanzetta, F., Gavignet, E., Nika, P., and Meunier, C. 2005. Microthermocouples for the simultaneous measurements of temperature/pressure and temperature/velocity in microsystems, *12th International Metrology Congress*, June 20–23, 2005, Lyon, France (CD-Rom, article No. 68, 6 pages).

62. Lanzetta, F. Gavignet, E., and Girardot, L. 2008. Microthermocouple anemometry versus laser Doppler anemometry for experimental velocity measurement in air jet, *ISFV13—13th International Symposium on Flow Visualization, FLUVISU12—12th French Congress on Visualization in Fluid Mechanics*, July 1–4, 2008, Nice, France (CD-Rom, ID#380, 11 pages).

63. Doms, M., Bekesch, A., and Mueller, J. 2005. A microfabricated Pirani pressure sensor operating near atmospheric pressure, *J. Micromech. Microeng.* 15:1504–1510.

64. Umrath, W. 1998. Fundamentals of vacuum technology. http://www.metrovac.eu/Publicacoes/assets/VacuumBook%20Fundamentals.pdf (2010/12/22).

65. Delchar, T.A. 1993. *Vacuum Physics and Techniques (Physics and Its Applications)*, Springer, New York.

4

Temperature Measurements: Resistive Sensors

Paulo Seleghim, Jr.

CONTENTS

4.1 Introduction

Temperature, as well as pressure, acceleration, and so on, is a variable that can be measured to acquire information about a physical process to be scientifically described and mastered in an engineering application. To do this, one must interact with the process through a measuring system embedding a physical phenomenon capable of translating the process variable into an "indicated" suitable signal, usually some electrical variable such as voltage, current, capacitance, etc. The indicated signal should be of electrical nature because further processing can be accomplished through analog circuits and digital microprocessors that are basically electronic devices. Possibly, in a near future, signal processing will be accomplished through photonic devices and our transducers will be based on physical phenomena through which the process variable modulates some light-related variable as, for example, the effect of temperature on a fiber Bragg grating sensor. Anyway, the fundamental and frequently overlooked concept here is that the measured or indicated variable is the response to the stimulus imposed by the process and, as such, it contains transformed rather than original information about the process. Thus, any measurement problem is actually an inverse problem (in the mathematical sense of the term) because one wants to recover the original information from the transformed information, that is, the process signal from the indicated signal. The question of if and how this is possible constitutes an important new research area.

One important type of sensor is the resistance temperature detector or resistive thermal device (RTD) whose working principle is based on the change in electrical resistance of some material with changing temperature. Several materials can be used, such as iron and copper. However, platinum is the most common resistance thermometer (PRT) because of its linearity with temperature and chemical stability. RTDs are gradually becoming predominant in industrial applications, particularly in applications under 600°C, due to their higher accuracy and repeatability, in addition to the simplicity of its conditioning electronics compared with thermocouples or other types of thermal sensors. More specifically, the RTD being essentially a resistor element, one can take advantage of a great number of standard electronic measurement techniques and integrated components suitable for measuring under myriads of practical condition.

Specifying the most adequate RTD to a particular application can be a difficult task, as it can be for all other types of sensor. To ensure the desired performance, one must consider a number of aspects such as thermochemical compatibility and materials, dimensions and size, temperature range and dynamics, accuracy, precision and errors, effects of lead wiring configuration, conditioning electronics, and nominal resistance and temperature coefficients. Some of these aspects are informed by the manufacturers of the sensor, electronic components, etc., and others are dictated by the specificities of the application. Platinum RTD (PRT) standards help defining a general frame of reference within which these issues can be addressed. The European standard DIN/IEC 60751, one of the most commonly adopted worldwide, requires that the RTD's electrical resistance has to be of $100.00\ \Omega$ at 0°C with a temperature coefficient of resistance of $0.00385\ \Omega/\Omega/°C$ between 0°C and 100°C. In DIN/IEC 60751, there are two classes of resistance tolerances: Class $A = 100.00 \pm 0.06\ \Omega$ @ 0°C and Class $B = 100.00 \pm 0.12\ \Omega$ @ 0°C. The combination of resistance tolerance and temperature coefficient defines the resistance/temperature characteristics of the sensor and, ultimately, an envelope around the nominal transduction equation within which lies the actual calibration curve of each particular sensor. (This point will be elaborated in the section dedicated to error analysis.) Consequently, the greater the sensor's resistance tolerance the more the calibration curve will deviate from the generalized curve and more variation there will be from sensor to sensor. Interchangeability is an important issue in applications where the sensor is expected to be replaced from time to time, particularly if the RTD's information is used for billing purposes, such as in custody transfer in the petroleum industry.

As mentioned above, some aspects to be considered when specifying an RTD are intrinsic to the sensor, and others are application dependent. Among the intrinsic aspects, probably the most important one is the necessary conditioning electronics and lead wiring. An RTD is intrinsically a two-wire resistance that must be connected to its conditioning electronics through lead wires, which introduce stray impedances to the circuit. Therefore, most applications are developed based on three- or four-wire circuitry to compensate for these stray effects producing a truer indication of the measured temperature. Figure 4.1 shows the corresponding diagrams. The three-wire circuit is based on the assumption that the lead wires have the same impedance that can be cancelled out by adding a third resistance to one of the adjoining arms. Due to its simplicity and the availability of high-quality connection cables, this is a very common choice in industrial applications in which the distance between the sensor and the conditioning electronics is less than 500 m. The four-wire Kelvin connection uses separate pairs of current-carrying and voltage-sensing cables, providing virtually full cancellation of stray impedances of up to $15\ \Omega$ cables. Due to its complexity, this configuration is commonly restricted to laboratory applications where very high accuracies are required.

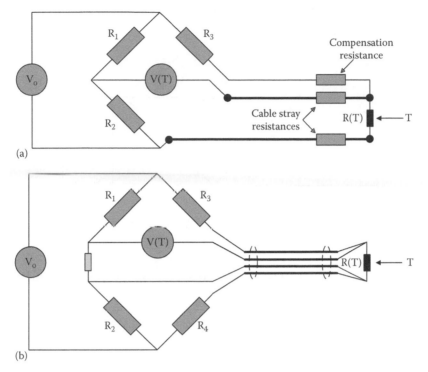

FIGURE 4.1
Three-wire (a) and four-wire (b) circuits used to compensate for stray impedances of lead cables.

4.2 Transduction Equation and Conditioning Electronics

A thermal measurement system interacts with the process and generates a response voltage, or other electrical variable, which is indicative of the stimulus temperature. This is generally done through a physical phenomenon that responds electrically to a thermal stimulus such as Seebeck's effect used in thermocouples, photosensitivity used in photodetectors, and Joule's effect used in RTDs. In addition to this, some electronic circuitry is always necessary to transform the transduced electrical variable into a more convenient one, usually by magnification, denoising, offset correction, etc. The overall relation between the stimulus and the indicated variables is described by a mathematical model embedding both physical transduction effects and the associated conditioning electronics. We will see this in more details, starting by analyzing usual electronic configurations.

One of the simplest electronic conditioning configurations is the Wheatstone bridge circuit shown in Figure 4.2. Generically speaking, a transduction operator is defined as the operator (F) that transforms the stimulus variable (T) into the response variable (V) that is in mathematical terms

$$V(T) = F[T] \tag{4.1}$$

Noting that V_o defines the excitation voltage, R_1, R_2, and R_3 are known resistances, $R(T)$ is the sensor's resistance when exposed to the temperature T, and $V(T)$ is the voltage

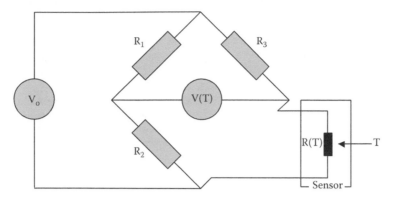

FIGURE 4.2
Conditioning electronics used to generate a voltage V as response to the stimulus temperature T.

difference across the bridge, the transduction operator can be written in the form of the following equation:

$$V(T) = V_o \left(\frac{R(T)}{R_3 + R(T)} - \frac{R_2}{R_1 + R_2} \right) \tag{4.2}$$

where the relation between the process temperature and the sensor's resistance is usually defined through a polynomial:

$$R(T) = R_{ref} \left[1 + a_1 \cdot T + a_2 \cdot T^2 + a_3 \cdot (T - 100)^3 \right] \tag{4.3}$$

By construction, R_{ref} is commonly standardized at 100 or 1000 Ω at 0°C, and the coefficients a_k (k = 1, 2, and 3) depend on the sensor material. For a PRT, these coefficients are

$$
\begin{aligned}
a_1 &= +3.9083 \times 10^{-3} {}^\circ C^{-1} \\
a_2 &= -5.7750 \times 10^{-7} {}^\circ C^{-2} \\
a_3 &= \begin{cases} 0 & \text{if } T < 0 \\ -4.1830 \times 10^{-12} {}^\circ C^{-3} & \text{if } 0 \le T \le 630°C \end{cases}
\end{aligned}
\tag{4.4}
$$

Because the coefficients a_2 and a_3 are small compared with a_1, the resistance of a PRT behaves almost linearly with temperature. Linearity is a very important characteristics, which, to be preserved at the transduction equation, requires a careful design of the conditioning electronics. Suppose, for example, that a $R_{ref} = 100\ \Omega$ PRT is supposed to measure temperatures ranging from 0°C to 200°C. Defining $R_1 = R_2 = 1\ k\Omega$, $R_3 = 100\ \Omega$, and $V_o = 7.27$ V, Equation 4.1 implies that the output voltage ranges from 0 to 1.00 V as shown in the graph of Figure 4.3, together with a linear regression model. Linearity is so important in many practical applications that the actual transduction equation is often replaced by its regression line with associate intrinsic errors and characterization parameters.

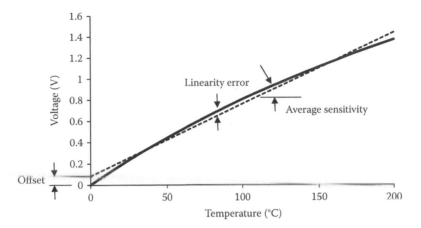

FIGURE 4.3
Graph of the transduction equation (4.1) (solid) with $R_{ref} = 100\ \Omega$, $R_1 = R_2 = 1\ k\Omega$, $R_3 = 100\ \Omega$, and $V_o = 7.27\ V$ and a least squares fitted line (dashed).

Some of these characterization parameters are as follows:

- Offset error—the output reading obtained when the input is set to $T = 0°C$ (in volts for this example)
- Linearity error—the difference between the transduction curve and its linear regression model (in volts for this example)
- Average sensitivity—the average inclination of the transduction curve defined through the inclination of its linear regression model (in volts/°C for this example)

If one cannot live with these errors, additional electronic conditioning circuitry must be designed, particularly to compensate offset and linearity errors. Dedicated microcontrollers are ideal for this task because they include in one single chip the input analog-to-digital converter, the floating point processor, and the output digital-to-analog converter, in addition to being cheap. Disregarding the necessity to convert to and from digital representation, the whole linearization and normalization procedure can be viewed mathematically as constructing the following transformation:

$$v(T) = a \cdot L[V(T)] + b \qquad (4.5)$$

where
a and b are normalization parameters
L is a linearization operator

Let us start by the construction of L. It is quite intuitive that according to Equation 4.1, if this operator is constructed such that $L \equiv F^{-1}$, the relation between v and T in (4.5) will be forcibly linear. However, this is rarely possible in a strict mathematical sense because of the successive nonlinear transformations between input and output variables. For the particular example above, although it is possible to explicitly obtain R in terms of V in Equation 4.1, the polynomial relation between the process temperature and the sensor's resistance in

Equation 4.2 makes it impractical to explicitly obtain T in terms of R. Instead, it is possible to identify an approximation of the transduction equation (4.1)

$$V_{model}(T) = F_{model}[T] \tag{4.6}$$

such that it is possible to explicitly obtain T from V_{model} in (4.6). This can be done by adjusting a suitable mathematical form to (4.1), for instance, Equations 4.2 and 4.3 combined in the example above. For instance, if T relates to V according to an "s-shaped" curve, it is usual to try to fit

$$V_{model}(T) = V_{\infty}(1 - e^{-\alpha \cdot T^{\beta}}) \tag{4.7}$$

where V_{∞}, α, and β are calculated to minimize an overall error such as

$$e(V_{\infty}, \alpha, \beta) = \int_{T_{min}}^{T_{max}} [V(T) - V_{model}(T)]^2 dT \tag{4.8}$$

Then, the linearization operator can be defined by

$$L[\cdot] = V_{model}^{-1}(\cdot) \tag{4.9}$$

which takes the following form, after assuming that $V \approx V_{model}$ and substituting (4.9), (4.7), and (4.6) into (4.5):

$$v(T) = a \cdot \left[\frac{1}{\alpha} \ln\left(\frac{V_{\infty}}{V_{\infty} - V(T)}\right)\right]^{(1/\beta)} + b \tag{4.10}$$

Finally, a and b are calculated by associating the intervals*

$$[T_{min}, T_{max}] \xrightarrow{\text{Equation 4.5}} [v_{min}, v_{max}] \tag{4.11}$$

resulting in

$$a = \frac{v_{max} - v_{min}}{T_{min} - T_{max}} \tag{4.12}$$

$$b = \frac{T_{max}v_{min} - T_{min}v_{max}}{T_{min} - T_{max}} \tag{4.13}$$

Figure 4.4 illustrates how this process works and the role of F_{model} is clearly that of allowing the inversion to restore linearity, otherwise not practical through F.

---

* Usually 0 to 10 V or 0 to 5 V because most of data acquisition boards work over this range.

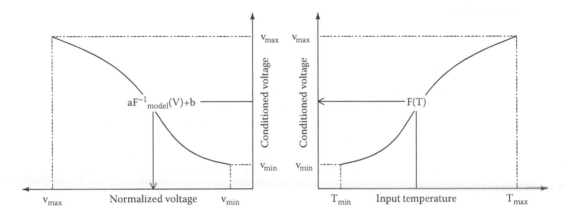

FIGURE 4.4
Schematic representation of the linearization procedure through an invertible model transduction equation.

The whole linearization procedure can thus be summarized in the following steps:

1. Fit an invertible analytic model to the transduction equation (4.1) by minimizing an error functional of the type given in (4.8). This can be done by some numerical optimization procedure such as Newton's method if derivatives of first and second order can be calculated, or a genetic algorithm if not.

2. Express the input variable (T) in function of the output variable (V_{model}) in (4.6). The resulting expression corresponds to the linearization operator as in Equation 4.9.

3. Substitute V_{model} for V in (4.9) and calculate coefficients a and b in (4.5) according to a previously defined mapping range as defined by (4.11), 0 to 10 V, for instance.

This three-step procedure applied to the example above gives very good results as shown in Figure 4.5. The linearity error is less than 0.2 V over 0 to 10 V, that is, less than 2% of the transducer's span.

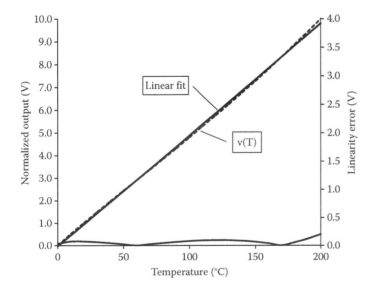

FIGURE 4.5
Linearized response (dotted) and its linear fit (solid) voltage for the example shown in Figure 4.3.

4.3 Propagation of Uncertainties—Error Analysis

Once one gets a reading from a transducer, two important questions arise: (1) How close is this reading from the true value (accuracy = degree of veracity)? (2) How repetitive is this reading if the same stimulus is applied (precision = degree of reproducibility)? Several sources of uncertainties and external influences contribute to create reading errors. For instance, the values of the electronic components of the conditioning circuit may change due to ambient temperature fluctuations. Or the circuit's wiring may work as an antenna adding electromagnetic noise to the output variable. Usually all these influences are neither predictable nor controllable, and, consequently, the transduction equation may deviate and fluctuate.

The deviations, which are closely related to a loss of accuracy, can be estimated by first recognizing that the response variable V, in addition to depending on the stimulus variable T through the transduction Equation 4.1, also depends on a number of intrinsic parameters denoted by x_k, that is

$$V(T) = F[T; x_1, x_2, \ldots, x_N] \tag{4.14}$$

Thus, it is possible to estimate a small deviation ΔV produced by errors $\Delta T, \Delta x_1, \Delta x_2, \ldots,$ Δx_N according to the following formula:

$$\Delta V \cong \Delta T \frac{\partial F}{\partial T} + \Delta x_1 \frac{\partial F}{\partial x_1} + \Delta x_2 \frac{\partial F}{\partial x_2} + \cdots + \Delta x_N \frac{\partial F}{\partial x_N} \tag{4.15}$$

To illustrate the application of this formula, consider the example given by Equation 4.2 with errors ΔR_1, ΔR_2, ΔR_3, and ΔR associated respectively to the circuit's and to the sensor's resistances, this last one due to random fluctuations between $\pm \Delta T$ of the stimulus temperature. The excitation voltage and the sensor represented by the parameters a_k in (4.3) are supposed to be free of errors for simplicity. The corresponding deviation is then

$$\Delta V \cong \Delta T \frac{\partial V}{\partial T} + \Delta R_1 \frac{\partial V}{\partial R_1} + \Delta R_2 \frac{\partial V}{\partial R_2} + \Delta R_3 \frac{\partial V}{\partial x_3} + \Delta R \frac{\partial V}{\partial x} \tag{4.16}$$

Equation 4.16 may be used to define an envelope containing all possible deviations of the original transduction equation (4.2), generated by all possible combinations of errors between the intervals $\pm \Delta R_{1,max}$, $\pm \Delta R_{2,max}$, $\pm \Delta R_{3,max}$, and $\pm \Delta R_{max}$, that is,

$$\Delta V_{max} \cong \left| \Delta T_{max} \frac{\partial V}{\partial T} \right| + \left| \Delta R_{1,max} \frac{\partial V}{\partial R_1} \right| + \left| \Delta R_{2,max} \frac{\partial V}{\partial R_2} \right| + \left| \Delta R_{3,max} \frac{\partial V}{\partial x_3} \right| + \left| \Delta R_{max} \frac{\partial V}{\partial x} \right| \tag{4.17}$$

This is shown in Figure 4.6 for 1% maximum variations on all parameters.

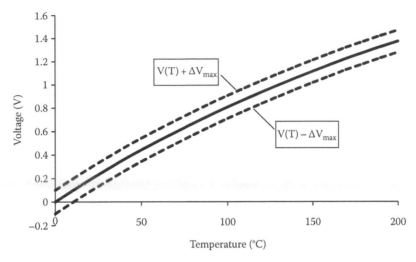

FIGURE 4.6
Graph of the transduction equation (4.1) (solid line) and its deviation envelopes according to (4.17) (dashed line) with 1% errors on $R_{ref} = 100\ \Omega$, $R_1 = R_2 = 1\ k\Omega$, $R_3 = 100\ \Omega$, and constant $V_o = 10.0\ V$.

Let us now investigate the aspects related to the reproducibility. A very precise transducer produces very close readings when the same stimulus is applied. Considering the more general transduction equation given by (4.14), when the same input variable T is applied, one gets very stable output readings V, which may or may not be close to the true output value depending on x_k having experienced some deviation or not. (The parameters x_k may vary due to ambient temperature changes, for example.) But why the readings are not exactly the same, regardless of being accurate or not? As already mentioned, the transducer is submitted to a number of unpredictable external influences making its readings fluctuate, even if the stimulus is kept rigorously constant. These random fluctuations are related to the transducer's precision and must be characterized according to a statistical approach.

Suppose the same stimulus temperature T is applied repeatedly to the transducer, producing response voltages V_k that vary randomly according to a probability histogram $p(V_k|T)$. In other words, a particular reading V_k is seen as a statistical variable or the outcome of exposing the transducer to the temperature T. Within this idea, accuracy, precision, and reproducibility can be quantified through parameters describing statistically $p(\ )$; respectively, its location, dispersion, and shape.

Let then $V_\mu(T)$ be the mean value of V_k, for instance, the arithmetic mean given by

$$V_\mu(T) = \sum_k p(V_k|T)V_k \tag{4.18}$$

Although the arithmetic mean is the most commonly adopted one, other types of statistical location parameter can also be used, such as geometric and harmonic means or median or mode averages. The location of a strongly asymmetrical distribution $p(\ )$ may be better characterized by its median value rather than the arithmetic mean, for example. A convenient definition for the mean value being established, and denoting $V_{true}(T)$ the true response voltage to the stimulus temperature T, a natural definition for the accuracy error is the following:

$$e_{accuracy}(T) = V_{true}(T) - V_\mu(T) \tag{4.19}$$

One advantage of a statistical approach to these issues is the possibility of discriminating and characterizing subtle aspects of reproducibility, which are related to how V_k spreads around V_μ. Let then $\sigma(T)$, $\gamma(T)$, and $\kappa(T)$ denote, respectively, the standard deviation, skewness, and kurtosis of $p(V_k|T)$ given by

$$\sigma^2(T) = \sum_k p(V_k|T)(V_k - V_\mu(T))^2 \tag{4.20}$$

$$\gamma^3(T) = \frac{1}{\sigma^3(T)} \sum_k p(V_k|T)(V_k - V_\mu(T))^3 \tag{4.21}$$

$$\kappa^4(T) = \frac{1}{\sigma^4(T)} \sum_k p(V_k|T)(V_k - V_\mu(T))^4 \tag{4.22}$$

The standard deviation is a measure of the dispersion with which the voltage readings V_k spread around V_μ at a given temperature. A low standard deviation indicates that the readings tend to be very close to the mean, whereas high standard deviation indicates that they spread out over a large range of values. In a situation of perfect reproducibility, all the readings are the same, p( ) becomes an infinitely concentrated Dirac distribution, and $\sigma(T)$ tends to zero. Conversely, if all possible readings V_k have the same probability of being observed, p( ) becomes uniform (equiprobable) and $\sigma(T)$ grows unbounded. This is coherent with the idea that between Dirac and equiprobable probability histograms, one ranges from a situation of perfect predictability to perfect unpredictability, or from full reproducibility to full lack of reproducibility.

The fact that the standard deviation has the same physical units that the readings allow to define a precision error, or simply precision. This error corresponds to an interval around V_k, which can be considered as the support of $p(V_k|T)$, and whose upper and lower bounds are defined by a previously defined interval of all possible reading. If the probability histogram is not known a priori, the precision error $e_{precision}$ must be determined by solving the equation

$$\int_{V_\mu - e_{precision}}^{V_\mu + e_{precision}} p(V_k|T)\, dV = E_\% \tag{4.23}$$

in which $E_\%$ represents the desired confidence level. However, if p( ) is known, $e_{precision}$ can be related to the corresponding standard deviation. For instance, for the normal or Gaussian distribution, $V_\mu \pm \sigma$ contains 68.2%, $V_\mu \pm 2\sigma$ contains 95.4%, and $V_\mu \pm 3\sigma$ contains 99.7% of the readings. Although Gaussian distributions are very common, other probability histograms may also be found in practice, particularly when nonlinear and hysteretic effects influence the transduction phenomenon. This is why the dispersion alone is not sufficient to characterize precision, and other shape parameters such as the skewness γ and the kurtosis κ must be used. Actually it is possible to demonstrate that if p( ) is Gaussian, γ and κ and all other higher order statistical moments are uniquely determined from V_μ and σ.

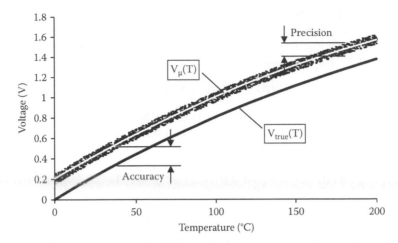

FIGURE 4.7
Schematic representation of the accuracy and precision associated to the example given in Figure 4.6.

Skewness, as defined in (4.21), is a measure of the asymmetry of the probability histogram, which is related to the reading bias. A negative skewness indicates that readings lower than V_μ tend to be more dispersed than readings greater than V_μ, which implies a tendency to underestimate. On the contrary, a positive skewness indicates that readings greater than V_μ tend to be more dispersed than readings lower than V_μ, which implies a tendency to overestimate. This approach can fail in multimodal distributions or in distributions where one tail is long but the other is heavy.

Kurtosis, as defined in (4.22), quantifies how peaked the probability histogram is and, therefore, also measures dispersion but with a different emphasis. A high kurtosis implies that p( ) has a sharper peak and longer, fatter tails, while a low kurtosis indicates that p( ) has a more rounded peak and shorter thinner tails. To interpret this, suppose two distinct probability histograms, but with the same standard deviation. The one with higher kurtosis will produce a greater amount of readings close to the mean V_μ, but a few will be more dispersed. On the other hand, the one with lower kurtosis will produce readings uniformly dispersed within the same precision. In other words, under equal precisions, high kurtosis means very good reproducibility with some highly dispersed readings, while low kurtosis implies that readings are uniformly reproducible.

These concepts are illustrated in Figure 4.7, corresponding to the example shown in Figure 4.6, with 10% uniform random error added to the voltage readings.

4.4 Temperature Measurements under Time Varying Conditions

In the previous sections, time varying conditions were not considered, implying that a response is obtained simultaneously with the application of the stimulus, independently of the state of the transducer before that. This situation can be reproduced in practice by applying the stimulus to the transducer and waiting a sufficient amount of time for all the transients to vanish before reading the response. Thus, the general transduction equation (4.14) actually expresses a static or steady state relation between stimulus and response. But what happens if one must measure under time varying conditions?

Before elaborating and elucidating this question, let us first better state the problem and illustrate with a practical example.

Mathematically speaking, if the stimulus temperature varies in time so will the response voltage and the general transduction equation should be rewritten as

$$V(t) = F[T(t); x_1, x_2, \ldots, x_N] \tag{4.24}$$

where the intrinsic parameters x_k may or may not vary in time. Consequently, the transduction operator F will probably involve derivatives and/or integrals of the T and V, most likely of the form

$$A_0 V(T) + A_1 \frac{dV(t)}{dt} + \cdots + A_P \frac{d^P V(t)}{dt^P} = T(t) + B_1 \frac{dT(t)}{dt} + \cdots + B_Q \frac{d^Q T(t)}{dt^Q} \tag{4.25}$$

where the coefficients A_k and B_k depend on the intrinsic parameters x_k and, also, $Q < P$ for stability. This obviously does not represent all possible dynamic transduction equations, but most practical applications can be cast into it with minor restrictive hypothesis. Let us see how this is so.

Consider the problem of monitoring fluidization patterns in a gas–solid fluidized bed reactor by measuring internal instantaneous temperature, as indicated in Figure 4.8.

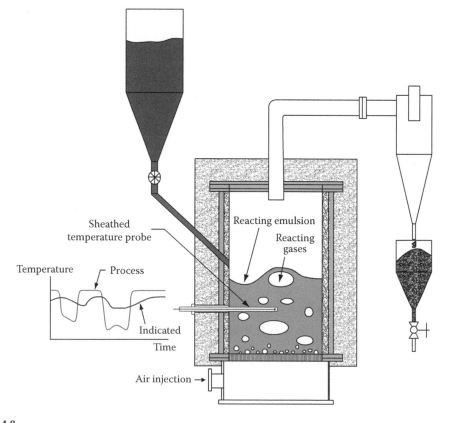

FIGURE 4.8
Dynamic temperature measurement in a fluidized bed reactor—oscillations are due to alternate passage of cold air bubbles through the probe. (From Oliveira, J. et al., *Powder Technol.*, 170, 123, 2006.)

Detecting gas bubbles with a thermal probe is based on the temperature difference between the hotter reacting emulsion phase and the colder gas bubbles. Due to the extremely severe measurement conditions (temperatures exceeding 600°C, material deterioration due to friction with particulate, chemical corrosion, presence of electrostatic charges, etc.), it is recommended to install the probe in a sheath or thermowell made of some resistant material such as stainless steel whose stray effects can be compensated by additional elaborate electronic conditioning.

The local instantaneous process temperature of the flow and the corresponding indicated temperature form the pair stimulus/response and will be respectively denoted by T_{proc} and T_{ind} by convenience. Thermal accumulation is characterized by the sheath's mass m (kg) and by its specific heat C_h (J/kg/K), while convective and radiative heat transfers through the area A (m^2) are accounted respectively by the convection coefficient h (W/m^2/K) and by the emissivity ε. Thus, neglecting the heat conduction through the sensor cable and admitting that the radiative medium completely involves the sensor tip, the governing equation relating T_{ind} and T_{proc} can be written as follows:

$$mC_h \frac{dT_{ind}}{dt} - hA(T_{proc} - T_{ind}) - \varepsilon\sigma A\left(T_\infty^4 - T_{ind}^4\right) = 0 \tag{4.26}$$

where T_∞ denotes the temperature at which radiative transfers occur and is given by the combustion temperature of the particulate in the case of a fluidized bed reactor. Equation 4.26 can be written in more appropriate terms by dividing both sides by hA and rearranging the powers of T_∞ and T_{ind}, which results in

$$\tau \frac{dT_{ind}}{dt} - (T_{proc} - T_{ind}) - \gamma(T_\infty - T_{ind}) = 0 \tag{4.27}$$

where

$$\tau = \frac{mC_h}{hA} \tag{4.28}$$

$$\gamma \cong \frac{4\varepsilon\sigma}{h}\left(\frac{T_\infty + T_{ind}}{2}\right)^3 \tag{4.29}$$

In these expressions, τ represents the probe's time constant, that is, the increase in temperature caused by heat accumulation over heat transferred by convection, while the radiation coefficient γ quantifies the intensity of radiative heat transfer in comparison with convective heat transfer. Equation 4.27, although embeds some restrictive hypotheses, represents a good cast into the general dynamic transduction equation given by (4.25), specially for practical applications.

Solving the inverse problem, that is, calculating T_{proc} from the measured values of T_{ind}, is certainly a difficult task because of its intrinsic ill-conditioned nature. In mathematical terms, the problem being inverse and intrinsically ill posed in the sense of Hadamard (1923), the solution may not exist or, if it exists, it may not be unique or not continuous with respect to the input data. In practice, this means that the solution process of T_{proc} from T_{ind}

will be strongly affected by the presence of experimental errors and noise. This effect has already been studied, and some techniques have been proposed such as Beck's function specification method (Beck et al., 1985) and Murio's mollification method (Murio, 1993). These techniques require relatively long computational codes and are not suited for online implementation. An appropriate solution technique for the online reconstruction of T_{proc} from T_{ind} was proposed by Oliveira et al. (2006) based on a modified version of the Savitzki–Golay filtering method (Savitzky and Golay, 1964).

The transduction equation can be discretized in time with the help of the finite difference method. By defining an adequate time step Δt, and a backward discretization scheme with indices n and (n − 1) indicating that the variable refers to times $t = n\Delta t$ and $t = (n − 1)\Delta t$ respectively, it is possible to obtain

$$\frac{\tau_n}{\Delta t}(T_{ind,n} - T_{ind,n-1}) - (T_{proc,n} - T_{ind,n}) - \gamma_n(T_\infty - T_{ind,n}) = 0 \qquad (4.30)$$

Thus, the direct and inverse problems are expressed as

$$T_{ind,n} = \frac{1}{(\tau_n/\Delta t) + 1 + \gamma_n}\left(T_{proc,n} + \gamma_n T_\infty + \frac{\tau}{\Delta t}T_{ind,n-1}\right) \qquad (4.31)$$

$$T_{proc,n} = \frac{\tau_n}{\Delta t}(T_{ind,n} - T_{ind,n-1}) + T_{ind,n} - \gamma_n(T_\infty - T_{ind,n}) \qquad (4.32)$$

A numerical experiment is effective to demonstrate the discrepancies introduced by thermal inertia, convection, and radiation, as well as the extreme sensitivities to the presence of noise when solving the inverse problem. Consider a reacting gas–solid bubbly flow whose temperature varies between characteristic levels around 900 and 1000 K. These temperature levels are respectively associated with the colder gas within the bubbles and with the hotter solid particles in the emulsion phase and, for simplicity, are assumed to vary according to a square wave. Thus, $T_\infty = 1000$ K and additional parameters were adopted representing typical experimental values: $m = 4.712 \times 10^{-6}$ kg, $C = 380$ J/kg/K, $h = 550$ W/m^2/K, $A = 3.142 \times 10^{-6}$ m^2, and $\varepsilon = 0.9$, which implies an average time constant of 1.1 s adopted in Equations 4.31 and 4.32 for simplicity ($\tau_n = \bar{\tau} = 1.1$ s). The synthetic measured signal was generated by solving Equation 4.31 with additive centered uniform noise with 0.01 K amplitude. The sampling period was set to 0.001 s and the recurrence on T_{ind} due to γ_n was handled by the Newton–Raphson method. The reconstruction of T_{proc} from the noisy values of T_{ind} was accomplished through Equation 4.32, and all these signals are shown in Figure 4.9. It can be seen that despite an extremely low and unrealistic noise level of 0.01 K over 950 K perturbing the input data, the error between the correct process signal and the reconstructed process signal has an average value of 0.091 K and a standard deviation of 8.503 K, which corresponds to a magnification of nearly 2500 times.

The online regularization technique proposed by Oliveira et al. (2006) is based on the smoothing procedure of Savitzki and Golay (1964) applied to problematical terms in Equation 4.27, among which the temporal derivative is surely the most important one.

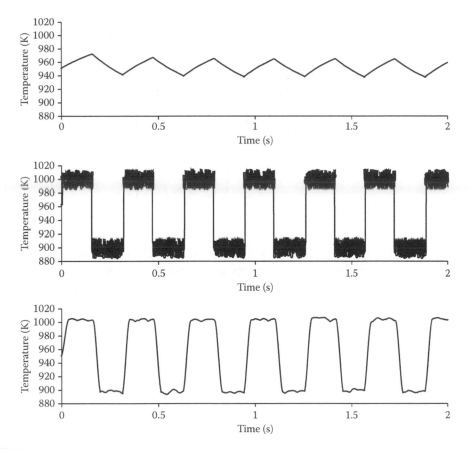

FIGURE 4.9
Process temperature, indicated temperature, and reconstructed process temperature obtained from Equation 4.7 without prior regularization. (The error level of the indicated temperature is 0.01 K.)

The basic idea is to fit a low-order polynomial of order N to the last $M + 1$ indicated temperatures and to replace dT_{ind}/dt and T_{ind} in Equation 4.27 by smoothed or regularized values obtained from this polynomial. According to the proposed method, Equation 4.31 will be transformed to

$$T_{proc,n} = -\tau_n \cdot a_{1,n} + a_{0,n} - \gamma_n \cdot (T_\infty - a_{0,n}) \tag{4.33}$$

where $a_{0,n}$ and $a_{1,n}$ are respectively the first and second coefficients of the smoothing polynomial replacing

$$T_{ind}(n\Delta t) \cong a_{0,n} \tag{4.34}$$

$$\frac{dT_{ind}}{dt}(n\Delta t) \cong -a_1 \tag{4.35}$$

in Equation 4.26. The index n was introduced to stress the fact that a_0 and a_1 refer to $t = n\Delta t$ and must be recalculated at all time steps by solving the associated least squares problem, that is,

$$
\begin{bmatrix}
\sum\limits_{k=0}^{M} k^0 & \sum\limits_{k=0}^{M} k^1 & \cdots & \sum\limits_{k=0}^{M} k^N \\[2mm]
\sum\limits_{k=0}^{M} k^1 & \sum\limits_{k=0}^{M} k^2 & \cdots & \sum\limits_{k=0}^{M} k^{N+1} \\[2mm]
\vdots & \vdots & \cdots & \vdots \\[2mm]
\sum\limits_{k=0}^{M} k^{N+1} & \sum\limits_{k=0}^{M} k^{N+2} & \cdots & \sum\limits_{k=0}^{M} k^{2N}
\end{bmatrix}
\cdot
\begin{pmatrix}
a_0 \cdot \Delta t^0 \\[2mm]
a_1 \cdot \Delta t^1 \\[2mm]
\vdots \\[2mm]
a_N \cdot \Delta t^N
\end{pmatrix}
=
\begin{pmatrix}
\sum\limits_{k=0}^{M} T_{ind,k} k^0 \\[2mm]
\sum\limits_{k=0}^{M} T_{ind,k} k^1 \\[2mm]
\vdots \\[2mm]
\sum\limits_{k=0}^{M} T_{ind,k} k^N
\end{pmatrix}
\tag{4.36}
$$

The main advantage of this approach is that Gram's matrix in Equation 4.36 does not depend on T_{ind} and, consequently, can be previously inverted and stored. The corresponding reconstructed temperature for numerical example above is also shown in Figure 4.9.

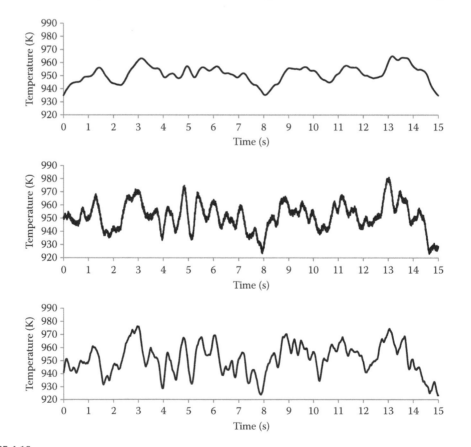

FIGURE 4.10
Flame temperature of a Bunsen burner measured with a sheathed PRT (T_{ind}), actual process temperature measured with a micro-thermocouple (T_{proc}), and reconstructed process temperature with regularization given by Equation 4.33.

Experimental tests were performed by Oliveira (2006) in which the flame temperature of a Bunsen burner was measured with a sheathed PRT and also with an exposed micro-thermocouple to determine the actual process temperature. The intrinsic parameters of the reconstruction procedure given by Equation 4.33 were optimized to better reproduce the actual process temperature probability density function. The corresponding signals are shown in Figure 4.10.

4.5 General Dynamic Behavior of a Temperature Probe

We have seen in the previous section that thermal accumulation, radiation, convection, and other thermal phenomena create dynamical effects that may significantly distort and delay the response with respect to the stimulus. This stimulus–response relation may become even more complex if elaborate conditioning electronics have to be designed to meet with performance requirements. This is the case when capacitive and/or inductive components are used, usually employed in analog filters, or the conditioning electronics contains feedback loops, which are common in constant temperature-sensing techniques. One important advantage of RTDs is that, being resistive in nature, their conditioning electronics tends to remain simple. Anyway, the general transduction equation (4.25) is useful for describing the majority of the transducers found in practice, including both sensor and its conditioning electronics. Consequently, it is of interest to characterize its behavior to generic dynamic stimuli.

This can be done very straightforwardly by using the Fourier transform which, defined for a generic signal s(t), takes the following form

$$\hat{s}(\omega) \overset{\text{Fourier}}{=} \int_{-\infty}^{+\infty} s(t) \cdot e^{-i\omega t} dt \tag{4.37}$$

The Fourier transform $\hat{s}(\omega)$ is an alternative representation to s(t), meaning that the features of the original signal are rearranged without loss of information. If one recognizes that (4.37) can be seen as scalar products with the analyzing harmonic signal $\exp(+i\omega t)$,* which is an orthogonal basis of the finite energy signal space (Hilbert space), $\hat{s}(\omega)$ can be interpreted as the frequency content or components with respect to the analyzing frequency ω. Thus, the inversion formula corresponding to (4.37), that is,

$$s(t) \overset{\text{Fourier}^{-1}}{=} \frac{1}{2\pi} \int_{-\infty}^{+\infty} \hat{s}(\omega) \cdot e^{+i\omega t} d\omega \tag{4.38}$$

has a very simple interpretation: s(t) is recreated from its frequency components by adding up harmonic signals $\exp(+i\omega t)$ weighted by $\hat{s}(\omega)$. Let us see an example of this: suppose s(t)

---

* The scalar product can be defined as $\langle x(t), y(t) \rangle = \int_{-\infty}^{+\infty} x(t) \cdot y^*(t) \, dt$.

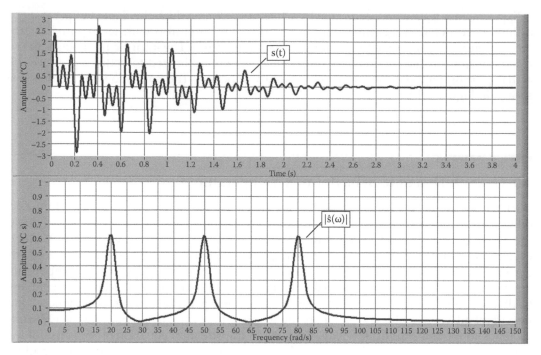

FIGURE 4.11
Graphs of the original signal given by Equation 4.40 for $\omega_1 = 20\,\text{rad/s}$, $\omega_2 = 50\,\text{rad/s}$, $\omega_3 = 80\,\text{rad/s}$, and $\alpha = 0.5\,\text{s}^{-2}$.

is generated by adding three sinuses of different frequencies and multiplying the result by a Gaussian window function

$$s(t) = [\sin(\omega_1 t) + \sin(\omega_2 t) + \sin(\omega_2 t)] \cdot \exp(-\alpha t^2) \qquad (4.39)$$

The graph of s(t) and of its Fourier transform is shown in Figure 4.11. It is clear that the amplitudes of $\hat{s}(\omega)$ are peaked under $\omega = \omega_1$, $\omega = \omega_2$, and $\omega = \omega_3$ and that these peaks are sharper as the essential duration of the Gaussian window increases because more oscillations are included in the analysis.

One very interesting mathematical property of the Fourier transform is transforming derivatives into polynomials in (iω). It can be demonstrated that

$$\int_{-\infty}^{+\infty} \frac{ds(t)}{dt} e^{-i\omega t} dt = i\omega \cdot \int_{-\infty}^{+\infty} s(t) e^{-i\omega t} dt = i\omega \cdot \hat{s}(\omega) \qquad (4.40)$$

which implies that

$$\int_{-\infty}^{+\infty} \frac{d^n s(t)}{dt^n} e^{-i\omega t} dt = (i\omega)^n \cdot \hat{s}(\omega) \qquad (4.41)$$

This is very convenient for the analysis of differential equations, particularly of the general transduction equation. Suppose, for simplicity, that A_k and B_k are constant in (4.25). Calculating its Fourier transform and considering (4.41) result in

$$A_0\hat{V}(\omega) + A_1(i\omega)\hat{V}(\omega) + \cdots + A_P(i\omega)^P\hat{V}(\omega) = \hat{T}(\omega) + B_1(i\omega)\hat{T}(\omega) + \cdots + B_Q(i\omega)^Q\hat{T}(\omega) \quad (4.42)$$

or, alternatively

$$\hat{V}(\omega) \cdot \left[\sum_{k=0}^{P} A_k(i\omega)^k\right] = \hat{T}(\omega) \cdot \left[1 + \sum_{k=0}^{Q} B_k(i\omega)^k\right] \quad (4.43)$$

Equation 4.43 gives us a way of expressing the intrinsic concept in (4.24) that the transduction process is a transformation of a stimulus signal into a response signal through a physical phenomenon. In other words, transducing the stimulus temperature T(t) into the response voltage V(t), which corresponds to solving (4.25), in the corresponding Fourier representations becomes a simple algebraic multiplication of the form

$$\hat{V}(\omega) = \hat{H}(\omega) \cdot \hat{T}(\omega) \quad (4.44)$$

where

$$\hat{H}(\omega) = \frac{1 + \sum_{k=0}^{Q} B_k(i\omega)^k}{\sum_{k=0}^{P} A_k(i\omega)^k} \quad (4.45)$$

The special function H characterizes the transduction equation, including both physics and conditioning electronics, since it depends exclusively on the parameters A_k and B_k, independently of the stimulus and the corresponding response. It is also called the "transfer function" associated to the linear time-invariant transduction system given by (4.25) and plays the role of a "dynamic calibration curve" from which any response can be determined by algebraic multiplication with the corresponding stimulus. This is more evident if Equation 4.44 is rewritten in terms of amplitude and phase

$$\hat{V}(\omega) = \rho_V(\omega) \cdot e^{i\varphi_V(\omega)} = \left[\rho_H(\omega) \cdot e^{i\varphi_H(\omega)}\right] \cdot \left[\rho_T(\omega) \cdot e^{i\varphi_T(\omega)}\right] = \hat{H}(\omega) \cdot \hat{T}(\omega) \quad (4.46)$$

from where it follows that

$$\rho_V(\omega) = \rho_H(\omega) \cdot \rho_T(\omega) \quad (4.47)$$

$$\phi_V(\omega) = \phi_H(\omega) + \phi_T(\omega) \quad (4.48)$$

Thus, in the Fourier representation, the response is determined by multiplying the amplitudes of the transfer function and of the stimulus, and by adding the corresponding phases.

Let us test this with the example of the previous section, defined by Equation 4.27, with $\gamma = 0$ for simplicity

$$\hat{H}(\omega) = \frac{1}{1 + i\omega\tau} = \rho_H(\omega) \cdot \exp\left[i\varphi_H(\omega)\right] \quad (4.49)$$

and the amplitude $\rho_H(\omega)$ and phase $\phi_H(\omega)$ given by

$$\rho_H(\omega) = \frac{1}{\sqrt{1 + (\omega\tau)^2}} \tag{4.50}$$

$$\phi_H(\omega) = -\omega\tau \tag{4.51}$$

Now, assuming that the stimulus temperature is a square pulse of the form

$$T(t) = \begin{cases} 1 & \text{if } 0 \le t \le \Delta t \\ 0 & \text{elsewhere} \end{cases} \tag{4.52}$$

whose Fourier transform is given by

$$\hat{T}(\omega) = \frac{1}{\omega}[\sin(\Delta t\omega) + i \cdot (\cos(\Delta t\omega) - 1)] \tag{4.53}$$

Calculating the amplitude of $\hat{T}(\omega)$ in (4.53) results in

$$\rho_T(\omega) = \frac{2 \cdot [1 - \cos(\Delta t\omega)]}{\omega} \tag{4.54}$$

and the amplitude of the response can determined accordindg to (4.47), that is,

$$\rho_V(\omega) = \frac{2 \cdot [1 - \cos(\Delta t\omega)]}{\omega \cdot \sqrt{1 + (\omega\tau)^2}} \tag{4.55}$$

shown in both time and frequency representation in Figure 4.12.

As already mentioned, H can be used as dynamic calibration curve from which the response to any possible stimulus temperatures can be calculated according to (4.46). Thus, it must be previously determined before using the associated transducer to measure the temperature of a dynamic process. But how can this be done? At first glance, Equation 4.44 suggests that any pair of stimulus/response would be enough for determining the transfer function by simply dividing the Fourier transform of the last by the Fourier transform of the first. Indeed, this can only work if the stimulus is capable of exciting all modes of the transfer function, which constitutes an important practical problem, and several techniques have been developed to solve it.

A very interesting one is based on a very obvious but insightful observation:

$$\text{if } \hat{T}(\omega) \equiv 1 \Rightarrow \hat{V}(\omega) = \hat{H}(\omega) \tag{4.56}$$

According to the interpretation of the Fourier transform given above that it reveals the frequency content of the analyzed signal, a unitary $\hat{T}(\omega)$ means that T(t) contains all possible frequencies in equal amounts, and, consequently, all possible modes of the

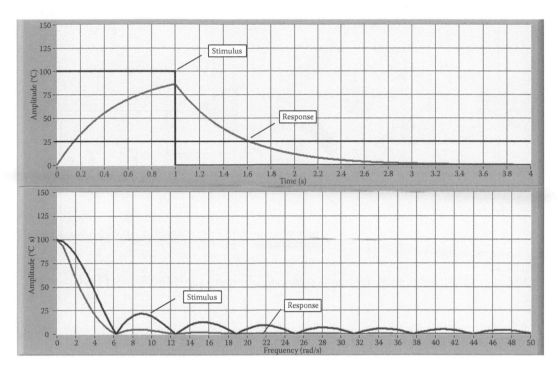

FIGURE 4.12
Stimulus temperature given by Equation 4.52 and the corresponding response voltage signals (top), together with the amplitudes of their Fourier transforms (bottom).

transfer function are equally excited. The temporal representation of such unitary stimulus is determined by calculating its inverse Fourier transform, as defined by Equation 4.39, resulting in the well-known Dirac distribution

$$T(t) = \int\limits_{-\infty}^{+\infty} [\hat{T}(\omega) = 1] \cdot \exp(-i\omega t)\, d\omega = \delta(t) \tag{4.57}$$

In practice, a Dirac stimulus can only be approximated by trying to reproduce its limiting function sequences, such as submitting the transducer to a very brief and intense temperature by plunging the thermal sensor into a hot bath for a short period of time. Other possibilities are to apply a random or chirp-like temperature signal, a sinusoid with slowly varying frequency, but generating these stimuli is a difficult task for temperature and heat flow. Let us see an example of how this works.

Suppose that a PRT is to be used to measure temperature fluctuations in a reacting turbulent flow and that, for some reason, only the components around a specific frequency are of interest. As already mentioned, the possibility of dealing with a thermal resistive sensor as a resistor element opens a wide range of solutions in terms of designing cheap and robust electronic conditioning circuits. Consider using the transduction circuit shown in Figure 4.1 to which a simple RLC filter and a follower amplifier are connected in order to read the bridge voltage, as shown in Figure 4.13.

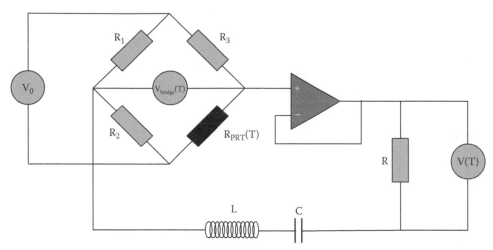

FIGURE 4.13
RCL band-pass filter used to measure temperature oscillations around a specific frequency of interest.

Applying Kirchhoff's mesh law, that is, the sum of the electric potential differences around any closed loop must be zero, we get the following equation:

$$V_{bridge}(T) = \frac{L}{R}\frac{dV}{dt} + V + \frac{1}{RC}\int_{-\infty}^{t} V(\tau)\,d\tau \qquad (4.58)$$

where
$V_{bridge}(T)$ is the normalized voltage generated across the bridge by the PRT
R, L, and C are respectively the resistance, inductance, and capacitance of the RLC filter
V is the response voltage

By applying the Fourier transform to (4.58) we get

$$\hat{V}_{bridge} = \left[1 + \frac{L}{R}i\omega + \frac{1}{RCi\omega}\right]\cdot\hat{V} \qquad (4.59)$$

or, in terms of the transfer function as defined in (4.45)

$$\hat{H}(\omega) = \frac{1}{1 + \dfrac{L}{R}i\omega + \dfrac{1}{RCi\omega}} \qquad (4.60)$$

which can be rewritten more conveniently as

$$\hat{H}(\omega) = \frac{1}{1 + iQ_f\left(\dfrac{\omega}{\omega_0} - \dfrac{\omega_0}{\omega}\right)} \qquad (4.61)$$

or, in polar representation

$$\hat{H}(\omega) = \frac{1}{\sqrt{1 + Q_f^2 \left(\dfrac{\omega}{\omega_0} - \dfrac{\omega_0}{\omega}\right)^2}} \cdot \exp\left[-iQ_f\left(\frac{\omega}{\omega_0} - \frac{\omega_0}{\omega}\right)\right] \tag{4.62}$$

where Q_f stands for the filter's quality factor and ω_0 is its central frequency given by

$$Q_f = \sqrt{\frac{L}{R^2C}} \tag{4.63}$$

$$\omega_0 = \frac{1}{\sqrt{LC}} \tag{4.64}$$

The filter is designed by defining R, L, and C so that the central frequency matches the frequency of interest and, also, to optimize the tradeoff between bandwidth and distortion caused by nonuniform delay. Arbitrating the quality factor to $Q_f = 5.0$ rad/s and the central frequency to $\omega_0 = 315.16$ rad/s (50 Hz) and fixing $R = 100\ \Omega$, by enforcing Equations 4.63 and 4.64 implies that $L = 1.59$ H and $C = 6.37\ \mu F$. We will now see how this transduction system, modeled by Equation 4.58 or 4.59, responds to two different temperature signals, starting by a Dirac stimulus.

As pointed out above, in practice, a Dirac stimulus can be represented by the limit of a sequence of unitary rectangular pulses, which is done by setting its amplitude equal to the reciprocal of its duration. In mathematical terms, this can be put into the following form:

$$V_{bridge}(t) = \begin{cases} \dfrac{1}{\varepsilon} & \text{if } 0 \leq t \leq \varepsilon \\ 0 & \text{elsewhere} \end{cases} \tag{4.65}$$

A finite difference discretization of Equation 4.58, as previously done in Equation 4.30, produces the following recurrence formula in which causality is already enforced:

$$V_n = \left[V_{bridge,n} + \frac{L}{R\Delta t}V_{n-1} - \frac{\Delta t}{RC}\sum_{k=0}^{n-1}V_k\right] \cdot \left[1 + \frac{L}{R\Delta t} + \frac{\Delta t}{RC}\right]^{-1} \tag{4.66}$$

As it can be seen in Figure 4.14, as the duration ε in (4.66) tends to zero, the Fourier transform of the stimulus $\hat{V}_{bridge}(\omega) \rightarrow 1$, indicating that all frequencies become equally present in the signal and, consequently, the Fourier transform of the response $\hat{V}(\omega) \rightarrow \hat{H}(\omega)$.

The band-pass filter's work can be illustrated by applying a stimulus temperature of the type defined in Equation 4.39 with $\omega_1 = 100$ rad/s, $\omega_2 = 315.16$ rad/s, $\omega_3 = 750$ rad/s, and $\alpha = 5$, with $\pm 5\%$ additive uniform noise to better mimic an actual experimental measurement condition, shown in Figure 4.14. It is clear that although there are three equal amplitude frequencies in the stimulus signal, the response signal is predominantly composed of the central frequency $\omega_2 = 315.16$ rad/s (50 Hz), ω_1 and ω_3 being attenuated by a factor of 0.01 approximately. Finally, another interesting effect is the improvement of the signal-to-noise ratio in the response signal. This is so because the frequency content of the added noise is uniform (white noise) and, consequently, the most part of it was attenuated by the filter (Figure 4.15).

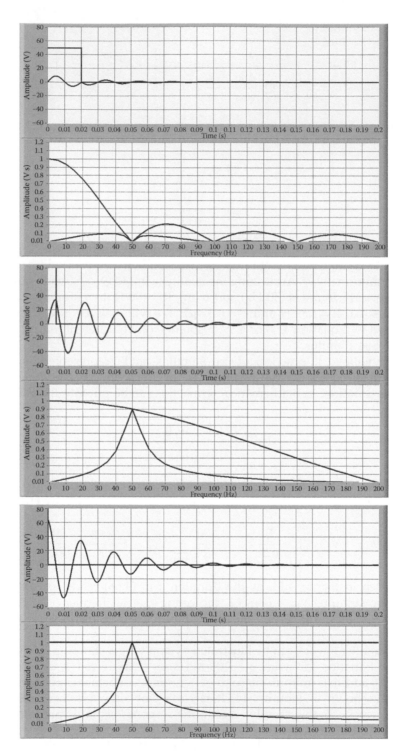

FIGURE 4.14

Identification of the transfer function of the RLC band-pass filter given by Equation 4.62 through application of rectangular pulses progressively approximating a Dirac stimulus ($\varepsilon = 0.02$, 0.005, and 0.0001 s).

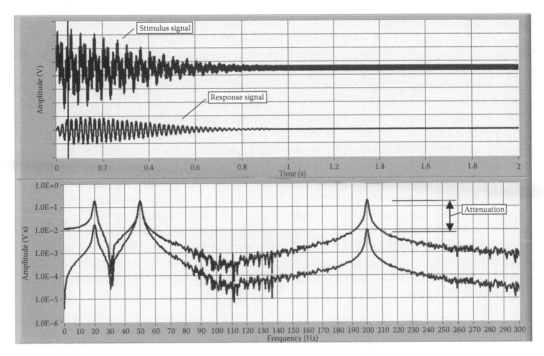

FIGURE 4.15

Response of the measurement system sketched in Figure 4.12 ($\omega_0 = 50$ Hz) to a stimulus temperature containing three frequencies, one of which matches the filters resonance frequency.

4.6 Conclusions

RTDs, or simply resistance thermometers, are accurate, robust, and cheap, which make them very attractive for problems involving severe experimental conditions involving temperature ranges between 0°C and 600°C, particularly in industrial applications. Another important advantage is their excellent interchangeability, a direct consequence of inherent long-term stability, and available well-accepted standards. Out of that range, thermocouples are preferable in applications involving very high (>1000°C) or very low temperatures (<−20°C), despite the necessity of more complicated conditioning electronics, particularly for cold temperature junction compensation and amplification. In general, designing an electronic circuit for an RTD is simpler because it can be treated as a resistor to which many robust electronic measurement techniques have been developed and tested in a great number of successful applications.

As any other transducer, an RTD can be used to measure under static or steady state conditions or to measure dynamic temperatures. Either way it is necessary to use the transduction equation to determine the stimulus variable from the measured variable. In static measurements, this procedure can be generally reduced to solving an algebraic equation through a calibration curve, and several characterization parameters can be defined, in particular those dealing with errors. For instance, accuracy, precision, and reproducibility are defined as statistical parameters, characterizing different aspects of the error histogram. Measuring under time varying conditions involves solving a more complex transduction equation, usually an integro-differential equation with related issues of existence, uniqueness, stability, etc. Under some circumstances (constant intrinsic parameters), the Fourier transform can be used to transform derivatives and integrals of the transduction equation into polynomials on a conjugate variable (frequency), which also reduces the solution process to solving an algebraic equation in the transformed domain. The quotient of the associated characteristic polynomials, also known as transfer function, works as a dynamic calibration curve: the response is obtained simply by multiplying the stimulus by the transfer function in the transformed representation.

Wrapping things up, the most important concept that we tried to tackle here is that all measurement problem is intrinsically an inverse problem, in the mathematical sense of the term. In other words, measurements being actually the response to stimuli imposed by the process, recovering the process variable from the indicated variable implies solving an ill-posed problem that, among other intrinsic difficulties, is extremely sensitive to perturbations in the input data such as experimental errors, electromagnetic noise, etc. This was clearly illustrated by the example in Figure 4.9, where uniform additive noise is 2500 times magnified by the non-regularized reconstruction algorithm. This justifies the importance of developing and applying adequate solution procedures, in addition to a good understanding of the physics involved in transduction together with the use of elaborate electronic conditioning circuitry.

Nomenclature

a, b	normalization parameters
a_k	polynomial coefficients
$a_{k,n}$	coefficients of the smoothing polynomial
A	area
A_k, B_k	ordinary differential equation model coefficients
C_h	specific heat
e( )	error functional
$e_{accuracy}$	accuracy error
$E_\%$	confidence level
F[ ]	transduction operator
h	convection coefficient
H	transfer function
L[ ]	linearization operator
m	mass of thermocouple sheath
M	number of indicated temperatures
n	integer time counter
N	order of smoothing polynomial
p( )	probability histogram
P, Q	order of differential operators
Q_f	filter quality factor
R, L, C	resistance, inductance, and capacitance, respectively
R_k	circuit resistance
R_{ref}	reference resistance
R(T)	sensor resistance
s(t)	generic temporal signal

t	time
T	stimulus temperature
T_{ind}	indicated temperature
T_{proc}	local instantaneous process temperature
T_∞	reference radiative temperature
v	normalized voltage
V	response voltage
V_k	random response voltages to T
V_o	excitation voltage
V_{true}	true response voltage to the stimulus temperature T
V_μ	mean value of V_k
V_∞, α, β	model parameters
x_k	intrinsic parameters in F[ ]

Greek Variables

$\delta(t)$	Dirac delta generalized function
Δ	deviation, error, difference
Δt	time step
ε	emissivity
γ	relative radiation coefficient
ω	frequency
ω_0	filter central frequency
ϕ	phase
ρ	amplitude
σ, γ, κ	standard deviation, skewness, and kurtosis, respectively
τ	time constant

References

Beck, J.V., Blackwell, B., and St-Clair, C.R., 1985, *Inverse Heat Conduction—Ill Posed Problems*, Wiley, Chichester, U.K., 320 p.

Hadamard, J., 1923, *Lecture on Cauchy's Problem in Linear Partial Differential Equations*, Yale University Press, New Haven, CT, 120 p.

Murio, D.A., 1993, *The Mollification Method and the Numerical Solution of Ill-Posed Problems*, Wiley-Interscience, New York, 320 p.

Oliveira, J., 2006, Development of an intelligent temperature sensor—Real time compensation of convective effects, radiation and thermal accumulation. PhD thesis, University of São Paulo, in Portuguese (available for download at www.teses.usp.br/teses/disponiveis/18/18135/tde-10062006–140116).

Oliveira, J., Santos, J.N., and Seleghim, P.J., 2006, Inverse measurement method for detecting bubbles in a fluidized bed reactor toward the development of an intelligent temperature sensor. *Powder Technol.*, 170, 123–135.

Savitzky, A. and Golay, M.J.E., 1964, Smoothing and differentiation of data by simplified least square procedures. *Anal. Chem.*, 36(8), 1627–1639.

5

Heat Flux Sensors

Saulo Güths

CONTENTS

5.1 Introduction

Despite advances in electronics and microelectronics (and even nanotechnology), thermo-couples are still widely used in industry and in scientific applications. The reason for this is the flexibility of this type of sensor. It can be in the form of wire or film of any thickness, resistant to high temperatures, acid, or alkaline media; it adapts to the shape of the measurement and can still be easily repaired. The inconvenient aspect is the low intensity signal, requiring high-quality measurement systems. The analysis of one-dimensional problems involving heat transfer needs two linearly independent boundary conditions. The heat flux boundary condition is present, but normally not used in experimental studies due to measurement difficulties. This chapter presents different types of heat fluxmeters, especially the "tangential gradient" type, and discusses methods of calibration and the error involved.

5.2 Thermocouples

In metals and semiconductors, the transport processes of charge (electric current) and energy are closely related and are due to the displacement of free electrons (conduction electrons). When the electrons of the electrosphere are weakly linked to their core and absorb sufficient energy from external sources, they can become free from their core (Kinzie 1973). At constant temperature, energy densities of free electrons in different materials are not necessarily the same. So when two different materials in thermal equilibrium are in contact, there is a tendency to occur diffusion of electrons through the interface. If the two materials are forming a closed circuit and the two junctions are at the same temperature, the resulting electric fields are opposite and there will be no flow of electrons. However, if the junctions are at different temperatures, there will be an electric current, as shown in Figure 5.1. If the circuit is broken at any point, a potential difference (V) can be measured, function of temperature difference of the two junctions and the type of material of the wires:

$$V = \alpha_{AB}(T_1 - T_2) \tag{5.1}$$

This voltage is called "Seebeck voltage or emf," a tribute to Thomas Seebeck who discovered this phenomenon in 1821 (Rowe 1995). The measurement of the Seebeck emf is made at zero current. Thus, the voltmeter must have low impedance (high internal resistance) to ensure this condition.

The law of intermediate metals says that the sum of the thermoelectric forces in a circuit composed of any number of different materials is zero if the entire circuit is at a uniform temperature. Thus, a homogeneous material can be added in a circuit and will not affect the emf as long as their ends are the same temperature (Figure 5.2). The thermocouple formed by materials A and B will not be affected by materials C or D, where $T_3 = T_4$ and $T_5 = T_6$.

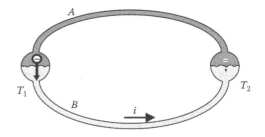

FIGURE 5.1
Diffusion of electrons in the material $A-B$ where $T_1 > T_2$.

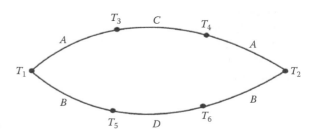

FIGURE 5.2
Circuit with intermediate metals.

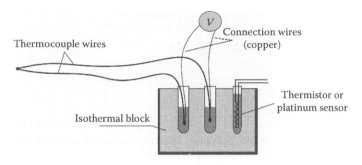

FIGURE 5.3
Thermocouple in open circuit configuration.

The thermocouple does not measure temperature directly, but a temperature difference between two bodies. It is necessary to know one of the temperatures, called the reference junction (or cold-junction). One of the reference junctions most commonly used is the ice-melting bath. It is preferable to use distilled water and a bath with finely crushed ice. For more precise work, the reference junction must be the triple point of water. It is recommended to immerse the connections in an oil or mercury bath. A simpler solution is to coat the wires with a layer of synthetic varnish or place them inside a synthetic glove. The law of intermediate metals allows the connection of a thermocouple in a configuration shown in Figure 5.3, called junction open reference. This situation is widely used because it preserves the thermocouple. Another widely used configuration, especially when there are several thermocouples, is to keep the reference junction at the same temperature as the environment, measuring the temperature reference with a bulb thermometer or a resistance thermometer. The connections can be at a liquid bath, or a metal block with large thermal inertia, and the thermocouples placed in holes filled with conductive material (mercury, mineral oil, or "thermal grease"). In electronic dataloggers, the reference junction is the connection terminals. However, these terminals are usually made from plastic (low thermal conductivity), and there is a risk of temperature gradient occurring, causing measurement error. Manufacturers usually recommend that the equipment be switched on in advance (around 1 h) to establish a uniform internal temperature, thus reducing measurement error. The reference temperature is measured by a thermistor, or a specific integrated circuit, named "electronic cold-junction."

The thermocouples are in fact nonlinear temperature transducers: the thermoelectric power varies with the temperature of the junctions. The thermocouple formed by copper/constantan has a thermoelectric power (α) of around 40 μV K^{-1} at temperatures near the environment, and $\alpha = 53$ μV K^{-1} at 200°C. Table 5.1 presents simplified equations for thermocouple type T (copper/constantan) and type K (chromel/alumel) with reference junction at 0°C. If a reference temperature different from 0°C is used, the emf of the reference junction temperature must first be added.

TABLE 5.1

Equations for Thermocouples Type T and Type K with Reference Junction at 0°C

Type	Temperature Range	emf (μV)	Temperature (°C)
T (copper/const)	−10°C to 100°C	$V = 39.011T + 0.0374T^2$	$T = -0.0259V - 7.11663 \times 10^{-7} V^2$ $+ 2.85872 \times 10^{-11} V^3$
K (chromel/alumel)	−10°C to 200°C	$V = 40.938T - 0.0008T^2$	$T = 0.0244V + 1.123 \times 10^{-8} V^2$

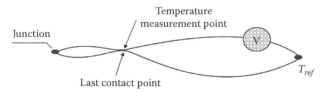

FIGURE 5.4
Measurement error by short circuit in thermocouple.

The extension cables are wires with lower purity than those defined by standards for the manufacturing of thermocouples. They are inserted between the measuring point and the reference junction, with the goal of reducing the cost. The presence of these wires can introduce uncertainties up to 2°C, but this can be greatly reduced if the system is calibrated with them, and the same temperature calibration is retained during use.

A simple electrical contact between the two wires is enough to build a thermocouple, because the flowing electrical current is very small. However, the oxidation may impede the passage of electrons. At low temperature, brazing with tin is sufficient. At higher temperatures, it becomes necessary to use acetylene or arc welding. However, the method of manufacturing a thermocouple differs depending on the need of use. When the measurement of fast transient phenomena is required, the thermocouple must be fine, and the junction should be as small as possible. Even when it is desired to measure an average temperature, this integration can be accomplished using a junction of large size, bearing in mind a possible influence of exchange by radiation. The measuring point of temperature of a thermocouple is the last region of contact between the two materials (Figure 5.4). A short circuit before the junction is a source of error.

5.2.1 Plated Thermocouple

The need to simplify the fabrication of thermoelectric circuits (eliminating the welding) led to the use of bimetallic circuits, made by electrolytic (or chemical) deposition of a high-conductivity metal layer (material 2, Figure 5.5) on a metal support with lower conductivity and different thermoelectric power (material 1, Figure 5.5). The thermoelectric power was defined by Hannay in 1959 as follows: "the power of a thermoelectric material is a measure of the tendency of free electrons to move from warm to cold regions. This shift results is a Seebeck difference of potential with an amplitude sufficient to offset the electrical current created by the displacement of loads in the circuit." To calculate the thermoelectric power at any point of a nonhomogeneous circuit, it is necessary to establish the relationship between the electric current at this point and the gradients of potential and temperature, and deduct the relationship to cancel the electric current.

5.2.1.1 Metal Homogeneous Region

In the section of the non-coated circuit in the presence of a thermal gradient, local Ohm's law is generalized in the form:

$$\mathbf{j} = \sigma E - \sigma \alpha \mathbf{T} \tag{5.2}$$

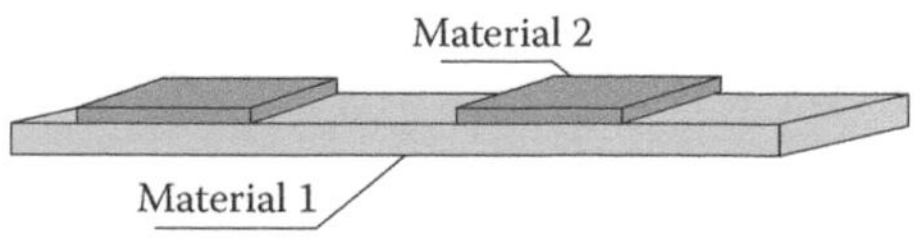

FIGURE 5.5
Bimetallic circuit.

To cancel the local current density, the gradient of electric potential must be proportional to the temperature gradient. The electrical current will be canceled if

$$\alpha = \frac{E}{\nabla \mathbf{T}} \tag{5.3}$$

which corresponds to the usual definition of the thermoelectric power.

5.2.1.2 Regions Coated by Metallic Deposit

The same method can be used to determine the thermoelectric power in the regions covered by the metallic deposit. If the temperature is constant in the transverse direction of the circuit, the electrical current flowing in the axial direction must be zero (Figure 5.6). The expressions of the currents I_1 and I_2 through the horizontal surfaces are as follows:

$$I_1 = \iint_{S_1} j_1 dS_1, \quad I_2 = \iint_{S_2} j_2 dS_2 \tag{5.4}$$

When the thickness of the deposit and substrate are constant and the streamlines are fully developed, the equations above reduce to

$$I_1 = S_1 j_1, \quad I_2 = S_2 j_2 \tag{5.5}$$

By definition, the current across the cross section of the bimetallic layer along the direction O-x must be nil, that is,

$$I = (\sigma_1 S_1 + \sigma_2 S_2) E_x - (\alpha_1 \sigma_1 S_1 + \alpha_2 \sigma_2 S_2) \Delta T_x = 0 \tag{5.6}$$

This relation can be identified as generalized Ohm's law applied to conductors showing an equivalent electrical conductivity (σ_{eq}).

$$I = \sigma_{eq}(S_1 + S_2) E_x - \sigma_{eq} \alpha_{eq}(S_1 + S_2) \Delta T_x = 0 \tag{5.7}$$

Comparing Equations 5.6 and 5.7, the linear conductivity can be expressed by

$$\sigma_{eq}(S_1 + S_2) = \sigma_1 S_1 + \sigma_2 S_2 \tag{5.8}$$

$$\sigma_{eq} = \frac{\sigma_1 S_1 + \sigma_2 S_2}{S_1 + S_2} \tag{5.9}$$

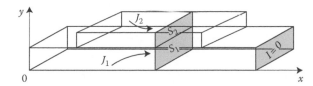

FIGURE 5.6
Definition of area integration areas of current densities.

which leads to an equivalent thermoelectric power (α_{eq}) given by

$$\alpha_{eq} = \frac{\alpha_1 \sigma_1 S_1 + \alpha_2 \sigma_2 S_2}{\sigma_1 S_1 + \sigma_2 S_2} \tag{5.10}$$

The thermoelectric equivalent power (α_{eq}) depends not only on the thermoelectric power of the materials involved but also on the electrical conductivities and cross section areas.

5.2.1.3 Seebeck Effect in Bimetallic Circuits

A thermocouple made by partial metallization of a wire or film generates an emf caused by the Seebeck effect, proportional to the temperature difference between the ends of the deposited electrodes (thermoelectric junctions). The potential difference between points A and B of the circuit (Figure 5.7) is obtained by integration of the gradient of electrical potential between these two points:

$$E = \alpha \Delta T \quad \text{or} \quad \frac{\partial V}{\partial x} = \alpha \frac{\partial T}{\partial x} \tag{5.11}$$

Integration from A to B leads to

$$V_B - V_A = \int_A^B \alpha_1 dT \tag{5.12}$$

Following the same method, the potential difference measured by the Seebeck effect (V) is obtained by integrating the potential gradient on the path AD:

$$V = \int_A^B \alpha_1 dT + \int_B^C \alpha_{eq} dT + \int_C^D \alpha_1 dT \tag{5.13}$$

and assuming that the temperatures at the ends of the circuit are the same ($T_A = T_D$), then

$$V = (\alpha_1 - \alpha_{eq})(T_B - T_C) \tag{5.14}$$

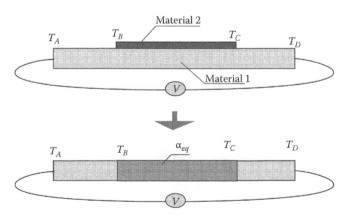

FIGURE 5.7
Plated thermocouple: equivalent circuit.

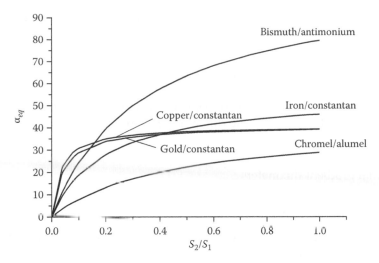

FIGURE 5.8
Equivalent thermoelectric power (α_{eq}) as a function of cross section area ratio (S_2/S_1).

Thus, the Seebeck emf is proportional to the difference in thermoelectric power between the substrate and the region with the metallic deposit. Figure 5.8 shows the difference in thermoelectric power of some pairs of materials as a function of cross section area ratio (S_2/S_1). It is possible to see that Bismuth deposited on a base of Antimony gives a high thermoelectric power difference, but it requires a thick deposit. The cause is a small difference in electrical conductivity of both materials. The fabrication of small thermopiles can be harmful because the high thermal conductivity leads to a "thermal short circuit" between the joints, decreasing the sensitivity of the device. The same phenomenon occurs in the iron/constantan pair. The copper/constantan pairs and gold/constantan pairs (constantan being the substrate), despite showing a regular difference of thermoelectric power, do not require a very thick deposit. The reason for this is the high contrast in electrical conductivity of materials. The use of the chromel/alumel pair is not practical because the deposit of this alloy is difficult (Delatorre et al. 2003).

5.3 Heat Fluxmeters

There are basically two types of heat fluxmeters: transient and stationary type. The transient type, also called calorimetric, correlates the increase in temperature of a body with mass (m) and specific heat (c) to the heat flux absorbed (Q):

$$Q = mc\frac{\partial T}{\partial t} \tag{5.15}$$

The second type, which is more widely used, is based on Fourier law, relating the heat flux (q) that crosses a body (called auxiliary wall) with the temperature difference (ΔT) between

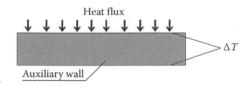

FIGURE 5.9
Measurement principle of stationary heat fluxmeter.

the faces, as shown in Figure 5.9. It is possible to distinguish two types of transducers by considering the way they measure the difference in temperature:

1. Transverse gradient fluxmeters
2. Tangential gradient fluxmeters

5.3.1 Transverse Gradient Fluxmeters

The temperature difference is measured in a direction transverse to the surface where the heat transfer rate is evaluated. The most common forms of measurement are presented below.

5.3.1.1 Welded Thermopile Sensor

In this configuration, the temperature difference is measured by a welded thermopile, and the resin is the auxiliary wall, as shown in Figure 5.10. The difficulty in this configuration consists in welding the thermocouples, requiring a large wall thickness (around 5 mm) (Philip 1961). The device has a high thermal resistance and significantly disturbs the measurement.

5.3.1.2 Plated Thermopile Sensor

The construction of the thermoelectric circuit can be simplified by using the electrolytic deposition of copper on a constantan wire (principle described in Section 5.3.3) to eliminate the production of a large number of thermoelectric welded joints (Figure 5.11). But the

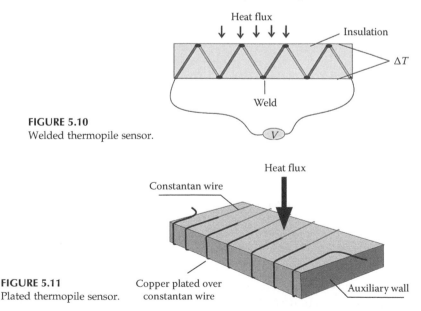

FIGURE 5.10
Welded thermopile sensor.

FIGURE 5.11
Plated thermopile sensor.

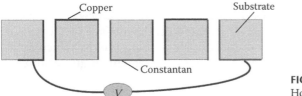

FIGURE 5.12
Hole-plated sensor.

transducer thickness is large (about 3 mm) which is a source of measurement error (Beasley and Figliola 1988).

5.3.1.3 Hole-Plated Sensor

In this configuration, the thermocouples are constructed by a photoetching technique and deposited in vacuum on a thin substrate (100 μm) (Figure 5.12). However, the high cost and difficulty of building sensors with large areas of measurement limit their use.

5.3.2 Tangential Gradient Fluxmeter

Here, the key is to modify the lines of heat flux to generate a temperature difference in a plane tangential to the plane of measurement (Güths 1994). The deviation of the flux lines is caused by a copper pin shown in Figure 5.13. The temperature differences are measured by the deposited thermocouples connected in series. Each thermocouple

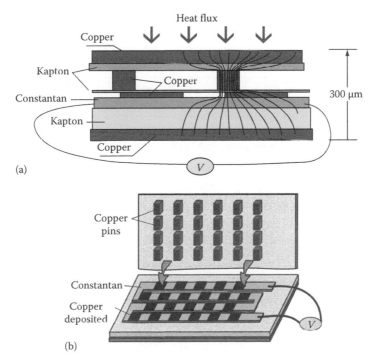

FIGURE 5.13
Tangential gradient heat fluxmeter: (a) cross section view and (b) open view.

converts a temperature difference in Seebeck emf. The emf produced is directly proportional to the number of thermoelements distributed on the surface of the sensor. This technique allows the manufacturing of thermocouples without welds, which facilitates the fabrication of transducers with large area of measurement, high sensitivity, and reduced thickness.

5.3.3 Calibration Methods

The accuracy of the calibration process defines the performance of transducers. The most common way to calibrate remains the use of a heater, which is considered as a standard procedure. This section shows two configurations for calibration: (a) the simultaneous method and (b) the "auxiliary transducer" method.

5.3.3.1 Simultaneous Method

One of the most standard and direct methods to calibrate using a heater is a simultaneous calibration of two transducers. Initially, the transducers are placed according to the configuration shown in Figure 5.14a. The same heat flux flows through the two transducers:

$$q_A = q_B \tag{5.16}$$

The insulation has the function of minimizing heat losses from the top surface of the heater. It does not play any active role in the process. The plate must be maintained at constant temperature. Afterwards, the heater is placed between the two transducers as shown in

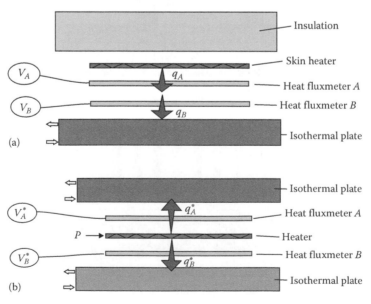

FIGURE 5.14
Simultaneous calibration: (a) first configuration and (b) second configuration.

Figure 5.14b. Assuming that the entire heat flux dissipated by the heater (P) crosses the surfaces (A) of the two transducers,

$$\frac{P}{A} = q_A^* + q_B^* \tag{5.17}$$

assuming a linear relationship between heat flux and the emf (V):

$$q = cV \tag{5.18}$$

Equations 5.16 and 5.17 are then written as

$$\frac{P}{A} = c_A V_A^* + c_B V_B^* \tag{5.19}$$

$$c_A V_A = c_B V_B \tag{5.20}$$

arriving at the following relations:

$$c_A = \frac{P/A}{V_A^* + (V_A/V_B)V_B^*} \quad \text{and} \quad c_B = \frac{P/A}{V_B^* + (V_B/V_A)V_A^*} \tag{5.21}$$

5.3.3.2 "Auxiliary Transducer" Method

The heat flux lost through insulation is measured by a transducer previously calibrated and subtracted from the value dissipated by the heater (Figure 5.15). This method is particularly interesting for "in situ" calibration.

The accuracy of calibration is directly dependent on the accuracy of the heat flux dissipated by the heater. It should be as thin as possible to minimize heat losses from the sides and have the same size, that is, the same area, as the fluxmeter. The heater-dissipated power should not also depend on its own temperature level. This can be obtained by the use of heaters in constantan, which have an electrical resistance that does not vary with temperature. To minimize uncertainties arising from the wires, it is recommended to measure the electrical resistance in the four-wire configuration and use the current measurement to calculate the power dissipation (De Ponte and Maccato 1980).

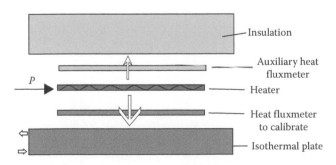

FIGURE 5.15
"Auxiliary transducer" calibration method.

5.4 Conclusion

Plated thermocouples facilitate the construction of thermopiles, eliminating the need of soldering. The thermopiles are the basic element of most heat fluxmeters, and the use of the technique of plating permits the development of new types of sensors, especially of "the tangential gradient" type, which combines small thickness and high sensitivity. The accuracy of the calibration process defines the performance of transducers, and the method of the "auxiliary transducer" has been shown to be more simple and with low uncertainty.

Nomenclature

a thermal diffusivity ($m^2\ s^{-1}$)
A surface area (m^2)
c calibration constant of heat fluxmeters ($W\ V^{-1}$)
E electric potential gradient vector ($V\ m^{-1}$)
I electrical current (A)
$\mathbf{j}$ current density vector ($A\ m^{-1}$)
k thermal conductivity ($W\ m^{-1}\ K^{-1}$)
m mass (kg)
P electrical power (W)
q heat transfer rate ($W\ m^{-2}$)
Q heat flow rate (W)
S cross section area (m^2)
T temperature (K)
V emf (V)
$\nabla \mathbf{T}$ temperature gradient vector ($K\ m^{-1}$)
α thermoelectric power ($V\ K^{-1}$)
σ electrical conductivity ($S\ m^{-1}$)

Superscripts

*   related to second configuration of calibration

Subscripts

A sensor A
B sensor B
AB differential properties of two elements
eq equivalent circuit
1   deposit material
2   substrate material

References

Beasley, D. and Figliola, R. S. 1988. A generalised analysis of a local heat flux probe. *J. Phys. E Sci. Instrum.* 21, 316–322.

Delatorre, R. G., Sartorelli, M. L., Schervernski, A. Q., Güths, S., and Pasa, A. A. 2003. Thermoelectric properties of electrodeposited CuNi alloys on Si. *J. Appl. Phys.* 93, 6154.

De Ponte, F. and Maccato, W. 1980. The calibration of heat flow meters. Thermal insulation performance. *ASTM STP* 718, 237–254.

Güths, S. 1994. Anémomètre a effet peltier et fluxmètre thermique: Conception et réalisation. application à l'etude de la convection naturelle. Thèse de Doctorat, Université d'Artois, FR.

Kinzie, P. A. 1973. *Thermocouple Temperature Measurement*. John Wiley & Sons, New York.

Philip, J. R. 1961. The theory of heat flux meters. *J. Geophys. Res.* 66, 571–579.

Rowe, D. M. 1995. *Handbook of Thermoelectrics*. CRC Press, Boca Raton, FL.

6

Radiative Measurements of Temperature

Jean-Claude Krapez

CONTENTS

6.1 Introduction

6.1.1 Basic Relations for Sensed Radiance in Radiative Temperature Measurement

Matter spontaneously emits electromagnetic radiation in a very broad spectrum enclosing ultraviolet (UV), visible light, infrared (IR), and microwaves. The emitted radiance from a surface in a given direction depends on wavelength, temperature, and the considered

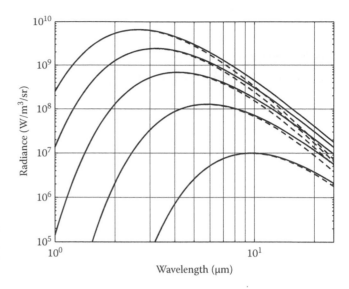

FIGURE 6.1
Blackbody radiance vs. wavelength for $T = 300, 500, 700, 900,$ and 1100 K (from bottom to top). Planck's law in continuous line and Wien's law in dashed line.

matter and direction. For a solid material, it also depends on the surface state: presence of corrosion and roughness. The maximum emitted radiance is given by Planck's law. It only depends on wavelength and temperature (Siegel and Howell 1972):

$$B(\lambda, T) = \frac{C_1}{\lambda^5} \frac{1}{\exp(C_2/\lambda T) - 1} \tag{6.1}$$

$B(\lambda, T)$ is expressed in W/m^3/sr, wavelength λ in m, and temperature T in K, with $C_1 = 1.191 \times 10^{-16}$ W m^2 and $C_2 = 1.439 \times 10^{-2}$ m K (see Figure 6.1).

$B(\lambda, T)$ is also called the blackbody radiance. A blackbody surface absorbs all incoming radiation, and no other surface, at the same temperature, emits more thermal radiation than it does. The blackbody is essentially a thermodynamic concept and it is difficult to find a material presenting such properties over the entire electromagnetic spectrum.

The blackbody radiance is described in Figure 6.1 for different temperature levels. The maximum emission is observed at a wavelength λ_{max} such that $\lambda_{max} T = 2898$ µm K, which is Wien's displacement law. The peak emissive intensity shifts to a shorter wavelength at a higher temperature in inverse proportion to T.

A common approximation to Plank's law is Wien's law, which is also plotted in Figure 6.1:

$$B_W(\lambda, T) = \frac{C_1}{\lambda^5} \exp\left(-\frac{C_2}{\lambda T}\right) \tag{6.2}$$

The approximation error increases with the wavelength. One can, however, consider that Wien's approximation is valid in the rising part of the radiance curve. As a matter of fact, the error is less than 1% provided $\lambda T < 3124$ µm K.

It is obvious that by measuring the thermal radiation emitted by the blackbody surface at a given wavelength and with reference to Planck's law, one can infer its temperature. This idea is at the origin of pyrometry, thermography, microwave radiometry, and more

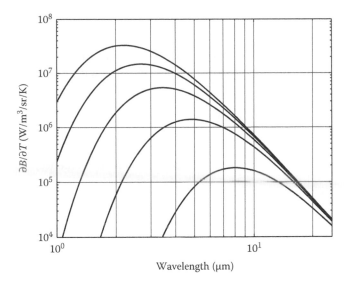

FIGURE 6.2
Absolute sensitivity of blackbody radiance to temperature for $T = 300$, 500, 700, 900, and 1100 K (from bottom to top).

generally all electromagnetic-based approaches that rely on the thermal radiation intensity measurement for temperature characterization.

The sensitivity of blackbody radiance to temperature, according to Planck's law, is plotted in Figures 6.2 and 6.3. Figure 6.2 refers to absolute sensitivity $\partial B/\partial T$ whereas Figure 6.3 refers to relative sensitivity $B^{-1}\partial B/\partial T$. The absolute sensitivity presents a maximum at a wavelength such that $\lambda T = 2410$ μm K. For a blackbody at 300 K, maximum radiance is observed at $\lambda = 9.65$ μm; however, the maximum sensitivity to temperature variations is observed at a shorter wavelength, namely, $\lambda = 8.03$ μm. On the other hand, the *relative* sensitivity is continuously decreasing (see Figure 6.3). The trend is like $1/\lambda$ at short wavelengths. The decreasing nature of relative sensitivity would favor short wavelengths for temperature measurement. Actually, one should consider all three aspects: radiance level, absolute sensitivity, and relative sensitivity, together with the spectral detectivity

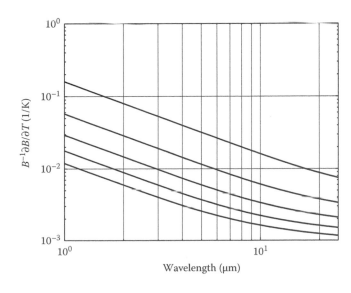

FIGURE 6.3
Relative sensitivity of blackbody radiance to temperature for $T = 300$, 500, 700, 900, and 1100 K (from top to bottom).

and thus the signal-to-noise ratio of the potential sensors, when selecting a wavelength or a spectral band for temperature measurement.

The ratio between $L(\lambda, T, \theta, \varphi)$, the radiance effectively emitted by a surface in the direction (θ, φ), and the blackbody radiance at same wavelength and same temperature is called the emissivity:

$$\varepsilon(\lambda, T, \theta, \varphi) = \frac{L(\lambda, T, \theta, \varphi)}{B(\lambda, T)} \leq 1 \tag{6.3}$$

The emissivity generally depends on the surface temperature but, just for convenience, we will drop the T dependency.

Second Kirchhoff's law states that the emissivity in a given direction is equal to the absorptance in the same direction:

$$\varepsilon(\lambda, \theta, \varphi) = \alpha(\lambda, \theta, \varphi) \tag{6.4}$$

The energy conservation law for an opaque material (i.e., the energy that is not absorbed by the surface is reflected in all directions) leads to the following relation between absorptance and directional hemispherical reflectance:

$$\alpha(\lambda, \theta, \varphi) + \rho'^{\cap}(\lambda, \theta, \varphi) = 1 \tag{6.5}$$

The radiation that leaves the surface $L(\lambda, T, \theta, \varphi)$ is the sum of the radiation emitted by the surface and the reflection by the surface of the radiation coming from the environment in all directions (θ_i, φ_i) of the upper hemisphere:

$$L(\lambda, T, \theta, \varphi) = \varepsilon(\lambda, \theta, \varphi)B(\lambda, T) + \int_{2\pi} \rho''(\lambda, \theta, \varphi, \theta_i, \varphi_i)L^{\downarrow}(\lambda, \theta_i, \varphi_i) \cos \theta_i d\Omega_i \tag{6.6}$$

where $\rho''(\lambda, \theta, \varphi, \theta_i, \varphi_i)$ is the bidirectional reflectance.

Let us now consider temperature measurement with an optical sensor. Depending on the application, the sensor is at a distance ranging from a fraction of a meter, in common industrial processes, to several kilometers in the case of airborne remote sensing and up to hundreds or even thousands of kilometers in the case of satellite remote sensing. Apart from the cases based on vacuum operation, the sensed thermal radiation is thus transmitted through an air layer ranging from a few centimeters to the whole atmosphere thickness (air layer thickness can be higher in the case of near-horizontal line of sight). Along this optical path, only a fraction of the radiation is transmitted (the corresponding fraction is defined by the transmission coefficient $\tau(\lambda, \theta, \varphi)$). The self-emitted radiation of the air layer between the surface and the sensor, $L^{\uparrow}(\lambda, \theta, \varphi)$, finally adds to the transmitted fraction to give the at-sensor radiance $L_s(\lambda, T, \theta, \varphi)$:

$$L_s(\lambda, T, \theta, \varphi) = \tau(\lambda, \theta, \varphi)L(\lambda, T, \theta, \varphi) + L^{\uparrow}(\lambda, \theta, \varphi) \tag{6.7}$$

A common approximation is to consider that the surface is Lambertian, i.e., its optical properties are direction independent. Equation 6.6 is then simplified as follows:

$$L(\lambda, T) = \varepsilon(\lambda)B(\lambda, T) + (1 - \varepsilon(\lambda))\frac{E^{\downarrow}(\lambda, T)}{\pi} \tag{6.8}$$

where the surface irradiance is given by

$$E^{\downarrow}(\lambda, T) = \int_{2\pi} L_{env}(\lambda, \theta_i, \varphi_i) \cos \theta_i d\Omega_i \qquad (6.9)$$

By introducing $L^{\downarrow}(\lambda, T) = E^{\downarrow}(\lambda, T)/\pi$, the equivalent isotropic environment radiance, one gets

$$L(\lambda, T) = \varepsilon(\lambda)B(\lambda, T) + (1 - \varepsilon(\lambda))L^{\downarrow}(\lambda) \qquad (6.10)$$

The influence of the air layer between the surface and the sensor was expressed through its transmission and its self-emission. The same approach can be applied to model the influence of the collecting optics of the sensor. Combining all together, a global transmission and a global self-emission can be defined therefrom.

This development has shown that, even for Lambertian surfaces, the sensed radiation depends on a series of additional variables: the surface emissivity, the irradiance from the environment, the path transmission, and the path self-emission. Therefore, in order to get the target temperature from the measured radiance, one also has to estimate these variables. Depending on the application, the difficulties they introduce are very different:

1. **Pyrometry of High-Temperature Surfaces**
   Generally the sensor is at a close range (the air path is on the order of 0.1–10 m). Therefore, by carefully selecting the wavelength(s), the air transmission can be very high. At the same time, the air self-emission can be negligible. In any case, a calibration can be performed for correcting the optical path transmission and its self-emission by aiming a blackbody which is put at the same distance from the sensor. This calibration is satisfactory as long as both air path contributions do not change. Regarding the reflection of the environment irradiance, the surrounding surfaces are usually much colder than the sensed surface; in that case, the reflection contribution is also negligible. For all these reasons, after a proper calibration of the optic instrument at each wavelength, one thus has access to the emitted radiance itself:

$$L(\lambda, T) = \varepsilon(\lambda)B(\lambda, T) \qquad (6.11)$$

2. **Airborne/Satellite Remote Sensing**
   Transmission and air layer self-emission cannot be discarded anymore. Furthermore, the aimed surface is most often in the same temperature range as the environment whose emitted radiation is reflected on the surface (the "environment" consists of the atmosphere layer itself and nearby solid surfaces in the case of "rough" scenes like urban scenes). The complete equation involving Equations 6.7 and 6.6 has thus to be considered. Generally, however, the terrestrial surfaces are considered as Lambertian surfaces. After proper calibration, one has access to the spectral at-sensor radiance:

$$L_s(\lambda, T, \theta, \varphi) = \tau(\lambda, \theta, \varphi)[\varepsilon(\lambda)B(\lambda, T) + (1 - \varepsilon(\lambda))L^{\downarrow}(\lambda)] + L^{\uparrow}(\lambda, \theta, \varphi) \qquad (6.12)$$

In both cases, we face the so-called *emissivity–temperature separation* (ETS) problem. In the second one, the atmosphere contributions are so important that a supplementary task of atmospheric compensation needs to be accomplished.

6.1.2 Relations and Databases for Spectral Emissivity

One could think that emissivity is a material-only related property and that it would be sufficient to refer to an emissivity database to solve the ETS problem. Some laws were indeed found for spectral emissivity but only for "ideal" materials. As an example, for pure metals, the Hagen–Rubens emissivity relation leads to

$$\varepsilon(T, \lambda) \approx 0.0221 \left(\frac{r_{273} T}{\lambda} \right)^{0.5} \tag{6.13}$$

where
   r_{273} is the resistivity at 273 K in $\Omega \cdot m$
   T is the temperature in K
   λ is the wavelength in m (the constant 0.0221 is in $(\Omega \cdot K)^{-1/2}$)

It was experimentally shown that this law is satisfactory only for $\lambda > 2\ \mu m$. Furthermore, it is not valid for corroded or rough surfaces. As stated by Siegel and Howell (1972), "these types of rules can be misleading because of the large property variations that can occur as a result of surface roughness, contamination, oxide coating, grain structure, and so forth. The presently available analytical procedures cannot account for all these factors so that it is not possible to directly predict radiative property values except for surfaces that approach ideal conditions of composition and finish."

For this reason, the emissivity of the considered material has to be evaluated in virtue of its surface specific state. An indirect approach consists in measuring the directional hemispherical reflectance using Equations 6.4 and 6.5 to infer directional emissivity. This requires the use of an additional radiation source and bringing close to the characterized surface an integrating hemisphere to collect all the reflected radiation. This approach was used to build several databases (see for example Touloukian and DeWitt 1970, Salisbury and d'Aria 1992, Baldridge et al. 2009), which give some hints on the emissivity range and spectral variations for a specific material (in Figures 6.4 and 6.5 some examples of emissivity spectra are reported in the 3–14 μm range).

6.1.3 Needs for Emissivity–Temperature Separation Methods

The indirect reflectance approach will not be dealt with in this chapter. We will rather review the approaches consisting in evaluating *simultaneously* temperature and emissivity, or which manage to get rid of emissivity in the temperature measurement procedure.

In the field of pyrometry, different methods were devised depending on the number of wavelengths or wavebands used for the measurement: monochromatic, bispectral to multi-wavelength pyrometry (MWP). They will be described in Sections 6.2 and 6.3.

In the field of remote sensing, the temperature range of common scenes (sea surface, rural and urban landscapes) is a few tens of degrees around 300 K. The atmospheric window corresponding to maximum radiance is the [8–14 μm] window. Fortunately in

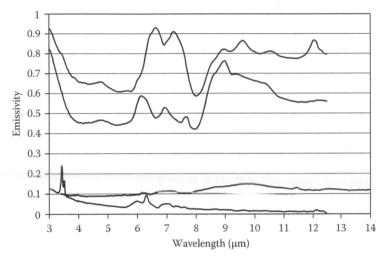

FIGURE 6.4
Spectral emissivity of a series of metals in the range of 3–14 μm (from top to bottom: oxidized galvanized steel, galvanized steel, brass, copper). (Data from reflectance spectra in ASTER spectral library, http://speclib.jpl.nasa. gov/. Copyright: Jet Propulsion Laboratory, California Institute of Technology, Pasadena, CA; see also Balridge, A.M. et al., *Remote Sens. Environ.*, 113, 711, 2009.)

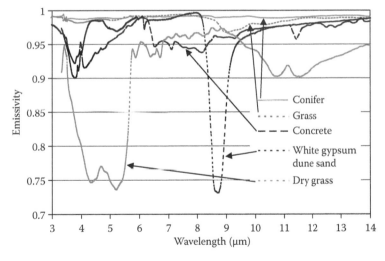

FIGURE 6.5
Spectral emissivity of a variety of natural and manmade materials in the range of 3–14 μm. (Data from reflectance spectra in ASTER spectral library, http://speclib.jpl.nasa.gov/. Copyright: Jet Propulsion Laboratory, California Institute of Technology, Pasadena, CA; see also Balridge, A.M. et al., *Remote Sens. Environ.*, 113, 711, 2009.)

this spectral range, natural surfaces have a high emissivity (see Figure 6.5). This property, together with the fact that several pixels in the IR image share the same atmospheric parameters (transmission, self-emission) allowed developing a series of efficient methods for ETS in the presence of participating atmosphere. A presentation of a few of these methods will be given in Section 6.4.

6.2 Single-Color and Two-Color Pyrometry

6.2.1 Single-Color Pyrometry

In single-color pyrometry one measures, in a given direction (θ, φ), the following radiance:

$$L(\lambda, \theta, \varphi, T) = \varepsilon(\lambda, \theta, \varphi)B(\lambda, T) \tag{6.14a}$$

However, from now on, we will not recall the angular dependency. Thus

$$L(\lambda, T) = \varepsilon(\lambda)B(\lambda, T) \tag{6.14b}$$

The raw signal also includes a multiplicative coefficient and an additive coefficient (assuming linearity between radiance and recorded signal). Nevertheless, by calibrating the sensor with a blackbody at two different temperatures, one can get rid of both coefficients. Such calibration is from now on assumed to have been applied.

Obviously, at this stage, it is necessary to know the spectral emissivity of the sensed surface $\varepsilon(\lambda)$ to infer the blackbody radiance and then the surface temperature. One has to refer to previous knowledge of the material optical properties, which, referring to the difficulties presented in Section 6.1.2 for establishing reliable emissivity databases, is prone to lead to substantial errors.

By differentiating Equation 6.14, one can evaluate the sensitivity of temperature to an error on emissivity:

$$\frac{dT}{T} = -\left(\frac{T}{B}\frac{dB}{dT}\right)^{-1}\frac{d\varepsilon}{\varepsilon} \tag{6.15}$$

The amplification factor $\left(\dfrac{T}{B}\dfrac{dB}{dT}\right)^{-1}$ can be deduced from the relative sensitivity $\dfrac{1}{B}\dfrac{dB}{dT}$ in Figure 6.3.

Also, with Wien's approximation, Equation 6.15 reduces to

$$\frac{dT}{T} = -\frac{\lambda T}{C_2}\frac{d\varepsilon}{\varepsilon} \tag{6.16}$$

The amplification factor is about 0.08 at 1 μm for a temperature of 1100 K. It reaches about 0.2 at 10 μm for a temperature of 300 K. A 10% underestimation of emissivity will lead to a 0.8% overestimation of temperature in the first case (i.e., 8 K) and 2% in the second case (i.e., 6 K). The advantage of working at short wavelength is evident from this perspective; as a matter of fact, the error amplification is proportional to λ. For this reason some authors promoted the use of visible pyrometry and even UV pyrometry (see for example Corwin and Rodenburgh 1994, Hervé and Sadou 2008, Pierre et al. 2008). However, although a given emissivity relative error has a lower impact on temperature evaluation at short wavelength, it should not occult the fact that a reasonable estimation of emissivity has nevertheless to be made. The retrieved temperature is unavoidably affected by this emissivity estimation (Duvaut et al. 1996). Apart from this, at short wavelength, both the signal and its *absolute* sensitivity to temperature decrease. The choice of the spectral range for pyrometry is thus always a compromise.

6.2.2 Two-Color Pyrometry

By performing a measurement at another wavelength, one adds new information, but unfortunately one also adds a new unknown, namely, the spectral emissivity at this supplementary wavelength:

$$\begin{cases} L(\lambda_1, T) = \varepsilon(\lambda_1)B(\lambda_1, T) \\ L(\lambda_2, T) = \varepsilon(\lambda_2)B(\lambda_2, T) \end{cases} \tag{6.17}$$

The most popular method consists in calculating the ratio of the spectral radiances:

$$R_{12} = \frac{L(\lambda_1, T)}{L(\lambda_2, T)} = \frac{\varepsilon(\lambda_1)}{\varepsilon(\lambda_2)} \frac{B(\lambda_1, T)}{B(\lambda_2, T)} = \frac{\varepsilon(\lambda_1)}{\varepsilon(\lambda_2)} \left(\frac{\lambda_2}{\lambda_1}\right)^5 \frac{\exp(C_2/\lambda_2 T) - 1}{\exp(C_2/\lambda_1 T) - 1} \tag{6.18}$$

which gives, with Wien's approximation,

$$R_{12} \approx \frac{\varepsilon(\lambda_1)}{\varepsilon(\lambda_2)} \left(\frac{\lambda_2}{\lambda_1}\right)^5 \exp\left(\frac{-C_2}{\lambda_{12}T}\right) = \frac{\varepsilon(\lambda_1)}{\varepsilon(\lambda_2)} \left(\frac{\lambda_2}{\lambda_1}\right)^5 \lambda_{12} \frac{1}{C_1} B_W(\lambda_{12}, T) \tag{6.19}$$

where the equivalent wavelength of the two-color sensor is defined by

$$\lambda_{12} = \frac{\lambda_1 \lambda_2}{\lambda_2 - \lambda_1} \tag{6.20}$$

Ratio-two-color pyrometry thus requires knowing the emissivity ratio $\varepsilon(\lambda_1)/\varepsilon(\lambda_2)$ in order to infer temperature from the radiance ratio R_{12} according to Equation 6.18 or to its approximation, Equation 6.19. One common assumption is that $\varepsilon(\lambda_1) = \varepsilon(\lambda_2)$ (for this purpose, the gray body assumption is often invoked; however, stating that $\varepsilon(\lambda_1) = \varepsilon(\lambda_2)$ is less restrictive than the gray body assumption that concerns the entire spectrum).

As for one-color pyrometry, it is easy to relate the temperature estimation error to the emissivity estimation error:

$$\frac{dT}{T} = -\frac{\lambda_{12}T}{C_2} \left(\frac{d\varepsilon_1}{\varepsilon_1} - \frac{d\varepsilon_2}{\varepsilon_2}\right) \tag{6.21}$$

Considering the examples ($\lambda_1 = 1\,\mu m$, $\lambda_2 = 1.5\,\mu m$) and $T = 1100\,K$ for the first one and ($\lambda_1 = 10\,\mu m$, $\lambda_2 = 12\,\mu m$) and $T = 300\,K$ for the second one, the amplification factor reaches, respectively, 0.22 and 1.2; these values are respectively three and six times higher than with single-color pyrometry as illustrated in Section 6.2.1.

The error on temperature can be lowered by reducing the equivalent wavelength, i.e., by increasing the higher wavelength λ_2 or decreasing the shorter one λ_1. The amplification factor will anyway be larger than with single-color pyrometry performed at the shortest wavelength. Reducing the equivalent wavelength also gives the opportunity to increase R_{12} sensitivity to temperature variations when approaching the $\lambda_{12}T = 2410\,\mu m\,K$ optimum product.

Anyway, a prior knowledge about the emissivity spectrum, more precisely the ratio $\varepsilon(\lambda_1)/\varepsilon(\lambda_2)$, is required to expect some success with ratio pyrometry. The advantage, however, as compared to one-color pyrometry is that, thanks to the ratioing, the method is insensitive to problems like a partial occultation of the line of sight, or an optical path transmission variation (provided that this transmission variation is the same in both spectral channels).

Bicolor pyrometry has been a matter of research for a long time and integrated instruments are now on the market. Some modifications to the basic approach were suggested in order to improve its performances. As an example the photothermal approach solves the problem of reflected fluxes (Loarer et al. 1990). Indeed when the sensed surface is not much hotter than the surrounding, the reflected radiance happens to be disturbing. The photothermal approach is an active method which, with the use of a modulated laser beam, allows the emitted flux to be rigorously separated from the reflected fluxes. The slight temperature modulation induced by the laser absorption gives rise to a modulated component in the signal whereas the reflected flux only contributes to the DC signal. A lock-in detection allows to separate them. Finally, by performing the measurement at two wavelengths, the ratio of the modulated signals is proportional to $\dfrac{\varepsilon(\lambda_1)\partial B/\partial T(\lambda_1, T)}{\varepsilon(\lambda_2)\partial B/\partial T(\lambda_2, T)}$ which is then used to infer the surface temperature (compare with Equation 6.18). A pulse laser can also be used, where the transient signal at both wavelengths leads to the same ratio as before (Loarer and Greffet 1992).

In some circumstances, it is possible to bring close to the characterized object a highly reflecting surface. By properly choosing its shape, one gets two benefits: the reflection fluxes from the environment are diminished and the apparent emissivity of the sensed surface is increased thanks to the multiple reflections of the emitted radiation between the surface and the mirror (Krapez et al. 1990). As a consequence, the temperature estimation error due to estimation errors on $\widehat{\varepsilon}(\lambda_1)/\widehat{\varepsilon}(\lambda_2)$ is diminished, where $\widehat{\varepsilon}$ is the apparent, actually amplified, emissivity.

6.3 Multiwavelength Pyrometry

With single-color pyrometry, we have at hand one radiance measurement and two unknowns: the monochromatic emissivity and the temperature. By performing a measurement at another wavelength we get an additional radiance value but at the same time we introduce an additional unknown: the emissivity at this new wavelength. The process can be repeated up to N wavelengths. Basically the problem of MWP is thus underdetermined: there are N values for the observable and $N+1$ unknown parameters. Furthermore, if the surface irradiance is significant and if the background radiation can be approximated by a blackbody radiation at temperature T_b, the number of unknowns reaches $N+2$ (the alternative is to evaluate independently this unknown background equivalent temperature).

MWP has been a subject of controversy for several decades (Gardner 1980, Coates, 1981, Hunter et al. 1985, 1986, Hiernault et al. 1986, Nordine 1986, DeWitt and Rondeau 1989, Tank and Dietl 1990, Gathers 1991, Khan et al. 1991a,b, Lindermeir et al. 1992, Duvaut et al. 1995, 1996, Chrzanowski and Szulim 1998a,b, 1999, Scharf et al. 2001, Cassady and Choueiri 2003, Mazikowski and Chrzanowski 2003, Sade and Katzir 2004, Wen and Mudawar 2004a,b, Uman and Katzir 2006, Duvaut 2008): some authors presented experimental results with various successes, sometimes with small temperature errors and at other times with unacceptably high errors, depending on the material, on its surface state, and on the chosen function for approximating the emissivity spectrum. Even the theoretical works do not agree on the advantage of using a large number of wavelengths (Gardner 1980, Coates 1981, Nordine 1986, Tank and Dietl 1990, Gathers 1991, Khan et al. 1991b, Lindermeir et al. 1992, Wen and Mudawar 2004a,b, Duvaut 2008).

6.3.1 Interpolation-Based Methods

In order to solve the underdetermined problem, a *potential* solution is to reduce by 1 the degree of freedom of the emissivity spectrum. A first approach consists in approximating $\varepsilon(\lambda)$ or $\ln[\varepsilon(\lambda)]$ by a polynomial of degree $N-2$. However, it was shown by Coates (1981), based on Wien's approximation and a polynomial approximation of $\ln[\varepsilon(\lambda)]$, that this method can rapidly lead to unrealistic temperature values as N increases.

As a matter of fact, by taking the logarithm of Equation 6.14 with Wien's approximation for blackbody radiance, one gets

$$\ln\left[\frac{L(\lambda_i, T)\lambda_i^5}{C_1}\right] = \ln(\varepsilon_i) - \frac{C_2}{\lambda_i T}, \quad i = 1, N \tag{6.22}$$

With a polynomial approximation of degree $N-2$ for $\ln[\varepsilon(\lambda)]$, a temperature T' is retrieved (it is actually extracted from the constant parameter of the polynomial of degree $N-1$ which interpolates the N values $\lambda_i \ln[L(\lambda_i, T)\lambda_i^5/C_1]$):

$$\ln\left[\frac{L(\lambda_i, T)\lambda_i^5}{C_1}\right] = \sum_{j=0}^{N-2} a_j \lambda_i^j - \frac{C_2}{\lambda_i T'}, \quad i = 1, N \tag{6.23}$$

It is then easy to see, by multiplying both equations by λ_i and subtracting them, that the temperature error expressed through $C_2(1/T - 1/T')$ (it is also called "temperature correction") corresponds to the constant parameter of the polynomial of degree $N-1$ passing through the N values $\lambda_i \ln[\varepsilon(\lambda_i)]$. Temperature corrections for $N = 1, 2, 3$ are (Coates 1981, Khan et al. 1991a):

$$\begin{aligned}
N = 1 \quad & C_2\left(\frac{1}{T} - \frac{1}{T'}\right) = \lambda_1 \ln(\varepsilon_1) \\[2mm]
N = 2 \quad & C_2\left(\frac{1}{T} - \frac{1}{T'}\right) = \frac{\lambda_1 \lambda_2}{\lambda_1 - \lambda_2} \ln\left(\frac{\varepsilon_2}{\varepsilon_1}\right) \\[2mm]
N = 3 \quad & C_2\left(\frac{1}{T} - \frac{1}{T'}\right) = \frac{\lambda_1 \lambda_2 \lambda_3}{(\lambda_2 - \lambda_1)(\lambda_3 - \lambda_1)(\lambda_3 - \lambda_2)} \\[2mm]
& \times \left(\lambda_1 \ln\left(\frac{\varepsilon_2}{\varepsilon_3}\right) + \lambda_2 \ln\left(\frac{\varepsilon_3}{\varepsilon_1}\right) + \lambda_3 \ln\left(\frac{\varepsilon_1}{\varepsilon_2}\right)\right)
\end{aligned} \tag{6.24}$$

With equidistant wavelengths, the temperature correction involves the ratio $\varepsilon_1\varepsilon_3/\varepsilon_2^2$ for three wavelengths and the ratio $\varepsilon_1\varepsilon_3^2/\varepsilon_2^3\varepsilon_4$ for four wavelengths (Khan et al. 1991a). Of course, one should estimate this ratio beforehand. Assigning arbitrarily a value of 1 to this ratio for different metals had the consequence that the temperature estimation error increased very rapidly with the number of wavelengths (Khan et al. 1991a).

It can be shown that the temperature correction limit for wavelength intervals decreasing to 0 is equal to $(-1)^{N-1}\lambda^N/(N-1)!\, d^{N-1}\ln[\varepsilon(\lambda)]/d\lambda^{N-1}$ (Nordine 1986). One can also recognize in the temperature correction the extrapolation error at $\lambda = 0$ of the $\lambda_i \ln[\varepsilon(\lambda_i)]$

polynomial interpolation. This finding can now be developed a little more. If, by chance, a polynomial of degree $N - 2$ could be found passing *exactly* through the N values $\ln[\varepsilon(\lambda_i)]$, the polynomial of degree $N - 1$ passing through the N values $\lambda_i \ln[\varepsilon(\lambda_i)]$ would have a 0 constant parameter—i.e., no extrapolation error—and the retrieved temperature would be the real one. Such an event is highly improbable and the result is tightly dependent on polynomial extrapolation properties. Unfortunately it is well known that an extrapolation based on polynomial interpolation leads to increasingly high errors as the polynomial degree rises. Furthermore, things get progressively worse as the extrapolation is performed far from the interpolation domain. This last point would actually advocate expanding the spectral range to the shortest possible wavelength, but it is a desperate remedy.

The potentially catastrophic errors described just before are systematic errors, i.e., method errors. They are obtained even when assuming errorless signal. To analyze the measurement error's influence, one can state, for simplicity, that the measurement error in channel i is described by a corresponding uncertainty of the apparent emissivity in the same channel, $d\varepsilon(\lambda_i)$. The interpolation of the $\lambda_i \ln[\varepsilon(\lambda_i) + d\varepsilon(\lambda_i)]$ values leads afterwards to the same extrapolation problem as described before and adds to it. The calculated temperature is thus increasingly sensitive to measurement errors as the number of channels increases.

The interpolation-based method error originates from *overfitting* of the experimental data. It was finally recognized that the interpolation-based method could be retained only for the simpler pyrometers, i.e., with two to three wavelengths at most (Coates 1981).

6.3.2 Regularization by Using a Low-Order Emissivity Model

6.3.2.1 Description of Emissivity Models

The overfitting shortcomings previously described can be alleviated by reducing the number of unknowns that are used for describing the emissivity spectrum. Different models were tested:

$$\varepsilon(\lambda_i) = \sum_{j=0}^{m} a_j \lambda_i^j, \quad i = 1, \ldots, N, \quad m < N - 2 \text{ (generally } m = 1 \text{ or } 2) \tag{6.25}$$

$$\ln[\varepsilon(\lambda_i)] = \sum_{j=0}^{m} a_j \lambda_i^j, \quad i = 1, \ldots, N, \quad m < N - 2 \text{ (generally } m = 1 \text{ or } 2) \tag{6.26}$$

$$\varepsilon(\lambda_i) = \frac{1}{(1 + a_0 \lambda_i^2)}, \quad i = 1, \ldots, N \tag{6.27}$$

Polynomials of $\lambda^{1/2}$ or $\lambda^{-1/2}$ for $\ln[\varepsilon(\lambda)]$ and functions involving the brightness temperature were also considered by Wen and Mudawar (2004a,b), a sinusoidal function of λ by Gardner (1980), and other more "physical" models like Maxwell, Hagen–Rubens, and Edwards models by Duvaut et al. (1995, 1996) and Duvaut (2008).

The gray-band model consists in separating the spectrum in a small number N_b of regions and assigning the same emissivity value to all channels of a given region (Tank and Dietl 1990). The bands can be narrowed down to three or even two channels as suggested by Lindermeir et al. (1992). In this way, the number of unknowns is reduced from $N + 1$ to $N/3 + 1$ or $N/2 + 1$. One can even go further by squeezing some bands to one channel. The extreme limit consists in $N - 1$ single-channel bands plus one dual-channel band. In that case, we face a problem with N measurements and N unknowns which is thus, *in principle*, invertible. We will see that it is actually very badly conditioned.

The concept of gray bands can be generalized by allowing that the channels chosen to share a common emissivity value are not necessarily close together: an iterative process was described by Barducci and Pippi (1996) where these wavelengths are each time reshuffled according to the pseudo-continuous emissivity spectrum, i.e., the one defined over the N wavelengths according to

$$\hat{\varepsilon}(\lambda_i, \hat{T}) = \frac{L(\lambda_i, T)}{B(\lambda_i, \hat{T})}, \quad i = 1, \dots, N \tag{6.28}$$

where
  $\hat{T}$ is the last estimation of temperature
  $\hat{\varepsilon}(\lambda_i, \hat{T})$ is sorted from lower to higher value
  the N_b groups of equal emissivity wavelengths are defined by cutting this vector into N_b parts

The unknown parameters of the emissivity function, together with temperature, are finally evaluated through least-squares minimization. By introducing Wien's approximation for radiance, a polynomial approximation for $\ln[\varepsilon(\lambda)]$, and by considering the observable $\ln[L(\lambda_i, T)\lambda_i^5/C_1]$, Equation 6.22 shows that the problem reduces to a *linear least-squares* problem (Gardner 1980, Hiernault et al. 1986, Cassady and Choueiri 2003, Mazikowski and Chrzanowski 2003). Otherwise, when considering the observable $L(\lambda_i, T)$ one faces a *non-linear least-squares* problem (Gardner et al. 1981, Hunter et al. 1985, 1986, DeWitt and Rondeau 1989, Tank and Dietl 1990, Gathers 1991, Khan et al. 1991b, Lindermeir et al. 1992, Duvaut et al. 1995, Chrzanowski and Szulim 1998a,b, Scharf et al. 2001, Cassady and Choueiri 2003, Sade and Katzir 2004, Wen and Mudawar 2004a,b, Uman and Katzir 2006, Duvaut 2008). Let us add that by rearranging the i equations as described in Equation 6.22 one could get rid of one parameter, either a constant parameter or the temperature (Gardner 1980, Hiernault et al. 1986, Cassady and Choueiri 2003). However, it is believed that no advantage in accuracy is obtained by manipulating the data to present the same information in a different form (Gardner 1980). As a matter of fact, in the case of linear fitting such a manipulation even increases the uncertainty of the identified parameters.

6.3.2.2 Least-Squares Solution of the Linearized ETS Problem

Let us take the logarithm of the measured radiance and adopt Wien's approximation. The chosen observable is

$$Y_i = \ln\left(\frac{L_i \lambda_i^5}{C_1}\right) + e_i, \quad i = 1, N \tag{6.29}$$

where e_i is the measurement error (noise) in channel i. We will assume that the e_i, $i = 1, N$ are uncorrelated random variables following a Gaussian distribution of uniform variance. Actually a noise of uniform variance is usually assumed for L_i, but, for ease, we will consider that this applies to its logarithm. This approximation is valid if the spectral range is not too wide with respect to the variations of $B(\lambda, T)$ and if the emissivity does not change too much.

According to Equation 6.22 where $\ln[\varepsilon(\lambda)]$ is approximated by a polynomial of degree m, the least-squares solution is

$$\hat{\mathbf{P}} = \begin{bmatrix} \hat{a}_0 & \cdots & \hat{a}_m & \hat{T} \end{bmatrix}^T = \arg\min_{a_j, T} \sum_{i=1}^{N} \left(Y_i - \left(\sum_{j=0}^{m} a_j \lambda_i^j - \frac{C_2}{\lambda_i T} \right) \right)^2 \tag{6.30}$$

For numerical purposes, it is preferable to replace the wavelength in the polynomial expression by its reduced and centered value so that $\lambda_i^* \in [-1, 1]$:

$$\lambda_i^* = 2\frac{\lambda_i - \lambda_{\min}}{\lambda_{\max} - \lambda_{\min}} - 1 \tag{6.31}$$

For the same reason, one can normalize T by T_{ref} so that $C_2/\lambda_i T_{\text{ref}}$ is on the order of 1. The associated unknown parameter is then $P_T^* = T_{\text{ref}}/T$. The parameter vector is

$$\mathbf{P}^* = \begin{bmatrix} a_0^* & \cdots & a_m^* & P_T^* \end{bmatrix}^T \tag{6.32}$$

where the parameters a_j^* are the coefficients of the polynomial in λ_i^*. The corresponding sensitivity matrix is

$$\mathbf{X} = \begin{bmatrix} 1 & \lambda_1^* & \lambda_1^{*2} & \cdots & \dfrac{-C_2}{\lambda_1 T_{\text{ref}}} \\ \cdots & \cdots & \cdots & \cdots & \cdots \\ 1 & \lambda_N^* & \lambda_N^{*2} & \cdots & \dfrac{-C_2}{\lambda_N T_{\text{ref}}} \end{bmatrix}_{N, m+2} \tag{6.33}$$

where the columns correspond to the sensitivity to successive parameters in vector $\mathbf{P}^*$ (i.e., the first derivative of the model functions relatively to each parameter).

The sensitivity to the parameters a_j^* and P_T^* is plotted vs. the reduced wavelength $\lambda_i' = \lambda_i/\lambda_{\min}$ in Figure 6.6 up to $j=2$ for the particular case of $\lambda_{\max}/\lambda_{\min} = 1.75$. The

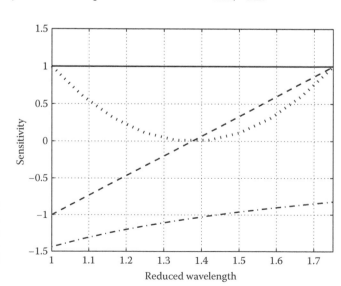

FIGURE 6.6
Sensitivity to the first three coefficients of the emissivity polynomial function (continuous, dashed, and dotted line) and to P_T, the inverse of normalized temperature (dashed–dotted line). Reduced wavelength is $\lambda' = \lambda/\lambda_{\min}$.

sensitivity to the temperature inverse is very smooth and close to linear. We thus expect a strong correlation between the parameters (near collinear sensitivity vectors).

An estimation of the parameter vector $\mathbf{P}^*$ in the least-squares sense is obtained by solving the linear system:

$$(\mathbf{X}^T\mathbf{X})\hat{\mathbf{P}}^* = \mathbf{X}^T\mathbf{Y} \tag{6.34}$$

The near-dependent sensitivities lead to an $\mathbf{X}^T\mathbf{X}$ matrix that is near singular. Indeed by computing the condition number of the matrix $\mathbf{X}^T\mathbf{X}$ one gets very high values, even for a low degree polynomial approximation (see Figure 6.7).

The condition number describes the rate at which the identified parameters will change with respect to a change in the observable. Thus, if the condition number is large, even a small error in the observable may cause a large error in the parameters (the condition number, however, only provides an upper bound). The condition number also reflects how a small change in the matrix $\mathbf{X}^T\mathbf{X}$ itself will affect the identified parameters. Such a change may be due to the measurement error of the equivalent wavelength corresponding to each spectral channel. From Figure 6.7, a first statement is that the regularization with a polynomial model of degree 2 and higher will not be efficient (the case of a polynomial of degree 1 would not be very stable either).

In the field of polymer regression, using orthogonal polynomials like Legendre polynomials instead of the monomial basis functions greatly helps for reducing the condition number. However, in present case, due to the smooth sensitivity of the temperature parameter, this does not help much.

A means of reducing the condition number would be to extend the spectral range. From the radiance curves in Figure 6.1 we notice that radiance is higher than 10% of its maximum over nearly one decade bandwidth. Assuming that the measurement is performed in different channels of such a large bandwidth, the condition number would decrease as shown in Figure 6.8. Unfortunately, due to technical reasons such as availability of sensors, spurious reflections from external sources (sun, ambient light, etc.), and

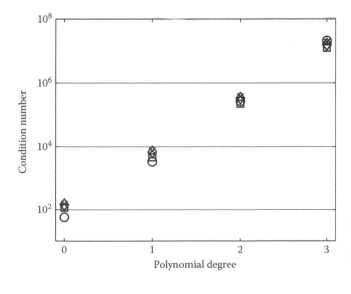

FIGURE 6.7
Condition number of the $\mathbf{X}^T\mathbf{X}$ matrix (O: $N=m+2$, □: $N=7$, ◇: $N=30$, ✕: $N=100$). Considered spectrum is such that $\lambda_{max}/\lambda_{min}=1.75$.

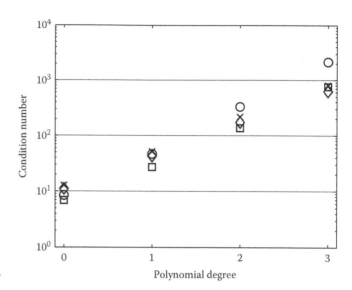

FIGURE 6.8
Same as in Figure 6.7 with $\lambda_{max}/\lambda_{min} = 10$.

presence of atmospheric absorption bands such a broadband temperature measurement remains hypothetical.

Instead of modeling $\ln[\varepsilon(\lambda)]$ by a polynomial function, one could use a staircase function (gray-band model). The sensitivity related to the emissivity assigned to a given band is a top-hat function. The condition number of the matrix $\mathbf{X}^T\mathbf{X}$ is represented in Figure 6.9. It slightly depends on the number of channels, but it rapidly rises with the number of bands N_b: the trend is roughly like N_b^3. This indicates that the ill-conditioned character of the identification problem becomes very critical if one looks at describing the emissivity profile with a high-resolution staircase function. It is expected that only rough approximations of the profile (surely with less than five to six bands) are likely to provide a safe characterization, i.e., with reasonably low parameter uncertainties.

The condition number is not all. It also depends on the choice of the reference temperature T_{ref}. Sometimes it could even be misleading because it only gives an upper bound of

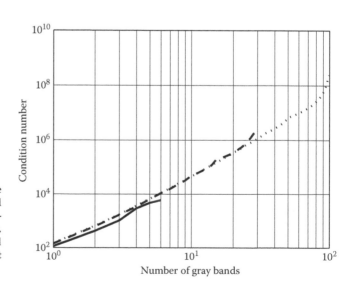

FIGURE 6.9
Condition number of $\mathbf{X}^T\mathbf{X}$ matrix in the case of a staircase emissivity model (gray-band model). Number of wavelengths is $N = 7$ (continuous line), $N = 30$ (dashed line), $N = 100$ (dotted line). Considered spectrum is such that $\lambda_{max}/\lambda_{min} = 1.75$.

the error propagation. It is better to analyze the diagonal values of the covariance matrix $(\mathbf{X}^T\mathbf{X})^{-1}$. They actually provide the variance amplification factor for each identified parameter $P*$:

$$\left[\sigma_{P*}^2\right] = \mathrm{diag}((\mathbf{X}^T\mathbf{X})^{-1})\sigma^2 \tag{6.35}$$

where σ^2 is the variance of the observable, i.e., $(\sigma_{L_i}/L_i)^2$, which is here assumed independent of the spectral channel i (if one assumed instead that the radiance variance $(\sigma_{L_i})^2$ is uniform, the result would be $\left[\sigma_{P*}^2\right] = \mathrm{diag}((\mathbf{X}^T\mathbf{\Psi}^{-1}\mathbf{X})^{-1})$ where $\mathbf{\Psi}$ is the inferred covariance matrix of the observable).

One should be aware that σ_{P*}^2 merely describes the error around the mean estimator value due to radiance error propagation to the parameters. If the mean estimator is biased, as is the case when the true emissivity profile is not well represented by the chosen model, one should add the square systematic error to get the root mean square (RMS) error which better represents the misfit to the true parameter value, either temperature or local emissivity (this will be described later through a Monte Carlo analysis of the inversion).

With the polynomial model, the mean standard relative error for emissivity, which is defined by

$$\frac{\sigma_\varepsilon}{\varepsilon} \equiv \sqrt{\frac{1}{N} \sum_{i=1}^{N} \frac{\sigma_{\varepsilon_i}^2}{\varepsilon_i^2}} \tag{6.36}$$

is related to the standard error of the retrieved polynomial coefficients through

$$\frac{\sigma_\varepsilon}{\varepsilon} = \sqrt{\frac{1}{N} \sum_{i=1}^{N} \left[X_{ij}^2\right]^T \left[\sigma_{a_j^*}^2\right]_{j=1,m}} \tag{6.37}$$

As such, it can be related to the uncertainty of the observable, which will be written as σ_L/L, through an error amplification factor K_ε:

$$\frac{\sigma_\varepsilon}{\varepsilon} = K_\varepsilon \frac{\sigma_L}{L} \tag{6.38}$$

With the gray-band model, the mean standard error and the amplification factor are defined according to

$$\frac{\sigma_\varepsilon}{\varepsilon} \equiv \sqrt{\frac{1}{N} \sum_{i=1}^{N_b} \left(\frac{\sigma_{\varepsilon_i}}{\varepsilon_i}\right)^2} = K_\varepsilon \frac{\sigma_L}{L} \tag{6.39}$$

From Wien's expression for radiance, it is clear that the standard relative error for temperature is proportional to temperature, to σ_L/L, and to a wavelength scale representative of the spectral window $\tilde{\lambda}$ (one can choose the geometric mean of the window limits: $\tilde{\lambda} \equiv \sqrt{\lambda_{min}\lambda_{max}}$). The error amplification factor for temperature, K_T, is thus defined according to

$$\frac{\sigma_T}{T} = K_T \tilde{\lambda} T \frac{\sigma_L}{L} \tag{6.40}$$

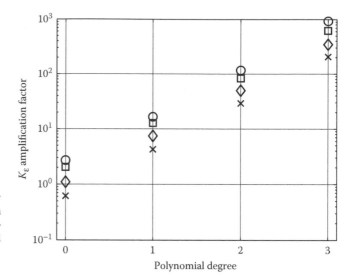

FIGURE 6.10
Error amplification factor of emissivity versus the polynomial degree m chosen for modeling $\ln[\varepsilon(\lambda)]$. Symbols correspond to different values for the total number of spectral channels: O: $N = m + 2$, □: $N = 7$, ◇: $N = 30$, ✕: $N = 100$.

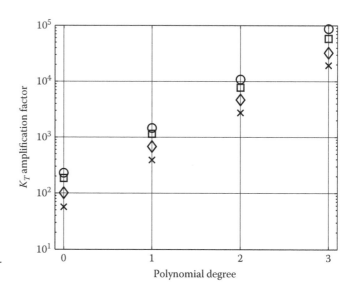

FIGURE 6.11
Same as in Figure 6.10 for the error amplification factor of temperature.

The error amplification factors K_T and K_ε are plotted in Figures 6.10 and 6.11 for the polynomial model and in Figures 6.12 and 6.13 for the gray-band model, assuming for both cases a relative bandwidth $\lambda_{max}/\lambda_{min}$ of 1.75 (this could correspond to the [8–14 μm] spectral interval, for example).

A first comment for the polynomial model is that the standard errors increase exponentially with the polynomial degree m, roughly like $\exp(2m)$. This increase can be slowed down by widening the spectral window. With the gray-band model, the standard errors increase nearly in proportion to the number of bands. In both cases, they decrease with the total number of channels, roughly like $N^{-1/2}$. Empirical relations can be found for the factors K_T and K_ε. They lead to the following error predictions for the particular case $\lambda_{max}/\lambda_{min} = 1.75$:

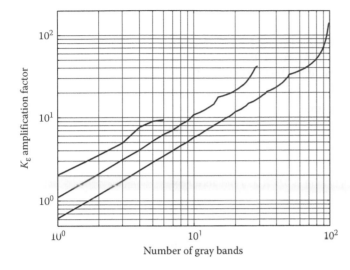

FIGURE 6.12
Error amplification factor of emissivity when using the gray-band model. The number of spectral channels is from top to bottom: $N = 7$, 30, 100. In each case, the number of gray bands ranges from 1 to $N - 1$.

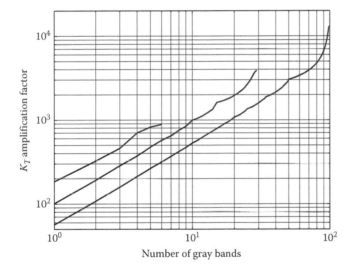

FIGURE 6.13
Same as in Figure 6.12 for the error amplification factor of temperature.

For the polynomial model:

$$\begin{cases} \dfrac{\sigma_T}{T} \cong 420 \dfrac{\exp(1.94m)}{\sqrt{N+4}} \tilde{\lambda} T \dfrac{\sigma_L}{L} \\[3mm] \dfrac{\sigma_\varepsilon}{\varepsilon} \cong 6.4 \dfrac{\exp(1.93m)}{\sqrt{N+4}} \dfrac{\sigma_L}{L} \end{cases} \tag{6.41}$$

For the gray-band model:

$$\begin{cases} \dfrac{\sigma_T}{T} \cong 410 \dfrac{N_b}{\sqrt{N+4}} \tilde{\lambda} T \dfrac{\sigma_L}{L} \\[3mm] \dfrac{\sigma_\varepsilon}{\varepsilon} \cong 5 \dfrac{N_b}{\sqrt{N+4}} \dfrac{\sigma_L}{L} \end{cases} \tag{6.42}$$

TABLE 6.1

Root-Mean-Square Error for the Estimated
Temperature and Emissivity Depending
on the Degree of the Polynomial Model
for Emissivity

Polynomial Degree	σ_T (K)	σ_ε
0	1.5	0.02
1	9.4	0.13
2	64	0.83

Target temperature is 320 K and radiance noise
is 1%.

Regarding the bandwidth influence, we notice that the relative error of temperature
depends both on λ_{min} and λ_{max} whereas the mean relative error of emissivity only depends
on the ratio $\lambda_{max}/\lambda_{min}$.

Assuming a target at 320 K, and 1% radiance noise, a pyrometer with seven wavelengths
between 8 and 14 μm will provide temperature and emissivity values with standard errors
as reported in Table 6.1, depending on the polynomial degree chosen for $\ln[\varepsilon(\lambda)]$.

The errors are rather high with a linear model for $\ln[\varepsilon(\lambda)]$ and they reach unacceptably
high values when using a degree 2 polynomial. These results seem to preclude using the
least squares linear regression approach together with a polynomial of degree 2 and more.
They were obtained with Wien's approximation. However, Planck's law is close to Wien's
approximation over a large spectrum; therefore, we expect that the general least squares
nonlinear regression will also face serious problems when using a polynomial model for
regularization.

Applying the gray-band model to the previous example leads to the standard errors
shown in Table 6.2 (the number of bands can be increased up to $N - 1 = 6$ for avoiding
underdetermination).

The errors increase with the number of bands, starting from the values corresponding to
a degree 0 polynomial and ending at values that are lower than those obtained with a

TABLE 6.2

Root-Mean-Square Error for the
Estimated Temperature and Emissivity
Depending on the Number of Bands When
Assuming a Gray-Band Model for Emissivity
and Seven Spectral Measurements

Number of Bands	σ_T (K)	σ_ε
1	1.5	0.020
2	2.6	0.035
3	3.7	0.049
4	5.7	0.076
5	6.7	0.090
6	7.2	0.094

Target temperature is 320 K and radiance noise
is 1%.

degree 1 polynomial. This is interesting in the sense that even with six bands, i.e., six degrees of freedom for emissivity, the errors do not "explode" as it was observed earlier by increasing the polynomial degree. The gray-band model, although not being smooth, could thus capture more easily rapid variations in the emissivity profile like peaks.

However, as previously stated, the standard errors that are here presented only reflect what happens when noise corrupts the radiance emitted by a surface which otherwise *perfectly follows the staircase model*. As an example, with the six-bands case, the emissivity should be *equal* in the two channels that were chosen to form the largest band.

6.3.2.3 A Look at the Solutions of the ETS Problem

Another way of presenting the ill posedness of the ETS problem and the difficulties in finding an appropriate regularization consists, like in Coates (1981), in exposing the multiple solutions to this underdetermined problem. For this purpose, we took two examples for the "true" emissivity profile: a linear profile and a polynomial of degree 6. These profiles are represented with bold lines in Figures 6.14 and 6.15 (let us mention that with reference to the measured spectra in Figures 6.4 and 6.5, the degree 6 polynomial spectrum in Figure 6.15 cannot be considered unreasonable in any way).

The emitted radiance was then calculated according to Planck's law assuming a 320 K temperature in both cases (for simplicity we discarded at this stage the eventual reflections; experimental noise was also discarded, but it will be added later). Then, from different temperature estimated values $\hat{T}$, one can infer the emissivity profile $\hat{\varepsilon}(\lambda, \hat{T})$, which *exactly* leads to the observed radiance. It is given by

$$\hat{\varepsilon}(\lambda, \hat{T}) = \frac{L(\lambda, T)}{B(\lambda, \hat{T})} = \varepsilon(\lambda) \frac{B(\lambda, T)}{B(\lambda, \hat{T})} \tag{6.43}$$

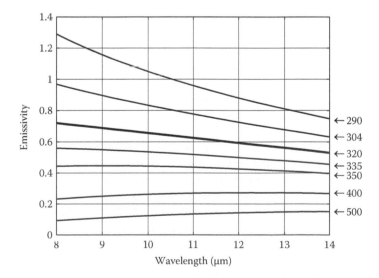

FIGURE 6.14

Emissivity profiles inferred by assuming a temperature $\hat{T}$ higher or lower than the "real" T temperature which is here 320 K. $\hat{T}$ values are indicated on the right. The "true" profile is in bold line (here assumed linear).

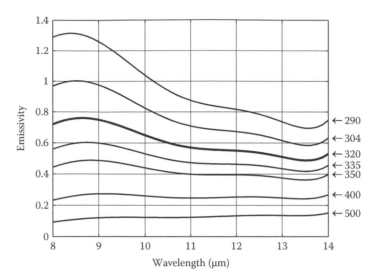

FIGURE 6.15
Same as in Figure 6.14 when the "true" profile is a degree 6 polynomial function (bold line).

Some profiles $\hat{\varepsilon}(\lambda, \hat{T})$ are reported in Figures 6.14 and 6.15 together with the corresponding estimated temperature $\hat{T}$. We must stress the point that these emissivity profiles are all *perfect solutions* to the problem, at least from the mathematical perspective. Of course one has to discard those presenting higher values than 1. With this constraint in mind, the admissible temperatures are from about 304 K up. Similarly, profiles that reach values less than, say, 0.02–0.03 can also be discarded if one has some prior information that the surface is not a very clean polished metal surface (refer to the examples in Figures 6.4 and 6.5).

The traditional way consists in looking for a solution of $\varepsilon(\lambda)$ in the form of a polynomial. Let us consider the case of a polynomial of degree 1. The problem can then be reformulated as follows: which profile in Figure 6.14, respectively in Figure 6.15, does fit a straight line at best, taking into account the weighting with the blackbody radiance? Of course, in Figure 6.14, the profile corresponding to $\hat{T} = 320$ K is the only one to be linear (the curvature of the profile changes on each side of $\hat{T} = 320$ K). Nevertheless, one has to admit that the profiles corresponding to an estimated temperature in the range 304 K $< \hat{T} < 350$ K are not far from a straight line. If one added some experimental noise, it is clear that the squared residuals after the linear fit would be in the same range for all profiles $\hat{\varepsilon}(\lambda, \hat{T})$ corresponding to this temperature range.

The case in Figure 6.15 is even worse: it is evident that, among all possible solutions, the "true" profile is not the straightest line. Evidently, in this example, the answer for optimal $\hat{T}$ will be a temperature much higher than the "true" value (lower profiles in the figure are indeed smoother than higher profiles). The final solution will thus present a bias. A bias would also be obtained for the case drawn in Figure 6.15 if the chosen emissivity model was a degree 0 polynomial instead of a degree 1 polynomial.

As often stated, when using LSMWP, it is necessary to choose an emissivity model that *exactly* corresponds to the true profile. The difficulty is that most often the profile shape is unknown. A misleading thought is that LSMWP performs a fit of the true profile with the chosen model (polynomial, exponential, etc.). Actually, as seen above, performing LSMWP comes to choosing among the different possible profile solutions $\hat{\varepsilon}(\lambda, \hat{T})$, the one that fits at best the model, in the least squares sense by weighting with the blackbody radiance (the fit deals with $\varepsilon(\lambda)$ if the observable is radiance and with $\ln(\varepsilon(\lambda))$ if it is its logarithm). This can lead to an emissivity profile of much higher or much lower mean value than the real one, together with an important temperature error. Actually, the problem with the present LSMWP is that it sticks to the emissivity *shape* rather than to its *magnitude*.

6.3.2.4 Least-Squares Solution of the Nonlinearized ETS Problem

When using Planck's law instead of Wien's approximation, LSMWP cannot be linearized anymore. The nonlinear least-squares problem can be tackled with the Levenberg–Marquardt method as provided, for example, by the *lsqnonlin* function from MATLAB® library. When choosing a linear model for emissivity and when the "true" emissivity profile is indeed linear this naturally leads to the right temperature and right emissivity profile (there is no systematic error when the simulated emissivity spectrum corresponds to the chosen model). On the contrary, when the "true" emissivity profile is not linear, the identification presents a bias. For a "true" emissivity profile corresponding to the bold line curve in Figure 6.15, the result is reported in Figures 6.16 and 6.17. For this example we assumed seven equidistant spectral measurements between 8 and 14 μm. The dots in Figure 6.16 correspond to the simulated measured radiance (no noise at this stage) and the line corresponds to the radiance calculated from $\hat{L}(\lambda, T) = \hat{\varepsilon}_{d1}(\lambda)B(\lambda, \hat{T})$ where $\hat{\varepsilon}_{d1}(\lambda)$ is the degree 1 polynomial solution of the LSMWP inversion. A perfect match for radiance is of course impossible: the low-order model chosen for emissivity (degree 1 polynomial) cannot explain the observed radiance variations. The least-squares procedure reveals that the $\hat{\varepsilon}(\lambda, \hat{T})$ profile in Figure 6.15 that fits at best a straight line, by taking into account the weighting with the blackbody radiance, is the one corresponding to 335.3 K. The seven dots in Figure 6.17 correspond to $\hat{\varepsilon}(\lambda, 335.3)$ and the dashed line is the best linear estimate for emissivity $\hat{\varepsilon}_{d1}(\lambda)$. The systematic error is thus +15 K for temperature and between −0.06 and −0.2 for emissivity.

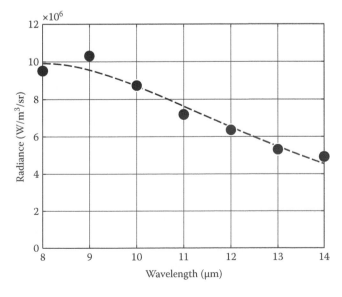

FIGURE 6.16

Inversion result for the degree 6 polynomial emissivity profile from Figure 6.15 when using a linear model. Dots represent the "true" noiseless radiance, and the dashed line is the emitted radiance according to the solution.

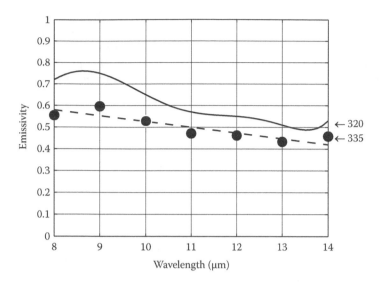

FIGURE 6.17
Inversion result for the degree 6 polynomial emissivity profile from Figure 6.15 ($T = 320$ K) when using a linear model. The "true" emissivity profile is labeled 320 K. The linear solution $\hat{\varepsilon}_{d1}(\lambda)$ that is associated with a temperature of 335.3 K is in dashed line. The profile $\hat{\varepsilon}(\lambda, 335.3)$ is represented with dots.

If the fitting happens to be too far from the $\hat{\varepsilon}(\lambda, \hat{T})$ profile, one should change the model. For this particular example, however, changing to a quadratic model leads to a complete failure: the profile in Figure 6.15 that is closest to a degree 2 polynomial is the one corresponding to 230 K and the retrieved (hypothetical) emissivity spectrum ranges between 2 and 6! Obviously, by imposing the constraint $\hat{\varepsilon}(\lambda, \hat{T}) < 1$, the acceptable solution would be the profile associated to $T = 304$ K, which means a 16 K underestimation.

Let us now analyze the influence of the measurement noise on the ETS performance. This can be easily performed by simulating experiments where the theoretical radiance is corrupted with artificial noise. The radiance is altered by adding values that are randomly generated with a predetermined probability density function. We assumed a Gaussian distribution with a spectrally uniform standard deviation. We fixed it to a value ranging from 0.2% to 6% of the maximum radiance (additive noise). The least-squares minimization was performed without constraint (i.e., without imposing $\varepsilon_i < 1$) in order to highlight the mathematical (poor) stability of the inversion procedure. A series of 200 radiance spectra were treated for each noise level and for both nominal emissivity profiles described in Figures 6.14 and 6.15 (polynomial functions of degree 1 or 6). As before, we assumed that the spectral measurements are performed at seven equidistant wavelengths between 8 and 14 μm. We chose a linear emissivity model for LSMWP inversion. The results for the maximum in the seven channels of the RMS emissivity error are plotted in Figure 6.18. Those for the RMS error of temperature are plotted in Figure 6.19. One can notice the following:

1. When the "true" profile is linear, the RMS error increase for temperature and for emissivity is roughly proportional to the radiance noise level (the temperature RMS error becomes somewhat erratic when noise is higher than about 3%). In particular, the RMS errors are 0.1 for emissivity and 8 K for temperature in the case of a 1% measurement noise.

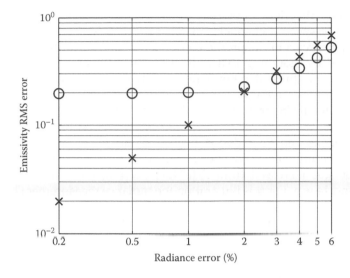

FIGURE 6.18
Statistic analysis (Monte Carlo sampling with 200 simulated experiments) of the measurement noise influence on the identified emissivity when using a linear emissivity model. The "true" emissivity was considered linear (crosses) or a degree 6 polynomial (circles). Seven channels between 8 and 14 μm.

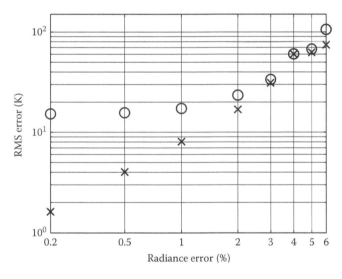

FIGURE 6.19
Same as in Figure 6.18 for the identified temperature.

2. When the "true" profile is a degree 6 polynomial, the RMS errors are first dominated by a systematic error, which corresponds to the model implementation error (the chosen model—degree 1 polynomial—is too crude for representing the "true" profile); statistic errors due to the measurement noise dominate only when noise is higher than 2%–3%.

Let us also add that the inversion leads to a systematic error as soon as the "true" profile departs from a straight line. The previous analysis allows us to evaluate the magnitude of this error when the deviation is small. Statistically, by considering several "true" profiles close to the nominal straight line in Figure 6.14, the RMS of the systematic errors would be equal to the RMS of the statistic errors obtained by adding the same amount of measurement noise. For this reason, a "true" profile departing by as little as 1% from a straight line leads to an emissivity bias whose RMS value is about 0.1. The temperature quadratic mean error is in this case about 8 K, which is far from negligible. This result highlights the considerable importance of choosing the right emissivity model. This impact can be reduced by increasing the number of spectral channels (the trend is like $N^{-1/2}$ as seen later), on the condition that the departure from the profile model is randomly distributed.

The same analysis was performed by assuming that both the "true" profile and the model are quadratic. The RMS errors (not presented here) are roughly proportional to the radiance noise level as when both profiles are linear, however, at a much higher level: in the case of a 1% measurement noise, the RMS errors reach 0.33 for emissivity and 49 K for temperature.

It is well known that statistic errors can be reduced by increasing the number of measurements, here by increasing the number of channels. This is confirmed in Figures 6.20 and 6.21 where this number was increased from 7 to 120, keeping the channels uniformly distributed between 8 and 14 μm. For this illustration the radiance measurement noise was fixed at 1%. One can notice that the RMS errors indeed decrease in the case of the linear "true" profile with a power-law trend, close to the $N^{-1/2}$ classical reduction. In the

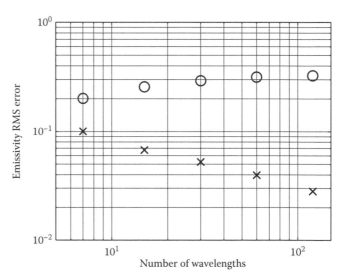

FIGURE 6.20
Same as in Figure 6.18 when assuming a 1% Gaussian noise on radiance, and increasing the number of channels from 7 to 120 (uniformly distributed between 8 and 14 μm).

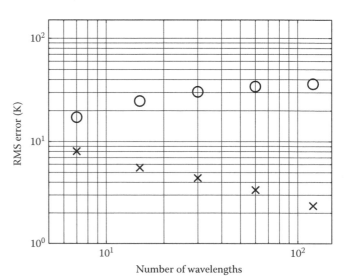

FIGURE 6.21
Same as in Figure 6.20 for the identified temperature.

case of the more complicated degree 6 polynomial "true" profile, there is no such reduction. As a matter of fact, systematic errors always dominate. There is even a progressive increase of the RMS errors with the number of wavelengths. The RMS errors of 0.2 for emissivity and 17 K for temperature that are observed with seven channels cannot be reduced by adding more channels.

As a conclusion we can state the following:

1. Even by reducing the number of unknowns, as was done here by modeling spectral emissivity with a polynomial of low degree, the problem remains badly conditioned; with a polynomial model (either for $\varepsilon(\lambda)$ or for $\ln \varepsilon(\lambda)$), reasonable inversion results are expected only up to degree 1.

2. Important systematic errors appear as soon as the real emissivity departs from the considered model: 1% departure from a straight line already leads to 8 K RMS error. More complicated spectral shapes lead to unpredictably high systematic errors (15 K for the considered example of a degree 6 polynomial).

3. Even if the real emissivity values at the sampled wavelengths ε_i $i=1, N$ perfectly fitted to a straight line, the demand on radiance measurement precision is very high: as a matter of fact, no more than 0.12% noise is allowed to get a 1 K RMS error near room temperature for a seven-band pyrometer between 8 and 14 μm.

The same analysis was performed by considering the gray-band model. From the degree 6 polynomial emissivity spectrum in Figure 6.15 (it will be called the "raw" profile), two "true" emissivity spectra were drawn. The first one was simply obtained by sampling the raw profile at the N wavelengths of the pyrometer. The second one was deduced from the latter one to be compliant with the gray-band model: the N emissivity values were averaged separately in each of the N_b gray bands. The second "true" profile thus *perfectly* fits to the gray-band model whereas the first one is more realistic.

A Monte Carlo approach was applied by adding Gaussian noise to the theoretical spectral radiance and performing the inversion on 300 such synthetic data. The results for emissivity and temperature RMS errors due to 1% RMS radiance noise are presented in Figures 6.22 and 6.23 when assuming that the measurement is made in $N=7$ channels between 8 and 14 μm and in Figures 6.24 and 6.25 when assuming $N=30$ channels. In each case, the number of bands was varied between 1 and $N-1$. The curves obtained with the model-compliant emissivity profile (crosses) are rising with the number of bands and are in agreement with those obtained with the covariance matrix (see Figure 6.12 for emissivity and Figure 6.13 for temperature). More interesting are the curves obtained with the model-not-compliant emissivity profile (circles): they are noticeably erratic and the RMS errors are more important, actually higher than 0.07 for emissivity and higher than 8 K for temperature. They are particularly important when the number of bands is either low or high with respect to the number of channels. In the first case, the number of bands is insufficient to describe correctly the true emissivity profile. In the second case, we again face a problem of overfitting. It thus appears to be better to choose intermediate values for the number of bands. One can notice that for some particular number of bands, the results are significantly better than with the linear emissivity model (compare with Figure 6.20 for the emissivity error and with Figure 6.21 for the temperature error). However, the results may vary by a factor of 2 by just changing the number of bands by one. This unpredictable behavior seems to preclude the gray-band model from leading to a safer and more efficient inversion than the linear emissivity model allows.

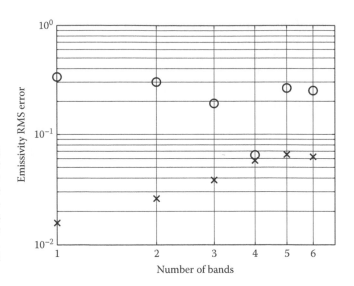

FIGURE 6.22

Statistic analysis (Monte Carlo sampling with 300 simulated experiments) of the measurement noise influence on the identified emissivity when using the gray-band model. Measurement is performed in seven spectral channels between 8 and 14 µm. Circles: emissivity spectrum is a degree 6 polynomial. Crosses: the previous profile is averaged in each band before inversion.

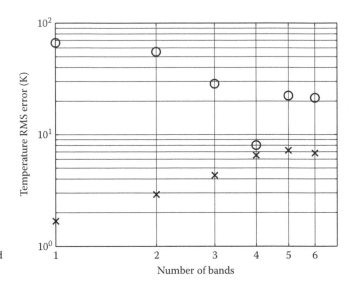

FIGURE 6.23

Same as in Figure 6.22 for the identified temperature.

Finally, LSMWP does not perform well for simultaneous evaluation of temperature and emissivity when using the emitted spectral radiance only. Reasonable RMS values can be obtained only when the emissivity spectrum perfectly matches with the implemented emissivity model (gray band or linear). Otherwise, important systematic errors are encountered. The problem is that, apart from a few exceptions, one does not know beforehand whether the emissivity of a tested material complies with such or another model.

As a conclusion, there is no valuable reason for implementing MWP instead of the simpler one-color or bispectral pyrometry. All methods need a priori information about emissivity. However, the requirements with one-color pyrometry (the knowledge of an emissivity level) or with bispectral pyrometry (the knowledge of the ratio of emissivity at two wavelengths) are less difficult to satisfy than the requirement with MWP, which is *a requirement of a strict shape conformity* of the emissivity profile with a given parametric function which, practically, is impossible to satisfy.

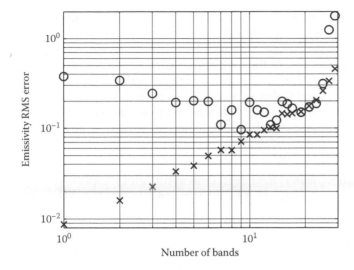

FIGURE 6.24

Same as in Figure 6.22 when the measurements are performed over spectral 30 channels (1 up to 29 gray bands are considered).

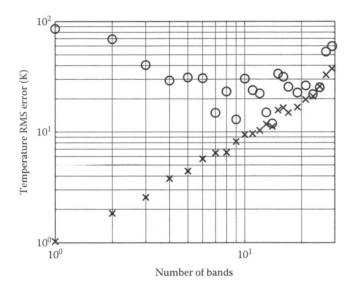

FIGURE 6.25

Same as in Figure 6.24 for the identified temperature.

6.4 Emissivity–Temperature Separation Methods in the Field of Remote Sensing

Optical remote sensing by airborne or satellite sensors presents the following characteristics:

1. The measurements are highly conditioned by the radiative properties of the atmosphere (transmission, upward emission along the optical path, downward emission and reflection on the earth surface, scattering, etc.). These properties depend on altitude, direction of sight, profiles of air humidity and temperature, aerosol type and size, etc. The wavelength selection for the optic sensor, be it broadband, multiwavelength, or hyperspectral, is of course dependent on the typical atmospheric conditions it will face during its mission.

2. The footprint is generally large: on the order of 10 cm for low-altitude airborne sensors to about 2 km for sensors aboard geostationary satellites. It may thus happen that different materials, possibly at different temperature levels, are present in the instantaneous field of view (i.e., the footprint corresponding to one pixel of the considered camera). Aggregation laws in the IR are generally complicated, especially if the surface is not flat and if there are reflections between different soil elements. A general approximation is to assimilate the integrated radiance to the one originating from a pure isothermal surface. The objective is then to evaluate the equivalent isothermal temperature. The disaggregation problem which consists in separating the different contributions will not be considered here.

3. Natural surfaces (soil, vegetation, water) have high emissivity values, especially in the longwave range, as for example in the [8–14 μm] atmospheric window, where they generally exceed 0.9 (see Figure 6.5). They are then close to blackbodies. Most often they are considered as Lambertian surfaces.

4. Since the 1970s, the radiometric and spectral performances of the optic sensors onboard satellites have continuously improved: once presenting a few bands in the [3–5 μm] and [8–14 μm] range, the sensors are now multispectral (some tens of bands) and even hyperspectral (about 100 bands). The spectral analysis is performed by dispersion or by Fourier transform interferometry.

The measured radiance at the sensor level can be written from Equations 6.7 and 6.8 as

$$L_s(\lambda, T, \theta, \varphi) = \tau(\lambda, \theta, \varphi)\left[\varepsilon(\lambda)B(\lambda, T) + (1 - \varepsilon(\lambda))\frac{E^{\downarrow}(\lambda)}{\pi}\right] + L^{\uparrow}(\lambda, \theta, \varphi) \qquad (6.44)$$

where
$\tau(\lambda, \theta, \varphi)$ is the path transmission
$E^{\downarrow}(\lambda)$ is the down-welling sky irradiance ($L^{\downarrow}(\lambda) = E^{\downarrow}(\lambda)/\pi$ is the corresponding isotropic equivalent radiance)
$L^{\uparrow}(\lambda, \theta, \varphi)$ is the up-welling path radiance emitted by the atmospheric constituents (the angle dependency (θ, φ) will not be recalled from now on)

To solve Equation 6.44 for emissivity and temperature we need to know the transmission and the upwelling and downwelling atmospheric radiances. Atmospheric radiative transfer models like MODTRAN (Berk et al. 1989) and MATISSE (Simoneau et al. 2009) can be used to compute the needed radiative parameters.

Figure 6.26 illustrates the relative importance of the various contributions to the at-sensor radiance in the longwave IR range. To achieve this, MODTRAN simulations were performed assuming a sensor flying at 1900 m altitude and aiming the ground at nadir. A mid-latitude summer atmospheric model was considered. For this illustration, we assumed that the ground is gray with 0.9 emissivity and 313 K temperature. A first observation is that the atmospheric influence is far from negligible even in the so-called atmospheric window 8–12 μm. Furthermore, the IR spectra of present gases (H_2O, CO_2, CH_4, O_3, etc.) print their characteristic features to all radiative contributions. Last, due to the low reflectance of the surface, the contribution from the downwelling atmospheric emission is very small and it is sometimes neglected (however, for a precise ETS, taking into account $L^{\downarrow}(\lambda)$ in the inversion process is of prime importance, especially with the SpSm approach; see Section 6.4.3.3).

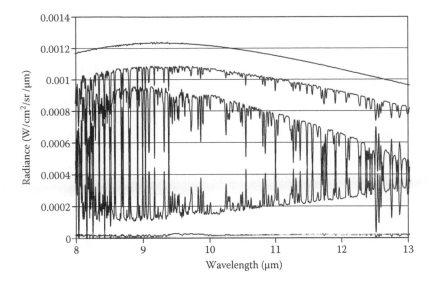

FIGURE 6.26
Remote sensing of a gray surface ($\varepsilon = 0.9$) at $T = 313$ K by a sensor at 1900 m altitude (MODTRAN simulations). From top to bottom: nominal blackbody radiance, total at-sensor radiance, soil contribution (after attenuation), atmospheric upwelling emission, down-welling sky radiance (after reflection and attenuation).

For computing the atmospheric transmission and emission with a radiation transfer simulation software, standard atmospheric models can be considered at first. They are defined by temperature and humidity profiles vs. altitude, depending on local climate and season, and they assume different aerosol types. However, for a higher fidelity of the radiation transfer calculation results, it is better to rely on radiosonde observation results obtained at the time and place of the remote sensing measurement. This solution is nevertheless expensive and radiosonde deployment is not always possible. Other means, either direct or indirect, were thus devised for the atmospheric compensation.

6.4.1 Combined Atmospheric Compensation and Emissivity–Temperature Separation

A series of methods, the dual channel and the split-window methods, were developed for evaluating sea surface temperature at the time the sensors had only a few spectral data in the IR (Nimbus, advanced very high resolution radiometer [AVHRR], etc.) (Barton 1983). They actually performed atmospheric compensation and, based on the fact that water emissivity is known and very high, they also performed the remaining ETS. The dual channel method is based on the differential absorption in adjacent IR bands. Regression laws of the following form were established between the true surface temperature T and the at-sensor brightness temperatures measured in two particular channels, one in the midwave range T_{R1} and the other in the longwave range T_{R2}:

$$T \approx a_1 T_{R1} + a_2 (T_{R2} - T_{R1}) + a_0 \tag{6.45}$$

The coefficients a_i were obtained from regression over simulated databases covering a large range of atmospheric conditions.

The split-window terminology is used when the two channels are in the same 8–13 μm band, more specifically between 10 and 12 μm (AVHRR has actually two channels, at

10.3–11.3 and 11.5–12.5 μm, thereby "splitting" the thermal infrared (TIR) spectral window). Reported typical errors (model errors) are less than 0.5 K.

A similar regression law was established when considering that the brightness temperatures are obtained in the same channel but when the sea is viewed from two different angles (Barton 1983).

The advantage of these differential approaches is that radiosonde atmospheric profiles are not required.

After the success of the split-window technique over sea surfaces, people extended it to land surfaces. The purpose was to apply it to geostationary satellites (SEVIRI on Meteosat-SG), to sun-synchronous satellites (MODIS and ASTER on TERRA), and to airborne remote sensing (DAIS imaging spectrometer, TIMS multispectral scanner, AHS hyperspectral scanner). However, over land, the unknown emissivities are a greater source of inaccuracy than atmospheric effects. Inaccuracy of only 0.01 in ε causes errors in T sometimes exceeding those due to imperfect atmospheric correction.

On land surfaces, the relationship between real temperature and at-sensor brightness temperatures is more complicated than the one expressed in Equation 6.43. Therefore more intricate regression laws were proposed which take into account a series of ancillary data. As a matter of fact, the emissivity at the two channels has to be estimated independently as shown below. The most recent relations are of the following type (Coll et al. 2003, Atitar and Sobrino 2009):

$$T \approx T_{R1} + a_1(T_{R2} - T_{R1}) + a_2(T_{R2} - T_{R1})^2 + a_3(1 - \varepsilon)$$
$$+ a_4W(1 - \varepsilon) + a_5\Delta\varepsilon + a_6W\Delta\varepsilon + a_0 \qquad (6.46)$$

where
ε is the mean emissivity $(\varepsilon_1 + \varepsilon_2)/2$
$\Delta\varepsilon$ is the emissivity difference $(\varepsilon_1 - \varepsilon_2)$
W is the columnar water vapor (CWV) in the direction of observation

The emissivity is approximated from vegetation and soil emissivities (ε_{vi} and ε_{si}), according to

$$\varepsilon_i = \varepsilon_{vi}f_v + \varepsilon_{si}(1 - f_v) + 4\langle d\varepsilon_i\rangle f_v(1 - f_v) \qquad (6.47)$$

where
f_v is the vegetation cover fraction
$\langle d\varepsilon_i\rangle$ is a term accounting for the cavity effect which depends on the surface geometry

Average values for vegetation and soil emissivity ε_{vi}, ε_{si} were computed from the spectral library built by Salisbury and d'Aria (1992). The vegetation cover fraction can be estimated from a vegetation index such as normalized difference vegetation index (NDVI). This index is equal to the normalized difference of the reflectance values measured on each side of the chlorophyll absorption band (i.e., at about 0.67 and 0.82 μm). Remote sensing in the visible range and in near IR is thus necessary to get the NDVI and therefrom an estimation of the vegetation cover fraction.

The CWV W itself can be obtained from the normalized difference of the IR signals at 10.8 and 12.0 μm (Sobrino and Romaguera 2008).

The a_i coefficients in Equation 6.46 are derived from a regression over a simulated database comprising global-scale atmospheric conditions (from MODTRAN simulations performed on thermodynamic initial guess retrieval (TIGR) database—Chevallier et al. (2006)) and a wide range of surface emissivities (Salisbury and d'Aria 1992). From the simulations, it is shown that land surface temperature can be obtained with an error of about 1.3 K.

6.4.2 Separate Evaluation of the Atmospheric Parameters

The in-scene atmospheric compensation (ISAC) method was developed to get the spectra of $\tau(\lambda)$ and $L^\uparrow(\lambda)$ from a hyperspectral datacube (i.e., the matrix obtained by piling the images obtained at all wavelengths) (Young et al. 2002). It simultaneously exploits the spectral and the spatial dimension of the IR images. The fundamental hypothesis is that the atmospheric parameters are constant over the whole scene. The method first consists in finding the (i, j) pixels corresponding to near-blackbodies in the scene. If the temperature of these pixels spans over a sufficiently broad range, a linear fit between their at-sensor radiance $L(i, j, \lambda)$ and their blackbody radiance $B(\lambda, T_{i,j})$ will provide the requested atmospheric parameters:

$$L_s(i, j, \lambda) = \tau(\lambda)B(\lambda, T_{i,j}) + L^\uparrow(\lambda) \quad (i, j) = \text{blackbody} \qquad (6.48)$$

By performing this fit at each wavelength one gets an estimation of transmittance and upwelling atmospheric radiance: $\hat{\tau}(\lambda)$ and $\hat{L}^\uparrow(\lambda)$.

The blackbody pixels are found from the $L_s(i, j, \lambda)$ vs. $B(\lambda, T_{i,j})$ scatterplot for the whole scene: the blackbody pixels are indeed those that stick to the upper part of the plot. Beforehand, the temperature $T_{i,j}$ has to be estimated at each pixel; this is done by selecting the wavelength λ_0 with the highest transmission ($\tau(\lambda_0) \approx 1, L^\uparrow(\lambda) \approx 0$), by assigning a common emissivity value (i.e., $\varepsilon_m = 0.95$) to all pixels, and by inverting Planck's law:

$$T_{i,j} = B^{-1}\left(\frac{\lambda_0, L_s(i, j, \lambda_0)}{\varepsilon_m}\right) \qquad (6.49)$$

Other techniques for correcting the acquired data from atmospheric effects are based on sounding techniques and neural networks. The sounding techniques use the opacity variations of the atmosphere in absorption bands to estimate atmospheric composition and temperature profiles. In the 4–12 μm spectral range the main molecules responsible for absorption and emission are H_2O around 6.7 μm and between 4.8 and 5.5 μm, and CO_2 around 4.3 μm. In the vicinity of these absorption bands the IR signal depends on a weighting function which is sensitive to both concentration profiles and temperature profiles. CO_2 concentration is relatively constant spatially and temporally; therefore sounding near 4.3 μm allows evaluating the temperature profile. From this knowledge, one can then analyze the IR signal in the vicinity of a water absorption band to retrieve the water vapor profile.

Estimating the mean temperature of the atmospheric layer and the CWV under the sensor is sufficient when the purpose in only to evaluate the ground-leaving radiance (Achard et al. 2007). Two artificial neural networks were trained to retrieve these parameters from radiances measured near the 4.3 μm CO_2 absorption band and in the 4.8–5.5 μm range (Achard et al. 2007). Training was performed on slightly noised radiance spectra obtained with MODTRAN simulations based on a 2000 atmospheric vertical profiles database representing polar to tropical climate. Without signal noise, the mean atmospheric temperature can be retrieved with about 0.3 K RMS error and the water vapor content with 0.12 g/cm^2 RMS error. These integrated values are then used to scale the mean vertical profiles in order to get the same mean temperature and CWV. Based on these scaled atmospheric profiles and a new MODTRAN simulation the atmospheric terms appearing in Equation 6.44 can finally be evaluated.

6.4.3 Emissivity–Temperature Separation Methods

We assume at this point that the atmospheric effects along the path from the ground to the sensor were properly estimated. One can thus estimate the ground-leaving radiance:

$$L(\lambda, T) = \frac{L_s(\lambda, T) - L^\uparrow(\lambda)}{\tau(\lambda)} = \varepsilon(\lambda)B(\lambda, T) + (1 - \varepsilon(\lambda))L^\downarrow(\lambda) \tag{6.50}$$

The separation methods that we will present are based on a priori information about the emissivity spectrum. One exception is the multi-temperature (or multitemporal) method which is based on recording the radiance field after the ground has reached different temperature levels.

A particular situation is when $L(\lambda, T) = L^\downarrow(\lambda)$. In this case, the emissivity cannot be determined and the temperature is obtained from

$$T(\lambda) = B^{-1}\lfloor L^\downarrow(\lambda)\rfloor \tag{6.51}$$

6.4.3.1 Normalized Emissivity Method

The maximum emissivity ε_{max} is assumed to be known, but at an unspecified wavelength. The problem can be tackled in two ways. The first approach is iterative: one selects a rather high temperature value $\hat{T}$ and one computes the emissivity-satisfying Equation 6.50 at each wavelength to get the estimated spectrum:

$$\hat{\varepsilon}(\lambda) = \frac{L(\lambda, T) - L^\downarrow(\lambda)}{B(\lambda, \hat{T}) - L^\downarrow(\lambda)} \tag{6.52}$$

The temperature estimate is then progressively decreased until the maximum of the spectrum $\hat{\varepsilon}(\lambda)$ reaches the target value ε_{max}. This process is easily understood by referring to Figure 6.15.

In the case where $L(\lambda, T) < L^\downarrow(\lambda)$, the first estimation $\hat{T}$ should be low and then progressively increased.

The second method is initiated by computing a brightness temperature assuming the surface to be a gray body with $\varepsilon(\lambda) = \varepsilon_{max}$:

$$T_R(\lambda) = B^{-1}\left[\frac{L(\lambda, T) - L^\downarrow(\lambda)}{\varepsilon_{max}} + L^\downarrow(\lambda)\right] \tag{6.53}$$

The temperature estimation is then simply $\hat{T} = \max[T_R(\lambda)]$ where $L(\lambda, T) > L^\downarrow(\lambda)$ and $\hat{T} = \min[T_R(\lambda)]$ where $L(\lambda, T) < L^\downarrow(\lambda)$. The emissivity spectrum is finally deduced from the application of Equation 6.52.

The emissivity to be assigned is often close to 0.96, which represents a reasonable average of maximum values in the thermal IR for exposed geologic surfaces. In the case of bare soils with rocks, this method is thus rather precise (0.02 emissivity uncertainty). The underlying hypothesis is, however, hard to satisfy when the sensed surface is totally unknown, even when it is composed of natural materials. Inaccuracies finally tend to be rather high (±3 K). The temperature–emissivity separation (TES) method was therefore proposed as an improvement of the normalized emissivity method for a broad application.

6.4.3.2 Temperature–Emissivity Separation (TES) Method

The TES method (Gillespie et al. 1998) was developed for land-surface temperature evaluation by the Advanced Space-borne Thermal Emission and Reflection Radiometer (ASTER) on board TERRA satellite, which includes a five-channel multispectral thermal-IR scanner.

TES is based on the observation that the relative spectrum defined by $\beta(\lambda) = \hat{\varepsilon}(\lambda)/\bar{\hat{\varepsilon}}$, where $\hat{\varepsilon}(\lambda)$ is obtained from Equation 6.52 with an estimation of $\hat{T}$, is rather insensitive to the temperature estimation error. The problem is then how to extract the *absolute* spectrum $\varepsilon(\lambda)$ from the *relative* spectrum $\beta(\lambda)$?

Gillespie et al. found out a correlation between ε_{min} and the minimum–maximum emissivity difference defined by $MMD = \beta_{max} - \beta_{min}$ (Gillespie et al. 1998):

$$\varepsilon_{min} \approx 0.994 - 0.687 MMD^{0.737} \tag{6.54}$$

The regression was based on 86 laboratory reflectance spectra from the ASTER spectral library (Salisbury and d'Aria 1992) for rocks, soils, vegetation, snow, and water between 10 and 14 μm. Ninety-five percent of the samples fall within 0.02 emissivity units of the regression line (see Figure 6.27). Nevertheless, this empirical relation is not "universal": data related to "artificial" materials like metals fall far below the regression line (see Figure 6.27).

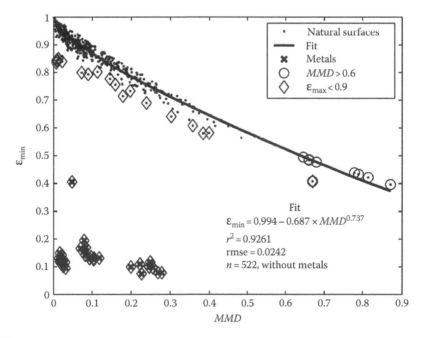

FIGURE 6.27
Correlation between ε_{min} and *MMD*. The line corresponds to the original regression law of Gillespie et al. (1998). Complementary data with low maximum emissivity (diamonds), surfaces with high contrast (circles) and metals (crosses). (From Payan, V. and Royer, A., *Int. J. Remote Sens.*, 25(1), 15, 2004. With permission.)

After evaluating ε_{min} from the regression law, one retrieves a new estimate of the emissivity spectrum through

$$\hat{\varepsilon}(\lambda) = \beta(\lambda)\frac{\varepsilon_{min}}{\beta_{min}} \tag{6.55}$$

A new temperature estimation is finally obtained by extracting the maximum emissivity value from the $\hat{\varepsilon}(\lambda)$ spectrum and inverting Planck's law at the corresponding wavelength:

$$\hat{T} = B^{-1}\left[\frac{L(\lambda_m, T) - L^{\downarrow}(\lambda_m)}{\hat{\varepsilon}_{max}} + L^{\downarrow}(\lambda_m)\right], \quad \lambda_m = \arg\max(\hat{\varepsilon}(\lambda)) \tag{6.56}$$

One or two iterations are sufficient for the convergence of the procedure.

To be effective, TES requires at least three or four bands of data; numerical simulations showed that uncertainties become larger as the number of bands is reduced further (for two bands, the products are only half as precise) (Gillespie et al. 1998).

TES would introduce a bias to emissivity in the case of near-gray materials (as a matter of fact, for all gray materials ε_{min} and thus $\hat{\varepsilon}(\lambda)$ would stick to 0.994). Instead of scaling the $\beta(\lambda)$ spectrum with ε_{min} for gray materials (those for which $MMD < 0.006$), it was decided to assign the emissivity of water to them.

The data dispersion in Figure 6.17 cannot be discarded. It can induce an error of 0.02 on ε_{min}. It is thus possible to have landscapes or terrains in which the TES regression will lead to systematic errors of that magnitude.

TES algorithm is presently used to calculate surface temperature and emissivity standard products for ASTER, which are predicted to be within +1.5 K and +0.015 of correct values, respectively. Validations performed on different sites demonstrated that TES generally performs within these limits (Sabol et al. 2009).

6.4.3.3 Spectral Smoothness Method (SpSm)

This relatively new method is specifically aimed at hyperspectral instruments (typically more than 100 channels are needed in the midwave or longwave IR bands) (Borel 1998, 2008, Knuteson et al. 2004, Kanani et al. 2004, Achard et al. 2007, Cheng et al. 2008). Its success is based on the statement that the emissivity spectra of natural surfaces are much *smoother* than the spectra of the atmospheric contributions (gas absorption bands print characteristic features in $L^{\downarrow}(\lambda)$, $L^{\uparrow}(\lambda)$, and $\tau(\lambda)$ spectra). If the estimated temperature $\hat{T}$ in Equation 6.52 is not equal to the real surface temperature, the retrieved emissivity spectrum $\hat{\varepsilon}(\lambda)$ will retain a part of the very detailed spectral features present in $L(\lambda, T)$ and in $L^{\downarrow}(\lambda)$, provided of course that the sensor has a high enough spectral resolution. When $\hat{T}$ comes closer to the real temperature, the apparent emissivity spectrum becomes progressively smoother. The solution is found when the computed spectrum is the smoothest, devoid of narrow atmospheric emission spectral lines. The ill-posed ETS problem is thus solved thanks to the fact that the detailed spectral features of the downwelling atmospheric radiance are "printed" in the leaving radiance.

Figure 6.28 illustrates this approach for the case of a concrete surface at $T = 313$ K and a typical atmospheric radiance (concrete spectrum was taken from ASTER spectral library—Baldridge et al. [2009]; see Figure 6.5, and the atmospheric radiance and transmittance were computed with MODTRAN code for the same conditions as for Figure 6.26). The estimated surface temperature was scanned from $T - 5$ to $T + 5$ K by 1 K steps. The smoothest apparent emissivity spectrum is indeed the one corresponding to a 313 K surface temperature.

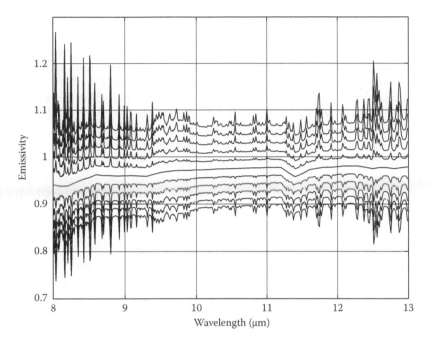

FIGURE 6.28

Illustration of the SpSm method. Case of a concrete surface at $T = 313$ K observed with an IR sensor at 1900 m altitude. Retrieved emissivity spectra for different estimated temperature values between $T - 5$ and $T + 5$ K with 1 K steps (from top to bottom). The smoothest profile corresponds to the true temperature value.

Suggested methods for optimizing the smoothness of the emissivity profile are minimizing the standard deviation of the difference between the computed emissivity $\hat{\varepsilon}(\lambda)$ and the corresponding three-point boxcar averaged emissivity profile (Borel 1998), minimizing the mean square derivative of $\hat{\varepsilon}(\lambda)$ (Knuteson et al. 2004), the integral of the absolute derivative (Kanani et al. 2004), or the correlation product between $L^{\downarrow}(\lambda)$ and $\hat{\varepsilon}(\lambda)$ (Cheng et al. 2008).

Borel (2008) noticed that there are potentially many atmospheres giving a smooth but physically incorrect emissivity. To be efficient in retrieving the true emissivity curve, the SpSm method requires the atmospheric compensation to be very precise. When applying the ISAC method, the initial hypotheses may bias the transmission spectrum to a detrimental level. A refinement for finding the best candidate atmosphere was presented by Borel (2008), and it consists in

1. Performing a large number of radiation transfer calculations assuming a well-populated distribution of atmospheric conditions in order to build a look-up table (LUT) of transmission profiles $\tau_{LUT}(\lambda)$

2. Calculating the cosine of the spectral angle between $\hat{\tau}(\lambda)$ and all $\tau_{LUT}(\lambda)$ from LUT and selecting some candidate atmospheres presenting a small spectral angle

3. Minimizing the smoothness criteria with the candidate atmospheres at a limited number of pixels and keeping the one leading to the best general minimization

4. Minimizing the smoothness criteria with the best candidate atmosphere on the whole image

Another requirement of the method is that the sensor should have a high spectral resolution in order to capture sufficient details of the atmospheric spectral features: it needs to have a resolution of 10 cm^{-1} or better to distinguish atmospheric spectral features from emissivity features. A direct consequence is that spectral calibration errors have a high impact on the output. Spectral shift and change of the width of the channels of as little as 1/20th of a wave center spacing, for the 128-channel SEBASS sensor in the 7.5–13.5 µm range, produce radiance artifacts that lead to incorrect choice of atmosphere, causing an offset in surface temperature of 1 K and emissivity of 0.04 to occur.

Assuming a measured radiance error of 0.5%, and combining neural network for evaluation of the atmospheric contribution and the SpSm method, leads to about 1.6 K RMS and 0.8 K bias for temperature and 0.023 RMS and 0.027 bias for emissivity in thermal IR (Achard et al. 2007). The emissivity errors in the 3–4.2 µm range are just slightly higher.

6.4.3.4 Multi-Temperature Method

We have seen that with one temperature measurement, the ETS problem is underdetermined because there are $N + 1$ unknowns and only N equations. In principle, by performing a measurement at another temperature level, the problem should be solved as one would then have $N + 2$ unknowns (one additional temperature unknown) and $2N$ equations. Of course, the surface emissivity and the environment contributions should not have changed in between. Actually two measurements at two wavelengths should suffice to evaluate the two temperature levels and then the emissivity. Things are, however, not as evident due to persistent correlations as it will be shown later.

The multi-temperature (or multitemporal) method is not very common in remote sensing, first because it seems to magnify greatly the measurement noise (unresolved ill posedness) and also because it requires a high-quality registration between the successive images.

In order to highlight the difficulties inherent to the multi-temperature method let us first consider the case of pyrometry where the environment radiation reflection on the sensed surface can be neglected. Let us also assume that the measured radiance is errorless. For this illustration, we assume that the spectral measurements are performed at two different temperatures.

The emissivity profile $\hat{\varepsilon}(\lambda, \hat{T}_1)$ is obtained from the first set of N measurements and from a temperature estimation $\hat{T}_1$:

$$\hat{\varepsilon}(\lambda, \hat{T}_1) = \frac{L(\lambda, T_1)}{B(\lambda, \hat{T}_1)} = \varepsilon(\lambda) \frac{B(\lambda, T_1)}{B(\lambda, \hat{T}_1)} \tag{6.57}$$

We obtain similarly the emissivity profile $\hat{\varepsilon}(\lambda, \hat{T}_2)$ from the second set of N measurements and an estimation $\hat{T}_2$ for the second temperature. The problem thus reduces to find $\hat{T}_1$ and $\hat{T}_2$ such that $\hat{\varepsilon}(\lambda, \hat{T}_1) = \hat{\varepsilon}(\lambda, \hat{T}_2)$ at each wavelength. It follows that

$$\frac{B(\lambda, T_1)}{B(\lambda, \hat{T}_1)} = \frac{B(\lambda, T_2)}{B(\lambda, \hat{T}_2)} \quad \forall \lambda \tag{6.58}$$

If we adopt Wien's approximation, this reduces to

$$\frac{\exp(C_2/\lambda \hat{T}_1)}{\exp(C_2/\lambda T_1)} = \frac{\exp(C_2/\lambda \hat{T}_2)}{\exp(C_2/\lambda T_2)} \quad \forall \lambda \tag{6.59}$$

It is easy to see that the problem is then degenerate and there is an infinity of solutions, defined by

$$\frac{1}{\hat{T}_2} - \frac{1}{\hat{T}_1} = \frac{1}{T_2} - \frac{1}{T_1} \tag{6.60}$$

The degeneracy can also be detected by analyzing the sensitivity matrix relatively to the N emissivity values and to the two temperatures T_1 and T_2. This matrix is used when solving the linear least-squares problem where the chosen observable is the logarithm of the emitted radiance:

$$\mathbf{X} = \begin{bmatrix} & \dfrac{-C_2}{\lambda_1 T_{\text{ref}}} & 0 \\ \mathbf{I}_N & \cdots & \cdots \\ & \dfrac{-C_2}{\lambda_N T_{\text{ref}}} & 0 \\ & 0 & \dfrac{-C_2}{\lambda_1 T_{\text{ref}}} \\ \mathbf{I}_N & \cdots & \cdots \\ & 0 & \dfrac{-C_2}{\lambda_N T_{\text{ref}}} \end{bmatrix}_{2N, N+2} \tag{6.61}$$

The determinant of $\mathbf{X}^T \mathbf{X}$ is 0 whatever the number of wavelengths. The sensitivities are thus correlated.

As a conclusion from this preliminary analysis, performing additional measurements at different temperature levels does not help for the ETS in pyrometry, at least in the spectral region where Wien's approximation is valid (the approximation is better than 1% when $\lambda T < 3124$ μm K). No new information is brought with these additional measurements.

When considering Planck's law instead of Wien's approximation, the identification problem becomes nonlinear, but it is reasonable to suspect that the ill posedness will be severe.

Let us now consider the nonlinear least-squares approach for identifying the N emissivities and the two temperatures. We here introduce a reflected flux from the environment:

$$[\varepsilon_i, T_1, T_2]^T = \arg\min_{\varepsilon_i, T_1, T_2} \sum_{i=1}^{N} \left(L(\lambda_i, T_1) - \left(\varepsilon_i B(\lambda_i, T_1) + (1 - \varepsilon_i) L^{\downarrow}(\lambda_i) \right) \right)^2$$
$$+ \left(L(\lambda_i, T_2) - \left(\varepsilon_i B(\lambda_i, T_2) + (1 - \varepsilon_i) L^{\downarrow}(\lambda_i) \right) \right)^2 \tag{6.62}$$

For the illustration, we will consider three cases:

1. A near-blackbody in a cold environment, i.e., without reflections from the environment (Figures 6.29 and 6.30).
2. A gray body ($\varepsilon_i = 0.9$) with reflections from a blackbody environment at 300 K (Figures 6.31 and 6.32).
3. Same as before but the radiation from the environment is weighted by a random spectral function with uniform probability distribution. The purpose is to simulate a spectrally rich environment radiance (Figures 6.33 and 6.34).

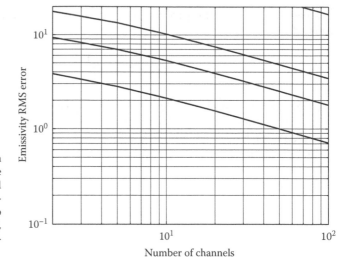

FIGURE 6.29
Multi-temperature pyrometry. Error on emissivity for a 1% error on radiance depending on the number of spectral channels between 8 and 14 μm. The temperature difference between the two experiments is from top to bottom: 1, 5, 10, 30 K. Case of a near-blackbody surface in a cold environment.

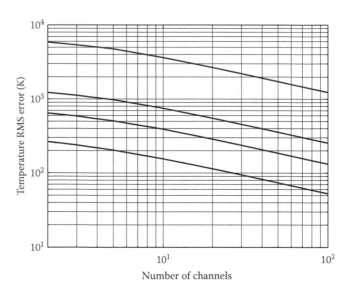

FIGURE 6.30
Same as in Figure 6.29 for temperature.

For all three cases we assume the surface to be at $T_1 = 320$ K for the first measurement, the second measurement is performed at a temperature of 1, 5, 10, or 30 K higher. IR detection is performed on a variable number of channels in the 8–14 μm bandwidth.

The objective is merely to present the expected standard errors on emissivity and temperature as obtained from the covariance matrix expressed at the solution. The sensitivity matrix is obtained by differentiating $(\varepsilon_i B(\lambda_i, T_1) + (1 - \varepsilon_i)L^{\downarrow}(\lambda_i))$ and $(\varepsilon_i B(\lambda_i, T_2) + (1 - \varepsilon_i)L^{\downarrow}(\lambda_i))$ according to each emissivity ε_i, and according to T_1 and T_2. We plotted in Figures 6.29 through 6.34 the standard errors for emissivity and temperature assuming a 1% measurement error).

A general comment is that the errors diminish when increasing the number of channels. Above some 10 channels, the decrease is roughly $N^{-1/2}$. An improvement is also obtained by increasing the temperature difference between the two experiments.

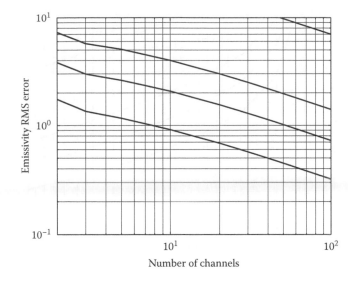

FIGURE 6.31
Same as in Figure 6.29 in the case of a gray body with $\varepsilon = 0.9$ and a blackbody environment radiation at $T_{env} = 300$ K.

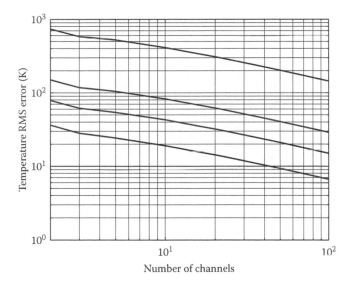

FIGURE 6.32
Same as in Figure 6.31 for temperature.

In the case of a near-blackbody surface without significant reflections, the standard errors are huge, although the assumed radiance noise is as little as 1%. This merely reflects that we are close to the ill-posed case expressed earlier with Wien's approximation. The inversion performance can be improved only when a reflected flux is present. This fact was already noticed by Kanani et al. (2004). It is also shown here that a definite advantage is obtained thanks to the high-frequency spectral features of the downwelling atmospheric radiation. When assuming a smooth environment radiation (Figures 6.31 and 6.32), the benefit is indeed not as important as when radiation contains detailed spectral features (Figures 6.33 and 6.34). As an example, by performing a measurement in 100 channels with a 1% radiance error and a 5 K temperature difference between the two experiments, the temperature standard error drops from 30 to 3 K when the environment radiation changes from smooth to spectrally rich. At the same time, the emissivity error drops from values higher than 1 to about 0.07. However, these errors are only estimates: they were extracted

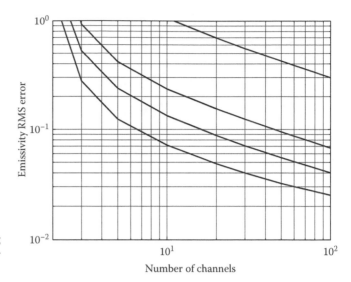

FIGURE 6.33
Same as in Figure 6.31 after introducing detailed spectral features to the down-welling environment radiation.

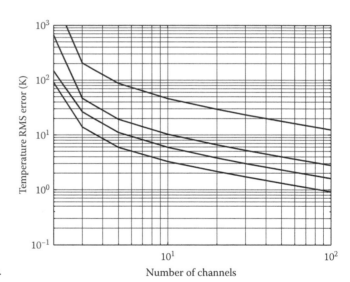

FIGURE 6.34
Same as in Figure 6.33 for temperature.

from the covariance matrix which is based on a local linearization of the nonlinear problem. For better estimates, one should simulate a large number of experiments through a Monte Carlo procedure.

6.5 Conclusion

Noncontact temperature measurement by radiative methods presents distinct advantages over contact methods but it always exposes the underdetermined problem of ETS. We describe some methods to solve this problem in the field of high-temperature MWP and in the field of airborne and satellite remote sensing. Different proposals were made in the past for ETS with various successes, sometimes by sticking to the myth that N equations should suffice to retrieve N unknowns! The work presented here aims to show that with a sensitivity analysis and by implementing simple linear and nonlinear tools, it is often possible to rapidly verify the advertised performances of the methods and avoid some traps.

Regarding MWP, one must finally admit that without knowledge about the emissivity *magnitude*, the temperature measurement cannot be very precise. Some vague intuition about the shape of the emissivity spectrum is not sufficient and adding more wavelengths does not help much. The blackbody spectrum is too regular, and therefore, introducing an emissivity polynomial model of a higher degree than 1 introduces high correlations and generally leads to poor results.

ETS in remote sensing is an active field of research as it is mingled with other problems like atmospheric compensation. However, the detailed spectral features of the downwelling atmospheric radiation offer an invaluable opportunity for succeeding in ETS as shown by the SpSm approach and the multi-temperature approach.

In remote sensing, when developing new methods, one always faces the difficulty of validating them on large, heterogeneous, and rough surfaces. The disaggregation problem which consists in evaluating the individual temperatures of a mixture pixel is also part of the inversion problems related to remote temperature measurement.

Nomenclature

B	blackbody spectral radiance ($W/m^3/sr$)
e_i	noise in spectral channel i
E	spectral irradiance (W/m^3)
L	spectral radiance ($W/m^3/sr$)
$L^{\uparrow}$	spectral radiance emitted by the atmosphere along the path from surface to sensor ($W/m^3/sr$)
$L^{\downarrow}$	spectral radiance emitted by the environment down to the surface ($W/m^3/sr$)
N	number of spectral channels
N_b	number of gray bands
P	scalar parameter
T	temperature (K)
Y	experimental observed data

Greek Variables

α	absorptance
ε	emissivity
θ	zenithal angle (rad)
λ	wavelength (m)
ρ'', $\rho'^{\cap}$	bidirectional reflectance and directional-hemispherical reflectance
σ	standard deviation
τ	transmission coefficient
φ	azimuthal angle (rad)

Superscripts

$\wedge$ estimated value
*   reduced and centered value (see Equation 6.31)
$'$ reduced value
T transpose of a matrix or a vector

Subscripts

s at-sensor (radiance)
W Wien's approximation of Planck's law

Abbreviations

ETS   emissivity–temperature separation
IR   infrared
LSMWP   least-squares multiwavelength pyrometry
MWP   multiwavelength pyrometry
SCP   single-color pyrometry
SpSm   spectral smoothness method
TCP   two-color pyrometry

References

Achard, V., Lesage, S., Poutier, L. 2007, Retrieval of atmospheric temperature and water vapor content from thermal infrared hyperspectral data in a purpose of atmospheric compensation, *Proc. SPIE* 6745:67451F.

Atitar, M., Sobrino, J.A. 2009, A split-window algorithm for estimating LST from Meteosat 9 Data: Test and comparison with *in situ* data and MODIS LSTs, *IEEE Geosci. Remote Sens. Lett.* 6(1): 122–126.

Baldridge, A.M., Hook, S.J., Grove, C.I., Rivera, G. 2009, The ASTER Spectral Library Version 2.0, *Remote Sens. Environ.* 113:711–715, http://speclib.jpl.nasa.gov/

Barducci, A., Pippi, I. 1996, Temperature and emissivity retrieval from remotely sensed images using the "grey body emissivity" method, *IEEE Trans. Geosci. Remote Sens.* 34(3):681–695.

Barton, I.J. 1983, Dual channel satellite measurements of sea surface temperature, *Quart. J. R. Meteor. Soc.* 109:365–378.

Berk, A., Bernstein, L.S., Robertson, D.C. 1989, *MODTRAN: A Moderate Resolution Model for LOW-TRAN7*, GL-TR-89-0122, Air Force Geophysic Lab., Hanscom AFB, MA.

Borel, C.C. 1998, Surface emissivity and temperature retrieval for a hyperspectral sensor, *Proceedings of the IEEE Conference on Geoscience and Remote Sensing*, Seattle, IGARSS'98, pp. 504–509.

Borel, C.C. 2008, Error analysis for a temperature and emissivity retrieval algorithm for hyperspectral imaging data, *Int. J. Remote Sens.* 29(17):5029–5045.

Cassady, L.D., Choueiri, E.Y. 2003, High accuracy multi-color pyrometry for high temperature surfaces, *IEPC-03-79 28th International Electric Propulsion Conference*, Toulouse, France, March 17–21, 2003.

Cheng, J., Liu, Q., Li, X., Xiao Q., Liu, Q., Du, Y. 2008, Correlation based temperature and emissivity separation algorithm, *Sci. China Ser. D: Earth Sci.* 51(3):357–369.

Chevallier, F., Morcrette, J.-J., Chédin, A., Cheruy, F. 2006, TIGR-like atmospheric-profile databases for accurate radiative-flux computation, *Quart. J. R. Meteor. Soc.* 126(563):777–785.

Chrzanowski, K., Szulim, M. 1998a, Measure of the influence of detector noise on temperature measurement accuracy for multiband infrared systems, *Appl. Opt.* 37(22):5051–5057.

Chrzanowski, K., Szulim, M. 1998b, Error on temperature measurement with multiband infrared systems, *Appl. Opt.* 38(10):1998–2006.

Chrzanowski, K., Szulim, M. 1999, Comparison of temperature resolution of single-band, dual-band and multiband infrared systems, *Appl. Opt.* 38(13):2820–2823.

Coates, P.B. 1981, Multiwavelength pyrometry, *Metrologia* 17:103–109.

Coll, C., Caselles, V., Valor, E., Rubio, E. 2003, Validation of temperature–emissivity separation and split-window methods from TIMS data and ground measurements, *Remote Sens. Environ.* 85:232–242.

Corwin, R.R., Rodenburgh, A. 1994, Temperature error in radiation thermometry caused by emissivity and reflectance measurement error, *Appl. Opt.* 33(10):1950–1957.

DeWitt, D.P., Rondeau, R.E. 1989, Measurement of surface temperatures and spectral emissivities during laser irradiation, *J. Thermophys.* 3(2):153–159.

Duvaut, T. 2008, Comparison between multiwavelength infrared and visible pyrometry: Application to metals, *Infrared Phys. Technol.* 51:292–299.

Duvaut, T., Georgeault, D., Beaudoin, J.L. 1995, Multiwavelength infrared pyrometry: Optimization and computer simulations, *Infrared Phys. Technol.* 36:1089–1103.

Duvaut, T., Georgeault, D., Beaudoin, J.L. 1996, Pyromètre multispectral infrarouge: Application aux métaux, *Rev. Gen. Therm.* 35:185–196.

Gardner, J.L. 1980, Computer modelling of a multiwavelength pyrometer for measuring true surface temperature, *High Temp. High Press.* 12:699–705.

Gardner, J.L., Jones, T.P., Davies, M.R. 1981, A six-wavelength radiation pyrometer, *High Temp. High Press.* 13:459–466.

Gathers, G.R. 1991, Analysis of multiwavelength pyrometry using nonlinear least square fits and Monte Carlo methods, *11th Symposium on Thermophysical Properties*, Boulder, CO, June 1991.

Gillespie, A., Rokugawa, S., Matsunaga, T., Cothern, J.S., Hook, S., Kahle, A.B. 1998, A temperature and emissivity separation algorithm for advanced spaceborne thermal emission and reflection radiometer (ASTER) images, *IEEE Trans. Geosci. Remote Sens.* 36(4):1113–1126.

Hervé, P., Sadou, A. 2008, Determination of the complex index of refractory metals at high temperatures: Application to the determination of thermo-optical properties, *Infrared Phys. Technol.* 51:249–255.

Hiernault, J.-P., Beukers, R., Heinz, W., Selfslag, R., Hoch, M., Ohse, R.W. 1986, Submillisecond six-wavelength pyrometer for high temperature measurements in the range 2000 K–5000 K, *High Temp. High Press.* 18:617–625.

Hunter, B., Allemand, C.D., Eager, T.W. 1985, Multiwavelength pyrometry: An improved method, *Opt. Eng.* 24(6):1081–1085.

Hunter, B., Allemand, C.D., Eager, T.W. 1986, Prototype device for multiwavelength pyrometry, *Opt. Eng.* 25(11):1222–1231.

Kanani, K., Poutier, L., Stoll, M.P., Nerry, F. 2004, Numerical assessment of spectral emissivity measurement methods using a field spectroradiometer in the 3–13 μm domain, *Proc. SPIE* 5232:479–488.

Khan, M.A., Allemand, C., Eager, T.W. 1991a, Noncontact temperature measurement. I. Interpolation based techniques, *Rev. Sci. Instrum.* 62(2):392–402.

Khan, M.A., Allemand, C., Eager, T.W. 1991b, Noncontact temperature measurement. II. Least square based techniques, *Rev. Sci. Instrum.* 62(2):403–409.

Knuteson, R.O., Best, F.A., DeSlover, D.H., Osborne, B.J., Revercomb, H.E., Smith, W.L. 2004, Infrared land surface remote sensing using high spectral resolution aircraft observations, *Adv. Space Res.* 33:1114–1119.

Krapez, J.-C., Bélanger, C., Cielo, P. 1990, A double-wedge reflector for emissivity enhanced pyrometry, *Meas. Sci. Technol.* 1:857–864.

Lindermeir, E., Tank, V., Hashberger, P. 1992, Contactless measurement of the spectral emissivity and temperature of surfaces with a Fourier transform infrared spectrometer, *Proc. SPIE* 1682:354–364.

Loarer, T., Greffet, J.-J. 1992, Application of the pulsed photothermal effect to fast surface temperature measurements, *Appl. Opt.* 31(25):5350–5358.

Loarer, T., Greffet, J.-J., Huetz-Aubert, M. 1990, Noncontact surface temperature measurement by means of a modulated photothermal effect, *Appl. Opt.* 29(7):979–987.

Mazikowski, A., Chrzanowski, K. 2003, Non-contact multiband method for emissivity measurement, *Infrared Phys. Technol.* 44:91–99.

Nordine, P.C. 1986, The accuracy of multicolour optical pyrometry, *High Temp. Sci.* 21:97–109.

Payan, V., Royer, A. 2004, Analysis of temperature emissivity separation (TES) algorithm applicability and sensitivity, *Int. J. Remote Sens.* 25(1):15–37.

Pierre, T., Rémy, B., Degiovanni, A. 2008, Microscale temperature measurement by the multispectral and statistic method in the ultraviolet-visible wavelengths, *J. Appl. Phys.* 103, 034904-1-10.

Sabol, D.E., Gillespie, A.R., Abbott, E., Yamada, G. 2009, Field validation of the ASTER temperature–emissivity separation algorithm, *Remote Sens. Environ.* 113:2328–2344.

Sade, S., Katzir, A. 2004, Multiband fiber optic radiometry for measuring the temperature and emissivity of gray bodies of low or high emissivity, *Appl. Opt.* 43(9):1799–1810.

Salisbury, J.W., d'Aria, D.M. 1992, Emissivity of terrestrial materials in the 8–14 μm atmospheric window, *Remote Sens. Environ.* 42:83–106

Scharf, V., Naftali, N., Eyal, O., Lipson, S.G., Katzir, A. 2001, Theoretical evaluation of a four-band fiber-optic radiometer, *Appl. Opt.* 40(1):104–111.

Siegel, R., Howell, J.R. 1972, *Thermal Radiation Heat Transfer*, McGraw Hill Ed., New York.

Simoneau, P., Caillault, K., Fauqueux, S., Huet, T., Labarre, L., Malherbe, C., Rosier, B. 2009, MATISSE-v1.5 and MATISSE-v2.0: New developments and comparison with MIRAMER measurements, *Proc. SPIE* 7300:73000L.

Sobrino, J.A., Romaguera, M. 2008, Water-vapor retrieval from Meteosat 8/SEVIRI observations, *Int. J. Remote Sens.* 29(3):741–754.

Tank, V., Dietl, H. 1990, Multispectral infrared pyrometer for temperature measurement with automatic correction of the influence of emissivity, *Infrared Phys.* 30(4):331–342.

Touloukian, Y.S., DeWitt, D.P. 1970, *Thermal Radiative Properties. Thermophysical Properties of Matter*, Plenum Corp., New York.

Uman, I., Katzir, A. 2006, Fiber-optic multiband radiometer for online measurements of near room temperature and emissivity, *Opt. Lett.* 31(3):326–328.

Wen, C.D., Mudawar, I. 2004a, Emissivity characteristics of roughened aluminium alloy surfaces and assessment of multispectral radiation thermometry (MRT) emissivity models, *Int. J. Heat Mass Transfer* 47:3591–3605.

Wen, C.D., Mudawar, I. 2004b, Emissivity characteristics of polished aluminium alloy surfaces and assessment of multispectral radiation thermometry (MRT) emissivity models, *Int. J. Heat Mass Transfer* 48:1316–1329.

Young, S.J., Johnson, B.R., Hackwell, J.A. 2002, An in-scene method for atmospheric compensation of thermal hyperspectral data, *J. Geophys. Res.* 107:D24–4774.

Part II

Inverse Heat Transfer Problems

7

Introduction to Linear Least Squares Estimation and Ill-Posed Problems for Experimental Data Processing

Olivier Fudym and Jean-Christophe Batsale

CONTENTS

7.1 Introduction

The experimental data collected from some measurement devices are often used according to a representative model in order to determine indirectly some properties of the system under study, characterize the environment of a sensor, change the output value of the sensor into another magnitude of interest, or even determine the calibration data of some instrument. Moreover, in many situations, the quantities that we wish to determine are different from the ones that we are able to measure directly, as could be seen in the first part of this book.

If our purpose is to determine the coefficients of some model, evaluate some physical properties, such as thermal conductivity or viscosity, or make the indirect measurement of some quantity (for instance, deducing some local heat flux density or convective heat transfer coefficient from the temperature measured by some sensor), it will be at the end necessary to make the comparison between some record of data points and a direct model conveniently chosen. The structure and consistency of the model to be used has been already discussed in depth in Chapter 1.

The errors attached to these measurement operations make unavailable the true value of the quantities in consideration. How to characterize and evaluate which effect will have these measurement errors on the magnitudes of interest will be a key point for the experimentalist approach for data processing. The definition and type of measurement errors are specified in the document, Definition and implementation International Vocabulary of Metrology (VIM), and the method for its evaluation in the so-called Guide to the Expression of Uncertainty (JCGM 200:2008) in Measurement (GUM) (JCGM 100:2008). As suggested in the VIM, the measurement errors may be split into two different types of errors:

1. The systematic measurement error is a "component of measurement error that in replicate measurements remains constant or varies in a predictable manner." The reference for this type of errors is a true quantity value, such as a standard of negligible measurement uncertainty.

2. The random measurement error is the "component of measurement error that in replicate measurements varies in an unpredictable manner." Random measurement errors of a set of replicate measurements form a distribution that can be summarized by its expectation, which is generally assumed to be zero, and its variance.

According to the VIM, the random measurement error equals the measurement error minus the systematic measurement error. The measurement bias is the estimate of a systematic measurement error. The systematic errors are often present in parametric estimation based on experimental measurements, due either to some constant error in the measuring process or to some error in the direct model, which may not be a perfect representation of the experiment—see Chapter 1 for instance. However, this kind of errors is considered to be out of the scope of this chapter, since for the sake of clarity, the models will be assumed to be errorless, in order to focus mainly on the effect of the measurement errors on the parameter estimation process. Moreover, systematic errors often simply appear as some additional parameters to be estimated. Hence, the measurement bias is not considered herein, but only the random measurement errors with expectation assumed to be zero. Thus, the measurement errors are considered as perturbations conveniently described by some random variables, which probability density function may be unknown (Lira 2002, Willink 2008).

The main goal of this chapter is to propose some methods to determine and quantify which would be the "best" way of estimating the parameters by processing the collected data. Hence, not only the resulting value of the retrieved parameters is of interest, but also the quantification of how the measurement errors propagate through the estimation procedure and may disturb the convenient interpretation of the experimental data. Moreover, in some cases, the estimation problem may be very sensitive to the errors and turn to be "ill-posed" (this term will be conveniently defined in Section 7.5). Inverse methods are devoted to solve such ill-posed problems. However, the techniques used to solve ill-posed problems are not within the scope of the present chapter, and will be discussed in the subsequent chapters of Part II.

As shown in Figure 7.1, the estimation problem yields the requirement for minimizing some "distance" (to be defined hereafter) between the collected data and the direct model. The model depends on some input variables, parameters, and independent variables. Generally, the structure of the model and the relationship between the input variables, parameters, and independent variables (such as time, or space variables) may be complex. Moreover, both the data and model outputs may be functions of continuous or discrete variables. However, since only a finite number of data values is generally recorded, and the occurrence of the outputs is computed in correspondence with these values, most of the problems can fall in finite-dimensional problems. Thus, only discrete variables are considered herein.

The topic of the present chapter is relative to linear estimation within a discrete framework. After a "Getting Started" section devoted to point out the main concepts and challenges of parameter estimation, the general ordinary least squares (OLS) method is derived and discussed. In Section 7.4, the more general case where some basic statistical assumptions regarding the errors are not satisfied (constant variance and uncorrelated), and various estimators are proposed. Section 7.5 is an introduction to ill-posed problems, envisioned through basic examples such as those currently faced when processing some experimental data. The anatomy of such linear transforms is analyzed through the singular value decomposition (SVD) tools and the corresponding spectral approach.

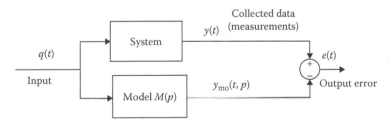

FIGURE 7.1
Fitting a model with collected data: the output error method.

Finally, Section 7.6 introduces the predictive model error which is obtained by building directly the model with the collected data and then turn a nonlinear estimation problem into a linear estimation problem.

7.2 Getting Started

In this section, the main characteristics of a parameter estimation problem are envisioned through some very simple examples. These basic cases are analyzed in order to make clear the main concepts and challenges of the OLS parameter estimation, and define the frame of an estimation model which is linear with respect to the parameters.

7.2.1 One Parameter, One Measurement

The most basic case that we can study is to estimate one parameter with a single measurement! For instance, let us track a target moving with a constant velocity on a one-dimensional trajectory (such as a free-falling body once constant velocity is achieved). You should like to deduce its velocity from a single position measurement y_1 triggered at a chosen time t_1. Then the corresponding model, for any position y and time t, is

$$y_{\mathrm{mo}}(t) = vt \tag{7.1}$$

It is assumed here that the initial position is perfectly known and chosen as the origin for y. With a single measurement, if t_1 is conveniently chosen to be not zero, the velocity can be estimated by matching exactly the measurement with the model given in Equation 7.1, which yields

$$\hat{v} = \frac{y_1}{t_1} \tag{7.2}$$

The measured position y_1 is here a dependent variable whose value is found by observation. The time is an independent variable. As suggested in Section 7.1, let us assume that the measurement of the position made by the camera is perturbed by a random measurement error (due to thermal noise, defect of focalization, or spatial resolution of the camera). Assuming that the error is additive, we may write:

$$y_1 = y_1^* + \varepsilon_1 \tag{7.3}$$

where the superscript "*" yields for the true (but unknown) value.

We assume that the expected value of the error is zero (the measurement of the position y is unbiased) and has a standard deviation (std) σ_y. We also assume that the independent variable t_1 is known without error. The meaning of Equation 7.3 is that the measurement y_1 is an outcome of a random variable Y_1. Equation 7.3 can now be substituted into Equation 7.2:

$$\hat{v} = \frac{y_1^*}{t_1} + \frac{\varepsilon_1}{t_1} = v^* + \frac{\varepsilon_1}{t_1} \tag{7.4}$$

The first very important result shown in Equation 7.4 is that the estimate of the velocity is also a random variable, since it depends directly on the random variable ε_1. Moreover, since the expected value of ε_1 is zero, the estimate of the velocity given by Equation 7.2 is unbiased, because its expected value, which can be deduced directly from Equation 7.4, is the true—but unknown—value v^*. Finally, the variance of the estimate can also be deduced directly from Equation 7.4 such as

$$\sigma_v^2 = \frac{\sigma_y^2}{t_1^2} \tag{7.5}$$

More generally, it is apparent in Equation 7.4 that the probability distribution of the estimate can be deduced from the probability distribution of the measurement error.

The last very significant result given by Equation 7.5 is that the variance of the estimate can grow drastically if the instant of measurement t_1 tends to zero, as well as it can be reduced when t_1 is increased.

Thus, a specific role of the independent variable t_1 appears through this discussion. In fact, the time variable has here another very important characteristic, which we may point out if we calculate X_v, the first derivative of the model with respect to velocity, computed at time t as

$$X_v(t) = \frac{\partial y_{\mathrm{mo}}(t)}{\partial v} = t \tag{7.6}$$

X_v is called the sensitivity coefficient of the model—that computes the position—with respect to the velocity—the parameter herein. In Equation 7.6, it is apparent that X_v is independent of the parameter: the model defined in Equation 7.1 makes the problem linear with respect to the parameter because the sensitivity coefficient is independent of the parameter.

A first-order expansion of the model is then written as

$$y_{\mathrm{mo}}(t, v + \Delta v) = y_{\mathrm{mo}}(t, v) + X_v(t)\Delta v + \vartheta(\Delta v^2) \tag{7.7}$$

Note that in the specific case considered herein where the model is given by Equation 7.1, $\vartheta(\Delta v^2) = 0$.

A low value of the sensitivity coefficient will not let discriminate low variations of the velocity. At the opposite, if the sensitivity coefficient has a high value, we may hope to discriminate very small variations of the velocity. Thus, the sensitivity coefficient plays an important role in the estimation as well as on the amplification of noise.

The std of the estimate as given in Equation 7.5 suggests that it is possible to get an estimation of the velocity as good as possible if the measurement is made at a sufficiently long time. Unfortunately, this is not easy to realize in practice, since the maximum time available for the experiment may not be long enough. Moreover, in some cases the measurement error may increase with the distance, the trajectory may turn to become nonlinear, etc. This is the reason why it is convenient to check how adding some measurement points to this problem may turn out the estimation to be much more efficient.

7.2.2 One Parameter, Two or More Data Points

Gathering m outcomes of the measurements y_i for the random variables Y_i corresponding to the independent measurements at times t_i (for $i = 1, 2, \ldots m$), and assuming that the zero value has not been chosen for any time, yields m independent equations such as

$$y_{\mathrm{mo}}(t_i) = vt_i, \quad i = 1, 2, \ldots, m \tag{7.8}$$

An exact matching solution would intend to solve simultaneously the m equations

$$y_i = vt_i, \quad i = 1, 2, \ldots, m \tag{7.9}$$

and a solution would be possible only if the outcomes y_i fit perfectly on a line, that is,

$$\hat{v} = \frac{y_1}{t_1} \stackrel{?}{=} \frac{y_2}{t_2} = \cdots \stackrel{?}{=} \frac{y_m}{t_m} \tag{7.10}$$

Equation 7.10 would mean that the model "hits" the data points. Due to the errors ε_i, this is generally not realistic, and Equation 7.9 yields an overdetermined linear system which has no solution, as shown in Figure 7.2, where the line computed with the model for the exact value of velocity v^* is drawn as a reference line. It is apparent how the two measurement points cannot fit together in the same line. Since there is no solution, we instead prefer to solve an alternative problem. The least squares approach consists in solving this problem by trying to make as small as possible the sum of squares of "differences" between the data points and the line generated by the model, that is, to find the minimum of the function,

$$S(v) = (y_1 - vt_1)^2 + (y_2 - vt_2)^2 + \cdots + (y_m - vt_m)^2 = e_1^2(v) + e_2^2(v) + \cdots + e_m^2(v) = \mathbf{e}^T(v)\mathbf{e}(v) \tag{7.11}$$

where the vector of differences is defined by $\mathbf{e} = [e_1 \, e_2 \cdots e_m]^T$.

The elements of $\mathbf{e}(v)$ are the differences between the model hits and the data points, and are not equal to the errors. However, $\mathbf{e}(v)$ is a random variable vector due to the presence of the errors in the y values. Hence, the objective function S also is a random variable. For the exact value of the velocity, such as $v = v^*$, then the distances $\mathbf{e}(v)$ are equal to the errors $\varepsilon = [\varepsilon_1 \, \varepsilon_2 \, \ldots \, \varepsilon_m]$, defined by $y_i = y_i^* + \varepsilon_i$, as shown in Figure 7.2 for two data points.

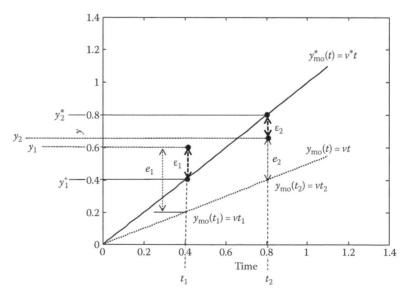

FIGURE 7.2
One parameter and two or more data points.

The minimization of S can be achieved by calculating the first derivative of S with respect to v, and then searching the estimate $\hat{v}$ as the value which cancels this derivative:

$$\frac{\partial S}{\partial v} = 2t_1(vt_1 - y_1) + 2t_2(vt_2 - y_2) + \cdots + 2t_m(vt_m - y_m) \quad \text{and} \quad \frac{\partial S}{\partial v}(\hat{v}) = 0 \qquad (7.12)$$

yields

$$\hat{v} = \frac{t_1 y_1 + t_2 y_2 + \cdots + t_m y_m}{t_1^2 + t_2^2 + \cdots + t_m^2} = \frac{\mathbf{x}_\mathbf{v}^T \mathbf{y}}{\| \mathbf{x}_\mathbf{v} \|^2} \qquad (7.13)$$

where the following vectorial notations are used:

$$\mathbf{x}_\mathbf{v} = [t_1 \quad t_2 \quad \cdots \quad t_m]^T; \quad \mathbf{y} = [y_1 \quad y_2 \quad \cdots \quad y_m]^T \quad \text{and} \quad \| \cdots \| \text{ is the } L_2 \text{ norm}$$

Since the data are collected with an error such as $y_i = y_i^* + \varepsilon_i$, the estimate is also a random variable. Let us also assume that the errors ε_i are independent with zero mean, have a constant variance (that is, $\sigma_Y^2 = \sigma_1^2 = \sigma_2^2 = \cdots = \sigma_m^2$) and are identically distributed with the same probability distribution function. According to Equation 7.13, we find a result similar to Equation 7.4 such as

$$\hat{v} = v^* + \frac{\mathbf{x}_\mathbf{v}^T \boldsymbol{\varepsilon}}{\| \mathbf{x}_\mathbf{v} \|^2} \quad \text{where } \boldsymbol{\varepsilon} = [\varepsilon_1 \quad \varepsilon_2 \quad \cdots \quad \varepsilon_m]^T \qquad (7.14)$$

This result is of great interest. The estimation of velocity was obtained by the minimization of the distance $\mathbf{e}$ between the model and the experimental data, but Equation 7.14 yields the estimation error, which is the distance between the solution (the estimated velocity) and its true (but unknown) value, involving the effect of the measurement error $\boldsymbol{\varepsilon}$ on the result.

Since the expected value of the errors is assumed to be zero, the estimator of the velocity is unbiased. Then, with these assumptions with respect to the errors, the variance of the estimate can also be deduced directly from Equation 7.14 such as

$$\sigma_v^2 = \frac{\sigma_Y^2}{\| \mathbf{x}_\mathbf{v} \|^2} \qquad (7.15)$$

Equation 7.14 means that the estimated variable is unbiased. Equation 7.15 gives a result about the "quality" of the estimation. It is apparent in Equation 7.15 that the variance of the estimated parameter will be lower if we design the experiment in order to increase the norm of the sensitivity vector. In that case, it is of interest to schedule the times where the measurements are recorded as long as possible, and also to increase the number of data points in consideration.

Some results of estimation of the velocity (in fact, the slope of a line) are shown in Figure 7.3. The trajectory y is computed, in order to give the "reference line." Then this vector is corrupted by an additive normal uncorrelated noise, in order to simulate some measurement points with errors corresponding to the previous assumptions. The simulated data y_i are used in order to estimate the parameter with Equation 7.13 for two different cases: estimation with all the data points and also estimation with only the first 18 data points.

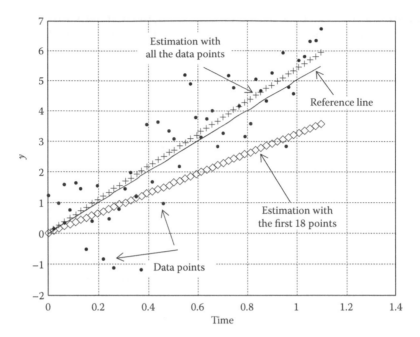

FIGURE 7.3
Estimation of the slope and retrieved lines with $\sigma_y = 1\,\text{s}$.

For this example, a very large noise level has been chosen, with $\sigma_y = 1\,\text{s}$. Then the retrieved lines generated with the corresponding estimated parameters are plotted in Figure 7.3.

The results of estimation obtained with Equations 7.13 and 7.15 are presented in Table 7.1. As expected, the estimation with the very first points yields a low norm of the sensitivity vector and is worse than the estimation made with the same number of data points but chosen at the end of the "experiment." Moreover, considering, on the same total range, a number of data points much higher (the last line of Table 7.1 simulates an experiment for which 1000 points would be collected) yields an almost perfect estimation of the velocity.

TABLE 7.1

Estimation Results for One Parameter and m Data Points ($m = 50$)

	Velocity, m/s	Standard Deviation, m/s
Reference value	5	—
Estimated with all the 50 points	5.38	0.24
Estimated with 18 points at the beginning	3.23	1.01
Estimated with 18 points at the end	5.38	0.27
Estimated when adding 150 points (on the same range)	4.97	0.11
Estimated with 1000 points (on the same range)	5.005	0.05

7.2.3 Statistical Assumptions Regarding the Errors

In the previous sections, there was no need to make any assumption regarding the random measurement errors to compute the estimated value of velocity with Equation 7.2 or Equation 7.13. However, if we are interested in evaluating the "quality" of the estimation, we must intend to predict which effect will have the measurement errors with its own probability distribution on the behavior of the estimates.

In Table 7.2, the main definitions related to the expectations, and used hereafter in the text, are written. For a discrete random variable, the expected value is the sum of the products of the possible values and their probability. The functional $E(\cdot)$ has the property of linearity. $V(X)$ is the variance, and the nonnegative root of $V(X)$ is called the standard deviation.

In Table 7.3 are listed the main statistical assumptions regarding the measurement errors which will be used hereafter in the text. The number of these assumptions will be used in our notation in order to validate some specific assumption. For instance "1237" means that only the assumptions no. 1-2-3 and 7 are considered to be satisfied. The term, "the standard assumptions" will be used for the case when all these nine assumptions are satisfied together, that is, the case "123456789."

Assuming that the standard assumptions are valid for the measurement errors corresponding to the cases of the previous examples given in the "Getting Started" section, we deduce that the estimators defined by Equations 7.4 and 7.14 are unbiased, since their expected value is the true value, and have a normal probability distribution. Moreover, the

TABLE 7.2

Some Definitions and Properties Related to the Expectations

Expected value	$E(X) = \int_{-\infty}^{\infty} x f_X(x) dx = \mu_X$ for continuous random variables
	$E(X) = \sum_x x P(X = x) = \mu_X$ for discrete random variables
Expected value of a function $g(X)$	$E(g(X)) = \sum_x g(x) P(X = x)$
$E(\cdot)$ is a linear functional	$E(aX + bY) = aE(X) + bE(Y)$
If X and Y are independent	$E(XY) = E(X)E(Y)$
Variance	$V(X) = E([X - E(X)]^2) = E(X^2) - [E(X)]^2 = \sigma_X^2$
The variance is not a linear functional	$V(aX) = a^2 V(X) = a^2 \sigma_X^2$
	$V(aX + bY) = a^2 V(X) + b^2 V(Y) + 2ab\, cov(X, Y)$
Covariance	$cov(X, Y) = E([X - \mu_X][Y - \mu_Y]) = E(X)E(Y) - \mu_X \mu_Y$
Correlation coefficient	$\rho_{X,Y} = \dfrac{cov(X, Y)}{\sigma_X \sigma_Y}, \quad \rho_{X,Y} = -1$ or $+1$ only if $Y = aX + b$
Expected value of a vector	$E(\mathbf{x}) = \begin{bmatrix} E(X_1) & E(X_2) & \dots & E(X_n) \end{bmatrix}^T$
Covariance matrix of a vector	$cov(\mathbf{y}) = E([\mathbf{y} - E(\mathbf{y})][\mathbf{y} - E(\mathbf{y})]^T)$
	$cov(\mathbf{y}) = \begin{bmatrix} var(y_1) & cov(y_1, y_2) & \cdots & cov(y_1, y_m) \\ & var(y_2) & \ddots & cov(y_2, y_m) \\ & & \ddots & \vdots \\ \text{symmetric} & & & var(y_m) \end{bmatrix}$
Covariance of a linear combination of random variables	$\mathbf{z} = \mathbf{G}\mathbf{y}$ where $\mathbf{G}$ is a not random matrix $cov(\mathbf{z}) = \mathbf{G}\, cov(\mathbf{y})\mathbf{G}^T$

TABLE 7.3

Statistical Assumptions Regarding the Measurement Errors

Number	Assumption	Explicitation
1	Additive errors	$y_i = y_i^* + \varepsilon_i$
2	Unbiased model	$E(\mathbf{y}|\mathbf{P}) = y_{mo}(\mathbf{P}) \Rightarrow y_{mo}(t_i, \mathbf{P}^*) = y_i^*$
3	Zero mean errors	$E(\varepsilon_i) = 0$
4	Constant variance	$var(\varepsilon_i) = \sigma_y^2$
5	Uncorrelated errors	$\forall i,j$ and $i \neq j$: $cov(\varepsilon_i, \varepsilon_j) = 0$
6	Normal probability distribution	
7	Known statistical parameters	
8	No error in the X_{ij}	$\mathbf{X}$ is not a random matrix
9	No prior information regarding the parameters	

variance of these estimators, such as calculated with Equations 7.5 and 7.15, respectively, yields important information about the quality of the retrieved estimated parameters. If the measurement points are equally spaced, with a time step Δt and a number of points N, then the variance computed with Equation 7.15 can easily be bounded by

$$\sigma_v^2 < \frac{1}{m\Delta t^2} \sigma_Y^2 \tag{7.16a}$$

which means that an increase in the number of points results in a decreasing variance.

The bias $b(\cdot)$ of a random variable is defined as the difference between the expected value and the true value. An unbiased estimator—such as the one given by Equation 7.14—is an estimator with a bias equal to zero. The root mean square error of an estimator (rms) is

$$rms(\hat{p}) = \sqrt{E((\hat{p} - p^*)^2)} \tag{7.16b}$$

and it can be shown that

$$rms(\hat{p})^2 = b(\hat{p})^2 + V(\hat{p}) \tag{7.16c}$$

One of the major goals in the estimation process will be to look out for an optimal estimator. The quality of an estimator can be specified by the values of its bias, variance, and root mean square error. Obviously, when possible, it is preferred to search an unbiased estimator and then minimize the variance. This optimal estimator can be searched only when some information regarding the statistic assumptions is available.

When the estimation is linear, and if the standard assumptions are fulfilled, except constant variance, uncorrelated errors, and normal probability distribution—"123789," the existence of a minimum variance unbiased (MVU) estimator can be proved (Serra and Besson 1999, Rao et al. 1999). The existence of the MVU estimator is a key point, since it yields the best estimator you can use. Of course, when nothing is known regarding the statistical assumptions, it is still possible to use some robust estimator such as the OLS estimator, assuming that some potentially better estimators are possibly discarded (Wolberg 2005, Kariya and Kurata 2004).

7.3 Ordinary Least Squares Estimation

In this section, the general case of OLS estimation is presented more deeply, since the main trends of OLS were envisioned previously. Only discrete variables are in consideration, thus all the variables involved may be written in a vector form. The vector of the m collected data is already defined in Equation 7.13. The model $\mathbf{y_{mo}}$ depends both on n parameters P_j to be estimated, and independent variables, that is, reduced to one variable t for the sake of clarity, but with no loss of consistency.

$$\mathbf{y_{mo}} = [y_{mo}(t_1) \quad y_{mo}(t_2) \quad \cdots \quad y_{mo}(t_m)]^T \quad \text{and} \quad \mathbf{P} = [P_1 \quad P_2 \quad \cdots \quad P_n]^T \tag{7.17a}$$

The general structure of the model may then be written as

$$y_{mo} = M(t, \mathbf{P}) \tag{7.17b}$$

7.3.1 Sensitivity Coefficients and Sensitivity Matrix

7.3.1.1 Sensitivity Coefficients

The very important role played by the sensitivity coefficients has already been suggested in Section 7.1. The definition given in the particular case presented for one single parameter is easily extended for each of the n parameters:

$$X_j(t, \mathbf{P}) = \frac{\partial y_{mo}}{\partial P_j}, \quad j = 1, 2, \ldots, n \tag{7.18a}$$

X_j is the sensitivity coefficient of the model with respect to the parameter P_j computed at time t. If the model is not easily derived with respect to the parameters, a simple method to get an approximate value of the sensitivity coefficients is to compute the model twice with a small variation of the parameter of interest, and compute the derivative with the central finite difference approximation, such as

$$X_j(t, \mathbf{P}) \approx \frac{y_{mo}(t, P_1, P_2, \ldots, P_j + \Delta P_j, \ldots, P_n) - y_{mo}(t, P_1, P_2, \ldots, P_j - \Delta P_j, \ldots, P_n)}{2\Delta P_j} \tag{7.18b}$$

7.3.1.2 Definition of Linear Estimation

In this text, we are only dealing with linear estimation. The basic cases analyzed in the "Getting started" section were linear in that sense. The model is linear with respect to the parameters if the following property is achieved:

$$y_{mo}(t, a_1 \mathbf{P_1} + a_2 \mathbf{P_2}) = a_1 y_{mo}(t, \mathbf{P_1}) + a_2 y_{mo}(t, \mathbf{P_2}) \quad \text{for any real value of } a_1 \text{ and } a_2 \tag{7.18c}$$

A very important property shown is that if the model is linear with respect to the parameters, the sensitivity coefficients do not depend on the parameters. Then the model can be written as

$$y_{mo}(t, \mathbf{P}) = \sum_{j=1}^{n} X_j(t) P_j \tag{7.18d}$$

Example 7.1

The property of some model to be linear with respect to the parameters does not imply that the sensitivity coefficients should be linear with respect to the independent variables (such as time in the previous examples). For instance, if we consider the three following examples:

$$\text{Case 1} \quad y_{mo} = at + b \Rightarrow X_a = t \quad \text{and} \quad X_b = 1$$

$$\text{Case 2} \quad y_{mo} = at\,\exp(\sin(t)) + b\sqrt{t} \Rightarrow X_a = t\,\exp(\sin(t)) \quad \text{and} \quad X_b = \sqrt{t}$$

These models are linear with respect to the parameters a and b. The latter is not linear with respect to time.

$$\text{Case 3} \quad y_{mo} = a\,\exp\left(-\left(\frac{t}{b}\right)\right) \Rightarrow X_a = \exp\left(-\left(\frac{t}{b}\right)\right) \quad \text{and} \quad X_b = \frac{a}{b^2 t}\,\exp\left(-\left(\frac{t}{b}\right)\right)$$

The model given in case 3 is not linear with respect to the parameters a and b, since the sensitivity coefficients do depend on the parameters.

7.3.1.3 Sensitivity Matrix

Since only discrete variables are in consideration herein, the sensitivity coefficients can be written for the discrete values of the independent variable t, that is, for the occurrences corresponding to the data points, which means that we compute the model only for the "observable" vector, represented by the data points vector $\mathbf{y}$. The sensitivity vector with respect to the jth parameter is

$$\mathbf{x_j} = \begin{bmatrix} X_j(t_1) & X_j(t_2) & \cdots & X_j(t_m) \end{bmatrix}^T \tag{7.19a}$$

The n sensitivity vectors have a length m and can be gathered in order to build the sensitivity matrix $\mathbf{X}$, which has m rows and n columns, such as

$$\mathbf{X} = \begin{bmatrix} \mathbf{x_1} & \mathbf{x_2} & \cdots & \mathbf{x_n} \end{bmatrix} = \begin{bmatrix} X_{11} & X_{12} & \cdots & X_{1n} \\ X_{21} & X_{22} & \cdots & X_{2n} \\ \cdots & \cdots & \cdots & \cdots \\ X_{m1} & X_{m2} & \cdots & X_{mn} \end{bmatrix} \tag{7.19b}$$

With these notations, if the model is assumed to be linear with respect to the parameters the discrete model involving both the model vector and the parameters vector, as defined in Equation 7.16, is then written as

$$\mathbf{y_{mo}} = \mathbf{XP} \tag{7.19c}$$

Equation 7.19c yields a system of m equations with n parameters.

7.3.2 The Normal Equations and the OLS Estimator

7.3.2.1 Deriving the Normal Equations

With the linear model of Equation 7.19c, the relationship involving the experimental data points, the errors, and the model can be expressed as

$$\mathbf{y} = \mathbf{y^*} + \boldsymbol{\varepsilon} = \mathbf{y_{mo}}(\mathbf{P^*}) + \boldsymbol{\varepsilon} = \mathbf{XP^*} + \boldsymbol{\varepsilon} \tag{7.20a}$$

where $\mathbf{P^*}$ is the exact-but-unknown value of the parameters vector.

It is apparent in Equation 7.20a that we assume the model to be exact and to describe correctly the experiment with no error of modeling: when we write Equation 7.20a, we make the implicit assumption that the model $\mathbf{y_{mo}}(\cdot)$ is able to match the exact value $\mathbf{y}^*$.

Due to the errors in the collected data, Equation 7.20a cannot be solved in order to find $\mathbf{P}^*$, and the system $\mathbf{y} = \mathbf{XP}$ has generally no solution, since it is overdetermined. Similarly to the example given in Section 7.2.2, the OLS objective function is chosen as the norm to be minimized, that is, the norm of the output error (as shown in Figure 7.1):

$$S_{OLS}(\mathbf{P}) = (\mathbf{y} - \mathbf{y_{mo}})^T(\mathbf{y} - \mathbf{y_{mo}}) = (\mathbf{y} - \mathbf{XP})^T(\mathbf{y} - \mathbf{XP}) = \mathbf{e}^T(\mathbf{P})\,\mathbf{e}(\mathbf{P}) \qquad (7.20b)$$

It is noteworthy that the output error (the vector of differences between the data and the model) to be minimized does not involve directly the true value $\mathbf{y}^*$. However, with $\boldsymbol{\varepsilon} = \mathbf{y} - \mathbf{y}^*$ and $\mathbf{e} = \mathbf{y} - \mathbf{y_{mo}}$, it is straightforward to deduce the difference between the model and the true value as $\boldsymbol{\varepsilon} - \mathbf{e} = \mathbf{y_{mo}} - \mathbf{y}^*$, which can be observed in Figure 7.2.

Again, the minimization is achieved by computing the derivatives of the objective function with respect to the parameters. We obtain n equations such as

$$\frac{\partial S_{OLS}}{\partial P_j} = -2\sum_{i=1}^{m} X_{ij}\left(y_i - \sum_{k=1}^{n} X_{ik}P_k\right) = -2\sum_{i=1}^{m} X_{ij}(y_i - \mathbf{l_i}\mathbf{P}) = -2\mathbf{x}_j^T(\mathbf{y} - \mathbf{l_i}\mathbf{P}) \qquad (7.20c)$$

where $\mathbf{l_i}$ is the ith line of $\mathbf{X}$, as shown in Figure 7.4.

Equaling to zero, these derivatives yield a system of $j = 1, 2, \ldots, n$ equations, applied to the optimum vector $\hat{\mathbf{P}}_{OLS}$, which can be written in a more compact matrix form such as

$$\mathbf{X}^T(\mathbf{y} - \mathbf{X}\hat{\mathbf{P}}_{OLS}) = 0 \Rightarrow (\mathbf{X}^T\mathbf{X})\hat{\mathbf{P}}_{OLS} = \mathbf{X}^T\mathbf{y} \qquad (7.20d)$$

Equations 7.20d are called "the normal equations" (the reason for this name is envisioned hereafter in Section 7.3.4). It is noteworthy that the matrix $\mathbf{X}^T\mathbf{X}$ is a square (n, n) matrix. It is important to note that the size of this matrix can be relatively small, according to the number n of parameters to be estimated.

7.3.2.2 The OLS Estimator

Solving the normal equations such as given in Equation 7.20d yields the estimated vector of parameters. If the matrix is $\mathbf{X}^T\mathbf{X}$ nonsingular, the formal solution of Equation 7.20d is straightforward, and can be obtained by direct inversion of the square matrix $\mathbf{X}^T\mathbf{X}$, such as

$$\hat{\mathbf{P}}_{OLS} = (\mathbf{X}^T\mathbf{X})^{-1}\mathbf{X}^T\mathbf{y} \qquad (7.21)$$

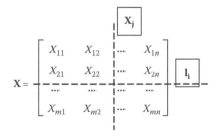

FIGURE 7.4
Structure of the sensitivity matrix.

Equation 7.21 yields the solution of the OLS minimization problem given by Equation 7.20b. The matrix $(\mathbf{X}^T\mathbf{X})^{-1}\mathbf{X}^T$ is the so-called Moore–Penrose matrix, also named as the pseudo-inverse of $\mathbf{X}$ (Campbell and Meyer 1991, Penrose 1955). Obviously, a necessary condition for $\mathbf{X}^T\mathbf{X}$ to be nonsingular is that the sensitivity coefficients are independent, and have a nonzero norm. This condition also requires that the number of measurements m be equal or greater than the number of parameters n to be estimated. Equation 7.21 yields a direct expression of the OLS estimator, which does not need any iterative procedure to be solved. Moreover, when the number of parameters to be estimated is small, a full analytical solution of Equation 7.21 is available. This option is quite suitable in image processing, such as thermal properties mapping from infrared (IR) images (Batsale et al. 2004).

Unfortunately, in some cases, it is not efficient to solve directly Equation 7.20d by direct inversion of the Moore–Penrose matrix, as suggested by Equation 7.21, due to the fact that either the number of parameters is huge or the pseudo-inverse of the sensitivity matrix is highly sensitive to the errors in the measurements. For instance, for geophysic applications such as gravity field computation (Baboulin 2006), "the computational task is quite challenging because of the huge quantity of daily accumulated data (about 90,000 parameters and several million observations) and because of the coupling of the parameters resulting in completely dense matrices." In other cases, the sensitivity matrix is ill-conditioned, resulting in an erratic inversion, which yields a hazardous amplification of the measurement errors toward very high values of the variance of the estimated parameters. Some examples of the latter will be proposed hereafter in Section 7.5 devoted to ill-posed problems, such as the case of thermal diffusivity mapping from IR images processing (Batsale et al. 2004).

Many methods are available for solving the normal equations which are detailed in Björck (1996), but giving a survey of these methods and of their numerical stability (Higham 2002) is out of the scope of this book. If the rank of $\mathbf{X}$ is n, then the matrix $\mathbf{X}^T\mathbf{X}$ is symmetric positive definite and can be decomposed using a Cholesky decomposition. A more reliable way of solving linear least squares problems consists in using orthogonal transformations. The commonly used QR factorization can be performed by using orthogonal transformations called Householders reflections (Golub and Van Loan 1996). QR factorization is less sensitive to ill-conditioned cases and is more numerically stable than the direct inversion of the normal equations. However, the floating-point operations involved in the Householder QR factorization have an order of magnitude equal to $2mn^2$, while it is mn^2 for the direct inversion of the normal equations (Golub and Van Loan 1996). An alternative decomposition of the sensitivity matrix is the SVD and the corresponding spectral analysis (Shenfelt et al. 2002). SVD is expensive in terms of computation time, but may be quite helpful for solving ill-conditioned systems (Bamford et al. 2008), and to analyze the structure of the linear transform, as seen hereafter in Section 7.5. The OLS estimator can also be calculated in a recursive form, by computing the pseudo-inverse with few measurements available, and then adding measurements sequentially (Borne et al. 1997, Ljung 1987).

7.3.2.3 Some Statistical Properties of the OLS Estimator

In this section, it is first assumed that the statistical assumptions "12389" are satisfied, according to Table 7.3, which means that the standard assumptions are fulfilled except normal, known statistical parameters, constant variance, and uncorrelated errors. Then some important statistical properties of the OLS estimator defined by Equation 7.21 can be evaluated.

First, since the error is additive, Equation 7.20a can be incorporated into the OLS estimator, Equation 7.21, such as to get

$$\hat{\mathbf{P}}_{OLS} = (\mathbf{X}^T\mathbf{X})^{-1}\mathbf{X}^T\mathbf{y} = \mathbf{Hy} = \mathbf{H}(\mathbf{y}^* + \boldsymbol{\varepsilon}) = \mathbf{H}(\mathbf{XP}^* + \boldsymbol{\varepsilon}) = \mathbf{P}^* + \mathbf{H}\boldsymbol{\varepsilon} \qquad (7.22a)$$

where the matrix $\mathbf{H} = (\mathbf{X}^T\mathbf{X})^{-1}\mathbf{X}^T$ has been introduced in order to facilitate the notations, and by using the fact that $\mathbf{HX} = \mathbf{I}_n$.

Then, the expected value of $\hat{\mathbf{P}}_{OLS}$ can be calculated by

$$E(\hat{\mathbf{P}}_{OLS}) = E(\mathbf{P}^* + \mathbf{H}\boldsymbol{\varepsilon}) = \mathbf{P}^* + \mathbf{H}E(\boldsymbol{\varepsilon}) = \mathbf{P}^* \qquad (7.22b)$$

since the expected value is a linear operator and the error is assumed to have a zero mean.

Hence Equation 7.22b yields the very important result that the OLS estimator is unbiased.

The covariance matrix of the parameter vector can also be calculated, by making use of Equation 7.22a, and the definition of the covariance matrix of a vector of random variables such as given in Table 7.2.

$$cov(\hat{\mathbf{P}}_{OLS}) = E\left(\left[\hat{\mathbf{P}}_{OLS} - \mathbf{P}^*\right]\left[\hat{\mathbf{P}}_{OLS} - \mathbf{P}^*\right]^T \right) = E(\mathbf{H}\boldsymbol{\varepsilon}[\mathbf{H}\boldsymbol{\varepsilon}]^T) = \mathbf{H}E(\boldsymbol{\varepsilon}\boldsymbol{\varepsilon}^T)\mathbf{H}^T = \mathbf{H}\,cov(\mathbf{y})\mathbf{H}^T$$

$$(7.22c)$$

With the additional assumptions that the errors have a constant variance and are uncorrelated—"1234589," this equation is turned into

$$cov(\hat{\mathbf{P}}_{OLS}) = (\mathbf{X}^T\mathbf{X})^{-1}\sigma_y^2 \qquad (7.22d)$$

This equation gives the very important result that the OLS estimator is the MVU estimator within the frame of assumptions "1234589," and has the "minimum" covariance matrix. It is said that the OLS estimator is efficient (Beck and Arnold 1977).

When the sensitivity vector tend to be correlated, the determinant of the matrix $\mathbf{X}^T\mathbf{X}$ tends to zero, resulting in the amplification of the variance of some specific parameters, as suggested by the form of Equation 7.22d, which means that the uncertainty of some parameters may turn to increase drastically. This important topic is discussed in Section 7.5.

Another important conclusion implied by Equation 7.22d is also that the parameters may be correlated, even if the observations are not, through the non-diagonal terms of the matrix $(\mathbf{X}^T\mathbf{X})^{-1}$. An illustration of this is given hereafter in Section 7.3.3, with an example relative to the estimation of two parameters.

Obviously, once the parameters have been estimated, it is also possible to find the regression model, obtained by substituting Equation 7.21 into Equation 7.19c, and also using Equation 7.22a, which yields

$$\hat{\mathbf{y}}_{OLS} = \mathbf{X}\hat{\mathbf{P}}_{OLS} = \mathbf{X}(\mathbf{X}^T\mathbf{X})^{-1}\mathbf{X}^T\mathbf{y} = \mathbf{y}^* + \mathbf{X}(\mathbf{X}^T\mathbf{X})^{-1}\mathbf{X}^T\boldsymbol{\varepsilon} \qquad (7.23a)$$

It is apparent in Equation 7.23a that the error on the regression model is evaluated by the projection matrix applied to the measurement error. It is also of great interest to define the vector of residuals, as the difference between the predicted values, computed with the estimated vector of parameters, and the original data, such as

$$\mathbf{r} = \hat{\mathbf{e}} = \mathbf{y} - \hat{\mathbf{y}}_{OLS} = \mathbf{y} - \mathbf{X}\hat{\mathbf{P}}_{OLS} = (\mathbf{I}_m - \mathbf{XH})\mathbf{y} \qquad (7.23b)$$

An important property of the vector of residuals can be deduced from Equation 7.23b. It is apparent, since the errors in $\mathbf{y}$ have been assumed to have a null expected value, that the vector of residuals also has a zero mean value, since we assume here that the matrices in Equation 7.23b are not random. Hence, it is of great interest to verify that the vector of residuals is close to have a zero mean value, in order to discard some bias in the model or some correlation effect. Also the expected value of the sum of squares of the residuals $S_{OLS}(\hat{\mathbf{P}}_{OLS}) = \mathbf{r}^T\mathbf{r}$ can be computed, and yields the following result:

$$E(S_{OLS}(\hat{\mathbf{P}}_{OLS})) = (m - n)\sigma_y^2 \qquad (7.23c)$$

Equation 7.23c can be used in order to evaluate a posteriori the variance of the noise associated to the signal $\mathbf{y}$.

If the assumption number 6 in Table 7.3, which states that the errors have a normal probability distribution, is assumed to be satisfied, it can be shown that $S_{OLS}(\hat{\mathbf{P}}_{OLS})/\sigma_Y^2$ has a $\chi^2(m-n)$ distribution and $\hat{\mathbf{P}}_{OLS} - \mathbf{P}^*$ has a normal distribution $N\left(\mathbf{0}, \sigma_Y^2(\mathbf{X}^t\mathbf{X})^{-1}\right)$.

The covariance matrix of the vector of residuals can also be computed from Equation 7.23b, by using the fact that the matrices $\mathbf{I}_m$, $\mathbf{X}$, and $\mathbf{H}$ are not random matrices, and making use of the covariance of a linear combination of vector random variables, such as shown in the last line of Table 7.2. Then the following result is obtained:

$$cov(\mathbf{r}) = (\mathbf{I}_m - \mathbf{XH})cov(\mathbf{y})(\mathbf{I}_m - \mathbf{XH})^T \qquad (7.23d)$$

Moreover, with the "1234589" assumptions in consideration herein, the covariance matrix of the residuals is reduced to

$$cov(\mathbf{r}) = (\mathbf{I}_m - \mathbf{XH})\sigma_y^2 \qquad (7.23e)$$

7.3.3 An Example with Two Parameters, *m* Data Points

The OLS estimation method presented in Section 7.3.2 is applied to the simple example of a straight-line regression, where both the slope and intercept are to be retrieved. The model is given by

$$y_{mo} = ax + b \qquad (7.24a)$$

Then the sensitivity coefficients are

$$X_a = x \quad \text{and} \quad X_b = 1 \qquad (7.24b)$$

Thus, the model is linear with respect to the parameters *a* and *b*.

The data are $\mathbf{y} = [y_1 \quad y_2 \quad \cdots \quad y_m]^T$ and the parameter vector is $\mathbf{P} = \begin{bmatrix} a \\ b \end{bmatrix}$

The sensitivity matrix is built with the sensitivity vectors, such as

$$\mathbf{X} = \begin{bmatrix} x_1 & 1 \\ x_2 & 1 \\ \cdots & \cdots \\ x_m & 1 \end{bmatrix} \tag{7.24c}$$

The two terms involved in the expression of the OLS estimator such as given by Equation 7.21 can be calculated explicitly:

$$\mathbf{X}^T\mathbf{y} = \begin{bmatrix} \sum x_i y_i \\ \sum y_i \end{bmatrix}, \quad \mathbf{X}^T\mathbf{X} = \begin{bmatrix} \sum x_i^2 & \sum x_i \\ \sum x_i & m \end{bmatrix}$$

The OLS estimation with Equation 7.21 yields

$$\hat{a} = \frac{S_{xy}^2}{S_{xx}^2}, \quad \hat{b} = \bar{y} - \hat{a}\bar{x} \tag{7.24d}$$

where

$$S_{xy}^2 = \frac{1}{m}\sum_{i=1}^{m}(x_i - \bar{x}) \cdot (y_i - \bar{y}), \quad S_{xx}^2 = \frac{1}{m}\sum_{i=1}^{m}(x_i - \bar{x})^2 \quad \text{and} \quad \bar{x} = \frac{1}{m}\sum_{i=1}^{m}x_i, \quad \bar{y} = \frac{1}{m}\sum_{i=1}^{m}y_i$$

For the standard assumptions regarding the measurement errors, the covariance matrix of the estimation error is obtained with Equation 7.22d such as

$$cov(\hat{\mathbf{P}}_{OLS}) = \begin{bmatrix} \sigma_a^2 & cov(\varepsilon_a, \varepsilon_b) \\ cov(\varepsilon_a, \varepsilon_b) & \sigma_b^2 \end{bmatrix} = \frac{\sigma_Y^2}{m^2 s_{xx}^2}\begin{bmatrix} m & -\sum x_i \\ -\sum x_i & \sum x_i^2 \end{bmatrix} \tag{7.24e}$$

which yields the std of the parameters estimators and the correlation coefficient r, such as

$$\sigma_a = \frac{\sigma_Y}{S_{xx}\sqrt{m}} \quad \text{and} \quad \sigma_b = \frac{\sigma_Y}{\sqrt{m}}\left(1 + \frac{\bar{x}^2}{S_{xx}^2}\right)^{1/2}$$

$$r = \frac{cov(\varepsilon_a, \varepsilon_b)}{\sigma_a \sigma_b} = -\frac{1}{(1 + S_{xx}^2/\bar{x}^2)^{1/2}}$$

The non-diagonal terms of the covariance matrix of the parameters indicate that the estimated parameters are correlated, even if the error in measurements is not.

These results confirm some usual characteristics of the statistical properties such as the following:

1. Increasing the number of data points m yields a better quality for the estimates, even if the measurements are quite noisy.
2. The quality of estimation of the slope a is better when the data points are chosen within a large dispersion (S_{xx} large).

3. The accurate estimation of the intercept needs both short times and large dispersion ($S_{xx}^2/\bar{x}^2$ is low).

4. The correlation coefficient r is negative, which means that the intercept is underestimated when the slope is overestimated. Moreover, the two parameters are more uncorrelated if the value of $S_{xx}^2/\bar{x}^2$ is large.

The present example is also used for the examination of the residuals, such as defined with Equation 7.23b. A normally distributed measurement error $N(0, \sigma_y = 1.89)$ is added to the output of the model of Equation 7.24a—computed with $a = 2$ and $b = 5$, and the corresponding simulated data are used for the OLS estimation of both the slope and intercept. The simulated measurement error is plotted in Figure 7.5, where its histogram is also shown. The residual obtained with the OLS estimator is plotted in Figure 7.6. As expected, the residuals have zero mean and are distributed like the measurement error (the mean is found to be 10^{-15} and $cov(r) \approx \sigma_y^2 \mathbf{I}_m$).

Then, the effect of an error in the model is tested. The OLS estimation is still made with the model given by Equation 7.24a, which means that the slope and intercept of a line are to be retrieved, but the experimental data are simulated now with a modified model, such as given in Equation 7.24f:

$$y_{\mathrm{mo}} = ax + b + g(x - x_0)^2 \tag{7.24f}$$

This situation would appear if the model which is intended for the estimation of parameters does not describe conveniently the experiment. In this case, we are implementing the estimation with an erroneous model. An example is plotted in Figure 7.7, with $g = 0.01$ and

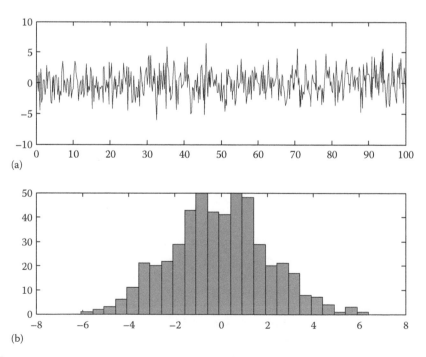

(a)

(b)

FIGURE 7.5

(a) Simulated error $N(0, \sigma_y = 1.89)$ and (b) histogram.

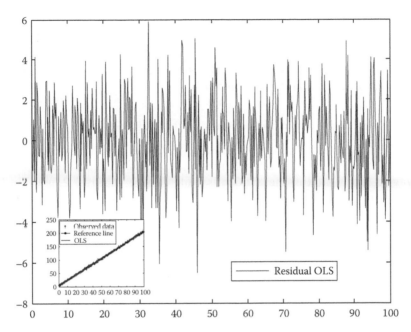

FIGURE 7.6
Residual for OLS estimation with the model given by Equation 7.24a, where the data have been simulated with the same model and the measurement error shown in Figure 7.5.

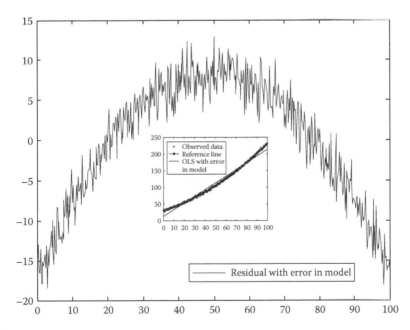

FIGURE 7.7
Residual for the OLS estimation of the slope and intercept of a line with the model given by Equation 7.24a, where the data have been simulated with an error in the model—Equation 7.24f—and the measurement error shown in Figure 7.5.

$x_0 = 50$. The curve of the residuals is not any more symmetric and centered on zero, but exhibits a "correlated" behavior, which is due to the error in the model. A MATLAB® program implemented to get these results is shown in Appendix 7.A.1. Hence, it is quite important to always examine the curves of residuals, since their form yields important information regarding the quality of estimation.

7.3.4 Another Look at the Normal Equations and the OLS Solution

With the previous example given in Section 7.1.2, we already noticed, as shown in Equation 7.10, that the model does not hit simultaneously all the measurement points. Another way to formulate this fact, which is quite clear in Figure 7.2, is to write that the data vector **y** is not included in the subspace $C(\mathbf{X})$ defined as the column space of **X**. It is then possible to define the difference vector **e** as the difference between the model and the data vectors, as shown in Figure 7.8a, and then minimize the norm of this vector. Obviously, the minimum norm of **e** is obtained when **e** is orthogonal to the subspace $C(\mathbf{X})$ (see Figure 7.8b). This is equivalent to searching the orthogonal projection of **y** onto the subspace spanned by the column vectors of **X**.

It seems quite reasonable to assume that the sensitivity vectors are independent, since otherwise the estimation problem is turned to be ill-posed: the information matrix would be singular and Equation 7.20d could not be solved.

If the sensitivity vectors are independent, they form a basis of **X**, and any vector of $C(\mathbf{X})$ can be written as a linear combination of these vectors. Hence, looking for a vector to be orthogonal to $C(\mathbf{X})$ is equivalent to searching a vector orthogonal to any linear combination of the sensitivity vectors, that is, which dot product would be zero with any sensitivity vector $\mathbf{x}_j$, such as

$$\mathbf{x}_j^T \hat{\mathbf{e}} = \mathbf{0}, \quad j = 1, 2, \dots, n \tag{7.25a}$$

which can be expressed in a more compact matrix form as

$$\mathbf{X}^T \hat{\mathbf{e}} = \mathbf{0} \tag{7.25b}$$

The orthogonal difference vector $\hat{\mathbf{e}} = \mathbf{r} = \mathbf{y} - \mathbf{X}\hat{\mathbf{P}}$ can be also recorded as the vector of residuals defined in Section 7.3.2.3 by Equation 7.23b. Substituting this expression into Equation 7.25b yields the result

$$\mathbf{X}^T (\mathbf{y} - \mathbf{X}\hat{\mathbf{P}}) = \mathbf{0} \tag{7.25c}$$

which is identical to the normal equation given in Equation 7.20d.

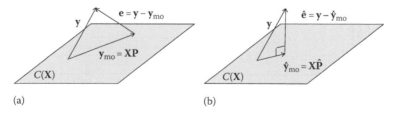

(a)   (b)

FIGURE 7.8
Projecting the data onto the subspace spanned by X: (a) non-orthogonal projection and (b) optimal orthogonal projection.

Thus, the normal equations are retrieved directly. It is noteworthy that the difference vector is orthogonal to the column space of $\mathbf{X}$.

As a result of this approach, we can also notice that the matrix $\mathbf{X}(\mathbf{X}^T\mathbf{X})^{-1}\mathbf{X}^T$ is the projection matrix that maps the data vectors in R^m onto the column space of the matrix $\mathbf{X}$. The regression model, as given by Equation 7.23b, which is the model predicted with the estimated parameters, can now be analyzed as the projection of the data vector onto the space spanned by the model. Moreover, Equation 7.23b, defining the vector of residuals—or as well the orthogonal difference vector in Figure 7.8—can be written in order to split the data into two orthogonal parts, such as

$$\mathbf{y} = \mathbf{r} + \hat{\mathbf{y}}_{\text{OLS}} = \mathbf{y}_{\perp} + \mathbf{y}_{/\!/} \tag{7.25d}$$

This kind of decomposition is usefully implemented for the SVD approaches, such as proposed in Bamford et al. (2009) and explained in Tan et al. (2006).

7.4 Nonconstant Variance Errors

In the previous developments, all the measurements errors are assumed to have the same precision. Obviously, this is not always the case, as the measurement errors may depend on the amplitude of the signal, be recorded by different sensors, or be expressed in different units.

7.4.1 Example and Motivation

In this section, we consider the problem of estimating a constant p, when two measurement points $y_i\ i = 1, 2$ are available. As previously, the first approach is obtained by minimizing the norm given by the objective function S, such as

$$S(p) = (p - y_1)^2 + (p - y_2)^2 \tag{7.26a}$$

which yields an unbiased estimation of p which is the arithmetic mean of the observed data, such as

$$\hat{p} = \bar{y} = \frac{y_1 + y_2}{2} \tag{7.26b}$$

However, this simple formulation may present two important drawbacks, since it is very sensitive both to a change of unit when the measurement data are not all expressed in the same units, and to a nonconstant level of precision of the collected data.

For instance, let us assume that y_2 is measured in a different unit than y_1, within a multiplication constant g, both for the measurement and the corresponding model, such as

$$S(p) = (p - y_1)^2 + (gp - gy_2)^2 \tag{7.26c}$$

Then the minimization of the square norm yields

$$\hat{p}_g = \frac{y_1 + g^2 y_2}{1 + g^2} \qquad (7.26d)$$

which gives a different solution according to the value of the scaling factor g.

Moreover, assuming that the statistical assumptions regarding the measurement errors are standard except for constant variance and normal: additive, zero mean, and uncorrelated errors, known statistical parameters, no error in the sensitivity matrix, and no prior information regarding the parameters, we may be looking for the MVU estimator being a linear combination of the observed data, and search the scalars a_1 and a_2 such as

$$\hat{p} = a_1 y_1 + a_2 y_2 \qquad (7.27a)$$

which has the expected value

$$E(\hat{p}) = E(a_1 y_1 + a_2 y_2) = a_1 E(y_1) + a_2 E(y_2) = (a_1 + a_2)y^* + E(\varepsilon_1) + E(\varepsilon_2) \qquad (7.27b)$$

Since the errors are assumed to have zero mean, a necessary condition to get an unbiased estimator is obviously written as

$$a_1 + a_2 = 1 \qquad (7.27c)$$

The variance of $\hat{p}$ is written as

$$var(\hat{p}) = E((\hat{p} - y^*)^2) = a_1^2 \sigma_1^2 + a_2^2 \sigma_2^2 = a_1^2 \sigma_1^2 + (1 - a_1)^2 \sigma_2^2 \qquad (7.27d)$$

and may be derived with respect to a_1 in order to find the minimum variance estimator, which yields

$$a_1 = \frac{\sigma_2^2}{\sigma_1^2 + \sigma_2^2} \Rightarrow a_2 = 1 - a_1 = \frac{\sigma_1^2}{\sigma_1^2 + \sigma_2^2} \qquad (7.28a)$$

The solution for the MVU estimator of p is then obtained as

$$\hat{p}_{MVU} = \frac{\dfrac{1}{\sigma_1^2}y_1 + \dfrac{1}{\sigma_2^2}y_2}{\dfrac{1}{\sigma_1^2} + \dfrac{1}{\sigma_2^2}} \qquad (7.28b)$$

In that case, it can be shown that the MVU has been obtained through the minimization of the following objective function:

$$S(p) = \frac{(p - y_1)^2}{\sigma_1^2} + \frac{(p - y_2)^2}{\sigma_2^2} \qquad (7.28c)$$

In Equation 7.28b, the new element with respect to the OLS estimation, as previously analyzed, is that the information relative to the precision of each individual measurement has been incorporated into the objective function to be minimized. It is apparent in Equation 7.28c that more weight is given to the measurement points with a lower variance while less weight is given to the data with a high level of uncertainty. It is also interesting to note that the variance of this estimator is

$$\frac{1}{var(\hat{p}_{MVU})} = \frac{1}{\sigma_1^2} + \frac{1}{\sigma_2^2} \tag{7.28d}$$

The form of Equation 7.28d confirms that the quality of the estimation is mostly determined by the data with the lower uncertainty: if $\sigma_1^2 \ll \sigma_2^2 \Rightarrow var(\hat{p}_{MVU}) \approx \sigma_1^2$.

7.4.2 Gauss–Markov Theorem

7.4.2.1 Gauss–Markov Estimator

In the example given in the previous section, the minimization of S does not depend in fact on the size of any individual variance σ_i^2 but only on their proportion. If the errors are not independent, it seems to be consistent to apply a weighting which involves both the covariances and the variances.

We consider herein the case where the statistical assumptions regarding the measurement errors are "123789," that is, the standard assumptions except normal, uncorrelated, and constant variances, and the covariance matrix is known within a multiplicative constant and is positive definite, such as

$$cov(\mathbf{y}) = \sigma^2 \mathbf{\Omega} \tag{7.29a}$$

With such assumptions regarding the errors, the Gauss–Markov theorem states the very important result that the MVU estimator of all estimators which are a linear combination of the observed data is the following estimator given as

$$\hat{\mathbf{P}}_{GM} = (\mathbf{X}^T \mathbf{\Omega}^{-1} \mathbf{X})^{-1} \mathbf{X}^T \mathbf{\Omega}^{-1} \mathbf{y} \tag{7.29b}$$

The Gauss–Markov estimator is obtained by minimizing the objective function

$$S_{GM} = \mathbf{e}^T \mathbf{e} = (\mathbf{y} - \mathbf{X}\mathbf{P}_{GM})^T \mathbf{\Omega}^{-1} (\mathbf{y} - \mathbf{X}\mathbf{P}_{GM}) \tag{7.29c}$$

and its covariance matrix is

$$cov(\hat{\mathbf{P}}_{GM}) = (\mathbf{X}^T \mathbf{\Omega}^{-1} \mathbf{X})^{-1} \sigma^2 \tag{7.29d}$$

In the case where the measurement errors are uncorrelated, then the covariance matrix is diagonal, and the objective function to be minimized, as seen with two data points in the previous example, is

$$S_{GM} = \sum_{i=1}^{m} \frac{\left(y_i - \sum_{j=1}^{n} X_{ij}P_j\right)^2}{\sigma_i^2} \tag{7.29e}$$

which means that each observation is simply weighted with the corresponding variance.

7.4.2.2 Example

The GM estimation method is applied to the simple example of a straight-line regression, where both the slope and intercept are to be retrieved, as seen with OLS with the model given in Equation 7.24a, but with nonconstant variances:

$$y_{mo} = 2t + 5 \tag{7.30a}$$

We assume that the covariance matrix is diagonal, known within a multiplicative constant, and that the variance is proportional to the square of the independent variable (the time here), $\sigma_y(t) = \alpha t$ such as

$$cov(\mathbf{y}) = \alpha^2 \mathbf{\Omega} = \alpha^2 \begin{bmatrix} t_1^2 & 0 & \dots & 0 \\ 0 & t_2^2 & 0 & \dots \\ \dots & \dots & \dots & 0 \\ 0 & \dots & 0 & t_n^2 \end{bmatrix} \tag{7.30b}$$

A MATLAB program implemented to get these results is shown in Appendix 7.A.2. Applying OLS and GM estimators to this problem gives the results shown in Figure 7.9, where the lines retrieved with the estimated parameters are plotted and compared to the reference line drawn with the exact value of the parameters. The line retrieved with OLS is far from the reference (exact) line, due to the variation of variance, while the GM estimator successfully achieves to match the reference line. The exact values for the slope and intercept and the estimated parameters are in that case respectively:

$$\mathbf{P}^* = \begin{bmatrix} 2 \\ 5 \end{bmatrix}; \quad \hat{\mathbf{P}}_{GM} = \begin{bmatrix} 2.01 \\ 4.99 \end{bmatrix}; \quad \hat{\mathbf{P}}_{OLS} = \begin{bmatrix} 2.10 \\ 3.52 \end{bmatrix}; \quad cov(\hat{\mathbf{P}}_{GM}) = \begin{bmatrix} 0.0013 & -0.0015 \\ -0.0015 & 0.0020 \end{bmatrix}$$

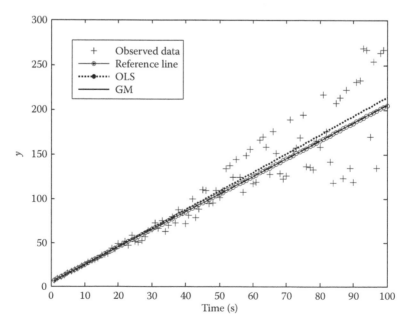

FIGURE 7.9
Estimation with nonconstant variance data.

The Gauss–Markov estimator is quite close to the reference values, since it takes into account the variation of uncertainty at each measurement point. The OLS estimator, which is applied but does not consider the knowledge about this statistical behavior of the errors, gives an "acceptable" result, but is far to be the best estimator in this case. Moreover, the GM estimator is in this case the MVU, but the OLS is not.

7.4.3 Maximum Likelihood Estimation

The method of maximum likelihood consists of choosing among the possible values for the parameter, the value that maximizes the probability of obtaining the sample of data that have been observed. The joint probability distribution $f(\mathbf{y}|\mathbf{P})$ defines the distribution of the measurement points for a given parameter vector, that is, associates a probability with each different outcome $\mathbf{y}$ for a fixed $\mathbf{P}$. On the other hand, if we are interested in finding the vector of parameters which maximizes the probability of getting the observation we got, then we are looking at $f(\mathbf{y}|\mathbf{P})$ as a function of $\mathbf{P}$ for a fixed set of measurement $\mathbf{y}$ already obtained. This function, similar in its form to $f(\mathbf{y}|\mathbf{P})$, is called the likelihood function $L(\mathbf{P}|\mathbf{y})$.

We consider the case where the assumptions are "1236789," that is, the standard assumptions are satisfied, except constant variance and uncorrelated errors. Note that the only difference with the Gauss–Markov theorem case is that the errors have a normal distribution.

The likelihood function is deduced directly from the probability density function as

$$L(\mathbf{P}|\mathbf{y}) = (2\pi)^{-m/2} |\psi|^{-1} \exp\left(\frac{-(\mathbf{y} - \mathbf{XP})^T \cdot \psi^{-1} \cdot (\mathbf{y} - \mathbf{XP})}{2} \right) \tag{7.31a}$$

where $\psi = cov(\mathbf{y})$.

Taking the natural logarithm of the likelihood function yields

$$\ln L(\mathbf{P}|\mathbf{y}) = -\frac{1}{2}[m \ln(2\pi) + \ln(|\psi|) + S_{\mathrm{ML}}] \tag{7.31b}$$

with

$$S_{\mathrm{ML}} = (\mathbf{y} - \mathbf{XP})^T \cdot \psi^{-1} \cdot (\mathbf{y} - \mathbf{XP}) \tag{7.31c}$$

Thus, maximizing the likelihood function is equivalent to minimizing the objective function S_{ML}. The ML estimator is then obtained as

$$\hat{\mathbf{P}}_{\mathrm{ML}} = (\mathbf{X}^T \psi^{-1} \mathbf{X})^{-1} \mathbf{X}^T \psi^{-1} \mathbf{y} \tag{7.32a}$$

The covariance matrix of $\hat{\mathbf{P}}_{\mathrm{ML}}$ is obtained by substituting the data by the model plus the error into Equation 7.32a, with Equation 7.20a, and is quite similar to the calculus in the OLS case, such as in Equation 7.22d. It yields

$$cov(\hat{\mathbf{P}}_{\mathrm{ML}}) = (\mathbf{X}^T \psi^{-1} \mathbf{X})^{-1} \tag{7.32b}$$

Despite the fact that there is more information available regarding the statistical assumptions regarding the errors for the ML than for the GM estimator, both give the same result for the case envisioned herein. Moreover, since the GM estimator produces the MVU for linear estimation, the ML does not yield more performance while it requires more a priori assumptions. Thus, the ML is in fact mostly useful for nonlinear estimator implementation, and is presented herein for the sake of general knowledge regarding the wide variety of approaches.

7.4.4 Weighted Least Squares

If the covariance matrix of the errors is not known, but some information is available that yields some specific variations regarding the errors, it might be of interest to include the effect of some symmetric weighting matrix $\mathbf{w}$ in the estimation method, even if this matrix is not necessarily equal to the inverse of the covariance matrix. This may be the case when some change in the observed variables is implemented. An example is encountered when the three-dimensional temperature field is separated into the multiplication of a two-dimensional in-plane temperature and an in-depth temperature in a semi-infinite medium, which is proportional to the square root of time. In that case, the measurement errors are also multiplied by the independent variable, and this information must be incorporated in the estimator.

The objective function which has to be minimized is then

$$S_{WLS} = (\mathbf{Y} - \mathbf{XP})^T \mathbf{w} (\mathbf{Y} - \mathbf{XP}) \tag{7.33a}$$

which yields the estimator

$$\hat{\mathbf{P}}_{WLS} = (\mathbf{X}^T \mathbf{w} \mathbf{X})^{-1} \mathbf{X}^T \mathbf{w} \mathbf{Y} \tag{7.33b}$$

If the standard assumptions are valid, except constant variance, uncorrelated, normal, and known statistical parameters (12389), the covariance matrix of the WLS estimator can be calculated as

$$cov(\hat{\mathbf{P}}_{WLS}) = (\mathbf{X}^T \mathbf{w} \mathbf{X})^{-1} \mathbf{X}^T \mathbf{w} \, cov(\mathbf{e}_Y) \mathbf{w} \mathbf{X} (\mathbf{X}^T \mathbf{w} \mathbf{X})^{-1} \tag{7.33c}$$

If $\mathbf{w}$ is chosen as the inverse of the covariance matrix of the errors, that is, $\mathbf{w} = cov(\mathbf{e}_Y)^{-1}$, then this expression collapses to

$$cov(\hat{\mathbf{P}}_{WLS}) = (\mathbf{X}^T \mathbf{w} \mathbf{X})^{-1} \tag{7.33d}$$

7.5 Introduction to Ill-Posed Problems

The parameter estimation problem, such as defined in Figure 7.1, is relative to the comparison of a model which computes the output from the knowledge of the input and some inner parameters which are used in the model. As seen previously in Chapter 1, the model structure may be different according to the kind of problem in consideration, the observation scale, and complexity.

The parameters to be recovered may be as well the passive structural parameters of the model (model identification), the parameters relative to the input variables, initial state, boundary conditions, some thermophysical properties, calibration, etc. For any of these cases in consideration, the output of the model can be properly computed if all the required information is available.

The problem is said to be well-posed, if, according to Hadamard (1923), three conditions are satisfied, such as

1. A solution exists.
2. The solution is unique.
3. The solution depends continuously in the data.

Problems that are not well-posed in the sense of Hadamard are said to be ill-posed problems. Note that the simple inversion of a well-posed problem may either be or not be a well-posed problem.

In this chapter, solving the direct problem in consideration by the discrete linear model defined by Equation 7.19c is a well-posed problem.

The example of searching the slope of a line with two or more data points, such as discussed in Section 7.2.2 and Equations 7.8 and 7.9 may be either a well-posed or an ill-posed problem:

1. A unique and stable solution exists if all the data points fit on the same line (no noise in the data), and the time zero has not been chosen for some noisy data point. In that very specific case, the problem of finding the slope is well-posed.
2. If, due to the noise in the measurement points, the data do not fit on the same line, a solution does not exist (as shown in Figure 7.2); the corresponding inverse problem of finding the slope is ill-posed.
3. If the values of time for taking the measurements are not properly chosen (mostly close to zero), it has been discussed with Equation 7.14 that the solution is unstable, since the errors in the measurement may increase drastically—see Table 7.1 and Figure 7.3; this inverse problem is ill-posed.

The parameter estimation problem stated by finding the vector of parameters by matching exactly the measurements to the model, and by making use of Equation 7.19c, is most often an ill-posed problem, since it is generally overdetermined (because the number of measurements m is greater than the number of parameters n), and has no solution because $\mathbf{y} \notin \mathrm{Im}(\mathbf{X})$. When the system is under-determined ($m < n$), it is also ill-posed because there is an infinity of solutions. Moreover, when $m = n$, the problem may be well-posed if it were stable, but may also be unstable due to the effect of noise in the data.

7.5.1 Examples and Motivation

In Section 7.2.2, the example given by Equation 7.10 was an ill-posed problem. Its conversion into the least squares problem, defined by minimizing the norm in Equation 7.11, was the approach chosen to turn this problem into a well-posed problem, and solve it with Equation 7.13. We already stated that if the instants of measurements were not properly chosen, the problem could turn to be unstable.

We give hereafter some typical examples of ill-posed problems, such as derivation, deconvolution, or extrapolation of the surface temperature by using some internal sensor measurement.

The first and second examples are typical of a parameterized function estimation. Instead of having a low number of parameters to be estimated with a high number of measurements, as for Example 3.3, of estimating the slope and intercept of a linear profile, the number of parameters to be estimated is very large and is quite of the same order as the number of observable data **y**, which makes the problem highly sensitive to noise. Unfortunately, in this case the inversion often also amplifies the measurement noise.

7.5.1.1 Derivation of a Signal

The derivation of a signal is often required for data processing. It is the case of time-dependent functions, for instance, when deriving the time evolution of the mass of a product during drying or deducing the velocity of a body from the measurement of its position. A usual case in heat transfer is the problem of estimating the heat flux $q(t)$ exchanged by a body with uniform temperature $T(t)$ and volumetric heat capacity C (lumped body approximation). The heat balance can be written as

$$C\frac{dT}{dt} = q(t) \text{ with the initial condition } t = 0 \quad T = 0$$

An inversion procedure for recovering an estimation of $q(t)$ from the measured temperature values $y(t_i)$, for different levels of the measurement noise, is based on the following steps:

a. Choose some heat flux function, such as $q(t) = 2t$ (arbitrarily chosen here).
b. Compute the corresponding analytical solution $T(t) = t^2/C$.
c. Add some random error, in order to simulate some experimental data, such as $y(t) = T(t) + \varepsilon(t)$.
d. Retrieve the estimation by discrete derivation of the signal $\hat{q}(t) = C\frac{\Delta y}{\Delta t} \approx C\frac{dT}{dt}$.
e. Repeat for different values of the signal-to-noise ratio (characterized by different levels of std).

The results are depicted in Figure 7.10, assuming that $C = 1$. When the std of the error is low, the heat flux is conveniently retrieved (Figure 7.10a). For case (b), the noise on the signal y remains very low, in the sense that it is still almost not visible in the corresponding curve. However, the heat flux is poorly computed. Increasing the level of noise, such as in Figure 7.10c, where the std is 0.9 K, results in a drastically poor computation of the heat flux. Thus, the derivation of an experimental signal is an ill-posed problem, due to its unstable nature. The numerical derivation yields the computation of the difference of successive measurements, divided by the time step. As previously seen in Equations 7.14 and 7.15, the ill-posed character of the problem is more important as the time step decreases.

The corresponding program can be found in Appendix 7.A.3.

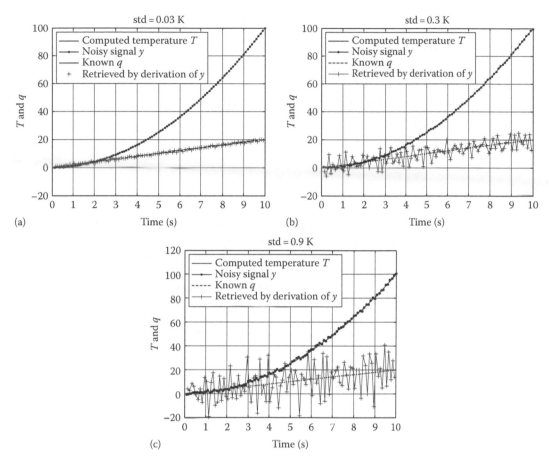

FIGURE 7.10
Derivation of an experimental signal: (a) std = 0.03 K, (b) std = 0.3 K, and (c) std = 0.9 K.

7.5.1.2 Deconvolution of a Signal

The deconvolution of a signal is also an operation often required when processing experimental data, for instance, when searching the transfer function of a system or sensor, in image processing, optics, geophysics, etc. We give again the heat transfer example of some heat capacity exchanging with convective heat losses with the surrounding medium, such as

$$C\frac{dT}{dt} = q(t) - hT \text{ with the initial condition } t = 0 \quad T = 0$$

We assume here that $C = 1, T_\infty = 0$ and that the boundary surface of the body is 1.

Solving this equation by using the Laplace transform of the temperature and heat flux and inverting yields the solution in the form of the following product of convolution:

$$T(t) = \int_0^t q(t - \tau)\exp(-h\tau)d\tau$$

The same approach as in the previous example is proposed herein, such as

a. Choose some heat flux function, such as $q(t)$.

b. Compute the corresponding analytical solution $T(t)$ as the convolution product above.

c. Add some random error, such as $y(t) = T(t) + \varepsilon(t)$.

d. Retrieve the heat flux by inverting the product this signal (deconvolution).

e. Repeat for different values of the signal-to-noise ratio (characterized by different levels of std).

The results are depicted in Figure 7.11. For a low std of the error (std = 0.03 K), the heat flux is conveniently retrieved by the deconvolution operation. When increasing the noise level (std = 0.3 K), the drastic amplification of the errors in the deconvolution operation makes the result absolutely inaccurate. The visual effect of the noise level in the curves where the temperature outputs are drawn shows that the increase of noise between the two situations, which makes the solution accurate or unavailable, is not significant. It is apparent with this example that the deconvolution of an experimental signal may be an ill-posed problem, depending on the functional form of the impulse response, due to its unstable nature.

The corresponding program can be found in Appendix 7.A.4.

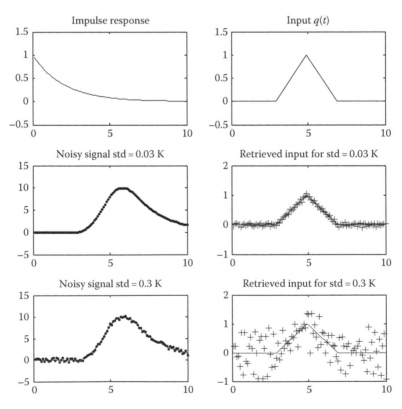

FIGURE 7.11
Effect of the noise level on the deconvolution of a signal.

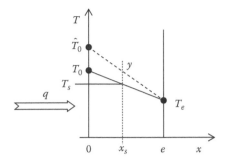

FIGURE 7.12
Inverse steady-state heat conduction problem.

7.5.1.3 Interpolation/Extrapolation of a Linear Profile

We consider the case of steady-state one-dimensional heat conduction in a homogeneous slab, such as depicted in Figure 7.12. If two boundary conditions and the thermal conductivity and thickness are known, the problem is obviously well-posed, and the resulting temperature profile is linear.

If the temperature at the left side of the wall is unknown, the problem is ill-posed, which makes it necessary to add some additional information in order to define a new well-posed problem. For instance, let us use the information of a thermal sensor inserted within the wall, at $x = x_s$, as shown in Figure 7.12. The inverse problem under consideration is to retrieve the temperature T_0 at the left wall from the measured temperature y.

The direct model, which computes the temperature at the sensor location, is written as

$$T_s = (1 - x_s^*)T_0 + x_s^*T_e \quad \text{where } x_s^* = \frac{x_s}{e} \tag{7.34a}$$

The error on the measured value given by the sensor is assumed to be additive, zero mean, and known std σ, such as

$$y = T_s + \varepsilon \tag{7.34b}$$

The estimator of T_0 is obtained by substituting the temperature of the sensor by the measured value y, which yields

$$\hat{T}_0 = \frac{1}{(1 - x_s^*)}y - \frac{x_s^*}{(1 - x_s^*)}T_e \tag{7.35a}$$

Thus, the error on the estimate can be calculated easily as

$$e_{T_0} = \hat{T}_0 - T_0 = \frac{y - T_s}{(1 - x_s^*)} = \frac{\varepsilon}{(1 - x_s^*)} \tag{7.35b}$$

Its expected value and std is then deduced as

$$E(e_{T_0}) = 0 \quad \text{and} \quad \sigma_{T_0} = \frac{\sigma}{(1 - x_s^*)} \tag{7.35c}$$

These results show that the estimation is unbiased, and that the std is an increasing function of the position of the sensor in the slab. The amplification of the error when the sensor is located toward the "fixed" point has already been noted in the previous examples on the estimation of the slope and intercept of a line, and is rather apparent in the Figure 7.12, when comparing the real temperature profile with the predicted profile. The predicted profile is obtained as

$$\hat{T}_x = \frac{1 - x^*}{(1 - x_s^*)} y + \frac{x^* - x_s^*}{(1 - x_s^*)} T_e \quad \text{where } x^* = \frac{x}{e} \tag{7.36a}$$

The error on the predicted model of the temperature at the position x can be calculated by the same approach as

$$e_{T_x} = \frac{1 - x^*}{(1 - x_s^*)} \varepsilon \tag{7.36b}$$

Its expected value and std are then deduced as

$$E(e_{T_x}) = 0 \quad \text{and} \quad \sigma_{T_x} = \frac{1 - x^*}{(1 - x_s^*)} \sigma \tag{7.36c}$$

Again, the estimator of T_x is unbiased, and the expression of the std suggests that two regions appear in the slab:

1. The layer located between the sensor and the right side of the slab $x_s \le x \le e$, where the boundary conditions are known exactly (T_s) or are approximated by the measurement (y), the profile is determined by a linear interpolation, and the std of the resulting error is reduced.
2. The layer located in front of the sensor $0 \le x < x_s$, where one boundary condition is unknown and the temperature is obtained by extrapolation of the temperature profile, the measurement noise is amplified. The inverse problem is ill-posed in this region, and more ill-posed while the sensor moves toward far from the front wall.

In the case of transient heat conduction, the extrapolation of the wall heat flux or wall temperature from the information collected by an inner thermal sensor is also an ill-posed problem, due to the amplification of the error in the inversion procedure. It is known as the inverse heat conduction problem (IHCP), and has been widely discussed (Beck et al. 1985, Ozisik and Orlande 2000, Alifanov 1994, Fudym et al. 2003, Petit and Maillet 2008).

7.5.2 Structure of the Linear Transform and Stability

7.5.2.1 Singular Value Decomposition

It was already discussed that the existence, unicity, and stability of the solution of the discrete linear parameter estimation problem, such as defined in this chapter, depend highly on the characteristics and structure of the rectangular matrix **X**. Moreover, when the overdetermined problem $\mathbf{y} = \mathbf{XP}$ is turned into the least squares problem given by the normal equations, such as Equation 7.20d, it appears that the structure of the square matrix $\mathbf{X}^T\mathbf{X}$ is also important for the propagation of the errors between the observed data and the parameters. The anatomy of such linear transformation is very clearly discussed in the text of Tan et al. (2006).

One approach of interest in order to analyze this problem is to consider the singular value decomposition (SVD) of $\mathbf{X}$. We assume herein that $m > n$ (overdetermined system; there is more data than parameters).

The square matrix $\mathbf{X}^T\mathbf{X}$ of size (n, n) and $\mathbf{X}\mathbf{X}^T$ of size (m, m) are symmetric and positive semi-definite and hence have eigenvalues that are real and nonnegative.

Let $\lambda_1^2, \lambda_2^2, \ldots, \lambda_n^2$ be the eigenvalues of $\mathbf{X}^T\mathbf{X}$ associated to the n eigenvectors $\mathbf{v_1}, \mathbf{v_2}, \mathbf{v_3}, \ldots, \mathbf{v_n}$ used to build the square matrix $\mathbf{V}^T$ of size (n, n).

The eigenvalues of $\mathbf{X}\mathbf{X}^T$, which are also $\lambda_1^2, \lambda_2^2, \ldots, \lambda_n^2$, are associated to the m eigenvectors $\mathbf{u_1}, \mathbf{u_2}, \mathbf{u_3}, \ldots, \mathbf{u_m}$ used to build the square matrix $\mathbf{U}$ of size (m, m).

The rank of the matrix is given by the number r of nonzero eigenvalues. The singular values are the square roots of the nonzero eigenvalues, written in decreasing order, such as $\lambda_1 \geq \lambda_2 \geq \cdots \geq \lambda_r > 0$.

The matrices $\mathbf{U}$ and $\mathbf{V}^T$ are orthogonal such as $\mathbf{V}\mathbf{V}^T = \mathbf{V}^T\mathbf{V} = \mathbf{I}_n$ and $\mathbf{U}\mathbf{U}^T = \mathbf{U}^T\mathbf{U} = \mathbf{I}_m$.

Let us define the rectangular diagonal matrix $\mathbf{S}$ of size (m, n), which is built with the singular values written in decreasing order—see Equation 7.37c.

The SVD of the matrix $\mathbf{X}$ is then written as

$$\mathbf{X} = \mathbf{U}\mathbf{S}\mathbf{V}^T \tag{7.37a}$$

Substituting the SVD of X in the linear relationship $\mathbf{y} = \mathbf{X}\mathbf{P}$ yields

$$\mathbf{U}\mathbf{S}\mathbf{V}^T\mathbf{P} = \mathbf{y} \;\Leftrightarrow\; \mathbf{U}^T\mathbf{U}\mathbf{S}\mathbf{V}^T\mathbf{P} = \mathbf{U}^T\mathbf{y} \;\Leftrightarrow\; \mathbf{S}\mathbf{V}^T\mathbf{P} = \mathbf{U}^T\mathbf{y} \tag{7.37b}$$

where the estimation problem can be considered now with the new parameter vector $\mathbf{b} = \mathbf{V}^T\mathbf{P}$ and a new observable vector: $\mathbf{z} = \mathbf{U}^T\mathbf{y}$, such as

$$\mathbf{S}\mathbf{b} = \mathbf{y} \quad \text{or} \quad \begin{bmatrix} \lambda_1 & 0 & \cdots & 0 & \cdots & 0 \\ 0 & \lambda_2 & & & & 0 \\ & & 0 & \cdots & & 0 \\ \cdots & & & \lambda_r & & \\ & & & \cdots & 0 & \cdots \\ & & & \cdots & \cdots & 0 \\ 0 & 0 & 0 & 0 & \cdots & 0 \end{bmatrix} \begin{bmatrix} b_1 \\ b_2 \\ \\ \cdots \\ b_r \\ \\ \cdots \\ b_n \end{bmatrix} = \begin{bmatrix} z_1 \\ z_2 \\ \cdots \\ z_r \\ \cdots \\ \cdots \\ z_m \end{bmatrix} \tag{7.37c}$$

The condition for the existence of a solution is that $z_i = 0$ for $i = r + 1, \ldots, m$, which is equivalent to a condition of orthogonality of the eigenvectors $\mathbf{U}_i^T$ for the values $i = r + 1, \ldots, n$ of the data y_i.

The unicity of the solution is true if $r = n$, which is possible only if $m \geq n$ (more data than parameters). When $r < n$, the matrix $\mathbf{X}$ does not have full rank, and the parameters to be estimated must be reduced, or some parameters must be determined in an arbitrary form.

The linear transform of $\mathbf{y}$ also yields a new covariance matrix associated to the observable measurement noise. Hopefully, we can note that this operation does not affect the variance of the error of the transformed signal $\mathbf{z}$:

$$cov(\mathbf{z}) = \mathbf{U}^T cov(\mathbf{y})\mathbf{U} = \sigma^2\mathbf{U}^T\mathbf{U} = \sigma^2\mathbf{I} \tag{7.38a}$$

Thus the statistical assumption for $\mathbf{b}$ is also verified.

The SVD can be used in the normal equations of Equation 7.20d in order to find the OLS estimator in the diagonal basis, which yields

$$(\mathbf{X}^T\mathbf{X})\hat{\mathbf{P}}_{OLS} = \mathbf{X}^T\mathbf{y} \Rightarrow \mathbf{S}^T\mathbf{S}\hat{\mathbf{b}}_{OLS} = \mathbf{S}^T\mathbf{z} \tag{7.38b}$$

Thus the OLS problem can be solved alternatively in the SVD basis. Moreover, this formal relationship will be useful to analyze the effect of the errors on the stability of the solution.

7.5.2.2 Condition Number and Stability

Assuming that the matrix has full rank ($r = n$), it is noteworthy that, from Equation 7.38b:

$$cov(\hat{\mathbf{b}}_{OLS}) = \sigma^2(\mathbf{S}^T\mathbf{S})^{-1} \quad \text{or} \quad cov(\hat{\mathbf{b}}_{OLS}) = \begin{bmatrix} \dfrac{\sigma^2}{\lambda_1^2} & \cdot & 0 \\ \cdot & \cdot & \cdot \\ 0 & \cdot & \dfrac{\sigma^2}{\lambda_n^2} \end{bmatrix} \tag{7.39}$$

The above equation shows that an effect of noise amplification appears due to the fact that the eigenvalues $\lambda_1^2, \lambda_2^2, \ldots, \lambda_n^2$ have a wide range of order of magnitude. It is of particular interest to note in Equation 7.39 that the covariance matrix of the estimator in the diagonal basis links the square of the singular values to the variance of noise, that is, to the level of uncertainty in the measurement errors.

A small pertubation applied to a single component k of $\mathbf{z}$, such as

$$\delta\mathbf{z} = \delta z_k \mathbf{U_k} \tag{7.40a}$$

yields the following variation to the OLS estimator:

$$\delta\hat{\mathbf{b}} = \frac{\delta z_k}{\lambda_k}\mathbf{V_k} \tag{7.40b}$$

which implies a relative variation corresponding to

$$\frac{\parallel \delta\hat{\mathbf{b}} \parallel}{\parallel \delta\mathbf{z} \parallel} = \frac{1}{\lambda_k} \tag{7.40c}$$

Thus, the singular values indicate how the same perturbation yields different effects on the components of the estimator. Moreover, this relative variation may increase drastically when the singular values are close to zero. It is apparent that the maximum relative variation factor is obtained between the first and the last component, such as

$$cond(\mathbf{X}) = \frac{\lambda_1}{\lambda_r} \tag{7.41}$$

where $cond(\mathbf{X})$ is the condition number of the matrix $\mathbf{X}$, defined as the ratio between the higher and the lower singular values. Hence, $cond(\mathbf{X}) \geq 1$, and $cond(\mathbf{X}) = 1$ occurs when $\mathbf{X}$ is similar to a multiple of the identity. If $cond(\mathbf{X})$ is not too large, the problem is said to be well-conditioned and the solution is stable with respect to small variations of the data. Otherwise the problem is said to be ill-conditioned. It is clear that the separation between well-conditioned and ill-conditioned problems is not very sharp and that the concept of well-conditioned problem is more vague than the concept of well-posed problem.

We can also note that

$$cond(\mathbf{X}^T\mathbf{X}) = \frac{\lambda_1^2}{\lambda_r^2} = (cond(\mathbf{X}))^2 \tag{7.42}$$

Equation 7.42 confirms the previous comments in Section 7.3.2.2, which stated that computing the normal equations by the direct inversion of the matrix $\mathbf{X}^T\mathbf{X}$ is not always the most efficient method, since it is highly sensitive to ill-conditioning.

Example 7.2

Example of an ill-conditioned matrix

$$\begin{bmatrix} 1 & 1 \\ 1 & 1.01 \end{bmatrix} \begin{bmatrix} p_1 \\ p_2 \end{bmatrix} = \begin{bmatrix} 1 \\ 1 \end{bmatrix} \text{ the inversion yields } \begin{bmatrix} p_1 \\ p_2 \end{bmatrix} = \begin{bmatrix} 1 \\ 0 \end{bmatrix}$$

Let us give a perturbation of 1% on the second data point, such as

$$\begin{bmatrix} 1 & 1 \\ 1 & 1.01 \end{bmatrix} \begin{bmatrix} p_1 \\ p_2 \end{bmatrix} = \begin{bmatrix} 1 \\ 1.01 \end{bmatrix} \text{ the inversion yields } \begin{bmatrix} p_1 \\ p_2 \end{bmatrix} = \begin{bmatrix} 0 \\ 1 \end{bmatrix}$$

Hence, the resulting perturbation on the solution of the matrix inversion is surprisingly as far as possible from the original solution. The solution is quite unstable. Note immediately that the determinant is close to zero.

The eigenvalues are (2.005, 0.005), and the condition number is 402 ≫ 1.

7.6 Minimization Based on a Predictive Model Error

The experimentalist is often faced with the problem of identifying directly the model of a system by processing the experimental data, or retrieving the transfer function of some system (Fudym et al. 2005). Moreover, the estimation of thermophysical properties or heat transfer parameters is often a nonlinear estimation problem, as will be more deeply discussed in Chapter 8. Nonlinear estimation yields the requirement for some iterative computation procedure for the minimization of the objective function. This point may be absolutely critical when dealing with the problem of retrieving the map of some physical properties, such as thermophysical properties mapping from thermal images provided by an IR camera, or more generally with global full fields methods where the field of some magnitude must be retrieved by the means of some imaging system. The amount of data to be processed and parameters to be estimated is in that case drastically increased. Solving the estimation problem in a nonlinear framework where the resolution is iterative yields in this case a prohibitive computation cost in terms of cpu time.

It is then of great interest to be able to turn this kind of problem into a similar but linear estimation problem, for the same parameters. It is shown in this section how this can be achieved by minimizing a predictive model error instead of an output model error (Borne et al. 1997, Ljung et al. 1987). This section is devoted to the implementation of the predictive model error minimization. First, a simple basic example is proposed for the sake of clarity. Then an application for thermal diffusivity mapping from thermal images processing is presented.

7.6.1 Minimization Based on a Predictive Model Error

In this section, let us consider first the example previously proposed in Section 7.5.1.2 of a body with uniform temperature and volumetric heat capacity C, exchanging heat at the boundary, both by an input heat flux and convection at the wall, in order to illustrate the nonlinear character of estimation. Moreover, we consider now that the input heat flux is constant, such as

$$C\frac{dT}{dt} = q - hT \text{ with the initial condition } t = 0 \quad T = 0 \tag{7.43a}$$

The analytical solution is, in that case, given by

$$y_{\text{mo}}(t) = T(t) = -\frac{q}{h}\exp\left(-\frac{h}{C}t\right) \tag{7.43b}$$

Assuming that C is known, the two parameters to be retrieved are q and h.

The sensitivity coefficients defined by Equation 7.18a are in that case such as

$$X_q(t, \mathbf{P}_{\text{mo}}) = \frac{\partial y_{\text{mo}}}{\partial q} = -\frac{1}{h}\exp\left(-\frac{h}{C}t\right) \tag{7.43c}$$

$$X_h(t, \mathbf{P}_{\text{mo}}) = \frac{\partial y_{\text{mo}}}{\partial h} = q\left(\frac{1}{h^2} + \frac{1}{C}\right)\exp\left(-\frac{h}{C}t\right) \tag{7.43d}$$

where $\mathbf{P}_{\text{mo}} = [q \quad h]^T$.

The subscript "mo" indicates that the vector of parameters is defined with respect to the model y_{mo}. If we want to implement the output error method, as depicted in previous sections, then we must compare the vector of the model $\mathbf{y_{mo}} = [T(t_1) \quad T(t_2) \quad \cdots \quad T(t_m)]^T$, computed for some discrete values of the independent variable t and written in the form given by Equation 7.17b, with the corresponding experimental measurements $\mathbf{y} = [y_1 \quad y_2 \quad \cdots \quad y_m]^T$.

Writing this problem in a matrix form for m discrete values of time, with the formulation corresponding to Equation 7.19c, the sensitivity matrix is

$$\mathbf{X} = \begin{bmatrix} \mathbf{x}_q & \mathbf{x}_h \end{bmatrix} \tag{7.44a}$$

with

$$\mathbf{x}_q = \begin{bmatrix} X_q(t_1, \mathbf{P}_{\text{mo}}) & X_q(t_2, \mathbf{P}_{\text{mo}}) & \cdots & X_q(t_m, \mathbf{P}_{\text{mo}}) \end{bmatrix}^T \text{ and}$$
$$\mathbf{x}_h = \begin{bmatrix} X_h(t_1, \mathbf{P}_{\text{mo}}) & X_h(t_2, \mathbf{P}_{\text{mo}}) & \cdots & X_h(t_m, \mathbf{P}_{\text{mo}}) \end{bmatrix}^T$$

Then Equation 7.19c is

$$\mathbf{y_{mo}} = \mathbf{X}_{(\text{P}_{\text{mo}})}\mathbf{P}_{\text{mo}} \tag{7.44b}$$

where the model $\mathbf{y_{mo}} = [T(t_1) \quad T(t_2) \quad \cdots \quad T(t_m)]^T$ is computed with Equation 7.43b.

Then the OLS objective function defined in Equation 7.20a is chosen to be minimized, which is the norm of the output error (as shown in Figure 7.1).

Obviously, the sensitivity coefficients are not independent of the parameters; hence, the estimation is nonlinear with respect to the parameters. Thus, the least squares approach applied to the minimization of the output model error, where the direct model is given by Equation 7.43b is a nonlinear estimation problem.

The predictive model approach yields from the very important observation that the temperature, which is given by Equation 7.43b, is obviously also required to satisfy Equation 7.43a for any value of time, which means that the governing equation is always fulfilled.

The predictive model is based on the idea that the experimental data may also be assumed to fulfill the governing equation as well. This idea is very close to the "exact matching" concept. The predictive model error is proposed to be built within the following steps:

1. First, consider that the finite difference discretization of Equation 7.43a can be used in order to predict the temperature at time t_i from the knowledge of temperature at time t_{i-1}, such as

$$T_i = \left(1 - \frac{h}{C\Delta t}\right)T_{i-1} + \frac{q}{C\Delta t} \qquad (7.45a)$$

2. Then consider the predictive model by computing the right side of Equation 7.45a with the experimental data points instead of the model, such as

$$\hat{y}_i = \left(1 - \frac{h}{C\Delta t}\right)y_{i-1} + \frac{q}{C\Delta t} \qquad (7.45b)$$

3. Write the predictive model in a matrix form:

$$\hat{\mathbf{y}} = \mathbf{X}_{(y)}\mathbf{P} \qquad (7.45c)$$

where

$$\hat{\mathbf{y}} = [\hat{y}_1 \quad \hat{y}_2 \quad \cdots \quad \hat{y}_m]^T; \quad \mathbf{P} = \left[1 - \frac{h}{C\Delta t} \quad \frac{q}{C\Delta t}\right]^T \quad \text{and} \quad \mathbf{X}_{(y)} = \begin{bmatrix} 0 & 1 \\ y_1 & 1 \\ \cdots & \cdots \\ y_{m-1} & 1 \end{bmatrix} \qquad (7.45d)$$

Note in Equation 7.45d that the sensitivity matrix depends now on the experimental data, but is independent of the parameters. Equation 7.45c yields a predictor which computes at any time step the predicted temperature, knowing the previous outputs of the experimental measurements. Obviously, the predictive model is linear with respect to the parameters. The sketch of the predictive model is drawn in Figure 7.13, which shows how the predictive model is directly built with the experimental data. Unfortunately, this kind of approach makes necessary to fill in the sensitivity matrix directly with some linear operations on the measured data.

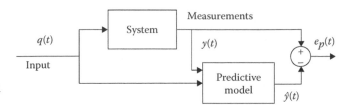

FIGURE 7.13
Minimization of the predictive model
error.

In this particular example, the vector of parameters is not exactly the same for the output model or the predictive model. This is absolutely not due to the method. In many cases, the vector of parameters may remain unchanged, as will be shown, for instance, in the example of Section 7.6.2.

4. Compute the predictive error and the norm to be minimized:

$$\mathbf{e_p} = \mathbf{y} - \hat{\mathbf{y}} = \mathbf{y} - \mathbf{X_{(y)}}\mathbf{P} \tag{7.46a}$$

$$S_{OLS}(\mathbf{P}) = \mathbf{e_p^T}\mathbf{e_p} = (\mathbf{y} - \hat{\mathbf{y}})^T(\mathbf{y} - \hat{\mathbf{y}}) = (\mathbf{y} - \mathbf{X_{(y)}}\mathbf{P})^T(\mathbf{y} - \mathbf{X_{(y)}}\mathbf{P}) \tag{7.46b}$$

The derivation of this norm can be made exactly as previously for the case of the output model error, which yields the OLS estimator of $\mathbf{P}$, such as

$$\hat{\mathbf{P}}_{OLS} = [(\mathbf{X}^T\mathbf{X})^{-1}\mathbf{X}^T]_{(y)}\mathbf{y} \tag{7.46c}$$

For the sake of clarity, the OLS estimator is considered herein, but according to the statistical assumptions regarding the measurement errors, another estimator such as a GM estimator could be used as well.

Various different approaches are available in order to take into account the effect of randomness of the sensitivity matrix, such as the Bayesian approach (Kaipio and Somersalo 2004), or the total least squares methods (Bamford et al. 2008, Van Huffel and Vandewalle 1991), which are not within the scope of the present chapter, but are discussed elsewhere in this book. Both sides of Equation 7.45c are computed with the experimental data. Hence, this equation presents some symmetry characteristic, and could be easily permuted. This symmetry is what is taken into account in the total least squares approach, where the effect of noise of both the observable and the sensitivity coefficients is considered.

However, assuming that the error is additive, such as defined with Equation 7.20a, it is apparent in Equation 7.46c that the stochastic nature of the sensitivity matrix may yield some bias in the OLS estimator, due to the correlation between the pseudo-inverse matrix and the errors. Thus, the OLS estimator is generally biased in this case.

7.6.2 Example: Thermal Diffusivity Mapping

We give herein an example relative to thermal diffusivity mapping of a heterogeneous thin plate, where the experimental data are recorded by an IR camera. Due both to the great number of pixels of the IR camera and the fast frame rate, the quantity of information to be processed is drastically increased (over 15 GB/s !), which makes it unrealistic to implement some global nonlinear estimation problems. In this case, the problem is addressed instead by using an approach based on the resolution of a linear estimation problem for each pixel, by matching the experimental data with the local discretized equations, which leads to a predictive error model. The parameter fields are retrieved by computing the local correlations between pixels. Moreover, in the case of the heat pulse response analysis, the signal-to-noise ratio is increased by applying a spatially random heat pulse heating (Batsale et al. 2004, Bamford et al. 2009).

The mathematical formulation for this problem is given by

$$C(x,y)\frac{\partial T}{\partial t} = \frac{\partial}{\partial x}\left[k(x,y)\frac{\partial T}{\partial x}\right] + \frac{\partial}{\partial y}\left[k(x,y)\frac{\partial T}{\partial y}\right] \tag{7.47a}$$

The equation is discretized using an explicit finite difference scheme, where each node (i,j) is considered as a sensor of the IR camera, such as

$$Y_{i,j}^{n+1} = L_{i,j}^n a_{i,j} + Dx_{i,j}^n \delta_{i,j}^x + Dy_{i,j}^n \delta_{i,j}^y \tag{7.47b}$$

where the subscripts pair (i,j) denotes the finite-difference node at $x_i = i\Delta x$, $i = 1\ldots n_i$, and $y_j = j\Delta y$, $j = 1\ldots n_j$, and the superscript n denotes the time $t_n = n\Delta t$, $n = 0\ldots(n_t-1)$. The other quantities appearing in Equation 7.47b are given by

$$Y_{i,j}^{n+1} = T_{i,j}^{n+1} - T_{i,j}^n \tag{7.48a}$$

$$L_{i,j}^n = \Delta t\left(\frac{T_{i-1,j}^n - 2T_{i,j}^n + T_{i+1,j}^n}{(\Delta x)^2} + \frac{T_{i,j-1}^n - 2T_{i,j}^n + T_{i,j+1}^n}{(\Delta y)^2}\right) \tag{7.48b}$$

$$Dx_{i,j}^n = \frac{\Delta t}{2\Delta x}\left(T_{i+1,j}^n - T_{i-1,j}^n\right) \quad \text{and} \quad Dy_{i,j}^n = \frac{\Delta t}{2\Delta y}\left(T_{i,j+1}^n - T_{i,j-1}^n\right) \tag{7.48c}$$

$$\delta_{i,j}^x = \frac{1}{C(x,y)}\frac{\partial k}{\partial x} \quad \text{and} \quad \delta_{i,j}^y = \frac{1}{C(x,y)}\frac{\partial k}{\partial y} \tag{7.48d}$$

Equation 7.48a defines the forward temperature difference in time. Equation 7.48b approximates the Laplacian of temperature at time t_n and node (i,j). Equations 7.48c and d are relative to the spatial derivatives of temperature and local thermal conductivity.

$a_{i,j}$, which is the local thermal diffusivity in m^2 s^{-1}.

$\delta_{i,j}^x$ and $\delta_{i,j}^y$, which are the local thermal conductivity gradients along the x and y directions, respectively, divided by the heat capacity.

We assume now that the temperature $T_{i,j}^n$ at any pixel and any time is obtained from the experimental images, and define the predictive model as

$$\hat{Y}_{i,j}^{n+1} = L_{i,j}^n(T)a_{i,j} + Dx_{i,j}^n(T)\delta_{i,j}^x + Dy_{i,j}^n(T)\delta_{i,j}^y \tag{7.49a}$$

where the variable "T" means that the operators are applied to the experimental data.

By writing Equation 7.49a for a given node (i,j) and all time steps, we obtain

$$\hat{\mathbf{Y}}_{ij} = \mathbf{J}_{ij}(T)\mathbf{P}_{ij} \tag{7.49b}$$

where

$$
\mathbf{J_{ij}}(T) = \begin{bmatrix} L_{i,j}^1 & Dx_{i,j}^1 & Dy_{i,j}^1 \\ L_{i,j}^2 & Dx_{i,j}^2 & Dy_{i,j}^2 \\ \vdots & \vdots & \vdots \\ L_{i,j}^{n_t} & Dx_{i,j}^{n_t} & Dy_{i,j}^{n_t} \end{bmatrix}(T); \quad \mathbf{Y_{ij}} = \begin{bmatrix} \hat{Y}_{i,j}^1 \\ \hat{Y}_{i,j}^2 \\ \vdots \\ \hat{Y}_{i,j}^{n_t} \end{bmatrix}; \quad \mathbf{P_{ij}} = \begin{bmatrix} a_{i,j} \\ \delta_{i,j}^x \\ \delta_{i,j}^y \end{bmatrix}
$$

where the different operators in the sensitivity matrix $\mathbf{J_{ij}}$ are computed directly with the experimental data obtained from the thermal images.

The predictive model $\hat{\mathbf{Y}}_{\mathbf{ij}}$ defined by Equation 7.49b is then used to compute the predictive model error by making the difference with the "observable data" $\mathbf{Y_{ij}}(y(t)$ in Figure 7.13), such as

$$
\mathbf{e_{pi,j}} = \mathbf{Y_{ij}} - \hat{\mathbf{Y}}_{\mathbf{ij}} = \mathbf{Y_{ij}} - \mathbf{J_{ij}P_{ij}} \quad \text{with } \mathbf{Y_{ij}} = \begin{bmatrix} Y_{i,j}^1 \\ Y_{i,j}^2 \\ \vdots \\ Y_{i,j}^{n_t} \end{bmatrix} \tag{7.50}
$$

The objective function to be minimized is the norm of the predictive error, such as

$$
S_{i,j} = \mathbf{e}_{\mathbf{pi,j}}^T \mathbf{e_{pi,j}} = (\mathbf{Y_{ij}} - \hat{\mathbf{Y}}_{\mathbf{ij}})^T(\mathbf{Y_{ij}} - \hat{\mathbf{Y}}_{\mathbf{ij}}) = (\mathbf{Y_{ij}} - \mathbf{J_{ij}P_{ij}})^T(\mathbf{Y_{ij}} - \mathbf{J_{ij}P_{ij}}) \tag{7.51}
$$

Equation 7.49b, when used as a predictor, yields a linear dependence of the system response with respect to the vector of parameters, based on the knowledge of the sensitivity matrix. With the spatial resolution and frequency of measurements made available by IR cameras, the sensitivity matrix can be approximately computed with the measurements. It is noteworthy that the choice is made to solve simultaneously as many local linear estimation problems as supplied by the measurement device (for instance, the number of pixels) instead of a single global huge problem.

In the case where the statistical assumptions regarding the errors are the standard assumptions, except no error in the sensitivity coefficients—12345679—we use the OLS estimator, such as written in Equation 7.21. The OLS estimator is the MVU when the sensitivity matrix is deterministic. We analyze now which is the effect of violation of this assumption when the sensitivity matrix is filled with measurements, and is then corrupted with random errors.

The plate under consideration is shown in Figure 7.14. It is obtained from a real IR image, which is binarized in order to show the structure of the medium, composed by some circles of material A in a matrix of material B. The thermal diffusivity map is drawn in Figure 7.14a. A numerical experiment is simulated with a level of noise such as $\sigma_T = 0.03\,\text{K}$, which is similar to the std of the IR camera sensors. The temperature field at the end of the experiment is shown in Figure 7.14b. Due to the spatially random photothermal pulse, two-dimensional heat transfer is achieved in the whole plate.

The retrieved maps of parameters estimated with the OLS estimator are depicted in Figures 7.15 and 7.16. The thermal diffusivity map is drawn in Figure 7.15. The whole

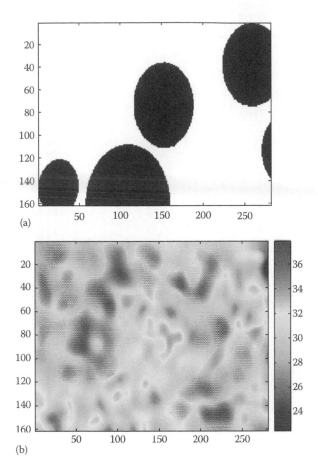

(a)

(b)

FIGURE 7.14
Heterogeneous plate: (a) thermal diffusivity material A (black): $a_A = 1.00 \times 10^{-7}$ m^2 s^{-1}; material B (white): $a_B = 2.53 \times 10^{-7}$ m^2 s^{-1} and (b) final temperature field ($t = 1.2$ s).

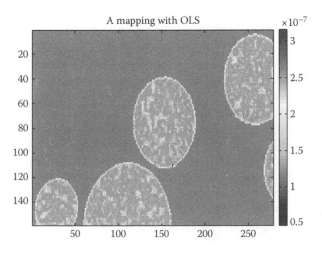

FIGURE 7.15
Thermal diffusivity mapping with OLS ($\sigma_T = 0.03$ K).

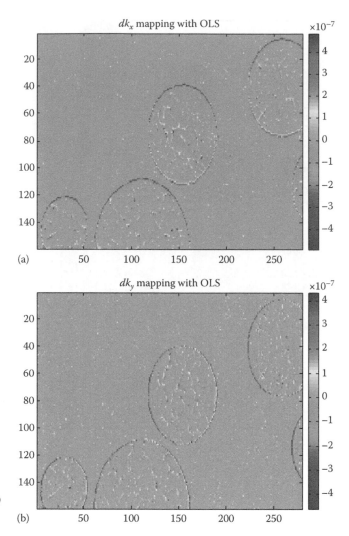

FIGURE 7.16
Thermal conductivity gradient maps: (a) $\delta_{i,j}^{x}$ and (b) $\delta_{i,j}^{y}$.

map is estimated conveniently; the root mean square error (rms) is found to be $rms = 1.7 \times 10^{-8}$ m^2 s^{-1}. In Figure 7.16a and b are drawn the retrieved maps of the thermal conductivity first derivatives in the x direction and y direction, respectively. As expected, for a medium with sharp interfaces between the two materials A and B, these parameters act as an edge filter.

The thermal diffusivity profiles corresponding to the line 60 of the plate are plotted in Figure 7.17. The results obtained with the predictor are compared with the OLS estimator which would be obtained when the sensitivity matrix is computed with the "exact" randomless temperature (without noise), while the observable data are still corrupted with measurement errors. As expected, the solution with the predicted model in X is more sensitive to noise, but no bias is present. Moreover, the residual between the predictor and the data is quite good, as shown in Figure 7.18, since it is similar to a white noise, with the same level of noise as the initial signal.

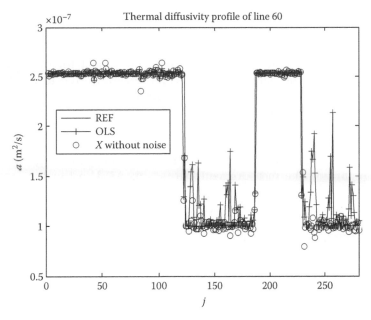

FIGURE 7.17
Thermal diffusivity profile of line 60. Comparison between OLS estimation with noise in X (full predictive model) and without noise in X.

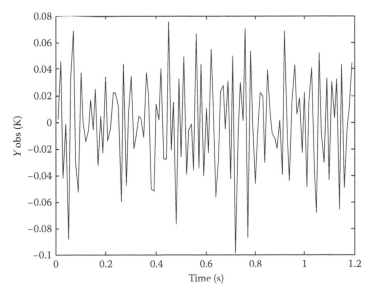

FIGURE 7.18
Residual $(\mathbf{Y_{ij}} - \mathbf{J_{ij}}(y)\mathbf{P_{ij}})$ for pixel $\mathbf{i} = 60$; $\mathbf{j} = 150$.

7.7 Conclusion

Linear least squares estimation is very efficient and robust for data processing in a discrete frame, when some parameters are to be retrieved, due to its straightforward implementation, since it is obtained directly, with no need for an iterative procedure, and may be even solved analytically, when the number of parameters to be estimated is low. Moreover, some statistical characteristics relative to the uncertainty on the estimated parameters can be derived. Another advantage of linearity is that when estimation is linear, the MVU always exists. According to the knowledge of the statistical assumptions regarding the measurement errors, this MVU estimator can be found or not. For the standard assumptions "1234589" previously discussed, the OLS estimator is the MVU. When the variance is not constant and the errors correlated, that is, "123789," the GM estimator is the MVU. When few information regarding the errors is available, the OLS estimator still may be used, but obviously with a loss of precision about its statistical behavior.

Hence, it is always of interest to turn the parameter estimation problem into a linear estimation problem whenever it is possible. This is absolutely a key point when processing an important quantity of data, and/or when the number of parameters is high. In that case, one way of turning the problem to linear is by minimizing the predictive model error instead of the output error. This approach is useful and efficient and is obtained by computing directly the model with the measured data, applying the governing equation to the measurements.

The inverse problem of retrieving some quantities by processing experimental data is in many cases an ill-posed problem (we gave herein the examples of derivation, deconvolution, extrapolation, and inverse heat conduction), or more specifically, in a linear estimation approach an ill-conditioned problem, where the solution is highly sensitive to experimental and/or modeling errors. The next chapters of this book are mostly devoted to methods that yield a better understanding and resolution of such problems.

Appendix 7.A.1: Example of Section 7.3.3

```
% Comparison of Residuals with or without error in the model
% y = ax + b
% yr = y + noise(x)
% yre = ye + noise(x)
% We estimate a and b with the noisy signal yr or yre
clear all;close all
n = 500;un = ones(size(1:n));x = (1:n)./5;y = 2*x + 5;
% Noise with normal distribution
gain = 2;noise = gain*randn(size(x));sigy = std(noise);
ye = y + 0.01*(x-50).^2;% With error in the model
yr = y + noise;yre = ye + noise; % Noisy signal
```

```
% OLS estimator with yr
X = [x' un'];% sensitivity matrix
bls = i nv(X'*X)*X'*yr'
covls = inv(X'*X).*sigy^2; % covariance matrix
yols = bls(1)*x + bls(2); % OLS predictive model
covr = (eye(n) - X*inv(X'*X)*X').*sigy^2; % cov matrix of residuals
% OLS estimator with yre
blse = inv(X'*X)*X'*yre'
yolse = blse(1)*x + blse(2); % OLS predictive model
figure(1),plot(x,yr, 'k+',x,y, 'k.-',x,yols, 'k-'),legend('Observed
   data', 'Reference line', OLS')
figure(2),plot(x,yols-yr, 'k'),legend('Residual OLS')
figure(3),plot(x,yre, 'k+',x,ye, 'k.-',x,yolse, 'k-'),legend('Observed
   data', 'Reference line', 'OLS with error in model')
figure(4),plot(x,yolse-yre, 'k'),legend('Residual with error in model')
figure(5),subplot(2,1,1),plot(x,noise, 'k'),subplot(2,1,2),
   hist(noise,25, 'k')
figure(6),imagesc(covr), colorbar
```

Appendix 7.A.2: Example of Section 7.4.2.2

```
% Gauss-Markov Estimator: Comparison with OLS
% y = ax + b
% yr = y + noise(x)
% We must estimate a and b with the noisy signal yr
clear all;close all
n = 100;un = ones(size(1:n));
x = (1:n);y = 2*x + 5;
% Noise with uniform distribution
gain = 0.02;amp = gain*(x.^2);noise = amp.*(0.5-rand(size(x)));
yr = y + noise; % Noisy signal
X = [x' un'];% sensitivity matrix
phi = diag(un./amp.^2);% inverse of covariance matrix of noise
% Gauss-Markov estimator and its covariance matrix (GM)
bgm = inv(X'*phi*X)*X'*phi*yr'
covgm = inv(X'*phi*X);
% OLS estimator and its covariance matrix
bls = inv(X'*X)*X'*yr'
covls = inv(X'*X);
ygm = bgm(1)*x + bgm(2); % GM predictive model
yols = bls(1)*x + bls(2); % OLS predictive model
figure(1), plot(x,yr, 'k+',x,y,x,yols,x,ygm), legend('Observed data',
   'Reference line', 'OLS', 'GM')
```

Appendix 7.A.3: Derivation of a Signal—Section 7.5.1.1

```
% dT/dt = q(t)
% T = t**2 ==> q(t) = 2t
% y = T + noise
clear;
dt = 0.1; nt = 100; time = dt*(1:nt); % define time
q = 2*time; T = time.^2; % define q and T
gain = 0.1; % adjust to change the std of noise
noise = gain*(0.5-1*rand(size(time))); % define noise
y = T + noise; % add noise to simulate experimental data
fi = diff(y)/dt; % derivation of the "experimental" signal
t1 = dt*(1:nt-1); tr = t1 + 0.5*dt*ones(size(t1)); % center time values
sigma = std(noise); % compute std of noise
% Plot of results
plot(time,T, 'k',time,y, 'k.-',time,q, 'k:',tr,fi, 'k+');
legend('computed temperature T,' 'noisy signal y,' 'known q,' 'retrieved by
   derivation of y');
xlabel('time (s)'); ylabel('T and q'); grid on
```

Appendix 7.A.4: Deconvolution of an Experimental Signal—Section 7.5.1.2

```
% dT/dt = q(t) -kT t=0 T=0
clear;
dt = 0.1; nt = 100; time = dt*(1:nt); % define time
% define input q(t)
un = ones(1,(2*nt-1)); q = zeros(size(time));
q(30:49) = 0.05*(1:20); q(50:69) = 1-0.05*(1:20);
h = 0.5; fc = exp(-h*time); % impulse response
y = conv(fc,q); % convolution product
gain = 0.1; noise = gain*(0.5*un-rand(size(un))); sigma = std(noise)
yr = y + noise; % noisy signal
fir = deconv(yr,fc); % deconvolution
gain = 1; noise = gain*(0.5*un-rand(size(un))); sigma = std(noise)
yr2 = y + noise; % noisy signal
fir2 = deconv(yr2,fc); % deconvolution
subplot(321),plot(time,fc, 'k',0,1.5,0,-0.5),title('Impulse
   response');
subplot(322),plot(time,q, 'k',0,1.5,0,-0.5),title('Input q(t)');
subplot(323),plot(time,y(1:nt), 'k',time,yr(1:nt), 'k.-',0,2),
   title('Noisy signal std= 0.03 K');
```

```
subplot(324),plot(time,q, 'k',time,fir, 'k+'),title('Retrieved input
  for std=0.03 K');
subplot(325),plot(time,y(1:nt), 'k',time,yr2(1:nt), 'k.-',0,2),
  title('Noisy signal std=0.3 K');
subplot(326),plot(time,q, 'k',time,fir2, 'k+'),title('Retrieved input
  for std=0.3 K');
```

Nomenclature

a	thermal diffusivity ($m^2 s^{-1}$)
$b(\cdot)$	bias of an estimated parameter vector: $b(\hat{\mathbf{P}}) = E(\hat{\mathbf{P}}) - \mathbf{P}^*$
$cov(\mathbf{u})$	covariance matrix of a random vector $\mathbf{u}$
C	volumetric heat capacity ($J\ m^{-3}\ K^{-1}$)
e_i	difference between measurement and model at time t_i
$\mathbf{e}$	vector of differences between measurement and model
$E(\cdot)$	expected value
$g(t)$	volumetric heat source ($W\ m^{-3}$)
k	thermal conductivity ($W\ m^{-1}\ K^{-1}$)
m	mass (kg)
p	scalar parameter
$\mathbf{P}$	vector of parameters
$q(t)$	heat flux density ($W\ m^{-2}$)
r	correlation coefficient
r_i	residual at time t_i
$\mathbf{r}$	vector of residuals
S	objective function
t	time (s)
T	temperature (K)
$\mathbf{T}$	temperature vector
u	input function
$\mathbf{u}$	input vector (components of $u_{param}(M, l)$)
$var(\cdot)$	variance of a scalar random quantity
$\mathbf{X}$	sensitivity matrix
X	sensitivity coefficient
y	experimental outcome of Y
Y	random variable of experimental observed data
$\mathbf{Y}$	experimental observed data vector
ε_i	measurement error at time t_i
ε	error vector
σ	standard deviation

Superscripts

$\wedge$	estimated value or estimator or predictive model
$*$	exact value or reduced value (for sensitivity functions and coefficients)
T	transpose of a matrix or a vector

Subscripts

GM   Gauss–Markov
OLS   ordinary least squares
ML   maximum likelihood
mo   model
p predictive model
WLS   weighted least squares

Abbreviations

MVU   minimum variance unbiased estimator
std   standard deviation
VIM   International Vocabulary of Metrology

References

Alifanov, O.M. 1994. *Inverse Heat Transfer Problems*. Springer-Verlag, New York.

Baboulin, M. 2006. Résolution de problèmes de moindres carrés linéaires denses de grande taille sur des calculateurs parallèles distribués. Application au calcul de champ de gravité terrestre. Thèse INP Toulouse.

Bamford, M., J.C. Batsale, D. Reungoat, and O. Fudym. 2008. Simultaneous velocity and diffusivity mapping in the case of 3-D transient heat diffusion: Heat pulse thermography and IR image sequence analysis. *QIRT Journal* 5(1):97–126. doi:10.3166/qirt.5.97–126.

Bamford, M., J.C. Batsale, and O. Fudym. 2009. Nodal and modal strategies for longitudinal thermal diffusivity profile estimation. Application to the non destructive evaluation of SiC/SiC composites under uniaxial tensile tests. *Infrared Physics & Technology* 52:1–13.

Batsale, J.C., J.L. Battaglia, and O. Fudym. 2004. Autoregressive algorithms and spatially random flash excitation for 3D non destructive evaluation with infrared cameras. *QIRT Journal* 1:5–20.

Beck, J.V. and K.J. Arnold. 1977. *Parameter Estimation in Engineering and Science*. John Wiley & Sons, New York.

Beck, J.V., B. Blackwell, and C. St. Clair. 1985. *Inverse Heat Conduction, Ill-Posed Problems*. Wiley-Interscience, New York.

Björck, Å. 1996. *Numerical Methods for Least Squares Problems*. SIAM, Philadelphia, PA.

Borne, P., G. Dauphin-Tanguy, J.P. Richard, F. Rotella, and L. Zambettakis. 1997. Modélisation et Identification des Processus, Technip.

Campbell, S.L. and C.D. Meyer, Jr. 1991. *Generalized Inverses of Linear Transformations*. Dover, New York.

Fudym, O., J.L. Battaglia, and J.C. Batsale. 2005. Measurement of thermophysical properties in semiinfinite media from random heating and fractional model identification. *Review of Scientific Instruments* 76(4):044902.

Fudym, O., C. Carrère-Gée, D. Lecomte, and B. Ladevie. 2003. Drying kinetics and heat flux in thin layer conductive drying. *International Communications on Heat and Mass Transfer* 30(3):335–349.

Golub, G.H. and C.F. Van Loan. 1996. *Matrix Computations*, 3rd edn. The Johns Hopkins University Press, Baltimore, MD.

Hadamard, J. 1923. *Lectures on Cauchy's Problem in Linear Differential Equations*. Yale University Press, New Haven, CT.

Higham, N.J. 2002. *Accuracy and Stability of Numerical Algorithms*, 2nd edn. SIAM, Philadelphia, PA.

Hui, Y. and S.D. Pagiatakis. 2004. *Geo-Spatial Information Science*, Vol. 7(4).

JCGM 100:2008. Evaluation of measurement data—Guide to the expression of uncertainty in measurement. GUM, Joint Committee for Guides in Metrology, http://www.bipm.org/utils/common/documents/jcgm/JCGM_100_2008_E.pdf

JCGM 200:2008. *International Vocabulary of Metrology—Basic and General Concepts and Associated Terms. VIM*, 3rd edn., Joint Committee for Guides in Metrology, http://www.bipm.org/utils/common/documents/jcgm/JCGM_200_2008.pdf

Kaipio, J. and E. Somersalo. 2004. *Statistical and Computational Inverse Problems*, Applied Mathematical Sciences, Vol. 160. Springer-Verlag, New York.

Kariya, T. and H. Kurata. 2004. *Generalized Least Squares*. Wiley, Chichester, U.K.

Lira, I. 2002. *Evaluating the Uncertainty of Measurement: Fundamentals and Practical Guidance*. Institute of Physics Publishing, Bristol, U.K. (243pp).

Ljung, L. 1987. *System Identification Theory for the User*. Prentice Hall, Englewood Cliffs, NJ.

Ozisik, M.N. and H.R.B. Orlande. 2000. *Inverse Heat Transfer: Fundamentals and Applications*. Taylor & Francis, New York.

Penrose, R. 1955. A generalized inverse for matrices. *Proceedings of the Cambridge Philosophical Society* 51:406–413.

Petit, D. and D. Maillet. 2008. Techniques inverses et estimation de paramètres, Techniques de l'Ingénieur, Volume Sciences Fondamentales, dossiers AF 4515:1–18 and AF 4516:1–24.

Rao, C.R., H. Toutenburg, A. Fieger, C. Heumann, T. Nittner, and S. Scheid. 1999. *Linear Models: Least Squares and Alternatives*, Springer Series in Statistics. Springer-Verlag, New York.

Serra, J.J. and O. Besson. 1999. Estimation de propriétés thermiques locales. Partie B: Introduction à l'estimation en traitement du signal. Métrologie Thermique et Techniques Inverses, Ecole d'hiver METTI, 25–30 janvier 1999, Odeillo. Presses Universitaires de Perpignan. 1:249–272.

Shenfelt, J.R., R. Luck, R.P. Taylor, and J.T. Berry. 2002. Solution to inverse heat conduction problems employing SVD and model reduction. *IJHMT* 45:67–74.

Tan, S., C. Fox, and G. Nicholls. 2006. *Inverse Problems*, Course Notes for Physics, Vol. 707, University of Auckland, New Zealand, 2006. http://www.math.auckland.ac.nz/%7Ephy707/

Van Huffel, S. and J. Vandewalle. 1991. *The Total Least Squares Problem—Computational Aspects and Analysis*. Society for Industrial and Applied Mathematics, Philadelphia, PA.

Willink R. 2008. Estimation and uncertainty in fitting straight lines to data: Different techniques. *Metrologia* 45(3):290–298.

Wolberg, J. 2005. *Data Analysis Using the Method of Least Squares: Extracting the Most Information from Experiments*. Springer, New York.

8

Inverse Problems and Regularization

Haroldo F. de Campos Velho

CONTENTS

8.1 Introduction

Looking at light propagation, we ask ourselves: What is light? How is light propagated? On observing the sea movements, we again wonder: How does the sea wave travel? And how about the dynamics of sky bodies?

The aim of science is to answer the above questions. More generally, the understanding of the natural and cultural phenomena is the challenge in science.

One historical point in the scientific development was in Isaac Newton's book, *Mathematical Principles of Natural Philosophy* (written in Latin: *Philosophiæ Naturalis Principia Mathematica*), usually called the *Principia*, the first edition of which was published in 1687. Actually, the *Principia* is made up of a set of three books: Book 1: *De motu corporum* (*On the Motion of Bodies*), Book 2: (*On the Motion of Bodies, but Motion through Resisting Mediums*), Book 3: *De systemate mundi* (*On the System of the World*). There are many contributions related to Newton's approach. The *Principia* has a lot of solved problems, and it establishes many aspects of modern scientific thought. For example, the physics applied for the planet Earth is the same as for the other celestial bodies. Moreover, natural science should be a quantitative issue, that is, the quantities involved are connected according to mathematical relations. From the development of differential and integral calculus, the quantitative relations are expressed as differential and/or integral equations.

Therefore, a long period was needed until we were able to describe the equations for fluid dynamics, for example, for understanding qualitatively and quantitatively the movement of a sea wave. The description of the mathematical equations, the material properties (constitutive equations), and the initial and boundary conditions constitute the *direct problem* (also called the *forward problem*).

However, estimating a property from a natural phenomenon, taking into account the quantity measured or desired, characterizes an *inverse problem*. The expression "inverse problem" is attributed to the Georgian astrophysicist Viktor Amazaspovich Ambartsumian (for more information about him, consult the Internet: http://www.phys-astro. sonoma.edu/brucemedalists/ambartsumian/). One definition for inverse problems is attributed to the eminent Russian scientist Oleg Mikailivitch Alifanov: "Solution of an inverse problem entails determining unknown *causes* based on observation of their *effects*" (see in the Internet: http://www.me.ua.edu/inverse/whatis.html).

In the first years of the twentieth century, a French mathematician, Jacques Hadamard, in his studies on differential equations, established the concept of a *well-posed problem*, which comprises the following: (1) the solution exists; (2) the solution is unique; (3) there is a continuous dependence on the input data. Hadamard has derived some examples that fail to follow one or more conditions cited. Such problems are called *ill-posed problems* (also known as *incorrectly posed* or *improperly posed problems*). However, Hadamard believed that ill-posed problems are curiosities, and problems from the physical reality should be well-posed problems, because nature *works* in a stable way.

Unfortunately, inverse problems belong to the class of ill-posed problems. This was a great motivation to change the conception that improperly posed problems are only a pathological mathematical curiosity. This is the motivation to study mathematical methods to deal with this kind of problems. Researchers like David L. Phillips (1962) and Sean A. Twomey (1963a, 1963b) deserve special mention, but it was the scientific work from Andrei Nikolaevich Tikhonov that was the starting point of the general formulation to the ill-posed problems—the regularization method. Professor Tikhonov was a prominent Russian mathematician from the famous Steklov Mathematical Institute. He worked mainly on topology, functional analysis, mathematical physics, and computational mathematics.

The regularization method is based on computing the smoothest approximated solution consistent with available data. The search for the smoothest (or *regular*) solution is the additional information added to the problem, moving the original problem from the ill-posed problem to the well-posed one. Figure 8.1 gives an outline of the idea behind the scheme. This is one of the most powerful techniques for computing inverse solutions.

One competitive strategy is to consider the function to be estimated as a realization of a stochastic process. The methods that employ only the statistical properties of the noise (a permanent feature in the measured quantities) are the maximum likelihood (ML) estimators. Another statistical approach is the Bayesian method, where an *a priori*

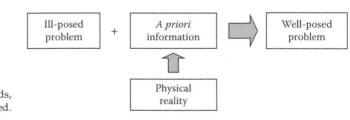

FIGURE 8.1
Principle of the regularization methods, where an additional information is used.

statistical density function is assumed to the unknown function. Some authors argue that a Bayesian scheme is a statistical justification to the regularization methods.

A standard reference on the regularization method is Tikhonov and Arsenin's book (1977). Another reference with some mathematical details is Engl et al. (1996). However, I would also like to suggest the books by Bertero and Boccacci (1998) and Aster et al. (2005). For statistical methods of inverse problems (IP), Tarantola (1987) and Kaipio and Somersalo (2005) are good references.

8.1.1 Estimating Initial Condition in Heat Transfer

Consider the problem of determining the initial condition of a linear conduction heat transfer in an insulated slab, isotropic and homogeneous, without heat sources in the domain. This problem can be mathematically formulated in a nondimensional form as follows:

$$\frac{\partial T}{\partial t} = \frac{\partial^2 T}{\partial x^2}, \quad t > 0, \quad x \in \Omega \equiv (0, 1), \tag{8.1a}$$

$$\frac{\partial T}{\partial x} = 0, \quad t > 0, \quad x = 0, \quad x = 1, \tag{8.1b}$$

$$T(x, 0) = f(x), \quad x \in [0, 1]. \tag{8.1c}$$

With convenient scaling, it is always possible to derive a nondimensional expression for system (8.1a) through (8.1c) (see Muniz et al. [1999]). The solution to the forward problem is explicitly obtained splitting the variables and applying the Fourier method (Fourier, 1940*; Özisik, 1980), $x \in (0, 1) \times \Re^+$ in the following equation:

$$T(x, t) = \sum_{m=0}^{\infty} e^{-\beta_m^2 t} \frac{1}{N(\beta_m)} X(\beta_m, x) \int_\Omega X(\beta_m, x') f(x') dx', \tag{8.2}$$

where
   $X(\beta_m, x)$ is the eigenfunction associated to the problem
   β_m are the eigenvalues
   $N(\beta_m)$ represents a *normalization* condition

For the Fourier method for the boundary condition (8.1b), we have

$$X(\beta_m, x) = \cos(\beta_m x), \quad \beta_m = m\pi, \quad N(\beta_m) = \int_0^1 X^2(\beta_m, x) dx, \quad m \in N \cup \{0\}. \tag{8.3}$$

Equation 8.2 can be written in an integral formulation:

$$T(x, t) = \int_0^1 K(x, x', t) f(x') dx', \tag{8.4}$$

---

* Poincaré (2001) said that Fourier has invented the Fourier series to deal with discontinuous functions, emerging from a physical problem: Heat conduction (Fourier, 1940). Poincaré also pointed out that our notion, about the fact that continuous functions are the only valid function, would remain for a longer time if there was no Fourier analysis.

where the kernel is defined by

$$K(x, x', t) = \sum_{m=0}^{\infty} e^{-\beta_m^2 t} \frac{1}{N(\beta_m)} X(\beta_m, x) X(\beta_m, x').$$ (8.5)

In Equation 8.4, the function $f(x)$ should be bounded following the boundary conditions.

If the temperature field is known at a time $t = \tau$, and using the orthogonality of the eigenfunction $X(\beta_m, x)$, the initial condition can be computed analytically as follows:

$$f(x) = \sum_{m=0}^{\infty} e^{\beta_m^2 \tau} \frac{1}{N(\beta_m)} X(\beta_m, x) \int_{\Omega} X(\beta_m, x') T(x', \tau) dx'.$$ (8.6)

The expression (8.6) could be verified employing an experiment. In a laboratory, one can measure the temperature by the sensors (e.g., thermal couple) at several points, for an insulated slab. The formula (8.6) can be written as

$$f(x) = \sum_{m=0}^{N_p} e^{\beta_m^2 \tau} \frac{1}{N(\beta_m)} X(\beta_m, x) \int_{\Omega} X(\beta_m, x') \theta^\delta(x', \tau) dx',$$ (8.7)

where $\theta^\delta(x_k, \tau)$, the observed temperature, differs from the true temperature $T(x_k, \tau)$, for all measured points by a small experimental error (with δ level of noise):

$$\theta^\delta(x', \tau) = T(x, \tau) + \delta.$$ (8.8)

For our test, we can consider the δ level of noise as a random variable with a 1% maximum deviation from the true temperature. The result of the inverse solution under this condition is shown in Figure 8.2. Clearly, the analytical inversion did not work. The inverse solution

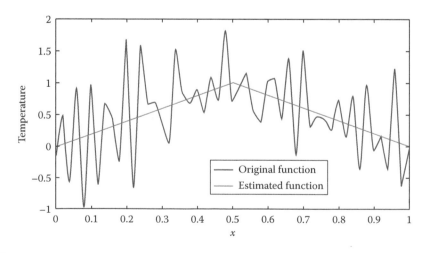

FIGURE 8.2
Initial condition: Triangular function is the true solution, while irregular function is the inverse solution computed by Equation 8.7.

exists, and it is unique. However, the solution (8.7) breaks the third Hadamard's condition, since it does not have a continuous dependency on input data (Hadamard, 1952).

Theorem 8.1

From the result (8.6), where $(f, T) \in L_2([0,1]) \times L_2([0,1])$, the computation of the initial condition is ill-posed under the L_2 norm.

Proof   Consider two initial conditions $f_1, f_2 \in L_2([0,1])$, such as $f_2(x) = f_1(x) + CX(\beta_m, x)$, with $C \in \Re - \{0\}$. The true temperatures are denoted by $T_1(x, \tau)$ and $T_2(x, \tau)$, for a fixed τ. From the linearity,

$$T_2(x, \tau) = T_1(x, \tau) + \sum_0^\infty e^{-\beta_m^2 \tau} \frac{1}{N(\beta_m)} X(\beta_m, x) \int_0^1 CX(\beta_n, x) X(\beta_m, x') dx'$$

$$= T_1(x, \tau) + Ce^{-\beta_m^2 \tau} X(\beta_n, x). \tag{8.9}$$

Therefore: $\|T_2 - T_1\|_2^2 = \int_0^1 [T_2(x, \tau) - T(x, \tau)]^2 dx = C^2 e^{-2\beta_n^2 \tau} N(\beta_n)$. For any arbitrary number C, the quantity $\|T_2 - T_1\|_2$ can be arbitrarily small for n sufficiently large.

Similarly, the difference between f_1 and f_2 can be evaluated on norm L_2, with $C \neq 0$

$$\|f_2 - f_1\|_2^2 = \int_0^1 C^2 X^2(\beta_n, x) dx = C^2 N(\beta_n) = \text{constant} > 0.$$

Summarizing, for arbitrary small differences between T_1 and T_2, one can select n and C in which the discrepancy between the corresponding inverse solutions could be arbitrary:

$$\|T_2 - T_1\|_2 \to 0, \quad \text{but} \quad \|f_2 - f_1\|_2 \to C^2 N(\beta_m) \text{ (arbitrary)}.$$

Actually, there is no surprise in the result from Theorem 8.1. A well-known property is the *Fredholm alternative*, asseverating a solution for integral equations of type (8.4). In a more technical statement, the Fredholm alternative is applied when the integral operator is a *compact operator* (Evans, 2000). In other words, from Fredholm's theory, smooth kernels of integral equations are compact operators. The spectrum of the compact operators is a subset (finite or infinity) of complex numbers, but countable, and it includes zero as a limit point (Groetsche, 1984). From the spectral theory of compact operators, it is possible to recognize that the inverse operator of Equation 8.4 will be ill defined for some points in the domain.*

In the next section, we describe some techniques to compute *good inversions*, avoiding the singularities in the inversion operation.

---

* Poincaré (2001) said that Fourier has invented the Fourier series to deal with discontinuous functions, emerging from a physical problem—heat conduction (Fourier, 1940). Our notion that continuous functions are only the valid functions could be retained for much more time without the Fourier analysis.

8.2 Regularization Method: Mathematical Formulation

We start this section by stressing the well-posed problem from Tikhonov's point of view. Following Tikhonov (Tikhonov and Arsenin, 1977), the problem $A(u) = f$, with $u \in X$ and $f \in F$, is well posed if the space X contains a subspace M such that

1. A solution exists and it belongs to the subspace M.
2. The solution in item-(1) is unique.
3. Small variations on f imply in small variations on the solution u, and the new solution remains in subspace M.

The well-posed problems from Tikhonov's point of view are conditionally well-posed problems. A pictorial representation is depicted in Figure 8.3. However, it is hard to determine the subspace M. Therefore, we need some additional condition to express what kind of solution in the space X we are searching. This is a selection process for indicating the type of properties we require from the candidate inverse solution. As indicated in Figure 8.1, this selection condition is taken from our previous knowledge of the physical problem—it is an *a priori* knowledge. For example, the requirement could allow only smooth or regular functions—this is an example of such constraint or selection condition.

The regularization procedure searches for solutions that display *global* regularity. In the mathematical formulation of the method, the inverse problem is expressed as optimization problem with constraint:

$$\min_{u \in X} \| A(u) - f^\delta \|_2^2, \quad \text{subject to } \Omega[u] \le \rho, \tag{8.10}$$

where
$A(u) = f^\delta$ represents the forward problem
$\Omega[u]$ is the regularization operator (Tikhonov and Arsenin, 1977)

The problem (8.10) can be written as an optimization problem without constrains using a Lagrange multiplier (penalty or regularization parameter)

$$\min_{u \in U} \left\{ \| A(u) - f^\delta \|_2^2 + \alpha \Omega[u] \right\}, \tag{8.11}$$

where α is the regularization parameter. The first term in the objective function (8.11) is the fidelity of the model with the observation data, while the second term expresses the regularity (or smoothness) required from the unknown quantity.

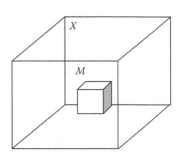

FIGURE 8.3
According to Tikhonov, it could be possible to find a subspace
$M \subset X$, where a well-posed problem can be defined.

Note that for $\alpha \to 0$, the fidelity term is overestimated; on the other hand, for $\alpha \to \infty$, all information in the mathematical model is lost.

In order to have a complete theory, the regularization operator should be known, and it is also necessary to have a scheme to compute the regularization parameter. A definition of a family of regularization operator is given in the following.

Definition 8.1 A family of continuous regularization operators R_α: $F \to U$ is called a regularization scheme for the inverse operation of $A(u) = f^\delta$, when

$$\lim_{\alpha \to 0} R_\alpha \{A(u)\} = u, \tag{8.12}$$

where $u \in U$ and α are regularization parameters.

The expression (8.11) is a practical implementation of Definition 8.1. Several regularizations operators have been derived from the pioneer works. Here, only three classes of these regularization operators will be described.

8.3 Determining the Regularization Parameter

Equations 8.11 and 8.12 introduce the regularization parameter. The best choice for this parameter is to indicate a good balance between the *fidelity term* (square difference between the mathematical model and the observations: $\|A(u) - f^\delta\|_2^2$) and the *smoothness term* (regularization term: $\Omega[u]$).

Several methods have been developed, for example (Bertero and Bocacci, 1998), Morosov's discrepancy criterion, Hansen's method (L-curve method), and generalized cross validation.

Morosov's criterion is based on the difference between data of the mathematical model and observations. It should have the same magnitude as measurement errors (Morozov and Stessin, 1992). Therefore, if δ is the error in the measure process, α is the root of the following equation:

$$\left\{ \|A(u) - f^\delta\|_2^2 \right\}_{\alpha^*} \approx \delta. \tag{8.13}$$

To determine N sensors in an inverse problem, assuming that measurement errors could be modeled by a Gaussian distribution with σ^2 variance, the discrepancy criterion for independent measures can be expressed as

$$\|A(u) - f^\delta\|_2^2 \approx N\sigma^2. \tag{8.14}$$

If the statistics on the observational data is not available, the generalized cross-validation method can be applied (Bertero and Bocacci, 1998; Aster et al., 2005). Considering now a linear forward problem ($A(u) = Au$, where A is a matrix), the goal of the cross-validation scheme is to minimize the *generalized cross-validation function* (Aster et al., 2005):

$$V(\alpha) = \frac{N \|A(u_\alpha) - f^\delta\|_2^2}{[\mathrm{Tr}\{I - B(\alpha)\}]^2}. \tag{8.15}$$

where Tr{C} is the trace of matrix C, and $B(\alpha)$ is the following matrix:

$$B(\alpha) \equiv AA^*(AA^* + \alpha I)^{-1}, \qquad (8.16)$$

where

A^* is the *adjoint matrix* $(Af,g) = (f,A^*g)$
I is the identity matrix

Another scheme to compute the regularization parameter is the L-curve method. The L-curve criterion is a geometrical approach suggested by Hansen (1992) (see Bertero and Bocacci [1998]; Aster et al. [2005]). The idea is to find the point of maximum curvature on the corner of the plot $\Omega[u_\alpha] \times \| A(u) - f^\delta \|_2^2$. In general, the plot smoothness $\times$ fidelity shows an L-shape curve type.

8.3.1 Generalized Discrepancy Principle

The Gaussian distribution assumption for modeling the noise in the measured data could be justified based on the central limit theorem* (Papoulis, 1984). However, there is a class of distribution where the second statistical value is not defined, and Morosov's criterion cannot be applied. There are many distributions in this class, for example, Cauchy and Lévy distributions or other distributions following a power law. Under the latter situation, the idea is to develop a generalization of Morosov's criterion (Morosov, 1984; Shiguemori et al., 2004a).

Let $\rho_f(x)$ be the distribution of the measure f^δ. Now, a *constrained variance* is defined as follows:

$$\bar{\sigma}^2 \equiv \int_{-d}^{d} x\rho_f(x)dx = 1. \qquad (8.17)$$

The generalization allows to compute the regularization parameter when the noise in the experimental data are given by Cauchy, t-Student, and Tsallis' distribution (with $q > 5/3$), producing good inverse solutions.

8.4 Tikhonov Regularization

Regularized inverse solutions search by *global* regularity for producing the smoothest reconstruction in agreement with the available measured (or desired) data.

The regularization operator in Equation 8.11 can be given by

$$\Omega[u] = \sum_{k=0}^{P} \mu_k \| u^{(k)} \|_2^2 . \qquad (8.18)$$

---

* The result can be expressed as the sum of a sequence of n independent and identically distributed (i.i.d.) random variables, with finite values for mean μ and variance σ^2: $S_n = X_1 + \cdots + X_n$, and considering the new random variable $Z_n \equiv (S_n - n\mu)/(\sigma\sqrt{n})$, the distribution of Z_n converges to the normal distribution $N(0,1)$, when $n \to \infty$.

Here, $u^{(k)}$ denotes the kth derivative (or difference, for discrete function) and $\mu_k \geq 0$. In general, $k = \delta_{kj}$ (Kronecker delta) and the operator becomes

$$\Omega[u] = \| u^{(k)} \|_2^2 . \tag{8.19}$$

This method is called the Tikhonov regularization of order-k, or Tikhonov-k. The effect of Tikhonov-0 regularization is to reduce the oscillations on function u, searching for smooth functions, in this case $u^{(0)} \equiv u \approx 0$. For the first-order regularization, the operator is $u^{(1)} \approx 0 \Rightarrow u \approx$ constant. Sometimes, the operator (8.18) is expressed as

$$\Omega[u] = \| L_k u \|_2^2 \tag{8.20}$$

with L_k the derivative or difference operators. For a discrete case, the Tikhonov operators of zero-, first-, and second-orders are given by

$$L_0 = I_{M \times M} = \begin{bmatrix} 1 & \cdots & 0 \\ \vdots & \ddots & \vdots \\ 0 & \cdots & 1 \end{bmatrix}_{M \times M} \quad L_1 = \begin{bmatrix} -1 & 1 & \\ & \ddots & \\ & -1 & 1 \end{bmatrix}_{(M-1) \times M} \quad L_2 = \begin{bmatrix} 2 & -1 & \\ -1 & 2 & -1 \\ & \vdots & \vdots \\ & & -1 & 2 \end{bmatrix}_{(M-2) \times M} . \tag{8.21}$$

The numerical experiment to determine the initial condition $f(x)$ in the system (8.1) is based on a triangular test function (Muniz et al., 1999):

$$f(x) = \begin{cases} 2x, & x \in [0, 0.5], \\ 2(1 - x), & x \in (0.5, 1]. \end{cases} \tag{8.22}$$

The synthetic experimental data (measured temperatures at time $\tau > 0$), which intrinsically contains errors, is obtained by adding a random perturbation to the exact solution of the direct problem, such that

$$\theta^\delta(x, \tau) = T(x, \tau) + \sigma \eta, \tag{8.23}$$

where
   σ is the standard deviation of the errors
   η is a random variable taken from a uniform distribution ($\eta \in [-1, 1]$)

The inverse solution is calculated using a discrete version of the system (8.1) with $N_x = 100$, and the experiment was performed at $\tau = 0.008$ for the level of noise $\sigma = 0.05$. According to Morozov and Stessin (1992), the optimal value for α is reached as $\| A(u_{\alpha^+}) - f^\delta \| \approx N_x \sigma^2 = 0.25$, where α^+ is the optimum regularization parameter (see Section 8.3 for more details).

   Regularized solutions for the inverse problem described in Section 8.1 are plotted in Figures 8.4 and 8.5. As pointed out in Section 8.2, a small regularization parameter yields oscillatory solutions, while $\alpha \to \infty$, the inverse solution tends to a fully uniform profile. By using the values estimated by the discrepancy criterion $\alpha = 0.073$ for zeroth-order

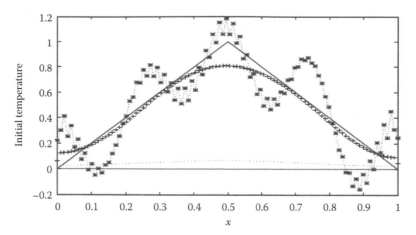

FIGURE 8.4
Inversion by Tikhonov-0, with "–": exact solution, "...": $\alpha = 10$, "*": $\alpha = 10^{-4}$, "+": $\alpha = 0.073$.

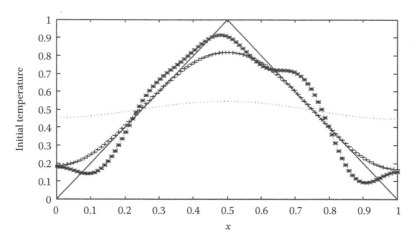

FIGURE 8.5
Inversion by Tikhonov-1, with "–": exact solution, "...": $\alpha = 1000$, "*": $\alpha = 10^{-2}$, "××": $\alpha = 34$.

Tikhonov regularization (Figure 8.6) and $\alpha = 0.34$ for first-order Tikhonov regularization, good estimations were obtained for the triangle test function. In real-world problems, the choice of a regularization parameter for a specific test function provides good results even when applied to other initial conditions.

8.5 Entropic Regularization

First proposed as a general inference procedure by Jaynes (1957), on the basis of Shannon's axiomatic characterization of the amount of information (Shannon and Weaver, 1949), the maximum entropy (MaxEnt) principle emerged at the end of the 1960s as a highly

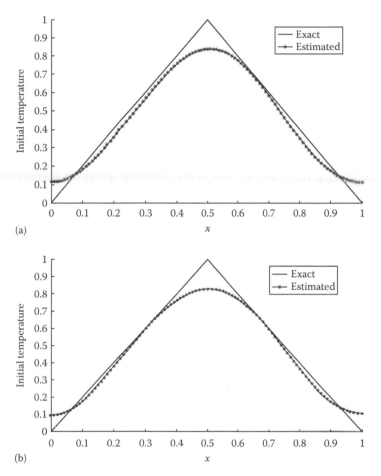

FIGURE 8.6
Initial condition estimation (see Section 8.1.1) using Tikhonov-0 regularization for Equation 8.11—see Section 8.4: (a) noise data following Cauchy distribution and (b) noise data following Tssalis' distribution ($q = 1.5$).

successful regularization technique. Since then, the MaxEnt principle has successfully been applied to a variety of fields—computerized tomography (Smith et al., 1991), nondestructive testing (Ramos and Giovannini, 1995), pattern recognition (Fleisher et al., 1990), and crystallography (de Boissieu et al., 1991).

As with other standard regularization techniques, MaxEnt searches for solutions that display *global* regularity. Thus, for a suitable choice of the penalty or regularization parameter, MaxEnt regularization yields the smoothest reconstructions, which are consistent with the available data. However, in spite of being very effective in preventing the solutions to be contaminated by artifacts, many times explicit penalizing roughness during the inversion procedure may not be the best approach to be followed. If, for instance, it is realistic to expect spikiness in the reconstruction of an image, or if there is prior evidence on the smoothness of, say, the second-derivatives of the true model, imposing an isotropic smoothing directly on the entire solution may lead to an unnecessary loss of resolution or to an unacceptable bias. In other words, the solution so obtained may no longer reflect the physical reality.

New entropic higher order regularization techniques have been introduced. They represent a generalization of the standard MaxEnt regularization method, and allow for a greater flexibility for introducing any prior information about the expected structure of the true physical model, or its derivatives, into the inversion procedure. One technique is based on the *minimization* of the entropy of the vector of *first-differences* of the unknown parameters. Adopting the standard terminology, it is called the minimum first-order entropy method (*MinEnt-1*). Unlike the classical maximum entropy formalism, this method constrains the class of possible solutions into a restricted set of low entropy models, constituted by locally smooth regions, separated by sharp discontinuities. The method MinEnt-1 was applied to the reconstruction of 2D geoelectric conductivity distributions from magnetotelluric data (Campos Velho and Ramos, 1997). A second-order entropic regularization is based on the *maximization* of the entropy of the vector of *second-differences* of the unknown parameters, and is denoted as the *MaxEnt-2* method. The MaxEnt-2 method was applied to the retrieval of vertical profiles of temperature in the atmosphere from remote sensing data (Ramos et al., 1999).

For the vector of parameters u_i with nonnegative components, the discrete entropy function S of vector u is defined by

$$S(u) = -\sum_{q=1}^{N} s_q \log(s_q), \quad \text{with} \quad \begin{cases} u = [u_1 \quad \cdots \quad u_N], \\ s_q = u_q / \sum_{q=1}^{N} u_q, \end{cases} \tag{8.24}$$

where $u_q = u(x_q)$. The (nonnegative) entropy function S attains its global maximum when all s_q are the same, which corresponds to a uniform distribution with a value of $S_{\max} = \log N$, while the lowest entropy level, $S_{\min} = 0$ is attained when all elements s_q but one are set to zero.

In practical situations, the definition of the maximal entropy regularizer can be extended for f not necessarily positive. In this case, assuming that $u_{\min} < u_i < u_{\max}$ $(i = 1, \ldots, N)$, the maximal entropy regularizer is redefined for the vector $p = [p_1 \quad \cdots \quad p_N]^T$, where $p_i = u_i - u_{\min} > 0$, $(i = 1, \ldots, N)$.

It is also possible to define higher order entropy functions, as in the Tikhonov regularization, defining new regularization procedures based on maximum entropy principle. These two approaches are based on the maximization of the entropy of the vector of *first-* and *second-differences* u. To this end, it is assumed that $u_{\min} < u_i < u_{\max}$ $(i = 1, \ldots, N)$. Under this last assumption, the methods of MaxEnt-1 and MaxEnt-2 are defined as follows (Muniz et al., 2000):

$$p_i = \begin{cases} u_i - u_{\min} + \varsigma & \text{(zeroth order)}, \\ u_{i+1} - u_i + (u_{\max} - u_{\min}) + \varsigma & \text{(first order)}, \\ u_{i+1} - 2u_i + u_{i-1} + 2(u_{\max} - u_{\min}) + \varsigma & \text{(second order)}, \end{cases} \tag{8.25}$$

where ς is a small parameter ($\varsigma = 10^{-10}$).

Maximum entropy of higher order is applied to obtain inverse solutions for problem described in Section 8.1. The results are shown in Figures 8.7 up to 8.9. For all entropic regularizations, Morosov's discrepancy principle was used to compute the regularization parameter for each entropic operator (Muniz et al., 2000) (Figures 8.8 and 8.9).

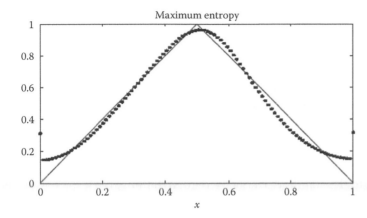

FIGURE 8.7
Regularized inverse solution for the zeroth-order entropy ($\alpha = 9$).

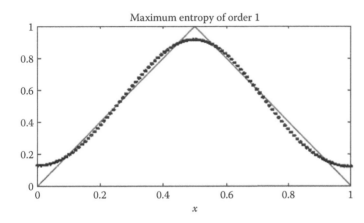

FIGURE 8.8
Regularized inverse solution for the first-order entropy ($\alpha = 7.9 \times 10^3$).

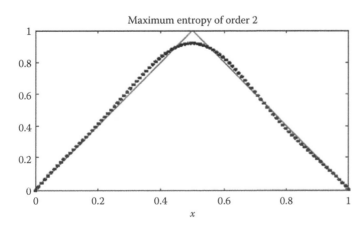

FIGURE 8.9
Regularized inverse solution for the second-order entropy ($\alpha = 1.5 \times 10^6$).

8.6 The Unified Regularization: The Principle of Maximum Nonextensive Regularization

A nonextensive formulation for the entropy has been proposed by Tsallis (1988) (see also: Tsallis [1999]). Recently, the nonextensive entropic was used to unify the Tikhonov and entropic regularizations (Campos Velho et al., 2006). The nonextensive parameter q plays a central role in Tsallis' thermostatistics, in which $q = 1$, the Boltzmann–Gibbs–Shannon's entropy is recovered. In the context of regularization theory, the nonextensive entropy includes another important particular case—when $q = 2$, the maximum nonextensive entropy principle is equivalent to the standard Tikhonov regularization.

Two methods were investigated for determining the regularization parameter for this new regularization operator: Morozov's discrepancy principle and the maximum curvature scheme of the curve relating smoothness versus fidelity, inspired by Hansen's geometrical criterion (Hansen, 1992).

A nonextensive form of entropy is given by the expression (Tsallis, 1988)

$$S_q(p) = \frac{k}{q-1} \left[1 - \sum_{i=1}^{N} p_i^q \right], \tag{8.26}$$

where

p_i is a probability
q is a free parameter (nonextensive parameter)

In thermodynamics the parameter k is known as Boltzmann's constant. Similarly, as in the mathematical theory of information, $k = 1$ is considered in the regularization theory. Tsallis' entropy reduces to the usual Boltzmann–Gibbs–Shanon formula, for the limit $q \to 1$.

Figure 8.10 shows the functional form for Tsallis' entropy for several values of q. For $q < 5/3$, the standard central limit theorem applies, implying that if p_i is written as a sum of

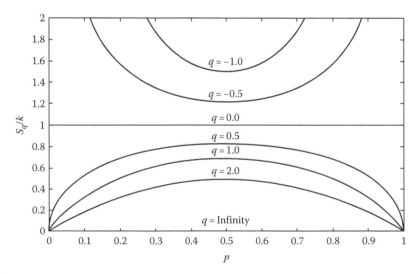

FIGURE 8.10
The behavior of the nonextensive entropy function for several values of q.

N random independent variables, in the limit case $N \to \infty$, the probability density function (PDF) for p_i in the distribution space is the *normal* (Gaussian) distribution (Tsallis, 1999). However, for $5/3 < q < 3$ the Levy–Gnedenko's central limit theorem applies, resulting for $N \to \infty$ in a Levy distribution as the PDF for the random variable p_i. The index in such Levy distribution is $\gamma = (3-1)/(q-1)$ (Tsallis, 1999).

The goal of this section is to describe formally the properties for this operator, taking into consideration the regularization purposes. Regularization property for entropy operator emerges from Jaynes' inference criterion—the maximum entropy principle—where all events have the same probability of occurring. Implying all parameters assume the same value $p_i = 1/N$, the following lemma extends this result for nonextensive entropy.

Lemma 8.1

The nonextensive function S_q is maximum as $p_i = 1/N$ for all i.

Proof The problem is to find the maximum of the function (8.26), with the following constrain:

$$\sum_{i=1}^{N} p_i = 1$$

since p_i represents a probability. Therefore, it is possible to define an objective function where the constraint can be added to the nonextensive function:

$$J(p) - S_q(p) + \alpha \left(\sum_{i=1}^{N} p_i - 1 \right),$$

where α is the Lagrange multiplier. The Lagrange multiplier, in this case, can be determined when a minimum for the objective function $J(p)$ is found, as follows:

$$\frac{\partial J(p_i)}{\partial p_i} = - \left[\frac{q}{q-1} p_i^{q-1} \right] + \alpha = 0 \Rightarrow p_i = \left[\frac{\alpha(q-1)}{q} \right]^{1/(q-1)}.$$

This result can be used to obtain the value of the p_i's that maximizes the function $J(p)$:

$$\sum_{i=1}^{N} p_i = \sum_{i=1}^{N} \left[\frac{\alpha(q-1)}{q} \right]^{1/(q-1)} = N \left[\frac{\alpha(q-1)}{q} \right]^{1/(q-1)} = 1 \Rightarrow p_i = \frac{1}{N},$$

i.e., if $p_i = 1/N$ for all $i = 1, \dots, N$, the nonextensive entropy function is maximum.

Theorem 8.2 shows that the extensive entropy and Thikhonov's regularizations are particular cases of the nonextensive entropy.

Theorem 8.2

For particular values for nonextensive entropy $q = 1$ and $q = 2$ are equivalents to the extensive entropy and Tikhonov regularizations, respectively.

Proof

(i) $q = 1$: taking the limit,

$$\lim_{q \to 1} S_q(u) = \lim_{q \to 1} \frac{1 - \sum_{i=1}^{N} p_i^q}{q - 1} = \lim_{q \to 1} \frac{1 - \sum_{i=1}^{N} \exp\left(q \log p_i\right)}{q - 1}$$

$$= \lim_{q \to 1} \frac{-\sum_{i=1}^{N} \log\left(p_i \exp\left(q \log p_i\right)\right)}{1} = -\sum_{i=1}^{N} p_i \, \log p_i.$$

(ii) $q = 2$: remembering that $\max\{S_2\}$ is equivalent to $\min\{-S_2\}$, yields

$$\max \ S_2(u) = \max\left\{1 - \sum_{i=1}^{N} p_i^2\right\} \Leftrightarrow \min\left\{-S_2(u)\right\} = \min\left\{\sum_{i=1}^{N} p_i^2 - 1\right\}.$$

Now, the maximum (minimum) value holds $\nabla_u S_2 = 0$, therefore

$$\nabla_u S_2(u) = \nabla_u \left(\sum_{i=1}^{N} p_i^2 - 1\right) = \nabla_u \left(\sum_{i=1}^{N} p_i^2\right) = \nabla_u \|u\|_2^2 .$$

In conclusion, $\max\{S_2(u)\} = \min\{\|u\|_2^2\}$, the zeroth-order Tikhonov regularization.

Table 8.1 shows the regularization parameters computed by the discrepancy principle and the L-curve method. Figures 8.11 and 8.12 show inverse solutions determined for the initial condition to the heat conduction problem (8.1). The L-curve scheme allows some oscillations for $q = 0.5$ and $q = 1.5$, while Morosov's technique has a stronger regularization for all values of the nonextensive parameters. It is important to note that boundary conditions for $x = 0$ and $x = 1$ are better identified by S_q with $q = 2.5$. Unfortunately, there is no theory until now to indicate the best value for the nonextensive parameter, and it is probably problem dependent.

TABLE 8.1

Regularization Parameters Computed by
Morosov's and Hansen's Criteria

q	α-Morosov	α-Hansen
0.5	0.0285	0.0011
1.5	0.0234	0.0008
2.0	0.0414	0.0040
2.5	0.0579	0.0040

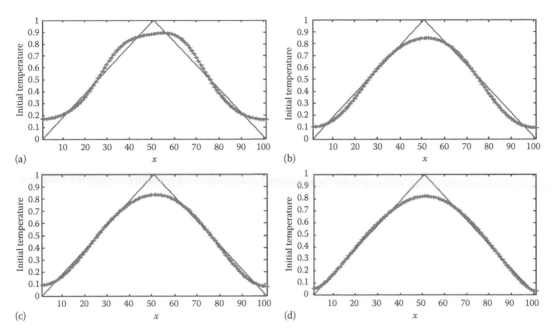

FIGURE 8.11
Reconstructions for triangular test function, with α determined by Morosov's criterion: (a) $q = 0.5$, (b) $q = 1.5$, (c) $q = 2.0$, and (d) $q = 2.5$.

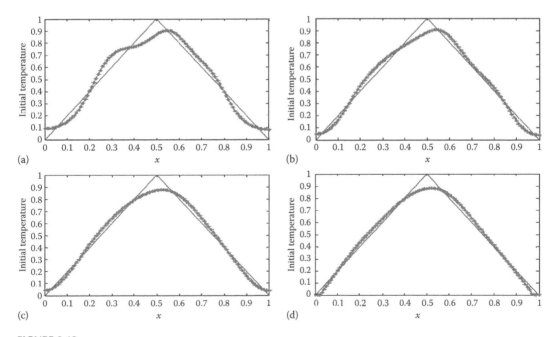

FIGURE 8.12
Reconstructions for triangular test function, with α determined by Hansen's criterion: (a) $q = 0.5$, (b) $q = 1.5$, (c) $q = 2.0$, and (d) $q = 2.5$.

8.7 Intrinsic Regularization

For solving the optimization problem (8.11), there are deterministic and stochastic techniques. In general, the advantage attributed to the stochastic techniques is to avoid local minima, while deterministic methods present a faster convergence (if the candidate solution is in the attractor basin). One example is the regularized inverse solution for the determination of initial condition (8.1) by genetic algorithm (GA) and epidemic genetic algorithm (EGA)—Figure 8.13 shows both solution, with regularization parameter computed by the discrepancy principle (Chiwiacowsky and Campos Velho, 2003).

Recently, a novelty was proposed by stochastic methods dealing with population, such as GA, extreme optimization, ant colony optimization (ACO), multiple particle collision, etc. The new strategy was named *intrinsic regularization*. Regularization is an operation to select some solutions with a specific characteristic, for example, smoothness. Therefore, the evaluation of the solution candidates (elements of the population from a stochastic method) can be

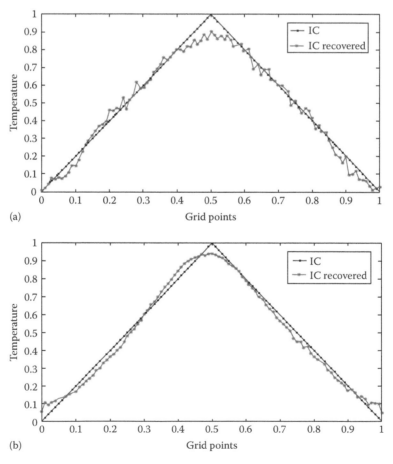

FIGURE 8.13
Reconstructions for triangular test function, for Tikhonov-0 regularization: (a) GA without epidemic and (b) epidemic GA.

done not only for the agreement with the experimental data, but also verifying the regularity of the solution.

From an *a priori* information about the smoothness of the solution profile, such knowledge is included in the generation of the candidate solutions. A larger number of elements of population is randomly generated, but only a subset of this population is selected, according to a smoothness criterion. Only the elements (in our example in heat conduction—possible profiles for the initial condition) of this subset are evaluated by the objective function:

$$J(u) = \| A(u) - f^\delta \|_2^2 , \tag{8.27}$$

requiring the solution of the direct problem. This preselection scheme can be viewed as a preregularization. For quantifying the smoothness any regularization norm can be applied.

The intrinsic regularization scheme was tested with radiation problems: estimating albedo in heat transfer (Stephany et al., 2010) and optical properties for hydrological optics (Souto et al., 2006; Carvalho et al., 2008). For these problems, a Tikhonov' second-order norm was applied to evaluate the smoothness

$$\Omega[u] = u_{k+1} - 2u_k + u_{k-1}, \quad k = 1, \ldots, N - 2. \tag{8.28}$$

For applications mentioned, the ACO was employed for minimizing the objective function (8.27). A list of advantages is as follows:

1. From the original population, only 15% of the smoothest solutions are preselected, reducing the computational cost for the inverse solution.

2. For this technique, the regularization is not explicit. Therefore, no schemes are necessary to compute a regularization parameter.

3. For the worked inverse problems in radiative transfer problems, better inverse solutions were obtained applying the intrinsic regularization operation, instead of using the standard regularization formulation.

8.8 Statistical Methods

The statistical estimation theory provides many methods for estimating properties and/or parameters such as the following: best linear unbiased estimator (BLUE), minimum variance unbiased estimator (MVUE), minimum mean squared error (MMSE), least-square estimator, maximum *a posteriori* (MAP), generalized method of moments, Markov chain Monte Carlo (MCMC), ML, Wiener filter, Kalman filter, ensemble Kalman filter (EnKF), particle filter, Bayes estimators, etc.

Some of statistical methods cited before have close relations, such as MCMC, EnKF, particle filter, and Bayes estimator, for example. It is not our goal here to investigate such relations. Our intention is to realize if some of these approaches present regularization properties, and to verify under which conditions the methods inspired on statistical considerations address similar inverse solutions (mathematical properties), as in the standard regularization strategy discussed in Section 8.2.

We only treat in this chapter on the ML and Bayesian methods. A brief description of these methods will be displayed, but for some definitions the previous definitions are necessary (see Chapter 12 for more details on the Bayesian approach).

The *expectation value* on one variable (or function) is defined by

$$E\{z\} \equiv \int_{-\infty}^{+\infty} w p_z(w) dw, \tag{8.29}$$

where $p_z(w)$ is the PDF of random variable z. In order to become clear in our application, the forward model is expressed in an *exact manner* as

$$f = A(u) + \delta, \tag{8.30}$$

where δ is a random variable with zero mean. If we have a similar model, but under different realizations, a random variable η can be written as

$$\eta = A(u) + \mu \tag{8.31}$$

with μ as a random variable with zero mean too. Denoting by $p_\eta(f)$ is the PDF of random variable η, our assumption is that the expected value for η should result in the same value by the function f, or, in other words,

$$E\{\eta\} = E\{f\} = A(u^0), \tag{8.32}$$

since μ and δ have zero mean. The expected value $E\{\eta\}$ is not known, but the function (vector, in discrete case) is a realization of the set $\eta_u = A(u) + \mu$. One issue to be pointed out is $p_\eta(f|u)$ is the conditional PDF of the random variable η_u. In addition, $E\{\eta_u\} = A(u)$; and finally, for $u = u^0$, $p_\eta(f|u^0)$ is the PDF of the random variable η_u, given by Equation 8.31, corresponding to the unknown function u^0.

Assuming that μ is an additive noise with zero expectation value (or, *zero mean*) and known PDF $p_\mu(\delta)$, the PDF of η_u is described by

$$p_\eta(f|u) = p_\mu(f - A(u)), \tag{8.33}$$

and condition 8.32 for $p_\mu(f - A(u))$ follows the previous statement.

We are looking for a criterion to select the best representation from the set of realizations η_u, considering a noisy observation f^δ. The technique to provide an estimate for u^0 is the procedure that we are looking for.

The procedure adopted by the ML method is appropriate to take into account the PDF for the random variable η, as expressed by the following equation:

$$L(u) = p_\eta(f|u). \tag{8.34}$$

ML estimation of u^0 is the function that maximizes $L(u)$ on its domain. In most of the cases, the Gaussian function is adopted as the likelihood function, but other PDF can also be used (Bertero and Bocaccio, 1998):

$$L(u) = p_\eta(f|u) = \frac{1}{\sigma_f \sqrt{2\pi}} \exp\left[-\frac{\|A(u) - f^\delta\|_2^2}{2\sigma_f^2}\right]. \tag{8.35}$$

To calculate the maximum of expression (8.35), one can employ an optimization solver (e.g., deriving Equation 8.35 related to u becoming equal to zero, and the new equation is solved by Newton method or other deterministic method; or applying a stochastic optimization scheme). For N independent measurements, the PDF of function η_u is a product of N functions. As the logarithm function is a monotonic function, the maximum of the logarithm of $L(u)$ will be the same result for the function $L(u)$ itself. However, the problem (8.35) is unstable (there is no continuous dependency on the data f^δ), similar to the problem in the deterministic approach (8.11) with $\alpha = 0$. One *a priori* condition on the unknown function is to assume that the distribution of the function u is a uniform distribution on a subset of domain X (see Figure 8.1):

$$p_\eta(u) = \begin{cases} \text{constant} \neq 0, & \text{for } u \in M, \\ 0, & \text{for } u \notin M. \end{cases} \tag{8.36}$$

Under this assumption, the likelihood function becomes

$$L(u) = p_\eta(f|u) = \frac{1}{\sigma_f \sqrt{2\pi}} \exp\left[-\frac{\|A(u) - f^\delta\|_F^2}{2\sigma_f^2}\right] \times \frac{1}{\sigma_u \sqrt{2\pi}} \exp\left[-\frac{\|u - u^+\|_X^2}{2\sigma_u^2}\right], \tag{8.37}$$

where u^+ is a reference value (our *a priori* assumption). In Equation 8.37, we point out the space that we are dealing in, as indicated by the norm used. The logarithm of the above function, as mentioned, has the same optimal solution:

$$l(u) \equiv \log L(u) = \log\left(\frac{1}{2\pi\sigma_f\sigma_u}\right) - \frac{1}{2\sigma_f^2}\left\{\|A(u) - f^\delta\|_F^2 + \alpha_\sigma \|u - u^+\|_X^2\right\}. \tag{8.38}$$

Here, $\alpha_\sigma \equiv \sigma_f^2/\sigma_u^2$ (note that the variance σ_u^2 is unknown). The result obtained by ML method has some similarity with the regularized optimization method. This result leads some authors to conclude that there is no obvious advantage of the statistical methods over the deterministic approaches (Berdichevsky and Dmitriev, 2002).

Bayesian methods are procedures based on the Bayes theorem for estimating an *a posteriori* PDF, computed from a former PDF (or *a priori* PDF). There are some controversies among scientists of the estimation theory. Some of them do not believe on scientific foundation of the maximum entropy method. Another team argues that Bayesian approach uses a subjective (nonscientific) information for the choice of the prior distribution. In this chapter, we are going to keep out of these discussions.

There are two fundamental features in the Bayesian approach. First, in the previous methods, the solution is a specific function (could be random), but in the Bayesian approach the solution is a distribution. Second, there is a natural way to introduce the *a priori* information. Such a feature is due to the approach pioneered by Thomas Bayes, an eighteenth-century researcher in probability theory.

Theorem 8.3

For two statistically independent events A and B, with probability $P(B) \neq 0$, Bayes' rule follows:

$$P(A|B) = \frac{P(B|A)P(A)}{P(B)}. \qquad (8.39a)$$

Proof   The proof is simple, and it can be found in several texts on probability theory (Papoulis, 1984). From the definition of conditional probability of an event A given event B, denoted by $P(A|B)$ as

$$P(A|B) \equiv \frac{P(A \cap B)}{P(B)} \quad \text{with } P(B) > 0,$$

where $P(A \cap B)$ is the joint probability of A and B. Similarly,

$$P(B|A) \equiv \frac{P(A \cap B)}{P(A)} \quad \text{with } P(A) > 0$$

is the conditional probability of an event B given event A. From the conditional probability defined above, we have

$$P(A \cap B) = P(A|B)P(B) = P(B|A)P(A).$$

From the last equation, it is possible to obtain Bayes' rule (8.38). The posterior distribution is computed using Bayes' theorem

$$p(u|f^\delta) = \frac{L(u\,|f^\delta)p(u)}{\int_{\text{all solutions}} L(u\,|f^\delta)p(u)du} = \frac{L(u\,|f^\delta)p(u)}{c}, \qquad (8.39b)$$

where $L(.)$ is the likelihood function. The relation between the likelihood function and the conditional probability is $L(u\,|f^\delta) = g(f^\delta|u)$. The constant c can be interpreted as a normalization condition for the posterior distribution. Equation 8.39b can be expressed as

$$p(u|f^\delta) \propto L(u|f^\delta)p(u). \qquad (8.40)$$

For independent measurements, the likelihood function is written as a product likelihood for each observation:

$$L(u|f^\delta) = g(f^\delta|u) = \prod_{k=1}^{N} g(f_k^\delta|u). \qquad (8.41)$$

Assuming observations independent and normally distributed f_k^δ with the expected value $A(u)_k$ and variance σ^2, the conditional probability will be

$$g(f^\delta|u) = \frac{1}{\sigma\sqrt{2\pi}} \exp\left[-\sum_{k=1}^{N} \frac{(A(u)_k - f_k^\delta)^2}{2\sigma^2} \right]. \tag{8.42}$$

Expressions (8.35) and (8.43) are valid for uncorrelated measurements. To maximize Equation 8.42, it is equivalent to minimize the negative of exponent

$$\min \sum_{k=1}^{N} \frac{(A(u)_k - f_k^\delta)^2}{2\sigma^2}. \tag{8.43}$$

In the multivariate situation, under Gaussian distribution assumption, covariances associated with the unknown parameters C_u and the observations C_f have important rules. The prior distribution will be

$$p(u) \propto \exp\left\{ -\frac{1}{2}(u - u^0)^\mathrm{T} C_u^{-1}(u - u^0) \right\} \tag{8.44}$$

and the likelihood function of the data given a profile u is

$$L(u) \propto \exp\left\{ -\frac{1}{2}(A(u) - f^\delta)^\mathrm{T} C_f^{-1}(A(u) - f^\delta) \right\}. \tag{8.45}$$

Finally, combining Equations 8.42 and 8.44, the posterior density function is given by

$$p(u) \propto \exp\left\{ -\frac{1}{2}(A(u) - f^\delta)^\mathrm{T} C_f^{-1}(A(u) - f^\delta) + (u - u^0)^\mathrm{T} C_u^{-1}(u - u^0) \right\}. \tag{8.46}$$

As before, maximizing Equation 8.46 is equivalent to minimizing the negative of exponent

$$\min\left\{ [A(u) - f^\delta]^\mathrm{T} C_f^{-1}[A(u) - f^\delta] + (u - u^0)^\mathrm{T} C_u^{-1}(u - u^0) \right\}. \tag{8.47}$$

The last term in Equation 8.47 can be understood as a regularization (Tikhonov-0). However, there are many implementations of the Bayesian formulation (Tarantola, 1987; Gordon et al., 1993; Kaipio and Somersalo, 2005; Fudym et al., 2008).

Related to the Bayesian approach, the lesson in this section is that the *a priori* distribution assumed for the unknown is a key issue to become the inverse process stable.

8.9 Regularized Neural Networks

After some initial controversy, artificial neural networks (ANNs) have become a well-established research field in artificial intelligence (AI). As a first approach (sometimes also called strong AI), the idea was to develop an artificial device able to emulate the ability of the human brain. The cognitive process is one component of intelligence. Information storage and the association ability allow us to make inference. The response of the brain under different situations is something linked to the experiences from the past for calibrating an extrapolation, or for adopting some previous strategy. Can the cognition emerge from a very complex connected system with unities with the ability for processing? We do not know. However, it is possible to believe that we are able to produce an answer (output) from inputs, after the establishment of the connection and/or reinforce some connections (the "configuration" of the neural network). This is a dynamical process with feedbacks, but a learning procedure is necessary to design a good connection configuration.

A simplified representation of a biological neuron is the artificial neuron—a weighted combination of the inputs is the value for the nonlinear activation function (in general, a sigmoid one). In ANNs, the identification of the connection weights is the learning (or training) phase.

Neural networks are a new technique to solve inverse problems. Different from other methodologies for computing inverse solutions, neural networks do not need the knowledge on forward problem. In other words, neural networks can be used as "inversion operator" without a mathematical model to describe the direct problem.

There are several architectures for the ANNs. But, related to the learning process, ANNs can be classified into two important classes: supervised NNs and unsupervised NNs. For the supervised strategy, the connection weights are selected to become the output from the ANN close to the target set, for example, by minimizing the functional of the square differences between the ANN output and the target values. Some authors have suggested the use of a Tikhonov's functional, as given by Equation 8.11 (Poggio and Girosi, 1990; Orr, 1995). The regularization suggested by Poggio and Girosi (1990) is applied on the output values of the NN

$$H[y] = \sum_{i=1}^{N} [y_i^{\text{Target}} - y_i(W)]^2 + \alpha \, \|Py\|, \tag{8.48}$$

where
  y is the NN output
  W is the connection weight matrix
  P is a linear operator

Poggio and Girosi (1990) use a radial base function (RBF) NN

$$y_i = \sum_{k=1}^{N} w_{ik} G(\|x - c_i\|), \tag{8.49}$$

$G(\|\cdot\|)$ being an RBF (Gaussian, for example).

Orr (1995) also employs an RBF-NN, but the regularized learning procedure is based on minimization of more standard regularization procedure (Tikhonov zeroth-order regularization):

$$\Omega[y] = \alpha \, \|W\|_2^2 \, . \tag{8.50}$$

Shiguemori et al. (2004b) employed three different architectures for the NNs, with a focus on the determination of the initial profile for the heat conduction problem in Section 8.1. The NNs used were multilayer perceptron, RBF-NN, and cascade correlation. The backpropagation algorithm was used for the learning process (without regularization).

Radial basis function networks are feedforward networks with only one hidden layer. They have been developed for data interpolation in multidimensional space. RBF nets can also learn arbitrary mappings. The primary difference between a backpropagation with one hidden layer and an RBF network is in the hidden layer units. RBF hidden layer units have a receptive field, which has a center, that is, a particular input value at which they have a maximal output. Their output tails off as the input moves away from this point. The most used function in an RBF network is a Gaussian (Figure 8.14).

RBF networks require the determination of the number of hidden units, the centers, and the sharpness (standard deviation) of their Gaussians. Generally, the centers and standard deviations are decided first by examining the vectors in the training data. The output layer weights are then trained using the delta rule.

The backpropagation training is a supervised learning algorithm that requires both input and output (desired) data. Such pairs permit the calculation of the error of the network as the difference between the calculated output and the desired vector ($\varepsilon^{1/2} = y^{\text{target}} - y(w)$). The weight adjustments are conducted by backpropagating such error to the network, governed by a change rule. The weights are changed by an amount proportional to the error at that unit, times the output of the unit feeding into the weight. Equation 8.51 shows the general weight correction according to the so-called delta rule

$$\Delta w_{ji} = \eta \frac{\partial \varepsilon}{\partial w_{ji}} y_i = \eta \delta_i y_i, \tag{8.51}$$

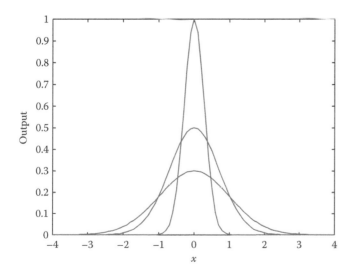

FIGURE 8.14

Gaussian for three different variances.

where
   δ_j is the local gradient
   y_i is the input signal of neuron j
   η is the learning rate parameter that controls the strength of change

For the NNs, the training sets are constituted by synthetic data obtained from the forward model, that is, profile of a *measure* points from probes spread in the space domain. Two different data sets were used. The first data set is the profiles obtained from 500 similar functions (see examples in Figure 8.15a). The second one is that obtained with 500 no-similar functions (Figure 8.15b). Similar functions are those belonging to the same class (linear function class, trigonometric function class, such as sine functions with different amplitude and/or phase, and so on). No-similar functions are those completely different, in which each one belongs to a distinct class.

The *activation* is a regular test used for checking out the NN performance, where a function belonging to the test function set is applied *to activate* (to run) the NN. Good activations were obtained for all three NNs for observational data with noise and noiseless data, for similar and nonsimilar test function sets (not shown). In the activation test the NN trained with similar data were systematically better than the training with nonsimilar functions (not shown too), with and without noise in the data.

Nevertheless, the activation test is an important procedure, indicating the performance of an NN, the effective test is defined using a function (initial condition) that did not belong to the training function set. This action is called the *generalization* of the NN.

An interesting remark is the result for the activation test (evaluation of the performance to produce the same answer with function in the data set used for training), where the training with similar functions produced better identification than nonsimilar function. However, reconstructions using nonsimilar functions were systematically better for the generalization, except in one case—the estimation of semitriangular function by RBF-NN with 5% of noise (not shown). Figure 8.16 shows the initial condition reconstruction for noisy experimental data.

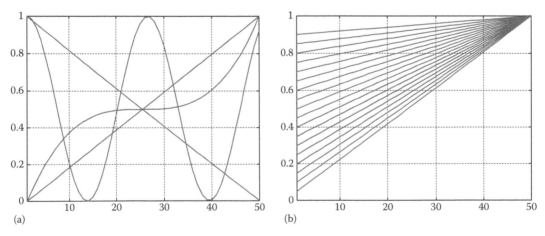

(a)   (b)

FIGURE 8.15
Sample of test functions for training: (a) nonsimilar functions and (b) similar functions.

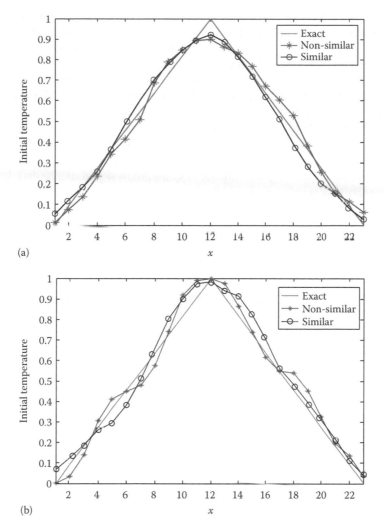

FIGURE 8.16
Reconstruction using NN with 5% of noise: (a) multilayer perceptron and (b) radial base function.

8.10 Conclusion

The regularization technique was probably the first general mathematical procedure to solve inverse problems. If we can elect two general schemes applied to inverse problems, for sure regularization and Bayesian approaches are most used and very strong procedures. This chapter is dedicated to the general aspects for regularization strategy. Of course, there are other techniques to solve inverse problems. Some missing methodologies for solving inverse problems that we can cite are as follows: variational approach, methods based on filtering properties (inverse Fourier filtering, Wiener filter, mollification method, splines), Kalman filter (and its variants), and iterative method (conjugate gradient, Landweber).

Another important issue that was not included here is related to the optimization methods applied to find the minimum of regularization functional (8.11). There are many deterministic (Newton, quasi-Newton, steepest descent, Levenberg–Marquardt, conjugate gradient, simplex method) and stochastic ones (simulated annealing, GA, ant colony system, particle swarm optimization, among others). A brief description of some of the methods is given by Campos Velho (2008).

A novelty in the field is the application of ANNs. There are few examples where the inverse methodology could be implemented on hardware devices. Two exceptions are methods based on fast Fourier transform (FFT) and neural networks (Campos Velho et al., 2007). If neural networks are good for many applications, some problems are not easy for this new method, for example, in the context of damage identification, with stiffness dependent on time; for systems with few degrees of freedom, neural networks are able to compute good solutions (Chiwiacowsky et al., 2008a), but for larger structures (see Chiwiacowsky et al. [2008b]), neural networks have not presented good performance.

Appendix 8.A: Some Properties for Nonextensive Thermostatics

For N_p microstates with probabilities $p_i \geq 0$, $i = 1, \ldots, N_p$:

A1: Nonextensive entropy:

$$S_q(p) = \frac{k}{q-1} \left(1 - \sum_{i=1}^{N_p} p_i^q \right).$$ (8.A.1)

A2: q-expectation of an observable:

$$O_q \equiv \langle O \rangle_q = \sum_{i=1}^{N_p} p_i^q o_i.$$ (8.A.2)

Properties

1. If $q \to 1$:

$$S_1 = k \sum_{i=1}^{N_p} p_i \ln p_i,$$ (8.A.3)

$$O_1 = \sum_{i=1}^{N_p} p_i O_i.$$ (8.A.4)

2. Nonextensive entropy is positive: $S_q \geq 0$.
3. Nonextensivity:

$$S_q(A + B) = S_q(A) + S_q(B) + (1 - q)S_q(A)S_q(B),$$ (8.A.5)

$$O_q(A + B) = O_q(A) + O_q(B) + (1 - q)[O_q(A)S_q(B) + O_q(B)S_q(A)].$$ (8.A.6)

4. Max S_q under constraint $O_q = \sum_i p_i^q \varepsilon_i$ (canonical ensemble):

$$p_i = \frac{1}{Z_q}[1 - \beta(1-q)\varepsilon_i]^{1/(1-q)}, \tag{8.A.7}$$

where

the ε_i is the energy of state i

$O_q = U_q$ is the nonextensive form to the internal energy

and the normalization factor Z_q (partition function), for $1 < q < 3$, is given by

$$Z_q = \left[\frac{\pi}{\beta(1-q)}\right]^{1/2} \frac{\Gamma[(3-q)/2(q-1)]}{\Gamma[1/(q-1)]}, \tag{8.A.8}$$

where $\Gamma(x)$ is the gamma function. For $q = 1$, yields

$$p_i = e^{-\beta\varepsilon_i}/Z_1. \tag{8.A.9}$$

Nomenclature

$E\{.\}$	expected value
$f(x)$	initial condition
$J(u)$	objective function
k	Boltzmann's constant
$K(x, x', t)$	kernel for integral solution of the heat transfer problem
$L(u)$	kikelihood function applied to function (vector) u
$N(\beta_m)$	"norm" for the forward heat transfer problem
p_i	probability of state i
R_α	family of regularization operators
$S(u)$	entropy of function u
S_q	Tsallis' entropy (nonextensity entropy, or q-entropy)
t	time variable
T	temperature
x	space variable
$X(\beta_m, x)$	eigenfunction for the forward heat transfer problem

Greek Variables

α	regularization parameter
β_m	eigenvalue for the forward heat transfer problem
δ	noise level
ε	square difference between target and the output from the neural network
η	learning ratio in delta rule for neural network
Ω	internal domain in partial differential equation
$\Omega(u)$	general representation for regularization operator applied to function u
σ	standard deviation
θ	temperature corrupted by noise

References

Aster, R. C., B. Borchers, C. H. Thuerber. *Parameter Estimation and Inverse Problems*, Elsevier, Burlington, VT, 2005.

Berdichevsky, M. N., V. I. Dmitriev, *Magnetotellurics in the Context of the Theory of Ill-Posed Problems*, Society of Exploration Geophysics, Tulsa, OK, 2002.

Bertero, M., P. Boccacci, *Introduction to Inverse Problems in Imaging*, Institute of Physics, London, U.K., 1998.

de Boissieu, J., R. J. Papoular, C. Janot, Maximum entropy method as applied in quasi-crystallography, *Europhysics Letters*, 16, 343–347, 1991.

Campos Velho, H. F., Inverse problems in space research, short-course, *National Congress on Computational and Applied Mathematics*, Belem, Brasil, 8–11 September, 2008 (in Portuguese).

Campos Velho, H. F., F. M. Ramos, Numerical inversion of two-dimensional geoelectric conductivity distributions from eletromagnetic ground data, *Brazilian Journal of Geophysics*, 15, 133–143, 1997, available in the internet: http://www.scielo.br/scielo.php?pid = S0102-261X1997000200003& script = sci_arttext (accessed at December 10, 2010).

Campos Velho, H. F., F. M. Ramos, E. H. Shiguemori, J. C. Carvalho, A unified regularization theory: The maximum non-extensive entropy principle, *Computational & Applied Mathematics*, 25, 307–330, 2006, available in the internet: http://www.scielo.br/scielo.php?script = sci_arttext&pid = S1807-03022006000200011 (accessed at December 10, 2010).

Campos Velho, H. F., J. D. S. Silva, E. H. Shiguemori, Hardware implementation for the atmospheric temperature retrieval from satellite data, *Inverse Problems, Design and Optimization Symposium IPDO-2007*, April 16–18, Miami, FL, 2007.

Carvalho, A. R., H. F. Campos Velho, S. Stephany, R. P. Souto, J. C. Becceneri, S. Sandri, Fuzzy ant colony optimization for estimating chlorophyll concentration profile in offshore sea water, *Inverse Problems in Science and Engineering*, 16, 705–715, 2008.

Chiwiacowsky, L. D., H. F. Campos Velho, Different approaches for the solution of a backward heat conduction problem, *Inverse Problems in Engineering*, 11, 471–494, 2003.

Chiwiacowsky, L. D., E. H. Shiguemori, H. F. Campos Velho, P. Gasbarri, J. D. S. Silva, A comparison of two different approaches for the damage identification problem, *Journal of Physics: Conference Series*, 124, 012017, 2008a, available in the internet: http://iopscience.iop.org/1742-6596/124/1/012017 (accessed at December 10, 2010).

Chiwiacowsky, L. D., P. Gasbarri, H. F. Campos Velho, Damage assessment of large space structures through the variational approach, *Acta Astronautica*, 62, 592–604, 2008b.

Cowie, J., *Climate Change: Biological and Human Aspects*, Cambridge University Press, Cambridge, U.K., 2007, p. 3.

Engl, H. W., M. Hanke, A. Neubauer, *Regularization of Inverse Problems: Mathematics and Its Applications*, Kluwer, Dordrecht, the Netherlands, 1996.

Evans, L. C., *Partial Differential Equations*, American Mathematical Society, Providence, RI, 2000.

Fleisher, M., U. Mahlab, J. Shamir, Entropy optimized filter for pattern recognition, *Applied Optics*, 29, 2091–2098, 1990.

Fourier, J., *Analytical Theory of Heat*, Cambridge University Press, Cambridge, U.K., 1940.

Fudym, O., H. R. B. Orlande, M. Bamford, J. C. Batsale, Bayesian approach for thermal diffusivity mapping from infrared images with spatially random heat pulse heating, *Journal of Physics: Conference Series* 135, 012042 (8 pp.), 2008.

Gordon, N., D. Salmond, A. Smith, Novel approach to nonlinear/non-Gaussian Bayesian state estimation, *IEE Proceedings*, 140, 107–113, 1993.

Groetsche, C. W., *The Theory of Tikhonov Regularization for Fredholm Integral Equation of the First Kind*, Pitman, London, U.K., 1984.

Hadamard, J., *Lectures on Cauchy's Problem in Linear Partial Differential Equations*, Dover, Mineola, NY, 1952.

Hansen, P. C., Analysis of discrete Ill-posed problems by means of the L-curve, *SIAM Review*, 34, 561–580, 1992.

Jaynes, E. T., Information theory and statistical mechanics, *Physical Review*, 106, 620–630, 1957.

Kaipio, J., E. Somersalo, *Statistical and Computational Inverse Problems*, Springer, New York, 2005.

Morosov, V. A., *Methods for Solving Incorrectly Posed Problems*, Springer, Verlag, 1984.

Morozov, V. A., M. Stessin, *Regularization Methods for Ill-Posed Problems*, CRC Press, Boca Raton, FL, 1992.

Muniz, W. B., H. F. Campos Velho, F. M. Ramos, A comparison of some inverse methods for estimating the initial condition of the heat equation, *Journal of Computational and Applied Mathematics*, 103, 145–163, 1999.

Muniz, W. B., F. M. Ramos, H. F. Campos Velho, Entropy- and Tikhonov-based regularization techniques applied to the backwards heat equation, *Computers & Mathematics with Applications*, 40, 1071–1084, 2000.

Orr, M. J. L., Regularisation in the selection of radial basis function centres, *Neural Computation*, 7, 606–623, 1995.

Özisik, M. N., *Heat Conduction*, Wiley Interscience, New York, 1980.

Papoulis, A., *Probability, Random Variables, and Stochastic Processes*, 2nd edn., McGraw-Hill, New York, 1984.

Phillips, D. L., A technique for the numerical solution of certain integral equations of the first kind, *Journal of the ACM*, 9(1), 84–97, 1962.

Poggio, T., F. Girosi, Networks for approximation and learning, *Proceedings of the IEEE*, 78, 1481–1497, 1990.

Poincaré, H., *The Value of Science*, Modern Library, 2001.

Ramos, F. M., A. Giovannini, Résolution d'un problème inverse multidimensionnel de diffusion de la chaleur par la méthode des eléments analytiques et par le principe de l'entropie maximale, *International Journal of Heat and Mass Transfer*, 38, 101–111, 1995.

Ramos, F. M., H. F. de Campos Velho, J. C. Carvalho, N. J. Ferreira, Novel approaches on entropic regularization, *Inverse Problems*, 15, 1139–1148, 1999.

Shannon, C. E., W. Weaver, *The Matemathical Theory of Communication*, University of Illinois Press, Champaign, IL, 1949.

Shiguemori, E. H., H. F. Campos Velho, J. D. S. da Silva, *Generalized Morozov's Principle, Inverse Problems, Design and Optimization Symposium (IPDO)*, March 17–19, Rio de Janeiro, Brazil, Proceedings in CD-Rom, paper code IPDO-079, Vol. 2, pp. 290–298, 2004a.

Shiguemori, E. H., H. F. Campos Velho, J. D. S. da Silva, Estimation of initial condition in heat conduction by neural network, *Inverse Problems in Science and Engineering*, 12, 317–328, 2004b.

Smith, R. T., C. K. Zoltani, G. J. Klem, M. W. Coleman, Reconstruction of tomographic images from sparse data sets by a new finite element maximum entropy approach, *Applied Optics*, 30, 573–582, 1991.

Souto, R. P., S. Stephany, H. F. de Campos Velho, Reconstruction vertical profiles of absorption and scattering coefficients from multispectral radiances, *Mathematics and Computers in Simulation*, 73(1–4), 255–267, 2006.

Stephany, S., J. C. Becceneri, R. P. Souto, H. F. Campos Velho, A. J. Silva Neto, A pre-regularization scheme for the reconstruction of a spatial dependent scattering albedo using a hybrid ant colony optimization implementation, *Applied Mathematical Modelling*, 34(3), 561–572, 2010.

Tarantola, A., *Inverse Problem Theory*, Elsevier, Amsterdam, the Netherlands, 1987.

Tikhonov, A. N., V. Y. Arsenin, *Solution of Ill-Posed Problems*, John Wiley & Sons, New York, 1977.

Tsallis, C., Possible generalization of Boltzmann-Gibbs statistics, *Journal of Statistical Physics*, 52, 479–487, 1988.

Tsallis, C., Nonextensive statistics: Theoretical, experimental and computational evidences and connections, *Brazilian Journal of Physics*, 29, 1–35, 1999.

Twomey, S. A., The determination of aerosol size distributions from diffusional decay measurements, *Journal of the Franklin Institute*, 275(2), 121–138, 1936a.

Twomey, S. A., On the numerical solution of Fredholm integral equations of the first kind by the inversion of the linear system produced by quadrature, *Journal of the ACM*, 10(1), 97–101, 1963b.

9

Nonlinear Estimation Problems

Benjamin Remy and Stéphane Andre

CONTENTS

9.1 Introduction

The experimental characterization of an insulating material will be studied first. This material is sandwiched between two layers of highly conductive materials. Hence this problem will be referred to as the three-layer thermal problem. The first part will be dedicated to a self-sufficient presentation of the basic tools we need in order to analyze a problem in view of data inversion for parameter estimation purposes. Some very common concepts such as sensitivity analysis and confidence bounds will be recalled and applied to our test case. The discussion will then be oriented to the analysis of "after-estimation" residuals. In this second part, recent developments that are still under progress will be presented and the use of information (or noninformation) contained in the residuals will be discussed. Especially, we will describe the tools that may be used to correct an identification procedure from the existence of some bias in the model. In order to master the proposed tools, the bias will be introduced by fixing the values of some parameters to their nominal value. Once validated, this procedure can be applied to a very general situation where the bias is not known (effect of model reduction or real experimental bias).

Regarding parameter estimation, the (philosophical) position of the authors is that nothing can be done in the case of an ill-conditioned problem except recognizing that the initial goal is in vain, or modifying the problem through physical thinking to make it well-posed or adequately conditioned. This position emerges from the well-known parsimony "principle" (see http://en.wikipedia.org/wiki/Parsimony), which in the field of science could be summarized by this sentence: "trying to perfectly recover reality is indeed very easy, when one adds parameters to each others so that it connects-the-dots." There is much more to learn and to retrieve from the distance maintained between a model and the observations it is supposed to match. As a consequence, and to return back to our subject, any minimization algorithm is a good one when the problem is well defined. For this reason, no discussion will be found in this chapter regarding the minimization algorithmic techniques (the authors will use the "basic" Levenberg–Marquardt algorithm for all the adjustments presented in the text). On the contrary, we want to put forward that a deeper understanding of how things go is always preferable.

9.2 A Parameter Estimation Problem in Thermal Transfer

A sample made of an insulating material of thickness e is considered. This material (ρ, c, λ) is sandwiched between two copper plates of very thin or very large thickness e_c. The characterization of the material (the measurement of some of its thermal properties) is attempted using the flash technique: a heat pulse (Dirac distribution) is produced to irradiate the so-called front face $(z = 0)$ of the sample. The heat transfer is experimentally made 1D (large aspect ratio). Front and rear surfaces $(z = 0$ and $z = e + 2e_c)$ can exchange heat with the surroundings through a global heat exchange conductance h.

Figure 9.1 gives the schematic principle and main variables of the problem.

For this study, the following typical values of the heat density Q and heat exchange coefficient h will be considered: $Q = 10,000\,\text{J m}^{-2}$ and $h = 10\,\text{W m}^{-2}\,\text{K}^{-1}$. These parameters have nothing to do with our metrological goals but are indivisibly present: Q for the experiment to be possible and h as an undesired nuisance parameter. Two test cases A

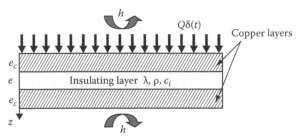

FIGURE 9.1

Principle of the flash experiment for thermal characterization of materials.

TABLE 9.1

Physical Properties for the Two Test Cases

Test Case A	Copper	Insulating Material A
Thickness (mm)	0.6	6
Thermal conductivity (SI)	385	0.02
Volumic heat capacity (SI)	3.6×10^6	5000
Test Case B	**Copper**	**Insulating Material B**
Thickness (mm)	2.0	0.200
Thermal conductivity (SI)	385	0.2
Volumic heat capacity (SI)	3.6×10^6	1.0×10^6

and B will be considered depending on the materials and thicknesses considered (refer to Table 9.1). In test case A, the thickness of the copper blades is very small and a highly insulating material of very low capacity is considered (aerogel substance). In test case B, the thickness of the copper blades is pretty much larger than that of the insulating material (polyvinyl chloride [PVC] or polymer glue). This case corresponds to the measurement of thermal (contact) resistances.

9.2.1 The Physical Direct Model

Modeling in physics and engineering is one of the prior tasks to carry on. The model is the basic instrument to approach reality either by anticipating the observation in order to see whether the predictions make sense physically, or by "comparing" experimentally obtained results with the model, supposed to be used in the same "conditions" (with the same entries). But there may be different ways of constructing a model, which may lead to different models (having different mathematical structures, for example). By different, we mean that they may not approach physical reality with the same precision. And by precision, we mean some acceptable distance that depends purely on the modeler decision, which basically relies on well-assigned objectives. An open field exists for the experimentalist–metrologist in order to realize some compromise between "sufficient" precision in the modeling and "optimal" estimation of parameters that he or she wants to measure. Our example illustrates this point.

A basic reduced direct model (referred to as RDM1 in the following) is developed here that, besides the thermal experiment design, integrates the fact that the material to be characterized presents a very low thermal conductivity and heat capacity and that the two layers used for technical constraints have been chosen to be highly conductive and capacitive. The assumption that can be made here is that the capacitive effect of the insulating material can be neglected when compared to that of the copper layers, and that the thermal resistance of these layers may be of negligible impact compared to the resistance of the well-defined insulating material. In other words, we assume that the diffusion time in the copper layers is extremely short, thus leading to the sole effect of the inertial behavior (lumped-body approximation), and that the diffusion time in material A or B is so long that it should mask their own inertial effect. With respect to the full parameter inversion problem, relying on the perfectly refined (or detailed) modeling of this experiment, this is in fact a first parameter model reduction which consists in neglecting the thermal conductivity of the copper layers and the heat capacity of the sandwiched material.

The model can be represented in the electrical analogy framework as a combination of capacities and resistances as depicted in Figure 9.2.

Capacity C corresponds to the heat capacity of a copper layer expressed in $J\,m^{-2}\,K^{-1}$, product of the mass of copper per unit surface m and the specific heat capacity of copper c_c: $C = mc_c = \rho e_c c_c$. Note that this means that the problem is free from any area definition and that symbol ϕ stands for a surface heat flux density. An identical heat exchange coefficient is assumed on both the front and rear faces of the three-layer sample. Subscripts $i = 1$ and $i = 2$ denote, respectively, the font and rear face vectors (θ_i, ϕ_i). In the Laplace space (all variables are now considered to be Laplace transformed, p being the Laplace variable), the solution of this problem can be put in the form of a quadrupole representation (Maillet et al. 2000), that is, in a matrix form. The exact correspondence of the electrical scheme above is given in Figure 9.3.

Note that the convective resistance $1/h$, instead of appearing on the upper right corner of the corresponding quadrupole matrix, has been located in the lower left corner in the form of impedance $(1/h)^{-1}$. These special quadrupoles connected in series lead finally to a direct

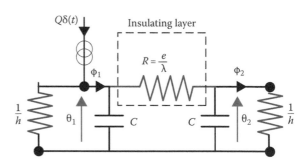

FIGURE 9.2
RDM1 as seen through electrical analogy (pure resistances and capacities).

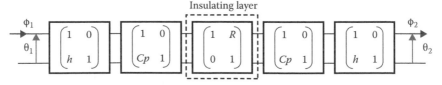

FIGURE 9.3
RDM1 in Laplace space (quadrupole matrix product formulation).

relationship between both front and rear face temperature-heat flux vectors. By a simple matrix multiplication, we have

$$
\begin{bmatrix} \theta_1 \\ \phi_1 \end{bmatrix} = \begin{bmatrix} A' & B' \\ C' & D' \end{bmatrix} \begin{bmatrix} \theta_2 \\ \phi_2 \end{bmatrix} \quad \text{where} \quad \begin{cases} A' = 1 + hR + RCp, \\ B' = R, \\ C' = h(2 + hR) + 2C(1 + hR)p + RC^2p^2, \\ D' = 1 + hR + CRp, \\ \phi_2 = 0, \\ \phi_1 = Q. \end{cases} \tag{9.1}
$$

The front face Laplace temperature θ_1 and the rear face Laplace temperature θ_2, the metrological signals we want to model in real-time space, are given by the following formulas:

$$
\begin{cases} \theta_1 = \dfrac{A'}{C'} Q \\ \theta_2 = \dfrac{Q}{C'} \end{cases} \quad \text{or} \quad \begin{cases} \theta_1 = \dfrac{Q}{C} \dfrac{(p + \alpha)}{(p + \alpha)^2 - \omega^2} \\ \theta_2 = \dfrac{Q}{C} \dfrac{\omega}{(p + \alpha)^2 - \omega^2} \end{cases} \quad \text{with} \quad \begin{cases} \alpha = \dfrac{1 + Rh}{RC}, \\ \omega = \dfrac{1}{RC}. \end{cases} \tag{9.2}
$$

The advantage of this (approximated) model in view of our further discussions is that it presents a comfortable analytical solution. Indeed, return to the temporal domain gives (Abramowitz and Stegun 1970)

$$
\begin{cases} T_1(t) = \dfrac{Q}{2C} \left[\exp\left(-\dfrac{h}{C} t \right) + \exp\left(-\left(\dfrac{2}{RC} + \dfrac{h}{C} \right) t \right) \right], \\ T_2(t) = \dfrac{Q}{2C} \left[\exp\left(-\dfrac{h}{C} t \right) - \exp\left(-\left(\dfrac{2}{RC} + \dfrac{h}{C} \right) t \right) \right]. \end{cases} \tag{9.3}
$$

These two model response equations give us a sound basis to deal with the Parameter Estimation Problem (PEP). We have responses made of two components, each of them being a time exponential. Therefore, two characteristic times control the phenomenon. They are made of the product of the two different resistances R and $1/h$, and the thermal capacity of a copper layer C.

Finally, this model gives rise to three parameters: $\beta_1 = Q/2C$, $\beta_2 = C/h$, and $\beta_3 = RC/2$. They depend on the three physical parameters (Q, h, λ), C and e being parameters assumed to be perfectly known. Regarding the PEP, this model is obviously NonLinear in the Parameters (NLP). It is evident, for example, that depending on the values of h and R, the two parameters β_2 and β_3 will be identifiable or not (if the two characteristic times of the exponentials are too close, these will be indistinguishable).

9.2.2 Direct Simulations for the Two Test Cases

Of course, a more refined model may be developed that takes into account the heat capacity of the insulating material (but still neglecting the diffusion transfer in the copper layers). This second Reduced Direct Model (RDM2) can be derived without effort since the insulating material can now be represented (still approximately) by two resistances

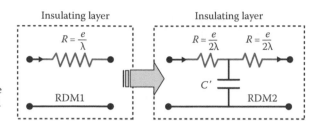

FIGURE 9.4
Two options for modeling the behavior of the insulating material which lead to either RDM1 or RDM2.

(whose sum is exactly the real resistance $R = e/\lambda$) and the capacity of the layer denoted $C' = C_i = \rho c e$ (Figure 9.4).

The model equations can be derived as previously, using the quadrupole representation in Laplace space. Computing the new matrix product in Laplace domain gives equations for $\theta_1(p)$ and $\theta_2(p)$ that differ from those given in Section 9.2.1, which do not lead to simple analytical expressions when returning to the real-time domain. A numerical Laplace inversion will be used in the computations.

Finally, the complete direct model (CDM) can be built for such a thermal problem, which would be applicable to any case (without any approximations). But since the objective of the chapter is to discuss NLE, RDM1 will be a nice object for such a purpose. Anyway, in order to prove the consistency of both reduced models RDM1 and RDM2 with, for example, the "perfect" model CDM, we have plotted in Figure 9.5 the rear and front face temperature responses obtained for the two test cases under consideration, for both the RDMs and the CDM. This latter can be obtained very easily through the quadrupole framework (Maillet et al. 2000). It is obvious from these curves that the agreement is nice. By "nice," we mean that although one may be able to distinguish some slight discrepancies, nothing has been obviously omitted regarding the phenomena involved in the problem. And the point is that with such an agreement, the bias induced by working with RDM1 instead of RDM2 or CDM is far more preferable simply because the latter involves, respectively, one or two additional parameters (thermal conductivity of the copper layers and heat capacity of the insulating layer). Taking these two additional parameters into account in our estimation problem (for the two test cases we have chosen) can be proved to be seriously dangerous because we simply do not need them (this will be demonstrated in

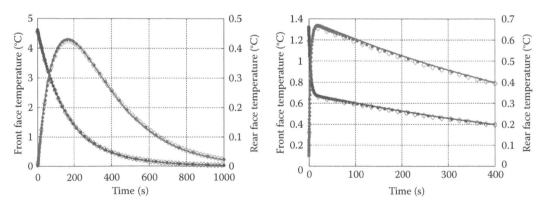

FIGURE 9.5
Comparison between the CMD (dots •) and the reduced models RDM1 (lines) and RDM2 ($\diamond$s) for test case A (left) and test case B (right).

Section 9.2.3). Mathematically speaking, they would either yield a very weak sensitivity to the global response or, alternatively, they would introduce some correlation between all parameters, which in both cases destroys the conditioning of the PEP and induces very dispersed estimations. The question is, is there and what is the impact of this reduction in parameter vector dimension on the measurements of the other parameters?

9.2.3 Inverse Analysis

9.2.3.1 Prerequisite Basic Tools

We recall rapidly all the mathematical ingredients necessary for conducting a stochastical approach of the analysis. The method we used for this PEP is based on the (unweighted) ordinary least square (OLS) criterion. The parameter vector $\boldsymbol{\beta}$ (dimension p) is found by minimization of the OLS sum:

$$S(\boldsymbol{\beta}) = \sum_{i=1}^{n} (Y_i - T(t_i, \boldsymbol{\beta}))^2, \tag{9.4}$$

where the signal is corrupted by an additive noise ε_i for each data point $i = 1:n$.

$$Y_i = T(t_i, \boldsymbol{\beta}) + \varepsilon_i. \tag{9.5}$$

This noise is a stochastic variable that we choose to consider of zero mean, constant standard deviation (std) σ, and uncorrelated along time (Beck and Arnold 1977); hence,

$$E(\varepsilon_i) = 0 \quad \mathrm{var}(\varepsilon_i) = \sigma^2 \quad \mathbf{cov}(\varepsilon_i) = \sigma^2 \mathbf{Id}. \tag{9.6}$$

The observable, the model variable, and the noise are considered as (discrete) vectorial data of dimension $n \times 1$ according to the number of considered experimental acquisition times. The least square sum (9.4) can be rewritten as

$$S(\boldsymbol{\beta}) = (\mathbf{Y} - \mathbf{T}(\boldsymbol{\beta}))^t \quad (\mathbf{Y} - \mathbf{T}(\boldsymbol{\beta})). \tag{9.7}$$

Its minimum is obtained when the jth equations $\partial S(t, \boldsymbol{\beta})/\partial \beta_j = 0$ ($j = 1 \ldots p$) are verified.

Since the model is NLP, the minimum is found through an iterative process (Gauss–Newton algorithm basically, Levenberg 1944) of the form

$$\hat{\boldsymbol{\beta}}^{(k+1)} = \hat{\boldsymbol{\beta}}^{(k)} + (\mathbf{X}^{(k)^t} \mathbf{X}^{(k)})^{-1} \mathbf{X}^{(k)^t} (\mathbf{Y} - \mathbf{T}(\boldsymbol{\beta}^{(k)})), \tag{9.8}$$

with $\mathbf{X}^{(k)} = \mathbf{X}(\hat{\boldsymbol{\beta}}^{(k)})$ and where the sensitivity matrix $\mathbf{X}$ gathers the sensitivity coefficients $X_{\beta_j} = \partial T(t, \boldsymbol{\beta})/\partial \beta_j$.

We recall that the iterative process (9.8) requires using the inverse of a matrix; therefore, it is clear that the parameters can be found only if the sensitivity coefficients are nonzero and linearly independent. Without any specialized and dedicated tools, this iterative process can be stopped when the residuals are of the same order of magnitude as the measurement noise, that is, when

$$S(\hat{\boldsymbol{\beta}}^{(k)}) \approx n\sigma^2. \tag{9.9}$$

At convergence, the standard deviation of the error made on the estimated parameters can be evaluated, thanks to the (symmetrical) *estimated* covariance matrix of the estimator. It characterizes the precision that can be reached on the estimated parameters (its inverse is sometimes named the precision matrix) and depends on the statistical assumptions that can be made on the data. We assume here the validity of the set 1111—11 for these assumptions (according to Beck's taxonomy, see Beck and Arnold 1977, p. 134 and Chapter VII), which means additive, zero mean, uncorrelated errors with constant variance, nonstochastic independent variable (time), and no prior information on the parameters.

This OLS method is of particular interest since it allows the estimation of the uncertainties on estimated values of parameters. From Equations 9.5 through 9.8, expectancy E and standard deviations $\sigma_{\hat{\beta}}$ of the estimated parameters can be evaluated.

9.2.3.1.1 *Expectancy of the Estimator*

$$E(\hat{\boldsymbol{\beta}}) = \boldsymbol{\beta} + E[(X^tX)^{-1}X^t\boldsymbol{\varepsilon}(t)] = \boldsymbol{\beta} + (X^tX)^{-1}X^tE[\boldsymbol{\varepsilon}(t)]. \qquad (9.10)$$

And, because of a zero mean assumption on the noise,

$$E(\hat{\boldsymbol{\beta}}) = \boldsymbol{\beta}. \qquad (9.11)$$

The expectancy shows that the estimated values are equal to the true values of parameters (unbiased estimator). This fully justifies the use of least squares method for the determination of unknown parameters.

9.2.3.1.2 *Standard Deviation of the Estimator (Variance–Covariance Matrix)*
By definition of the covariance matrix,

$$\text{cov}(\hat{\boldsymbol{\beta}}) = E[(\hat{\boldsymbol{\beta}} - E(\hat{\boldsymbol{\beta}}))(\hat{\boldsymbol{\beta}} - E(\hat{\boldsymbol{\beta}}))^t] = E[(\hat{\boldsymbol{\beta}} - \boldsymbol{\beta})(\hat{\boldsymbol{\beta}} - \boldsymbol{\beta})^t]. \qquad (9.12)$$

Substituting $\hat{\boldsymbol{\beta}}$ by its expression

$$\text{cov}(\hat{\boldsymbol{\beta}}) = E[(X^tX)^{-1}X^t\boldsymbol{\varepsilon}\boldsymbol{\varepsilon}^tX(X^tX)^{-1}] = (X^tX)^{-1}X^tE(\boldsymbol{\varepsilon}\boldsymbol{\varepsilon}^t)X(X^tX)^{-1}. \qquad (9.13)$$

As $\boldsymbol{\varepsilon}$ is assumed to be noncorrelated, the matrix $E(\boldsymbol{\varepsilon}\boldsymbol{\varepsilon}^t)$ is diagonal. Assuming the standard deviation of noise to be constant, this matrix is spherical, $E(\boldsymbol{\varepsilon}\boldsymbol{\varepsilon}^t) = \sigma^2.\text{Id}$. We finally obtain

$$\text{cov}(\hat{\boldsymbol{\beta}}) = \sigma_b^2(X^tX)^{-1}. \qquad (9.14)$$

This can be also written as

$$\text{cov}(\hat{\boldsymbol{\beta}}) \approx \begin{bmatrix} \text{var}(\hat{\beta}_i) & \text{cov}(\hat{\beta}_i, \hat{\beta}_j) & \cdots \\ \text{cov}(\hat{\beta}_i, \hat{\beta}_j) & \text{var}(\hat{\beta}_j) & \cdots \\ \vdots & \vdots & \ddots \end{bmatrix} = \sigma^2(X^t(\hat{\boldsymbol{\beta}})X(\hat{\boldsymbol{\beta}}))^{-1}. \qquad (9.15)$$

It depends obviously on the level of the signal-to-noise ratio (SNR) and brings into play the inverse of the (X^tX) matrix, already pointed as a decisive operation for a troubleless estimation. It also depends on the number of points and of their distribution along the estimation interval, which, by the way, may be optimized if necessary (Beck et al. 1985). The diagonal terms represent the square of the estimated standard deviation of each parameter $\sigma_{\hat{\beta}_i}^2$. They quantify the error that one can expect through inverse estimation. This is true if the above assumption made for the noise is consistent within the experiment. The problem being NLP, retrieving these optimum bounds through a statistical analysis may depend on the starting guesses made to initialize the estimation algorithm. This matrix can also be an indicator for detecting possible correlations between the parameters. Estimation of the correlation matrix is calculated according to

$$\mathbf{cor}(\hat{\boldsymbol{\beta}}) \approx \begin{bmatrix} 1 & \rho_{ij} & \cdots \\ \rho_{ij} & 1 & \cdots \\ \vdots & \vdots & \ddots \end{bmatrix} \quad \text{all terms being the result of } \rho_{ij} = \frac{\mathrm{cov}(\hat{\beta}_i, \hat{\beta}_j)}{\sqrt{\sigma_{\hat{\beta}_i}^2 \sigma_{\hat{\beta}_j}^2}}. \tag{9.16}$$

The correlation coefficients (off-diagonal terms) correspond to some measure of the correlation existing between the two parameters β_i and β_j. They vary between -1 and 1. They are global quantities (in some sense, "averaged" over the considered identification interval, the whole $[0, t]$ here). Gallant (1975) suggested that difficulty in computation may be encountered when the common logarithm of the ratio of the largest to smallest eigenvalues of **cor** exceeds one-half the number of significant decimal digits used by the computer. A more practical hybrid matrix representation **Vcor** can be constructed. It gathers the diagonal terms of the covariance matrix (more precisely their square root, normalized by the value of the estimated parameter) and the off-diagonal terms of the correlation matrix:

$$\mathbf{Vcor}(\hat{\boldsymbol{\beta}}) \approx \begin{bmatrix} \sqrt{\mathrm{var}(\hat{\beta}_i)}/\hat{\beta}_i & \rho_{ij} & \cdots \\ \rho_{ij} & \sqrt{\mathrm{var}(\hat{\beta}_j)}/\hat{\beta}_j & \cdots \\ \vdots & \vdots & \ddots \end{bmatrix}. \tag{9.17}$$

The OLS estimator can be proved to be unbiased, which means that the statistical mean of multiple estimated values $\hat{\boldsymbol{\beta}}$ is equal to the exact parameter vector $\boldsymbol{\beta}$.

Now the inverse analysis can start. The first step is to compute the sensitivity coefficients (this can be made analytically here in the case of the RDM1, but it is generally made through centered finite differences approximations). Remember that if the (deterministic) minimization numerical algorithm is based on exact sensitivity coefficients, the plot of these coefficients and the conclusions you may extract from it (if any) are more pertinent in the nondimensional form. We will then compute the reduced sensitivity coefficients $X_{\beta_i}^* = \beta_i(\partial T(t, \boldsymbol{\beta})/\partial \beta_i)$ and plot them as a function of the explanatory variable t. Regarding the synthetic noise introduced in the present study for stochastic and statistical analysis, a histogram plot is given in Figure 9.6.

9.2.3.1.3 Expectancy of the Residuals Curve
One way to validate the estimation is to compare the experimental curve with the theoretical curve given by the analytical model, using the estimated values of the unknown parameters by calculating the residuals denoted **r**, which is defined as the difference

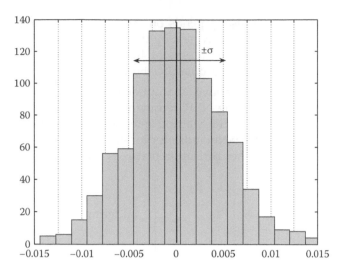

FIGURE 9.6
Synthetic noise histogram used for statistical analysis (type: Gaussian, std $\sigma = 0.005$, SNR $= 100$, $n = 1000$ data points, equally distributed).

between the experimental and theoretical curves. If Equation 9.5 is checked, then we can easily show that the expectancy of residuals curves is equal to a null function:

$$\mathbf{E(r)} = \mathbf{E[Y(t, \boldsymbol{\beta}) - F(t, \hat{\boldsymbol{\beta}})]} = \mathbf{E[X(\boldsymbol{\beta} - \hat{\boldsymbol{\beta}})]} = \mathbf{E[-X(X^tX)^{-1}X^t\boldsymbol{\varepsilon}(t)]}$$
$$= -\mathbf{X(X^tX)^{-1}X^tE[\boldsymbol{\varepsilon}(t)]}. \tag{9.18}$$

Since $\mathbf{E[\boldsymbol{\varepsilon}(t)] = 0}$,

$$\mathbf{E(r) = 0}. \tag{9.19}$$

So, if the model we used for describing the experiment is adapted, the residuals curve is "unsigned" and allows to prove the theoretical model used for the estimations is unbiased.

9.2.3.2 Case of Null Sensitivity to C_i

The sensitivity and identifiability study starts by considering RDM2 (taking into account the capacity of the insulating layer). The reduced sensitivities are plotted for both test cases in Figures 9.7 through 9.10 for both the front and rear faces. The four parameters $\boldsymbol{\beta} = [Q \, h \, \lambda \, C_i]^t$ are considered at their nominal values given in Table 9.1.

Note first that the sensitivity to parameter C_i, the capacity ($C' = C_i$) of the sandwiched insulated layer, is very weak (nearly null) in both test cases. This means that with its nominal value, this parameter plays no role in the calculations or has no influence on the experimental signal. As a consequence, it can be assumed to be a known parameter and then, RDM1 would be the most appropriate model. Note also that on these figures, the reduced sensitivity X_Q^* appears equal to the signal itself, $T_i (i = 1, 2)$. It is evident from Equations 9.3 of RDM1 that the pre-factor role of Q is responsible for the fact that $X_Q^* = T_2$ or T_1.

Before going on to the analysis of RDM1, we want to show here the effect of conserving a parameter of poor sensitivity in the estimation process. Considering only test case A, the following sensitive curves are obtained and corresponding stochastic results (information from matrix Vcor) have been reported in Table 9.2.

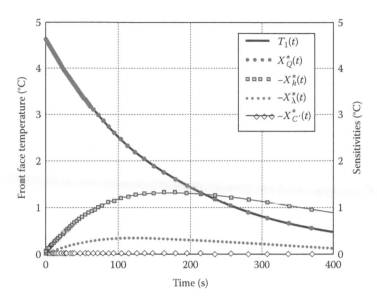

FIGURE 9.7

Test case A: Front face temperature signals along with the reduced sensitivities to the set of parameters—RDM2: $\beta = [Q = 10,000 \ h = 10 \ \lambda = 0.02 \ C_i = 30]^t$.

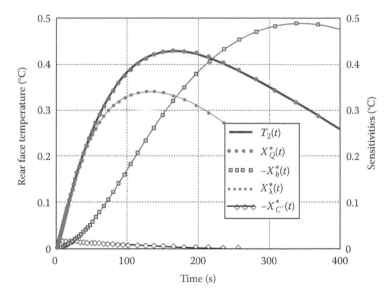

FIGURE 9.8

Test case A: Rear face temperature signals along with the reduced sensitivities to the set of parameters—RDM2: $\beta = [Q = 10,000 \ h = 10 \ \lambda = 0.02 \ C_i = 30]^t$.

The following remarks can be made. For the rear face, three of the correlation coefficients calculated for the nominal β vector are extremely high. This means that the PEP exhibits strong correlations between three parameters and this will make the estimation difficult. Note that Beck and Arnold (1977, p. 379) give a condition according to which, if *all* correlation coefficients are greater than 0.9, then the PEP will fail. But the fact that in the present case, the

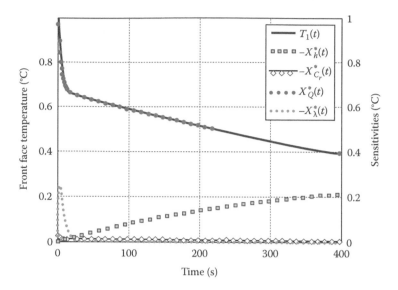

FIGURE 9.9
Test case B: Front face temperature signals along with the reduced sensitivities to the set of parameters—RDM2:
$\beta = [Q = 10,000\, h = 10\, \lambda = 0.2\, C_i = 200]^t$.

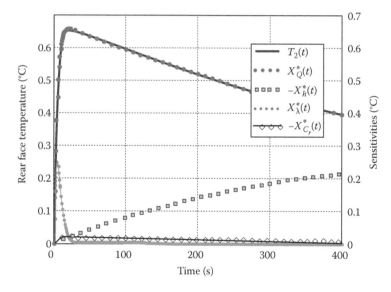

FIGURE 9.10
Test case B: Rear face temperature signals along with the reduced sensitivities to the set of parameters—RDM2:
$\beta = [Q = 10,000\, h = 10\, \lambda = 0.2\, C_i = 200]^t$.

sensitivity to one of the parameter is zero (or nearly so) deserves to be discussed. Anyway, if one looks at the estimated variances of the parameters, we observe that they are quite large (the estimated values cannot be considered as reliable). Parameters Q, λ, and C_i would be estimated with an error level of, respectively, 3%, 3.5%, and 10%. This means that if a 95% confidence interval is given for approximately $2\sigma_\beta$ (in a perfect world!), the capacity of the

TABLE 9.2

Stochastic Data for Test Case A Using RDM2 (Errors on Parameter Estimates Given for $\sigma = 0.005$)

	Rear Face T_2				Front Face T_1		
det$(\mathbf{X^tX})$	5.5×10^{-11}				2.3×10^{-5}		
Cond$(\mathbf{X^tX})^{-1}$	2.7×10^{14}				1.5×10^{10}		
Gallant number	8.3				3.8		
$\rho_{Qh}, \rho_{Q\lambda}, \rho_{QC'}$	0.99_4	-0.99_9	-0.74	-0.34	-0.61	0.4	
$\rho_{h\lambda}, \rho_{hC'}, \rho_{\lambda C'}$	-0.98	-0.66	0.77	-0.92	0.014	0.11	
$\sigma_{\hat{\beta}_i}/\beta_i$ for $\hat{Q}, \hat{h}, \hat{\lambda}, \hat{C}'$	0.03	0.007	0.035	0.10	0.0002	0.0008	0.004	0.015

insulated layer would be identified with around ±20% error. It is interesting to note that the parameter with quasi-null sensitivity presents the highest variance in its estimate. Clearly, this PEP appears badly conditioned. Note nevertheless that the (more) global information given by the determinant of matrix $(\mathbf{X^tX})$, the conditioning number of matrix $(\mathbf{X^tX})^{-1}$, or the Gallant number $Ga = 2\log_{10}(\lambda^{max}/\lambda^{min})_{cor}$—where λ is the singular value of the correlation matrix—all indicate that computer precision will not be responsible for the estimation algorithm failure (for this nominal value!). The Gallant number gives the minimum number of significant digits required to overcome the bad conditioning of the problem.

We must also note that on the contrary, working with the front face observable appears to be the good solution here.

A statistical analysis (random noise of $\sigma = 0.005$ added to the simulated signal and random initial parameter vector selected within a $[-50\%; +50\%]$ range around the nominal values) performed over 200 estimations gives estimators with no bias, except for C_i (5%) and reduced standard deviations $\sigma_{\hat{\beta}_i}/\hat{\beta}_i$ of 0.032, 0.010, 0.035, and 0.40, respectively, for Q, h, λ, and C_i. One must note that these values corroborate in a perfect manner the prediction of the stochastic analysis (Table 9.2) except for C_i. As said earlier, these estimations of the errors made on the parameters are optimum values. Choosing initial parameters in a larger set around the nominal values, for example, leads to larger errors.

Considering the residuals obtained for one of the runs of the repeated estimations, one obtains (Figure 9.11) perfect "unsigned" residuals with a standard deviation exactly equal to the standard deviation given as input to the normal synthetic random noise. This means that based on the observation of the residuals, the fitting procedure has been successful. This is absolutely true. But this has been made possible because of the overdetermined character of the model (too many degrees of freedom) which has allowed the values of the parameters to adjust in such a way that the OLS criterion is satisfied. The presence of parameter C_i is responsible for this. Regarding the metrological aspect (measuring parameters precisely), the conclusion is nevertheless that the procedure has failed. For the figure below, the following values have been identified: $Q = 10,443$, $h = 10.08$, $\lambda = 0.019$, and $C_i = 24.53$!

As a conclusion to this section, we can see that RDM2 is not an appropriate model for our metrological purpose basically because one of the parameters in the unknown set has a very poor sensitivity. It is of course out of question to use the CDM in these conditions. The decision that must be made at this point is to retrieve this (unnecessary) parameter from the model and make a new sensitivity analysis for model RDM1. One can also recall the following principle:

Perfectly uncorrelated and zero-mean experimental residuals should be regarded with suspicion. It is not a sign of good experimental design! It may signify that the PEP is overdetermined.

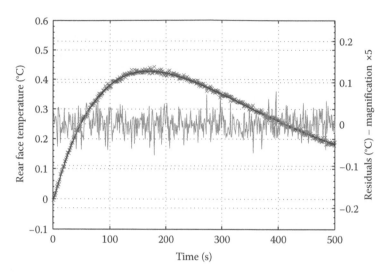

FIGURE 9.11
Simulated identification performed for test case A, observable T_2, with RDM$_2$.

9.2.3.3 Inverse Analysis with RDM1

The reduced sensitivities for the new parameter vector $\boldsymbol{\beta} = [Q \; h \; \lambda]^t$ (C_i being set to its nominal value) are still observable in Figures 9.7 through 9.10 (they do not change in a visible manner). What can be said from these figures is that the sensitivities do not apparently exhibit evident correlations. This is true for test case B (Figures 9.9 and 9.10). This is also true for test case A. Note that the front face response (Figure 9.7) shows that the sensitivity to λ is weaker than the other two. Regarding the rear face response, the three sensitivities exhibit the same behavior, but shifted in time. Direct correlation $\mathbf{X}^*_{\beta_1} = K \mathbf{X}^*_{\beta_2}$ would be manifested by some linear relationship between the sensitivities. Therefore, one can plot in the same figure the curves $\mathbf{X}^*_{\beta_i} = f(\pm \mathbf{X}^*_{\beta_j})$ for $i, j = 1, 2, 3$. This has been done in Figure 9.12 for the rear face response, which seems more interesting in order to identify all three parameters. One can see that all three parameters seem to be uncorrelated when

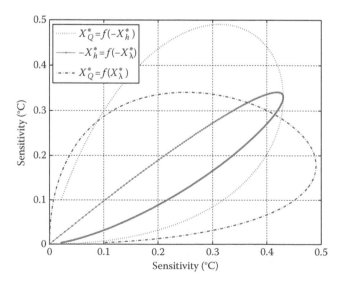

FIGURE 9.12
Sensitivities plotted by pairs.

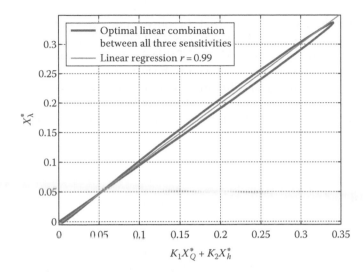

FIGURE 9.13
Evidence of linear combination between all three parameters.

compared by pair. The plot $X_h^* = f(-X_\lambda^*)$ in Figure 9.12 is the most suspect, as it draws a narrow loop whose orientation axis passes close to the origin.

But trouble in the identification process can also come from a more indirect correlation, like some linear combination between all three parameters of the type $X_\lambda^* = K_1 X_Q^* + K_2 X_h^*$. Attempting to find the optimal coefficients K_i through a least-square minimization process lead in our case to find $K_1 = 0.939$ and $K_2 = 0.225$, with the result plotted in Figure 9.13. As can be seen, this linear combination is quasi-perfect (correlation coefficient of 0.99 for the line adjusted on the loop). For RDM1 and test case A, the PEP appears badly conditioned because of this evident correlation between parameters. Let us look at all the major data that can be obtained from the stochastic and statistical analysis and reported in Table 9.3. All correlation coefficients for observable T_2 are greater than 0.99. This confirms that all the three parameters are correlated. According to Beck and Arnold (1977, p. 380), the fact that all the correlation coefficients are greater than 0.9 in absolute value leads to inaccurate estimates. But all other indicators reveal that the estimation is possible, especially the Gallant criterion. The variances of the parameter estimates show that for a standard deviation of 0.005 on the noise, the confidence bounds on the thermal conductivity will be roughly of the order of $\pm 2\sigma_\lambda/\lambda = \pm 6\%$. This table reveals that some errors are also made on the parameters; there is no problem in estimating the parameters from the rear face temperature (the algorithm converges through the unbiased values). Of course, one

TABLE 9.3

Stochastic and Statistical Data for Test Case A Using RDM1 (Errors on Parameter Estimates Given for $\sigma = 0.005$)

	Rear Face T_2			Front Face T_1		
$\det(\mathbf{X^t X})$	1.14×10^{-5}			0.18		
Cond $(\mathbf{X^t X})^{-1}$	2×10^{14}			1.3×10^{10}		
Gallant number	7.8			3.8		
$\rho_{Qh}, \rho_{Q\lambda}, \rho_{h\lambda}$	0.99_4	-0.99_9	-0.98_9	-0.38	0.63	-0.93
$\sigma_{\hat{\beta}_i}/\beta_i$ for $\hat{Q}, \hat{h}, \hat{\lambda}$	0.0275	0.0066	0.029	0.0002	0.0008	0.0042
Statistical result over 200 estimations						
$\sigma_{\hat{\beta}_i}/\beta_i$ for $\hat{Q}, \hat{h}, \hat{\lambda}$	0.026_4	0.0062	0.028_1	0.0002	0.0006	0.0037

must be very careful with these results which assume that the observable does not carry any experimental bias, which is not true in general (see Section 9.2.3.5).

What this table also reveals is that, compared to RDM2 and Table 9.2, the removal of the parameter with null sensitivity has diminished the variances on the remaining parameters. This must be retained as a general rule in nonlinear estimation (and in linear cases as well). In our case, the effect is not very spectacular because in this problem, the correlation coefficients between C_i and the remaining three parameters are not very strong. But in other cases, the effect can be drastic.

9.2.3.4 About a Change in the Parameterization

It has been suggested earlier that some change of parameterization would allow to overcome parameter estimation difficulties such as in the case of high correlation coefficients inducing high variances for the estimated parameters, for example. Here, we want to come back to this discussion to give, very briefly, some precisions and our conclusions.

First, and taking experience of what has been shown previously, if a change of parameterization is made that results in the production of a new parameter of null sensitivity, this new parameterization will have a benefit effect to properly estimate the remaining ones. Note that it is the object of dimensional analysis to help make such reparameterization efficient.

Second, if all the parameters of the problem have nonnegligible sensitivities but appear correlated, the question is, is it possible to find a new set of parameters defined from the initial one, to enhance the quality of the estimation process?

The answer is no. It can be demonstrated (see Annex in Remy and Degiovanni [2005]) that the sensitivities to a new set of parameters can be expressed from the sensitivities of the current set (using the Jacobian of the transformation). The same is true for the variance–covariance matrix. What these relations reveal is that

- If two parameters appear correlated in a given set of parameters, two parameters of a new set, recombined from the previous ones, will also be correlated.

- If the sensitivity of a parameter is changed with a new parameterization (for example, it is enhanced), this will not change its variance in fine.

9.2.3.5 About the Presence of Some Bias

Here, a bias means that there exists a systematic and generally unknown inconsistency between the model and the experimental data. A statistical analysis can be performed in the above case in the same way as before to see whether the presence of some bias in the data may change our results. This analysis will be made through simulations of estimation, with synthetic data produced in the same way as before (corruption with a noise of standard deviation $= 0.005$). The initial values of the parameters are generated in a random manner within a $[-100\%; +100\%]$ range about the nominal values. The influence of a bias when using RDM1 will be checked in two different manners:

- First, we use RDM2 to generate the synthetic data (with C_i set to its true value of 30 for test case A). RDM1 is then used to perform the identifications where the capacity C_i of the insulating layer does not exist! Note that compared to the capacity of the copper layers, this means that an error of $30/2160 = 1.4\%$ is introduced in the total heat capacity of the system. We then use a model which appears biased with respect to the data (or on the contrary, it is a matter of point of view). This bias comes from the structure of the model itself and, more precisely, from the fact that one of the parameters is assumed to be known. A total of 200 estimations have been performed. The populations of estimated parameters, the mean estimated value of each parameter, and its calculated standard deviation are given in Figure 9.14a through c. This shows first that the bias existing between the data and the model has totally biased the estimated parameters. The average errors introduced in the estimated values are of 32%, 5%, and 27%, respectively, for Q, h, and λ. The relative errors $\sigma_{\hat{\beta}}/\hat{\beta}$ are of the order of 0.05, 0.0086, and 0.005. Contrary to what has been shown when RDM1 was used to identify parameters through RDM1 (Table 9.3), no general conclusion can be predicted. Q and h have increased and λ has decreased. The residuals for one of the run are plotted in Figure 9.14d. They present now a "signed" character (oscillation around zero with a much smaller frequency than the noise). This signed character has been made visible in Figure 9.14d by superposing the residuals obtained for the same estimation in the absence of noise (solid dark line) and the residuals obtained with the noise (gray line). This signed character is typical of the presence of some bias which in real life is generally not known and can originate from different sources. It will be the object of Section 9.3 to demonstrate that the residuals can be used with benefit to predict the bias that has been made on the estimated parameters. This bias can also be calculated rigorously from the knowledge of the sensitivities to the remaining

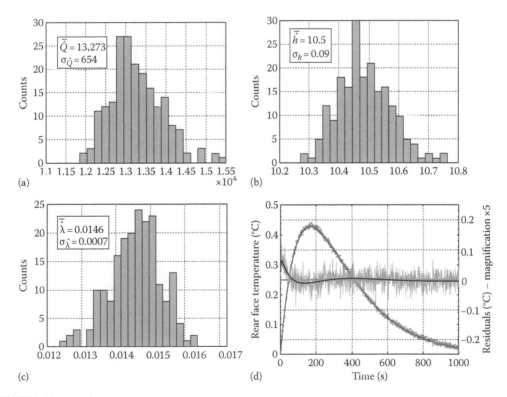

FIGURE 9.14
Influence of a bias coming from the model. (a, b, c) Histograms of the estimated parameters. (d) Noised thermogram, identified modeling curve, magnified residuals.

and known parameters. It will be shown further on that this bias is a systematic error due to the error made on the parameters supposed known, to which adds up a stochastic component (due to the noise). It must be retained that the bias introduced on the estimated parameters caused by giving fixed nominal value to the others (with some error) depends on the degree of correlation existing between estimated and fixed parameters.

- Second, we use RDM1 to generate the synthetic data and add to it a bias supposed to represent a real perturbation (a drift) of experimental origin: we choose a linear increase supposed to occur when the temperature signal is not at equilibrium before the flash excitation. Now the bias comes from the experiment, and the structure of both the identification and direct models is conserved. It is very small, as can be seen in Figure 9.15d, when compared to the temperature response. The populations of estimated parameters, the mean estimated value of each parameter, and its calculated standard deviation are given in Figure 9.15a through c. Also in this case, one must recognize that this small bias has nevertheless drastically corrupted the identification.

One possibility for the experimentalist to check if its estimations may be biased is to observe the output of the inversion process for varying identification ranges of the independent variable.

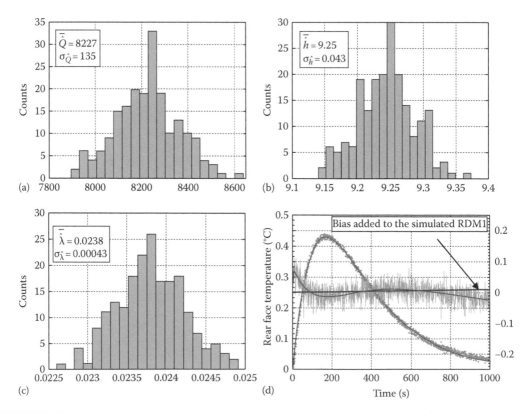

FIGURE 9.15
Influence of a bias obtained from the experiment. (a, b, c) Histograms of the estimated parameters. (d) Noised thermogram, identified modeling curve, magnified residuals.

TABLE 9.4

Influence of the Existence of Some Bias on the Parameter Estimates
for a Badly Conditioned Problem RDM1—Test Case A

Time Interval	70 s	150 s	300 s
a (m^2 s^{-1})	3.76×10^{-6}	3.22×10^{-6}	2.21×10^{-6}
λ (W m^{-1} °C^{-1})	0.031	0.064	0.084

Thermal diffusivity and conductivity estimates from real thermograms obtained
on an aerogel medium sandwiched between two copper layers.

For example, we can vary the time identification interval. If a bias affects the data when
compared to the modeling, then the estimations will vary, depending on the selected
identification interval. This can be observed in Table 9.4 where three identifications have
been performed for three different time intervals [0–70 s], [0–150 s], and [0–300 s]. The
thermal diffusivity and conductivity estimated from RDM1 depend strongly on the identification intervals. The values can change within a factor of 60% or 170% in such a case.

As a conclusion to this section, the sensitivity coefficients must be carefully analyzed in
order to identify possible correlations between them. It must be retained that in case of
high correlations for a set of nominal values representative of the metrological problem
under concern, the most important task for the "inverse man" is to track the presence of
some bias in the data, or in the model (especially through the parameters that are assumed
to be known). Even a small bias may lead the PEP to fail.

Of course, the demonstration would be complete by considering either the front face
response of test case A or any of the front or rear face responses for test case B. Because this
model (RDM1) is in fact NLP, the sensitivity and Vcor matrix analysis shows that the PEP
is not critical for those cases. It can be seen that the variances calculated after statistical
analysis (repetitions of identification) correspond exactly to the ones predicted by the
stochastic approach. One must also say that the statistical data have been obtained with
initial values of the parameters selected randomly in a +100% range of deviation from the
nominal values, which is a broad range. It has been observed that the results may slightly
depend on the algorithm used. This illustrates the robustness of this case. These results also
confirm, for example, that the estimator is unbiased. In order to save space in this text, no
additional results or comments are given, but the interested reader can develop its own
code with all the equations given above to verify this by himself or herself. As a benchmark, we give in Table 9.5 the stochastic data obtained for test case B.

TABLE 9.5

Stochastic and Statistical Data for Test Case B Using RDM1 (Errors on Parameter Estimates
Given for $\sigma = 0.005$)

	Rear Face T_2			Front Face T_1		
det $(\mathbf{X^t X})$	2.9×10^{-6}			2.9×10^{-6}		
Cond $(\mathbf{X^t X})^{-1}$	1.8×10^{7}			1.8×10^{7}		
Gallant number	1.9			1.9		
$\rho_{Qh}, \rho_{Q\lambda}, \rho_{h\lambda}$	0.78_5	-0.24	-0.18_6	0.78_8	0.38	0.30
$\sigma_{\hat{\beta}_i}/\beta_i$ for $\hat{Q}, \hat{h}, \hat{\lambda}$	0.0006	0.0012	0.0079	0.0006	0.0012	0.0083
Statistical result over 200 estimations						
$\sigma_{\hat{\beta}_i}/\beta_i$ for $\hat{Q}, \hat{h}, \hat{\lambda}$	0.0006	0.0011	0.0075	0.0006	0.0011	0.0079

9.2.4 Conclusion of the Three-Layer Characterization Problem

With this simple physical model and PEP, we do have an NLP problem. We have shown that

1. An appropriate reduced model must be chosen that allows correct estimation of the parameters (problem with a parameter having a too low sensitivity).
2. The reduced model, which is still NLP, exhibits different behaviors with respect to the estimation problem.
3. These different behaviors are caused by possible correlations between the remaining parameters and the presence of some bias in the experimental data.

In test case A, we can expect troubles in the estimation procedure using the rear face temperature signal. If we look at the physical reasons behind this phenomenon, it is interesting to note that the physical model is made of the sum (or difference) of two exponentials. Among the two characteristic times, $\tau_1 = \beta_2$ is directly proportional to parameter $1/h$. The second time constant $\tau_2 = (1/\beta_3 + 1/\beta_2)^{-1} = (1/\beta_3 + 1/\tau_1)^{-1}$ depends on parameter λ. Table 9.6 gives the calculated time constants for both test cases. It can be seen that for test case A, these time constants are in a ratio of 1.66, which is too low to discriminate between two exponentials (a ratio of 3 is generally the required order of magnitude). In test case B, the ratio $\tau_1/\tau_2 = 200$ is large enough to distinguish the contributions of both exponentials.

In test case B, the experimentalist has advantage in diminishing the thickness of the thin layer that he or she wants to characterize: makes the layer you want to characterize disappear! Indeed, for a given very large τ_1, determined by the capacity of the copper layers and the heat exchange coefficient, the best conditions for the identification of λ (β_3) are obtained when τ_2 is made as small as possible. This implies to make $\beta_3 = RC/2$ as small as possible and hence to decrease the thickness of the intermediate layer.

Finally, we want to convince the reader that there is no general rule in non-linear PEP. However, a toolbox of mathematical instruments exists: it can be used to analyze each metrological problem as a particular case. A last example is mentioned below.

The three-layer problem considered in the first section (test case B) is again considered but material B is now a layer of liquid* (water) and its properties are given in Table 9.7. This problem has led to the development of a special apparatus dedicated to the

TABLE 9.6

Time Constants Analysis for RDM1 and the Two Test Cases A and B

	β_2 (s)	β_3 (s)	τ_1 (s)	τ_2 (s)	τ_1/τ_2
Test case A	216	324	216	129.6	1.66
Test case B	720	3.6	720	3.58	200

---

* The aspect ratio of the measurement cell (height/thickness) is very large compared to 1, which suppress any natural convection contribution to heat transfer.

TABLE 9.7

Physical Properties for the Flash Equipment Developed
for Thermal Characterization of Liquids

Test Case: Water	Copper	Material B: Water
Thickness (mm)	2.0	4.5
Thermal conductivity (SI)	385	0.597
Volumic heat capacity (SI)	3.6×10^6	4.175×10^6

measurement of both the thermal conductivity and heat capacity of liquids using the flash method (Remy and Degiovanni 2005). The heat capacity of the layer $(\rho Ce)_W$ is 80 times larger and the thermal resistance about 7 times larger compared to test case B. As a consequence, with this new set of nominal parameters, the PEP has been moved to a different "point" in the parameter space. The heat capacity of the layer cannot be neglected anymore. RDM2 is now the expected, more pertinent, reduced model we have to consider for the PEP. Furthermore, a four-parameter identification must be considered, which in this study involves $\beta_1 = a_W$, $\beta_2 = (\rho C)_W$, $\beta_3 = Q$, and $\beta_4 = h$.

As done earlier (see Section 9.2.3.2), the results of the stochastic analysis applied to this new PEP are summarized through the Vcor matrix. In Table 9.8, one can see that although some (but not all) of the correlation coefficients are very high, the variances for the four parameters are nevertheless small enough to guarantee a correct identification. In Table 9.8, right column, the stochastic analysis has been made by assuming that one of the four parameters is fixed to its assumed correct value. One can once again verify that although a bias can be introduced, this has the consequence of diminishing the variances on the remaining parameters: in the previous four-parameter identification, the relative errors (within 1 standard deviation and in %) were 0.5% for the thermal diffusivity and 1.6% for the heat capacity.

In this experiment, the bias has been shown to occur exclusively from an error made due to a difficult-to-achieve perfect concentricity between the two copper cylinders. The negative effect of this bias has been avoided (and suppressed) by introducing four thermocouples located at a 90° angle from each other and taking the average as observable. The gap between the cylinders can then be precisely calculated and ascribed to the thickness of the liquid layer.

TABLE 9.8

Vcor Matrix: Diagonal Terms = Relative Errors on Parameters β—Off-Diagonal Terms = Correlation Coefficients (Std. of the Noise $\sigma = 0.005\,K$)

Test Case W—RDM2 Identification		Test Case W—RDM2 Identification Parameter β_2 Is Fixed	

$$\text{Vcor}(\hat{\beta}) \approx \begin{bmatrix} 0.005 & -0.989 & -0.995 & -0.115 \\ \bullet & 0.016 & 0.998 & -0.004 \\ \bullet & \bullet & 0.011 & 0.006 \\ \bullet & \bullet & \bullet & 0.011 \end{bmatrix} \qquad \text{Vcor}(\hat{\beta}) \approx \begin{bmatrix} 0.0008 & -0.861 & -0.81 \\ \bullet & 0.0007 & 0.978 \\ \bullet & \bullet & 0.011 \end{bmatrix}$$

9.3 Calculation of the Standard Deviations of Estimated Parameters in the Case of a Biased Model

In the ideal case of a complete or detailed model (unbiased model) as that considered in Section 9.2.3.1, where the model is assumed to be corrupted by an additive noise in the signal with classical assumptions, we have previously shown that the OLS method through an iterative procedure (in the case of a nonlinear problem) is a very interesting tool for estimating values of the unknown parameters because the estimator is unbiased, $\mathbf{E}(\hat{\boldsymbol{\beta}}) = \boldsymbol{\beta}$, and the errors on the estimated parameters are only stochastic, $\mathbf{cov}(\hat{\boldsymbol{\beta}}) = \sigma_b^2 (\mathbf{X}^t \mathbf{X})^{-1}$. This means that the covariance matrix shows that errors in the estimated parameters are small if the measurement noise is small.

For a given measurement noise level, errors in the estimated parameters can become large if the determinant of the sensitivity matrix product tends to zero as, for instance, if the sensitivity of the model to a given parameter is small or if two parameters are correlated.

As shown in Section 9.2.3, optimization of the parameters' estimation consists in reducing the standard deviations or variances of the parameters by reducing the number of parameters in the detailed model by setting different parameters to their nominal values. To describe the rear face response of a sample submitted to heat pulse stimulation in its front face as in the case of a Flash method, one can use, for example, a reduced model like the adiabatic model in which heat loss effects can be neglected, or a model in which heat losses are fixed to their nominal values.

These two kinds of strategies allow reducing the stochastic errors caused by noise, but as illustrated in Section 9.2.3.5, a systematic error or bias may appear for the estimated parameters. In the next section, we will see how it is possible to evaluate this bias from known quantities such as the residuals curve and the sensitivities to the parameters involved in the reduced model and by working with time variable intervals. For more clarity, we decided to work in a more simple case than those used in Section 9.2, the "Flash" method on a single layer material. The reason is that this model is more simple and the number of unknown parameters is reduced both in the unbiased and biased models.

9.3.1 Evaluation of the Bias on the Estimated Parameters of a Reduced Model

As explained in the introduction of Section 9.2 and demonstrated in Section 9.2.3.5, setting different parameters to their nominal values or using a reduced model that neglects different physical aspects (a particular case where the nominal value is null) leads to a bias in the parameters that are estimated. In this section, we will show how it can be evaluated.

The biased (reduced) model can be formally written from the unbiased model (exact model):

$$\tilde{\mathbf{T}}(t, \hat{\boldsymbol{\beta}}_r, \boldsymbol{\beta}_{c_{nom}}) = \mathbf{T}(t, \boldsymbol{\beta}_r, \boldsymbol{\beta}_c) + \mathbf{X}_r|_{\boldsymbol{\beta}_r}(\hat{\boldsymbol{\beta}}_r - \boldsymbol{\beta}_r) + \mathbf{X}_c|_{\boldsymbol{\beta}_c}(\boldsymbol{\beta}_{c_{nom}} - \boldsymbol{\beta}_c) \tag{9.20}$$

with

- $\boldsymbol{\beta}_r$ and $\boldsymbol{\beta}_c$: true values of the "unknown" and "known" parameters, respectively
- $\hat{\boldsymbol{\beta}}_r$: current estimated values of "unknown" parameters
- $\boldsymbol{\beta}_{c_{nom}}$: nominal values of fixed of "known" parameters

The sign "$\sim$" on $\mathbf{T}$ indicates that the model we used for the estimation is a biased model (reduced model). In the vicinity of the true solution, $\tilde{\mathbf{T}}$, the only function of $\boldsymbol{\beta}_r$, can be written from the detailed $\mathbf{T}$ model which is the function of $(\boldsymbol{\beta}_r, \boldsymbol{\beta}_c)$.

The error on the fixed parameters (assumed to be "known" parameters) is defined by

$$\mathbf{e}_{\boldsymbol{\beta}_c} = \boldsymbol{\beta}_{c_{nom}} - \boldsymbol{\beta}_c. \tag{9.21}$$

The estimator vector $\boldsymbol{\beta}$ can be split into two components: an "unknown" component $\boldsymbol{\beta}_r$, composed of the unknown parameters, and a "known" component $\boldsymbol{\beta}_c$, composed of parameters fixed to their nominal value—$\boldsymbol{\beta} = (\boldsymbol{\beta}_r|\boldsymbol{\beta}_c)$. The sensitivity matrix is composed of the sensitivities of the model to the "unknown" and "known" parameters: $\mathbf{X} = (\mathbf{X}_r|\mathbf{X}_c)$. Equation 9.20 shows that $\tilde{\mathbf{X}}_r^t(\hat{\boldsymbol{\beta}}_r) = (\partial\tilde{\mathbf{T}}/\partial\hat{\boldsymbol{\beta}}_r) = \mathbf{X}_r^t(\boldsymbol{\beta}_r)$. The linear system that has to be solved only depends on the sensitivities $\tilde{\mathbf{X}}_r^t$ of the "unknown" parameters of the biased model. As $\tilde{\mathbf{X}}_r^t(\hat{\boldsymbol{\beta}}_r) = \mathbf{X}_r^t(\boldsymbol{\beta}_r)$,

$$\mathbf{X}_r^t(\mathbf{Y} - \tilde{\mathbf{T}}) = 0. \tag{9.22}$$

Expression (9.20) can be written as

$$\tilde{\mathbf{T}}\left(t, \hat{\boldsymbol{\beta}}_r, \boldsymbol{\beta}_{c_{nom}}\right) = \mathbf{T}(t, \boldsymbol{\beta}_r, \boldsymbol{\beta}_c) + (\mathbf{X}_r|\mathbf{X}_c)\left(\begin{array}{c}\hat{\boldsymbol{\beta}}_r - \boldsymbol{\beta}_c \\ \mathbf{e}_{\boldsymbol{\beta}_c}\end{array}\right). \tag{9.23}$$

The experimental curve $\mathbf{Y}$ depends on the true values of both "unknown" $\boldsymbol{\beta}_r$ and "known" $\boldsymbol{\beta}_c$ parameters and is given by the detailed model $\mathbf{T}$ with an assumed additive noise:

$$\mathbf{Y}(t, \boldsymbol{\beta}_r, \boldsymbol{\beta}_c) = \mathbf{T}(t, \boldsymbol{\beta}_r, \boldsymbol{\beta}_c) + \boldsymbol{\varepsilon}(t). \tag{9.24}$$

Substituting $\mathbf{Y}$ and $\mathbf{T}$ by their expressions (9.23) and (9.24) in Equation 9.22 leads to

$$\hat{\boldsymbol{\beta}}_r = \boldsymbol{\beta}_r + \underbrace{\left(\mathbf{X}_r^t\mathbf{X}_r\right)^{-1}\mathbf{X}_r^t\boldsymbol{\varepsilon}}_{\mathbf{e}_\varepsilon} \underbrace{-\left(\mathbf{X}_r^t\mathbf{X}_r\right)^{-1}\mathbf{X}_r^t\mathbf{X}_c\mathbf{e}_{\boldsymbol{\beta}_c}}_{\mathbf{b}_{\boldsymbol{\beta}_r}}. \tag{9.25}$$

As expected, the estimated parameters $\hat{\boldsymbol{\beta}}_r$ are composed of two additive errors: a stochastic error $\mathbf{e}_\varepsilon$ as in the unbiased model and a deterministic error $\mathbf{b}_{\boldsymbol{\beta}_r}$ defined by

$$\mathbf{b}_{\boldsymbol{\beta}_r} = -\left(\mathbf{X}_r^t\mathbf{X}_r\right)^{-1}\mathbf{X}_r^t\mathbf{X}_c\mathbf{e}_{\boldsymbol{\beta}_c}. \tag{9.26}$$

As for the detailed model, Equation 9.25 allows the calculation of the expectancy and standard deviation of the estimated values of the "unknown" parameter $\boldsymbol{\beta}_r$.

9.3.1.1 Expectancy of the Estimator

$$\mathbf{E}(\hat{\boldsymbol{\beta}}_r) = \boldsymbol{\beta}_r + \left(\mathbf{X}_r^t\mathbf{X}_r\right)^{-1}\mathbf{X}_r^t\mathbf{E}(\boldsymbol{\varepsilon}) - \left(\mathbf{X}_r^t\mathbf{X}_r\right)^{-1}\mathbf{X}_r^t\mathbf{X}_c\mathbf{e}_{\boldsymbol{\beta}_c}. \tag{9.27}$$

Since the mean value of noise (expectancy) is null, we finally obtain

$$\mathbf{E}(\hat{\boldsymbol{\beta}}_r) = \boldsymbol{\beta}_r - \left(\mathbf{X}_r^t\mathbf{X}_r\right)^{-1}\mathbf{X}_r^t\mathbf{X}_c\mathbf{e}_{\boldsymbol{\beta}_c} = \boldsymbol{\beta}_r + \mathbf{b}_{\boldsymbol{\beta}_r}. \tag{9.28}$$

In this case, the estimated parameters are biased. This error $\mathbf{b}_{\beta_r}$ or "bias" is linearly linked to the errors on the "known" parameters $\mathbf{e}_{\beta_c}$.

We can notice that this bias is small if

1. The sensitivities of the detailed model to the fixed parameters are small.
2. The nominal values of fixed parameters are close to the true values of these parameters.
3. The "known" and "unknown" parameters are independent (i.e., $\mathbf{X}_r^t \mathbf{X}_c = 0$).

Furthermore, if the sensitivity matrix to "unknown" parameters $\mathbf{X}_r^t \mathbf{X}_r$ is well conditioned, then the effect of the errors of "known" parameters $\mathbf{e}_{\beta_c}$ on estimated parameters $\hat{\boldsymbol{\beta}}_r$ will be reduced.

9.3.1.2 Standard Deviation of the Estimator (Variance–Covariance Matrix)

By definition

$$\mathrm{cov}(\hat{\boldsymbol{\beta}}_r) = \mathbf{E}\left[(\hat{\boldsymbol{\beta}}_r - \mathbf{E}(\hat{\boldsymbol{\beta}}_r))(\hat{\boldsymbol{\beta}}_r - \mathbf{E}(\hat{\boldsymbol{\beta}}_r))^t\right] = \mathbf{E}\left[(\hat{\boldsymbol{\beta}}_r - \boldsymbol{\beta}_r - \mathbf{b}_{\beta_c})(\hat{\boldsymbol{\beta}}_r - \boldsymbol{\beta}_r - \mathbf{b}_{\beta_c})^t\right]. \tag{9.29}$$

Using Equations 9.25 and 9.26,

$$\mathrm{cov}(\hat{\boldsymbol{\beta}}_r) = \mathbf{E}\left[(\mathbf{X}_r^t \mathbf{X}_r)^{-1} \mathbf{X}_r^t \boldsymbol{\varepsilon}\boldsymbol{\varepsilon}^t \mathbf{X}_r (\mathbf{X}_r^t \mathbf{X}_r)^{-1}\right] = \sigma_b^2 (\mathbf{X}_r^t \mathbf{X}_r)^{-1}, \tag{9.30}$$

$$\mathrm{cov}(\hat{\boldsymbol{\beta}}_r) = \sigma_b^2 (\mathbf{X}_r^t \mathbf{X}_r)^{-1}. \tag{9.31}$$

The expression of the covariance matrix remains unchanged and is only a function of the standard deviation of the noise and of the sensitivity matrix to "unknown" parameters. Consequently, the dispersion is lower than in the case of detailed model because the size of the sensitivity matrix is smaller. Nevertheless, it is important to remember that this covariance matrix has been evaluated from the biased values of "unknown" parameters (expectancy of the estimated parameter): $\mathbf{E}(\hat{\boldsymbol{\beta}}_r) = \boldsymbol{\beta}_r + \mathbf{b}_{\beta_r}$.

9.3.1.3 Modified Standard Deviation of the Estimator (Mean Squared Error)

A more realistic approach for the standard deviation evaluation of "unknown" parameters consists in evaluating the covariance matrix not from the estimated and biased values of "unknown" parameters $\mathbf{E}(\hat{\boldsymbol{\beta}}_r) = \boldsymbol{\beta}_r + \mathbf{b}_{\beta_r}$ but from their true values $\boldsymbol{\beta}_r$, that is,

$$\mathrm{cov}_m(\hat{\boldsymbol{\beta}}_r) = \mathbf{E}[(\hat{\boldsymbol{\beta}}_r - \boldsymbol{\beta}_r)(\hat{\boldsymbol{\beta}}_r - \boldsymbol{\beta}_r)^t]. \tag{9.32}$$

To distinguish the variance calculated from the expectancy of "unknown" parameters from this new variance, this latter is denoted cov_m for "modified" parameters. Using Equation 9.25,

$$\mathrm{cov}_m(\hat{\boldsymbol{\beta}}_r) = \mathbf{E}\left[\left((\mathbf{X}_r^t \mathbf{X}_r)^{-1} \mathbf{X}_r^t \boldsymbol{\varepsilon} + \mathbf{b}_{\beta_r}\right)\left((\mathbf{X}_r^t \mathbf{X}_r)^{-1} \mathbf{X}_r^t \boldsymbol{\varepsilon} + \mathbf{b}_{\beta_r}\right)^t\right]. \tag{9.33}$$

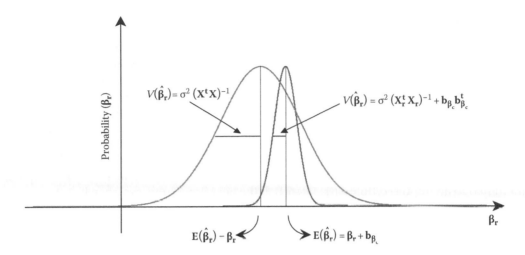

FIGURE 9.16
Comparison between the standard deviation of parameters estimated in the case of detailed and reduced models.

As the noise is a stochastic quantity and the bias a deterministic quantity, this yields

$$\mathbf{cov_m(\hat{\beta}_r)} = \sigma_b^2 (\mathbf{X_r^t X_r})^{-1} + \mathbf{b_{\beta_r} b_{\beta_r}^t}. \tag{9.34}$$

Thus, the errors on estimated values of "unknown" parameters are composed of two components: "stochastic" and "systematic" errors. In order to present an interest, the use of a reduced model or the reduction of the number of "unknown" parameters must lead to smaller errors.

The illustration of the two consequences yielded by the above demonstration, namely, the error introduced in the expected value of the unknown parameters and the reduction of their variance, is given in Figure 9.16. Application on a basic example, the classical "Flash" method, will be made in the next section.

9.3.1.4 Expectancy of the Residuals Curve

The expectancy of residuals in the case of a reduced model is given by

$$E(\mathbf{r}) = E[Y(t, \boldsymbol{\beta}) - F(t, \hat{\boldsymbol{\beta}})] = E[\boldsymbol{\varepsilon}(t) - \mathbf{X_r}(\hat{\boldsymbol{\beta}}_r - \boldsymbol{\beta}_r) - \mathbf{X_c e_{\beta_c}}]$$
$$= -\mathbf{X_r} E\left[(\mathbf{X_r^t X_r})^{-1} \mathbf{X_r^t}\boldsymbol{\varepsilon}(t) - (\mathbf{X_r^t X_r})^{-1} \mathbf{X_r^t X_c e_{\beta_c}}\right] - \mathbf{X_c e_{\beta_c}} \tag{9.35}$$

and finally,

$$E(\mathbf{r}) = \left(\mathbf{X_r}(\mathbf{X_r^t X_r})^{-1} \mathbf{X_r^t} - \mathbf{I}\right) \mathbf{X_c e_{\beta_c}}. \tag{9.36}$$

This relation shows that the residuals are "signed" in the case of a biased model.

9.3.2 Application: Case of the Classical "Flash" Method

For validating the theoretical results established in Sections 9.2.3.1 and 9.3.1, Monte Carlo simulations have been performed on a simulated Flash experiment (see Figure 9.17) (100 runs with different random noise have been generated). The expectancy and standard deviations on estimated parameters have been then calculated using biased and unbiased models.

The Flash experiment consists in applying a uniform space heat pulse stimulation of a very short duration (Dirac) onto a sample. Measurement of its rear face temperature evolution allows the estimation of the thermal diffusivity. Different nonideal aspects of the experiment, related to heat losses with the surrounding environment, will be considered. The sample is assumed cylindrical with a thickness e and a radius R. This sample is submitted to an impulsed flux: $\phi(t) = \phi_0 \delta(t)$. If the heat flux is uniform, then heat transfer is one dimensional. The heat transfer equation in 1D is given by

$$\frac{\partial^2 T}{\partial x^2} = \frac{1}{a}\frac{\partial T}{\partial t} \tag{9.37}$$

with the following boundary conditions:

$$\begin{cases} \text{at } t = 0, & T = 0, \\[2mm] \text{in } x = 0, & \lambda\dfrac{\partial T}{\partial x} = hT_0 - \phi(t), \\[2mm] \text{in } x = e, & -\lambda\dfrac{\partial T}{\partial x} = hT_e. \end{cases} \tag{9.38}$$

The solution can be easily obtained using a Laplace transform and is a function of three independent parameters:

$$T(t, \boldsymbol{\beta}) = f\left(\frac{Q}{\rho c e}, \frac{he}{\lambda}, \frac{a}{e^2}, t\right). \tag{9.39}$$

Working on the reduced thermogram normalized by its maximum, the problem is reduced to two "unknown" parameters (this normalized thermogram T^* will be noted T in the followings for more clarity):

$$T^*(t, \boldsymbol{\beta}) = \frac{T(t, \boldsymbol{\beta})}{T_{\max}} = f\left(\frac{he}{\lambda}, \frac{a}{e^2}, t\right). \tag{9.40}$$

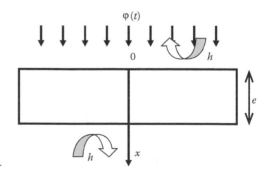

FIGURE 9.17
Principle of the Flash method.

- The Fourier number (inverse of the characteristic time): $Fo = (a/e^2)(t_c = 1/Fo)$
- The Biot number (heat losses): $Bi = (he/\lambda)$

The "unknown parameters" vector is defined by $\boldsymbol{\beta} = \begin{pmatrix} 1/Fo \\ Bi \end{pmatrix}$.

Nominal values:

$$Fo = 10,$$
$$Bi = 0.05.$$

For validating the theoretical relations we obtained, Monte Carlo simulations have been carried out. Estimations of a theoretical signal composed of 1000 points in time and corrupted by an additive "white" noise with a reduced standard deviation σ of 1% (100 runs have been considered) have been carried out.

To make the difference between the theoretical expectancy $\mathbf{E}$ and standard deviation $\mathbf{cov} = \boldsymbol{\sigma}^2$ and the results obtained in practice by Monte Carlo simulations, the practical quantities are denoted by the superscript "^." Estimations have been performed at first without considering a bias and second using a biased model by fixing the heat losses to a nominal value.

9.3.2.1 Estimation without Bias

Figure 9.18 gives an example of estimation and compares the estimated thermogram calculated from the estimated values of the "unknown" parameters given by a Levenberg–Marquardt algorithm (OLS method). The "oscillating" curve represents the residuals curve (difference between direct model corrupted by noise and estimated thermogram).

The following estimated values were obtained:

- For the expectancy, $\hat{\mathbf{E}}(\hat{\boldsymbol{\beta}}) = \begin{pmatrix} 0.099998 \\ 0.049987 \end{pmatrix}$.

  Since $\boldsymbol{\beta} = \begin{pmatrix} 0.10 \\ 0.05 \end{pmatrix}$, an error (of stochastic origin) exists but is small $\begin{pmatrix} 0.002\% \\ 0.026\% \end{pmatrix}$.

  This error can be reduced through a repetition of the experiments and averaging of estimations since $\mathbf{E}(\hat{\boldsymbol{\beta}}) = \boldsymbol{\beta}$.

- For the standard deviations on parameters (diagonal terms of $\mathbf{cov}$ matrix),

$$\begin{pmatrix} \hat{\sigma}_{\hat{\beta}_1}^2 \\ \hat{\sigma}_{\hat{\beta}_2}^2 \end{pmatrix} = \begin{pmatrix} 0.0242 \times 10^{-9} \\ 0.9070 \times 10^{-9} \end{pmatrix}, \quad \text{since} \quad \begin{pmatrix} \sigma_{\beta_1}^2 \\ \sigma_{\beta_2}^2 \end{pmatrix} = \text{diag} \ (\sigma_b^2.(\mathbf{X^t X})^{-1}) = \begin{pmatrix} 0.0243 \times 10^{-9} \\ 0.9003 \times 10^{-9} \end{pmatrix}.$$

- For expectancy of residuals curves, we obtain $\hat{\mathbf{E}}(\mathbf{r}) \simeq 0$.

Thus, theoretical results of Section 9.2.3.1 have been checked.

9.3.2.2 Estimation with a Bias

For generating an artificial bias, we now reduce intentionally the number of "unknown" parameters involved in the theoretical model (two parameters in the model considered here) by setting some parameters to their nominal values (only one parameter in this case).

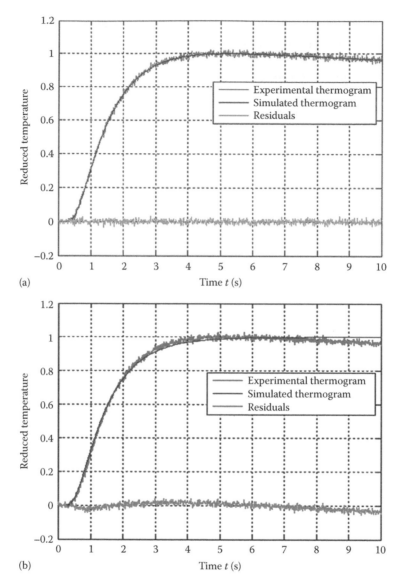

FIGURE 9.18
Estimation from a thermogram corrupted by noise. (a) With an unbiased model. (b) With a biased or reduced model.

For instance, we decide to fix the heat loss coefficient by setting the Biot number to $\beta_c = 0.04$. This corresponds to an error $e_{\beta_c} = -0.01$ on this parameter.

In this case, the Biot number is like an assumed "known" parameter and plays the role of parameter vector β_c. The only "unknown" parameter in this case corresponds to the characteristic time and is denoted β_r.

- For the expectancy, we can check that the expectancy of estimated parameters is biased. We found $\hat{E}(\hat{\beta}_r) = 0.101095$ instead of $\beta_r = 0.1$. We can calculate the bias (difference between the expectancy of estimated parameter $\hat{\beta}_r$ and its true value β_r)

$\hat{E}(\hat{\beta}_r) - \beta_r = 0.001095$ and compare this result with the theoretical bias given by Equation 9.26, that is, $b_{\beta_r} = 0.001096$. This result is very close and consequently allows us to check Equation 9.28: $E(\hat{\beta}_r) = \beta_r + b_{\beta_r}$.

- For the variance, we obtained $\hat{\sigma}_{\beta_1}^2 = 1.3311 \times 10^{-11}$ (standard deviation calculated with 100 runs) compared to the theoretical result $\sigma_{\beta_1}^2 = \sigma_b^2 (X_r^t X_r)^{-1} = 1.3968 \times 10^{-11}$. Within an error of 4.7%, we verified Equation 9.31 $\text{cov}(\hat{\beta}_r) = \sigma_b^2 (X_r^t X_r)^{-1}$.

- For the "modified" variance (9.34) or mean squared error, calculated from 100 runs and from true values, we find

$$\hat{\text{cov}}_m(\hat{\beta}_r) = 1.2031 \times 10^{-6} \quad \text{to compare with } \sigma_b^2 (X_r^t X_r)^{-1} + b_{\beta_r} b_{\beta_r}^t = 1.2031 \times 10^{-6}.$$

- It is interesting to notice that the "modified" variance (standard deviation between the estimated value and the true value) is larger than the stochastic error. In this case, the bias is larger than the stochastic error.

- Two residuals curves are compared in Figure 9.19. The calculated ones (resulting from the difference between the input thermogram and the after-estimation recalculated thermogram) and the theoretical ones obtained from the theoretical relation correspond to $E(r)$ given by Equation 9.36. A good agreement between these two curves can be observed. The difference corresponds to the noise that appears in the residuals but not in its theoretical expectancy.

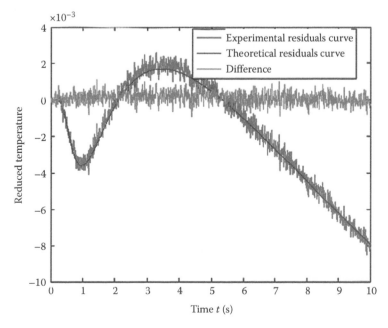

FIGURE 9.19
Residuals curves in the case of a biased model.

9.3.3 Can Residuals Curve Help to Determine Information Regarding Estimation of Bias?

As illustrated in the previous section, in the ideal case, when we use a detailed model that perfectly describes the experiment (experimental curve is equal to the theoretical model with an additive random noise), the residuals curve is "unsigned" and the standard deviations in estimated parameters are of stochastic origin and only depend on the random noise in the signal (see Sections 9.2.3 and 9.3.2.1).

In practice, the detailed model that must perfectly describe the experiment is not known because to obtain the theoretical model, different simplifying assumptions are made. Depending on these assumptions, we only have a more or less reduced and/or biased model. In this case, we can observe a signature in the residuals curve and in its expectancy (see Sections 9.2.3.5 and 9.3.2.2 for instance). We call of "signed" residuals. The standard deviations on parameters are composed of two components: a stochastic error and a systematic bias $\mathbf{b}_{\beta_r}$. The evaluation of this bias is important because it can lead, as previously shown, to some errors larger than the stochastic one.

As shown in the previous section (see Equation 9.26), the bias $\mathbf{b}_{\beta_r}$ on the estimated parameters is a function of the sensitivities of the detailed model and errors on fixed parameters through $\mathbf{X}_c \mathbf{e}_{\beta_c}$, which are unknown in practical case.

The problem is: How can we evaluate the bias on the estimated parameters of the reduced model if the detailed (unbiased) model is unknown?

We will show that it is possible to estimate the unknown quantity $\mathbf{X}_c \mathbf{e}_{\beta_c}$ from the residuals curve $\mathbf{r}$, more precisely from its expectancy $\mathbf{E}(\mathbf{r})$. The unknown quantity $\mathbf{X}_c \mathbf{e}_{\beta_c}$ can be formally eliminated by combining Equations 9.26 and 9.36:

$$\mathbf{b}_{\beta_r} = -\left(\mathbf{X}_r^t \mathbf{X}_r\right)^{-1} \mathbf{X}_r^t \left[\left(\mathbf{X}_r \left(\mathbf{X}_r^t \mathbf{X}_r\right)^{-1} \mathbf{X}_r^t - \mathbf{I} \right)^{-1} \mathbf{E}(\mathbf{r}) \right]. \tag{9.41}$$

Consequently, a direct relation between the bias $\mathbf{b}_{\beta_r}$ on "unknown" parameters and the residuals curve $\mathbf{E}(\mathbf{r})$ exists and allows the calculation of modified standard deviation on "unknown" parameters $\boldsymbol{\sigma}_m(\hat{\boldsymbol{\beta}}_r)$ through Equation 9.34.

Remark

In practice and as usually done with the sensitivity coefficients in a nonlinear case where sensitivity to real value of parameters $\mathbf{X}(\mathbf{t}, \boldsymbol{\beta})$ is assumed close to the sensitivity coefficients to estimated values $\mathbf{X}(\mathbf{t}, \hat{\boldsymbol{\beta}})$, we can consider that a good evaluation of the residuals expectancy is given by the residual curves of a given estimation, that is, $\mathbf{E}(\mathbf{r}) \simeq \hat{\mathbf{E}}(\mathbf{r}) \simeq \mathbf{r}$.

9.3.3.1 Estimation of the Sensitivity of the Detailed Model: $\mathbf{X}_c \mathbf{e}_{\beta_c}$

To obtain $\mathbf{X}_c \mathbf{e}_{\beta_c}$ allowing the calculation of bias (Equation 9.26), we have to solve a linear problem "$A.X = B$" (see Equation 9.36):

$$\left(\mathbf{X}_r \left(\mathbf{X}_r^t \mathbf{X}_r\right)^{-1} \mathbf{X}_r^t - \mathbf{I} \right) \cdot \mathbf{X}_c \mathbf{e}_{\beta_c} = \mathbf{E}(\mathbf{r}) \rightarrow A \cdot X = B \tag{9.42}$$

with

- $A = \left(\mathbf{X}_r(\mathbf{X}_r^t\mathbf{X}_r)^{-1}\mathbf{X}_r^t - \mathbf{I}\right)$
- $X = \mathbf{X}_c\mathbf{e}_{\boldsymbol{\beta}_c}$
- $B = \mathbf{E}(\mathbf{r})$

A is $(n \times n)$ matrix (n being the number of points in time) that is perfectly known since it only depends on the sensitivities of the reduced model $\mathbf{X}_r$. (Let us note that this matrix has no unit.) Vector B corresponds to the expectancy of residuals and X is the unknown vector that we are seeking for.

Unfortunately, we can show that the matrix $\mathbf{A}$ is singular (determinant equal to zero). This means that different detailed models that lead to the same residuals curve can exist.

A solution consists in finding an approximate solution using the "pseudo-inverse" of this matrix using a singular value decomposition (SVD):

$$\mathbf{A} = \mathbf{U}\mathbf{S}\mathbf{V}^t. \tag{9.43}$$

$\mathbf{U}$ and $\mathbf{V}$ are two orthogonal matrices so that $\mathbf{U}^t \cdot \mathbf{U} = \mathbf{I}$ and $\mathbf{V}^t \cdot \mathbf{V} = \mathbf{I}$. $\mathbf{S}$ is a diagonal matrix that contains the singular values of A sorted in decreasing order.

In the case of our matrix $\mathbf{A} = \left(\mathbf{X}_r(\mathbf{X}_r^t\mathbf{X}_r)^{-1}\mathbf{X}_r^t - \mathbf{I}\right)$, we can show that it can be decomposed under the following form $\mathbf{A} = \mathbf{U}\mathbf{S}\mathbf{V}^t$ with

$$USV^t = \begin{bmatrix} U_{1,1} & \cdots & U_{n,1} \\ \vdots & \ddots & \vdots \\ U_{1,n} & \cdots & U_{n,n} \end{bmatrix} \begin{bmatrix} 1 & \cdots & 0 \\ \vdots & 1 & \vdots \\ 0 & \cdots & 0 \end{bmatrix} \begin{bmatrix} V_{1,1} & \cdots & V_{n,1} \\ \vdots & \ddots & \vdots \\ V_{1,n} & \cdots & V_{n,n} \end{bmatrix}. \tag{9.44}$$

The diagonal matrix is only composed of two different singular values 0 (order 1) and 1 (order $n - 1$). The matrix $\mathbf{A}$ is a projector. As only one null singular value exists, the rank of this matrix is thus equal to $n - 1$. Two approximated solutions can be obtained either by using a direct truncature method which consists of reducing the size of matrix $\mathbf{A}$ (the truncature method consists in deleting one row and one column of matrix $\mathbf{A}$, the last ones for instance), or by using the "pseudo-inverse" solution given by deleting the null singular value (last row and column of the matrix $\mathbf{S}$), the last column of matrix $\mathbf{U}$ and the last row of the matrix $\mathbf{V}$ (SVD solution).

To validate the calculation of $\mathbf{X}_c\mathbf{e}_{\boldsymbol{\beta}_c}$ by these two techniques, these two solutions are compared in the case of the Flash "method" with the "true" function of $\mathbf{X}_c\mathbf{e}_{\boldsymbol{\beta}_c}$. The results are drawn in Figure 9.20:

- First, if we compare truncature and SVD solutions, they are in good agreement with small discrepancies due to the truncature in the SVD. This means that the solution obtained by the truncature of the matrix $\mathbf{A}$ and the solution given by SVD leads to the same results. If we now compare these two results with the true expression of $\mathbf{X}_c\mathbf{e}_{\boldsymbol{\beta}_c}$, we can observe that these two solutions are different from $\mathbf{X}_c\mathbf{e}_{\boldsymbol{\beta}_c}$.

- Nevertheless, in all cases, as shown in Figure 9.21, if we compute the residuals (using Equation 9.36 and the solution obtained by SVD) and compare this curve with the true residuals curve, we find the same curve. This means that the solution of our problem is not unique.

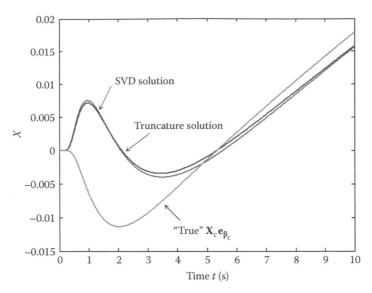

FIGURE 9.20
Solution obtained by SVD and comparison with true function $X = \mathbf{X}_c \mathbf{e}_{\beta_c}$.

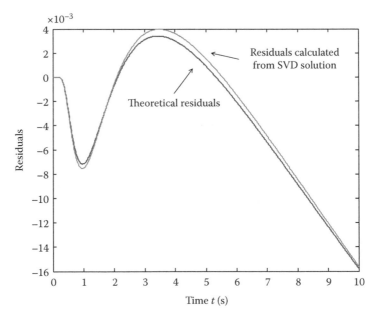

FIGURE 9.21
Comparison of the true residuals and the residuals calculated from $X = \mathbf{X}_c \mathbf{e}_{\beta_c}$ obtained by SVD.

This result can be explained by the fact that $\mathbf{A}$ is a projector and, consequently, that different vectors can lead to the same projection.

It is clear that the SVD procedure yields a particular solution $\mathbf{X}_{\text{SVD}}$ of the linear system $\mathbf{AX} = \mathbf{B}$. The true solution of our problem is composed of a second term, which corresponds to the eigenvector associated with the null eigenvalue (Kernel solution) that has been removed in the "pseudo-inverse" calculation by SVD. This vector corresponds to the direction of projection.

Thus, to obtain the general solution, we have also to take into account the solution associated to the null eigenvalue denoted X_0 that satisfies the system $AX_0 = 0$ (Kernel solution):

$$\left(X_r\left(X_r^t X_r\right)^{-1} X_r^t - Id\right) X_0 = 0 \tag{9.45}$$

or

$$\left(X_r\left(X_r^t X_r\right)^{-1} X_r^t\right) X_0 = 1^* X_0. \tag{9.46}$$

The solution X_0 can be obtained by solving an eigenvalue problem (9.46). It is plotted in Figure 9.22.

Since X_{SVD} satisfies $AX_{SVD} = B$ and X_0 satisfies $AX_0 = 0$, it is clear that $X = X_{SVD} + \alpha X_0$ is also a solution of the system $AX = B$ for any values of the coefficient α.

The previous decomposition is fully justified from the expressions of the expectancy of residuals and bias:

$$\begin{cases} E(r) = \left(X_r\left(X_r^t X_r\right)^{-1} X_r^t - Id\right) X_c e_{\beta_c}, \\ b_{\beta_r} = -\left(X_r^t X_r\right)^{-1} X_r^t X_c e_{\beta_c}. \end{cases} \tag{9.47}$$

Thus,

$$E(r) = X_r\left(X_r^t X_r\right)^{-1} X_r^t X_c e_{\beta_c} - X_c e_{\beta_c} \tag{9.48}$$

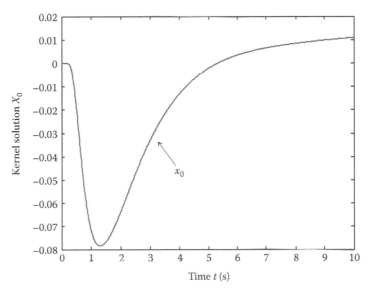

FIGURE 9.22
Solution X_0 obtained by solving an eigenvalue problem (Kernel solution).

and consequently,

$$E(r) = -X_r b_{\beta_r} - X_c e_{\beta_c}. \tag{9.49}$$

We find that

$$X_c e_{\beta_c} = -E(r) + X_r^t b_{\beta_r}. \tag{9.50}$$

By identification with $X_c e_{\beta_c} = X_{SVD} + \alpha X_0$, we find that

$$\begin{cases} X_{SVD} = -E(r), \\ \alpha X_0 = -X_r b_{\beta_r}. \end{cases} \tag{9.51}$$

We show that an infinite number of solutions to our problem exist, which appears as the sum of the SVD solution (pseudo-inverse) X_{SVD} and the Kernel solution X_0, with a weighting coefficient α. The problem now stays in the determination of α, because only one value of this coefficient allows us to calculate the real values of the bias.

As shown here, we cannot determine the coefficient α through only one estimation without any complementary information.

The idea is to work with variable time intervals. Indeed, with an unbiased model, the bias is null for any time intervals. On the contrary, in the case of a reduced or biased model, bias on parameters varies with the time interval as illustrated in Table 9.4 (see also Section 9.2.3.5).

In the next section, we will show how it is possible to use the variations of the bias with respect to time in order to estimate it.

9.3.3.2 Estimation of the Bias Using a Variable Time Interval

We will show in this section that it is possible to have an approximation of the bias in the estimated parameters if we are working with at least two different time intervals. For more simplicity, we make the assumption that time intervals are equally spaced $\Delta t = C^{ste}$, which is not necessary. The maximum time interval length is noted by $t_{max} = n \cdot \Delta t$ (n is the maximum number of time steps). We used 1000 points in practice.

The experimental curve is simulated using the unbiased model T but with no added noise. The sensitivities of the unbiased models T to the "unknown" β_r and "known" β_c parameters are defined by

$$X_r = \frac{\partial T(t, \beta_r, \beta_c)}{\partial \beta_r} \quad \text{and} \quad X_c = \frac{\partial T(t, \beta_r, \beta_c)}{\partial \beta_c}. \tag{9.52}$$

The sensitivities of the biased model $\tilde{T}$ to the "unknown" β_r denoted $\tilde{X}_{\beta_r}$ are defined by

$$\tilde{X}_r = \frac{\partial \tilde{T}(t, \hat{\beta}_r, \beta_{c_{nom}})}{\partial \hat{\beta}_r} = X_r. \tag{9.53}$$

Equation 9.20 clearly shows that in first approximation, sensitivities to the detailed model X_r usually unknown are close to the sensitivities of the reduced model $\tilde{X}_r$. We decided to keep these two quantities in the following equations to remember what type of sensitivity should be rigorously considered.

The estimated parameters $\boldsymbol{\beta}_r$ of the biased model $\tilde{\mathbf{T}}$ are estimated by an OLS method that consists in minimizing the scalar product between $\tilde{\mathbf{X}}_r$ and the residuals curve $\mathbf{r}$ (residuals vector orthogonal with each sensitivity coefficient of the biased model $\tilde{\mathbf{X}}_r$ to "unknown" parameters). The residuals are given here by $\mathbf{r} = \mathbf{T} - \tilde{\mathbf{T}}$ (no noise):

$$\tilde{\mathbf{X}}_r^t \cdot \mathbf{r} = 0. \tag{9.54}$$

Since $\mathbf{r} = \mathbf{T}(t, \boldsymbol{\beta}_r, \boldsymbol{\beta}_c) - \hat{\mathbf{T}}(t, \hat{\boldsymbol{\beta}}_r, \boldsymbol{\beta}_{c_{nom}}) = -\mathbf{X}_r \mathbf{b}_{\boldsymbol{\beta}_r} - \mathbf{X}_c \mathbf{e}_{b_c}$ from (9.20) and $\tilde{\mathbf{X}}_r^t \mathbf{r} = 0$, we obtain

$$\tilde{\mathbf{X}}_r^t \cdot \mathbf{r} = 0 = -\tilde{\mathbf{X}}_r^t \mathbf{X}_r \cdot \mathbf{b}_{\boldsymbol{\beta}_r} - \tilde{\mathbf{X}}_r^t \mathbf{X}_c \cdot \mathbf{e}_{b_c}, \tag{9.55}$$

and finally,

$$\mathbf{b}_{\boldsymbol{\beta}_r} - -\left(\tilde{\mathbf{X}}_r^t \mathbf{X}_r\right)^{-1} \tilde{\mathbf{X}}_i^t \mathbf{X}_c \mathbf{e}_{b_c}, \tag{9.56}$$

In practice, it is not possible to evaluate the bias from this relation because the sensitivities $\mathbf{X}_r$ and $\mathbf{X}_c$ of the unbiased model (ideal model) are unknown. As previously shown, a good approximation of $\mathbf{X}_r$ can be given by $\tilde{\mathbf{X}}_r$ but $\mathbf{X}_c \mathbf{e}_{\boldsymbol{\beta}_c}$ remains unknown.

The expression of $\mathbf{X}_c \mathbf{e}_{\boldsymbol{\beta}_c}$ can be calculated from the residuals by $\mathbf{X}_c \mathbf{e}_{b_c} = -\mathbf{r} - \mathbf{X}_r \mathbf{b}_{\boldsymbol{\beta}_r}$ (see (9.20)) and gives

$$\mathbf{r} = \left(\mathbf{X}_r (\tilde{\mathbf{X}}_r^t \mathbf{X}_r)^{-1} \tilde{\mathbf{X}}_r^t - \mathbf{Id}\right) \mathbf{X}_c \mathbf{e}_{b_c}. \tag{9.57}$$

The difficulty then comes from the fact that this matrix is singular and leads to an infinite number of solutions.

Let consider now two different time intervals $[0 - t_1 = n_1 \cdot \Delta t]$ and $[0 - t_2 = n_2 \cdot \Delta t]$, with $n_1 \leq n_2 \leq n$. Truncated vector quantities (thermograms and sensitivity curves are truncated to their n_1 and n_2 first points) related to these two time intervals and the corresponding values of the "unknown" parameters $\boldsymbol{\beta}_r$ are denoted with the subscripts 1 or 2 ($\hat{\boldsymbol{\beta}}_{r_1}$ and $\hat{\boldsymbol{\beta}}_{r_2}$).

The OLS minimization allows us to write (see Equation 9.54)

$$\tilde{\mathbf{X}}_{r1} \cdot \mathbf{r}_1 = 0 \quad \text{and} \quad \tilde{\mathbf{X}}_{r_2} \cdot \mathbf{r}_2 = 0. \tag{9.58}$$

From bias expression (Equation 9.56), we obtain

$$\begin{aligned} \mathbf{b}_{\boldsymbol{\beta}_{r1}} = \hat{\boldsymbol{\beta}}_{r_1} - \boldsymbol{\beta}_r = -\left(\tilde{\mathbf{X}}_{r_1}^t \mathbf{X}_{r_1}\right)^{-1} \tilde{\mathbf{X}}_{r_1}^t \mathbf{X}_{c_1} \mathbf{e}_{\boldsymbol{\beta}_c}, \\ \mathbf{b}_{\boldsymbol{\beta}_{r2}} = \hat{\boldsymbol{\beta}}_{r_2} - \boldsymbol{\beta}_r = -\left(\tilde{\mathbf{X}}_{r_2}^t \mathbf{X}_{r_2}\right)^{-1} \tilde{\mathbf{X}}_{r_2}^t \mathbf{X}_{c_2} \mathbf{e}_{\boldsymbol{\beta}_c}. \end{aligned} \tag{9.59}$$

$\hat{\boldsymbol{\beta}}_{r_1}$ and $\hat{\boldsymbol{\beta}}_{r_2}$ are estimated by an OLS procedure with two different time intervals and are thus perfectly known. So it is for $\tilde{\mathbf{X}}_{r_1}$ and $\tilde{\mathbf{X}}_{r_2}$. If we take the difference of these two relations, we obtain the bias variations $\Delta \mathbf{b}_{\boldsymbol{\beta}_{r2-1}} = \mathbf{b}_{\boldsymbol{\beta}_{r2}} - \mathbf{b}_{\boldsymbol{\beta}_{r1}}$:

$$\Delta \mathbf{b}_{\boldsymbol{\beta}_{r2-1}} = \hat{\boldsymbol{\beta}}_{r_2} - \hat{\boldsymbol{\beta}}_{r_1} = -\left(\tilde{\mathbf{X}}_{r_2}^t \mathbf{X}_{r_2}\right)^{-1} \tilde{\mathbf{X}}_{r_2}^t \mathbf{X}_{c_2} \mathbf{e}_{\boldsymbol{\beta}_c} + \left(\tilde{\mathbf{X}}_{r_1}^t \mathbf{X}_{r_1}\right)^{-1} \tilde{\mathbf{X}}_{r_1}^t \mathbf{X}_{c_1} \mathbf{e}_{\boldsymbol{\beta}_c}. \tag{9.60}$$

From this expression, it is possible to obtain information on $\mathbf{X}_c \mathbf{e}_{\boldsymbol{\beta}_c}$.

If we assume that $t_2 \rightarrow t_n$ and $t_1 \rightarrow t_2$, then we have in a first approximation for the cumulative sum $\tilde{X}^t_{r_2} X_{r_2}$:

$$\tilde{X}^t_{r_2} X_{r_2} = \tilde{X}^t_{r_1} X_{r_1} + \sum_{i=n_1+1}^{n_2} \tilde{X}^t_{ri} X_{ri} \simeq \underset{t_1 \rightarrow t_2/t_2 \rightarrow t_{\max}}{\tilde{X}^t_{r_1} X_{r_1}}, \tag{9.61}$$

with

$$\Delta b_{\beta_{r2-1}} = \hat{\beta}_{r_2} - \hat{\beta}_{r_1} = -\left(\tilde{X}^t_{r_1} X_{r_1}\right)^{-1}\left[\tilde{X}^t_{r_2} X_{c_2} e_{\beta_c} - \tilde{X}^t_{r_1} X_{c_1} e_{\beta_c}\right]. \tag{9.62}$$

We also have

$$\tilde{X}^t_{r_2} X_{c_2} e_{b_c} = \tilde{X}^t_{r_1} X_{c_1} e_{b_c} + \sum_{i=n_1+1}^{n_2} \tilde{X}^t_{ri} X_{ci} e_{\beta_c}. \tag{9.63}$$

This leads to

$$\Delta b_{\beta_{r2-1}} = \hat{\beta}_{r_2} - \hat{\beta}_{r_1} = -\left(\tilde{X}^t_{r_1} X_{r_1}\right)^{-1}\left[\sum_{i=n_1+1}^{n_2} \tilde{X}^t_{ri} X_{ci} . e_{\beta_c}\right]. \tag{9.64}$$

If $t_2 = t_1 + \Delta t$ ($n_2 = n_1 + 1$), then

$$\Delta b_{\beta_{r2-1}} = \hat{\beta}_{r_2} - \hat{\beta}_{r_1} = -\left(\tilde{X}^t_{r_1} X_{r_1}\right)^{-1}\left[\tilde{X}^t_r(t_2) X_c(t_2) \cdot e_{\beta_c}\right]. \tag{9.65}$$

If t_1 and $t_2(t_m = (t_1 + t_2)/2 \rightarrow m = (n_1 + n_2)/2)$ are close and assuming that the function $X_c(t)e_{\beta_c}$ is nearly constant in the interval $[t_1 - t_2]$ and equal to the value of the function in the middle of the interval $X_c(t_m)e_{\beta_c}$, then

$$\Delta b_{\beta_{r2-1}} = \hat{\beta}_{r_2} - \hat{\beta}_{r_1} = -\left(\tilde{X}^t_{r_1} X_{r_1}\right)^{-1}\left[\tilde{X}^t_r(t_m) X_c(t_m) e_{\beta_c}(n_2 - n_1)\right]. \tag{9.66}$$

We finally obtain

$$\tilde{X}^t_r(t_m) X_c(t_m) e_{\beta_c} = -\frac{\left(\tilde{X}^t_{r_1} X_{r_1}\right)\Delta b_{\beta_{r2-1}}}{(n_2 - n_1)}. \tag{9.67}$$

From the value of residuals $r(t_m) = -X_r(t_m)b_{\beta_r} - X_c(t_m)e_{b_c}$ at time t_m, the value of the bias can be calculated by

$$b_{\beta_r} = \left(\tilde{X}^t_r(t_m) X_r(t_m)\right)^{-1}\left(-\tilde{X}^t_r(t_m) r(t_m) - \tilde{X}^t_r(t_m) X_c(t_m) e_{b_c}\right). \tag{9.68}$$

So, we have shown that two biased estimations on two different intervals $[0 - t_1]$ and $[0 - t_2]$ can give access to the estimation of bias. Of course, the main assumption is about the approximation of the cumulative sum (Equation 9.61) that can be only checked if

$t_1 \rightarrow t_2$ and $t_2 \rightarrow t_n$. An illustration of this method is presented in the next and last section. Finally, knowing the value of bias $\mathbf{b}_{\beta_r}$, we can obtain the function $\mathbf{X}_c \mathbf{e}_{b_c}$ from the residuals and $\mathbf{X}_r$ by

$$\mathbf{X}_c \mathbf{e}_{b_c} = -\mathbf{r} - \mathbf{X}_r \mathbf{b}_{\beta_r}. \tag{9.69}$$

9.3.3.3 Application in the Case of the "Flash" Method

To illustrate these results, we consider the case of the Flash method. As previously shown, this problem is a function of two parameters: the characteristic time (thermal diffusivity of the material) and the Biot number (heat losses).

For the simulation of an experimental result, the following nominal values were considered:

- Fourier number: $Fo = \dfrac{a}{e^2} = 0.1$
- Biot number: $Bi = \dfrac{he}{\lambda} = 0.05$

To simulate a bias on this fine model, we consider the heat losses as a "known" parameter. The value of the Biot number is fixed to a nominal value $Bi = 0.03$ (relative error of -40%). To simplify, we will consider the signal without noise (only the systemic error is considered).

Theoretical thermogram (fine model) and sensitivities to "unknown" $\mathbf{X}_r$ and assumed "known" $\mathbf{X}_c$ parameters are plotted in Figure 9.23. The sensitivity curves show that thermogram is sensitive to $\mathbf{X}_r$. This parameter has been estimated by an OLS method with the biased model. The solution is presented in Figure 9.24. It is clear that the estimated value is different from the input nominal value.

If we then try to perform this estimation for different time interval lengths, we can observe a variation of bias as illustrated in Figure 9.25. This means, as explained before,

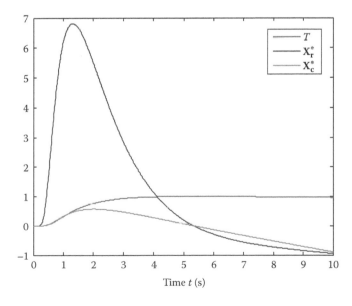

FIGURE 9.23
Thermogram with heat loss and sensitivities.

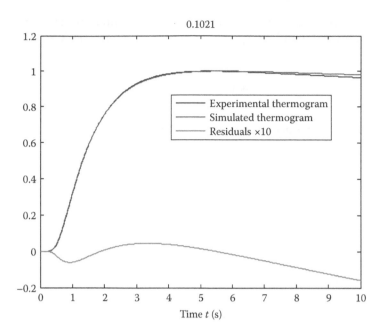

FIGURE 9.24
Estimated thermogram $\hat{\beta}_r = 0.1021$ and $\beta_c = 0.03$—residuals curve.

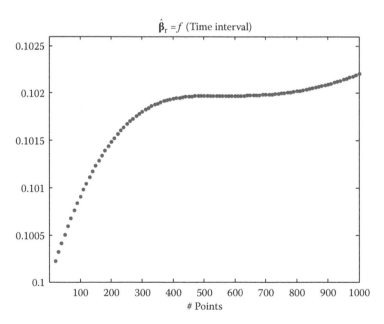

FIGURE 9.25
Evolution of bias versus time interval length.

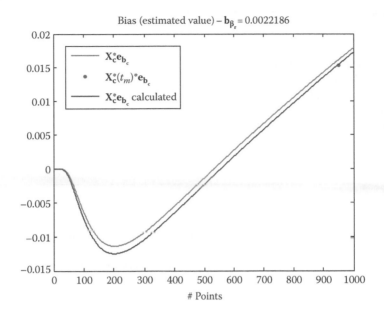

FIGURE 9.26
Comparison between theoretical and estimated $X_c e_{\beta_c}$ curves.

that a bias on the estimated parameter exists. Figure 9.25 clearly shows that the estimated values are a function of the time interval length. This information can be used for the determination of $X_c e_{\beta_c}$. Only one point in time is required to determine the value of the bias on the estimated parameter.

In Figure 9.26, the estimated function $X_c e_{\beta_c}$ obtained using (9.69) is compared with the theoretical one. These two curves are indeed very close. The bias we calculate is equal to 0.0022186 and is very close to the theoretical one.

9.4 Conclusion

To conclude, we have presented different relations giving an estimation of the bias or "systematic error" introduced in the estimated parameters when they are estimated from a biased model or a reduced model. These relations have been checked in the case of the Flash method. We have shown that it is possible to estimate the bias even if the detailed model is unknown. This bias can be calculated from the residuals curve and sensitivity to the "unknown" parameters. This method must be considered as the first track for estimating bias. Of course, more investigations are needed to improve, for instance, the choice of the number of temporal points that can be used for its determination.

Abbreviations

NLP Nonlinear in the parameters
OLS Ordinary least squares
PEP Parameter estimation problem
SNR Signal-to-noise ratio
SVD Singular value decomposition

References

Abramowitz, M. and Stegun, I.A. 1970. *Handbook of Mathematical Functions*, Dover Publications, Inc., New York.

Beck, J.V. and Arnold, K.J. 1977. *Parameter Estimation in Engineering and Science*, John Wiley & Sons, New York.

Beck, J.V., Blackwell, B., and St. Clair, C.R. 1985. *Inverse Heat Conduction: Ill-Posed Problems*, John Wiley & Sons, New York.

Gallant, A.R. 1975. Nonlinear regression. *Am. Stat.* 29:73–81.

Levenberg, K. 1944. A method for the solution of certain problems in least squares. *Quart. Appl. Math.* 2:164–168.

Maillet, D., Andre, S., Batsale, J.C, Degiovanni, A., and Moyne, C. 2000. *Thermal Quadrupoles: Solving the Heat Equation through Integral Transforms*, John Wiley & Sons, New York.

Remy, B. and Degiovanni, A. 2005. Parameters estimation and measurement of thermophysical properties of liquids. *Int. J. Heat Mass Transf.* 48(19–20):4103–4120.

10

A Survey of Basic Deterministic, Heuristic, and Hybrid Methods for Single-Objective Optimization and Response Surface Generation

Marcelo J. Colaço and George S. Dulikravich

CONTENTS

10.1 Introduction

Inverse problems usually involve the minimization of some objective function as part of their formulation. Such minimization procedures require the use of an optimization technique. Thus, in this chapter, we address solution methodologies for single-objective optimization problems, based on minimization techniques. Several gradient-based and non-gradient-based (stochastic) techniques are discussed, together with their basic implementation steps and algorithms. We present some deterministic methods, such as the conjugate gradient method, the Newton Method, and the Davidon–Fletcher–Powell (DFP) method (Levenberg, 1944; Hestenes and Stiefel, 1952; Davidon, 1959; Fletcher and Powell, 1963; Marquardt, 1963; Fletcher and Reeves, 1964; Broyden, 1965, 1967; Daniel, 1971; Polak, 1971; Beale, 1972; Bard, 1974; Beck and Arnold, 1977; Moré, 1977; Powell, 1977; Tikhonov and Arsenin, 1977; Dennis and Schnabel, 1983; Beck et al., 1985; Stoecker, 1989; Murio, 1993; Alifanov, 1994; Alifanov et al., 1995; Kurpisz and Nowak, 1995; Dulikravich and Martin, 1996; Trujillo and Busby, 1997; Jaluria, 1998; Beck, 1999; Belegundu and Chandrupatla, 1999; Colaço and Orlande, 1999, 2001a,b, 2004; Fletcher, 2000; Ozisik and Orlande, 2000; Woodbury, 2002). In addition, we present some of the stochastic approaches, such as the simulated annealing method (Corana et al., 1987; Goffe et al., 1994), the differential evolutionary method (Storn and Price, 1996), genetic algorithms (Goldberg, 1989; Deb, 2002), and the particle swarm method (Kennedy and Eberhart, 1995; Kennedy, 1999; Eberhart et al., 2001; Naka et al., 2001). Deterministic methods are, in general, computationally faster (they require fewer objective function evaluations in case of problems with low number of design variables) than stochastic methods, although they can converge to a local minima or maxima, instead of the global one. On the other hand, stochastic algorithms can ideally converge to a global maxima or minima, although they are computationally slower (for problems with relatively low number of design variables) than the deterministic ones. Indeed, the stochastic algorithms can require thousands of evaluations of the objective functions and, in some cases, become nonpractical. In order to overcome these difficulties, we will also discuss the so-called hybrid algorithms that take advantage of the robustness of the stochastic methods and the fast convergence of the deterministic methods (Dulikravich et al., 1999, 2003, 2004, 2008; Colaço and Orlande, 2001a,b; Colaço et al., 2004, 2005b, 2006, 2008; Colaço and Dulikravich, 2006, 2007; Dulikravich and Colaço, 2006; Wellele et al., 2006; Silva et al., 2007; Padilha et al., 2009). Each technique provides a unique approach with varying degrees of convergence, reliability, and robustness at different stages during the iterative minimization process. A set of analytically formulated rules and switching criteria can be coded into the program to automatically switch back and forth among the different algorithms as the iterative process advances (Dulikravich et al., 1999; Colaço et al., 2005b, 2008).

In many optimization problems, evaluation of the objective function is extremely expensive and time consuming. For example, optimizing chemical concentrations of each of the alloying elements in a multicomponent alloy requires manufacturing each candidate alloy and evaluating its properties using classical experimental techniques. Even with the most efficient optimization algorithms (Dulikravich et al., 2008), this means that often thousands of alloys having different chemical concentrations of their constitutive elements would have to be manufactured and tested. This is understandably too expensive to be economically acceptable. Similar is the situation when attempting to optimize three-dimensional aerodynamic shapes. Aerodynamics of thousands of different shapes needs to be analyzed using computational fluid dynamics software, which would be unacceptably time consuming.

For problems where objective function evaluations are already expensive and where the number of design variables is large thus requiring many such objective function evaluations, the only economically viable approach to optimization is to use an inexpensive and as accurate as possible surrogate model (a metamodel or a response surface) instead of the actual high fidelity analysis method. Such surrogate models are often known as response surfaces (Colaço et al., 2007, 2008). In the case of more than three design variables, a response surface becomes a high-dimensional hypersurface that needs to be fitted through the available (often small) set of high fidelity values of the objective function. Once the response surface (hypersurface) is created using an appropriate analytic formulation, it is very easy and fast to search such a surface for its minima given a set of values of design variables supporting such a response surface. Therefore, we also present in this chapter some basic concepts related to the response surface generation methodology.

10.2 Basic Concepts

10.2.1 Objective Function

The first step in establishing a procedure for the solution of either inverse problems or optimization problems is the definition of an *objective function*. The objective function is the mathematical representation of an aspect under evaluation, which must be minimized (or maximized). The objective function can be mathematically stated as

$$S = S(\mathbf{P}); \quad \mathbf{P} = \{P_1, P_2, \ldots, P_N\} \tag{10.1}$$

where $P_1, P_2, \ldots, P_N$ are the variables of the problem under consideration, which can be modified in order to find the minimum value of the function S.

The relationship between S and $\mathbf{P}$ can, most of the time, be expressed by a physical/mathematical model. However, in some cases, this relationship is impractical or even impossible and the variation of S with respect to $\mathbf{P}$ must be determined by experiments.

10.2.2 Unimodal versus Multimodal Objective Functions

Some of the methods that will be discussed here are only applicable to certain types of functions, namely unimodal, which are those having only one maximum (or minimum) inside the range of parameters being analyzed. This does not mean that the function must be continuous, as one can see from the Figure 10.1, where the first two functions are unimodals. The third function is unimodal in the interval $0 < P < 3\pi/2$ and the forth function is multimodal.

For unimodal functions, it is extremely easy to eliminate parts of the domain being analyzed in order to find the place of the maximum or minimum. Consider, for example, the first function of Figure 10.1: if we are looking for the maximum value of the function, and we know that $S(P = 1)$ is less than $S(P = 2)$, we can immediately eliminate the region to the left of $P = 1$, since the function is monotonically increasing its value. This is not true for multimodal functions, sketched as the fourth function in Figure 10.1.

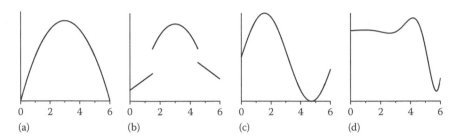

FIGURE 10.1
Some examples of functions S (ordinate) of a single design variable P (abscissa). (a) Continuous unimodal function, (b) discontinuous unimodal function, (c) and (d) multimodal functions.

10.2.3 Single- and Multi-Objective Functions

This chapter will deal only with single-objective functions. However, it is interesting to introduce the reader to the multi-objective optimization problems (Deb, 2002) since their applications in industry are very important. Consider, for example, the project of development of an automobile. Usually, we are not interested in only minimizing or maximizing a single function (e.g., fuel consumption), but extremizing a large number of objective functions as, for example: fuel consumption, automobile weight, final price, performance, etc. This problem is called a multi-objective optimization and it is more complex than the case of a single-objective optimization.

In an aero-thermo-elasticity problem, for example, several disciplines are involved with various (often conflicting) objective functions to be optimized simultaneously. This case can be illustrated by the Figure 10.2.

10.2.4 Constraints

Usually, the variables $P_1, P_2, \ldots, P_N$, which appear in the objective function formulation, are only allowed to vary within some prespecified ranges. Such *constraints* are, for example, due to physical or economical limitations.

We can have two types of constraints. The first one is the *equality constraint*, which can be represented by

$$G = G(\mathbf{P}) = 0 \tag{10.2}$$

This kind of constraint can represent, for example, the prespecified power of an automobile.

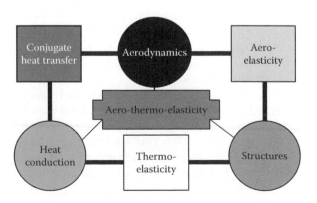

FIGURE 10.2
An example of a multi-objective design optimization problem.

The second type of constraint is called *inequality constraint*, and it is represented by

$$Q = Q(\mathbf{P}) < 0 \qquad (10.3)$$

This can represent, for example, the maximum temperature allowed in a gas turbine engine.

10.2.5 Optimization Problems

Inverse problems are mathematically classified as *ill-posed*, whereas standard heat transfer problems are *well-posed*. The solution of a well-posed problem must satisfy the conditions of existence, uniqueness, and stability with respect to the input data (Hadamard, 1923). The existence of a solution for an inverse heat transfer problem may be assured by physical reasoning. On the other hand, the uniqueness of the solution of inverse problems can be mathematically proved only for some special cases. Also, the inverse problem is very sensitive to random errors in the measured input data, thus requiring special techniques for its solution in order to satisfy the stability condition.

Successful solution of an inverse problem generally involves its reformulation as an approximate well-posed problem and makes use of some kind of regularization (stabilization) technique. Although the solution techniques for inverse problems do not necessarily make use of optimization techniques, many popular methods are based on them.

Despite their similarities, inverse and optimization problems are conceptually different. *Inverse problems are concerned with the identification of unknown quantities appearing in the mathematical formulation of physical problems, by using measurements of the system response.* On the other hand, *optimization problems generally deal with the minimization or maximization of a certain objective or cost function, in order to find design variables that will result in extreme value of the objective function.* In addition, inverse and optimization problems involve other different concepts. For example, the solution technique for an inverse problem is required to cope with instabilities resulting from the noisy measured input data, while for an optimization problem, the input data is given by the desired response(s) of the system. In contrast to inverse problems, the solution uniqueness may not be an important issue for optimization problems, as long as the solution obtained is physically feasible and can be practically implemented. Engineering applications of optimization techniques are very often concerned with the minimization or maximization of different quantities, such as minimum weight (e.g., lighter airplanes), minimum fuel consumption (e.g., more economic cars), maximum autonomy (e.g., longer range airplanes), etc. The necessity of finding the maximum or minimum values of some parameters (or functions) can be governed by economic factors, as in the case of fuel consumption, or design characteristics, as in the case of maximum autonomy of an airplane. Sometimes, however, the decision is more subjective, as in the case of choosing a car model. In general, different designs can be idealized for a given application, but only a few of them will be economically viable.

For optimization problems, the objective function S can be, for example, the fuel consumption of an automobile and the variables $P_1, P_2, \ldots, P_N$ can be the aerodynamic profile of the car, the material of the engine, the type of wheels used, the distance from the floor, etc.

In this chapter, we present deterministic and stochastic techniques for the minimization of an objective function $S(\mathbf{P})$ and the identification of the parameters $P_1, P_2, \ldots, P_N$, which appear in the objective function formulation. This type of minimization problem is solved in a space of finite dimension N, which is the number of unknown parameters. For many minimization problems, the unknowns cannot be recast in the form of a finite number of parameters and the minimization needs to be performed in an infinite dimensional space of functions (Hadamard, 1923; Daniel, 1971; Beck and Arnold, 1977; Tikhonov and Arsenin, 1977; Sabatier, 1978; Morozov, 1984; Beck et al., 1985; Hensel, 1991; Murio, 1993; Alifanov, 1994; Alifanov et al., 1995; Kurpisz and Nowak, 1995; Dulikravich and Martin, 1996; Kirsch, 1996; Trujillo and Busby, 1997; Isakov, 1998; Beck, 1999; Denisov, 1999; Yagola et al., 1999; Zubelli, 1999; Ozisik and Orlande, 2000; Ramm et al., 2000; Woodbury, 2002).

10.3 Deterministic Methods

In this section, some deterministic methods like the steepest descent method, the conjugate gradient method, the Newton–Raphson, and the quasi-Newton methods will be discussed. Some practical aspects and limitations of such methods will be addressed.

These types of methods, as applied to nonlinear minimization problems, generally rely on establishing an iterative procedure, which, after a certain number of iterations, will hopefully converge to the minimum of the objective function. The iterative procedure can be written in the following general form (Bard, 1974; Beck and Arnold, 1977; Dennis and Schnabel, 1983; Stoecker, 1989; Alifanov, 1994; Alifanov et al., 1995; Jaluria, 1998; Belegundu and Chandrupatla, 1999; Fletcher, 2000; Fox, 1971):

$$\mathbf{P}^{k+1} = \mathbf{P}^k + \alpha^k \mathbf{d}^k \tag{10.4}$$

where
    $\mathbf{P}$ is the vector of design variables
    α is the search step size
    $\mathbf{d}$ is the direction of descent
    k is the iteration number

An iteration step is *acceptable* if $S^{k+1} < S^k$. The direction of descent $\mathbf{d}$ will generate an acceptable step if and only if there exists a positive definite matrix $\mathbf{R}$, such that $\mathbf{d} = -\mathbf{R}\nabla S$ (Bard, 1974).

Such requirement results in directions of descent that form an angle greater than 90° with the gradient direction. A minimization method in which the directions are obtained in this manner is called an *acceptable gradient method* (Bard, 1974).

A *stationary point* of the objective function is one at which $\nabla S = 0$. The most that we can hope for any gradient-based method is that it converges to a stationary point. Convergence to the true minimum can be guaranteed only if it can be shown that the objective function has no other stationary points. In practice, however, one usually reaches the local minimum in the valley where the initial guess for the iterative procedure was located (Bard, 1974).

10.3.1 Steepest Descent Method

The most basic gradient-based method is the steepest descent method (Daniel, 1971; Stoecker, 1989; Jaluria, 1998; Belegundu and Chandrupatla, 1999). Some of the concepts developed here will be used in the next sections, where we will discuss more advanced methods. The basic idea of this method is to "walk" in the opposite direction of the locally highest variation of the objective function, in order to locate the minimum value of it. This can be exemplified in Figure 10.3.

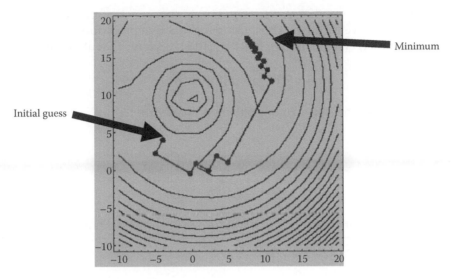

FIGURE 10.3
Convergence history for the steepest descent method.

The objective function can be mathematically stated as

$$S = S(\mathbf{P}); \quad \mathbf{P} = \{P_1, P_2, \ldots, P_N\} \tag{10.5}$$

The direction in which the objective function S varies most rapidly is the direction of gradient of S. For example, for the case with two variables (Figure 10.3), the gradient is

$$\nabla S = \frac{\partial S}{\partial P_1}\mathbf{i}_1 + \frac{\partial S}{\partial P_2}\mathbf{i}_2 \tag{10.6}$$

The iterative process for finding the minimum value of the objective function can be written in the most general terms as

$$\mathbf{P}^{k+1} = \mathbf{P}^k - \alpha^k \nabla S(\mathbf{P}^k) \tag{10.7}$$

where
  $\mathbf{P}$ is the vector of variables being optimized
  α is the search step size
  k is a counter for the iterations

Comparing Equations 10.4 and 10.7, one can check that for the steepest descent method, the direction of descent $\mathbf{d}$ is given by

$$\mathbf{d}^k = -\nabla S(\mathbf{P}^k) \tag{10.8}$$

In spite of this being the natural choice for the direction of descent, it is not very efficient as can be seen in Figure 10.3. Usually, the method starts with large variations in the objective function. As the minimum of the objective function is being approached, the convergence rate of this method becomes very low.

The optimum choice for the search step size is the one that causes the maximum variation in the objective function. Thus, using the iterative procedure given by Equation 10.7 and the definition of the objective function (10.1), we have that at iteration level $k+1$,

$$S(\mathbf{P}^{k+1}) = S(\mathbf{P}^k + \alpha^k \mathbf{d}^k) \tag{10.9}$$

The optimum value of the step size α is obtained by solving

$$\frac{dS(\mathbf{P}^{k+1})}{d\alpha^k} = 0 \tag{10.10}$$

Using the chain rule,

$$\frac{dS(\mathbf{P}^{k+1})}{d\alpha^k} = \frac{dS(P_1^{k+1})}{dP_1^{k+1}}\frac{dP_1^{k+1}}{d\alpha^k} + \frac{dS(P_2^{k+1})}{dP_2^{k+1}}\frac{dP_2^{k+1}}{d\alpha^k} + \cdots + \frac{dS(P_N^{k+1})}{dP_N^{k+1}}\frac{dP_N^{k+1}}{d\alpha^k} \tag{10.11}$$

Or

$$\frac{dS(\mathbf{P}^{k+1})}{d\alpha^k} = \left\langle [\nabla S(\mathbf{P}^{k+1})]^T, \frac{d\mathbf{P}^{k+1}}{d\alpha^k} \right\rangle \tag{10.12}$$

However, from Equations 10.7 and 10.8, it follows that

$$\frac{d\mathbf{P}^{k+1}}{d\alpha^k} = \mathbf{d}^k = -\nabla S(\mathbf{P}^k) \tag{10.13}$$

Substituting Equation 10.13 into (10.12) and (10.10), it follows that for steepest descent (Figure 10.4)

$$\left\langle [\nabla S(\mathbf{P}^{k+1})]^T, \nabla S(\mathbf{P}^k) \right\rangle = 0 \tag{10.14}$$

Thus, the optimum value of the search step size is the one that makes the gradients of the objective function at two successive iterations mutually orthogonal (Figure 10.3).

In "real life" applications, it is not possible to use Equation 10.14 to evaluate the search step size, α. Thus, some univariate search methods need to be employed in order to find the best value of the search step size at each iteration. In the case of a unimodal function, some classical procedures can be used, such as the dichotomous search (Stoecker, 1989; Jaluria, 1998), Fibonacci search (Stoecker, 1989; Jaluria, 1998), golden search (Stoecker, 1989; Jaluria, 1998), and cubic spline interpolation (de Boor, 1978), among others. However, for some realistic cases, the variation of the objective function with the search step size is not unimodal and then, more robust techniques are presented. The first one is the exhaustive search method and the second one is a technique based on exhaustive interpolation.

10.3.1.1 Exhaustive Search

This method (Stoecker, 1989; Jaluria, 1998) is one of the less efficient search methods available for sequential computation (which means not parallel computation). However,

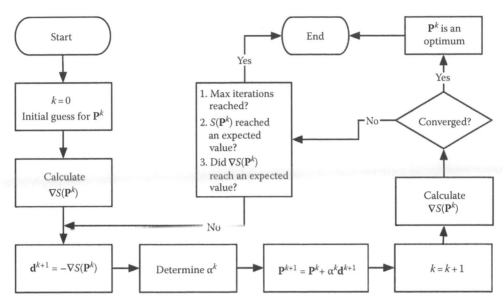

FIGURE 10.4
Iterative procedure for the steepest descent method.

it is a very good approach for parallel computing. Let us suppose, for example, that we are on a highway searching for a gas station with the lowest price of gasoline within an interval of 5 miles. If we do not have a newspaper or a telephone, the best way to do this is to go to each gas station and check the price and then determine the lowest value. This is the basis of the exhaustive search method. This method serves as an introduction to the next method, which is based on splines.

The basic idea consists in uniformly dividing the domain that we are interested in (the initial uncertainty region), and finding the region where the maximum or minimum value are located. Let us call this domain I_0. Let us suppose, for instance, the situation shown in Figure 10.5, where an uncertainty interval I_0 was divided into eight subregions, which are not necessarily the same size.

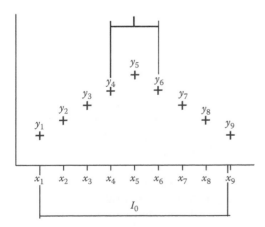

FIGURE 10.5
Exhaustive search method.

The objective function is evaluated at each of the nine points shown in the previous figure. From this analysis, we obtain the following:

$$
y_1 < y_2 < y_3 < y_4 < y_5
$$
$$
y_5 > y_6 > y_7 > y_8 > y_9
$$

$$(10.15)$$

Thus, the maximum point must be located between x_4 and x_6. Notice that we cannot say that the optimum is located between x_4 and x_5, nor between x_5 and x_6, since only a more refined grid could indicate this.

Thus, the final uncertainty interval I is $(x_6 - x_4)$ and the optimum point is located somewhere inside this interval. It can be shown (Stoecker, 1989; Jaluria, 1998) that I is given by

$$
I = \frac{2I_0}{n+1}
$$

$$(10.16)$$

where n is the number of objective functions evaluated. Notice that, once I is found, the process can be restarted making $I_0 = I$ and a more precise location for the maximum can be found. However, its precise location can never be reached.

In terms of sequential computation, this method is very inefficient. However, if we have a hypothetically large number of computers, all objective functions at each point in I_0 can be evaluated at the same time. Thus, for the example shown in Figure 10.5, for $n = 9$, if we can assign the task of calculating the objective function at each point to an individual computer, the initial uncertainty region is reduced by five times within the time needed to just perform one calculation of the entire region using a single computer. Other more sophisticated methods, such as the Fibonacci method, for example, need sequential evaluations of the objective function. The Fibonacci method, for example, requires four objective function evaluations for the same reduction of the uncertainty region. Thus, in spite of its lack of efficiency in single processor applications, the exhaustive search method may be very efficient in parallel computing applications. A typical parallel computing arrangement is where one computer is the master and the other computers perform the evaluations of the objective function at each of the locations. A typical arrangement for the case depicted in Figure 10.5 is presented in Figure 10.6 where there are 10 computers; one of them being the master and the other nine performing the evaluations of the objective functions at the nine locations shown on Figure 10.5.

10.3.1.2 Exhaustive Interpolation Search

This method is an improvement over the previous one, in that it requires fewer calculations to find the location of the minima. The method starts as the previous one, where domain is divided into several regions, where the objective functions are evaluated. The objective function is evaluated at a number of points in this domain. Next, a large number of points needs to be generated inside this domain and the objective function at these new points is estimated by spline fitting at the original points and interpolating at the new points using cubic splines (Dulikravich and Martin, 1994), B-splines (de Boor, 1978), kriging (Oliver and Webster, 1990), or other interpolants. Interrogating these interpolated values, we can find

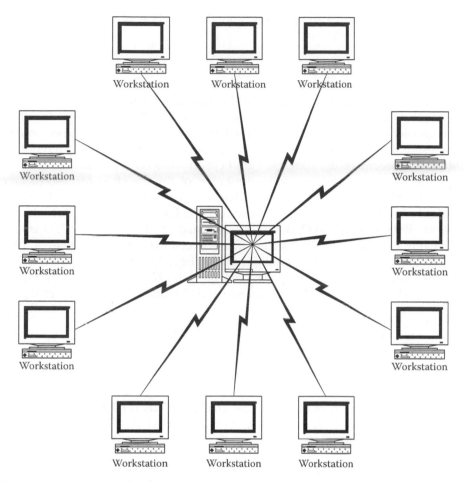

FIGURE 10.6
Typical setup for a parallel computing.

the region where the maximum or minimum values are located. The process can be repeated until a sufficiently small interval of uncertainty is obtained.

10.3.2 Conjugate Gradient Method

The steepest descent method, in general, converges slowly for non-quadratic functions, since optimum search step sizes produce orthogonal gradients between two successive iterations. The conjugate gradient method (Hestenes and Stiefel, 1952; Fletcher and Reeves, 1964; Daniel, 1971; Polak, 1971; Beale, 1972; Alifanov, 1974, 1994; Powell, 1977; Stoecker, 1989; Jarny et al., 1991; Artyukhin, 1993; Truffart et al., 1993; Dantas and Orlande, 1996; Huang and Tsai, 1997; Machado and Orlande, 1997; Orlande et al., 1997; Alencar Jr. et al., 1998; Colaço and Orlande, 1998; Jaluria, 1998; Belegundu and Chandrupatla, 1999; Colaço and Orlande, 2000, 2001a,b, 2002) tries to improve the convergence rate of the steepest descent method by choosing the directions of descent that reach the minimum value of the objective function faster. The iterative process for this method is given by the same general equation used in the steepest descent method, Equation 10.4. The difference is in the formulation for the direction of descent, which, for the conjugate gradient method, is given as a conjugation of the gradient and the direction of descent of the previous iteration, given as

$$\mathbf{d}^{k+1} = -\nabla S(\mathbf{P}^k) + \gamma^k \mathbf{d}^{k-1} + \psi^k \mathbf{d}^q \tag{10.17}$$

where γ^k and ψ^k are conjugation coefficients. The superscript q in Equation 10.17 denotes the iteration number where a restarting strategy is applied to the iterative procedure of the conjugate gradient method. Restarting strategies for the conjugate gradient method of parameter estimation were suggested by Powell (1977) in order to improve its convergence rate. Different versions of the conjugate gradient method can be found in the literature depending on the form used for the computation of the direction of descent given by Equation 10.17 (Hestenes and Stiefel, 1952; Fletcher and Reeves, 1964; Daniel, 1971; Polak, 1971; Beale, 1972; Alifanov, 1974, 1994; Powell, 1977; Jarny et al., 1991; Artyukhin, 1993; Truffart et al., 1993; Dantas and Orlande, 1996; Machado and Orlande, 1997; Orlande et al., 1997; Alencar Jr. et al., 1998). In the Fletcher–Reeves version (Fletcher and Reeves, 1964), the conjugation coefficients γ^k and ψ^k are obtained from the following expressions (Fletcher and Reeves, 1964; Daniel, 1971; Alifanov, 1974, 1994; Powell, 1977; Jarny et al., 1991; Dantas and Orlande, 1996; Huang and Tsai, 1997; Machado and Orlande, 1997; Orlande et al., 1997):

$$\gamma^k = \frac{\|\nabla S(\mathbf{P}^k)\|^2}{\|\nabla S(\mathbf{P}^{k-1})\|^2}, \quad \text{with } \gamma^0 = 0 \quad \text{for } k = 0 \tag{10.18a}$$

$$\psi^k = 0, \quad \text{for } k = 0, 1, 2 \tag{10.18b}$$

In the Polak–Ribiere version of the conjugate gradient method (Daniel, 1971; Polak, 1971; Powell, 1977; Jarny et al., 1991; Artyukhin, 1993; Truffart et al., 1993; Alifanov, 1994), the conjugation coefficients are given by

$$\gamma^k = \frac{[\nabla S(\mathbf{P}^k)]^T [\nabla S(\mathbf{P}^k) - \nabla S(\mathbf{P}^{k-1})]}{\|\nabla S(\mathbf{P}^{k-1})\|^2}, \quad \text{with } \gamma^0 = 0 \quad \text{for } k = 0 \tag{10.19a}$$

$$\psi^k = 0, \quad \text{for } k = 0, 1, 2, \ldots \tag{10.19b}$$

Based on a previous work by Beale (1972), Powell (1977) suggested the following expressions for the conjugation coefficients, which gives the so-called Powell–Beale's version of the conjugate gradient method (Beale, 1972; Alifanov, 1974; Powell, 1977):

$$\gamma^k = \frac{[\nabla S(\mathbf{P}^k)]^T [\nabla S(\mathbf{P}^k) - \nabla S(\mathbf{P}^{k-1})]}{[\mathbf{d}^{k-1}]^T [\nabla S(\mathbf{P}^k) - \nabla S(\mathbf{P}^{k-1})]}, \quad \text{with } \gamma^0 = 0 \quad \text{for } k = 0 \tag{10.20a}$$

$$\psi^k = \frac{[\nabla S(\mathbf{P}^k)]^T [\nabla S(\mathbf{P}^{q+1}) - \nabla S(\mathbf{P}^q)]}{[\mathbf{d}^q]^T [\nabla S(\mathbf{P}^{q+1}) - \nabla S(\mathbf{P}^q)]}, \quad \text{with } \gamma^0 = 0 \quad \text{for } k = 0 \tag{10.20b}$$

In accordance with Powell (1977), the application of the conjugate gradient method with the conjugation coefficients given by Equations 10.20 requires restarting when gradients at successive iterations tend to be non-orthogonal (which is a measure of the local nonlinearity of the problem) and when the direction of descent is not sufficiently downhill. Restarting is performed by making $\psi^k = 0$ in Equation 10.17.

The non-orthogonality of gradients at successive iterations is tested by the following equation:

$$\text{ABS}([\nabla S(\mathbf{P}^{k-1})]^T \nabla S(\mathbf{P}^k)) \geq 0.2 \, \| \nabla S(\mathbf{P}^k) \|^2 \tag{10.21a}$$

where $\text{ABS}(\cdot)$ denotes the absolute value.

A non-sufficiently downhill direction of descent (i.e., the angle between the direction of descent and the negative gradient direction is too large) is identified if either of the following inequalities is satisfied:

$$[\mathbf{d}^k]^T \nabla S(\mathbf{P}^k) \leq -1.2 \, \| \nabla S(\mathbf{P}^k) \|^2 \tag{10.21b}$$

$$[\mathbf{d}^k]^T \nabla S(\mathbf{P}^k) \geq -0.8 \, \| \nabla S(\mathbf{P}^k) \|^2 \tag{10.21c}$$

We note that the coefficients 0.2, 1.2, and 0.8 appearing in Equations 10.21a through c are empirically determined and are the same values used by Powell (1977).

In Powell–Beale's version of the conjugate gradient method, the direction of descent given by Equation 10.17 is computed in accordance with the following algorithm for $k \geq 1$ (Powell, 1977):

Step 1: Test the inequality (10.21a). If it is true, set $q = k - 1$.

Step 2: Compute γ^k using Equation 10.20a.

Step 3: If $k = q + 1$, set $\psi^k = 0$. If $k \neq q + 1$, compute ψ^k using Equation 10.20b.

Step 4: Compute the search direction $\mathbf{d}^{k+1}$ using Equation 10.17.

Step 5: If $k \neq q + 1$, test the inequalities (10.21b and c). If either one of them is satisfied, set $q = k - 1$ and $\psi^k = 0$. Then, recompute the search direction using Equation 10.17.

The steepest descent method, with the direction of descent given by the negative gradient equation, would be recovered with $\gamma^k = \psi^k = 0$ for any k in Equation 10.17. We note that the conjugation coefficients γ^k given by Equations 10.18a, 10.19a, and 10.20a are equivalent for quadratic functions, because the gradients at different iterations are mutually orthogonal (Daniel, 1971; Powell, 1977).

The same procedures used for the evaluation of the search step size in the steepest descent method can be employed here. Figure 10.7 illustrates the convergence history for the Fletcher–Reeves version of the conjugate gradient method for the same function presented in Figure 10.3. One can see that the conjugate gradient method is faster than the steepest descent. It is worth noting that the gradients between two successive iterations are no longer mutually orthogonal.

Colaço and Orlande (1999) presented a comparison of Fletcher–Reeves', Polak–Ribiere's, and Powell–Beale's versions of the conjugate gradient method, as applied to the estimation of the heat transfer coefficient at the surface of a plate. This inverse problem was solved as a function estimation approach, by assuming that no information was available regarding the functional form of the unknown. Among the three versions tested for the conjugate gradient method, the method suggested by Powell and Beale appeared to be the best, as applied to the cases examined in that paper. This algorithm did not present the anomalous increase of the functional as observed with the other versions, and its average rates of reduction of the functional were the largest. As a result, generally, the smallest values for

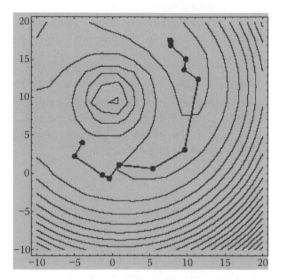

FIGURE 10.7
Convergence history for the Fletcher–Reeves version of
the conjugate gradient method.

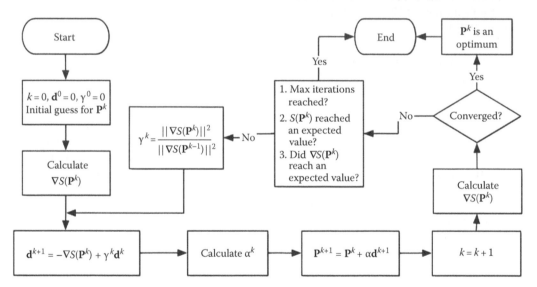

FIGURE 10.8
Iterative procedure for the Fletcher–Reeves version of the conjugate gradient method.

the root mean square (RMS) error of the estimated functions were obtained with Powell–
Beale's version of the conjugate gradient method.

Figure 10.8 shows the iterative procedure for the Fletcher–Reeves version (Fletcher and
Reeves, 1964) of the conjugate gradient method.

10.3.3 Newton–Raphson Method

While the steepest descent and the conjugate gradient methods use gradients of the
objective function in their iterative procedures, the Newton–Raphson method (Daniel,
1971; Stoecker, 1989; Jaluria, 1998; Belegundu and Chandrupatla, 1999) uses information
of the second derivative of the objective function in order to achieve a faster convergence
rate (which does not necessarily mean a shorter computing time).

Let us consider a function $S(\mathbf{P})$, which is at least twice differentiable. The Taylor expansion of $S(\mathbf{P})$ around a vector $\mathbf{h}$ is given by

$$S(\mathbf{P} + \mathbf{h}) = S(\mathbf{P}) + \nabla S(\mathbf{P})^T \mathbf{h} + \frac{1}{2}\mathbf{h}^T D^2 S(\mathbf{P})\mathbf{h} + O(\mathbf{h}^3) \qquad (10.22)$$

where
$\nabla S(\mathbf{P})$ is the gradient (vector of first-order derivatives)
$D^2 S(\mathbf{P})$ is the Hessian (matrix of second-order derivatives)

If the objective function $S(\mathbf{P})$ is twice differentiable, then the Hessian is always symmetrical, and we can write

$$\nabla S(\mathbf{P} + \mathbf{h}) \cong \nabla S(\mathbf{P}) + D^2 S(\mathbf{P})\mathbf{h} \qquad (10.23)$$

The optimum is obtained when the left side of Equation 10.23 vanishes. Thus, we have

$$\mathbf{h}_{\text{optimum}} \cong -[D^2 S(\mathbf{P})]^{-1}\nabla S(\mathbf{P}) \qquad (10.24)$$

and the vector that optimizes the function $S(\mathbf{P})$ is

$$(\mathbf{P} + \mathbf{h}_{\text{optimum}}) \cong \mathbf{P} - [D^2 S(\mathbf{P})]^{-1}\nabla S(\mathbf{P}) \qquad (10.25)$$

Thus, introducing a search step size, which can be used to control the rate of convergence of the method, we can rewrite the Newton–Raphson method in the form of the Equation 10.4 where the direction of descent is given by

$$\mathbf{d}^{k+1} = -[D^2 S(\mathbf{P}^k)]^{-1}\nabla S(\mathbf{P}^k) \qquad (10.26)$$

The Newton–Raphson method is faster than the conjugate gradient method as demonstrated in Figure 10.9. However, the calculation of the Hessian matrix coefficients takes a long time. Figure 10.10 shows the iterative procedure for the Newton–Raphson method.

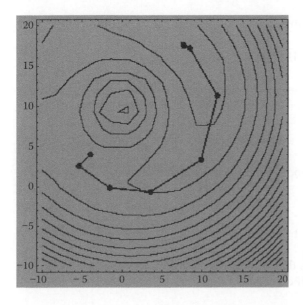

FIGURE 10.9
Convergence history for the Newton–Raphson method.

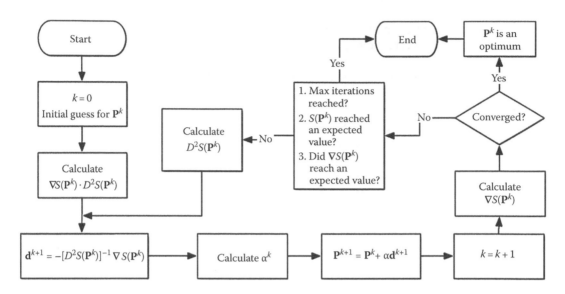

FIGURE 10.10
Iterative procedure for the basic Newton–Raphson method implementation.

Some other methods, which do not require second-order derivatives, so-called quasi-Newton methods, will be addressed in the next section.

10.3.4 Quasi-Newton Methods

The quasi-Newton methods (Daniel, 1971; Stoecker, 1989; Jaluria, 1998; Belegundu and Chandrupatla, 1999) try to calculate the Hessian appearing in the Newton–Raphson method in a manner that does not involve second-order derivatives. Usually, they employ approximation for the Hessian based only on first-order derivatives. Thus, they have a slower convergence rate than the Newton–Raphson method, but they are overall computationally faster.

Let us define a new matrix $\mathbf{H}$, which is an approximation to the inverse of the Hessian as

$$\mathbf{H}^k = [D^2 S(\mathbf{P}^k)]^{-1} \tag{10.27}$$

Thus, the quasi-Newton methods follow the general iterative procedure given by Equation 10.4, where the direction of descent is given by

$$\mathbf{d}^{k+1} = -\mathbf{H}^k \nabla S(\mathbf{P}^k) \tag{10.28}$$

The matrix $\mathbf{H}$ for the quasi-Newton methods is iteratively calculated as

$$\mathbf{H}^k = \mathbf{H}^{k-1} + \mathbf{M}^{k-1} + \mathbf{N}^{k-1} \quad \text{for } k = 1, 2, \ldots \tag{10.29a}$$

$$\mathbf{H}^k = \mathbf{I} \quad \text{for } k = 0 \tag{10.29b}$$

where $\mathbf{I}$ is the identity matrix. This means that during the first iteration, the quasi-Newton method starts as the steepest descent method.

Different quasi-Newton methods can be found depending on the choice for the matrices **M** and **N**. For the DFP method (Davidon, 1959; Fletcher and Powell, 1963), such matrices are given by

$$\mathbf{M}^{k-1} = \alpha^{k-1} \frac{\mathbf{d}^{k-1}(\mathbf{d}^{k-1})^T}{(\mathbf{d}^{k-1})^T \mathbf{Y}^{k-1}} \tag{10.30a}$$

$$\mathbf{N}^{k-1} = -\frac{(\mathbf{H}^{k-1}\mathbf{Y}^{k-1})(\mathbf{H}^{k-1}\mathbf{Y}^{k-1})^T}{(\mathbf{Y}^{k-1})^T \mathbf{H}^{k-1}\mathbf{Y}^{k-1}} \tag{10.30b}$$

where

$$\mathbf{Y}^{k-1} = \nabla S(\mathbf{P}^k) - \nabla S(\mathbf{P}^{k-1}) \tag{10.30c}$$

Figure 10.11 shows the results for the minimization of the objective function shown before, using the DFP method. One can see that its convergence rate is between the conjugate gradient method and the Newton–Raphson method.

Note that, since the matrix **H** is iteratively calculated, some errors can be propagated and, in general, the method needs to be restarted after certain number of iterations (Colaço et al., 2006). Also, since the matrix **M** depends on the choice of the search step size α, the method is very sensitive to its value.

A variation of the DFP method is the Broyden–Fletcher–Goldfarb–Shanno (BFGS) method (Davidon, 1959; Fletcher and Powell, 1963; Broyden, 1965, 1967), which is less sensitive to the choice of the search step size. For this method, the matrices **M** and **N** are calculated as

$$\mathbf{M}^{k-1} = \left(\frac{1 + (\mathbf{Y}^{k-1})^T \mathbf{H}^{k-1}\mathbf{Y}^{k-1}}{(\mathbf{Y}^{k-1})^T \mathbf{d}^{k-1}} \right) \frac{\mathbf{d}^{k-1}(\mathbf{d}^{k-1})^T}{(\mathbf{d}^{k-1})^T \mathbf{Y}^{k-1}} \tag{10.31a}$$

$$\mathbf{N}^{k-1} = -\frac{\mathbf{d}^{k-1}(\mathbf{Y}^{k-1})^T \mathbf{H}^{k-1} + \mathbf{H}^{k-1}\mathbf{Y}^{k-1}(\mathbf{d}^{k-1})^T}{(\mathbf{Y}^{k-1})^T \mathbf{d}^{k-1}} \tag{10.31b}$$

Figure 10.12 shows the iterative procedure for the BFGS method.

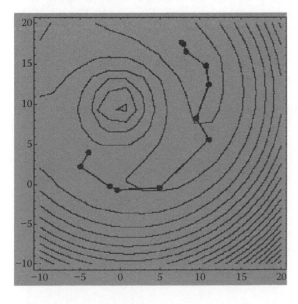

FIGURE 10.11
Convergence history for the DFP method.

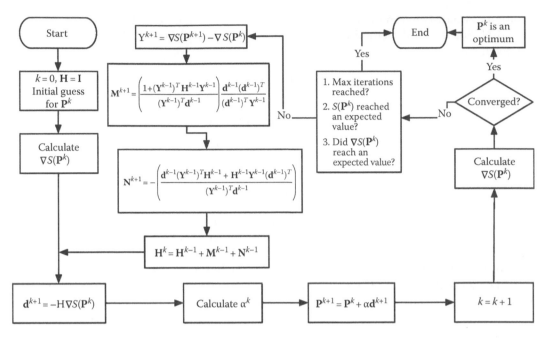

FIGURE 10.12
Iterative procedure for the BFGS method.

At this point, it is of interest to explore the influence on the initial guess for the four methods introduced thus far. Usually, all these methods quickly converge to the minimum value if it is close to the initial guess. The Newton–Raphson method, however, without the search step size, moves to the extreme point closest to the initial guess, irregardless if it is a maximum, minimum, or a saddle point. This is the reason why we introduce a search step size in Equation 10.25. The search step size prevents the method from jumping to a maximum value when we look for a minimum and vice versa. Figures 10.13 and 10.14 show the influence of the initial guess for all four methods for a Rosenbrock "banana-shape" function (More et al., 1981).

It should be pointed out that in real-life situations, topology of the objective function space is not smooth and second derivatives of the objective function cannot be evaluated with any degree of confidence. Thus, all gradient-based and second-derivative-based search optimization algorithms have serious issue with robustness and reliability of their applications to realistic problems.

10.3.5 Levenberg–Marquardt Method

The Levenberg–Marquardt method was first derived by Levenberg (1944), by modifying the ordinary least squares norm. Later, in 1963, Marquardt (1963) derived basically the same technique by using a different approach. Marquardt's intention was to obtain a method that would tend to the Gauss method in the neighborhood of the minimum of

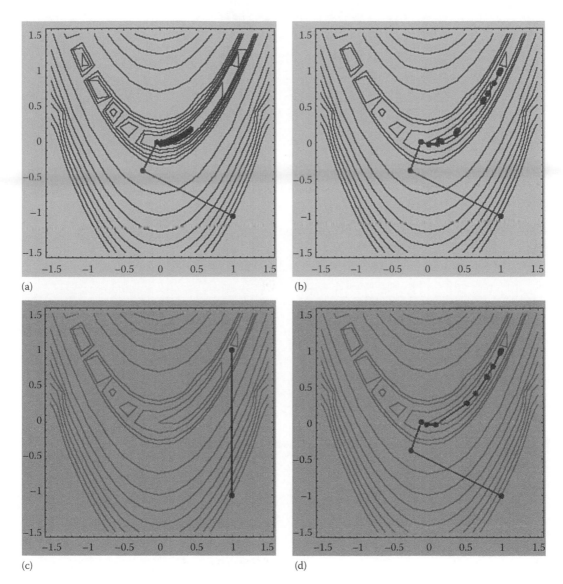

FIGURE 10.13
First initial guess for the (a) steepest descent, (b) conjugate gradient, (c) Newton–Raphson, and (d) DFP methods.

the ordinary least squares norm, and would tend to the steepest descent method in the neighborhood of the initial guess used for the iterative procedure. This method actually converts a matrix that approximates the Hessian into a positive definite one, so that the direction of descent is acceptable.

The method rests on the observation that if $\mathbf{J}$ is a positive definite matrix, then $\mathbf{A} + \lambda\mathbf{J}$ is positive definite for sufficiently large λ. If $\mathbf{A}$ is an approximation for the Hessian, we can

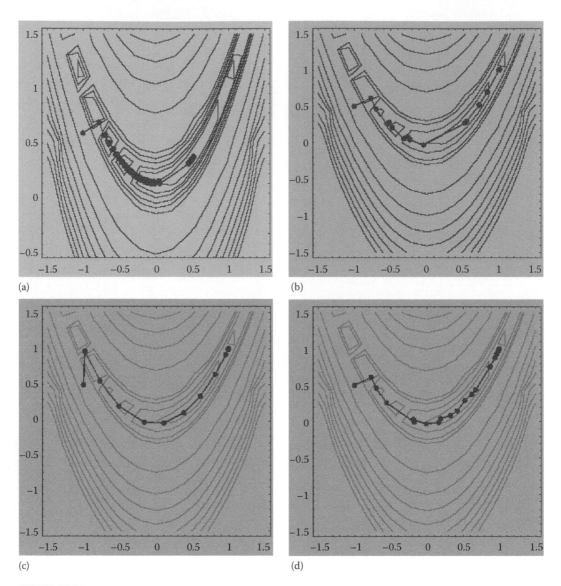

FIGURE 10.14
Second initial guess for the (a) steepest descent, (b) conjugate gradient, (c) Newton–Raphson, and (d) DFP methods.

choose **J** as a diagonal matrix whose elements coincide with the absolute values of the diagonal elements of **A** (Bard, 1974).

The direction of descent for the Levenberg–Marquardt method is given by (Bard, 1974)

$$\mathbf{d}^k = -(\mathbf{A}^k + \lambda^k \mathbf{J}^k)^{-1} \nabla S(\mathbf{P}^k) \tag{10.32}$$

and the step size is taken as $\alpha^k = 1$. Note that for large values of λ^k, a small step is taken along the negative gradient direction. On the other hand, as λ^k tends to zero, the

Levenberg–Marquardt method tends to an approximation of Newton's method based on the matrix $\mathbf{A}$. Usually, the matrix $\mathbf{A}$ is taken as that for the Gauss method (Bard, 1974; Beck and Arnold, 1977; Ozisik and Orlande, 2000).

10.4 Evolutionary and Stochastic Methods

In this section, some evolutionary and stochastic methods like genetic algorithm, differential evolution, particle swarm, and simulated annealing will be discussed. Evolutionary methods, in contrast to the deterministic methods, do not rely, in general, on strong mathematical basis and do not make use of the gradient nor second derivative of the objective function as a direction of descent. The evolutionary optimization algorithms attempt to mimic nature in order to find the minimum of the objective function.

10.4.1 Genetic Algorithms

Genetic algorithms (Goldberg, 1989) are heuristic global optimization methods that are based on the process of natural selection. Starting from a randomly generated population of candidate designs, the optimizer seeks to produce improved designs from one generation to the next. This is accomplished by exchanging genetic information between designs in the current population, in what is referred to as the crossover operation. Hopefully, this crossover produces improved designs, which are then used to populate the next generation (Goldberg, 1989; Deb, 2002).

The basic genetic algorithm works with a collection or population of candidate solutions to the optimization problem. The algorithm works in an iterative manner. At each iteration, also called generation, three operators are applied to the entire population of designs. These operators are selection, crossover, and mutation. For the operators to be effective, each candidate solution or design must be represented as a collection of finite parameters, also called genes. Each design must have a unique sequence of these parameters that define it. This collection of genes is often called the chromosome. The genes themselves are often encoded as binary strings, though they can be represented as real numbers. The length of the binary string determines how precisely the value, also known as the allele, of the gene is represented.

The genetic algorithm applied to an optimization problem proceeds as follows. The process begins with an initial population of random designs. Each gene is generated by randomly generating 0's and 1's. The chromosome strings are then formed by combining the genes together. This chromosome string defines the design. The objective function is evaluated for each design in the population. Each design is assigned a fitness value, which corresponds to the value of the objective function for that design. In the minimization case, a higher fitness is assigned to designs with lower values of the objective function.

Next, the population members are selected for reproduction, based upon their fitness. The selection operator is applied to each member of the population. The selection operator chooses pairs of individuals from population who will mate and produce an offspring. In the tournament selection scheme, random pairs are selected from the population and the individual with the higher fitness of each pair is allowed to mate.

Once a mating pair is selected, the crossover operator is applied. The crossover operator essentially produces new designs or offspring by combining the genes from the parent designs in a stochastic manner. In the uniform crossover scheme, it is possible to obtain any combination of the two parent's chromosomes. Each bit in each gene in the chromosome is assigned a probability that crossover will occur (e.g., 50% for all genes). A random number between 0 and 1 is generated for each bit in each gene. If a number greater than 0.5 is generated, then that bit is replaced by the corresponding bit in the gene from the other parent. If it is less than 0.5, the original bit in the gene remains unchanged. This process is repeated for the entire chromosome for each of the parents. When complete, two offsprings are generated, which may replace the parents in the population.

The mutation process follows next. When the crossover procedure is complete and a new population is formed, the mutation operator is applied. Each bit in each gene in the design is subjected to a chance for a change from 0 to 1, or vice versa. The chance is known as the mutation probability, which is usually small. This introduces additional randomness into the process, which helps to avoid local minima. Completion of the mutation process signals the end of a design cycle. Many cycles may be needed before the method converges to an optimum design.

For more details or for the numerical implementation of genetic algorithms, the reader is referred to Goldberg (1989) and Deb (2002).

10.4.2 Differential Evolution

The differential evolution method (Storn and Price, 1996) is an evolutionary method based on Darwin's theory of evolution of the species (Darwin, 1859). This non-gradient-based optimization method was created in 1995 (Storn and Price, 1996) as an alternative to the genetic algorithm methods. Following Darwin's theory, the strongest members of a population will be more capable of surviving in a certain environmental condition. During the mating process, the chromosomes of two individuals of the population are combined in a process called crossover. During this process, mutations can occur, which can be good (individual with a better objective function) or bad (individual with a worse objective function). The mutations are used as a way to escape from local minima. However, their excessive usage can lead to a non-convergence of the method.

The method starts with a randomly generated population in the domain of interest. Thus, successive combinations of chromosomes and mutations are performed, creating new generations until an optimum value is found.

The iterative process is given by (Figure 10.15)

$$\mathbf{P}_i^{k+1} = \delta_1 \mathbf{P}_i^k + \delta_2[\boldsymbol{\alpha} + F(\boldsymbol{\beta} - \boldsymbol{\gamma})] \tag{10.33}$$

where
  $\mathbf{P}_i$ is the ith individual of the vector of parameters
  $\boldsymbol{\alpha}$, $\boldsymbol{\beta}$, and $\boldsymbol{\gamma}$ are three members of population matrix $\mathbf{\Pi}$, randomly choosen
  F is a weight constant, which defines the mutation ($0.5 < F < 1$)
  k is a counter for the generations
  δ_1 and δ_2 are two functions that define the mutation

In this minimization process, if $S(\mathbf{P}^{k+1}) < S(\mathbf{P}^k)$, then $\mathbf{P}^{k+1}$ replaces $\mathbf{P}^k$ in the population matrix $\mathbf{\Pi}$. Otherwise, $\mathbf{P}^k$ is kept in the population matrix.

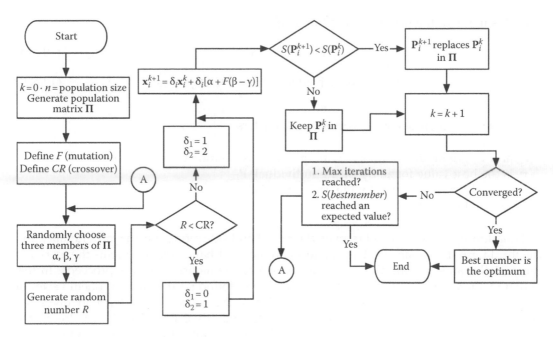

FIGURE 10.15
Iterative procedure for the differential evolution method.

The binomial crossover is given as

$$\delta_1 = 0, \quad \text{if } R < CR$$
$$1, \quad \text{if } R > CR \tag{10.34a}$$

$$\delta_2 = 1, \quad \text{if } R < CR$$
$$0, \quad \text{if } R > CR \tag{10.34b}$$

where
CR is a factor that defines the crossover $(0.5 < CR < 1)$
R is a random number with uniform distribution between 0 and 1

10.4.3 Particle Swarm

This non-gradient-based optimization method was created in 1995 by an electrical engineer (Russel Eberhart) and a social psychologist (James Kennedy) (Kennedy and Eberhart, 1995; Kennedy, 1999; Eberhart et al., 2001; Naka et al., 2001) as an alternative to the genetic algorithm methods. This method is based on the social behavior of various species and tries to equilibrate the individuality and sociability of the individuals in order to locate the optimum of interest. The original idea of Kennedy and Eberhart came from the observation of birds looking for a nesting place. When the individuality is increased, the search for alternative places for nesting is also increased. However, if the individuality becomes too high, the individual might never find the best place. In other words, when the sociability is increased, the individual learns more from their neighbor's experience. However, if the sociability becomes too high, all the individuals might converge to the first place found (possibly a local minima).

In this method, the iterative procedure is given by

$$\mathbf{P}_i^{k+1} = \mathbf{P}_i^k + \mathbf{v}_i^{k+1} \tag{10.35a}$$

$$\mathbf{v}_i^{k+1} = \alpha \mathbf{v}_i^k + \beta \mathbf{r}_{1i}(\boldsymbol{\pi}_i - \mathbf{P}_i^k) + \beta \mathbf{r}_{2i}(\boldsymbol{\pi}_g - \mathbf{P}_i^k) \tag{10.35b}$$

where
$\mathbf{P}_i$ is the ith individual of the vector of parameters
$\mathbf{v}_i = 0$, for $k = 0$
$\mathbf{r}_{1i}$ and $\mathbf{r}_{2i}$ are random numbers with uniform distribution between 0 and 1
$\boldsymbol{\pi}_i$ is the best value found by the ith individual, $\mathbf{P}_i$
$\boldsymbol{\pi}_g$ is the best value found by the entire population
$0 < \alpha < 1; 1 < \beta < 2$

In Equation 10.35b, the second term on the right-hand side represents the individuality and the third term the sociability. The first term on the right-hand side represents the inertia of the particles and, in general, must be decreased as the iterative process proceeds. In this equation, the vector $\boldsymbol{\pi}_i$ represents the best value ever found for the ith component vector of parameters $\mathbf{P}_i$ during the iterative process. Thus, the individuality term involves the comparison between the current value of the ith individual $\mathbf{P}_i$ and its best value in the past. The vector $\boldsymbol{\pi}_g$ is the best value ever found for the entire population of parameters (not only the ith individual). Thus, the sociability term compares $\mathbf{P}_i$ with the best value of the entire population in the past.

Figure 10.16 shows the iterative procedure for the particle swarm method.

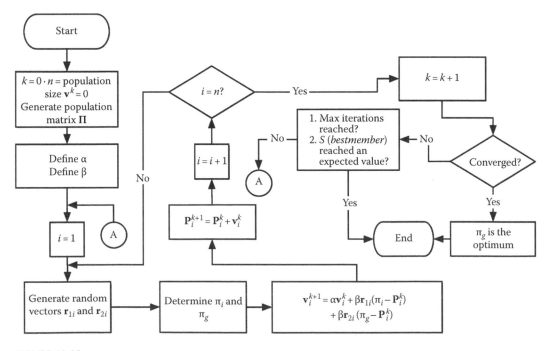

FIGURE 10.16
Iterative procedure for the particle swarm method.

10.4.4 Simulated Annealing

The simulated annealing method (Goffet et al., 1994; Corana et al., 1987) is based on the thermodynamics of the cooling of a material from a liquid to a solid phase. If a liquid material (e.g., liquid metal) is slowly cooled and left for a sufficiently long time close to the phase change temperature, a perfect crystal will be created, which has the lowest internal energy state.

On the other hand, if the liquid material is not left for a sufficient long time close to the phase change temperature, or, if the cooling process is not sufficiently slow, the final crystal will have several defects and a high internal energy state. This phenomena is similar to the quenching process used in metallurgical applications.

The gradient-based methods move in directions that successively lower the objective function value when minimizing the value of a certain function or in directions that successively raise the objective function value in the process of finding the maximum value of a certain function. The simulated annealing method can move in any direction at any point in the optimization process, thus escaping from possible local minimum or local maximum values.

We can say that gradient-based methods "cool down too fast," going rapidly to an optimum location which, in most cases, is not the global, but a local one. As opposed to gradient-based methods, nature works in a different way. Consider, for example, the Boltzmann probability function given as

$$\text{Prob}(E) \propto e^{(-E/KT)} \tag{10.36}$$

This equation expresses the idea that a system in thermal equilibrium has its energy distributed probabilistically among different energy states E, where K is the Boltzmann constant. Equation 10.36 tells us that even at low temperatures, there is a chance, although small, that the system is at a high energy level, as illustrated in Figure 10.17. Thus, there is a chance that the system could get out of this local minimum and continue looking for another one, possibly the global minimum.

Figure 10.18 shows the iterative procedure for the simulated annealing method. The procedure starts generating a population of individuals of the same size as the number of variables ($n = m$), in such a way that the population matrix is a square matrix. Then, the initial temperature (T), the reducing ratio (RT), the number of cycles (N_s), and the number of iterations of the annealing process (N_{it}) are selected. After $N_s^* n$ function evaluations, each element of the step length V is adjusted so that approximately half of all function evaluations are accepted. The suggested value for the number of cycles is 20. After $N_{it}^* N_s^* n$

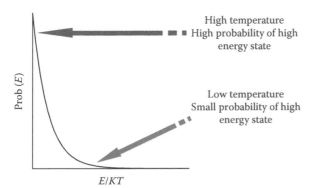

FIGURE 10.17
Schematic representation of Equation 10.36.

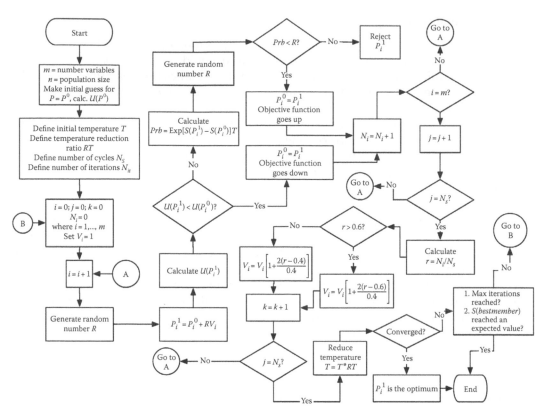

FIGURE 10.18
Iterative procedure for the simulated annealing method.

function evaluations, the temperature (T) is changed by the factor RT. The value suggested for the number of iterations by Corana et al. (1987) is MAX(100, 5*n).

The iterative process follows the equation

$$P_i^1 = P_i^0 + RV_i \tag{10.37}$$

Here, R is a random number with a uniform distribution between 0 and 1 and V is a step size, which is continuously adjusted.

Initially, it randomly chooses a trial point within the step length V (a vector of length n) of the user-selected starting point. The function is evaluated at this trial point (P_i^1) and its value is compared to its value at the initial point (P_i^0). In a minimization problem, all downhill moves are accepted and the algorithm continues from that trial point. Uphill moves may also be accepted; the decision is made by the Metropolis (Corana et al., 1987) criteria. It uses T (temperature) and the size of the downhill move in a probabilistic manner

$$P = e^{[S(P_i^1) - S(P_i^0)]/T} \tag{10.38}$$

The smaller T and the size of the uphill move are, the more likely that move will be accepted. If the trial is accepted, the algorithm moves on from that point. If it is rejected, another point is chosen for a trial evaluation.

Each element of V is periodically adjusted, so that half of all function evaluations in that direction are accepted. The number of accepted function evaluations is represented by the variable N^i. Thus, the variable r represents the ratio of accepted over total function evaluations for an entire cycle N_s and it is used to adjust the step length V.

A decrease in T is imposed upon the system with the RT variable by using

$$T(i+1) = RT*T(i) \tag{10.39}$$

where i is the ith iteration. Thus, as T declines, uphill moves are less likely to be accepted and the percentage of rejections rises. Given the scheme for the selection for V, V falls. Thus, as T declines, V falls and simulated annealing focuses upon the most promising area for optimization.

The parameter T is crucial in using simulated annealing successfully. It influences V, the step length over which the algorithm searches for optima. For a small initial T, the step length may be too small; thus not enough function evaluations will be performed to find the global optima. To determine the starting temperature that is consistent with optimizing a function, it is worthwhile to run a trial run first. The user should set $RT = 1.5$ and $T = 1.0$. With $RT > 1.0$, the temperature increases and V rises as well. Then, the value of T must be selected, which produces a large enough V.

10.5 Hybrid Optimization Methods

The hybrid optimization methods (Dulikravich et al., 1999, 2003, 2004, 2008; Colaço and Orlande, 2001a,b; Colaço et al., 2004, 2005b, 2006, 2008; Colaço and Dulikravich, 2006, 2007; Dulikravich and Colaço, 2006; Wellele et al., 2006; Silva et al., 2007; Padilha et al., 2009) are not more than a combination of the deterministic and the evolutionary/stochastic methods, in the sense that they try to use the advantages of each of these methods. The hybrid optimization method usually employs an evolutionary/stochastic method to locate a region where the global extreme point is located and then automatically switches to a deterministic method to get to the exact point faster (Dulikravich et al., 1999).

One of the possible hybrid optimization methods encountered in the literature (Dulikravich et al., 1999, 2003, 2004, 2008; Colaço and Orlande, 2001a,b; Colaço et al., 2004, 2005b, 2006, 2008; Colaço and Dulikravich, 2006, 2007; Dulikravich and Colaço, 2006; Wellele et al., 2006; Silva et al., 2007; Padilha et al., 2009), called in this chapter *H1*, is illustrated in Figure 10.19 (Colaço et al., 2005b). The driven module is very often the particle swarm method, which performs most of the optimization task. When a certain percent of the particles find a minima (let us say, some birds already found their best nesting place), the algorithm switches automatically to the differential evolution method and the particles (birds) are forced to breed. If there is an improvement in the objective function, the algorithm returns to the particle swarm method, meaning that some other region is more prone to having a global minimum. If there is no improvement on the objective function, this can indicate that this region already contains the global value expected and the algorithm automatically switches to the BFGS method in order to find its location more precisely. In Figure 10.19, the algorithm returns to the particle swarm method in order to check if there are no changes in this location and the entire procedure repeats itself. After some maximum number of iterations is performed (e.g., five), the process stops.

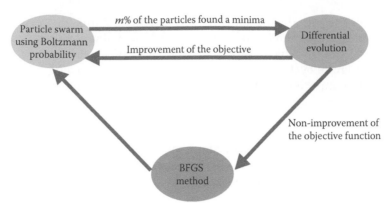

FIGURE 10.19
Global procedure for the hybrid optimization method *H1*.

In the particle swarm method, the probability test of the simulated annealing is performed in order to allow the particles (birds) to escape from local minima, although this procedure most often does not make any noticeable improvement in the method.

10.6 Response Surfaces

From the viewpoint of kernel interpolation/approximation techniques, many response surface methods are based on linear and nonlinear regression and other variants of the least square technique. This group of mesh-free methods has been successfully applied to many practical, but difficult, problems in engineering that are to be solved by the traditional mesh-based methods.

One of the most popular mesh-free kernel approximation techniques is the one that uses radial basis functions (RBFs). Initially, RBFs were developed for multivariate data and function interpolation. It was found that RBFs were able to construct an interpolation scheme with favorable properties such as high efficiency, good quality, and capability of dealing with scattered data, especially for higher dimension problems. A convincing comparison (Colaço et al., 2007) of an RBF-based response surface method and a wavelet-based artificial neural network method (Sahoo and Dulikravich, 2006) demonstrated superiority of RBF-based methods especially for high-dimensionality response surfaces.

The use of RBFs followed by collocation, a technique first proposed by Kansa (1990), after the work of Hardy (1971) on multivariate approximation, is now becoming an established approach. Various applications to problems in mechanics have been made in recent years—see, for example, Leitão (2001, 2004).

Kansa's method (or asymmetric collocation) starts by building an approximation to the field of interest (normally displacement components) from the superposition of RBFs (globally or compactly supported) conveniently placed at points in the domain and/or at the boundary.

The unknowns (which are the coefficients of each RBF) are obtained from the approximate enforcement of the boundary conditions as well as the governing equations by means of collocation. Usually, this approximation only considers regular RBFs, such as the globally supported multiquadrics or the compactly supported Wendland functions (Wendland, 1998).

There are several other methods for automatically constructing multidimensional response surfaces. Notably, a classical book by Lancaster and Salkauskas (1986) offers a variety of methods for fitting hypersurfaces of a relatively small dimensionality. Kauffman et al. (1996) obtained reasonably accurate fits of data by using second-order polynomials. Ivakhnenko and his team in Ukraine (Madala and Ivakhnenko, 1994) have published an exceptionally robust method for fitting non-smooth data points in multidimensional spaces. Their method is based on a self-assembly approach where the analytical description of a hypersurface is a multilevel graph of the type "polynomial-of-a-polynomial-of-a-polynomial-of-a-..." and the basis functions are very simple polynomials (Moral and Dulikravich, 2008). This approach has been used in indirect optimization based upon self-organization (IOSO) (IOSO, 2003) commercial optimization software that has been known for its extraordinary speed and robustness.

10.6.1 RBF Model Used in This Chapter

Let us suppose that we have a function of L variables P_i, $i = 1, \ldots, L$. The RBF model used in this work has the following form:

$$S(\mathbf{P}) \cong \xi(\mathbf{P}) = \sum_{j=1}^{N} \alpha_j \phi(|\mathbf{P} - \mathbf{P}_j|) + \sum_{k=1}^{M} \sum_{i=1}^{L} \beta_{i,k} q_k(P_i) + \beta_0 \tag{10.40}$$

where
$\mathbf{P} = (P_1, \ldots, P_i, \ldots, P_L)$
$S(\mathbf{P})$ is known for a series of points $\mathbf{P}$

Here, $q_k(P_i)$ is one of the M terms of a given basis of polynomials (Buhmann, 2003). This approximation $\xi(\mathbf{P})$ is solved for the α_j and $\beta_{i,k}$ unknowns from the system of N linear equations, subject

$$\sum_{j=1}^{N} \alpha_j q_k(P_1) = 0$$

$$\vdots \tag{10.41}$$

$$\sum_{j=1}^{N} \alpha_j q_k(P_L) = 0$$

$$\sum_{j=1}^{N} \alpha_j = 0 \tag{10.42}$$

In this chapter, the polynomial part of Equation 10.40 was taken as

$$q_k(P_i) = P_i^k \tag{10.43}$$

and the RBFs are selected among the following:

$$\text{Multiquadrics:} \quad \phi(|P_i - P_j|) = \sqrt{(P_i - P_j)^2 + c_j^2} \tag{10.44a}$$

$$\text{Gaussian:} \quad \phi(|P_i - P_j|) = \exp\left[-c_j^2(P_i - P_j)^2\right] \tag{10.44b}$$

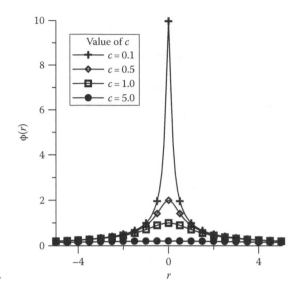

FIGURE 10.20
Influence of the shape parameter.

$$\text{Squared multiquadrics:} \quad \phi(|P_i - P_j|) = (P_i - P_j)^2 + c_j^2 \tag{10.44c}$$

$$\text{Cubical multiquadrics:} \quad \phi(|P_i - P_j|) = \left[\sqrt{(P_i - P_j)^2 + c_j^2}\right]^3 \tag{10.44d}$$

with the shape parameter c_j kept constant as $1/N$. The shape parameter is used to control the smoothness of the RBF. Figure 10.20 shows the influence on its choice for the multiquadrics RBF. From Equation 10.40, one can notice that a polynomial of order M is added to the RBF. M was limited to an upper value of 6. After inspecting Equations 10.40 through 10.43, one can easily check that the final linear system has $[(N + M*L) + 1]$ equations. Some tests were made using the cross product polynomials $(P_i\,P_j\,P_k\ldots)$, but the improvements of the results were irrelevant. Also, other types of RBFs were used, but no improvement of the interpolation was observed.

The choice of which polynomial order and which RBF are the best to a specific function, was made based on a cross-validation procedure. Let us suppose that we have N_{TR} training points, which are the locations on the multidimensional space where the values of the function are known. Such set of training points is equally subdivided into two subsets of points, named N_{TR1} and N_{TR2}. Equations 10.40 through 10.42 are solved for a polynomial of order zero and for the RBF expression given by Equations 10.44 using the subset N_{TR1}. Then, the value of the interpolated function is checked against the known value of the function for the subset N_{TR2} and the error is recorded as

$$RMS_{N_{TR1},M=0,RBF_1} = \sum_{i=1}^{N_{TR2}} [S(P_i) - \xi(P_i)]^2 \tag{10.45}$$

Then, the same procedure is made, using the subset N_{TR2} to solve Equations 10.40 through 10.42 and the subset N_{TR1} to calculate the error as

$$RMS_{N_{TR2},M=0,RBF_1} = \sum_{i=1}^{N_{TR1}} [S(P_i) - \xi(P_i)]^2 \tag{10.46}$$

Finally, the total error for the polynomial of order zero and the RBF expression given by Equations 10.44 is obtained as

$$RMS_{M=0,RBF_1} = \sqrt{RMS_{N_{TR1},M=0,RBF_1} + RMS_{N_{TR2},M=0,RBF_1}} \tag{10.47}$$

This procedure is repeated for all polynomial orders, up to $M=6$ and for each one of the RBF expressions given by Equations 10.44. The best combination is the one that returns the lowest value of the RMS error. Although this cross-validation procedure is quite simple, it worked very well for all test cases analyzed in this chapter.

10.6.2 Performance Measurements

In accordance with having multiple metamodeling criteria, the performance of each meta-modeling technique is measured from the following aspects (Jin et al., 2000):

- Accuracy—The capability of predicting the system response over the design space of interest.
- Robustness—The capability of achieving good accuracy for different problem types and sample sizes.
- Efficiency—The computational effort required for constructing the metamodel and for predicting the response for a set of new points by metamodels.
- Transparency—The capability of illustrating explicit relationships between input variables and responses.
- Conceptual simplicity—Ease of implementation. Simple methods should require minimum user input and be easily adapted to each problem.

For accuracy, the goodness of fit obtained from "training" data is not sufficient to assess the accuracy of newly predicted points. For this reason, additional confirmation samples are used to verify the accuracy of the metamodels. To provide a more complete picture of metamodel accuracy, three different metrics are used: R square (R^2), relative average absolute error ($RAAE$), and relative maximum absolute error ($RMAE$) (Jin et al., 2000).

10.6.2.1 R Square

$$R^2 = 1 - \frac{\sum_{i=1}^{n} (y_i - \hat{y}_i)^2}{\sum_{i=1}^{n} (y_i - \bar{y})^2} = 1 - \frac{MSE}{\text{variance}} \tag{10.48}$$

where
$\hat{y}_i$ is the corresponding predicted value for the observed value y_i
$\bar{y}$ is the mean of the observed values

While mean square error (MSE) represents the departure of the metamodel from the real simulation model, the variance captures how irregular the problem is. The larger the value of R^2, the more accurate the metamodel.

10.6.2.2 Relative Average Absolute Error

$$RAAE = \frac{\sum_{i=1}^{n} |y_i - \hat{y}_i|}{n * STD} \tag{10.49}$$

where *STD* stands for standard deviation. The smaller the value of *RAAE*, the more accurate the metamodel.

10.6.2.3 Relative Maximum Absolute Error

$$RMAE = \frac{\max\left(|y_1 - \hat{y}_1|, |y_2 - \hat{y}_2|, \ldots, |y_n - \hat{y}_n|\right)}{STD} \tag{10.50}$$

Large *RMAE* indicates large error in one region of the design space even though the overall accuracy indicated by R^2 and *RAAE* can be very good. Therefore, a small *RMAE* is preferred. However, since this metric cannot show the overall performance in the design space, it is not as important as R^2 and *RAAE*.

Although the R^2, *RAAE*, and *RMAE* are useful to ascertain the accuracy of the interpolation, they can fail in some cases. For the R^2 metric, for example, if one of the testing points has a huge deviation of the exact value, such discrepancy might affect the entire sum appearing on Equation 10.48 and, even if all the other testing points are accurately interpolated. Similarly, the R^2 result can be very bad. For this reason, we also calculate the percentage deviation of the exact value of each testing point. Such deviations are collected according to six ranges of errors: 0%–10%; 10%–20%; 20%–50%; 50%–100%; 100%–200%; >200%. Thus, an interpolation that has all testing points within the interval of 0%–10% of relative error might be considered good in comparison to another one where the points are all spread along the intervals from 10% to 200%.

10.6.3 Response Surface Test Cases

In order to show the accuracy of the RBF model presented, 296 test cases were used, representing linear and nonlinear problems with up to 100 variables. Such problems were selected from a collection of 395 problems (actually 296 test cases), proposed by Hock and Schittkowski (1981) and Schittkowski (1987). Figure 10.21 shows the number of variables of each one of the problems. Note that there are 395 problems, but some of them were not used.

Three methodologies were used to solve the linear algebraic system resulting from Equations 10.40 through 10.42: LU decomposition, singular value decomposition (SVD), and the generalized minimum residual (GMRES) iterative solver. When the number of equations was small (less than 40), the LU solver was used. However, when the number of variables increased over 40, the resulting matrix becomes too ill-conditioned and the SVD solver had to be used. For more than 80 variables, the SVD solver became too slow. Thus, the GMRES iterative method with the Jacobi preconditioner was used for all test cases.

In order to verify the accuracy of the interpolation over a different number of training points, three sets were defined. Also, the number of testing points varied, according to the number of training points. Table 10.1 presents these three sets, based on the number of dimensions (variables) *L* of the problem.

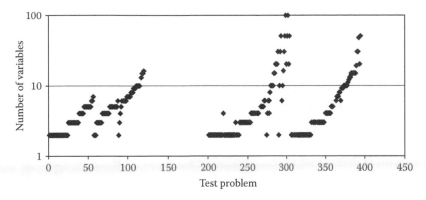

FIGURE 10.21

Number of variables for each problem considered.

TABLE 10.1

Number of Training and Testing Points

	Number of Training Points	Number of Testing Points
Scarce set	$3L$	$300L$
Small set	$10L$	$1000L$
Medium set	$50L$	$5000L$

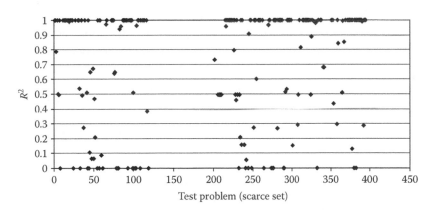

FIGURE 10.22

R^2 metric for the scarce set of training points.

Figure 10.22 shows the R^2 metric for all test cases, using the scarce set of training points. It can be noticed that the results are all spread from 0 (completely inadequate interpolation) to 1 (very accurate interpolation). However, even for this very small number of training points, most cases have an excellent interpolation, with $R^2 = 1$.

Figure 10.23 shows the CPU time required to interpolate each test function, using the scarce set of training points. For most of the cases, the CPU time was less than 1 s, using an AMD Opteron 1.6 GHz processor and 1GB registered ECC DDR PC-2700 RAM. In fact,

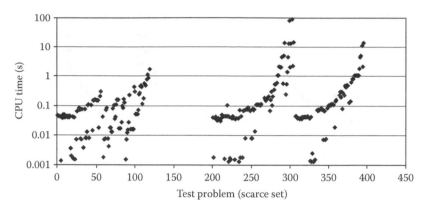

FIGURE 10.23
CPU time for the scarce set of training points.

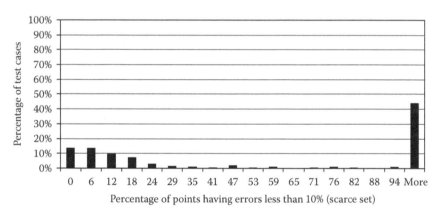

FIGURE 10.24
Testing points with less than 10% error, for the scarce set of training points.

the highest dimensional test cases, which had 100 variables, required only 100 s to be interpolated.

Although the R^2 might indicate some performance behavior of the interpolation function, we decided to use a different measure of accuracy. Figure 10.24 shows the percentage of testing points having errors less than 10%, against the percentage of all 296 test cases, for the scarce set of testing points. Thus, from this figure, it can be noticed that for more than 40% of all test functions, the relative errors were less than 10%. This is a very good result, considering the extremely small number of training points used in the scarce set.

Figure 10.25 shows the R^2 metric for the small set of training points. Compared to Figure 10.22, it can be seen that the points move toward the value of $R^2 = 1.0$, showing that the accuracy of the interpolation gets better when the number of training points increase.

Figure 10.26 shows the CPU time required for all test cases, when the small number of training points is used. Although the test case with 100 variables requires almost 1000 s, in almost all test cases, the CPU time is low.

Figure 10.27 shows the percentage of points having errors lower than 10%. Comparing with Figure 10.24, one can see that increasing the number of training points from 3 L

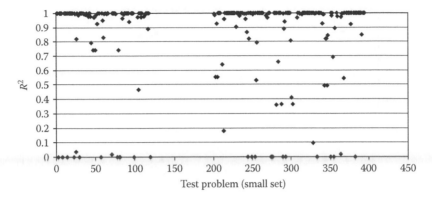

FIGURE 10.25
R^2 metric for the small set of training points.

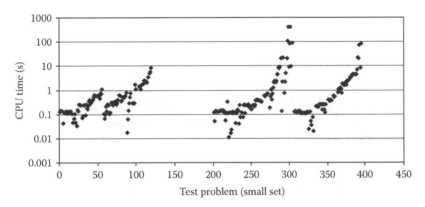

FIGURE 10.26
CPU time for the small set of training points.

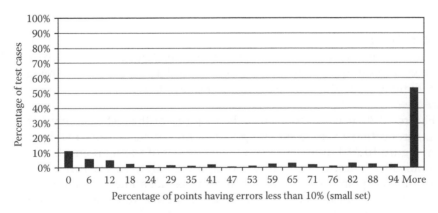

FIGURE 10.27
Testing points with less than 10% error, for the small set of training points.

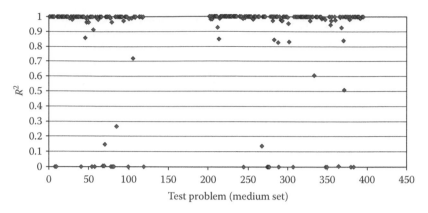

FIGURE 10.28
R^2 metric for the medium set of training points.

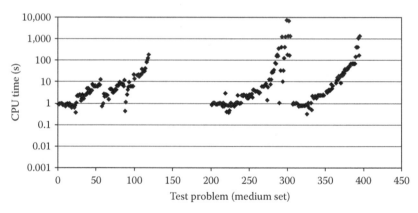

FIGURE 10.29
CPU time for the medium set of training points.

(scarce set) to 10 L (small set), the number of testing points having less than 10% of relative error for all 296 test cases increases from approximately 45% to approximately 55%, showing a very good interpolation, even for a not so large number of training points.

Finally, Figures 10.28 through 10.30 show the results when a medium set of training points are used.

From Figure 10.28, one can notice that the majority of the test cases have the R^2 metric close to 1.0, indicating a very good interpolation, for a not so large CPU time, as it can be verified at Figure 10.29. From Figure 10.30, the number of testing points having errors less than 10% for all 296 test cases increases to approximately 75% when a medium (50 L) number of training points is used. This indicates that such interpolation can be used as a metamodel in an optimization task, where the objective function takes too long to be calculated. Thus, instead of optimizing the original function, an interpolation can be used, significantly reducing the computational time.

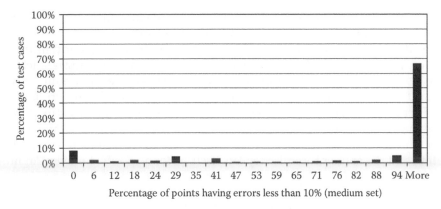

FIGURE 10.30
Testing points with less than 10% error, for the medium set of training points.

10.7 Hybrid Methods with Response Surfaces and Examples

Once the response surface methodology and the hybrid optimizer idea were presented, we will combine both of the sections. This method, called hybrid optimizer *H2* (Colaço and Dulikravich, 2007), is quite similar to the *H1* presented in Section 10.5, except for the fact that it uses a response surface method at some point of the optimization task. The global procedure is illustrated in Figure 10.31. It can be seen from this figure that after a certain number of objective functions were calculated, all this information was used to obtain a response surface. Such a response surface is then optimized using the same proposed hybrid code defined in the *H1* optimizer so that it fits the calculated values of the objective

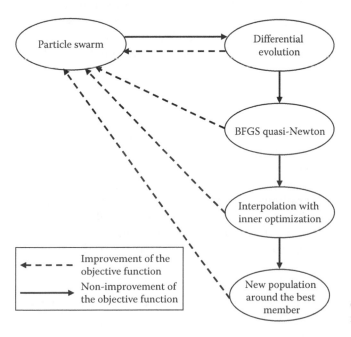

FIGURE 10.31
Global procedure for the hybrid optimization method *H2*.

function as closely as possible. New values of the objective function are then obtained very cheaply by interpolating their values from the response surface.

In Figure 10.31, if the BFGS cannot find any better solution, the algorithm uses an RBF interpolation scheme to obtain a response surface and then optimizes such response surface using the same hybrid algorithm proposed. When the minimum value of this response surface is found, the algorithm checks to see if it is also a solution of the original problem. Then, if there is no improvement of the objective function, the entire population is eliminated and a new population is generated around the best value obtained so far. The algorithm returns to the particle swarm method in order to check if there are no changes in this location and the entire procedure repeats itself. After a specified maximum number of iterations is performed (e.g., five), the process stops.

An even more efficient algorithm, which will be called *H3*, is an extension of the previous ones. The global procedure is enumerated in the following:

1. Generate an initial population, using the real function (not the interpolated one) $f(\mathbf{P})$. Call this population $\mathbf{\Pi}_{real}$.

2. Determine the individual that has the minimum value of the objective function, over the entire population $\mathbf{\Pi}_{real}$ and call this individual $\mathbf{P}_{best}$.

3. Determine the individual that is more distant from the $\mathbf{P}_{best}$, over the entire population $\mathbf{\Pi}_{real}$. Call this individual $\mathbf{P}_{far}$.

4. Generate a response surface, with the methodology at Section 10.6, using the entire population $\mathbf{\Pi}_{real}$ as training points. Call this function $g(\mathbf{P})$.

5. Optimize the interpolated function $g(\mathbf{P})$ using the hybrid optimizer *H1*, defined in Section 10.5, and call the optimum variable of the interpolated function as $\mathbf{P}_{int}$. During the generation of the internal population to be used in the *H1* optimizer, consider the upper and lower bounds limits as the minimum and maximum values of the population $\mathbf{\Pi}_{real}$ in order to not extrapolate the response surface.

6. If the real objective function $f(\mathbf{P}_{int})$ is better than all objective functions of the population $\mathbf{\Pi}_{real}$, replace $\mathbf{P}_{far}$ by $\mathbf{P}_{int}$. Otherwise, generate a new individual, using Sobol's pseudorandom number sequence generator (Sobol and Levitan, 1976) within the upper and lower bounds of the variables, and replace $\mathbf{P}_{far}$ by this new individual.

7. If the optimum is achieved, stop the procedure. Otherwise, return to step 2.

From the sequence above, one can notice that the number of times that the real objective function $f(\mathbf{P})$ is called is very small. Also, from step 6, one can see that the space of search is reduced at each iteration. When the response surface $g(\mathbf{P})$ is no longer capable to find a minimum, a new call to the real function $f(\mathbf{P})$ is made to generate a new point to be included in the interpolation. Since the CPU time to calculate the interpolated function is very small, the maximum number of iterations of the *H1* optimizer can be very large (e.g., 1000 iterations).

The hybrid optimizer *H3* was compared against the optimizer *H1*, *H2*, and the commercial code IOSO 2.0 for some standard test functions. The first test function was the Levy #9 function (Sandgren, 1977), which has 625 local minima and 4 variables. Such function is defined as

$$S(\mathbf{P}) = \sin^2(\pi - z_1) + \sum_{i=1}^{n-1}(z_i - 1)^2\left[1 + 10\sin^2(\pi z_{i+1})\right] + (z_4 - 1)^2 \tag{10.51}$$

where

$$z_i = 1 + \frac{P_i - 1}{4} \quad (i = 1, 4) \tag{10.52}$$

The function is defined within the interval $-10 \leq \mathbf{P} \leq 10$ and its minimum is $S(\mathbf{P}) = 0$ for $\mathbf{P} = 1$. Figure 10.32 shows the optimization history of the IOSO, *H1*, *H2*, and *H3* optimizers. Since the *H1*, *H2*, and *H3* optimizers are based on random number generators (because the particle swarm module), we present the best and worst estimatives for these three optimizers.

From Figure 10.32, it can be seen that the performance of the *H3* optimizer is very close to the IOSO commercial code. The *H1* code is the worst and the *H2* optimizer also has a

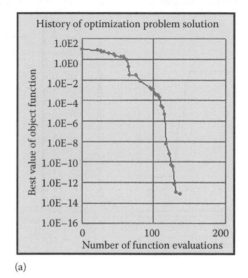

(a)

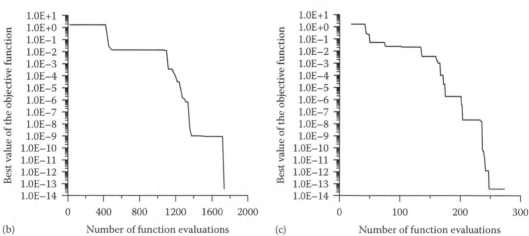

(b)   (c)

FIGURE 10.32
Optimization history of the Levy #9 function for the (a) IOSO, (b) *H1*-best, (c) *H2*-best,

(continued)

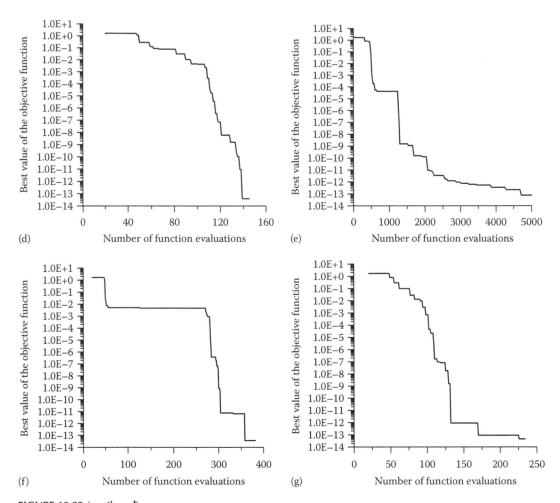

FIGURE 10.32 (continued)
(d) *H3*-best, (e) *H1*-worst, (f) *H2*-worst, and (g) *H3*-worst optimizers.

reasonably good performance. It is interesting to note that the *H1* code is the only one that does not have a response surface model implemented.

The second function tested was the Griewank function (Sandgren, 1977), which is defined as

$$S(\mathbf{P}) = \sum_{i=1}^{n} \frac{P_i^2}{4000} - \prod_{i-1}^{n} \cos\left(\frac{P_i}{\sqrt{i}}\right) + 1; \quad P_i \in \,]-600, 600[\quad (i = 1, 2) \tag{10.53}$$

The global minima for this function is located at $\mathbf{P} = 0$ and is $S(\mathbf{P}) = 0$. This function has an extremely large number of local minima, making the optimization task quite difficult.

Figure 10.33 shows the optimization history of the IOSO, *H1*, *H2*, and *H3* optimizers. Again, the best and worst results for *H1*, *H2*, and *H3* are presented.

From this figure, it is clear that the *H1*, *H2*, and *H3* optimizers are much better than the IOSO commercial code. The *H1* code was the best, while the *H2* sometimes stopped at some local minima. The worst result of the *H3* optimizer was, however, better than the result obtained by IOSO. It is worth pointing out that, with more iterations, the *H3* code could reach the minimum of the objective function, even for the worst result.

The next test function implemented was the Rosenbrook function (More et al., 1981), which is defined as

$$S(P_1, P_2) = 100(P_2 - P_1^2)^2 + (1 - P_1)^2 \qquad (10.54)$$

The function is defined within the interval $-10 \leq \mathbf{P} \leq 10$ and its minimum is $S(\mathbf{P}) = 0$ for $\mathbf{P} = 1$. Figure 10.34 shows the optimization history of the IOSO, *H1, H2*, and *H3* optimizers.

For this function, which is almost flat close to the global minima, the IOSO code was the one with the best performance, followed by the *H3* optimizer. The *H2* performed very

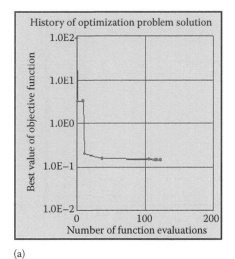

(a)

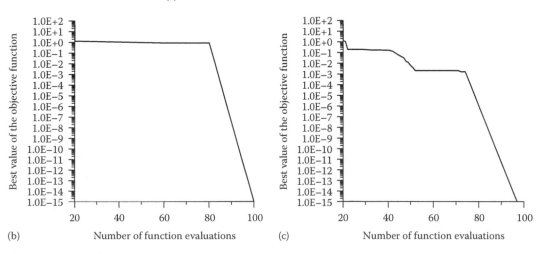

(b)   (c)

FIGURE 10.33
Optimization history of the Griewank function for the (a) IOSO, (b) *H1*-best, (c) *H2*-best,

(*continued*)

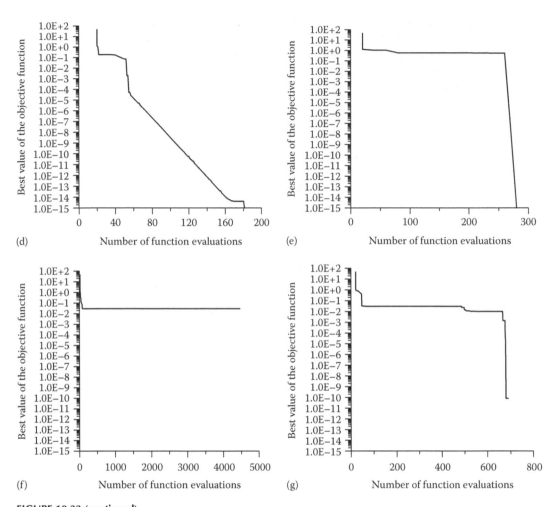

FIGURE 10.33 (continued)
(d) *H3*-best, (e) *H1*-worst, (f) *H2*-worst, and (g) *H3*-worst optimizers.

inadequately and the *H1* was able to get close to the minimum, but with a huge number of objective function calculations. When looking at the *H3* results, the final value of the objective function differed by some orders of magnitude. However, the optimum solution obtained with this new optimizer was $P_1 = 0.9996$ and $P_2 = 0.9992$, while the IOSO obtained $P_1 = 1.0000$ and $P_2 = 1.0000$. Thus, the relative error among the variables was less than 0.01%, indicating that despite the discrepancy among the final value of the objective function, the *H3* code was able to recover the value of the optimum variables with a neglectable relative error.

The last test function analyzed was the Mielle–Cantrel function (Miele and Cantrell, 1969), which is defined as

$$S(\mathbf{P}) = \left[\exp^{(P_1 - P_2)}\right]^4 + 100(P_2 - P_3)^6 + \arctan^4(P_3 - P_4) + P_1^2 \tag{10.55}$$

The function is defined within the interval $-10 \leq \mathbf{P} \leq 10$ and its minimum is $S(\mathbf{P})=0$ for $P_1=0$ and $P_2=P_3=P_4=1$. Figure 10.35 shows the optimization history of the IOSO, *H1*, *H2*, and *H3* optimizers. Again, the best and worst results for *H1*, *H2*, and *H3* are presented.

For this function, the IOSO code was the best, followed by the *H3*. The *H2* code performed very inadequately again. The *H1* was able to get to the global minimum after a huge number of objective function calculations. As occurred with the Rosenbrook function, in spite of the fact that *H3* results for the objective function differ from the IOSO code, the final values of the variables were $P_1=4.0981 \times 10^{-8}$, $P_2=0.9864$,

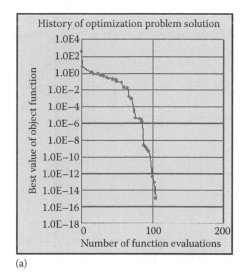

(a)

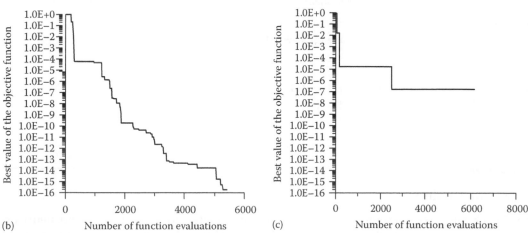

(b)   Number of function evaluations   (c)   Number of function evaluations

FIGURE 10.34
Optimization history of the Rosenbrook function for the (a) IOSO, (b) *H1*-best, (c) *H2*-best,

(continued)

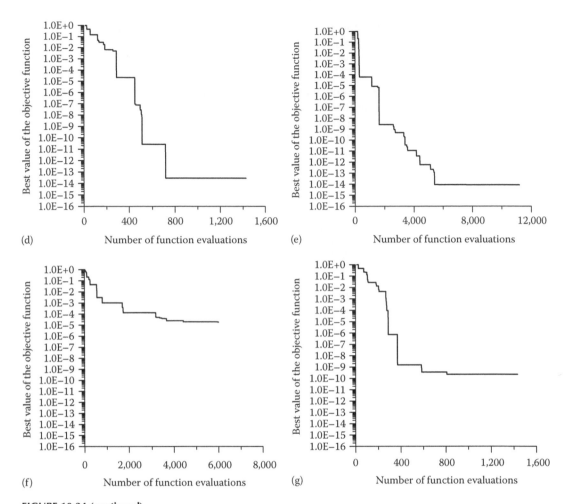

FIGURE 10.34 (continued)
(d) *H3*-best, (e) *H1*-worst, (f) *H2*-worst, and (g) *H3*-worst optimizers.

$P_3 = 0.9688$, and $P_4 = 0.9626$ for the **H3** optimizer and $P_1 = -0.1216 \times 10^{-5}$, $P_2 = 1.002$, $P_3 = 0.9957$, and $P_4 = 0.9962$ for the IOSO code.

10.8 Conclusion

In this chapter, we presented some basic concepts related to deterministic and heuristic methods, applied to single-objective optimization. Three different hybrid methods were also presented, as well as a powerful response surface methodology. The combination of the techniques presented here can be used in very complex engineering problems, which demand thousands of objective function calculations.

Acknowledgments

This work was partially funded by CNPq, CAPES (agencies for the fostering of science from the Brazilian Ministry of Science and Education), and FAPERJ (agency for the fostering of science from the Rio de Janeiro State). The authors are also very thankful to Professor Alain J. Kassab from the University of Central Florida for his suggestions (during IPDO-2007 in Miami) on how to choose the best shape parameter for RBF approximations. The first author is very grateful for the hospitality of George and Ellen Dulikravich during his stay in Miami for several periods of time during the years 2006, 2007, 2008, and 2009.

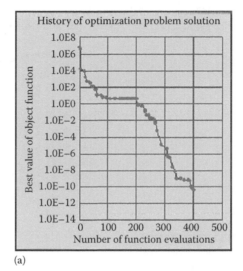

(a)

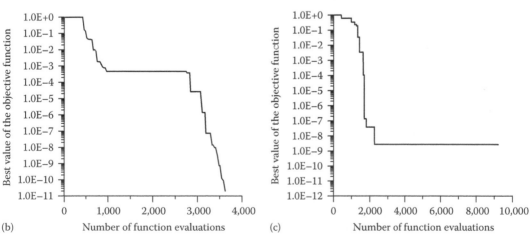

(b)   (c)

FIGURE 10.35

Optimization history of the Griewank function for the (a) IOSO, (b) *H1*-best, (c) *H2*-best,

(*continued*)

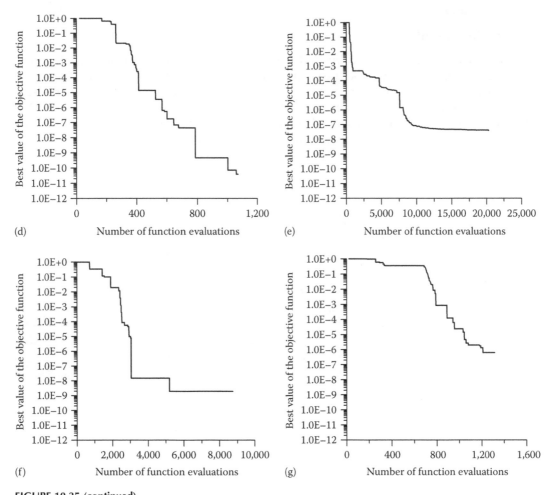

FIGURE 10.35 (continued)
(d) *H3*-best, (e) *H1*-worst, (f) *H2*-worst, and (g) *H3*-worst optimizers.

Nomenclature

A approximation of the Hessian
CR crossover constant
d direction of descent
E energy state
F weight constant which defines the mutation
G equality constraint
H approximation for the inverse of the Hessian
I uncertainty interval
I identity matrix
J matrix composed by the diagonal elements of **A**
k counter for the number of iterations
K Boltzmann constant

M, N	auxiliary matrices for the quasi-Newton methods
N	number of parameters (variables)
P	vector of parameters (variables) of the objective function S
q	iteration number for the restraint strategy in the Conjugate Gradient Method
Q	inequality constraint
$\mathbf{r}_1, \mathbf{r}_2$	random number vectors
S	objective function
T	temperature
Y	auxiliary vector for the quasi-Newton methods

Greeks

α	search step size
$\boldsymbol{\alpha}, \boldsymbol{\beta}, \boldsymbol{\gamma}$	vectors of parameters used in the differential evolution method
δ	delta Dirac function
γ, ψ	conjugation coefficients
λ	auxiliary parameter for the Levenberg-Marquardt method
π_I	best value of some individual
π_g	best value of the population
Π	population matrix

References

Alencar Jr., J.P., Orlande, H.R.B., and Ozisik, M.N. 1998. A generalized coordinates approach for the solution of inverse heat conduction problems, in: *11th International Heat Transfer Conference*, Korea, Vol. 7, pp. 53–58, August.

Alifanov, O.M. 1974. Solution of an inverse problem of heat conduction by iteration methods. *Journal of Engineering Physics*, 26(4):471–475.

Alifanov, O.M. 1994. *Inverse Heat Transfer Problems*, Springer Verlag, New York.

Alifanov, O.M., Artyukhin, E., and Rumyantsev, A. 1995. *Extreme Methods for Solving Ill-Posed Problems with Applications to Inverse Heat Transfer Problems*, Begell House, New York.

Artyukhin, E.A. 1993. Iterative algorithms for estimating temperature-dependent thermophysical characteristics, in: *First International Conference on Inverse Problems in Engineering: Theory and Practice*, eds. N. Zabaras, K. Woodburry, and M. Raynaud, pp. 101–108, Palm Coast, FL, June.

Bard, Y.B. 1974. *Nonlinear Parameter Estimation*, Academic Press, New York.

Beale, E.M.L. 1972. A derivation of conjugate gradients, in: *Numerical Methods for Nonlinear Optimization*, ed. F.A. Lootsma, pp. 39–43, Academic Press, New York.

Beck, J.V. 1999. Sequential methods in parameter estimation, in: *Third International Conference on Inverse Problems in Engineering*, Tutorial Session, Port Ludlow, WA.

Beck, J.V. and Arnold, K.J. 1977. *Parameter Estimation in Engineering and Science*, Wiley Interscience, New York.

Beck, J.V., Blackwell, B., and St. Clair, C.R. 1985. *Inverse Heat Conduction: Ill-Posed Problems*, Wiley Interscience, New York.

Belegundu, A.D. and Chandrupatla, T.R. 1999. *Optimization Concepts and Applications in Engineering*, Prentice Hall, Denver, CO.

Broyden, C.G. 1965. A class of methods for solving nonlinear simultaneous equations. *Mathematics of Computation*, 19:577–593.

Broyden, C.G. 1967. Quasi-Newton methods and their applications to function minimization. *Mathematics of Computation*, 21:368–380.

Buhmann, M.D. 2003. Radial basis functions on grids and beyond, in: *International Workshop on Meshfree Methods*, Lisbon, Portugal.

Colaço, M.J. and Dulikravich, G.S. 2006. A multilevel hybrid optimization of magnetohydrodynamic problems in double-diffusive fluid flow. *Journal of Physics and Chemistry of Solids*, 67:1965–1972.

Colaço, M.J. and Dulikravich, G.S. 2007. Solidifcation of double-diffusive flows using thermo-magneto-hydrodynamics and optimization. *Materials and Manufacturing Processes*, 22:594–606.

Colaço, M.J., Dulikravich, G.S., and Martin, T.J. 2004. Optimization of wall electrodes for electro-hydrodynamic control of natural convection during solidification. *Materials and Manufacturing Processes*, 19(4):719–736.

Colaço, M.J., Dulikravich, G.S., and Martin, T.J. 2005a. Control of unsteady solidification via optimized magnetic fields. *Materials and Manufacturing Processes*, 20(3):435–458.

Colaço, M.J., Dulikravich, G.S., Orlande, H.R.B., and Martin, T.J. 2005b. Hybrid optimization with automatic switching among optimization algorithms, in: *Evolutionary Algorithms and Intelligent Tools in Engineering Optimization*, eds. W. Annicchiarico, J. Periaux, M. Cerrolaza, and G. Winter, pp. 92–118, WIT Press/Computational Mechanics, Southampton, U.K.

Colaço, M.J., Dulikravich, G.S., and Sahoo, D. 2007. A comparison of two methods for fitting high dimensional response surfaces, in: *Proceedings of International Symposium on Inverse Problems, Design and Optimization (IPDO-2007)*, eds. G.S. Dulikravich, H.R.B. Orlande, M. Tanaka, and M. J. Colaço, Miami Beach, FL, April 16–18.

Colaço, M.J., Dulikravich, G.S., and Sahoo, D. 2008. A response surface method based hybrid optimizer. *Inverse Problems in Science and Engineering*, 16:717–742.

Colaço, M.J. and Orlande, H.R.B. 1998. Estimation of the heat transfer coefficient at the surface of a plate by using the conjugate gradient method, in: *VII Brazilian National Meeting of Thermal Sciences— ENCIT*, Rio de Janeiro, Brazil, Vol. 1, pp. 189–194.

Colaço, M.J. and Orlande, H.R.B. 1999. A comparison of different versions of the conjugate gradient method of function estimation. *Numerical Heat Transfer, Part A*, 36(2):229–249.

Colaço, M.J. and Orlande, H.R.B. 2000. A function estimation approach for the identification of the transient inlet profile in parallel plate channels, in: *International Symposium on Inverse Problems in Engineering Mechanics*, eds. M. Tanaka and G. S. Dulikravich, pp. 409–418, Nagano, Japan.

Colaço, M.J. and Orlande, H.R.B. 2001a. Inverse problem of simultaneous estimation of two boundary heat fluxes in parallel plate channels. *Journal of the Brazilian Society of Mechanical Engineering*, XXIII(2):201–215.

Colaço, M.J. and Orlande, H.R.B. 2001b. Inverse forced convection problem of simultaneous estimation of two boundary heat fluxes in irregularly shaped channels. *Numerical Heat Transfer, Part A*, 39:737–760.

Colaço, M.J. and Orlande, H.R.B. 2002. A natural convection inverse problem of simultaneous identification of two boundary heat fluxes in rectangular cavities, in: *12th International Heat Transfer Conference*, Grenoble, France.

Colaço, M.J. and Orlande, H.R.B. 2004. Inverse natural convection problem of simultaneous estimation of two boundary heat fluxes in irregular cavities. *International Journal of Heat and Mass Transfer*, 47:1201–1215.

Colaço, M.J., Orlande, H.R.B., and Dulikravich, G.S. 2006. Inverse and optimization problems in heat transfer. *Journal of the Brazilian Society of Mechanical Sciences and Engineering*, 28(1):1–24.

Corana, A., Marchesi, M., Martini, C., and Ridella, S. 1987. Minimizing multimodal functions of continuous variables with the 'simulated annealing algorithm.' *ACM Transactions on Mathematical Software*, 13:262–280.

Daniel, J.W. 1971. *The Approximate Minimization of Functionals*, Prentice-Hall, Englewood Cliffs, NJ.

Dantas, L.B. and Orlande, H.R.B. 1996. A function estimation approach for determining temperature-dependent thermophysical properties. *Inverse Problems in Engineering*, 3:261–279.

Darwin, C. 1859. *On the Origin of Species*, John Murray, London, U.K.

Davidon, W.C. 1959. *Variable Metric Method for Minimization*, Argonne National Laboratory, ANL-5990, Argonne, IL.

Deb, K. 2002. *Multi-Objective Optimization Using Evolutionary Algorithms*, John Wiley & Sons, New York.

de Boor, C. 1978. *A Practical Guide to Splines*, Springer Verlag, New York.

Denisov, A.M. 1999. *Elements of the Theory of Inverse Problems*, VSP, the Netherlands.

Dennis, J. and Schnabel, R. 1983. *Numerical Methods for Unconstrained Optimization and Nonlinear Equations*, Prentice Hall, New York.

Dulikravich, G.S. and Colaço, M.J. 2006. Convective heat transfer control using magnetic and electric fields. *Journal of Enhanced Heat Transfer*, 13(2):139–155.

Dulikravich, G.S., Colaço, M.J., Dennis, B.H., Marting, T.J., Egorov, I.N., and Lee, S. 2004. Optimization of intensities, and orientations of magnets controlling melt flow during solidification. *Materials and Manufacturing Processes*, 19(4):695–718.

Dulikravich, G.S., Colaço, M.J., Martin, T.J., and Lee, S. 2003. An inverse method allowing user-specified layout of magnetized micro-fibers in solidifying composites. *Journal of Composite Materials, UK*, 37(15):1351–1365.

Dulikravich, G.S., Egorov, I.N., and Colaço, M.J. 2008. Optimizing chemistry of bulk metallic glasses for improved thermal stability. *Modelling and Simulation in Materials Science and Engineering*, 16:075010.

Dulikravich, G.S. and Martin, T.J. 1994. Inverse design of super-elliptic cooling passages in coated turbine blade airfoils. *AIAA Journal of Thermophysics and Heat Transfer*, 8(2):288–294.

Dulikravich, G.S. and Martin, T.J. 1996. Inverse shape and boundary condition problems and optimization in heat conduction, in: *Advances in Numerical Heat Transfer*, eds. W.J. Minkowycz and E. M. Sparrow, Chapter 10, pp. 381–426, CRC Press, Boca Raton, FL.

Dulikravich, G.S., Martin, T.J., Dennis, B.H., and Foster, N.F. 1999. Multidisciplinary hybrid constrained GA optimization, Invited lecture, in: *EUROGEN'99—Evolutionary Algorithms in Engineering and Computer Science: Recent Advances and Industrial Applications*, eds. K. Miettinen, M. M. Makela, P. Neittaanmaki, and J. Periaux, Chapter 12, pp. 231–260, John Wiley & Sons, Jyvaskyla, Finland, May 30–June 3.

Eberhart, R., Shi, Y., and Kennedy, J. 2001. *Swarm Intelligence*, Morgan Kaufmann, San Francisco, CA.

Fletcher, R. 2000. *Practical Methods of Optimization*, John Wiley & Sons, New York.

Fletcher, R. and Powell, M.J.D. 1963. A rapidly convergent descent method for minimization. *Computer Journal*, 6:163–168.

Fletcher, R. and Reeves, C.M. 1964. Function minimization by conjugate gradients. *Computer Journal*, 7:149–154.

Fox, R.L. 1971. *Optimization Methods for Engineering Design*, Addison-Wesley Publishing Company, Reading, MA.

Goffe, W.L., Ferrier, G.D., and Rogers, J. 1994. Global optimization of statistical functions with simulated annealing. *Journal of Econometrics*, 60:65–99.

Goldberg, D.E. 1989. *Genetic Algorithms in Search, Optimization, and Machine Learning*, Addison-Wesley Publishing Company, Reading, MA.

Hadamard, J. 1923. *Lectures on Cauchy's Problem in Linear Differential Equations*, Yale University Press, New Haven, CT.

Hardy, R.L. 1971. Multiquadric equations of topography and other irregular surfaces. *Journal of Geophysics Research*, 176:1905–1915.

Hensel, E. 1991. *Inverse Theory and Applications for Engineers*, Prentice-Hall, Englewood Cliffs, NJ.

Hestenes, M.R. and Stiefel, E. 1952. Method of conjugate gradients for solving linear systems. *Journal of Research of the National Bureau of Standards. Section B*, 49:409–436.

Hock, W. and Schittkowski, K. 1981. *Test Examples for Nonlinear Programming Codes, Lecture Notes in Economics and Mathematical Systems, 187*, Springer Verlag, Berlin/Heidelberg/New York.

Huang, C.H. and Tsai, C.H. 1997. A shape identification problem in estimating time-dependent irregular boundary configurations, in: *National Heat Transfer Conference*, ASME HTD-340, Vol. 2, eds. G.S. Dulikravich and K.A. Woodbury, pp. 41–48.

IOSO NM Version 1.0, User's Guide. 2003. IOSO Technology Center, Moscow, Russia.

Isakov, V. 1998. *Inverse Problems for Partial Differential Equations, Applied Mathematical Sciences*, Vol. 127, Springer, New York.

Jaluria, Y. 1998. *Design and Optimization of Thermal Systems*, McGraw Hill, New York.

Jarny, Y., Ozisik, M.N., and Bardon, J.P. 1991. A general optimization method using adjoint equation for solving multidimensional inverse heat conduction. *International Journal of Heat and Mass Transfer*, 34(11):2911–2919.

Jin, R., Chen, W., and Simpson, T.W. 2000. Comparative studies of metamodeling techniques under multiple modeling criteria, in: *Proceedings of the 8th AIAA/USAF/NASA/ISSMO Multidisciplinary Analysis & Optimization Symposium*, AIAA 2000-4801, Long Beach, CA, September 6–8.

Kansa, E.J. 1990. Multiquadrics—A scattered data approximation scheme with applications to computational fluid dynamics—II: Solutions to parabolic, hyperbolic and elliptic partial differential equations. *Computers and Mathematics with Applications*, 19:149–161.

Kaufman, M., Balabanov, V., Burgee, S.L., Giunta, A.A., Grossman, B., Mason, W.H., and Watson, L.T. 1996. Variable complexity response surface approximations for wing structural weight in HSCT design, AIAA Paper 96-0089, in: *Proceedings of the 34th Aerospace Sciences Meeting and Exhibit*, Reno, NV.

Kennedy, J. 1999. Small worlds and mega-minds: Effects of neighborhood topology on particle swarm performance, in: *Proceedings of the 1999 Congress of Evolutionary Computation*, IEEE Press, Vol. 3, pp. 1931–1938.

Kennedy, J. and Eberhart, R.C. 1995. Particle swarm optimization, in: *Proceedings of the 1995 IEEE International Conference on Neural Networks*, Perth, Australia, Vol. 4, pp. 1942–1948.

Kirsch, A. 1996. *An Introduction to the Mathematical Theory of Inverse Problems, Applied Mathematical Sciences*, Vol. 120, Springer, New York.

Kurpisz, K. and Nowak, A.J. 1995. *Inverse Thermal Problems*, WIT Press, Southampton, U.K.

Lancaster, P. and Salkauskas, K. 1986. *Curve and Surface Fitting: An Introduction*, Academic Press, Harcourt Brace Jovanovic, New York.

Leitão, V.M.A. 2001. A meshless method for Kirchhoff plate bending problems. *International Journal of Numerical Methods in Engineering*, 52:1107–1130.

Leitão, V.M.A. 2004. RBF-based meshless methods for 2D elastostatic problems. *Engineering Analysis with Boundary Elements*, 28:1271–1281.

Levenberg, K. 1944. A method for the solution of certain non-linear problems in least squares. *Quarterly of Applied Mathematics*, 2:164–168.

Machado, H.A. and Orlande, H.R.B. 1997. Inverse analysis for estimating the timewise and spacewise variation of the wall heat flux in a parallel plate channel. *International Journal for Numerical Methods for Heat & Fluid Flow*, 7(7):696–710.

Madala, H.R. and Ivakhnenko, A.G. 1994. *Inductive Learning Algorithms for Complex Systems Modeling*, CRC Press, Boca Raton, FL.

Marquardt, D.W. 1963. An algorithm for least squares estimation of nonlinear parameters. *Journal of the Society for Industrial and Applied Mathematics*, 11:431–441.

Miele, A. and Cantrell, J.W. 1969. Study on a memory gradient method for the minimization of functions. *Journal of Optimization, Theory and Applications*, 3(6):459–470.

Moral, R.J. and Dulikravich, G.S. 2008. A hybrid self-organizing response surface methodology, Paper AIAA-2008–5891, in: *12th AIAA/ISSMO Multidisciplinary Analysis and Optimization Conference*, Victoria, BC, Canada, September 10–12.

Moré, J.J. 1977. The Levenberg-Marquardt algorithm: Implementation and theory, in: *Numerical Analysis*, Lecture Notes in Mathematics, Vol. 630, ed. G.A. Watson, pp. 105–116, Springer Verlag, Berlin.

More, J.J., Gabow, B.S., and Hillstrom, K.E. 1981. Testing unconstrained optimization software. *ACM Transactions on Mathematical Software*, 7(1):17–41.

Morozov, V.A. 1984. *Methods for Solving Incorrectly Posed Problems*, Springer Verlag, New York.

Murio, D.A. 1993. *The Mollification Method and the Numerical Solution of Ill-Posed Problems*, Wiley Interscience, New York.

Naka, S., Yura, T.G., and Fukuyama, T. 2001. Practical distribution state estimation using hybrid particle swarm optimization, in: *Proceedings IEEE Power Engineering Society*, Winter Meeting, Columbus, OH, January 28–February 1.

Oliver, M.A. and Webster, R. 1990. Kriging: A method of interpolation for geographical information system. *International Journal of Geographical Information Systems*, 4(3):313–332.

Orlande, H.R.B., Colaço, M.J., and Malta, A.A. 1997. Estimation of the heat transfer coefficient in the spray cooling of continuously cast slabs, in: *National Heat Transfer Conference*, ASME HTD-340, Vol. 2, eds. G.S. Dulikravich and K.A. Woodbury, pp. 109–116, June.

Ozisik, M.N. and Orlande, H.R.B. 2000. *Inverse Heat Transfer: Fundamentals and Applications*, Taylor & Francis, New York.

Padilha, R.S., Santos, H.F.S., Colaço, M.J., and Cruz, M.E.C. 2009. Single and multi-objective optimization of a cogeneration system using hybrid algorithms. *Heat Transfer Engineering*, 30:261–271.

Polak, E. 1971. *Computational Methods in Optimization*, Academic Press, New York.

Powell, M.J.D. 1977. Restart procedures for the conjugate gradient method. *Mathematical Programming*, 12:241–254.

Ramm, A.G., Shivakumar, P.N., and Strauss, A.V. (eds.). 2000. *Operator Theory and Applications*, American Mathematical Society, Providence, RI.

Sabatier, P.C. (ed.). 1978. *Applied Inverse Problems*, Springer Verlag, Hamburg.

Sahoo, D. and Dulikravich, G.S. 2006. Evolutionary wavelet neural network for large scale function estimation in optimization, in: *11th AIAA/ISSMO Multidisciplinary Analysis and Optimization Conference*, AIAA Paper AIAA-2006–6955, Portsmouth, VA, September 6–8.

Sandgren, E. 1977. The utility of nonlinear programming algorithms. PhD thesis, Purdue University, IN.

Schittkowski, K. 1987. *More Test Examples for Nonlinear Programming, Lecture Notes in Economics and Mathematical Systems*, 282, Springer Verlag, New York.

Silva, P.M.P., Orlande, H.R.B., Colaço, M.J., Shiakolas, P.S., and Dulikravich, G.S. 2007. Identification and design of a source term in a two-region heat conduction problem. *Inverse Problems in Science and Engineering*, 15:661–677.

Sobol, I. and Levitan, Y.L. 1976. The production of points uniformly distributed in a multidimensional cube, Preprint *IPM Akad. Nauk SSSR*, No. 40, Moscow, Russia.

Stoecker, W.F. 1989. *Design of Thermal Systems*, McGraw Hill, New York.

Storn, R. and Price, K.V. 1996. Minimizing the real function of the ICEC'96 contest by differential evolution, in: *IEEE Conference on Evolutionary Computation*, Nagoya, Japan, pp. 842–844.

Tikhonov, A.N. and Arsenin, V.Y. 1977. *Solution of Ill-Posed Problems*, Winston & Sons, Washington, DC.

Truffart, B., Jarny, J., and Delaunay, D. 1993. A general optimization algorithm to solve 2-d boundary inverse heat conduction problems using finite elements, in: *First International Conference on Inverse Problems in Engineering: Theory and Practice*, eds. N. Zabaras, K. Woodburry, and M. Raynaud, pp. 53–60, Palm Coast, FL, June.

Trujillo, D.M. and Busby, H.R. 1997. *Practical Inverse Analysis in Engineering*, CRC Press, Boca Raton, FL.

Wellele, O., Orlande, H.R.B., Ruberti, N.J., Colaço, M.J., and Delmas, A. 2006. Identification of the thermophysical properties of orthotropic semi-transparent materials, in: *13th International Heat Transfer Conference*, Sydney, Australia.

Wendland, H. 1998. Error estimates for interpolation by compactly supported radial basis functions of minimal degree. *Journal of Approximation Theory*, 93:258–272.

Woodbury, K. 2002. *Inverse Engineering Handbook*, CRC Press, Boca Raton, FL.

Yagola, A.G., Kochikov, I.V., Kuramshina, G.M., and Pentin, Y.A. 1999. *Inverse Problems of Vibrational Spectroscopy*, VSP, the Netherlands.

Zubelli, J.P. 1999. *An Introduction to Inverse Problems: Examples, Methods and Questions*, Institute of Pure and Applied Mathematics, Rio de Janeiro, Brazil.

11

Adjoint Methods

Yvon Jarny and Helcio R. Barreto Orlande

CONTENTS

11.1 Introduction

The numerical solution of inverse heat transfer problems dealing with experimental data processing for estimating unknown functions, like spatial distribution and/or time history of heat sources, thermal properties versus temperature, etc., is known to be ill conditioned. Regularization methods then have to be used for solving such problems, in order to build numerical stable solutions and to avoid the amplification of data and/or model errors.

In this chapter, a general presentation of the conjugate gradient method (CGM), combined with the Lagrange multiplier technique, is given for estimating unknown functions, classically found in heat transfer modeling. The method is iterative, and the regularization parameter is given in the form of the stopping criterion used to select the number of iterations, which is chosen according to the discrepancy principle proposed by Morozov (1984).

The presentation is organized as follows: First, a general formulation of inverse problems is given and illustrated with two basic examples of inverse heat source problems. Then, the principle of the CGM is developed in the simplest situation of linear model equations, and the Lagrangian technique is briefly presented. The CGM is then applied to a nonlinear estimation problem.

11.2 Formulation of Inverse Problems

To formulate an inverse problem of experimental data processing, two main components are required, namely,

1. A set of equations (algebraic, differential, integral, etc.), the solution of which describes the relationship between the unknown function to be estimated, denoted by u, and the observable variables or model output, denoted by y
2. A set of experimental data, denoted by $\tilde{y}$, used to determine u by matching in some sense with the output predicted through the model equations

A generic formulation for inverse problems thus consists in

1. Introducing the metric spaces U, Y for functions u and y, respectively
2. Considering the functional relation, denoted by X, which connects the input $u \in U_{ad} \subset U$ to the output $y \in Y$ through the model equations
3. Solving the equation $\tilde{y} = X(u)$

In practice and more often, no *exact* solution $u = X^{-1}(\tilde{y})$ exists for inverse problems, because the inverse operator X^{-1} cannot be defined or because of its high sensitivity to measurement and/or model errors. However, elements $u \in U_{ad}$ for which the deviation between the predicted output $X(u)$ and the measured output $\tilde{y}$ remains less than some tolerance ε can be acceptable. When the model is assumed to perfectly represent the physical problem under analysis, this tolerance is specified only by the level of the measurement error, that is, the variance of the random noise of the input signal.

Therefore, the solution of the inverse problem aims at selecting $u \in U_{ad}$ such that

$$\|X(u) - \tilde{y}\|_Y^2 \leq \varepsilon^2 \tag{11.1}$$

The inverse problem is then formulated in the least square sense, that is, it consists in finding $u^* \in U_{ad}$, which minimizes the least square criterion

$$S(u) = \|X(u) - \tilde{y}\|_Y^2 \tag{11.2}$$

It can be shown that this approach overcomes the question of existence of a solution but not that of uniqueness or stability under data perturbations.

We will see how a numerical solution can be computed iteratively by using the CGM in order to minimize $S(u)$ and we illustrate, according to the discrepancy principle (Morozov 1984), how the stopping rule $S(u_{nf}) = \varepsilon^2$ allows to get a regularized solution. The final iteration number nf is then the regularizing parameter.

11.3 Illustrative Examples: Linear Estimation

11.3.1 Inverse 1-D Boundary Heat Source Problem

A semi-infinite heat conducting body is submitted to a boundary heat flux density $u(t)$, which is to be determined from the observed output $y(t)$ given by the temperature history at the sensor location x_s, on the time interval $(0, t_f)$. Hence, we can write the state and output (observation) equations, respectively, as

State equations:

$$\frac{\partial T}{\partial t}(x, t) = \alpha \frac{\partial^2 T}{\partial x^2}(x, t), \quad x > 0, \quad 0 < t < t_f \tag{11.3a}$$

$$-k \frac{\partial T}{\partial x}(0, t) = u(t), \quad 0 < t < t_f \tag{11.3b}$$

$$T(x, 0) = 0, \quad x > 0 \tag{11.3c}$$

Output equation:

$$y(t) = T(x_s, t), \quad 0 < t < t_f \tag{11.3d}$$

The problem may be put under the generic form by introducing the functional spaces $U = Y = L^2(0, t_f)$, with the norm defined as

$$\|u\|_U = \left[\int_0^{t_f} u^2(t)dt \right]^{1/2} \quad \text{and} \quad \| \cdots \|_U = \| \cdots \|_Y \tag{11.4}$$

and the linear operator $X(.)$ is defined by the convolution integral

$$y(t) = X^* u(t) = \int_0^t f_s(t - \tau)u(\tau)d\tau, \quad t \in (0, t_f) \tag{11.5}$$

where
$f_s(t) = K(\sqrt{t_c/t}) \exp(-t_c/t)$ is the impulse response
$t_c = x_s^2/4\alpha$ is the characteristic time of the body at the sensor location
$K = 2/\rho c x_s \sqrt{\pi}$

A finite dimension approximation for Equation 11.5 is straightforward. The time interval $(0, t_f)$ is divided into $n = m$ subintervals $]t_{i-1}, t_i[$, of length Δt. The linear convolution operator may be approximated by a matrix operator. In practice, this is not the usual way to solve the inverse problem, but it is used here in order to illustrate its ill-posed character:

$$y(t_i) = \sum_j f_s(t_i - \tau_j)u(\tau_j)\Delta\tau = \sum_{j=1}^{i} X_{ij}u_j \tag{11.6}$$

$$X_{ij} = \begin{cases} \Delta t f_s((i - j + 1)\Delta t) & \text{for } j = 1 \text{ to } i \quad \text{and} \quad i = 1 \text{ to } m \\ 0 & \text{elsewhere} \end{cases} \tag{11.7}$$

The problem is then reformulated by introducing the new vector spaces $\mathbf{U} = \mathbf{Y} = \mathbf{R}^n$ and the model output is now written as

$$\mathbf{y} = \mathbf{Xu} \tag{11.8}$$

The square matrix $\mathbf{X}$ is a Toeplitz matrix, lower triangular, and the components are given by $z_i = \Delta t\, f_s(t_i)$, $i = 1, \ldots, n$, which go to zero when $\Delta t \to 0$.

$$\mathbf{X} = \begin{bmatrix} z_1 & 0 & & & \\ z_2 & z_1 & 0 & & 0 \\ z_3 & z_2 & z_1 & & \\ z_4 & z_3 & z_2 & \ddots & 0 \\ & & & \ddots & z_1 & 0 \\ z_m & z_{m-1} & z_{m-2} & & z_2 & z_1 \end{bmatrix} \tag{11.9}$$

This inverse boundary heat source problem is clearly a time deconvolution problem, and the solution is known to be very sensitive to data errors. A numerical solution for this problem, computed with the CGM, is presented in Section 11.4.

To illustrate the ill posedness of this problem, let us consider the solution given by

$$\hat{\mathbf{u}} = \mathbf{X}^{-1}\tilde{\mathbf{y}} \tag{11.10}$$

A numerical experiment is performed with the following numerical data, in the time interval $0 < t < t_f = 20$ s : $k = 1$ W m^{-1} K^{-1}; $x_s = 2$ mm; $\alpha = 10^{-6}$ m^2 s^{-1}; $\Delta \tau = 1$ s, $\rho c = 10^6$ J m^{-3} K^{-1}; $K = 0.564 \times 10^{-3}$ K m^2 J^{-1}.

Figure 11.1a shows the exact applied heat flux and the corresponding output of the problem, y. The solutions $\hat{u}(t)$ are presented in Figure 11.1b, computed with the time step $\Delta t = 0.5$ s, for different levels of noise on the simulated measured data $\tilde{y}(t)$. Figure 11.1b shows that, as the standard deviation of the measurement errors ($\sigma = 0.0$; 0.002 and 0.005 K) increases, the solution clearly becomes unstable.

It must be noted that by decreasing the time step Δt, the numerical inversion process becomes more ill conditioned. In fact, although the accuracy of the direct problem solution increases, the stability condition of the inverse problem solution decreases. There is thus a question of selecting the solution between accuracy and stability.

11.3.2 Inverse 2-D Boundary Heat Source Problem

Let us consider the solution of a 2-D steady-state heat conduction process (see Figure 11.2a), described by the following set of equations:

Steady-state equations:

$$k\Delta T = 0, \quad \text{in } \Omega \tag{11.11a}$$

$$T|_{\Gamma_1} = T_1 \tag{11.11b}$$

$$\left.\frac{\partial T}{\partial n}\right|_{\Gamma_2 \cup \Gamma_3} = 0 \tag{11.11c}$$

$$-k\left.\frac{\partial T}{\partial n}\right|_{\Gamma_4} = u \tag{11.11d}$$

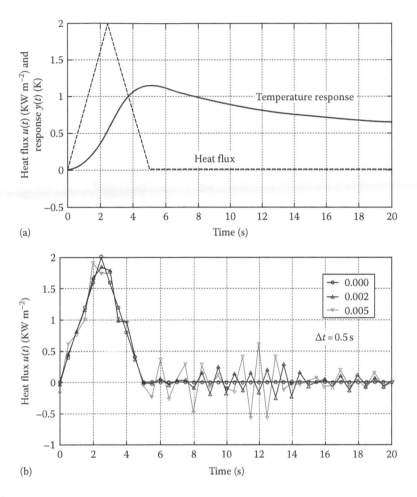

FIGURE 11.1

(a) Heat flux $u(t)$ (dashed line) and temperature output $y(t)$ (continuous line). (b) Solutions for different levels of noise.

Output equation:

$$y = T|_{\Gamma_3} \tag{11.12}$$

Numerical results of the direct problem are presented in Figure 11.2b and c; they were computed with the following input data:

$$T_1 = 0.5 \tag{11.13a}$$

$$u = q(\xi) = q_0[\sin(\pi\xi/2) - 1], \quad 0 < \xi < 1 \tag{11.13b}$$

where ξ denotes the coordinate along Γ_4. The heat flux and the temperature at surface Γ_4 are presented in Figure 11.2d, while Figure 11.2e presents the temperature at surface Γ_3.

Now, suppose that the boundary heat flux u is unknown at Γ_4 and that the output y is the observed temperature at Γ_3. This example of a 2-D inverse boundary heat source problem

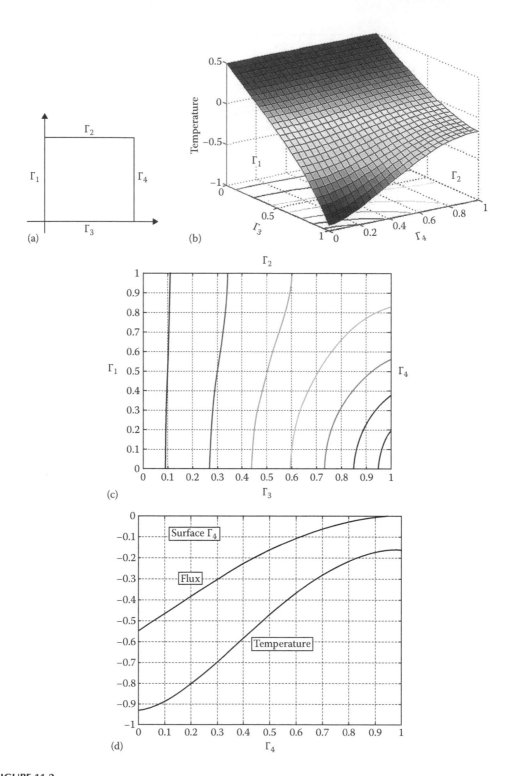

FIGURE 11.2
(a) Spatial domain. (b) Temperature field, solution of the direct heat conduction problem. (c) Contour plots of the direct problem solution. (d) Boundary condition on Γ_4.

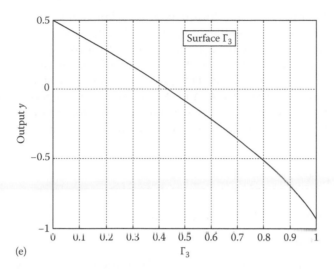

(e)

FIGURE 11.2 (continued)
(e) Observed output on Γ_3.

aims at the determination of the boundary heat flux u. The problem may be formulated after the discretization of the spatial variable within the spatial domain, which leads to the following state and output equations, respectively:

$$\mathbf{AT} = \mathbf{Bu} + \mathbf{b}T_1 \tag{11.14a}$$

$$\mathbf{y} = \mathbf{CT} \tag{11.14b}$$

where
  $\mathbf{T}$ is the state (temperature) vector
  $\mathbf{u} \in R^n$ is the heat source vector on Γ_4
  T_1 is the fixed temperature on Γ_1
  $\mathbf{y} \in R^m$ is the observed output vector
  $\mathbf{C}$ is the output matrix

Finally, the linear operator between the inputs (the heat flux $\mathbf{u}$ and the fixed temperature T_1) and the observed output takes the form of the following single matrix equation:

$$\mathbf{y} = \mathbf{CA}^{-1}(\mathbf{Bu} + \mathbf{b}T_1) \tag{11.15}$$

The sensitivity analysis (see Chapter 7 for more details) is an important step for evaluating the ill posedness of the inversion process. It is based on the sensitivity equations

$$\delta\mathbf{y} = \mathbf{CA}^{-1}\mathbf{B}\delta\mathbf{u} \tag{11.16}$$

$$\mathbf{X} = \mathbf{CA}^{-1}\mathbf{B} \tag{11.17}$$

where the elements of the sensitivity matrix are $X_{ij} = (\partial y_i / \partial u_j)$, $i = 1, \ldots, n$ and $j = 1, \ldots, m$.

This example illustrates how the formulation of a linear inverse heat source problem leads to the construction of a (possible) large size matrix operator to be inverted, here the **X** matrix. The size of the state vector, e.g., the number N of free nodes of the spatial grid, (except the nodes on the boundary Γ_1) defines the size of the square matrix **A**. It is clear in this example that the solution of the inverse problem requires the inversion of the matrix **X** and that the size of the matrices **A** and **X** are different. Generally, the matrix **X** is rectangular, and its size depends on two variables—the size m of the observed output data vector and the size n of the heat flux vector to be determined.

11.4 The Conjugate Gradient Algorithm

11.4.1 Gradient Methods

Iterative methods for computing the values of the minimum of the least square criterion $S(u)$, which are based on the computation of the first derivatives $\nabla_u(S)$, are called *gradient methods*. In these methods, an initial guess u^0 is given and the new estimates u^k are computed with the following general expression:

$$u^{k+1} = u^k + \rho^k w^k \tag{11.18}$$

until a termination condition is achieved. Here, ρ^k is a positive scalar, and w^k has the same size of u^k. Both ρ^k and w^k have to be determined at each iteration k. The modification $\rho^k w^k$ is chosen in order to decrease $S(u)$. Different choices are available to determine $\rho^k w^k$ (see, e.g., Chapter 10), but, in any case, such a choice is expected to satisfy

$$S(u^{k+1}) - S(u^k) < 0, \quad k = 0, 1, \ldots \tag{11.19}$$

It is convenient to normalize the w^k, that is, $\|w^k\| = 1$. Then, the step size in the direction w^k is equal to the scalar $\rho^k = \|u^{k+1} - u^k\|$.

By using a linear approximation of the function $S(u^{k+1})$, one obtains

$$S(u^{k+1}) - S(u^k) = \rho^k \langle \nabla S^k, w^k \rangle_U + \cdots \tag{11.20}$$

Assuming that the scalar ρ^k is positive, a necessary condition to decrease S, called the *descent condition*, is

$$\langle \nabla S^k, w^k \rangle_U < 0 \tag{11.21}$$

The directions given by w^k that satisfy this condition are called *descent directions*.

11.4.2 The Conjugate Gradient Method

In the CGM, the descent direction is generally computed as

$$w^k = -\nabla S^k, \quad k = 0 \tag{11.22a}$$

$$w^k = -\nabla S^k + \gamma^k w^{k-1}, \quad k > 0 \tag{11.22b}$$

Different versions of the CGM can be found in the literature, depending on the form used for the computation of the conjugation parameter γ^k (see Chapter 10). In the Fletcher–Reeves (1964) version, γ is given by

$$\gamma^k = \frac{\|\nabla S^k\|^2}{\|\nabla S^{k-1}\|^2} \tag{11.23}$$

In the Polak–Ribiere (Polak 1985) version of the CGM, the conjugation coefficient is given by

$$\gamma^k = \frac{\langle \nabla S^k - \nabla S^{k-1}, \nabla S^k \rangle_U}{\|\nabla S^{k-1}\|^2} \tag{11.24}$$

For the Hestenes–Stiefel (1952) version of the CGM, we have

$$\gamma^k = \frac{\langle \nabla S^k - \nabla S^{k-1}, \nabla S^k \rangle_U}{\langle \nabla S^k - \nabla S^{k-1}, w^{k-1} \rangle_U} \tag{11.25}$$

At each iteration, a line search is then performed along the direction w^k to determine the optimal step size, ρ^k. The line search consists in minimizing the single-variable function with respect to ρ^k, that is,

$$\phi(\rho) = S(u^{k+1}) = S(u^k + \rho^k w^k) \tag{11.26}$$

Equating the derivative $\phi'(\rho)$ to zero gives the optimal descent length ρ^k, but it also leads to the useful following property which holds at the termination of a line search

$$\phi'(r) = \langle \nabla S^{k+1}, w^k \rangle_U = 0 \tag{11.27}$$

The basic steps for the application of the CGM can be summarized as (Jarny 2002):

Step 1: Choose an initial guess u^0 at $k = 0$.

Repeat steps 2 to 6 until a termination condition is satisfied.

Step 2: Compute the LS-criterion S^k.

Step 3: Compute the gradient ∇S^k and determine the descent direction:

$$\text{if } k = 0, \quad w^0 = -\nabla S^0$$
$$\text{else} \quad w^k = -\nabla S^k + \gamma^k w^{k-1}$$

Step 4: Perform a line search in the direction w^k:

$$\rho^k = \arg\min_{\rho > 0} S(u^k + \rho w^k)$$

Step 5: Compute the new iterate

$$u^{k+1} = u^k + \rho^k w^k$$

Step 6: Make $k \leftarrow k + 1$.

A version of the CGM with the descent direction given by a more general form than that of Equation 11.22b was proposed by Powell (1977) as

$$w^k = -\nabla S^k + \gamma^k w^{k-1} + \Gamma^k w^q \tag{11.28}$$

where γ^k and Γ^k are conjugation coefficients given by

$$\gamma^k = \frac{\langle \nabla S^k - \nabla S^{k-1}, \nabla S^k \rangle_U}{\langle \nabla S^k - \nabla S^{k-1}, w^{k-1} \rangle_U} \quad \text{with } \gamma^0 = 0 \quad \text{for } k = 0 \tag{11.29a}$$

$$\Gamma^k = \frac{\langle \nabla S^{q+1} - \nabla S^q, \nabla S^k \rangle_U}{\langle \nabla S^{q+1} - \nabla S^q, w^q \rangle_U} \quad \text{with } \Gamma^0 = 0 \quad \text{for } k = 0 \tag{11.29b}$$

The superscript q in Equations 11.28 and 11.29 denotes the iteration number where a *restarting strategy* is applied to the iterative procedure of the CGM. Restarting strategies were suggested for the CGM of parameter estimation in order to improve its convergence rate (Powell 1977). In accordance with Powell (1977), the application of the CGM with the conjugation coefficients given by Equations 11.29 requires restarting when gradients at successive iterations tend to be nonorthogonal (which is a measure of the local nonlinearity of the problem) and when the direction of descent is not sufficiently downhill. Restarting is performed by making $\Gamma^k = 0$ in Equation 11.28.

The nonorthogonality of gradients at successive iterations is tested by using

$$\text{ABS}(\langle \nabla S^{k-1}, \nabla S^k \rangle_U) \geq 0.2 \langle \nabla S^k, \nabla S^k \rangle_U \tag{11.30a}$$

where ABS($\cdot$) denotes the absolute value.

A nonsufficiently downhill direction of descent (i.e., the angle between the direction of descent and the negative gradient direction is too large) is identified if either of the following inequalities are satisfied:

$$\langle \nabla S^k, w^k \rangle_U \leq -1.2 \langle \nabla S^k, \nabla S^k \rangle_U \tag{11.30b}$$

or

$$\langle \nabla S^k, w^k \rangle_U \geq -0.8 \langle \nabla S^k, \nabla S^k \rangle_U \tag{11.30c}$$

In Powell–Beale's version of the CGM, the direction of descent given by Equation 11.28 is computed in accordance with the following algorithm for $k \geq 1$:

Step 1: Test the inequality (11.30a). If it is true, set $q = k - 1$.

Step 2: Compute γ^k with Equation 11.29a.

Step 3: If $k = q + 1$, set $\Gamma^k = 0$. If $k \neq q + 1$ compute Γ^k with Equation 11.29b.

Step 4: Compute the search direction w^k with Equation 11.28.

Step 5: If $k \neq q + 1$ test the inequalities with Equations 11.30b and c. If either one of them is satisfied, set $q = k - 1$ and $\Gamma^k = 0$. Then recompute the search direction with Equation 11.28.

11.4.3 Application of the CGM to the Solution of Linear Inverse Problems under Matrix Form

For linear inverse problems like those introduced above in Section 11.3, the model equations can be put in the standard matrix form $\mathbf{y} = \mathbf{Xu}$, where $\mathbf{X}$ is a rectangular matrix. Then the least square criterion takes the generic quadratic form:

$$S(u) = \|\tilde{\mathbf{y}} - \mathbf{Xu}\|_Y^2 \tag{11.31}$$

When the elements of the matrix $\mathbf{X}$ (sensitivity matrix) have been computed, the application of the conjugate gradient algorithm is straightforward. According to the gradient equation

$$\nabla S = -2\mathbf{X}^T(\tilde{\mathbf{y}} - \mathbf{Xu}) \tag{11.32}$$

the descent directions w^k are then easily computed, and the descent step

$$\rho^k = \arg\min_{\rho > 0} S(\mathbf{u}^k + \rho\mathbf{w}^k) \tag{11.33}$$

is determined without a line search, by minimizing the scalar function:

$$\phi(\rho) = \|\tilde{\mathbf{y}} - \mathbf{X}(\mathbf{u}^k + \rho\mathbf{w}^k)\|^2 = \|\tilde{\mathbf{y}} - \mathbf{Xu}^k\|_Y^2 - 2\rho\langle\mathbf{X}^T(\tilde{\mathbf{y}} - \mathbf{Xu}^k), \mathbf{w}^k\rangle_U + \rho^2\|\mathbf{Xw}^k\|_Y^2 \tag{11.34}$$

It results in

$$\phi'(\rho) = 0 \Rightarrow \rho^k = -\frac{1}{2}\frac{\langle\nabla S^k, \mathbf{w}^k\rangle_U}{\|\mathbf{Xw}^k\|_Y^2} \tag{11.35}$$

For such a quadratic minimization problem, it can be shown that the iterative process is achieved under n iterations, with $n = \text{size}(u)$. In practice, the iterative procedure is stopped according to the discrepancy principle at the iteration nf such that $S(u^{nf}) = \varepsilon^2$.

11.4.4 The Lagrangian Technique to Compute the Gradient of the LS-Criterion

More often, the modeling equations are not explicit between the input u to be determined and the output y, like in the linear examples above. Only an implicit relationship is available. Then the Lagrangian technique can be used to compute the gradient of the LS-criterion:

$$S(u) = \|\tilde{y} - y(u)\|_Y^2 \tag{11.36}$$

In this section, for simplicity, we consider a mathematical model defined by a set of n algebraic equations, denoted by

$$R[\mathbf{y}, \mathbf{u}] = 0 \tag{11.37}$$

When **y** is the solution of the modeling equations, it is noted as **y(u)**. For both examples of linear problems introduced in Section 11.3, we have obviously $R[\mathbf{y}, \mathbf{u}] = \mathbf{y} - \mathbf{Xu}$ and $\mathbf{y(u)} = \mathbf{Xu}$. But the method is not restricted to linear equations. For nonlinear model equations, it is assumed that the following linear approximation is valid:

$$[\nabla_y R^T]^T \delta \mathbf{y} + [\nabla_u R^T]^T \delta \mathbf{u} = 0 \Rightarrow \delta \mathbf{y} = -\{[\nabla_y R^T]^T\}^{-1} [\nabla_u R^T]^T \delta \mathbf{u} = \mathbf{X} \delta \mathbf{u} \qquad (11.38)$$

where the superscript T denotes transpose. In the next sections, other kinds of modeling equations will be considered.

The Lagrange multiplier ψ and the Lagrangian $L(\mathbf{y}, \mathbf{u}, \psi)$ are introduced as

$$L(\mathbf{y}, \mathbf{u}, \psi) = \|\tilde{\mathbf{y}} - \mathbf{y}\|^2 - \langle \psi, R[\mathbf{y}, \mathbf{u}] \rangle_Y \qquad (11.39)$$

If $\mathbf{y} = \mathbf{y(u)}$, the Lagrangian is equal to the LS-criterion for any multiplier. It means that the constraints are satisfied, that is,

$$L(\mathbf{y(u)}, \mathbf{u}, \psi) = \|\tilde{\mathbf{y}} - \mathbf{y(u)}\|^2 - \langle \psi, R[\mathbf{y(u)}, \mathbf{u}] \rangle_Y = S(\mathbf{u}) \qquad (11.40)$$

Assuming that ψ is fixed, independent of **u**, then the differential of the Lagrangian takes the forms

$$dL = \langle \nabla_y S, \delta \mathbf{y} \rangle - \langle \psi, [\nabla_y R^T]^T \delta \mathbf{y} + [\nabla_u R^T]^T \delta \mathbf{u} \rangle_Y \qquad (11.41a)$$

$$dL = \langle \nabla_y S - [\nabla_y R^T]^T \psi, \delta \mathbf{y} \rangle_Y - \langle [\nabla_u R^T]^T \psi, \delta \mathbf{u} \rangle_U \qquad (11.41b)$$

The choice of the Lagrange multiplier is not restricted. For convenience, let us choose ψ to be the solution of the linear equation

$$\nabla_y S - [\nabla_y R^T]^T \psi = 0, \quad \text{or} \quad [\nabla_y R^T] \psi = -2(\tilde{\mathbf{y}} - \mathbf{y}) \qquad (11.42)$$

Hence, when ψ is a solution of this last equation, the stationary condition of the Lagrangian, $dL = 0$, reduces to

$$dL = \langle [\nabla_u R^T] \psi, \delta \mathbf{u} \rangle_U = 0 \qquad (11.43)$$

And finally, by taking $\mathbf{y} = \mathbf{y(u)}$, the gradient of the LS-criterion $\nabla S(\mathbf{u})$ can be determined by

$$\nabla S(\mathbf{u}) = -[\nabla_u R^T(\mathbf{y}, \mathbf{u})] \psi \qquad (11.44)$$

One of the main advantages in introducing the Lagrange multiplier ψ is clearly to allow the use of gradient algorithms to minimize the LS-criterion, when no explicit relationship is available between the model output y and the variable u to be determined. It must be noted that the Jacobian matrix $\mathbf{X} = [\nabla_u \mathbf{y}^T]^T$ is never used in the development above.

The Lagrange multiplier ψ is also called the *adjoint variable* (Jarny et al. 1991; Alifanov 1994; Alifanov et al. 1995). Combined with the Fletcher–Reeves version of the CGM (Fletcher and Reeves 1964), the Lagrangian technique leads to the following algorithm:

Step 1: Choose an initial guess u^0, $k=0$.

Repeat steps 2 to 7 until a termination condition is satisfied.

Step 2: Compute $\mathbf{y}^k = \mathbf{y}(\mathbf{u}^k)$ as a solution of $R[\mathbf{y}(\mathbf{u}), \mathbf{u}] = 0$, and compute $S(\mathbf{u}^k)$.

Step 3: Compute ψ^k as a solution of $[\nabla_y R^T]\psi^k = -2(\tilde{\mathbf{y}} - \mathbf{y}^k)$ and the gradient $\nabla S(\mathbf{u}^k) = -[\nabla_u R^T]\psi^k$.

Step 4: Determine the descent direction

$$\text{if } k = 0, \quad w^0 = -\nabla S^0$$

$$\text{else } \quad \gamma^k = \frac{\|\nabla S^k\|^2}{\|\nabla S^{k-1}\|^2}, \quad \text{and} \quad \mathbf{w}^k = -\nabla S^k + \gamma^k \mathbf{w}^{k-1}$$

Step 5: Perform a line search in the direction $\mathbf{w}^k$ to compute $\rho^k = \arg\min \phi_k(\rho)$, which minimizes the scalar function $\phi_k(\rho) = S(\mathbf{u}^k + \rho\mathbf{w}^k)$.

Step 6: Compute the new estimate: $\mathbf{u}^{k+1} = \mathbf{u}^k + \rho^k\mathbf{w}^k$.

Step 7: Make $k \leftarrow k+1$.

This algorithm is now illustrated with the solution of the previous linear examples discussed in Section 11.3.

11.4.5 Solution of the 1-D Boundary Heat Source Problem

The 1-D transient example presented in Section 11.3 is now revisited. When the surface of a semi-infinite heat conducting body is submitted to a time varying heat flux $u(t)$, the resulting temperature response $y(t; u)$ inside the body is modeled with the integral equation

$$y(t; u) = \int_0^t f_s(t - \tau)u(\tau)d\tau \tag{11.45}$$

The function f_s is Green's function for the pulse at $x = 0$

$$f_s(t) = \frac{1}{\sqrt{t}} \exp\left(-\frac{1}{\sqrt{t}}\right) \tag{11.46}$$

where the constants τ and K have been taken equal to one without loss of generality.

The model output y, computed for a triangular variation of the function $u(t)$, is shown in Figure 11.3a.

The inverse heat flux problem aims at determining the function u over the time interval $[0, t_f = 5]$, from the noisy output data $\tilde{y}$, knowing f_s, and without computing the sensitivity matrix $\mathbf{X}$.

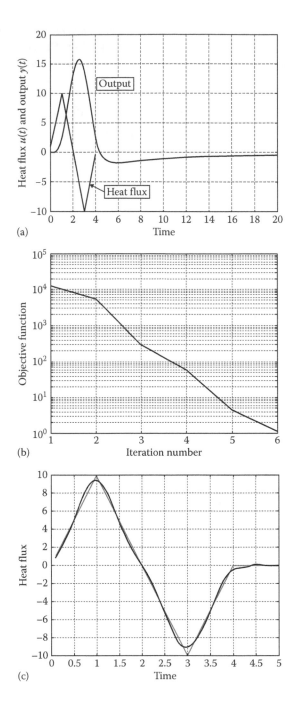

FIGURE 11.3
(a) 1-D boundary heat source data. (b) 1-D inverse heat source problem: the LS criterion versus the iteration number. (c) 1-D inverse heat source problem: exact heat flux (gray line) and estimated heat flux with $\varepsilon^2 = \|\delta Y\|^2 = 2.12$ (black line).

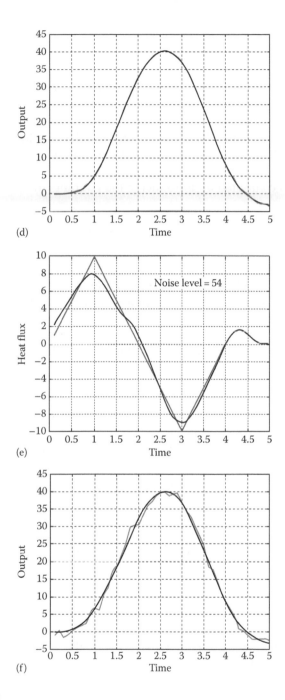

(d)

(e)

(f)

FIGURE 11.3 (continued)
(d) 1-D inverse heat source problem: predicted (black line) and measured (gray line) output with $\varepsilon^2 = \|\delta Y\|^2 = 2.12$. (Part figures b-d from Jarny, Y., 2003, *Inverse Engineering Handbook*, CRC Press, Boca Raton, FL. With permission.) (e) 1-D inverse heat source problem: exact heat flux (gray line) and estimated heat flux with $\varepsilon^2 = \|\delta Y\|^2 = 53.3$ (black line). (f) 1-D inverse heat source problem: predicted (black line) and measured output (gray line) with $\varepsilon^2 = \|\delta Y\|^2 = 53.30$.

The LS-criterion is introduced as

$$S(u) = \frac{1}{2} \int_0^{t_f} [\tilde{y}(t) - y(t; u)]^2 \, dt = \frac{1}{2} \|\tilde{y} - y(u)\|_Y^2, \quad \text{with } Y = L^2(0, t_f) \tag{11.47}$$

The solution that minimizes the LS-criterion is computed according to the adjoint method and the conjugate gradient algorithm. Notice that the modeling integral equation is linear, hence the LS-criterion is a quadratic form.

The gradient ∇S_u is computed by introducing the Lagrangian:

$$L(y, u, \psi) = \frac{1}{2} \|\tilde{y} - y\|_Y^2 - \langle \psi, y - f_s^* u \rangle_Y \tag{11.48}$$

where
$f_s^* u$ is the convolution product of the functions f and u
$\langle \psi, g \rangle_Y = \int_0^{t_f} \psi(t)g(t)dt$ is the scalar product of the function ψ with any function g over the time interval $(0, t_f)$

When the Lagrange multiplier ψ is fixed, the differential of the Lagrangian is

$$dL = \langle \tilde{y} - y, \delta y \rangle_Y - \langle \psi, \delta y \rangle_Y + \langle \psi, f_s^* \delta u \rangle_Y \tag{11.49a}$$

$$dL = \langle \tilde{y} - y - \psi, \delta y \rangle_Y + \langle \psi, f_s^* \delta u \rangle_Y \tag{11.49b}$$

By choosing ψ as the solution of the (simplest) adjoint equation

$$\psi = \tilde{y} - y \tag{11.50}$$

the stationary condition of the Lagrangian reduces to

$$dL = \langle \psi, f_s^* \delta u \rangle_Y = 0, \quad \forall \delta u \tag{11.51}$$

Hence, the gradient equation is obtained from

$$\langle \nabla S, \delta u \rangle_U = \langle \psi, f_s^* \delta u \rangle_Y, \quad \forall \delta u \tag{11.52}$$

To develop this gradient equation, if we start from the definitions of the scalar and convolution products, we have

$$\langle \nabla S, u \rangle_U = \int_0^{t_f} \nabla S(x)u(x)dx \tag{11.53}$$

$$\langle \psi, f^* u \rangle_Y = \int_0^{t_f} \psi(t) \int_0^t f(t - x)u(x)dx \, dt \tag{11.54}$$

Using the property of Green's function f

$$x > t \Rightarrow t - x < 0 \Rightarrow f(t - x) = 0 \tag{11.55}$$

the last double integral becomes

$$\langle \psi, f^*u \rangle_Y = \int_0^{t_f} u(x) \int_0^{t_f} \psi(t)f(t - x)dt\,dx \tag{11.56}$$

Then the first integral implies that

$$\nabla S(x) = \int_0^{t_f} \psi(t)f(t - x)dt, \quad 0 < x < t_f \tag{11.57a}$$

$$\nabla S(x) = \int_x^{t_f} \psi(t)f(t - x)dt, \quad 0 < x < t_f \tag{11.57b}$$

or, by denoting x as the dummy integration variable instead of t

$$\nabla S(t) = \int_t^{t_f} \psi(x)f(x - t)dx, \quad 0 < t < t_f \tag{11.58}$$

Finally, from the solution of the adjoint equation, a new integral equation for the gradient $\nabla S(t; u)$ of the LS-criterion results:

$$\nabla S(t; u) = \int_t^{t_f} [\tilde{y}(\xi) - y(\xi; u)]f(\xi - t)d\xi, \quad 0 < t < t_f \tag{11.59}$$

Then, the gradient can be computed, and the conjugate gradient algorithm can be performed.

To compute $y(t; u)$ and $\nabla S(t; u)$, the time variable is discretized as

$$t_k = k\Delta t, \quad \Delta t = \frac{t_f}{nt} \tag{11.60}$$

where nt is the number of time steps and the integral equations are put in standard algebraic forms

$$y_k = y(t_k) = \sum_{i=1}^{k} f_{k-i} u_i, \quad k = 1, \ldots, nt \tag{11.61a}$$

$$\nabla S_k = \nabla S(t_k) = \sum_{i=k+1}^{nf} f_{i-k} \psi_i, \quad k = 1, \ldots, nt \tag{11.61b}$$

with

$$f_n = f(t_n) = f(n\Delta t) = \frac{1}{\sqrt{t_n}} \exp\left(-\frac{1}{\sqrt{t_n}}\right) \Delta t, \quad n = 1, \ldots, nt \tag{11.62}$$

The solution is computed over the time interval $[0, t_f = 5]$, with $\Delta t = 0.1$ and $nf = 50$. The initial guess is taken equal to zero: $u_i^0 = 0, i = 1, \ldots, nt$. A normally distributed noise is added to the output vector data y, and the noise level is $\varepsilon^2 = \|\delta Y\|^2 = 2.12$. The iterative regularizing principle is adopted to stop the conjugate gradient algorithm, which produces the results presented in Figure 11.3b through d.

After five iterations, the computed solution is compared to the exact solution in Figure 11.3c and d. The noise level is now amplified to $\varepsilon^2 = \|\delta Y\|^2 = 53.30$ and the algorithm is stopped after four iterations. The results obtained are presented in Figure 11.3e and f.

This example shows how to use the adjoint method to compute the solution of an inverse input problem with the integral equation model. Due to the linearity of the modeling equations, the LS criterion is quadratic, and the conjugate gradient algorithm is very well adapted. The iterative regularization principle is an efficient way to avoid the amplification of data errors and is easy to implement.

Note that the output model equation $y(t) = \int_0^t f_s(t - \tau)u(\tau)d\tau$ admits the equivalent form $y(t) = \int_0^t u(t - \tau)f_s(\tau)d\tau$.

The numerical solution of the inverse problem, which consists of the determination of the parameter function $f_s(t)$ from the input and output measurements, could be performed with the same algorithm.

11.4.6 Solution of the Inverse 2-D Boundary Heat Source Problem

The stationary inverse 2-D heat conduction problem considered in Section 11.3 is now solved with the conjugate gradient algorithm presented above. The state variable T is the solution of the stationary heat conduction equations

$$\lambda \Delta T = 0, \quad \text{in } \Omega \tag{11.63a}$$

$$T|_{\Gamma_1} = T_1 \tag{11.63b}$$

$$\left. \frac{\partial T}{\partial n} \right|_{\Gamma_2 \cup \Gamma_3} = 0 \tag{11.63c}$$

$$-\lambda \left. \frac{\partial T}{\partial n} \right|_{\Gamma_4} = u \tag{11.63d}$$

where
   Ω is the spatial domain
   u is the spatial heat flux distribution to be determined on the boundary Γ_4 (see Figure 11.2a)

When u is fixed, the solution is denoted as $T(u)$.

The output variable is related to the state T through the output operator C

$$y = CT \tag{11.64}$$

Here, we take as in Section 11.3, the observation on the boundary Γ_3, $y = CT = T|_{\Gamma_3}$.

The LS-criterion to be minimized is $S(u) = \|\tilde{y} - CT(u)\|_{\Gamma_3}^2$, where $\tilde{y}$ is the noisy output data and the norm of a general function g in Γ_3 is given by $\|g\|_{\Gamma_3}^2 = \int_{\Gamma_3} g^2(s)ds$.

The Lagrangian takes the form

$$L(T, u, \psi) = \frac{1}{2}\|\tilde{y} - CT\|_{\Gamma_3}^2 - \langle \psi, \lambda\Delta T \rangle_Y \tag{11.65}$$

where the adjoint variable ψ is a function of the space variable in Ω, as the state T, and the scalar product

$$\langle \psi, v \rangle_Y = \int_\Omega \psi v d\Omega \tag{11.66}$$

Integration by parts on the spatial domain Ω, gives

$$\int_\Omega \psi \Delta T \, d\Omega = \int_\Omega T\Delta\psi \, d\Omega + \int_\Gamma \left[\psi\frac{\partial T}{\partial n} - T\frac{\partial \psi}{\partial n} \right] d\Gamma \tag{11.67}$$

For any function T that satisfies the boundary conditions, we get

$$\lambda \int_\Gamma \left[\psi\frac{\partial T}{\partial n} - T\frac{\partial \psi}{\partial n} \right] d\Gamma = \lambda \int_{\Gamma_1} \left[\psi\frac{\partial T}{\partial n} - T_1\frac{\partial \psi}{\partial n} \right] d\Gamma + \lambda \int_{\Gamma_1 \cup \Gamma_3} \left[\psi\frac{\partial T}{\partial n} - T\frac{\partial \psi}{\partial n} \right] d\Gamma + \int_{\Gamma_4} \left[-\psi u - \lambda T\frac{\partial \psi}{\partial n} \right] d\Gamma \tag{11.68}$$

When the adjoint variable ψ is fixed, the differential of the Lagrangian becomes

$$dL = -\langle \lambda\Delta\psi, \delta T \rangle + \lambda \int_{\Gamma_1} \frac{\partial T}{\partial n}\delta\psi \, d\Gamma - \lambda \int_{\Gamma_2 \cup \Gamma_4} \frac{\partial \psi}{\partial n}\delta T d\Gamma$$
$$+ \int_{\Gamma_3} \left[-\lambda\frac{\partial \psi}{\partial n} + (\tilde{y} - CT|_{\Gamma_3}) \right] \delta T \, d\Gamma - \int_{\Gamma_4} \psi\delta u d\Gamma \tag{11.69}$$

The choice of the adjoint variable ψ is not restricted. When ψ is taken as the solution of the adjoint equations

$$\lambda\Delta\psi = 0, \quad \text{in } \Omega \tag{11.70a}$$

$$\psi|_{\Gamma_1} = 0 \tag{11.70b}$$

$$\frac{\partial \psi}{\partial n}\bigg|_{\Gamma_2 \cup \Gamma_4} = 0 \tag{11.70c}$$

$$\lambda \frac{\partial \psi}{\partial n}\bigg|_{\Gamma_3} = \tilde{y} - CT|_{\Gamma_3} \tag{11.70d}$$

the differential of the Lagrangian reduces to

$$dL = \int_{\Gamma_4} \psi \delta u \, d\Gamma \tag{11.71}$$

and the gradient equations takes the form

$$\nabla_u S(s) = \psi(s), \quad s \in \Gamma_4 \tag{11.72}$$

11.4.7 A Nonlinear Problem

In this section, we illustrate the function estimation approach described above, as applied to the solution of an inverse problem involving the following diffusion equation:

$$C^*(\mathbf{r}^*)\frac{\partial T^*(\mathbf{r}^*, t^*)}{\partial t^*} = \nabla \cdot [D^*(\mathbf{r}^*)\nabla T^*] + \mu^*(\mathbf{r}^*)T^* \tag{11.73}$$

where
  $\mathbf{r}^*$ denotes the vector of coordinates
  the superscript * denotes dimensional quantities

Equation 11.73 can be used for the modeling of several physical phenomena, such as heat conduction, groundwater flow, and tomography. We focus our attention here to a 1-D version of Equation 11.73 written in the dimensionless form as (Rodrigues et al. 2004):

$$\frac{\partial T}{\partial t} = \frac{\partial}{\partial x}\left(D(x)\frac{\partial T}{\partial x}\right) + \mu(x)T \quad \text{in } 0 < x < 1 \quad \text{for } t > 0 \tag{11.74a}$$

and subject to the following boundary and initial conditions:

$$\frac{\partial T}{\partial x} = 0 \quad \text{at } x = 0 \quad \text{for } t > 0 \tag{11.74b}$$

$$D(x)\frac{\partial T}{\partial x} = 1 \quad \text{at } x = 1 \quad \text{for } t > 0 \tag{11.74c}$$

$$T = 0 \quad \text{for } t = 0 \quad \text{in } 0 < x < 1 \tag{11.74d}$$

Notice that in the *direct problem*, the diffusion coefficient function $D(x)$ and the source term distribution function $\mu(x)$ are regarded as known quantities, so that a direct problem is concerned with the computation of $T(x, t)$.

For the *inverse problem* of interest here, the functions $D(x)$ and $\mu(x)$ are regarded as unknown. Such functions will be simultaneously estimated by using measurements of $T(x, t)$ taken at appropriate locations in the medium or on its boundaries. These measurement errors are assumed to be uncorrelated, additive, normally distributed, with zero mean, and with a known constant standard deviation.

Practical applications of this inverse problem include the identification of nonhomogeneities in the medium, such as inclusions, obstacles or cracks, the determination of thermal diffusion coefficients, and the distribution of heat sources, groundwater flow, and tomography physical problems, in which both $D(x)$ and $\mu(x)$ vary.

For the simultaneous estimation of the functions $D(x)$ and $\mu(x)$, with the adjoint problem of the CGM (Jarny et al. 1991; Alifanov 1994; Alifanov et al. 1995; Ozisik and Orlande 2000; Jarny 2002), we consider the minimization of the following objective functional:

$$S[D(x), \mu(x)] = \frac{1}{2} \int_{t=0}^{t_f} \sum_{m=1}^{M} \{y[x_m, t; D(x), \mu(x)] - \tilde{y}_m(t)\}^2 dt \qquad (11.75)$$

where $\tilde{y}_m(t)$ are the transient measurements taken at the positions x_m, $m = 1, \ldots, M$. The estimated dependent variable $y[x_m, t; D(x), \mu(x)]$ is obtained from the solution of the direct problem (11.74a through 11.74d) at the measurement positions x_m, $m = 1, \ldots, M$, with estimates for $D(x)$ and $\mu(x)$. We note in Equation 11.75 that, for simplicity in the analytical analysis developed below, the measurements $\tilde{y}_m(t)$ are assumed to be continuous in the time domain.

We present below the use of the CGM for the minimization of the objective functional (11.75), by using two auxiliary problems, known as *sensitivity and adjoint problems*, for the computation of the step size and gradient directions, respectively.

The *sensitivity function*, the solution of the sensitivity problem, is defined as the directional derivative of $y(x, t)$ in the direction of the perturbation of the unknown function (Jarny et al. 1991; Alifanov 1994; Alifanov et al. 1995; Ozisik and Orlande 2000; Jarny 2002). Since the present problem involves two unknown functions, two sensitivity problems are required for the estimation procedure, resulting from perturbations in $D(x)$ and $\mu(x)$.

The sensitivity problem for $\Delta T_D(x, t)$ is obtained by assuming that the dependent variable $T(x, t)$ is perturbed by $\varepsilon \Delta T_D(x, t)$ when the diffusion coefficient $D(x)$ is perturbed by $\varepsilon \Delta D(x)$, where ε is a real number. The sensitivity problem resulting from perturbations in $D(x)$ is then obtained by applying the following limiting process:

$$\lim_{\varepsilon \to 0} \frac{L_\varepsilon(D_\varepsilon) - L(D)}{\varepsilon} = 0 \qquad (11.76)$$

where $L_\varepsilon(D_\varepsilon)$ and $L(D)$ are the direct problem formulations written in the operator form for perturbed and unperturbed quantities, respectively. The application of the limiting process given by Equation 11.76 results in the following sensitivity problem:

$$\frac{\partial \Delta T_D}{\partial t} = \frac{\partial}{\partial x} \left(D(x) \frac{\partial \Delta T_D}{\partial x} + \Delta D(x) \frac{\partial T}{\partial x} \right) + \mu(x) \Delta T_D \quad \text{for } t > 0 \quad \text{in } 0 < x < 1 \qquad (11.77a)$$

$$\frac{\partial \Delta T_D}{\partial x} = 0 \quad \text{at } x = 0 \quad \text{for } t > 0 \qquad (11.77b)$$

$$\Delta D(x) \frac{\partial T}{\partial x} + D(x) \frac{\partial \Delta T_D}{\partial x} = 0 \quad \text{at } x = 1 \quad \text{for } t > 0 \qquad (11.77c)$$

$$\Delta T_D = 0 \quad \text{in } 0 \le x \le 1 \quad \text{for } t = 0 \qquad (11.77d)$$

A limiting process analogous to that given by Equation 11.76, obtained from the perturbation $\varepsilon\Delta\mu(x)$, results in the following sensitivity problem for $\Delta T_\mu(x,t)$:

$$\frac{\partial \Delta T_\mu}{\partial t} = \frac{\partial}{\partial x}\left(D(x)\frac{\partial \Delta T_\mu}{\partial x}\right) + \mu(x)\Delta T_\mu + \Delta\mu(x)T \quad \text{in } 0 < x < 1 \quad \text{for } t > 0 \tag{11.78a}$$

$$\frac{\partial \Delta T_\mu}{\partial x} = 0 \quad \text{at } x = 0 \quad \text{and} \quad x = 1 \quad \text{for } t > 0 \tag{11.78b,c}$$

$$\Delta T_\mu = 0 \quad \text{in } 0 \leq x \leq 1; \quad \text{for } t = 0 \tag{11.78d}$$

We note in Equations 11.77a through d and 11.78a through d that the sensitivity problems depend on the unknown functions $D(x)$ and $\mu(x)$. Therefore, the present estimation problem is nonlinear and the objective functional nonquadratic, despite the fact that the direct problem (11.74a through d) is linear.

A Lagrange multiplier $\psi(x,t)$ is utilized in the minimization of the functional (11.75) because the estimated dependent variable $T[x_m, t; D(x), \mu(x)]$ appearing in such functional needs to satisfy a constraint, which is the solution of the direct problem. Such a Lagrange multiplier, needed for the computation of the gradient equations (as will be apparent below), is obtained through the solution of problems *adjoint* to the sensitivity problems, given by Equations 11.77a through d and 11.78a through d (Jarny et al. 1991; Alifanov 1994; Alifanov et al. 1995; Ozisik and Orlande 2000; Jarny 2002). Despite the fact that the present inverse problem involves the estimation of two unknown functions, thus resulting in two sensitivity problems as discussed above, one single problem, adjoint to problems (11.77a through d) and (11.78a through d), is obtained.

In order to derive the adjoint problem, the governing equation of the direct problem, Equation 11.74a, is multiplied by the Lagrange multiplier $\psi(x,t)$, integrated in the space and time domains of interest and added to the original functional (11.75). The following Lagrangian is obtained:

$$L[D(x), \mu(x)] = \frac{1}{2}\int_{x=0}^{1}\int_{t=0}^{t_f}\sum_{m=1}^{M}[y - \tilde{y}]^2\delta(x - x_m)dt\,dx$$

$$+ \int_{x=0}^{1}\int_{t=0}^{t_f}\left[\frac{\partial T}{\partial t} - \frac{\partial}{\partial x}\left(D(x)\frac{\partial T}{\partial x}\right) - \mu(x)T\right]\psi(x,t)dt\,dx \tag{11.79}$$

where δ is the Dirac delta function.

The directional derivatives of $L[D(x), \mu(x)]$, in the directions of perturbations in $D(x)$ and $\mu(x)$, are, respectively, defined by

$$\Delta L_D[D, \mu] = \lim_{\varepsilon \to 0}\frac{L[D_\varepsilon, \mu] - L[D, \mu]}{\varepsilon} \tag{11.80a}$$

$$\Delta L_\mu[D, \mu] = \lim_{\varepsilon \to 0}\frac{L[D, \mu_\varepsilon] - L[D, \mu]}{\varepsilon} \tag{11.80b}$$

where $L[D_\varepsilon, \mu]$ and $L[D, \mu_\varepsilon]$ denote the Lagrangian (11.79) written for perturbed $D(x)$ and $\mu(x)$, respectively.

After letting the above directional derivatives of $L[D(x), \mu(x)]$ go to zero, which is a necessary condition for the minimization of the Lagrangian (11.79), and after performing some lengthy but straightforward manipulations (Jarny et al. 1991; Alifanov 1994; Alifanov et al. 1995; Ozisik and Orlande 2000; Jarny 2002), the following adjoint problem for the Lagrange multiplier $\psi(x, t)$ is obtained:

$$-\frac{\partial \psi}{\partial t} - \frac{\partial}{\partial x}\left(D(x)\frac{\partial \psi}{\partial x}\right) - \mu(x)\psi + \sum_{m=1}^{M}[T - Y]\delta(x - x_m) = 0 \quad \text{in } 0 < x < 1, \quad \text{for } t > 0$$

(11.81a)

$$\frac{\partial \psi}{\partial x} = 0 \quad \text{at } x = 0 \quad \text{and} \quad x = 1 \quad \text{for } t > 0 \tag{11.81b,c}$$

$$\psi = 0 \quad \text{in } 0 \leq x \leq 1 \quad \text{for } t = t_f \tag{11.81d}$$

During the limiting processes used to obtain the adjoint problem, applied to the directional derivatives of $L[D(x), \mu(x)]$ in the directions of perturbations in $D(x)$ and $\mu(x)$, the following integral terms are, respectively, obtained:

$$\Delta L_D[D, \mu] = \int_{x=0}^{1} \int_{t=0}^{t_f} \Delta D(x)\frac{\partial T}{\partial x}\frac{\partial \psi}{\partial x} dt\, dx \tag{11.82a}$$

$$\Delta L_\mu[D, \mu] = -\int_{x=0}^{1} \int_{t=0}^{t_f} \Delta\mu(x)\psi(x, t)T(x, t)dt\, dx \tag{11.82b}$$

By invoking the hypotheses that $D(x)$ and $\mu(x)$ belong to the Hilbert space of square integrable functions in the domain $0 < x < 1$, it is possible to write (Jarny et al. 1991; Alifanov 1994; Alifanov et al. 1995; Ozisik and Orlande 2000; Jarny 2002):

$$\Delta L_D[D, \mu] = \int_{x=0}^{1} \nabla S[D(x)]\Delta D(x)dx \tag{11.83a}$$

$$\Delta L_\mu[D, \mu] = \int_{x=0}^{1} \nabla S[\mu(x)]\Delta\mu(x)dx \tag{11.83b}$$

Hence, by comparing Equations 11.83a and b and 11.82a and b, we obtain the gradient components of $S[D, \mu]$ with respect to $D(x)$ and $\mu(x)$, respectively, as

$$\nabla S[D(x)] = \int_{t=0}^{t_f} \frac{\partial T}{\partial x}\frac{\partial \psi}{\partial x} dt \tag{11.84a}$$

$$\nabla S[\mu(x)] = -\int_{t=0}^{t_f} \psi(x, t)T(x, t)dt \tag{11.84b}$$

An analysis of Equations 11.84a and 11.81b through 11.81c reveals that the gradient component with respect to $D(x)$ is null at $x = 0$ and at $x = 1$. As a result, the initial guess used for $D(x)$ is never changed by the iterative procedure of the CGM at such points, which can create instabilities in the inverse problem solution in their neighborhoods.

For the simultaneous estimation of $D(x)$ and $\mu(x)$, the iterative procedure of the CGM is written, respectively, as (Rodrigues et al. 2004):

$$D^{k+1}(x) = D^k(x) + \rho_D^k w_D^k(x) \tag{11.85a}$$

$$\mu^{k+1}(x) = \mu^k(x) + \rho_\mu^k w_\mu^k(x) \tag{11.85b}$$

where

$w_D^k(x)$ and $w_\mu^k(x)$ are the directions of descent for $D(x)$ and $\mu(x)$, respectively
ρ_D^k and ρ_μ^k are the search step sizes for $D(x)$ and $\mu(x)$, respectively
k is the number of iterations

For the iterative procedure for each unknown function, the direction of descent is obtained as a linear combination of the gradient direction with directions of descent of previous iterations. The directions of descent for the CGM for $D(x)$ and $\mu(x)$ can be written in a general form, respectively, as

$$d_D^k(x) = -\nabla S[D^k(x)] + \gamma_D^k w_D^{k-1}(x) + \Gamma_D^k w_D^{qD} \tag{11.86a}$$

$$d_\mu^k(x) = -\nabla S[\mu^k(x)] + \gamma_\mu^k w_\mu^{k-1}(x) + \Gamma_\mu^k w_\mu^{q\mu} \tag{11.86b}$$

where γ_D^k, γ_μ^k, Γ_D^k, and Γ_μ^k are the conjugation coefficients.

In this example, we use Powell–Beale's version of the CGM because of its superior robustness and convergence rate in nonlinear problems (Powell 1977). The conjugation coefficients for this version of the CGM are given by

$$\gamma_D^k = \frac{\int_{x=0}^1 \{\nabla S[D^k(x)] - \nabla S[D^{k-1}(x)]\}\nabla S[D^k(x)]dx}{\int_{x=0}^1 \{\nabla S[D^k(x)] - \nabla S[D^{k-1}(x)]\}w_D^{k-1}(x)dx} \tag{11.87a}$$

$$\gamma_\mu^k = \frac{\int_{x=0}^1 \{\nabla S[\mu^k(x)] - \nabla S[\mu^{k-1}(x)]\}\nabla S[\mu^k(x)]dx}{\int_{x=0}^1 \{\nabla S[\mu^k(x)] - \nabla S[\mu^{k-1}(x)]\}w_\mu^{k-1}(x)dx} \tag{11.87b}$$

$$\Gamma_D^k = \frac{\int_{x=0}^1 \{\nabla S[D^{qD+1}(x)] - \nabla S[D^{qD}(x)]\}\nabla S[D^k(x)]dx}{\int_{x=0}^1 \{\nabla S[D^{qD+1}(x)] - \nabla S[D^{qD}(x)]\}w_D^{qD}(x)dx} \tag{11.87c}$$

$$\Gamma_\mu^k = \frac{\int_{x=0}^1 \{\nabla S[\mu^{q\mu+1}(x)] - \nabla S[\mu^{q\mu}(x)]\}\nabla S[\mu^k(x)]dx}{\int_{x=0}^1 \{\nabla S[\mu^{q\mu+1}(x)] - \nabla S[\mu^{q\mu}(x)]\}w_\mu^{q\mu}(x)dx} \tag{11.87d}$$

where $\gamma_D^k = \gamma_\mu^k = \Gamma_D^k = \Gamma_\mu^k = 0$, for $k = 0$.

The search step sizes ρ_D^k and ρ_μ^k, appearing in the expressions of the iterative procedures for the estimation of $D(x)$ and $\mu(x)$, Equations 11.86a and b, respectively, are obtained by minimizing the objective functional at each iteration along the specified directions of descent.

If the objective functional given by Equation 11.75 is linearized with respect to ρ_D^k and ρ_μ^k, closed form expressions can be obtained for such quantities as follows:

$$\rho_d^k = \frac{F_1 A_{22} - F_2 A_{12}}{A_{11} A_{22} - A_{12}^2} \tag{11.88a}$$

$$\rho_\mu^k = \frac{F_2 A_{11} - F_1 A_{12}}{A_{11} A_{22} - A_{12}^2} \tag{11.88b}$$

where

$$A_{11} = \int_{t=0}^{t_f} \sum_{m=1}^{M} [\Delta T_D^k(x_m, t)]^2 dt \tag{11.89a}$$

$$A_{22} = \int_{t=0}^{t_f} \sum_{m=1}^{M} [\Delta T_\mu^k(x_m, t)]^2 dt \tag{11.89b}$$

$$A_{12} = \int_{t=0}^{t_f} \sum_{m=1}^{M} \Delta T_D^k(x_m, t) \Delta T_\mu^k(x_m, t) dt \tag{11.89c}$$

$$F_1 = \int_{t=0}^{t_f} \sum_{m=1}^{M} [Y_m^k - T^k(x_m, t)][\Delta T_D^k(x_m, t)] dt \tag{11.89d}$$

$$F_2 = \int_{t=0}^{t_f} \sum_{m=1}^{M} [Y_m^k - T^k(x_m, t)][\Delta T_\mu^k(x_m, t)] dt \tag{11.89e}$$

In Equations 11.89a through e, $\Delta T_D^k(x, t)$ and $\Delta T_\mu^k(x, t)$ are the solutions of the sensitivity problems given by Equations 11.77a through d and 11.78a through d, respectively, obtained by setting $\Delta D^k(x) = w_D^k(x)$ and $\Delta \mu^k(x) = w_\mu^k(x)$.

The test cases examined below in dimensionless form are physically associated with a heat conduction problem in a homogeneous steel bar of length 0.050 m. The diffusion coefficient and the spatial distribution of the source term are supposed to vary from the base values of $D(x) = 54$ W mK^{-1} and $\mu(x) = 10^5$ W m^{-3} K, respectively. The base values for the diffusion coefficient and source term distribution are associated with solid–solid phase transformations in steels. The final time is assumed to be 60 s, resulting in a dimensionless value of $t_f = 0.36$, and 50 measurements are supposed available per sensor.

Figure 11.4 shows the results obtained with the measurements of two nonintrusive sensors, for a step variation of $D(x)$ and for a constant $\mu(x)$. The results presented in Figure 11.4 were obtained with Powell–Beale's version of the CGM. The simulated measurements in this case contained random errors with standard deviation $\sigma = 0.01 y_{max}$, where y_{max} is the maximum absolute value of the measured variable. The initial guesses used for the iterative procedure of the CGM were $D(x) = 0.9$ and $\mu(x) = 4.5$. We note in Figure 11.4 that quite accurate results were obtained for such a strict test case, involving a discontinuous variation for $D(x)$, and only nonintrusive measurements. Although some blurring is

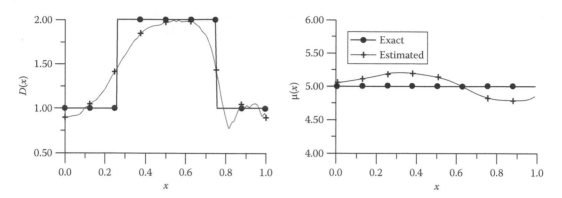

FIGURE 11.4
Simultaneous estimation of $\mu(x)$ and $D(x)$ obtained with measurements of two nonintrusive sensors with standard deviation $\sigma = 0.01y_{max}$.

observed near the discontinuity of $D(x)$ at $x = 0.25$, the locations of the discontinuities and the maximum value of the function are quite accurately estimated. Furthermore, the estimated function for $\mu(x)$ oscillates about the constant exact one with an amplitude smaller than the original distance of the initial guess to the exact function. The accuracy of the estimated functions improves when measurements of more sensors are used in the inverse analysis, as illustrated in Figure 11.5, which was obtained with measurements containing random errors ($\sigma = 0.01y_{max}$) of 10 sensors evenly located inside the medium.

Figure 11.6 illustrates the results obtained for the simultaneous estimation of $D(x)$ and $\mu(x)$, for a constant exact functional form for $D(x)$ and a triangular variation for $\mu(x)$. The results presented in Figure 11.6 were obtained with measurements containing the random errors ($\sigma = 0.01y_{max}$) of 10 sensors evenly located inside the medium, by using Powell–Beale's version of the CGM. Differently from the results shown above in Figure 11.5, we note in Figure 11.6 that the present solution approach fails to estimate the peak value of the exact triangular function for $\mu(x)$. The locations of the discontinuities on the first derivative of the exact function, which characterize the change of $\mu(x)$ from its base

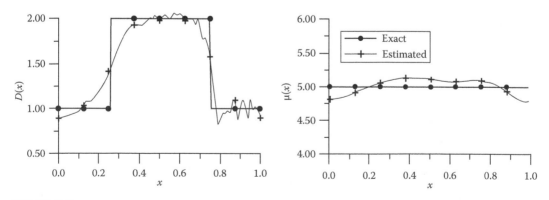

FIGURE 11.5
Simultaneous estimation of $\mu(x)$ and $D(x)$ obtained with measurements of 10 sensors equally spaced in the medium with standard deviation $\sigma = 0.01y_{max}$.

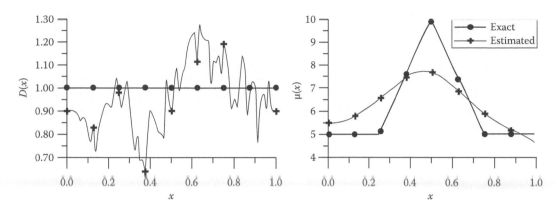

FIGURE 11.6
Simultaneous estimation of $\mu(x)$ and $D(x)$ obtained with measurements of 10 sensors equally spaced in the medium with standard deviation $\sigma = 0.01 y_{max}$.

value, could not be accurately estimated. Furthermore, the function estimated for $D(x)$ is characterized by large oscillations. We note that, generally, results similar to those presented in Figure 11.6 were obtained whenever the exact function for $\mu(x)$ was not constant. This is due to the lower sensitivity of the measured variable with respect to $\mu(x)$ as compared to the sensitivity with respect to $D(x)$.

Figure 11.7 presents the reduction of the objective functional with respect to the number of iterations obtained with Powell–Beale's, Polak–Ribiere's, and Fletcher–Reeves' versions of the conjugate gradient. The results presented in Figure 11.7 correspond to the test case shown in Figure 11.6, involving a constant functional form for $D(x)$ and a triangular variation for $\mu(x)$. Figure 11.7 shows that the prescribed tolerance for the iterative procedure of the CGM was reached only with Powell–Beale's version; the other two versions did not effectively reduce the objective functional, and the iterative procedure was stopped

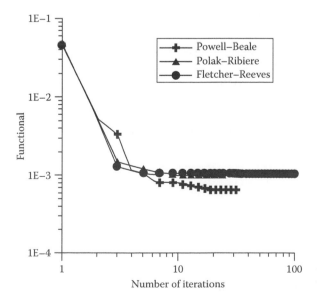

FIGURE 11.7
Comparison of different versions of the CGM for the simultaneous estimation of $\mu(x)$ and $D(x)$.

when the specified maximum number of iterations (100) was reached. The results presented in Figure 11.7 are representative of the other test cases examined, that is, Powell–Beale's version of the CGM resulted in the largest rate of reduction of the objective functional, so that the tolerance prescribed for the stopping criterion was reached in the smallest number of iterations. However, for some test cases, the use of Polak–Ribiere's version of the CGM resulted in reduction rates for the functional comparable to those obtained with Powell–Beale's version, but unexpected oscillations were observed on the values of the functional.

11.5 Conclusions

In this chapter, we introduced the adjoint method for the computation of the gradient of the objective functional, thus avoiding the use of computationally expensive techniques such as finite differences. The adjoint method is generally used together with the CGM and the discrepancy principle in order to obtain regularized solutions for inverse problems. This iterative regularization technique was illustrated with examples involving inverse problems in vector and function spaces. Furthermore, linear and nonlinear inverse problems have been addressed.

Appendix 11.A: Conjugate Directions and Conjugation Parameter

The convergence of the CGM to the minimum of the quadratic form in a finite number n iterations is due to the choice of the conjugate directions $\mathbf{w}^k$. Let us put the least square criterion under the form:

$$S(\mathbf{u}) = \langle \mathbf{C}\mathbf{u}, \mathbf{u} \rangle_U - 2\langle \mathbf{X}^T \tilde{\mathbf{y}}, u \rangle_U + \|\tilde{\mathbf{y}}\|^2$$

where $\mathbf{C} = \mathbf{X}^T \mathbf{X}$ is a symmetric matrix. Then by definition, two linear independent vectors $\mathbf{w}^1$, $\mathbf{w}^2$ are said to be C-conjugate if $\langle \mathbf{C}\mathbf{w}^1, \mathbf{w}^2 \rangle = 0$.

Now let us prove that the following linear combination

$$\mathbf{w}^2 = -\nabla S(\mathbf{u}^2) + \gamma \mathbf{w}^1 \quad \text{with the scalar } \gamma = \frac{\|\nabla S^2\|^2}{\|\nabla S^1\|^2}$$

produces a direction $\mathbf{w}^2$ C-conjugate to $\mathbf{w}^1$.

Proof

1. The gradient satisfies the equation

$$\nabla S^2 - \nabla S^1 = \mathbf{C}(\mathbf{u}^2 - \mathbf{u}^1)$$

2. The iteration prescription is

$$(\mathbf{u}^2 - \mathbf{u}^1) = \rho \mathbf{w}^1$$

3. Then, the conjugation equation can be rewritten as

$$\rho \langle \mathbf{C}\mathbf{w}^1, \mathbf{w}^2 \rangle = 0$$

$$\langle \mathbf{C}(\mathbf{u}^2 - \mathbf{u}^1), \mathbf{w}^2 \rangle = 0$$

$$\langle (\nabla S^2 - \nabla S^1), \mathbf{w}^2 \rangle = 0$$

$$\langle (\nabla S^2 - \nabla S^1), \nabla S^2 + \gamma \nabla S^1 \rangle = 0$$

4. Expansion produces

$$\langle (\nabla S^2 - \nabla S^1), \nabla S^2 \rangle + \gamma \langle (\nabla S^2 - \nabla S^1), \nabla S^1 \rangle = 0$$

5. The termination condition implies

$$\langle \nabla S^1, \nabla S^2 \rangle = 0$$

Then, finally we have

$$\langle \nabla S^2, \nabla S^2 \rangle - \gamma \langle \nabla S^1, \nabla S^1 \rangle = 0$$

which completes the proof.

Nomenclature

$S(u)$ objective function
X_{ij} sensitivity coefficients
t time
u unknown function to be estimated
w direction of descent
X functional relation that connects the input to the output through the model equations
y observable variables or model output
$\tilde{y}$ experimental data

Greek Variables

γ, Γ conjugation coefficients
ε tolerance for the stopping criterion
ρ search step size

σ　　　　standard deviation
ψ　　　　Lagrange multiplier
$\nabla_u(S)$　　gradient of the objective function

Superscripts

k　　　　iteration number
q　　　　iteration number where a restarting strategy is applied in Equations 11.28 and 11.29

References

Alifanov, O. M., 1994, *Inverse Heat Transfer Problems*, Springer Verlag, New York.

Alifanov, O. M., Artyukhin, E., and Rumyantsev, A., 1995, *Extreme Methods for Solving Ill-Posed Problems with Applications to Inverse Heat Transfer Problems*, Begell House, New York.

Fletcher, R. and Reeves, C. M., 1964, Function minimization by conjugate gradients, *Computer J.*, 7, 149–154.

Hestenes, M. R. and Stiefel, E., 1952, Method of conjugate gradients for solving linear systems, *J. Res. Nat. Bur. Stand.*, 49, 409–436.

Jarny, Y., 2002, The adjoint method to compute the numerical solutions of inverse problems, in K. A. Woodbury (Ed.), *Inverse Engineering Handbook*, CRC Press, Boca Raton, FL.

Jarny, Y., 2003, *Inverse Engineering Handbook*, CRC Press, Boca Raton, FL.

Jarny, Y., Özisik, M. N., and Bardon, J. P., 1991, A general optimization method using adjoint equation for solving multidimensional inverse heat conduction, *Int. J. Heat Mass Transf.*, 34, 2911–2929.

Morozov, V. A., 1984, *Methods for Solving Incorrectly Posed Problems*, Springer Verlag, New York.

Ozisik, M. N. and Orlande, H. R. B., 2000, *Inverse Heat Transfer: Fundamentals and Applications*, Taylor & Francis, New York.

Polak, E., 1985, *Computational Methods in Optimization*, Academic Press, New York.

Powell, M. J. D., 1977, Restart procedures for the conjugate gradient method, *Math. Prog.*, 12, 241–254.

Rodrigues, F. A., Orlande, H. R. B., and Dulikravich, G. S., 2004, Simultaneous estimation of spatially-dependent diffusion coefficient and source-term in a nonlinear 1D diffusion problem, *Math. Comput. Simul.*, 66, 409–424.

12

Bayesian Approaches for the Solution of Inverse Problems

Marina Silva Paez

CONTENTS

12.1 Introduction

Statistical inference is the process of drawing conclusions from statistical samples, and there are a number of different ways to perform it. In this chapter, we present the Bayesian approach to inference as an alternative to the more traditional approach (frequentist), which was used in Chapters 7 and 9. Under the frequentist approach, the unknown model parameters are considered to be fixed (cannot be represented probabilistically), and they are estimated by appropriately chosen statistics (estimators, which are functions of the observed data). Confidence intervals and hypothesis tests can be obtained through the sample distribution of the estimators, and predictions can be made when considering the estimated parameters as their true values. Note, however, that this kind of prediction procedure does not take into consideration the uncertainty about the model parameters.

Under the Bayesian approach, all unknown quantities, including unknown parameters, missing data, future observation, etc., can be represented through a probabilistic model—the prior distribution. Our interest is to obtain the posterior distribution of these unknown quantities, which is the conditional distribution of the parameter after observing a related data set. Once the posterior distributions are obtained, any desired inference about the unknown quantities can be performed, as all the information available is described through these distributions. The greatest advantage of the Bayesian approach to inference is that it takes into consideration all the uncertainty about the unknown quantities in a model, unlike the frequentist approach.

Analytically obtaining the posterior distribution can, however, be a hard task (or even impossible), when dealing with complicated models. Numerical methods have been proposed to sample these distributions and solve this problem, and, with the rapid advance of computers in recent years, they have gained a lot of popularity. In the last decade, the most popular among all the methods to estimate Bayesian models was the Markov chain Monte Carlo (MCMC). As a consequence, Bayesian inference has also gained popularity, and it is being extensively used nowadays to solve statistical problems in a wide range of applications. O'Hagan (2003) gives an overview on the subject and mentions a few applications in which he was involved, such as cost-effectiveness of medicines, terrestrial carbon dynamics, auditing, radiocarbon dating, setting water quality standards, and monitoring environmental pollution.

12.2 Bayesian Inference

In the Bayesian approach to inference, all the information available about the unknown parameters before the observation of a data set must be represented by a prior distribution. The prior distribution should be specified either through subjective knowledge about the parameter or through the use of information obtained from previous experiments. An indirect procedure is the specification through functional forms of parametric densities. The parameters of these densities, known as hyper-parameters, are chosen in a subjective way, according to the information available.

A systematic procedure involves choosing the functional form of the prior distribution such that the prior and posterior distributions belong to the same family of distributions, which are the so-called conjugate families. The advantage of conjugacy is that it makes the analysis easier and also permits exploiting the sequential aspect of the Bayesian paradigm.

When no information is available about the unknown parameters, noninformative priors or reference priors can be assigned. A noninformative prior can be obtained through a conjugate prior when setting the variance of this distribution to a relatively high value.

Suppose $\boldsymbol{\theta}$ is a vector of the unknown parameters in a given model and $p(\boldsymbol{\theta})$ represents its prior distribution. Once a data set Y is observed from that model, this prior distribution must be combined to the information given by the likelihood function $p(Y|\boldsymbol{\theta})$. This combination results in the posterior distribution of $\boldsymbol{\theta}$, represented by $p(\boldsymbol{\theta}|Y)$, which can be obtained via the Bayes theorem, given by

$$p(\boldsymbol{\theta}|Y) = \frac{p(Y|\boldsymbol{\theta})p(\boldsymbol{\theta})}{p(Y)},\tag{12.1}$$

with

$$p(Y) = \int p(Y|\boldsymbol{\theta})p(\boldsymbol{\theta})d\boldsymbol{\theta}.\tag{12.2}$$

The more informative the prior information, the more it influences the posterior. The bigger the number of observations in the data set, the more the likelihood that it will influence the posterior.

All the information available about the parameters $\boldsymbol{\theta}$ is contained in their posterior distribution. When the posterior distribution of $\boldsymbol{\theta}$ is known, all kinds of inferences can be performed, such as the punctual estimation (such as mean, median, and mode) and the construction of credibility intervals.

Note that any unknown quantity can be seen as a parameter under the Bayesian approach to inference. That way, performing prediction for the future (in time series) or interpolation (in spatial models), for example, can be easily performed under this approach. A good review on Bayesian statistics can be found in Migon and Gamerman (1999).

12.3 MCMC Methods

The posterior density $p(\boldsymbol{\theta}|Y)$ can be too complex and impossible to be obtained directly. With the use of the MCMC methods it is possible to sample from a Markov chain, which has the desired posterior as an equilibrium distribution. That way, after the chain converges to the equilibrium, the sampled values form a sample from this distribution, which can be used for Monte Carlo calculations. In this section, we present two MCMC algorithms to sample from the posterior of a set of parameters $\boldsymbol{\theta} = (\theta_1, \ldots, \theta_d)$: Gibbs sampling and Metropolis–Hastings. These and other methods can be seen in detail in Gamerman and Lopes (2006).

12.3.1 Gibbs Sampling

With the objective of obtaining a sample from the posterior $p(\theta_1, \ldots, \theta_d|Y)$, Gibbs sampling (Gelfand and Smith 1990) samples from the full conditional distributions of each parameter given the other parameters of the model, or in other words, it samples values of θ_i from $p(\theta_i|\boldsymbol{\theta}_{-i}, Y)$, $i, \ldots, d$, where $\boldsymbol{\theta}_{-i} = (\theta_1, \ldots, \theta_{i-1}, \theta_{i+1}, \ldots, \theta_d)'$. It is assumed that these distributions are known explicitly, and samples from them can be obtained through simple algorithms.

The steps of this sampling scheme are

1. Do $j = 1$ and give initial values to the parameters $\boldsymbol{\theta}^{(0)} = (\theta_1^{(0)}, \ldots, \theta_d^{(0)})'$
2. Obtain a new value $\boldsymbol{\theta}^{(j)} = (\theta_1^{(j)}, \ldots, \theta_d^{(j)})'$ through the successive sampling of values

$$
\begin{aligned}
\theta_1^{(j)} &\sim \pi(\theta_1|\theta_2^{(j-1)}, \ldots, \theta_d^{(j-1)}, Y), \\
\theta_2^{(j)} &\sim \pi(\theta_2|\theta_1^{(j)}, \theta_3^{(j-1)}, \ldots, \theta_d^{(j-1)}, Y), \\
&\vdots \\
\theta_d^{(j)} &\sim \pi(\theta_d|\theta_1^{(j)}, \ldots, \theta_{d-1}^{(j)}, Y);
\end{aligned}
\tag{12.3}
$$

3. Do $j = j + 1$ and return to step 2 until convergence is achieved

As the number of interactions increase, the chain gets closer to its equilibrium state. For inference purposes, the period before convergence (called the "burn-in") must be discarded. The sample obtained after the "burn-in" period can be seen as a sample from the posterior distribution of $\boldsymbol{\theta}$.

In many cases, the sampled Markov chains present an autocorrelation structure and therefore we do not observe an independent sample of values from the posterior. Working with this autocorrelated chain can lead to the underestimation of the variance. To reduce this problem, it is advised to work with a systematic subsample of the sampled values (after convergence), for example, at every $k > 1$ iteration.

12.3.2 Metropolis–Hastings

The idea behind the Metropolis–Hastings algorithm (Metropolis et al. 1953; Hastings 1970) is to sample candidate values for $\boldsymbol{\theta}$ from a proposal density $q(x|y)$. The proposal density depends on the current state y to generate a new proposed sample x, which can be accepted or not, according to a certain probability specified in a way to preserve the equilibrium distribution of interest.

The steps of this sampling scheme are

1. Do $j = 1$ and give initial values to the parameters $\boldsymbol{\theta}^{(0)} = (\theta_1^{(0)}, \ldots, \theta_d^{(0)})'$.
2. Sample a new proposed value ϕ for $\boldsymbol{\theta}$ from the distribution $q(\phi|\boldsymbol{\theta}^{(j-1)})$.
3. Accept this new value with probability:

$$\alpha(\boldsymbol{\theta}^{(j-1)}, \phi) = \min\left\{1, \frac{p(\phi|Y)q(\boldsymbol{\theta}^{(j-1)}|\phi)}{p(\boldsymbol{\theta}^{(j-1)}|Y)q(\phi|\boldsymbol{\theta}^{(j-1)})}\right\}. \tag{12.4}$$

If the new value is accepted, $\boldsymbol{\theta}^{(j)} = \phi$; otherwise, $\boldsymbol{\theta}^{(j)} = \boldsymbol{\theta}^{(j-1)}$.

4. Do $j + 1$ and return to step 2 until convergence is achieved.

After the chain converges to its equilibrium state, say, in iteration J, the values $\boldsymbol{\theta}^{(J)}, \ldots, \boldsymbol{\theta}^{(M)}$ form a sample from the posterior distribution of $\boldsymbol{\theta}$. As in the Gibbs sampling scheme, the obtained sample can be autocorrelated.

In general, the acceptance rate of the new values is adjusted to be around 50%. This adjustment can normally be made by altering the variance of the proposal density.

12.3.3 Convergence of the Chain

In theory, the Markov chain will eventually converge to the equilibrium distribution if we perform a sufficiently large number of iterations. This number, however, will vary according to the application in consideration, and therefore it is necessary to verify if and when the convergence was reached in every particular case.

An informal way of verifying the convergence is by analyzing the trajectory of at least two independently generated chains (starting at different points of the parametric space) and visually verifying if they all converge to the same place. Formal methods were also proposed to verify the convergence of the chains. The convergence diagnostics of Geweke (1992), Gelman and Rubin (1992), and Raftery and Lewis (1992) are the most popular among the statistical community, and a brief description of these methods are as follows:

- Geweke (1992): The idea behind this criteria is that after reaching convergence, if we take nonoverlapping parts (say the first n_1 iterations and the last n_2 iterations) of the Markov chain, both should constitute samples of the stationary distribution. The means of both parts are then compared through a difference of a means test.

- Gelman and Rubin (1992): To calculate the Gelman and Rubin statistic, $k \geq$ two chains must be run in parallel starting at different points in the parametric space. The idea is that after convergence, all chains will reach the stationary distribution, and therefore, the within-chain and the between-chain variances should be similar. The statistic R is calculated as a rate of the estimated variance (which is a weighted average of these two variances) and the within-chain variance. Once convergence is reached, R should be approximately equal to one.

- Raftery and Lewis (1992): This method is intended both to detect convergence to the stationary distribution and to provide a way of bounding the variance of estimates of quantiles of functions of parameters. If we define some acceptable tolerance for the desired quantile and a probability of being within that tolerance, the Raftery and Lewis diagnostic will calculate the minimum number of iterations n and the number of burn-ins m that would be needed to estimate the specified quantile to the desired precision if the samples in the chain were independent. Positive autocorrelation will increase the required sample size above these minimum values. An estimate I (the "dependence factor") of the extent to which autocorrelation inflates the required sample size is provided. Values of I larger than 5 are worrisome and can indicate influential starting values, high correlations between coefficients, or poor mixing.

Details about these methods can be found in Gamerman and Lopes (2006) and Cowles and Carlin (1996). Cowles and Carlin (1996) also provide an expository review of 10 other convergence diagnostics, describing the theoretical basis and practical implementation of each.

12.4 Applications in Different Statistical Problems

In this section, different applications of statistical models will be presented with the use of MCMC methods to obtain samples of the desired posterior distributions. The first application presents a normal hierarchical model for the weights of rats, and it was implemented in the user-friendly software WinBugs (Bayesian Using Gibbs Sampling for Windows; Thomas et al. 1992); the second application concerns a Poisson count change point problem to model the number of accidents in coal mines in Great Britain, and it was implemented in the software R (R Development Core Team 2005); in the third application, a space-time model is presented to model a pollutant set measured in the metropolitan region of Rio de Janeiro, and it was also implemented in the software WinBugs.

12.4.1 Rats: A Normal Hierarchical Model

This example is taken from Gelfand et al. (1990) and presented in the WinBugs user manual (Spiegelhalter et al. 2000). It concerns 30 young rats whose weights were measured weekly for 5 weeks. We denote by Y_{ij} the weight of the ith rat measured at age x_j.

A plot of the 30 growth curves suggests some evidence of downward curvature, and the proposed model is essentially a random effects linear growth curve given by

$$
\begin{aligned}
Y_{ij} &\sim N(\alpha_i + \beta_i(x_j - \bar{x}), \tau_c^{-1}), \\
\alpha_i &\sim N(\alpha_c, \tau_\alpha^{-1}), \\
\beta_i &\sim N(\beta_c, \tau_\beta^{-1}) \quad \text{for } i, j = 1, \ldots, 30,
\end{aligned}
\tag{12.5}
$$

where $N(\mu, \sigma^2)$ denotes the normal distribution with mean μ and variance σ^2, and therefore, τ_c, τ_α, and τ_β represent precisions. $\bar{x} = 22$ is the mean weight of the rats, and $x_j's$ were standardized around their mean to reduce dependence between α_i and β_i in their likelihood.

The unknown quantities in this model are α_i, β_i, $i = 1, \ldots, 30$, and τ_c, and the hyperparameters α_c, τ_α, β_c, and τ_β. The prior distributions of the parameters are set to be independent noninformative conjugate priors.

This application was implemented in the software WinBugs, which is a user friendly software that permits the specification of various types of models, and estimates parameters via MCMC. The model, including the priors, must be specified by the user, as well as the number of chains and iterations. As a result, the sampled values can be displayed, as well as various different graphics, showing, for example, the trajectory of the chains or histograms of the posterior distributions. When the model is simple enough, the user can design a doodle representation of the model and the code is automatically generated by the program. For this example, a doodle representation is provided in Figure 12.1.

Two chains starting at different points of the parametric space were obtained with 500 iterations each. Convergence was achieved very fast, and Figure 12.2 shows, as an example, the trajectory of the two chains obtained for τ_α and τ_β. The "burn-in" was chosen to be of 100 observations.

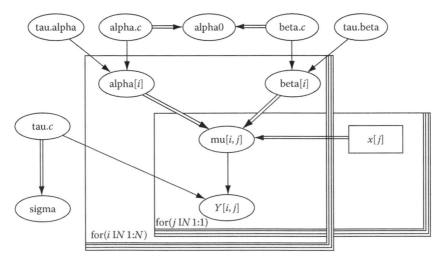

FIGURE 12.1
Graphical model for the example of rats.

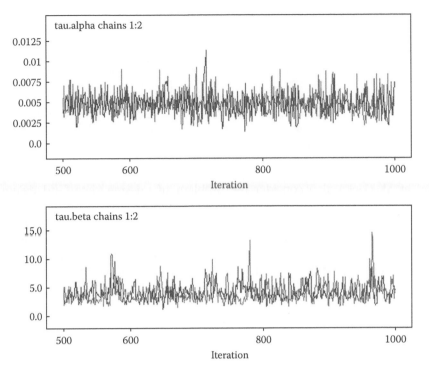

FIGURE 12.2
Trajectory of the two chains obtained for τ_α and τ_β.

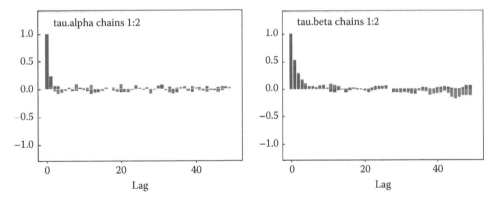

FIGURE 12.3
Autocorrelation in the chains for τ_α and τ_β.

One important aspect that must be observed is the autocorrelation in the chains. Figure 12.3 shows the autocorrelation graphics obtained for the parameters τ_α and τ_β. It can be seen that there is significant autocorrelation until leg 5 for the τ_β chain. The autocorrelation in the chains for the other parameters was also investigated, and it was found reasonable to select values at every five iterations in order to obtain independent samples.

Table 12.1 presents descriptive statistics of the sampled values from the posterior distributions of τ_c and the hyper-parameters α_c, τ_α, β_c, and τ_β.

TABLE 12.1

Descriptive Statistics of the Posterior Samples for α_c, τ_α, β_c, τ_β, and τ_c

Parameter	Mean	S.D.	MC Error	2.5%	Median	97.5%
α_c	242.5	2.638	0.1733	236.8	242.6	247.4
τ_α	0.004867	0.001391	1.171E$-$4	0.002611	0.004646	0.008008
β_c	6.173	0.1009	0.007608	5.97	6.182	6.396
τ_β	3.951	1.38	0.07741	1.861	3.685	7.429
τ_c	0.02743	0.004608	3.985E$-$4	0.01945	0.02736	0.03704

Gelfand et al. (1990) also consider the problem of missing data, and delete the last observations of some of the rats. The appropriate data file is obtained by simply replacing data values by NA. The model specification is unchanged, since the distinction between observed and unobserved quantities is made in the data file and not in the model specification. These unobserved data are seen as parameters of the model and their posterior (predictive) distributions can be automatically obtained by the program.

12.4.2 Poisson Count Change Point Problem

This example was presented by Dellaportas and Roberts (2001), and it was originally taken from Carlin and Louis (2000, p. 185). The data set consists of a series relating to the number of British coal mining disasters per year, from 1851 to 1962. The total number of observed years is $T = 112$.

It is clear from Figure 12.4, which shows a plot of the data, that the rate of disasters has reduced over the years.

Carlin and Louis (2000) propose that the number of disasters per year follow a Poisson distribution, but the rate of the Poisson changes at a certain (unknown) year. The model can be written as

$$
\begin{aligned}
Y_t &\sim \text{Poisson}(\theta), \quad t = 1, \ldots, k, \\
Y_t &\sim \text{Poisson}(\lambda), \quad t = k+1, \ldots, T,
\end{aligned}
\tag{12.6}
$$

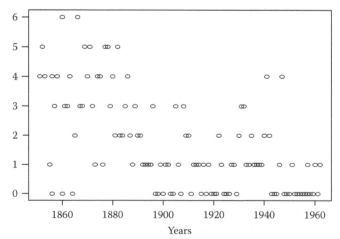

FIGURE 12.4

Counts of coal mining disasters in Great Britain.

where

Y_t is the number of disasters in year t

θ is the rate up to the kth year

λ is the rate thereafter

The Bayesian specification of the model is completed with a hierarchical framework: $\theta \sim$ Gamma(a_1, b_1), $\lambda \sim$ Gamma(a_2, b_2), k follows a discrete uniform over $\{1,\ldots,T\}$, each independent of one another, and finally $b_1 \sim$ Gamma(c_1, d_1) and $b_2 \sim$ Gamma(c_2, d_2).

Defining $\boldsymbol{Y} = \{Y_1,\ldots,Y_T\}$, this choice of specification leads to the following full conditional distributions:

$$\theta|\boldsymbol{Y}, \lambda, b_1, b_2, k \sim \text{Gamma}\left(a_1 + \sum_{i=1}^{k} Y_i, k + b_1\right),$$

$$\lambda|\boldsymbol{Y}, \theta, b_1, b_2, k \sim \text{Gamma}\left(a_2 + \sum_{i=k+1}^{T} Y_i, T - k + b_2\right), \tag{12.7}$$

$$b_1|\boldsymbol{Y}, \theta, \lambda, b_2, k \sim \text{Gamma}(a_1 + c_1, \theta + d_1),$$

$$b_2|\boldsymbol{Y}, \theta, \lambda, b_1, k \sim \text{Gamma}(a_2 + c_2, \lambda + d_2),$$

and

$$p(k|\boldsymbol{Y}, \theta, \lambda, b_1, b_2) = \frac{\pi(Y|k, \theta, \lambda)}{\sum_{j=1}^{T} \pi(Y|j, \theta, \lambda)}, \tag{12.8}$$

where

$$\pi(y|k, \theta, \lambda) = \exp\{k(\lambda - \theta)\}\left(\frac{\theta}{\lambda}\right)\sum_{i=1}^{k} Y_i. \tag{12.9}$$

Note that as all the full conditional distributions could be obtained, the Gibbs sampling algorithm can be used to sample the posterior distributions of the parameters θ, λ, b_1, b_2, and k. The hyper-parameters a_1, a_2, c_1, c_2, d_1, and d_2 are set to be equal to 0.001, such that prior distributions of b_1 and b_2 are flat with mean 1 and variance 1000. The computation was programmed in R, which is a free software environment for statistical computing and graphics.

Figure 12.5 shows the histograms of the samples obtained from the posterior distributions for the rate up to the kth year (θ), the rate after the kth year (λ), and the change point k. It is clear by these results that θ is significantly higher than λ. The higher posterior probability for k is when $k = 41$, which corresponds to the year 1891.

12.4.3 Atmospheric Pollution in Rio de Janeiro

In this example, we present a data set consisting of measurements of pollutants with less than 10 $\mu g/m^3$ of diameter (PM$_{10}$). This data set was obtained from a monitoring campaign in 1999, where concentrations of PM$_{10}$ were observed in 16 monitoring stations spread over the metropolitan region of Rio de Janeiro. The observations were made from January to

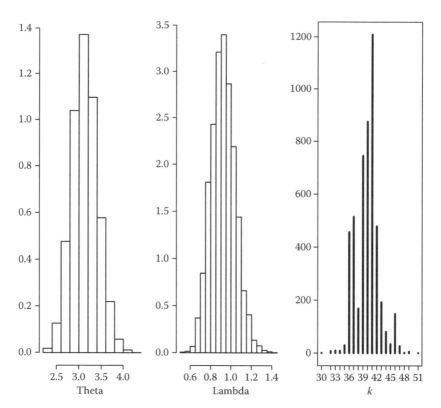

FIGURE 12.5
Posterior distribution of the parameters θ, λ, and k.

December, every 6 days, but there is a great amount of missing data. A more detailed description of this example can be found in Paez and Gamerman (2003).

Figure 12.6 shows the map of the metropolitan region of Rio de Janeiro with the location of the monitoring stations. The rectangle defines the area with more information regarding the PM_{10} concentrations, and where 15 of the 16 monitoring stations are located.

Figure 12.7 shows the 16 time series, each one corresponding to a monitoring station. It can be observed that the series tend to have similar behavior through time. That clearly shows the dependence of the PM_{10} concentrations to the day they are measured. The big amount of missing data is also evident by looking at Figure 12.7.

Other important question is how the location of observation influences the observed concentration of pollution. To have a preliminary idea about this, Figure 12.8 presents the means of PM_{10} concentrations per monitoring station. Lower concentrations are observed in the south and east of the area of study, which can be explained by the fact that this region is mostly residential, and suffers less with emissions from industries.

A preliminary analysis performed by Paez and Gamerman (2003) shows that the maximum daily temperature observed in the region of study and indicators of the day of the week can be used to explain part of the variation of the PM_{10} concentrations. Besides that,

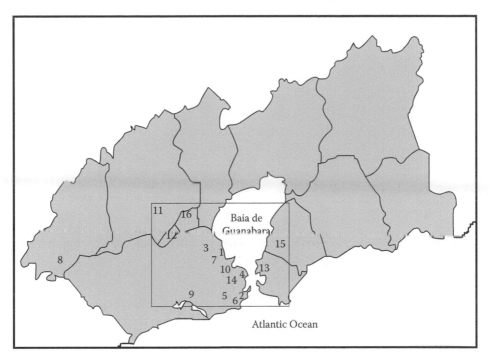

FIGURE 12.6
Map of the metropolitan region of Rio de Janeiro.

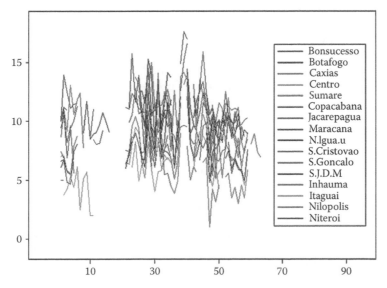

FIGURE 12.7
Concentration of PM_{10} through time.

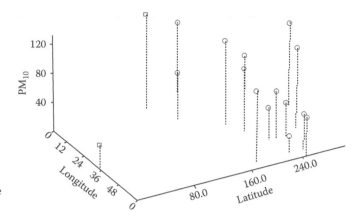

FIGURE 12.8
Means of PM_{10} concentrations by the
monitoring station.

it is suggested that the effects of the maximum temperature could be varying in space. The
model specified by Paez et al. (2005) is given by

$$Y_t(s_i) \sim N(\mu_t(s_i), \sigma_e^2),$$

$$\mu_t(s_i) = \theta_0(s_i) + \theta_1(s_i)\text{TEMP}_t + \mathbf{X}_t'\theta + \phi_t, \qquad (12.10)$$

$$\theta_j(\cdot) \sim GP(\gamma_j, \sigma_j^2 \rho_j(\cdot; \lambda_j)), \quad j = 0, 1,$$

where
  $Y_t(s_i)$ is the square root of the PM_{10} concentration levels, $s_i \in S, i = 1, \ldots, 16$ and period of
    time $t = 1, \ldots, 59$
  TEMP_t is the maximum temperature in day t
  $X_t = (\text{Monday, Tuesday, Wednesday, Thursday, Friday, Saturday})'_t$ are indicators of the
    day of the week

It is assumed that $\{\theta_j(s_i), s_i \in S\}, j = 1, 2$ are Gaussian processes (GP) (multivariate normal
distribution) with mean $\gamma_j \in R$ and covariance function $\sigma_j^2 \rho_j(s, u; \lambda_j), \sigma_j^2 > 0, \lambda_j \in \Lambda_j \subset R^+$.
The correlation functions $\rho_j(\cdot; \lambda_j), j = 1, 2$ are specified as independent exponential correl-
ation functions. ϕ is modeled through an AR(1) process given by

$$\phi_t = \delta\phi_{t-1} + \omega_t, \ \omega_t|\sigma_\phi^2 \overset{\text{ind}}{\sim} N(0, \sigma_\phi^2), \qquad (12.11)$$

with $\delta \in [0, 1), \sigma_\phi^2 > 0$ and $t \in Z$.
  To complete the Bayesian specification, the prior distributions for the unknown parameters in the model need to be specified. Paez et al. (2005) specify noninformative priors for
$\theta_2, \ldots, \theta_7, \sigma_\phi^{-2}, \gamma_0, \gamma_1,$ and δ; and informative priors for $\sigma_e^{-2}, \sigma_0^{-2}, \sigma_1^{-2}, \lambda_1$ and λ_0, with parameters chosen by preliminary analysis.
  Samples from the posterior distributions were obtained via WinBugs. A summary of the
results is presented in Table 12.2. It is interesting to note that the concentrations of PM_{10}
tend to be higher on Thursdays and Fridays than during the rest of the week. The
estimated precision $\hat{\sigma}_0^{-2}$ of the intercept process is lower than the estimated precisions
from the error and temporal component $\hat{\sigma}_e^{-2}$ and $\hat{\sigma}_\phi^{-2}$, showing that the variation in the
intercept is important to explain the variability of the PM_{10} concentrations. The estimated
precision $\hat{\sigma}_1^{-2}$ is also a lot higher than $\hat{\sigma}_0^{-2}$.

TABLE 12.2

Descriptive Statistics from the Posterior Samples
of the Model Parameters

Parameter	2.5%	97.5%	Mean	S.D.
θ_{MON}	−0.140	1.827	0.901	0.501
θ_{TUE}	−0.959	1.261	0.157	0.583
θ_{WED}	−0.477	2.224	0.907	0.688
θ_{THU}	0.526	3.378	1.876	0.720
θ_{FRI}	0.605	2.953	1.824	0.604
θ_{SAT}	−0.537	1.300	0.382	0.477
γ_0	1.413	9.345	5.860	2.040
γ_1	0.065	0.152	0.112	0.023
λ_0	0.0178	0.596	0.224	0.131
λ_1	0.0111	1.468	0.318	0.409
σ_0^{-2}	0.0567	0.532	0.234	0.124
σ_1^{-2}	109.5	1196.0	495.1	279.0
Δ	0.289	0.927	0.636	0.169
σ_ϕ^{-2}	0.441	1.246	0.770	0.204
σ_e^{-2}	0.638	0.831	0.731	0.035

Figure 12.9 shows the histograms of the posterior distribution of the parameters $\sigma_e^{-2}, \sigma_0^{-2}, \sigma_1^{-2}$, and λ_1. All the parameters are highly concentrated in small intervals with the exception of σ_0^{-2}.

One of the main interests in this kind of application is to be able to do interpolation in space and prediction for the future. In this application, Paez et al. (2005) perform interpolation in a 50×50 rectangular grid of points, covering part of the metropolitan region of Rio de Janeiro, which corresponds to the rectangle in Figure 12.6.

Once a sample of the posterior of $Y_{59}(\cdot)$ is obtained, a sample in the original scale of the PM_{10} concentrations can be obtained applying the quadratic transformation to every sampled value. Figure 12.10 shows the interpolated surface of the response process $(Y_t(\cdot))^2$ for time $t = 59$, showing that the levels of PM_{10} tend to be lower in the region that corresponds to the south zone of the city of Rio de Janeiro.

12.5 Bayesian Inference for the Solution of Inverse Problems

Collecting data is a way of obtaining information about a physical system or phenomenon of interest. In many situations, however, the quantities that we are able to measure are different from the ones we actually wish to determine, and the measurements can only give us some information about the desired quantities. The problem of trying to reconstruct the quantities that we really want is called an inverse problem. Typical inverse problems include, among others, inverse heat transfer problems, computer axial tomography, model fitting, image analysis, numerical analysis, and geophysics. In this chapter, we emphasize the solution of inverse heat transfer problems.

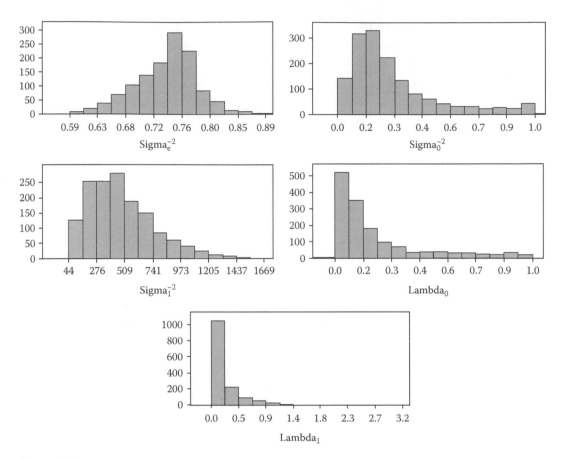

FIGURE 12.9
Histograms of the posterior distribution of the parameters $\sigma_e^{-2}, \sigma_0^{-2}, \sigma_1^{-2}, \lambda_0$, and λ_1.

With the fast advance in computational power and with the growth of interest in robustness and reliability, optimization under uncertainty has become the center of current engineering identification, design, and control research. Very recently, a sequence of methods have been proposed to solve stochastic inverse heat transfer problems, including sensitivity analysis, the extended maximum likelihood estimator approach, the spectral stochastic method, and the Bayesian inference method.

More specifically, the Bayesian statistical inference method has many advantages over other methods of solving inverse heat transfer problems. It provides not only point estimates but also the probability distribution of unknown quantities conditional on available data. Also, it explores uncertainty in the polluted data, which is rather critical because solutions to inverse problems are extremely sensitive to data errors. Applications of Bayesian inference to inverse heat transfer problems can be found, for example, in Wang and Zabaras (2004, 2005a, 2005b, 2006).

In this section, we present an example of the application of Bayesian statistics and MCMC methods for the solution of an inverse heat transfer problem, presented in Naveira et al. (2008). In this application, a simulated data set is obtained based on an experiment where one side of a flat surface, which has an initial temperature of T_0, is suddenly

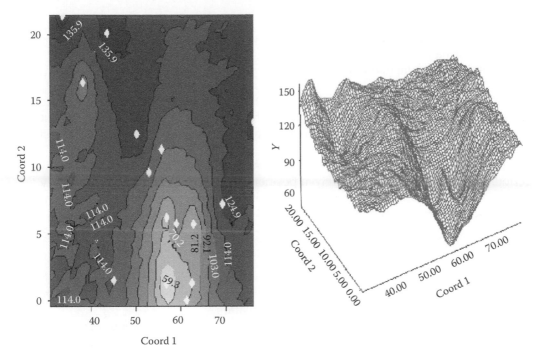

FIGURE 12.10
Interpolated surface of the response process $(Y_t(\cdot))^2$ for time $t = 59$.

submitted to a temperature of 0°C. The considered problem was modeled considering the following hypothesis:

- The surface was treated as a half-infinite and its thermal diffusivity was assumed unknown and independent from the temperature.
- The heat transfer was considered to be 1D.
- The initial temperature T_0 was considered uniform and unknown.

That way, the mathematical formulation of the considered physical model is given by

$$
\begin{aligned}
\frac{\partial T(x,t)}{\partial t} &= \alpha \frac{\partial^2 T(x,t)}{\partial x^2}, \quad 0 < x < \infty, \quad t > 0, \\
T(x,0) &= T_0, \quad 0 < x < \infty, \\
T(0,t) &= 0, \quad t > 0.
\end{aligned}
\tag{12.12}
$$

The partial differential equation given above can be solved analytically leading to the following solution:

$$
T(x,t) = T_0 \operatorname{erf}\left(\frac{x}{\sqrt{4\alpha t}}\right),
\tag{12.13}
$$

where

    $T(x, t)$ is the temperature at a location x of the surface in time t

    α is its thermal diffusivity

    $\text{erf}(\cdot)$ is the error function defined by $\text{erf}(x) = (1/\sqrt{\pi}) \int_0^x \exp\{-t^2\}\, dt$

A simulation of the problem described above was performed to obtain temperatures in a fixed location of the surface. After that, an exercise was made to simultaneously estimate the parameters of the initial temperature (T_0) and the thermal diffusivity (α) of the surface using MCMC methods.

Temperatures $y_1, y_2, \ldots, y_T$ were simulated, independently, for a set of $T = 250$ different periods of time $t_1, t_2, \ldots, t_{250}$ for a given position $x_{\text{sensor}} = 0.01$ m of the surface. They were simulated from a normal distribution with their mean given by the direct solution given above and variance given by $\sigma^2 = 2$, which will be considered known.

$$y_i \overset{\text{ind}}{\sim} N\left(T_0\, \text{erf}\left(\frac{x_{\text{sensor}}}{\sqrt{4\alpha t_i}}\right), \sigma^2\right), \quad i = 1, \ldots, T. \tag{12.14}$$

Note that the mean of y_i depends on the time it was measured t_i. The parameters were set to $T_0 = 50°C$ and $\alpha = 4.9 \times 10^{-7}\,\text{m}^2/\text{s}$ for the simulation.

Given the temperatures, our aim in this application is to obtain posterior distributions for the model parameters, T_0 and α, and compare these distributions to their real values.

Assuming that the measurements are independent, the likelihood is given by

$$p(Y|\theta) = \frac{1}{(2\pi\sigma^2)^{n/2}} \exp\left\{\frac{-(Y - E[Y|\theta])^T (Y - E[Y|\theta])}{2\sigma^2}\right\}, \tag{12.15}$$

where

    θ is the set of unknown parameters $\theta = \{T_0, \alpha\}$

    Y is the vector of measured temperatures $Y = \{y_1, y_2, \ldots, y_{250}\}$

    $E(Y|\theta)$ denotes the expectation of Y given the model parameters θ

For a given temperature y_i, $i = 1, \ldots, 250$, $E(y_i|\theta) = T(x_{\text{sensor}}, t_i)$.

We adopt uniform distributions as prior distributions for the parameters T_0 and α in the intervals $[45°C, 60°C]$ and $[10^{-7}, 10^{-5}\,\text{m}^2/\text{s}]$, respectively.

Three different MCMC algorithms were proposed for the estimation of the model parameters. Note that in theory, all MCMC algorithms should give equal results, as they must provide samples from the posterior distribution of the unknown parameters of the model. Different algorithms can, however, have different convergence speeds and different computational running times, and therefore this kind of comparison is of interest. The three different cases are

- Case 1: α and T_0 are sampled through Metropolis–Hastings steps. The proposal distribution for each parameter is normally centered on its previous value in the Markov chain, and with a conveniently chosen variance. The acceptance of the move is tested jointly for the two parameters.

- Case 2: α and T_0 are sampled through Metropolis–Hastings steps. The proposal distribution for each parameter is normally centered on its previous value in the Markov chain, and with a conveniently chosen variance. The acceptance of the move is tested separately for each parameter.

- Case 3: α is sampled through Metropolis–Hastings steps and T_0 is sampled by Gibbs sampling. The proposal distribution for α is normally centered on its previous value in the Markov chain, and with a conveniently chosen variance.

Note that it is not possible to obtain a closed form for the full conditional of α. However, it is possible to obtain a closed form for the full conditional of T_0, which is easy to sample from. It is a normal distribution given by

$$T_0|Y,\alpha \sim N\left(\frac{\sum_i y_i \operatorname{erf}\left(-1/\sqrt{4\alpha t_i}\right)}{\sum_i \operatorname{erf}\left(-1/\sqrt{4\alpha t_i}\right)^2}, \frac{\sigma^2}{\sum_i \operatorname{erf}\left(-1/\sqrt{4\alpha t_i}\right)^2}\right). \tag{12.16}$$

That way, it is possible to use Gibbs Sampling to sample from T_0 (as in case 3) but not from α.

The algorithms were implemented in the software MATLAB®, a numerical computing environment and a fourth-generation programming language developed by The Math-Works (see Gilat [2004] for an introduction to the software). Convergence is achieved after around 1000 iterations for the algorithms in cases 2 and 3, and it takes around 2000 iterations for the chains to converge in the algorithm in case 1, showing a possible disadvantage in updating T_0 and α jointly. The burn-in period was set to 10,000 iterations for case 1 and 5000 to cases 2 and 3. After the burn-in period, 50,000 iterations were sampled for each case. Computational times were 178, 316, and 188 s under cases 1, 2, and 3 respectively. That way, even though it took longer to converge, the algorithm in case 1 was faster than the other algorithms, making its use worthwhile.

The results obtained under the three algorithms were similar, as expected. Tables 12.3 and 12.4 show the posterior mean, median, standard deviation, and 95% credibility intervals for T_0 and α, respectively. Both parameters were well estimated, with their posterior means being very close to their real values.

TABLE 12.3

Posterior Results for the Parameter T_0

Case	Mean	Median	S.D.	95% CI
1	50.5146	50.4907	0.249757	[50.0582, 51.0544]
2	49.9987	50.005	0.274711	[49.4564, 50.5368]
3	50.0935	50.0916	0.208832	[49.6887, 50.5323]

TABLE 12.4

Posterior Results for the Parameter α

Case	Mean	Median	S.D.	95% CI
1	5.026×10^{-7}	5.019×10^{-7}	8.269×10^{-9}	$[4.872, 5.199] \times 10^{-7}$
2	4.935×10^{-7}	4.938×10^{-7}	8.781×10^{-9}	$[4.767, 5.112] \times 10^{-7}$
3	4.950×10^{-7}	4.948×10^{-7}	8.653×10^{-9}	$[4.803, 5.122] \times 10^{-7}$

12.6 Final Remarks

In this chapter, we presented a review on Bayesian statistics and MCMC methods to solve statistical problems. The methodology described here can have many advantages over other methods of solving inverse problems. An application was made to solve a simple inverse heat transfer problem where we compared the performance of different MCMC algorithms. It is important to note that Bayesian statistics and MCMC methods can also be applied to more complex inverse problems.

Nomenclature

$E[\cdot]$	expected value
exp	exponential function
Gamma(α, β)	Gamma distribution with mean α/β and variance α/β^2
GP(μ, $\sigma^2\rho(u)$)	Gaussian process with mean μ, variance σ^2, and correlation function $\rho(u)$
$N(\mu, \sigma^2)$	normal distribution with mean μ and variance σ^2
$p(\boldsymbol{\theta})$	density function of the prior distribution of $\boldsymbol{\theta}$
$p(\boldsymbol{\theta}|Y)$	density function of the posterior distribution of $\boldsymbol{\theta}$
$p(Y|\boldsymbol{\theta})$	likelihood function of $\boldsymbol{\theta}$
Poisson(λ)	Poisson distribution with mean λ
Y	vector of observations

Greek Variables

θ	scalar parameter
$\boldsymbol{\theta}$	vector of parameters

Superscripts

$\wedge$	estimated value
$-$	mean value
T	transpose of a matrix or a vector

References

Carlin, B. P. and T. A. Louis. 2000. *Bayes and Empirical Bayes Methods for Data Analysis*, New York: Chapman and Hall.

Cowles, M. K. and B. P. Carlin. 1996. Markov chain Monte Carlo convergence diagnostics: A comparative review. *Journal of the American Statistical Association* 434:883–904.

Dellaportas, P. and G. O. Roberts. 2001. Introduction to MCMC. Notes for the summerschool on spatial statistics and computational methods, Aalborg University, Denmark.

Gamerman, D. and H. F. Lopes. 2006. *Markov Chain Monte Carlo: Stochastic Simulation for Bayesian Inference*, 2nd edn., New York: Chapman and Hall.

Gelfand, A. E., S. E. Hills, A. Racine-Poon, and A. F. M. Smith. 1990. Illustration of Bayesian inference in normal data models using Gibbs sampling. *Journal of the American Statistical Association* 85:972–985.

Gelfand, A. E. and A. M. F. Smith. 1990. Sampling-based approaches to calculating marginal densities. *Journal of the American Statistical Association* 85:398–409.

Gelman, A. and D. Rubin. 1992. Inference from iterative simulation using multiple sequences. *Statistical Science* 7:457–511.

Geweke, J. 1992. Evaluating the accuracy of sampling-based approaches to the calculation of the posterior moments (with discussion). *Bayesian Statistics 4* (eds. J. M. Bernardo, J. O. Berger, A. P. Dawid, and A. F. M. Smith), Oxford, U.K.: Oxford University Press, pp. 169–193.

Gilat, A. 2004. *MATLAB: An Introduction with Applications*, 2nd edn., New York: John Wiley & Sons.

Hastings, W. K. 1970. Monte Carlo sampling methods using Markov chains and their applications. *Biometrika* 57:97–109.

Metropolis, N., A. W. Rosenbluth, M. N. Rosenbluth, A. H. Teller, and E. Teller. 1953. Equation of state calculations by fast computing machine. *Journal of Chemical Physics* 21:1087–1091.

Migon, H. S. and D. Gamerman. 1999, *Statistical Inference: An Integrated Approach*, London, U.K.: Arnold.

Naveira, C. P., H. M. Fonseca, M. S. Paez, and H. R. B. Orlande. 2008. Bayesian approach for parameter estimation in heat transfer. Extended abstract published in the *8th World Congress on Computational Mechanics (WCCM8)/5th European Congress on Computational Methods in Applied Sciences and Engineering (ECCO-MAS)*, Venice.

O'Hagan, T. 2003. Bayesian statistics: Principles and benefits. In *Bayesian Statistics and Quality Modelling in the Agro-Food Production Chain* (eds. MAJS van Boekal, A. Stein, and AHC van Bruggen), Wageningen, the Netherlands, Kluwer, Dordecht, 35–45. *Proceedings of the Frontis Workshop on Bayesian Statistics and Quality Modelling in the Agro-Food Production Chain*, Wageningen, the Netherlands.

Paez, M. S. and D. Gamerman. 2003. Study of the space-time in the concentration of airborne pollutants in the Metropolitan area of Rio de Janeiro. *Environmetrics* 14:387–408.

Paez, M. S., D. Gamerman, and V. de Oliveira. 2005. Interpolation performance of a spatiotemporal model with spatially varying coefficients: Application to PM10 concentrations in Rio de Janeiro. *Environmental and Ecological Statistics* 12:169–193.

Raftery, A. E. and S. M. Lewis. 1992. One long run with diagnostics: Implementation strategies for Markov chain Monte Carlo. *Statistical Science* 7:493–497.

R Development Core Team. 2005. R: A language and environment for statistical computing. http://www.R-project.org (Accessed on 16 May 2010).

Spiegelhalter, D. J., A. Thomas, and N. Best. 2000. *WinBUGS Version 1.3 User Manual*, Imperial College, London, U.K.

Thomas, A., D. J. Spiegelhalter, and W. R. Gilks. 1992. BUGS: A program to perform Bayesian inference using Gibbs samples. *Bayesian Statistics 4* (eds. J. M. Bernardo, J. O. Berger, A. P. Dawid, and A. F. M. Smith), Oxford, U.K.: Oxford University Press, pp. 837–842.

Wang, J. and N. Zabaras. 2004. A Bayesian inference approach to the inverse heat conduction problem. *International Journal of Heat and Mass Transfer* 47:3927–3941.

Wang, J. and N. Zabaras. 2005a. Hierarchical Bayesian models for inverse problems in heat conduction. *Inverse Problems* 21:183–206.

Wang, J. and N. Zabaras. 2005b. Using Bayesian statistics in the estimation of heat sources in radiation. *International Journal of Heat and Mass Transfer* 48:15–29.

Wang, J. and N. Zabaras. 2006. A Markov random field model of contamination source identification in porous media flow. *International Journal of Heat and Mass Transfer* 49:939–950.

13

Identification of Low-Order Models and Their Use for Solving Inverse Boundary Problems

Manuel Girault, Daniel Petit, and Etienne Videcoq

CONTENTS

13.1 Introduction

Thermal and fluid mechanics problems are governed by partial differential equations. The numerical resolution of such systems usually requires a fine spatial discretization of the corresponding domain. This yields to a large number N of degrees of freedom and consequently to a high CPU time consumption. When the objective of the modeling is the inversion of experimental data or the real-time control of a process, this drawback may become an impossible task to accomplish. In order to circumvent this issue, a helpful approach is to find another model that reproduces accurately enough the behavior of the system, but with a much lower number of degrees of freedom $n(n \ll N)$. Model reduction methods provide interesting solutions to obtain such low-order models.

13.2 Main Ideas of Model Reduction

Model reduction may be understood in at least two different ways:

The first one, that may be obvious, is to model the physical behavior of the system with a very few number of parameters, as low as possible: this is the principle of parsimony, which realizes a good compromise between required accuracy and model complexity.

The second one is a "mathematical point of view" where the user reduces, in his modeling, the number of equations to solve. Indeed, as soon as a spatial discretization is chosen in a domain, it corresponds to a model reduction from an infinite dimension to a finite one. But the resulting number of degrees of freedom is usually large. Among ways to reduce it, one consists in splitting an original boundary value problem into smaller ones on different subdomains. This approach is particularly suitable for parallel computing

(Han and Yin 2003, Langer et al. 2008). Another approach to reduce the size of the system is the use of the boundary element method (BEM) (Wrobel 2002): in this method, thanks to Green's functions analytically pre-calculated according to the PDE under consideration, only a boundary mesh of the domain is needed. That allows to replace a 3D (or 2D) problem with a 2D (1D) formulation, and it decreases strongly the size of the final system of equations to solve.

A widely used approach consists in writing down the model in a specific basis, different of the original physical basis of state variables, and operating a truncation or a selection of dominant modes to obtain a reduced set of equations able to adequately reproduce the systems dynamics. Some modal reduction methods are developed in Chapter 14. One may cite the approaches developed for linear systems by Marshall (Marshall 1966), Litz (Litz 1981), or Moore (Moore 1981), for example. A comparison between some reduction methods has been proposed in Ben Jaafar et al. (1990). The state space representation presented in the next section is frequently the starting point for such an approach.

13.3 State Space Representation of a Linear System and Its Modal Formulation

13.3.1 General Formulation

The state space representation (Rowell 2002), introduced in automatics for control theory, is also very much used in model reduction methods. The main characteristic of this dynamical representation is to underline the relationship between an input vector $\mathbf{U}(t)$ and an output vector $\mathbf{Y}(t)$ through a state vector $\mathbf{X}(t)$, as shown in Figure 13.1.

The linear form of the state equations is the following:

$$\dot{\mathbf{X}}(t) = \mathbf{A}\mathbf{X}(t) + \mathbf{B}\mathbf{U}(t) \tag{13.1}$$

$$\mathbf{Y}(t) = \mathbf{C}\mathbf{X}(t) \tag{13.2}$$

where
  $\mathbf{X}(t)$ and $\dot{\mathbf{X}}(t) \in \mathbb{R}^N$ are the state vector and its derivative with respect to time
  $\mathbf{U}(t) \in \mathbb{R}^p$ is the input vector
  $\mathbf{Y}(t) \in \mathbb{R}^q$ is the output vector
  $\mathbf{A} \in \mathbb{R}^{N \times N}$ is the state matrix
  $\mathbf{B} \in \mathbb{R}^{N \times p}$ is the input matrix
  $\mathbf{C} \in \mathbb{R}^{q \times N}$ is the output matrix that allows to select some components that interest more specially the user

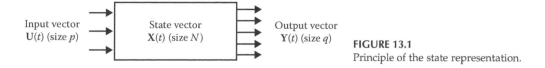

Input vector
U(t) (size p)

State vector
X(t) (size N)

Output vector
Y(t) (size q)

FIGURE 13.1
Principle of the state representation.

Note that another output equation can be found in literature:

$$\mathbf{Y}(t) = \mathbf{C}\mathbf{X}(t) + \mathbf{D}\mathbf{U}(t)$$

where $\mathbf{D} \in \mathbb{R}^{q \times p}$ is called the feedforward matrix.

If necessary, the change of variables $\mathbf{Y}'(t) = \mathbf{Y}(t) - \mathbf{D}\mathbf{U}(t)$ can be written to keep formulation (13.2).

The analytical solution of Equation 13.1 is as follows:

$$\mathbf{X}(t) = e^{\mathbf{A}(t-t_0)}\mathbf{X}(t_0) + \int_{t_0}^{t} e^{\mathbf{A}(t-\tau)}\mathbf{B}\mathbf{U}(\tau)\,d\tau \tag{13.3}$$

where
$\mathbf{X}(t_0)$ is the initial state vector
$e^{\mathbf{A}(t-t_0)} \in \mathbb{R}^{N \times N}$ is a matrix exponential

The matrix exponential is defined as follows:

$$e^{\mathbf{M}} = \sum_{k=1}^{\infty} \frac{\mathbf{M}^k}{k!}$$

where $\mathbf{M} \in \mathbb{R}^{N \times N}$. Of course, most of the time, this matrix is not easy to compute and a discrete time formulation (see Section 13.3.3) is needed.

The first part in the solution (13.3) is the free response to the initial state $\mathbf{X}(t_0)$; the second part is the forced response due to the input vector $\mathbf{U}(t)$ applied from time t_0 to time t.

13.3.2 State Space Representation for the Linear Heat Conduction Problem

The aim is here to show how a partial differential equation is transformed into a state space formulation ready to be reduced. A linear thermal system on a domain Ω is governed by the energy equation

$$\rho C_p \frac{\partial T}{\partial t} = k\nabla^2 T + q_v \tag{13.4}$$

where
q_v is the volumetric heat source (W m^{-3})
k is the thermal conductivity (W m^{-1} K^{-1})
C_p is the specific heat (J kg^{-1} K^{-1})
ρ is the density (kg m^{-3})

This equation is associated to the boundary conditions and the initial condition. Whatever the dimension of the problem, its geometry and the spatial discretization method, which is used (finite differences, finite volumes, finite elements, ...), this energy equation and associated boundary conditions can be written under the matrix form

$$\mathbf{C_a}\dot{\mathbf{T}}(t) = \mathbf{KT}(t) + \mathbf{P}(t) \tag{13.5}$$

where

$\mathbf{T}(t) \in \mathbb{R}^N$ is the vector function of temperatures at the N discretization nodes
$\dot{\mathbf{T}}(t)$ is its derivative with respect to time t
$\mathbf{C_a} \in \mathbb{R}^{N \times N}$ is the matrix of thermal capacities
$\mathbf{K} \in \mathbb{R}^{N \times N}$ is the matrix of thermal conductances
$\mathbf{P}(t) \in \mathbb{R}^N$ is the vector function containing thermal inputs (boundary conditions and/or internal heat sources) for each node of discretization

Writing $\mathbf{A} = \mathbf{C_a}^{-1}\mathbf{K}$ and $\mathbf{BU}(t) - \mathbf{C_a}^{-1}\mathbf{P}(t)$, Equation 13.5 becomes

$$\dot{\mathbf{T}}(t) = \mathbf{AT}(t) + \mathbf{BU}(t) \tag{13.6}$$

This equation is the same as Equation 13.1 with $\mathbf{T} = \mathbf{X}$: the state matrix $\mathbf{A}$ links temperature values at discretization nodes, and the input matrix $\mathbf{B}$ links discretization nodes to thermal inputs gathered in vector $\mathbf{U}$. An output matrix $\mathbf{C} \in \mathbb{R}^{q \times N}$ allows to select q temperatures in the whole temperature field $\mathbf{T}$ and to store them in vector $\mathbf{Y}$:

$$\mathbf{Y}(t) = \mathbf{CT}(t) \tag{13.7}$$

An application of such a model and its reduction is given in Section 13.8.

13.3.3 Example of Time Discretization of State Equations

Let us call Δt the time step. Assuming that the input vector $\mathbf{U}(t)$ is constant between time t_k and time t_{k+1} and equal to $\mathbf{U}(t_{k+1}) = \mathbf{U_{k+1}}$ and using a fully implicit Euler scheme, Equation 13.1 for t_{k+1} is written as follows:

$$\dot{\mathbf{X}} = \frac{\mathbf{X_{k+1}} - \mathbf{X_k}}{\Delta t} = \mathbf{AX_{k+1}} + \mathbf{BU_{k+1}}$$

which leads to

$$[\mathbf{I} - \mathbf{A}\Delta t]\mathbf{X_{k+1}} = \mathbf{X_k} + \mathbf{B}\Delta t \mathbf{U_{k+1}} \tag{13.8}$$

Equation 13.2 for t_{k+1} is written as follows:

$$\mathbf{Y_{k+1}} = \mathbf{CX_{k+1}} \tag{13.9}$$

From these equations, we can obtain an example of the state representation under a discrete form

$$\mathbf{X_{k+1}} = \mathcal{A}\mathbf{X_k} + \mathcal{B}\mathbf{U_{k+1}} \tag{13.10}$$

$$\mathbf{Y_{k+1}} = \mathbf{CX_{k+1}} \tag{13.11}$$

with $\mathcal{A} = [\mathbf{I} - \mathbf{A}\Delta t]^{-1}$ and $\mathcal{B} = [\mathbf{I} - \mathbf{A}\Delta t]^{-1}\mathbf{B}\Delta t = \mathcal{A}\mathbf{B}\Delta t$.

These equations allow computing the forward problem. They can also be used in an inverse problem consisting in the sequential estimation of the input vector $\mathbf{U_{k+1}}$ from the knowledge of the output vector $\mathbf{Y_{k+1}}$ (see Section 13.7).

Another discrete formulation, more accurate but with exponential diagonal matrices, is given in Section 13.6.

13.3.4 Principle of Model Reduction Using the State Space Equations

Using a similar representation, the aim of model reduction with a state space representation is then to find an equivalent form such as

$$\dot{\mathbf{X}}_r(t) = \mathbf{A}_r\mathbf{X}_r(t) + \mathbf{B}_r\mathbf{U}(t) \tag{13.12}$$

$$\mathbf{Y_{RM}}(t) = \mathbf{C}_r\mathbf{X}_r(t) \tag{13.13}$$

where

$\mathbf{X}_r(t)$ and $\dot{\mathbf{X}}_r(t) \in \mathbb{R}^n$ are the reduced state vector and its derivative, with $n \ll N$

matrices $\mathbf{A}_r \in \mathbb{R}^{n \times n}$, $\mathbf{B}_r \in \mathbb{R}^{n \times p}$, and $\mathbf{C}_r \in \mathbb{R}^{q \times n}$ are the new state, input, and output matrices, respectively

Of course, the output vector $\mathbf{Y_{RM}}(t)$ of the reduced model (RM) must be as close as possible to the output $\mathbf{Y}(t)$ of the original model for any $\mathbf{U}(t)$.

In several methods, a feedforward matrix $\mathbf{D}_r \in \mathbb{R}^{q \times p}$ arises in the RM output equation whereas it did not exist in the original model:

$$\mathbf{Y_{RM}}(t) = \mathbf{C}_r\mathbf{X}_r(t) + \mathbf{D}_r\mathbf{U}(t) \tag{13.14}$$

Most of the time, the role of this matrix is to preserve the stationary states between both models (see Section 13.4.2).

Figure 13.2 summarizes the principle of reduced state space representation.

13.3.5 State Space Representation in Modal Form

The state space representation is not unique. Some changes of variables are particularly interesting. From the original Equation 13.1, suppose now that $\mathbf{A}$ has N distinct eigenvalues $(\lambda_1, \lambda_2, \ldots, \lambda_N)$. They are calculated through the resolution of the spectral problem:

$$\mathbf{AV} = \lambda\mathbf{V} \tag{13.15}$$

where $\mathbf{V} \in \mathbb{R}^N$.

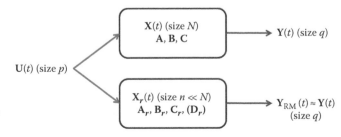

FIGURE 13.2
Principle of the reduced state representation.

That leads to the resolution of

$$\det [\mathbf{A} - \lambda \mathbf{I}] = 0$$

These eigenvalues generate N eigenvectors $\mathbf{V}_i$ linearly independent that can be stored in the modal matrix $\mathbf{M}$:

$$\mathbf{M} = [\mathbf{V}_1 \quad \mathbf{V}_2 \quad \cdots \quad \mathbf{V}_N] \tag{13.16}$$

Let us call $\tilde{\mathbf{F}}$ the diagonal matrix containing the eigenvalues of matrix $\mathbf{A}$. One has therefore $\tilde{\mathbf{F}} = \mathbf{M}^{-1}\mathbf{A}\mathbf{M}$.

We call $\tilde{\mathbf{G}} = \mathbf{M}^{-1}\mathbf{B}$ and $\tilde{\mathbf{H}} = \mathbf{C}\mathbf{M}$.

The change of variable $\tilde{\mathbf{X}} = \mathbf{M}\mathbf{X}$ in Equation 13.1 then yields

$$\dot{\tilde{\mathbf{X}}}(t) = \tilde{\mathbf{F}}\tilde{\mathbf{X}}(t) + \tilde{\mathbf{G}}\mathbf{U}(t) \tag{13.17}$$

$$\mathbf{Y}(t) = \tilde{\mathbf{H}}\tilde{\mathbf{X}}(t) \tag{13.18}$$

$\tilde{\mathbf{X}}(t) \in \mathbb{R}^N$ is the new state vector.

Several advantages appear in this formulation:

- As the differential equation is diagonal, all the state variables in Equation 13.17 are uncoupled, then easy to integrate.
- If the system is supposed to be stable, all the real part of the eigenvalues $\tilde{F}_i$ are real negative.
- If the input matrix $\tilde{\mathbf{G}}$ has no null line, all the components of $\tilde{\mathbf{X}}$ can be reached: the system is fully controllable.
- If the output matrix $\tilde{\mathbf{H}}$ has no null column, all the components of $\tilde{\mathbf{X}}$ can be seen: the system is fully observable.
- Many reduction methods use a truncation of this modal form (see Section 13.4).

13.4 Truncation of a State Space Representation in Modal Form

13.4.1 Principle

Among reduction methods, truncation and selection through the modal base of the state equation has been used quite a lot with several approaches. We just explain the principle of these methods. The matter is to split Equations 13.17 and 13.18 into two parts under the following form:

$$\begin{vmatrix} \dot{\tilde{\mathbf{X}}}_1(t) \\ \dot{\tilde{\mathbf{X}}}_2(t) \end{vmatrix} = \begin{vmatrix} \tilde{\mathbf{F}}_1 & 0 \\ 0 & \tilde{\mathbf{F}}_2 \end{vmatrix} \begin{vmatrix} \tilde{\mathbf{X}}_1(t) \\ \tilde{\mathbf{X}}_2(t) \end{vmatrix} + \begin{vmatrix} \tilde{\mathbf{G}}_1 \\ \tilde{\mathbf{G}}_2 \end{vmatrix} \mathbf{U}(t) \tag{13.19}$$

$$\mathbf{Y}(t) = [\tilde{\mathbf{H}}_1 \quad \tilde{\mathbf{H}}_2] \begin{bmatrix} \tilde{\mathbf{X}}_1(t) \\ \tilde{\mathbf{X}}_2(t) \end{bmatrix} \tag{13.20}$$

To make this partition, n "dominant eigenmodes" are selected among the N original eigenvalues of $\tilde{\mathbf{F}}$ ($n \ll N$). They are stored in $\tilde{\mathbf{F}}_1$. The RM consists in keeping in the system of ODE to solve, only the part relative to $\tilde{\mathbf{F}}_1$. Therefore, the RM has the form of Equations 13.12 and 13.14:

$$\dot{\tilde{\mathbf{X}}}_1(t) = \tilde{\mathbf{F}}_1\tilde{\mathbf{X}}_1(t) + \tilde{\mathbf{G}}_1\mathbf{U}(t) \tag{13.21}$$

$$\mathbf{Y}_{RM}(t) = \mathbf{H}_r\tilde{\mathbf{X}}_1(t) + \mathbf{D}_r\mathbf{U}(t) \tag{13.22}$$

The differences between the methods appear

- In the criterion used to select the modes in $\tilde{\mathbf{F}}_1$
- In the way to obtain matrices $\mathbf{H}_r$ and $\mathbf{D}_r$

The main advantages of an RM under a modal form are as follows:

- The system of ODE (13.21) is low dimensioned and uncoupled. Its integration is easy and very fast.
- The size q of the output vector $\mathbf{Y}_{RM}$ is independent of the order n of the RM.

13.4.2 Simple Classical Truncation: The Marshall Method

As an example, Marshall, a pioneer researcher in that field, proposed to retain only then greatest eigenvalues in $\tilde{\mathbf{F}}_1$ (Marshall 1966). As they are all with a real negative part (the system is supposed to be stable), the truncation corresponds to the n greatest time constants of the system. The contribution of the remaining $(N - n)$ eigenmodes is therefore assumed to be negligible. In this former approach, $\mathbf{H}_r = \tilde{\mathbf{H}}_1$ and $\mathbf{D}_r = 0$ in Equation 13.22.

However, such a practice usually leads to errors in static solutions. A way to circumvent this shortcoming is to search for a matrix $\mathbf{D}_r$. This can be made by writing that the RM with n modes should have exactly the same static outputs as the original detailed model (DM) with N modes. According to Equations 13.19 and 13.20, static outputs of DM are $\mathbf{Y}^s = \tilde{\mathbf{H}}_1\tilde{\mathbf{X}}_1^s + \tilde{\mathbf{H}}_2\tilde{\mathbf{X}}_2^s$ where $\tilde{\mathbf{X}}_1^s = -\tilde{\mathbf{F}}_1^{-1}\tilde{\mathbf{G}}_1\mathbf{U}^s (\dot{\tilde{\mathbf{X}}}_1^s = 0)$ and $\tilde{\mathbf{X}}_2^s = -\tilde{\mathbf{F}}_2^{-1}\tilde{\mathbf{G}}_2\mathbf{U}^s (\dot{\tilde{\mathbf{X}}}_2^s = 0)$. According to Equations 13.21 and 13.22, static outputs of RM are $\mathbf{Y}_{RM}^s = \mathbf{H}_r\tilde{\mathbf{X}}_1^s + \mathbf{D}_r\mathbf{U}^s$. Hence, we write

$$\mathbf{Y}_{RM}^s = -\mathbf{H}_r\tilde{\mathbf{F}}_1^{-1}\tilde{\mathbf{G}}_1\mathbf{U}^s + \mathbf{D}_r\mathbf{U}^s = \mathbf{Y}^s = -\tilde{\mathbf{H}}_1\tilde{\mathbf{F}}_1^{-1}\tilde{\mathbf{G}}_1\mathbf{U}^s - \tilde{\mathbf{H}}_2\tilde{\mathbf{F}}_2^{-1}\tilde{\mathbf{G}}_2\mathbf{U}^s \tag{13.23}$$

By identification of terms in both $\mathbf{Y}_{RM}^s$ and $\mathbf{Y}^s$, matrices $\mathbf{H}_r$ and $\mathbf{D}_r$ are then $\mathbf{H}_r = \tilde{\mathbf{H}}_1$ and $\mathbf{D}_r = -\tilde{\mathbf{H}}_2\tilde{\mathbf{F}}_2^{-1}\tilde{\mathbf{G}}_2$, respectively.

Remark

According to Equation 13.22, the dynamical outputs of RM are now $\mathbf{Y}_{RM}(t) = \tilde{\mathbf{H}}_1\mathbf{X}_1(t) - \tilde{\mathbf{H}}_2\tilde{\mathbf{F}}_2^{-1}\tilde{\mathbf{G}}_2\mathbf{U}(t)$. By comparing with Equation 13.20 of the DM, this means that dynamical modes $\tilde{\mathbf{X}}_2(t)$ are such as $\tilde{\mathbf{X}}_2(t) = -\tilde{\mathbf{F}}_2^{-1}\tilde{\mathbf{G}}_2\mathbf{U}(t)$, that is, $\tilde{\mathbf{F}}_2\tilde{\mathbf{X}}_2(t) + \tilde{\mathbf{G}}_2\mathbf{U}(t) = 0$. According to Equation 13.19, this means that modes $\tilde{\mathbf{X}}_2(t)$ do satisfy $\dot{\tilde{\mathbf{X}}}_2(t) = 0$. The nondominant $(N - n)$ eigenmodes are therefore supposed to reach instantaneously their asymptotic values.

The main drawback of the Marshall method is that a mode associated to a short time constant can be eliminated whereas it may be associated to a high energy level (in the signal sense).

Numerous other methods, more competitive, originate from the Marshall method. They use other criteria for selecting modes, taking into account the level of controllability and observability of the mode (see, e.g., [Litz 1981]). The balanced representation (Moore 1981) also takes into account both controllability and observability, but the truncation is made in a specific basis, different of the modal basis of the heat transfer operator.

13.5 Modal Identification Method for Linear Systems

In this section, we will develop the principles and some applications of the modal identification method (MIM) in the linear case. The handling of nonlinearities will be presented in Sections 13.9 and 13.10. The MIM has been developed for many years: its use to solve inverse problems has already been tested in several cases, showing the benefits obtained: 2D and 3D heat conduction (Videcoq and Petit 2001, Girault et al. 2003), 2D forced convection (Girault and Petit 2004), and 3D forced convection (Girault et al. 2006, 2008).

The main principle is to obtain a modal form as Equations 13.21 and 13.22 but without having any spectral problem to solve. Indeed, all the parameters of the model are identified through an optimization problem.

13.5.1 General Form of the Reduced Model in MIM

We have mentioned in Section 13.3.2 that the linear conduction Equation 13.4:

$$\rho C_p \frac{\partial T}{\partial t} = k \nabla^2 T + q_v$$

and associated boundary conditions can be written under the discrete form on a spatial mesh involving N *nodes*.

As seen in Section 13.3.5, this discrete formulation may be transformed to a modal form with N *modes*, as defined by Equations 13.17 and 13.18.

The MIM aims at building a low-order model with the same structure as Equations 13.17 and 13.18 but containing only $n \ll N$ eigenmodes that reproduce the essentials of the system dynamics. RMs that will be identified through MIM hence write

$$\dot{X}(t) = FX(t) + GU(t) \tag{13.24}$$

$$Y(t) = HX(t) \tag{13.25}$$

$X \in \mathbb{R}^n$ is the RM state vector. $U \in \mathbb{R}^p$ is the command vector, the same as in the DM, under the classic from (Equation 13.1) or under the modal form (Equation 13.17), taking into account boundary conditions and heat sources. $F = \mathrm{diag}(F_i) \in \mathbb{R}^{n \times n}$ contains $n \ll N$ eigenvalues, $G \in \mathbb{R}^{n \times p}$ is the RM command matrix, $H \in \mathbb{R}^{q \times n}$ is the RM output matrix, and $\mathbf{Y} \in \mathbb{R}^q$ is the RM output vector, which we want to be close to the DM output vector $\mathbf{Y_{DM}} \in \mathbb{R}^q$.

Note: From now on, as reduced models will be used more frequently than detailed ones, we will call the RM output vector $\mathbf{Y}$ instead of $\mathbf{Y_{RM}}$ and the DM output vector $\mathbf{Y_{DM}}$ instead of $\mathbf{Y}$. Moreover, the subscript r of the RM matrices and vectors will be omitted to lighten notations. As a consequence, $\mathbf{X}$ will now denote the RM state vector of size n rather than the original state vector of size N, which will be denoted according to the original variables (for instance, $\mathbf{T}$ for discrete temperature vector, $\mathbf{V}$ for discrete velocity vector).

13.5.2 Methodology for a Single Input

The analysis is made now for a single input: $\mathbf{U}(t)$ is restricted to a single scalar $u(t)$. The shape of the RM is then the same as Equations 13.24 and 13.25 with $\mathbf{G}u(t)$ instead of $\mathbf{G}\mathbf{U}(t)$ and $\mathbf{G} \in \mathbb{R}^n$ instead of $\mathbf{G} \in \mathbb{R}^{n \times p}$.

If the order n of RM is fixed, the aim is then to identify the matrices $\mathbf{F}, \mathbf{G}$, and $\mathbf{H}$. It is easy to prove that the product $\mathbf{GH}$ is constant when a change of variable $\mathbf{X}' = \mathbf{PX}$ is made ($\mathbf{P}$ is square and regular). Hence, as the matrices are going to be identified, we can fix vector $\mathbf{G}$ and then identify the corresponding matrix $\mathbf{H}$. For instance, we may choose $\mathbf{G} = \mathbf{1}$ where $\mathbf{1}$ is the vector of size n whose all components are equal to 1. The model is then written as follows:

$$\dot{\mathbf{X}}(t) = \mathbf{FX}(t) + \mathbf{1}u(t) \tag{13.26}$$

$$\mathbf{Y}(t) = \mathbf{HX}(t) \tag{13.27}$$

In the MIM, the reduction procedure is cast in a parameter estimation problem. Unknown components of matrices $\mathbf{F}$ and $\mathbf{H}$ are identified through the minimization of a quadratic criterion $\mathcal{J}_{\text{red}}$ built on an output error:

$$\mathcal{J}_{\text{red}}(\mathbf{F}, \mathbf{H}) = \| \mathbf{Y}(t; \mathbf{F}, \mathbf{H}) - \mathbf{Y^{data}}(t) \|^2_{L_2} \tag{13.28}$$

where
   $\mathbf{Y}(t; \mathbf{F}, \mathbf{H})$ is the RM output vector depending on $\mathbf{F}$ and $\mathbf{H}$
   $\mathbf{Y^{data}}(t)$ is a data output vector, which is either an output vector $\mathbf{Y}^*_{\mathbf{DM}}(t)$ of simulations made with the DM or the output vector $\mathbf{Y^{m}}^*(t)$ of in situ measurements recorded on the real system

Both $\mathbf{Y}(t; \mathbf{F}, \mathbf{H})$ and $\mathbf{Y^{data}}(t)$ correspond to the same known input signal $u(t)$ applied to both the RM and the DM (or the real system). As the system is linear, we will apply a Heaviside signal on $u(t)$.

Important remark: As matrices $\mathbf{F}$ and $\mathbf{H}$ are identified, there is no spectral problem to solve as opposed to the approach presented in Section 13.4. The reduction is not obtained from the matrices of the DM under the modal form.

Practically, data $\mathbf{Y^{data}}$ used for the RM identification are recorded for a discrete number of time steps N_t^{id}, so that

$$\mathcal{J}_{\text{red}}^{(n)}((\mathbf{F})_{n \times n}, (\mathbf{H})_{q \times n}) = \sum_{i=1}^{q} \sum_{j=1}^{N_t^{id}} \left(Y_i(t_j; (\mathbf{F})_{n \times n}, (\mathbf{H})_{q \times n}) - Y_i^{\text{data}}(t_j) \right)^2 \tag{13.29}$$

Let us define $\sigma_Y^{id,(n)}$ as the mean quadratic discrepancy (i.e., root mean square [rms] of the residues) between data $\mathbf{Y}^{\mathbf{data}}$ to be fitted and RM outputs $Y(\mathbf{F}, \mathbf{H})$:

$$
\sigma_Y^{id,(n)} = \sqrt{\frac{\mathcal{J}_{\text{red}}^{(n)}((\mathbf{F})_{n \times n}, (\mathbf{H})_{q \times n})}{q \times N_t^{id}}}
$$

$$
= \sqrt{\sum_{i=1}^{q} \sum_{j=1}^{N_t^{id}} \left(Y_i(t_j; (\mathbf{F})_{n \times n}, (\mathbf{H})_{q \times n}) - Y_i^{\text{data}}(t_j) \right)^2 \Big/ (q \times N_t^{id})} \tag{13.30}
$$

The objective functional $\mathcal{J}_{\text{red}}^{(n)}$ is first minimized for $n = 1$, corresponding to the identification of components of diagonal matrix $\mathbf{F}$ of size $(1, 1)$ (a single scalar) and matrix $\mathbf{H}$ of size $(q, 1)$. Then, the minimization is performed for $n = 2$ so that new matrices $\mathbf{F}$ of size $(2, 2)$ and $\mathbf{H}$ of size $(q, 2)$ are identified. This procedure is repeated until a stopping criterion is satisfied:

1. $n \leftarrow 1$
2. Minimization of $\mathcal{J}_{\text{red}}^{(1)}((\mathbf{F})_{1 \times 1}, (\mathbf{H})_{q \times 1}) \Rightarrow$ identification of $(\mathbf{F})_{1 \times 1}$ and $(\mathbf{H})_{q \times 1}$
3. $n \leftarrow n + 1$
4. Minimization of $\mathcal{J}_{\text{red}}^{(n)}((\mathbf{F})_{n \times n}, (\mathbf{H})_{q \times n}) \Rightarrow$ identification of $(\mathbf{F})_{n \times n}$ and $(\mathbf{H})_{q \times n}$
5. Test of stopping criterion: three possibilities:
   a. If $\sigma_Y^{id,(n+1)} \approx \sigma_Y^{id,(n)}$, then STOP else go to 3 (mainly for numerical applications)
   b. If $\sigma_Y^{id,(n+1)} \leq \sigma_Y^m$ where σ_Y^m is the standard deviation of measurement errors, then STOP else go to 3 (in the case of experimental data)
   c. If $\sigma_Y^{id,(n+1)} \leq \sigma_Y^a$ where σ_Y^a is the standard deviation corresponding to the accuracy wished by the user, then STOP else go to 3

According to Equations 13.26 and 13.27, $\mathbf{Y}(t)$ is nonlinear with respect to $\mathbf{F}$ whereas Equation 13.27 shows that $\mathbf{Y}(t)$ is linear with respect to $\mathbf{H}$. As a consequence, two types of optimization methods are used for the minimization of $\mathcal{J}_{\text{red}}^{(n)}((\mathbf{F})_{n \times n}, (\mathbf{H})_{q \times n})$:

1. A nonlinear iterative method is employed for the estimation of $\mathbf{F}$. It may be a deterministic method such as a conjugate gradient (Press et al. 2007) or quasi-Newton method (Gill et al. 1992), for instance, or a stochastic method (particle swarm optimization (Clerc 2005), genetic algorithm, etc). An initial guess for $\mathbf{F}$ is, therefore, required.
2. Ordinary (linear) least squares (OLS) are used for the estimation of $\mathbf{H}$ at each iteration of the above mentioned nonlinear iterative algorithm. In fact, at each iteration, the reduced state vector $\mathbf{X}$ is computed at all times $t_j, j = 1, \ldots, N_t^{id}$ by solving Equation 13.26 with the current matrix $\mathbf{F}$, the input signal $u(t)$ being known. A matrix $\mathbb{X} \in \mathbb{R}^{n \times N_t^{id}}$ is then formed:

$$
\mathbb{X} = \begin{bmatrix} \mathbf{X}(t_1) & \cdots & \mathbf{X}(t_{N_t^{id}}) \end{bmatrix} \tag{13.31}
$$

Similarly, a matrix $\mathbb{Y}^{\mathbf{data}} \in \mathbb{R}^{q \times N_t^{id}}$ is formed with data output vector $\mathbf{Y}^{\mathbf{data}}$ at all times:

$$
\mathbb{Y}^{\mathbf{data}} = \begin{bmatrix} \mathbf{Y}^{\mathbf{data}}(t_1) & \cdots & \mathbf{Y}^{\mathbf{data}}(t_{N_t^{id}}) \end{bmatrix} \tag{13.32}
$$

As Equation 13.27 may be written for each time $t_j, j = 1, \ldots, N_t^{id}$, one has searches $\mathbf{H}$ verifying:

$$\mathbb{Y}^{\mathbf{data}} = \mathbf{H}\mathbb{X} \tag{13.33}$$

which also writes

$$\mathbb{X}^{\mathbf{T}}\mathbf{H}^{\mathbf{T}} = \mathbb{Y}^{\mathbf{data}^{\mathbf{T}}} \tag{13.34}$$

Matrices $\mathbb{X}$ and $\mathbb{Y}^{\mathbf{data}}$ being known, and under the condition $N_t^{id} \geq n$ (easily obtained in practice since n is ranging from 1 to about 10), matrix $\mathbf{H}^{\mathbf{T}}$ may be estimated using OLS:

$$\mathbf{H}^{\mathbf{T}} \approx \hat{\mathbf{H}}^{\mathbf{T}} = [\mathbb{X}\,\mathbb{X}^{\mathbf{T}}]^{-1}\mathbb{X}\,\mathbb{Y}^{\mathbf{data}^{\mathbf{T}}} \tag{13.35}$$

No initial guess is hence needed for the estimation of $\mathbf{H}$. This is a good feature since $\mathbf{H}$ depends on the size q of the chosen output vector, $\mathbf{Y}$ and may therefore be of large size if q is large.

Note that the OLS approach is widely discussed in Chapter 7.

The n components of diagonal matrix $(\mathbf{F})_{n \times n}$ obtained by minimizing $\mathcal{J}_{\mathrm{red}}^{(n)}$ are used as initial guesses for the n first components of diagonal matrix $(\mathbf{F})_{n+1 \times n+1}$, which is identified through the minimization of $\mathcal{J}_{\mathrm{red}}^{(n+1)}$.

13.5.3 General Case of Several Independent Inputs

Up to now, this analysis has been made for a scalar input $u(t)$. We now consider the general case of an input vector $\mathbf{U}(t)$ with p components, $U_k(t), k = 1, \ldots, p$. According to the linearity of the system, an elementary reduced model (ERM) can be built for each independent input $U_k, k = 1, \ldots, p$, the $p - 1$ other inputs $U_{j \neq k}$ being inactive ($U_{j \neq k} = 0$) for the considered ERM. The global RM for $\mathbf{U}(t)$ can then be formed by assembling ERMs with respect to the superposition principle. Using the above presented algorithm, an ERM of order $n_{(k)}$ can be identified for U_k:

$$\dot{\mathbf{X}}_{(k)}(t) = \mathbf{F}_{(k)}\mathbf{X}_{(k)}(t) + \mathbf{1}_{(k)}U_{(k)}(t) \tag{13.36}$$

$$\mathbf{Y}_{(k)}(t) = \mathbf{H}_{(k)}\mathbf{X}_{(k)}(t) \tag{13.37}$$

where
   $\mathbf{X}_{(k)} \in \mathbb{R}^{n_{(k)}}$ is the state vector
   $\mathbf{F}_{(k)} \in \mathbb{R}^{n_{(k)} \times n_{(k)}}$ is diagonal
   $\mathbf{H}_{(k)} \in \mathbb{R}^{q \times n_{(k)}}$
   $\mathbf{1}_{(k)} \in \mathbb{R}^{n_{(k)}}$ has all its components equal to 1
   $\mathbf{Y}_{(k)} \in \mathbb{R}^q$ is the contribution of input component U_k to the global output vector $\mathbf{Y}$

The general scheme of the MIM is shown in Figure 13.3.

Once built, the p ERMs defined by Equations 13.36 and 13.37, $k = 1, \ldots, p$, may be assembled.

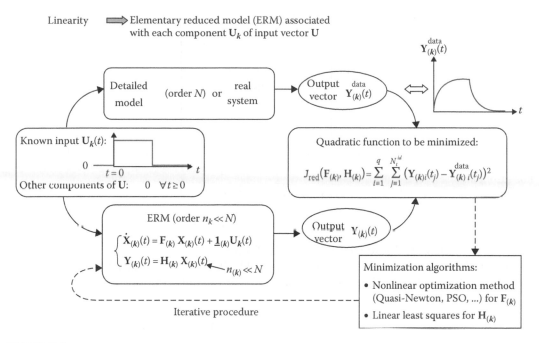

FIGURE 13.3
General scheme of MIM for the identification of an ERM relative to an independent input.

The superposition principle is thus applied:

$$\mathbf{Y}(t) = \sum_{k=1}^{p} \mathbf{Y}_{(k)}(t) = \sum_{k=1}^{p} \mathbf{H}_{(k)}\mathbf{X}_{(k)}(t) \tag{13.38}$$

Matrices and vectors of the global RM (Equations 13.24 and 13.25) are therefore defined as follows:

The global state vector $\mathbf{X}(t)$ of size $n = \sum_{k=1}^{p} n_{(k)}$ is

$$\mathbf{X}(t) = [\mathbf{X}_{(1)}(t) \quad \cdots \quad \mathbf{X}_{(k)}(t) \quad \cdots \quad \mathbf{X}_{(p)}(t)]^{\mathrm{T}} \tag{13.39}$$

The global output matrix $\mathbf{H}$ of size (q, n) is

$$\mathbf{H} = [\mathbf{H}_{(1)} \quad \cdots \quad \mathbf{H}_{(k)} \quad \cdots \quad \mathbf{H}_{(p)}] \tag{13.40}$$

The global diagonal state matrix $\mathbf{F}$ of size (n, n) is

$$\mathbf{F} = \begin{bmatrix} \mathbf{F}_{(1)} & & & & \\ & \ddots & & & \\ & & \mathbf{F}_{(k)} & & \\ & & & \ddots & \\ & & & & \mathbf{F}_{(p)} \end{bmatrix} \tag{13.41}$$

The global input matrix $\mathbf{G}$ of size (n, p) is

$$\mathbf{G} = \begin{bmatrix} \underline{\mathbf{1}}_{(1)} & & & & \\ & \ddots & & & \\ & & \underline{\mathbf{1}}_{(k)} & & \\ & & & \ddots & \\ & & & & \underline{\mathbf{1}}_{(p)} \end{bmatrix} \tag{13.42}$$

which multiplies the global input vector $\mathbf{U}(t) = [U_1(t) \quad \cdots \quad U_k(t) \quad \cdots \quad U_p(t)]^{\mathsf{T}}$.

For example:
We consider a thermal system, submitted to $p = 3$ independent thermal loads:

$$\mathbf{U}(t) = [U_1(t) \quad U_2(t) \quad U_3(t)]^{\mathsf{T}}$$

The spatial domain is discretized with $N = 1000$ nodes.
 The whole temperature vector is hence $\mathbf{T}(t) = [T_1(t) \quad \cdots \quad T_{N=1000}(t)]^{\mathsf{T}}$.
 We are interested in $q = 20$ output temperatures at distinct locations.
 For instance,

$$\mathbf{Y}(t) = [Y_1(t) \quad \cdots \quad Y_{q=20}(t)]^{\mathsf{T}} = [T_6(t) \quad \cdots \quad T_{82}(t) \quad \cdots \quad T_{324}(t) \quad \cdots \quad T_{857}(t)]^{\mathsf{T}}$$

Suppose that $p = 3$ ERMs have been built and that the final order of each ERM has been found to be

- $n_{(1)} = 3$ for the ERM relative to U_1, so that the number of identified parameters was $n_{(1)} = 3$ components of $\mathbf{F}_{(1)}$ through the nonlinear optimization algorithm and $q \times n_{(1)} = 60$ components of $\mathbf{H}_{(1)}$ through OLS.
- $n_{(2)} = 4$ for the ERM relative to U_2, so that the number of identified parameters was $n_{(2)} = 4$ components of $\mathbf{F}_{(2)}$ and $q \times n_{(2)} = 80$ components of $\mathbf{H}_{(2)}$.
- $n_{(3)} = 2$ for the ERM relative to U_3, so that the number of identified parameters was $n_{(3)} = 2$ components of $\mathbf{F}_{(3)}$ and $q \times n_{(3)} = 40$ components of $\mathbf{H}_{(3)}$.

The order of the global RM is therefore $n = 9$, which is also the size of the global state vector:

$$\mathbf{X} = [X_{(1)_1} \ X_{(1)_2} \ X_{(1)_3} \ X_{(2)_1} \ X_{(2)_2} \ X_{(2)_3} \ X_{(2)_4} \ X_{(3)_1} \ X_{(3)_2}]^{\mathsf{T}}$$

Matrices $\mathbf{F}$, $\mathbf{G}$, and $\mathbf{H}$ are respectively of size $(9, 9)$, $(9, 3)$, and $(20, 9)$.
 One has

$$\mathbf{F} = \begin{bmatrix} F_{(1)_1} & & & & & & & & \\ & F_{(1)_2} & & & & & & & \\ & & F_{(1)_3} & & & & & & \\ & & & F_{(2)_1} & & & & & \\ & & & & F_{(2)_2} & & & & \\ & & & & & F_{(2)_3} & & & \\ & & & & & & F_{(2)_4} & & \\ & & & & & & & F_{(3)_1} & \\ & & & & & & & & F_{(3)_2} \end{bmatrix}, \quad \mathbf{G} = \begin{bmatrix} 1 & 0 & 0 \\ 1 & 0 & 0 \\ 1 & 0 & 0 \\ 0 & 1 & 0 \\ 0 & 1 & 0 \\ 0 & 1 & 0 \\ 0 & 1 & 0 \\ 0 & 0 & 1 \\ 0 & 0 & 1 \end{bmatrix}$$

and

$$
\mathbf{H} = \begin{bmatrix} H_{(1)_{1,1}} & H_{(1)_{1,2}} & H_{(1)_{1,3}} & H_{(2)_{1,1}} & H_{(2)_{1,2}} & H_{(2)_{1,3}} & H_{(2)_{1,4}} & H_{(3)_{1,1}} & H_{(3)_{1,2}} \\ \vdots & \vdots & \vdots & \vdots & \vdots & \vdots & \vdots & \vdots & \vdots \\ H_{(1)_{q,1}} & H_{(1)_{q,2}} & H_{(1)_{q,3}} & H_{(2)_{q,1}} & H_{(2)_{q,2}} & H_{(2)_{q,3}} & H_{(2)_{q,4}} & H_{(3)_{q,1}} & H_{(3)_{q,2}} \end{bmatrix}
$$

with $q = 20$.

Remark

- The method allows building the RM only on the outputs that are interesting for the user. If the whole field is necessary, we have of course $q = N$.
- The knowledge of the DM matrices is not required to identify the RM. Only simulation results or measurements are needed to produce data included in $\mathbf{Y}^{\text{data}}$. This is the main advantage of the MIM compared to other methods for which matrices of the DM have to be computed and transformed (e.g., modal form) before performing the reduction by selection or truncation in the modes spectrum. With the MIM, an RM can be built from any simulation made by a commercial software, for example (Girault et al. 2006). Experimental data may also be used to identify a low-order model without DM simulations (Videcoq et al. 2003, Girault et al. 2008). This corresponds to an "experimental modeling." Of course, in the identification phase, it is needed to measure the outputs and the inputs. It is hence a kind of empirical calibration between inputs and outputs.

13.6 Time Discretization of the Reduced Model under the Modal Form

The time discretization of the RM under modal form defined by Equations 13.24 and 13.25, using an Euler implicit scheme such as in Section 13.3.3, leads to

$$
\mathbf{X}_{k+1} = \mathcal{F}\mathbf{X}_k + \mathcal{G}\mathbf{U}_{k+1} \tag{13.43}
$$

$$
\mathbf{Y}_{k+1} = \mathbf{H}\mathbf{X}_{k+1} \tag{13.44}
$$

with

$$
\mathcal{F} = [\mathbf{I} - \mathbf{F}\Delta t]^{-1} \tag{13.45}
$$

$$
\mathcal{G} = [\mathbf{I} - \mathbf{F}\Delta t]^{-1} \mathbf{G}\Delta t = \mathcal{F}\mathbf{G}\Delta t \tag{13.46}
$$

Note that as matrix $[\mathbf{I} - \mathbf{F}\Delta t]$ is a low rank (n) diagonal matrix, it is straightforward to invert it and obtain $\mathcal{F}$.

We propose also another time discretization: We take benefit of the modal form of RM where the matrix exponential of the state matrix is easy to compute because it is diagonal.

It is easy to prove, by using the analytical solution (13.3) between time t_k and time t_{k+1}, that the discretization of the modal form (13.24) gives

$$X_{k+1} = \mathcal{F}'X_k + \mathcal{G}'U_{k+1} \tag{13.47}$$

with $\mathcal{F}' = e^{F\Delta t}$ and $\mathcal{G}' = -F^{-1}[I - e^{F\Delta t}]G$.

13.7 Inverse Problem Using a Low-Order Model under the Modal Form

13.7.1 Sequential Formulation of the Inverse Problem

We now consider the following inverse problem: finding an estimation $\hat{U}$ of the input vector U from the knowledge of measured temperatures Y^{me} corresponding to the output vector Y of the RM defined by Equations 13.24 and 13.25. In the following, we will write U_k for $U(t_k)$, X_k for $X(t_k)$, Y_k for $Y(t_k)$, etc. We adopt a sequential estimation method: from the knowledge of a vector of measured temperatures Y_{k+1}^{me} and an estimate $\hat{U}_k$ of U_k, one looks for an estimate $\hat{U}_{k+1}$ at time t_{k+1}. The inverse problem aims at finding $\hat{U}_{k+1}$, minimizing the squared norm of the residual vector e_{k+1} between the measured data Y_{k+1}^{me} and the output vector Y_{k+1} of the model

$$\|e_{k+1}\|_{L_2}^2 = \|Y_{k+1} - Y_{k+1}^{me}\|_{L_2}^2 = \sum_{i=1}^{q} (e_i(t_{k+1}))^2 = \sum_{i=1}^{q} \left(Y_i(t_{k+1}) - Y_i^{me}(t_{k+1})^2\right) \tag{13.48}$$

Let us start from Equation 13.43:

$$X_{k+1} = \mathcal{F}X_k + \mathcal{G}U_{k+1} = \mathcal{F}[X_k + G\Delta t U_{k+1}] \tag{13.49}$$

Then, according to Equation 13.44:

$$Y_{k+1} = H\mathcal{F}[X_k + G\Delta t U_{k+1}] \tag{13.50}$$

Inserting Equation 13.50 in Equation 13.48 and letting

$$\mathcal{S} = H\mathcal{F}G\Delta t \tag{13.51}$$

and

$$\ell_k = H\mathcal{F}X_k \tag{13.52}$$

The squared norm to be minimized becomes

$$\|e_{k+1}\|_{L_2}^2 = \|\mathcal{S}U_{k+1} - (Y_{k+1}^{me} - \ell_k)\|_{L_2}^2 \tag{13.53}$$

meaning that a solution $\hat{\mathbf{U}}_{k+1}$ is sought for the system:

$$\mathcal{S}\mathbf{U}_{k+1} = \mathbf{Y}^{me}_{k+1} - \boldsymbol{\ell}_k \tag{13.54}$$

Equation 13.54 is a linear system where $\mathcal{S}$ is a dynamic sensitivity matrix, which does not vary with time, but depends on the time step Δt. High values for Δt make matrix $\mathcal{S}$ closer to the static sensitivity matrix $-\mathbf{HF}^{-1}\mathbf{G}$, making inversion easier.

Assuming that we have at least as many sensors as unknowns, that is, $q \geq p$, the solution is given by OLS:

$$\hat{\mathbf{U}}_{k+1} = [\mathcal{S}^{\mathrm{T}}\mathcal{S}]^{-1}\mathcal{S}^{\mathrm{T}}\left(\mathbf{Y}^{me}_{k+1} - \hat{\boldsymbol{\ell}}_k\right) \tag{13.55}$$

where $\hat{\boldsymbol{\ell}}_k = \mathbf{H}\mathcal{F}\hat{\mathbf{X}}_k$ is the estimation of $\boldsymbol{\ell}_k$ at the previous time step k, obtained from Equation 13.52 written for $\hat{\mathbf{X}}_k$, where $\hat{\mathbf{X}}_k$ is computed with Equation 13.49 written for $\hat{\mathbf{U}}_k$ and $\hat{\mathbf{X}}_{k-1}$ (both those vectors are known at time step $k+1$).

Two main advantages provided by the use of the RM under the modal form should be underlined here:

1. $\mathbf{F} \in \mathbb{R}^{n \times n}$ being a low rank diagonal matrix, it is easy to compute $\mathcal{F} = [\mathbf{I} - \mathbf{F}\Delta t]^{-1}$ and hence to obtain $\mathcal{S}$ (Equation 13.51) and $\hat{\boldsymbol{\ell}}_k = \mathbf{H}\mathcal{F}\hat{\mathbf{X}}_k$ (Equation 13.52 for $\hat{\mathbf{X}}_k$). In contrast, when using a DM, it would be needed to calculate the inverse of the non-diagonal matrix $\mathcal{A} = [\mathbf{I} - \mathbf{A}\Delta t]^{-1} \in \mathbb{R}^{N \times N}$.

2. At each time step, only the estimated reduced state vector $\hat{\mathbf{X}}_k \in \mathbb{R}^n$ has to be computed, instead of the estimated whole field $\hat{\mathbf{T}} \in \mathbb{R}^N$ of the original state variable (here temperature for instance) with a DM.

13.7.2 Use of Future Time Steps with Beck's Function Specification Method

In thermal diffusion, a variation of the input vector at time t_{k+1} does not instantaneously affect the sensors, because of the lagging and damping effects of diffusion. Thus, if $\mathbf{U}_{k+1}$ does not affect $\mathbf{Y}_{k+1}$, but $\mathbf{Y}_{k+2}$ or $\mathbf{Y}_{k+3}$, for instance, it is illusory to hope that $\mathbf{U}_{k+1}$ could be estimated from the knowledge of $\mathbf{Y}^{me}_{k+1}$ only. In order to take into account lagging and damping effects, it is useful to employ data measured at posterior times or future time steps (FTS) (Beck et al. 1996, Osman et al. 1997, Blanc et al. 1998), that is, to use sensors information $\mathbf{Y}^{me}_{k+2}, \mathbf{Y}^{me}_{k+3}, \ldots$, at times $t_{k+2}, t_{k+3}, \ldots$, to correctly estimate $\mathbf{U}_{k+1}$. Moreover, data $\mathbf{Y}^{me}$ are usually affected by measurement errors. FTS allow using information (variation of temperature) higher than the noise level. Thus, the extra information acts as a regularization procedure, which stabilizes the solution. If n_f is the number of FTS to be used, then for $0 \leq f \leq n_f$, Equation 13.54 is written for $\mathbf{U}_{k+1+f}$ and $\mathbf{Y}^{me}_{k+1+f}$. A temporary approximation of $\mathbf{U}_{k+1+f}$ is needed to search an estimate for $\mathbf{U}_{k+1}$. The *function specification method* proposed by Beck gives, for constant specification,

$$\mathbf{U}_{k+1+f} = \mathbf{U}_{k+1}, \quad 1 \leq f \leq n_f \tag{13.56}$$

Under the temporary assumption (13.56) (i.e., used for the current time t_{k+1} only), it is possible to rearrange the n_{f+1} Equation 13.54 to obtain

$$\mathbb{S}\mathbf{U}_{k+1} = \mathbb{Y}^{me} - \mathbb{B}_k \tag{13.57}$$

with

$$
\mathbb{S} = \begin{bmatrix} \mathbf{H}\mathcal{F}\mathbf{G}\Delta t \\ \mathbf{H}(\mathcal{F} + \mathcal{F}^2)\mathbf{G}\Delta t \\ \vdots \\ \mathbf{H}\left(\sum_{j=1}^{f+1} \mathcal{F}^j\right)\mathbf{G}\Delta t \\ \vdots \\ \mathbf{H}\left(\sum_{j=1}^{n_f+1} \mathcal{F}^j\right)\mathbf{G}\Delta t \end{bmatrix} \in \mathbb{R}^{q.(n_f+1)\times p}, \quad \mathbb{Y}^{\mathbf{me}} = \begin{bmatrix} \mathbb{Y}^{\mathbf{me}}_{\mathbf{k}+1} \\ \mathbb{Y}^{\mathbf{me}}_{\mathbf{k}+2} \\ \vdots \\ \mathbb{Y}^{\mathbf{me}}_{\mathbf{k}+1+f} \\ \vdots \\ \mathbb{Y}^{\mathbf{me}}_{\mathbf{k}+1+n_f} \end{bmatrix} \in \mathbb{R}^{q.(n_f+1)}
$$

and

$$
\mathbb{B}_{\mathbf{k}} = \begin{bmatrix} \mathbf{H}\mathcal{F}\mathbf{X}_{\mathbf{k}} \\ \mathbf{H}\mathcal{F}^2\mathbf{X}_{\mathbf{k}} \\ \vdots \\ \mathbf{H}\mathcal{F}^{f+1}\mathbf{X}_{\mathbf{k}} \\ \vdots \\ \mathbf{H}\mathcal{F}^{n_f+1}\mathbf{X}_{\mathbf{k}} \end{bmatrix} \in \mathbb{R}^{q.(n_f+1)}
$$

The estimate $\hat{\mathbf{U}}_{\mathbf{k}+1}$ given by OLS is then

$$
\hat{\mathbf{U}}_{\mathbf{k}+1} = [\mathbb{S}^{\mathbf{T}}\mathbb{S}]^{-1}\mathbb{S}^{\mathbf{T}}\left(\mathbb{Y}^{\mathbf{me}} - \hat{\mathbb{B}}_{\mathbf{k}}\right) \tag{13.58}
$$

where $\hat{\mathbb{B}}_{\mathbf{k}}$ is the estimation of $\mathbb{B}_{\mathbf{k}}$ at the previous time step k, hence computed with $\hat{\mathbf{X}}_{\mathbf{k}}$.

13.8 Simple Numerical 2D Transient Heat Transfer Problem

13.8.1 Description of the System and Its Modeling

The system under investigation is a square slab ABCD (AB = 0.1 m) composed of stainless steel. Figure 13.4 shows the geometry with its boundary conditions. Heat transfer is governed by Equation 13.4 with $q_v = 0$. The properties of the sample are $k = 16$ W m^{-1} K^{-1}, $C_p = 510$ J kg^{-1} K^{-1}, and $\rho = 7900$ kg m^{-3}. The spatial discretization is made with the control volume method using a regular mesh of $N = 121$ nodes (11×11). Temperatures at these 121 nodes form the temperature vector $\mathbf{T}$ and the corresponding detailed model is written under the formulation of Equation 13.6.

In the first step, an RM (Equations 13.24 and 13.25) is identified through the MIM. The second stage consists in using this RM to estimate heat flux densities $\varphi_1(t)$ and $\varphi_2(t)$ gathered in $\mathbf{U}(t)$ ($p = 2$) with only two sensors: one located in the middle of side BC and the other in the middle of side CD. These temperatures are stored in the output vector $\mathbf{Y}(t)$ ($q = 2$).

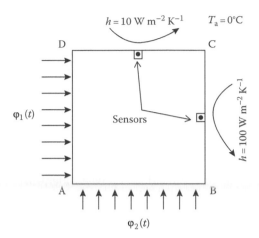

$h = 10 \ \text{W m}^{-2} \ \text{K}^{-1}$ $T_a = 0°\text{C}$

$\varphi_1(t)$

Sensors

$h = 100 \ \text{W m}^{-2} \ \text{K}^{-1}$

$\varphi_2(t)$

FIGURE 13.4
2D linear thermal system.

13.8.2 Identification and Validation of RM

An RM relative to both outputs is built following the method described in Section 13.5. This global RM is formed of two identified ERMs, each one relative to one of both heat flux densities. ERMs of order 3 have been found to be accurate enough, so that a global RM made of $n = 6$ dynamical uncoupled linear state equations is going to be used. This low number of degrees of freedom is very small compared to the original size of the DM ($N = 121$ equations).

13.8.3 Simulated Temperature Measurements and Mean Quadratic Errors

In order to simulate measurement errors, temperatures Y^{exact} computed with DM using φ^{exact} as test heat flux densities are altered with an additive Gaussian noise. The simulated noisy data Y^{sim} are expressed as

$$Y^{\text{sim}} = Y^{\text{exact}} + \omega\sigma$$

where
  σ is the standard deviation of the measurement errors that is supposed to be the same for all measurements
  ω is a random variable such as $-2.576 \le \omega \le 2.576$ that corresponds to the 99% confidence bounds for the temperature measurement

For all test cases analyzed here, we consider $\sigma = 0°\text{C}$ (errorless measurements), $\sigma = 0.1°\text{C}$ and $\sigma = 0.2°\text{C}$ (noisy data).
  Let us define σ_Y as the mean quadratic discrepancy between simulated data for inversion $Y_i^{\text{sim}}, i = 1, 2$ and corresponding outputs $\hat{Y}_i$ computed by the RM with the estimated inputs $\hat{\varphi}_i, i = 1, 2$:

$$\sigma_Y = \sqrt{\sum_{i=1}^{q} \sum_{j=1}^{N_t - n_f} \left(\hat{Y}_i(t_j) - Y_i^{\text{sim}}(t_j)\right)^2 \Big/ (q \times (N_t - n_f))} \tag{13.59}$$

where
  N_t is the number of time steps
  n_f is the number of FTS

Here we have $q = 2$ output data for inversion. Note that σ_Y is the rms of the residues between the model $\hat{Y}_i$ and the data Y_i^{sim}.

In this numerical example, the exact applied heat flux densities $\varphi_i^{\text{exact}}, i = 1, 2$, used as test signals to be retrieved, are of course perfectly known. In order to assess the quality of the inversion results, we may define σ_U as the mean quadratic discrepancy between exact inputs $\varphi_i^{\text{exact}}, i = 1, 2$, and estimated ones $\hat{\varphi}_i$:

$$\sigma_U = \sqrt{\sum_{i=1}^{p} \sum_{j=1}^{N_t - n_f} \left(\hat{\varphi}_i(t_j) - \varphi_i^{\text{exact}}(t_j) \right)^2 / (p \times (N_t - n_f))} \tag{13.60}$$

Here we have $p = 2$ inputs. Note that σ_U is the rms of the residues between identified values $\hat{\varphi}_i$ and exact values φ_i^{exact}. Of course, in real applications, σ_U is not known and σ_Y is the only available quantity.

13.8.4 Inversion Results

The RM of order 6 is used for the inversion. The heat flux densities $\varphi_i^{\text{exact}}, i = 1, 2$, used as test functions for the inverse problem, are shown in Figure 13.5. Temperatures computed by DM are depicted in Figure 13.6.

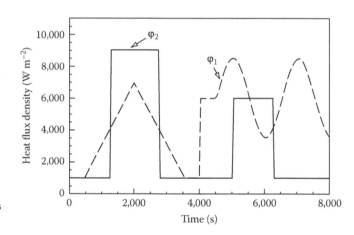

FIGURE 13.5
Heat flux densities used as test signals to be retrieved in the inverse problem.

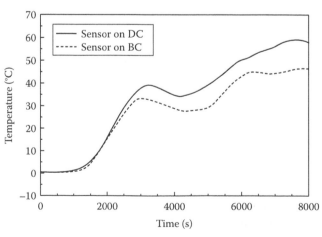

FIGURE 13.6
Temperature at sensors computed by DM.

TABLE 13.1

Inversion Results for Various Values of n_f

n_f	0	1	2	3	4	5
σ_Y (°C)	∞	1.12×10^{-2}	3.94×10^{-2}	6.68×10^{-2}	**0.104**	0.154
σ_U (W·m^{-2})	∞	1520	741	613	**609**	645

When no noise is added to computed temperatures ($\sigma = 0$°C), one future time step is needed ($n_f = 1$) to take into account lagging and damping effects of heat diffusion and to obtain physical results: $\sigma_Y = 2.04 \times 10^{-3}$°C and $\sigma_U = 452$ W m^{-2}.

The effect of noise on data is then studied. Temperatures computed with DM are altered with an additive noise ($\sigma - 0.1$°C). Table 13.1 shows values of σ_Y and σ_U for various values of n_f. For $n_f = 0$, $\sigma_Y = \infty$, meaning that the sensors sensitivities are not large enough to obtain physical results. Adding 1 future time step leads to $\sigma_Y \approx 10^{-2}$°C. One should not be misled by such an apparently good result: $\sigma_Y \ll \sigma$ in fact means that the noise on data signals has been fitted through the inversion process, which is not desirable. By increasing n_f to 2 and then 3, σ_Y grows but stays lower than the noise level σ. With $n_f = 4$, one obtains $\sigma_Y - 0.104$°C $\approx \sigma$, which is the goal we have to reach in order to get estimations in accordance with the noise level. This is the so-called *discrepancy principle*. Using $n_f = 5$ gives $\sigma_Y = 0.154$°C $> \sigma$, which is not the best we can aim for. The optimal value $n_f = 4$ is confirmed by the value of $\sigma_U = 609$ W m^{-2}, which is the lowest. Estimated heat flux densities for $\sigma = 0.1$°C and $n_f = 4$ are shown in Figure 13.7, along with the exact signals. Results are indeed quite satisfying. Thanks to the RM, the contribution of each input to the sensor temperature has been restored, although the time-dependent heat flux densities vary in a very different manner.

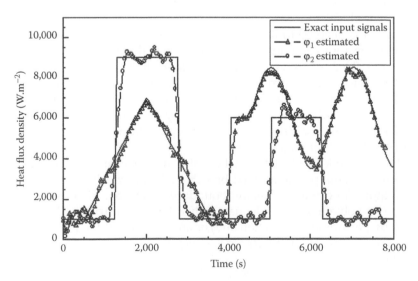

FIGURE 13.7
Exact and estimated heat flux densities. $\sigma = 0.1$°C, $n_f = 4$.

13.9 Modal Identification Method for Nonlinear Problems

13.9.1 Some Reduction Methods for Nonlinear Problems

Although some thermal systems may be described by a linear model, at least for a range of temperatures, many of them are governed by nonlinear equations. Model reduction techniques that can be applied to nonlinear problems are not as numerous as those developed for linear problems.

Among them, a commonly used approach in fluid mechanics is the proper orthogonal decomposition method coupled with a Galerkin projection of the partial differential equations (POD-G). POD is a powerful method, also known as Karhunen–Loève decomposition, and was introduced in fluid mechanics by Lumley in 1967 for the identification of coherent structures in turbulent flows. So far, the method has been widely used in fluid mechanics and heat transfer ((Holmes et al. 1997, Bergmann and Cordier 2008, Alonso et al. 2009), for instance, to cite but a few). The principle of the method consists in obtaining some orthogonal modes from the analysis of covariance matrices built with some data covering space and time domains. The data signal is then expanded on a truncation of the basis formed by these spatial POD modes. The time-varying coefficients of this decomposition carry the system dynamics. Classically, POD requires solving an eigenvalue problem relative to a two-point spatial correlation matrix, with time averaging (under the assumption of ergodic processes). However, for applications involving a large number of degrees of freedom, the spatial correlation matrix may become very large, and, hence, the computation of POD modes may become infeasible. To avoid this difficulty, Sirovich introduced the method of "snapshots," which has proved to be a powerful tool for the computation of the eigenmodes. The snapshot POD consists in solving an eigenvalue problem relative to a two-point temporal correlation matrix, with space averaging, before computing the spatial eigenmodes. For more details, the reader can see Sirovich (1987). Other approaches exist like the balanced POD (Willcox and Pereire 2002), which combines the POD with concepts from balanced realization. Note that a comparison of RMs built with the POD-Galerkin method and the MIM in the case of a 3D academic nonlinear diffusive system has been made in Balima et al. (2006).

Note: Chapter 14 is devoted to Karhunen–Loève decomposition methods.

The reduced basis approximation method based on Galerkin projection onto a Lagrange space (Rozza et al. 2008) relies on an underlying large-sized (N degrees of freedom) finite element model, which is an approximation to the considered infinite-dimensional PDEs, and on samples of solutions of this reference model for different occurrences of parameters of interest, such as boundary conditions, loads, physical properties, etc. A Galerkin projection onto an n-dimensional ($n \ll N$) Lagrange space associated with sampling parameters leads to an RM with n degrees of freedom.

The branch eigenmodes reduction method also uses a change of basis. It consists of solving a specific spectral problem called Branch problem, which concerns the advection–diffusion operator along with a Steklov boundary condition. This particular boundary condition allows to obtain a basis adequate to handle nonlinearities. The reduced basis is then obtained by performing a selection and/or an amalgam of the most dominant modes according to a particular criterion (temporal, energetic, . . .). The branch reduction method has notably been used in nonlinear heat diffusion, especially for solving inverse heat transfer problems (Videcoq et al. 2008) and also for building low-order models for advection–diffusion problems (Joly et al. 2008).

A goal-oriented, model-constrained optimization approach for model reduction that is able to target a particular output functional, span an applicable range of dynamic and parametric inputs, and respect the underlying governing equations of the system has been proposed in Bui-Thanh et al. (2007). The approach has therefore similarities with the MIM whose extension to nonlinear problems is presented in Section 13.9.2.

13.9.2 Extension of Modal Identification Method to Nonlinear Problems

The main idea is to work with RMs under the state space representation, as in the case of linear problems, but with extra nonlinear terms able to reproduce the nonlinearities in the local equations assumed to describe the behavior of the system. Hence, low-order models that will be identified take the following general form:

$$\dot{\mathbf{X}}(t) = \mathbf{F}\mathbf{X}(t) + \mathbf{C} + \mathbf{\Omega}\mathbf{\Psi}(\mathbf{X}(t)) + \mathbf{G}\mathbf{U}(t) \tag{13.61}$$

$$\mathbf{Y}(t) = \mathbf{H}\mathbf{X}(t) \tag{13.62}$$

where

- $\mathbf{X}(t) \in \mathbb{R}^n$ is a low-dimensioned state vector (typically, n will be ranging from 1 to 10).
- $\mathbf{F}\mathbf{X}(t)$ is a linear term. $\mathbf{F} \in \mathbb{R}^{n \times n}$ is a diagonal matrix, as the model is expressed under the modal form.
- $\mathbf{C} \in \mathbb{R}^n$ is a constant vector.
- $\mathbf{\Omega}\mathbf{\Psi}(\mathbf{X}(t))$ is a nonlinear term $\mathbf{\Psi}(\mathbf{X}(t)) \in \mathbb{R}^{n_\Psi}$ is the vector of nonlinearities, composed of nonlinear combinations of states $X_i(t)$, $1 \le i \le n$. The form of $\mathbf{\Psi}(\mathbf{X}(t))$ depends on the type of nonlinearities in the original equations assumed to be governing heat transfer in the actual system. The size n_Ψ of $\mathbf{\Psi}(\mathbf{X}(t))$ depends on both the kind of nonlinearities and the size n of state vector $\mathbf{X}(t)$, that is, the RM order. Matrix $\mathbf{\Omega} \in \mathbb{R}^{n \times n_\Psi}$ allows to distribute the contribution of $\mathbf{\Psi}$ on each one of the n state equations.
- $\mathbf{G}\mathbf{U}(t)$ is an input term, gathering the action of all thermal loads on the system: boundary conditions and internal heat sources. The input vector $\mathbf{U}(t) \in \mathbb{R}^p$ gathers thermal loads: prescribed heat flux densities (Neumann BCs), prescribed temperatures (Dirichlet BCs), ambient temperature, or temperature of external sources or sinks (convective or radiative exchanges), volumetric heat power of internal heat sources, whereas command matrix $\mathbf{G} \in \mathbb{R}^{n \times p}$ links these thermal loads to the state vector $\mathbf{X}(t)$.
- Matrix $\mathbf{H} \in \mathbb{R}^{q \times p}$ is an observation or output matrix allowing to link the outputs of interest, stored in vector $\mathbf{Y}(t) \in \mathbb{R}^q$, to the state vector $\mathbf{X}(t)$.

As for linear problems, the MIM for nonlinear problems aims at identifying low-order models through optimization techniques. Unknown components of matrices and vectors $\mathbf{F}, \mathbf{C}, \mathbf{\Omega}, \mathbf{G}, \mathbf{H}$ are estimated by solving a parameter estimation problem defined by an objective functional $\mathcal{J}_{\text{red}}$ based on the discrepancy between an output vector $\mathbf{Y}^{\text{data}}(t)$ of the system on the one hand and the corresponding output vector $\mathbf{Y}(t)$ of RM defined by Equations 13.61 and 13.62 on the other hand, when a specific input vector $\mathbf{U}^*(t)$ is applied:

$$\mathcal{J}_{\text{red}}(\mathbf{F}, \mathbf{C}, \mathbf{\Omega}, \mathbf{G}, \mathbf{H}) = \| \mathbf{Y}(t; \mathbf{F}, \mathbf{C}, \mathbf{\Omega}, \mathbf{G}, \mathbf{H}) - \mathbf{Y}^{\text{data}}(t) \|_{L_2}^2 \tag{13.63}$$

The objective functional $\mathcal{J}_{\text{red}}$ may take several different forms according to the involved heat transfer modes (diffusion, forced or natural convection, presence of radiative effects), the time-dependency of the problem (steady or unsteady heat transfer), and the number of independent thermal inputs. In the next section, we present the issues that arise in the general case of unsteady thermal diffusion with several independent inputs.

The input signals used to generate data for RM identification, gathered in vector $\mathbf{U}^*(t)$, must allow the system to react in large ranges of temperature levels and frequencies, in order, for RM, to adequately reproduce the system's nonlinearities when any other input signal $\mathbf{U}(t)$ is applied.

A good way to proceed is to use successive steps covering the considered working range, each one allowing reaching a steady state, in order to be able to identify the static characteristics of the system. For each steady state, a zero mean Gaussian white noise is added to ensure the data dynamical richness. Such signals have been used in numerical studies such as Balima et al. (2006) to identify an RM of 3D academic nonlinear diffusive system.

Such kind of signal is easy to use in a numerical framework; it may however be difficult to apply it from a practical point of view. In order to keep the main features discussed above, it is possible to replace random parts by sinusoidal functions with different magnitudes and frequencies (Girault et al. 2010).

Moreover, the RM must be able to return the proper resulting temperatures when both sources are heating simultaneously and independently, whatever the respective evolution of their intensities. It has been underlined in Section 13.5.3 that in the case of linear systems, the linear relationship between inputs and outputs allows to build an ERM for each input separately and then to use the superposition principle to get a global RM for the whole set of inputs (Videcoq and Petit 2001, Girault et al. 2003, 2006, 2008, Videcoq et al. 2003, Girault and Petit 2004). It is no longer possible in the case of nonlinear systems. The RM related to all inputs has to be built in a single step. A possible way to handle this issue is to use, in the case of p independent inputs, at least $p + 1$ data sets, the first p ones corresponding to each input acting separately, the others remaining inactive, and the last one relative to all inputs acting simultaneously (Girault et al. 2010). The functional $\mathcal{J}_{\text{red}}$ defined in Equation 13.63 therefore writes

$$\mathcal{J}_{\text{red}}(\mathbf{F}, \mathbf{C}, \mathbf{\Omega}, \mathbf{G}, \mathbf{H}) = \sum_{k=1}^{p+1} \sum_{i=1}^{q} \sum_{j=1}^{N_{t,k}^{id}} \left(Y_{ik}(t_j; \mathbf{F}, \mathbf{C}, \mathbf{\Omega}, \mathbf{G}, \mathbf{H}) - Y_{ik}^{\text{data}}(t_j) \right)^2 \qquad (13.64)$$

where the $p + 1$ data sets are thus defined as mentioned above:

$$\text{If } k = 1, \ldots, p: \begin{cases} U_{k,k} = U_{k,k}^* \neq 0 \\ U_{\ell,k} = 0, \quad \ell \neq k \end{cases}$$

$$\text{If } k = p + 1: U_{\ell,p+1} = U_{\ell,p+1}^* \neq 0 \quad \forall \ell = 1, \ldots, p$$

$N_{t,k}^{id}$ is the number of time steps in the kth data set, the t_j are the discretization times, and the $U_{\ell,k}^*$ are specific input signals such as those proposed previously, for the $\ell = 1, \ldots, p$ independent inputs and for the $k = 1, \ldots, p + 1$, data sets. Figure 13.8 summarizes the MIM for nonlinear problems when the objective functional is defined by Equation 13.64.

As in the MIM for linear problems, the objective functional $\mathcal{J}_{\text{red}}$ is first minimized for $n = 1$, leading to the identification of an RM of order 1. The value of n is then increased, and the minimization of $\mathcal{J}_{\text{red}}$, involving more unknown parameters, leads to RMs of higher

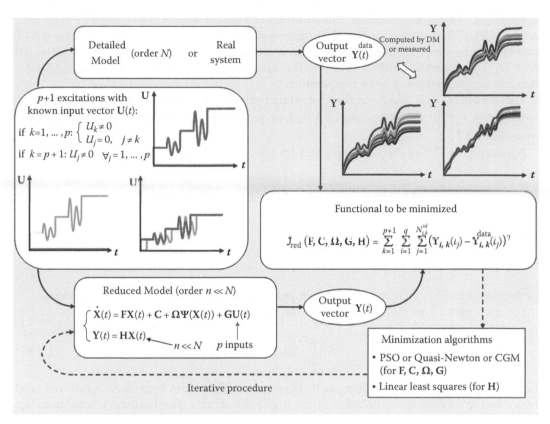

FIGURE 13.8
General scheme of MIM for nonlinear problems.

order. As in the linear case, two optimization techniques are used for the minimization of $\mathcal{J}_{red}$. $\mathbf{Y}(t)$ being nonlinear with respect to $\mathbf{F}$, $\mathbf{C}$, $\mathbf{\Omega}$, $\mathbf{G}$, an iterative method (conjugate gradient, quasi-Newton method, particle swarm optimization, genetic algorithm,...) is required to identify their components. $\mathbf{Y}(t)$ being linear with respect to matrix $\mathbf{H}$, components of $\mathbf{H}$ are obtained using OLS *at each iteration*.

13.10 Examples of MIM for Nonlinear Problems

13.10.1 Heat Diffusion with Radiative and Convective Boundary Conditions: Experimental Example

13.10.1.1 Local Governing Equations

The local equation governing heat diffusion with internal heat sources can be written as

$$\rho(M)C_p(M)\frac{\partial T}{\partial t}(M, t) = \vec{\nabla} \cdot (k(M)\vec{\nabla}T(M, t)) + \sum_{j=1}^{n_Q}\frac{Q_j(t)}{\text{Vol}_j}\chi_j(M) \tag{13.65}$$

where $T(M, t)$ is the temperature. The summation term corresponds to a number n_Q of heat sources. The jth source is uniformly distributed over a domain Ω_j whose volume is Vol$_j$ and has a time-varying strength $Q_j(t)$. $\chi_j(M)$ is defined as $\chi_j(M) = 1$ if $M \in \Omega_j$ and $\chi_j(M) = 0$ if $M \notin \Omega_j$. The thermal conductivity $k(M)$, the density $\rho(M)$, and the specific heat $C_p(M)$ are assumed to be independent of temperature.

We consider that both convective and radiative boundary conditions occur on some parts Γ_i^{CR} of the boundary and that prescribed heat flux densities occur on some other parts Γ_i^{Φ}.

For each part Γ_i^{CR} of the boundary and for any point M on Γ_i^{CR},

$$k(M)\frac{\partial T}{\partial n}(M, t) = h_i(T_a(t) - T(M, t)) + \varepsilon_i \sigma \left(T_a^4(t) - T^4(M, t)\right) \tag{13.66}$$

where
  h_i is the convective exchange coefficient
  ε_i is the emissivity

both related to Γ_i^{CR}. T_a is the ambient temperature, which can a priori vary with time.

For each part Γ_i^{Φ} of the boundary and for any point M on Γ_i^{Φ},

$$k(M)\frac{\partial T}{\partial n}(M, t) = \Phi_i(M, t) \tag{13.67}$$

Equations 13.66 and 13.67 allow us to keep generality in the boundary conditions and hence in the following formulation of the low-order model. The boundary conditions for the presented example will be given in Section 13.10.1.4.

13.10.1.2 State Space Representation

A formulation for a DM will be given here, from which the RM structure will then be defined.

Whatever the dimension of the problem, its geometry and the spatial discretization method (finite differences, finite volumes, finite elements,...), the energy Equation 13.65, and the associated boundary conditions (13.66) and (13.67) can be written as a set of nonlinear equations with constant coefficients matrices:

$$\dot{\mathbf{T}}(t) = \mathbf{A}\mathbf{T}(t) + \mathbf{B}_\mathbf{c}T_a(t) + \mathbf{R}[\mathbf{T}^4](t) + \mathbf{B}_\mathbf{r}T_a^4(t) + \mathbf{B}_\Phi\Phi(t) + \mathbf{B}_\mathbf{Q}\mathbf{Q}(t) \tag{13.68}$$

- $\mathbf{T}(t) \in \mathbb{R}^N$ is the vector of temperatures, function of time t, at the N discretization nodes, $\dot{\mathbf{T}}(t) \in \mathbb{R}^N$ its derivative with respect to time t.
- Matrix $\mathbf{A} \in \mathbb{R}^{N \times N}$ is the non-diagonal state matrix that connects temperatures at discretization nodes and contains diffusion terms as well as terms related to convective boundary conditions.
- Vector $\mathbf{B}_\mathbf{c} \in \mathbb{R}^N$ is associated with convective boundary conditions and links discretization nodes to the ambient temperature T_a.
- Matrix $\mathbf{R} \in \mathbb{R}^{N \times N}$ is associated with radiative boundary conditions and multiplies the vector $[\mathbf{T}^4](t) \in \mathbb{R}^N$ gathering temperatures at fourth power.

- Vector $\mathbf{B_r} \in \mathbb{R}^N$ is associated with radiative boundary conditions and links discretization nodes to the ambient temperature T_a at fourth power.

- Matrix $\mathbf{B_\Phi} \in \mathbb{R}^{N \times n_\Phi}$ links discretization nodes to prescribed heat flux densities gathered in vector function $\mathbf{\Phi}(t) = [\Phi_1(t) \cdots \Phi_{n_\Phi}(t)]^T \in \mathbb{R}^{n_\Phi}$.

- Matrix $\mathbf{B_Q} \in \mathbb{R}^{N \times n_Q}$ links discretization nodes to the internal heat sources gathered in vector function $\mathbf{Q}(t) = [Q_1(t) \cdots Q_{n_Q}(t)]^T \in \mathbb{R}^{n_Q}$.

- An output or observation matrix $\mathbf{C_{obs}} \in \mathbb{R}^{q_o \times N}$ allows to select q_o temperatures in the whole temperature field $\mathbf{T}(t)$ and to store them in vector function $\mathbf{Y_o^{DM}}(t) \in \mathbb{R}^{q_o}$:

$$\mathbf{Y_o^{DM}}(t) = \mathbf{C_{obs}}\mathbf{T}(t) \tag{13.69}$$

Let us call $p = n_\Phi + n_Q$ the total number of thermal inputs.

Let be $\mathbf{U}(t) = [\mathbf{\Phi}(t) \quad \mathbf{Q}(t)]^T$. Vector $\mathbf{U} \in \mathbb{R}^p$ is the input or command vector.

Let also be $\mathbf{B} = [\mathbf{B_\Phi} \ \mathbf{B_Q}]$. Matrix $\mathbf{B} \in \mathbb{R}^{N \times p}$ is called the input or command matrix.

In the following, we will consider constant ambient temperature T_a. We define the constant vector $\mathbf{V_c} \in \mathbb{R}^N$ such as $\mathbf{V_c} = \mathbf{B_c}T_a + \mathbf{B_r}T_a^4$. Then, the state space representation takes the following form:

$$\dot{\mathbf{T}}(t) = \mathbf{A}\mathbf{T}(t) + \mathbf{V_c} + \mathbf{R}[\mathbf{T^4}](t) + \mathbf{B}\mathbf{U}(t) \tag{13.70}$$

$$\mathbf{Y_o^{DM}}(t) = \mathbf{C_{obs}}\mathbf{T}(t) \tag{13.71}$$

Equations 13.70 and 13.71 constitute the structure of a DM of the system, called *state space representation*.

Note that the previous form is given for a better understanding of the following developments, but no DM or no solution of a DM has been used in the example described in Section 13.10.1.

13.10.1.3 Structure of Low-Order Model Equations

Let us consider the eigenvalue problem associated with matrix $\mathbf{A}$ of Equation 13.70. Let us call $\tilde{\mathbf{F}} \in \mathbb{R}^{N \times N}$ the diagonal matrix whose components are the N eigenvalues of $\mathbf{A}$, and $\mathbf{M} \in \mathbb{R}^{N \times N}$ the matrix whose columns are eigenvectors of $\mathbf{A}$.

Let us suppose that we operate a change of variables:

$$\mathbf{T}(t) = \mathbf{M}\tilde{\mathbf{X}}(t) \tag{13.72}$$

Then,

$$\mathbf{R}[\mathbf{T^4}](t) = \mathbf{R}\left[T_1^4(t) \cdots T_N^4(t)\right]^T = \mathbf{R}\left[\left(\sum_{j=1}^N M_{1j}\tilde{X}_j(t)\right)^4 \cdots \left(\sum_{j=1}^N M_{Nj}\tilde{X}_j(t)\right)^4\right]^T$$

$$= \mathbf{R}\mathbf{E}\mathbf{\Psi}(\tilde{\mathbf{X}}(t))$$

where

- $\mathbf{E} \in \mathbb{R}^{N \times N_\Psi}$ is a matrix.
- $\mathbf{\Psi}(\tilde{\mathbf{X}}(t)) \in \mathbb{R}^{N_\Psi}$ is a vector gathering nonlinear terms $\tilde{X}_i(t) \times \tilde{X}_j(t) \times \tilde{X}_k(t) \times \tilde{X}_l(t)$, $1 \leq i \leq j \leq k \leq l \leq N$ and whose dimension is $N_\Psi = N(N+1)(N+2)(N+3)/24$.

One has $\tilde{\mathbf{F}} = \mathbf{M}^{-1}\mathbf{AM}$. In addition, let us call $\tilde{\mathbf{C}} = \mathbf{M}^{-1}\mathbf{V_c}, \tilde{\boldsymbol{\Omega}} = \mathbf{M}^{-1}\mathbf{RE}, \tilde{\mathbf{G}} = \mathbf{M}^{-1}\mathbf{B}$, and $\tilde{\mathbf{H}}_\mathbf{o} = \mathbf{C_{obs}}\mathbf{M}$. We get

$$\dot{\tilde{\mathbf{X}}}(t) = \tilde{\mathbf{F}}\tilde{\mathbf{X}}(t) + \tilde{\mathbf{C}} + \tilde{\boldsymbol{\Omega}}\boldsymbol{\Psi}(\tilde{\mathbf{X}}(t)) + \tilde{\mathbf{G}}\mathbf{U}(t) \tag{13.73}$$

$$\mathbf{Y}_\mathbf{o}^{\mathrm{DM}}(t) = \tilde{\mathbf{H}}_\mathbf{o}\tilde{\mathbf{X}}(t) \tag{13.74}$$

Of course, such a model is not convenient at all because of the size ($N \times N_\Psi$) of matrix $\tilde{\boldsymbol{\Omega}}$.

The idea is now to keep the same structure and to search for a model with reduced state vector $\mathbf{X}$ of size $n \ll N$ and reduced associated matrices, so that the low-order model will be written as

$$\dot{\mathbf{X}}(t) = \mathbf{FX}(t) + \mathbf{C} + \boldsymbol{\Omega}\boldsymbol{\Psi}(\mathbf{X}(t)) + \mathbf{GU}(t) \tag{13.75}$$

$$\mathbf{Y}_\mathbf{o}(t) = \mathbf{H}_\mathbf{o}\mathbf{X}(t) \tag{13.76}$$

where matrix $\mathbf{F} \in \mathbb{R}^{n \times n}$ is a diagonal matrix, vector $\mathbf{C} \in \mathbb{R}^n$, matrix $\boldsymbol{\Omega} \in \mathbb{R}^{n \times n_\Psi}$, vector $\boldsymbol{\Psi}(\mathbf{X}(t)) \in \mathbb{R}^{n_\Psi}$ gathering nonlinear terms $X_i(t) \times X_j(t) \times X_k(t) \times X_l(t)\ 1 \leq i \leq j \leq k \leq l \leq n$, is of size $n_\Psi = n(n+1)(n+2)(n+3)/24$, command matrix $\mathbf{G} \in \mathbb{R}^{n \times p}$, and observation matrix $\mathbf{H}_\mathbf{o} \in \mathbb{R}^{q_o \times n}$.

Of course, $\mathbf{Y}_\mathbf{o}(t) \in \mathbb{R}^q$ should be as close as possible to $\mathbf{Y}_\mathbf{o}^{\mathrm{DM}}(t)$ whatever the input vector $\mathbf{U}(t)$.

13.10.1.4 Experimental Apparatus

The aim of the setup is to show the ability of RMs built by MIM to be used in a practical application involving two heat sources and many industrial constraints: nonlinear behavior, 3D configuration, high cooling (Girault et al. 2010).

The heterogeneous studied system is a parallelepiped block of steel (0.164 m × 0.098 m × 0.098 m). The block is drilled in its length by two circular ducts (0.016 m in diameter) as shown in Figure 13.9. The external surfaces of the block are insulated with ceramic sheets (0.024 m thick), except the largest vertical face Γ_1, painted in black in order to increase the

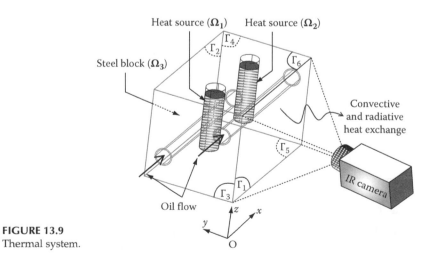

FIGURE 13.9
Thermal system.

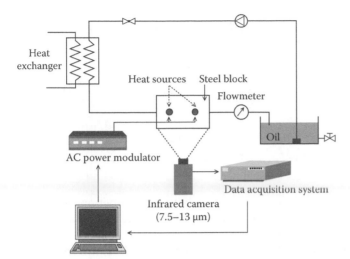

FIGURE 13.10
Schematic view of the experiment.

radiation heat transfer. Two cylindrical heat sources (20 mm in diameter and 60 mm long) are placed in the block. The heat sources are driven by a power modulator, and the heat extraction is realized with a high-rate oil flow in a closed circuit with a water/oil heat exchanger (cf. Figure 13.10). The two circular ducts, acting as coolers, are made of copper. Hence, the four annular cross sections corresponding to their inlet and outlet are forming thermal bridges.

Boundary conditions are described in the following:

- On the black painted surface Γ_1, which is not insulated, both radiative and convective exchanges do occur.
- The five other sides are theoretically insulated, but in practice, there are convective heat losses.

In fact, heat transfer in the steel block can be modeled by local Equation 13.65, and all boundary conditions can be cast in the form of Equation 13.66. When building a classical model based on the discretization of heat transfer balance equations, an important issue is to estimate proper values of heat exchange coefficients, contact resistances, and emissivities appearing in (13.66). One of the main interesting features of "experimental modeling" is to embed such quantities in the model without requiring the definition of proper values. In fact, when looking at terms of the right hand side of boundary Equation 13.66 and their counterparts in the DM state Equation 13.70 and RM state Equation 13.75, it appears that for each boundary,

- All terms of the form $h_i T_a + \varepsilon_i \sigma T_a^4$, which would appear explicitly in constant vector $\mathbf{V_c}$ of Equation 13.70, will be embedded in the constant vector $\mathbf{C}$ of Equation 13.75.
- All terms of the form $h_i T(M, t)$ that would appear in $\mathbf{AT}(t)$ will be embedded in the linear term $\mathbf{FX}(t)$.
- All terms of the form $\varepsilon_i \sigma T^4(M, t)$ that would appear in $\mathbf{R(T^4)}(t)$ will be embedded in the nonlinear term $\mathbf{\Omega\Psi(X}(t))$.

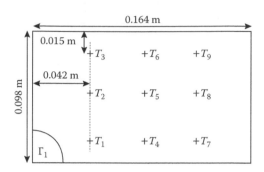

FIGURE 13.11
Temperature data on boundary surface Γ_1.

Moreover, thermal conductivities and contact resistances (between copper and steel, for instance) are also embedded in the RM. Note that the nonlinear term $\Omega\Psi(\mathbf{X}(t))$ also takes into account minor nonlinearities in such quantities.

Surface temperatures T_1 to T_9, whose location is shown in Figure 13.11, are measured using an infrared camera positioned 1 m away from Γ_1. Thanks to preliminary steady-state experiments, the standard deviation of the measurement errors has been estimated at $\sigma^m \approx 0.1$ K for the range 273–773 K. The nine pixels corresponding to $T_1 - T_9$ are extracted from the infrared picture at each time step. Three K-type thermocouples are placed inside the block at 10 mm from the boundary Γ_6 for the measurement of three internal temperatures T_{10}, T_{11}, and T_{12} as depicted in Figure 13.12. We wish to build a low-order model for the temperature vector $\mathbf{Y_o} = [T_1 \cdots T_{12}]^T (q_o = 12)$.

13.10.1.5 Low-Order Model Identification

Three input–output data sets are used in the RM identification procedure:

a. Source 1 is the only active source.
b. Source 2 is the only active source.
c. Both sources are active.

The input signals used as heating powers are typically those shown in Figure 13.8. Resulting temperatures at points 1–12 are recorded by the infrared camera. The number of time steps $N_{t,k}^{id}$, $k = 1, 2, 3$, is equal to 1078, 1075, and 1171 for data set (a), (b), and (c), respectively. The time step is $\Delta t^{id} = 20$ s.

A series of RMs (from order $n = 1$ to 4) has been constructed using MIM.

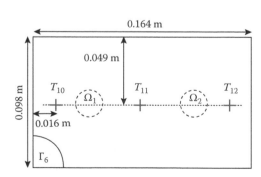

FIGURE 13.12
Temperature data inside the block.

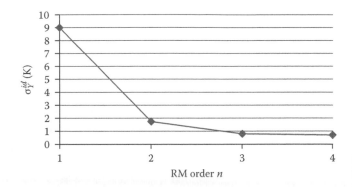

FIGURE 13.13

MIM results: σ_Y^{id} function of RM order n.

We define σ_Y^{id} as the mean quadratic discrepancy (i.e., rms of the residues) between measured temperatures Y_o^* and those computed with the identified RM of order n, Y_o:

$$
\sigma_Y^{id} = \sqrt{\mathcal{J}_{\text{red}} \Big/ \Big(q_0 \times \sum_{k=1}^{p+1} N_{t,k}^{id} \Big)}
$$

$$
= \sqrt{\sum_{k=1}^{p+1} \sum_{i=1}^{q_0} \sum_{j=1}^{N_{t,k}^{id}} \Big(Y_{o,i,k}(t_j) - Y_{o,i,k}^*(t_i) \Big)^2 \Big/ \Big(q_0 \times \sum_{k=1}^{p+1} N_{t,k}^{id} \Big)} \qquad (13.77)
$$

Figure 13.13 shows that σ_Y^{id} decreases with the RM order n. RMs of order 3 and 4 are almost equivalent. The mean quadratic value (i.e., rms value) of the temperature variation in the data is

$$
\sqrt{\sum_{k=1}^{p+1} \sum_{i=1}^{q_0} \sum_{j=1}^{N_{t,k}^{id}} \Big(Y_{o,i,k}^*(t_j) - Y_{o,i,k}^*(t_j) \Big)^2 \Big/ \Big(q_0 \times \sum_{k=1}^{p+1} N_{t,k}^{id} \Big)}
$$

This quantity is here equal to about 101 K.

The value $\sigma_Y^{id} \approx 0.8\,\text{K}$ obtained with RMs of order 3 and 4 is very small compared to it.

13.10.2 Coupling of Reduced Models in Forced Heat Convection: Numerical Example

13.10.2.1 Problem Description

In this section, an advection–diffusion problem is studied: forced heat convection is considered with an incompressible, stationary, laminar 2D flow over a backward-facing step with a time-varying heat flux density applied upstream of the step (Rouizi et al. 2010), as shown in Figure 13.14. The upstream height is $h = 1$ cm, and the step height is also $h = 1$ cm. The fluid, assumed to be Newtonian, is air with the following constant properties: dynamic viscosity $\mu = 1.81 \times 10^{-5}$ kg $(\text{ms})^{-1}$, density $\rho = 1.205$ kg m^{-3}, specific heat $C_p = 1005$ J $(\text{kg K})^{-1}$, and thermal conductivity $k = 0.0262$ W $(\text{mK})^{-1}$. u_1 and u_2 are respectively the streamwise and transverse velocity components.

The Reynolds number Re is defined as $Re = U_\infty 2h/v$ where U_∞ is the inlet mean velocity and $v = \mu/\rho$ is the kinematic viscosity of the fluid. The Reynolds number is based upon the hydraulic diameter $2h$ of the inlet channel. The flow entering the channel is assumed to be

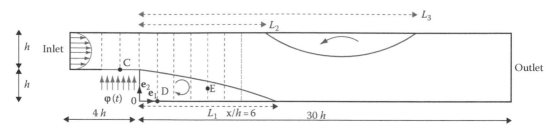

FIGURE 13.14
Backward-facing step.

laminar, fully developed, and is given by a parabola for the u_1 velocity component. An outflow boundary condition is used, assuming that the flow is fully developed at the outlet section. On all other boundaries, a null velocity is prescribed. At the entrance of the pipe, the fluid temperature is $T_\infty = 300$ K. A uniform heat flux varying in time $\varphi(t)$ is located at $x_1 \in [-2h; 0]$ and $x_2 = h$, while the other walls are treated as adiabatic surfaces. The thermal initial condition is $T(t=0) = 300$ K. The range [100;800] for the Reynolds number has been chosen to make sure the flow remains stable (Rouizi et al. 2010). The computational fluid dynamics (CFD) software (Fluent® 6.3.26) has been used to carry out computations. This is our DM. Grid independence tests have been performed using several grid densities, and the reattachment location on the stepped wall has been used as the criterion. In the present study, 144,247 nodes have been used for the computations. References may be found in Rouizi et al. (2010), especially those concerning the validation of our Fluent model against experimental and numerical studies. For all Reynolds numbers used, there is a main recirculation region, whose length L_1 increases with the Reynolds number. For a Reynolds number equal to 400, a second recirculation bubble appears attached to the upper wall of the channel between L_2 and L_3. The required CPU time for flow simulations was about 2 h on a dual-core bi-processor AMD Opteron 2.2 GHz with 3 GB of RAM on a HP DL 145G2 data processing server.

13.10.2.2 Local Equations and State Space Representation

The governing equations of the problem are the continuity, momentum, and energy equations:

$$\vec{\nabla} \cdot \vec{V} = 0 \tag{13.78}$$

$$\frac{\partial \vec{V}}{\partial t} + (\vec{V} \cdot \vec{\nabla})\vec{V} = \frac{\vec{\nabla}p}{\rho} + \nu \Delta \vec{V} \tag{13.79}$$

$$\frac{\partial T}{\partial t} + \vec{V} \cdot \vec{\nabla}T = \alpha \Delta T \tag{13.80}$$

where
  $\vec{V}$ is the velocity vector
  p is the pressure
  T is the temperature
  $\alpha = k/\rho C_p$ is the thermal diffusivity

We now present a matrix formulation of the discrete problem arising from Equations 13.78, 13.79, 13.80 and associated boundary conditions. We consider spatial meshes for velocity components, temperature and pressure, and call:

- $\mathbf{V} = [(V_i), i = 1, \ldots, 2N_V]^T = [(u_{1i} u_{2i}), i = 1, \ldots, N_V]^T$ the vector of size $2N_V$ containing the velocity components u_1 and u_2 at the discretization points
- $\mathbf{T} = [(T_i), i = 1, \ldots, N_T]^T$, the vector of size N_T containing the temperature value at the discretization points
- $\mathbf{P} = [(P_i), i = 1, \ldots, N_P]^T$, the vector of size N_P containing the pressure value at the discretization points

where N_V, N_T, and N_P depend on the chosen discretization method and on the mesh used.

Let us consider the case of a spatial discretization scheme involving, for any equation written for a specific node, all the other nodes in the domain. In such case, the discrete form of the nonlinear term $(\vec{V} \cdot \vec{\nabla})\vec{V}$ uses all possible products $V_i V_j, 1 \le i \le j \le 2N_V$. The state space equation relative to Equations 13.78 and 13.79 will write

$$\dot{\mathbf{V}} = \mathbf{A_V V} + \mathbf{Q_V \Psi(V)} + \mathbf{B_V} Re \tag{13.81}$$

where

- $\mathbf{A_V} \in \mathbb{R}^{2N_V \times 2N_V}$. $\mathbf{A_V V}$ is the discrete form of the linear diffusion term $\nu \Delta \vec{V}$ of Equation 13.79.
- $\mathbf{\Psi(V)} \in \mathbb{R}^{N_\Psi}$ is a vector of size $N_\Psi = 2N_V(2N_V + 1)/2$ gathering all possible products $V_i V_j, 1 \le i \le j \le 2N_V$.
- $\mathbf{Q_V} \in \mathbb{R}^{2N_V \times N_\Psi}$.
  In $\mathbf{Q_V \Psi(V)}$ are included the discrete form of the nonlinear advection term $(\vec{V} \cdot \vec{\nabla})\vec{V}$, which uses all possible quadratic terms of the velocity components products $V_i V_j, 1 \le i \le j \le 2N_V$. The discrete form of the pressure term $-\vec{\nabla}p/\rho$ in Equation 13.79, taking into account the continuity Equation 13.78, may also be expressed as a quadratic function of velocity components and is also included in the term $\mathbf{Q_V \Psi(V)}$.
- $\mathbf{B_V} \in \mathbb{R}^{2N_V}$. $\mathbf{B_V} Re$ is the discrete form of the flow boundary condition.

Although the components of matrices $\mathbf{A_V}$, $\mathbf{Q_V}$, and $\mathbf{B_V}$ depend on the chosen discretization method and mesh, only the structure of Equation 13.81 is of interest for our low-order model identification method.

In a same manner, the space discretization of the energy Equation 13.80 with associated boundary conditions leads to the following state space equation:

$$\dot{\mathbf{T}} = \mathbf{A_T T} + \mathbf{Q_T \Pi(V, T)} + \mathbf{B_T} \varphi(t) \tag{13.82}$$

where

- $\mathbf{A_T} \in \mathbb{R}^{N_T \times N_T}$. $\mathbf{A_T T}$ corresponds to the discrete form of the diffusion term $\alpha \Delta T$.
- $\mathbf{\Pi(V, T)} \in \mathbb{R}^{N_\Pi}$ where $N_\Pi = 2N_V N_T$ gathers all possible products $V_i T_j$, $1 \le i \le 2N_V, 1 \le j \le N_T$.
- $\mathbf{Q_T} \in \mathbb{R}^{N_T \times N_\Pi}$. $\mathbf{Q_T \Pi(V, T)}$ contains the discrete form of the transport term $\vec{V} \cdot \vec{\nabla} T$ of Equation 13.80.
- $\mathbf{B_T} \in \mathbb{R}^{N_T}$ is the input vector that allows to apply the heat flux density $\varphi(t)$.

Two kinds of RMs are developed in the following:

- The fluid reduced model (FRM) relative to steady fluid mechanics only, dedicated to the computation of the velocity field, for any constant Reynolds number Re in the range [100; 800].
- The thermal coupled reduced model (TCRM) designed to calculate temperature for a time-varying flux $\varphi(t)$ and for different values of $Re \in [300; 800]$. The TCRM is therefore weakly coupled to the FRM.

13.10.2.3 Form of the Fluid Reduced Model

We consider q_V points among the N_V ones of the velocity mesh. An output vector $\mathbf{Y}_V^{DM} \in \mathbb{R}^{2q_V}$ is introduced, allowing us to observe the velocity components u_1 and u_2 at each one of the q_V selected points:

$$\mathbf{Y}_V^{DM} = \mathbf{C}_V \mathbf{V} \tag{13.83}$$

where $\mathbf{C}_V \in \mathbb{R}^{2q_V \times 2N_V}$ is an output matrix.

Let us consider the eigenvalue problem associated with matrix $\mathbf{A}_V$. We call $\tilde{\mathbf{F}}_V \in \mathbb{R}^{2N_V \times 2N_V}$ the diagonal matrix containing the eigenvalues of $\mathbf{A}_V$ and $\mathbf{M}_V \in \mathbb{R}^{2N_V \times 2N_V}$ the matrix whose columns are eigenvectors of $\mathbf{A}_V$. Hence, $\tilde{\mathbf{F}}_V = \mathbf{M}_V^{-1} \mathbf{A}_V \mathbf{M}_V$. We now introduce in (13.81) and (13.83) the change of basis:

$$\mathbf{V} = \mathbf{M}_V \tilde{\mathbf{Z}} \tag{13.84}$$

Then $\quad \boldsymbol{\Psi}(\mathbf{V}) = \boldsymbol{\Psi}(\mathbf{M}_V \tilde{\mathbf{Z}}) = \mathbf{L}_V \boldsymbol{\Psi}(\tilde{\mathbf{Z}}) \quad$ where $\quad \mathbf{L}_V \in \mathbb{R}^{N_\Psi \times N_\Psi}$. Defining $\quad \tilde{\boldsymbol{\Gamma}}_V = \mathbf{M}_V^{-1} \mathbf{Q}_V \mathbf{L}_V$, $\tilde{\mathbf{G}}_V = \mathbf{M}_V^{-1} \mathbf{B}_V, \tilde{\mathbf{H}}_V = \mathbf{C}_V \mathbf{M}_V$, we get

$$\dot{\tilde{\mathbf{Z}}} = \tilde{\mathbf{F}}_V \tilde{\mathbf{Z}} + \tilde{\boldsymbol{\Gamma}}_V \boldsymbol{\Psi}(\tilde{\mathbf{Z}}) + \tilde{\mathbf{G}}_V Re \tag{13.85}$$

$$\mathbf{Y}_V^{DM} = \tilde{\mathbf{H}}_V \tilde{\mathbf{Z}} \tag{13.86}$$

The FRM that we are going to identify has a similar form, but with a new state vector $\mathbf{Z}$, whose size n_V is much lower than the one in the original model ($n_V \ll 2N_V$). Hence, the FRM has the following structure, where the steady flow is considered ($\mathbf{F}_V^{-1}$ can then be embedded in both $\boldsymbol{\Gamma}_V$ and $\mathbf{G}_V$):

$$0 = \mathbf{Z} + \boldsymbol{\Gamma}_V \boldsymbol{\Psi}(\mathbf{Z}) + \mathbf{G}_V Re \tag{13.87}$$

$$\mathbf{Y}_V = \mathbf{H}_V \mathbf{Z} \tag{13.88}$$

$\boldsymbol{\Psi}(\mathbf{Z}) \in \mathbb{R}^{n_\Psi}$ where $n_\Psi = n_V(n_V + 1)/2$ gathers all possible products $Z_i Z_j, 1 \leq i \leq j \leq n_V$. $\mathbf{Y}_V$ is the FRM output vector, which we want to be a good approximation of the DM output vector $\mathbf{Y}_V^{DM}$. Low-order matrices and vectors $\boldsymbol{\Gamma}_V \in \mathbb{R}^{n_V \times n_\Psi}$, $\mathbf{G}_V \in \mathbb{R}^{n_V}$, and $\mathbf{H}_V \in \mathbb{R}^{2q_V \times n_V}$ will be then identified.

13.10.2.4 Form of the Thermal Coupled Reduced Model

Forced convection is now considered, Equation 13.82 being weakly coupled to the Equation 13.81.

An output vector $\mathbf{Y_T^{DM}} \in \mathbb{R}^{q_T}$ allows the selection of some temperatures among the N_T contained in $\mathbf{T}$:

$$\mathbf{Y_T^{DM}} = \mathbf{C_T T} \tag{13.89}$$

where $\mathbf{C_T} \in \mathbb{R}^{q_T \times N_T}$ is the output matrix.

In addition to the change of basis (13.84) introduced in both (13.81) and (13.82), we consider the eigenvalue problem associated with matrix $\mathbf{A_T}$. We call $\tilde{\mathbf{F}}_T \in \mathbb{R}^{N_T \times N_T}$ the diagonal matrix containing the eigenvalues of $\mathbf{A_T}$ and $\mathbf{M_T} \in \mathbb{R}^{N_T \times N_T}$ the matrix whose columns are eigenvectors of $\mathbf{A_T}$. Hence, $\tilde{\mathbf{F}}_T = \mathbf{M_T}^{-1} \mathbf{A_T M_T}$.

We now introduce in (13.82) and (13.89) the change of basis:

$$\mathbf{T} = \mathbf{M_T} \tilde{\mathbf{X}} \tag{13.90}$$

Then $\mathbf{\Pi(V, T)} = \mathbf{\Pi(M_V \tilde{Z}, M_T \tilde{X})} = \mathbf{L_T \Pi(\tilde{Z}, \tilde{X})}$ where $\mathbf{L_T} \in \mathbb{R}^{N_\Pi \times N_\Pi}$. Defining $\tilde{\mathbf{\Gamma}}_T = \mathbf{M_T}^{-1} \mathbf{Q_T L_T}$, $\tilde{\mathbf{G}}_T = \mathbf{M_T}^{-1} \mathbf{B_T}, \tilde{\mathbf{H}}_T = \mathbf{C_T M_T}$, we get

$$\dot{\tilde{\mathbf{X}}} = \tilde{\mathbf{F}}_T \tilde{\mathbf{X}} + \tilde{\mathbf{\Gamma}}_T \mathbf{\Pi(\tilde{Z}, \tilde{X})} + \tilde{\mathbf{G}}_T \varphi(t) \tag{13.91}$$

$$\mathbf{Y_T^{DM}} = \tilde{\mathbf{H}}_T \tilde{\mathbf{X}} \tag{13.92}$$

The TCRM that we are going to identify has a similar form, but with a thermal state vector $\mathbf{X}$ whose size n_T is much lower than the one in the original model ($n_T \ll N_T$), and a fluid state vector $\mathbf{Z}$ of size $n_V \ll 2N_V$:

$$\dot{\mathbf{X}} = \mathbf{F_T X} + \mathbf{\Gamma_T \Pi(Z, X)} + \mathbf{G_T} \varphi(t) \tag{13.93}$$

$$\mathbf{Y_T} = \mathbf{H_T X} \tag{13.94}$$

$\mathbf{\Pi(Z, X)} \in \mathbb{R}^{n_\Pi}$ where $n_\Pi = n_V n_T$ gathers all possible products $Z_i X_j$, $1 \leq i \leq n_V$, $1 \leq j \leq n_T$. $\mathbf{Y_T}$ is the TCRM output vector, which we want to be a good approximation of the DM output vector $\mathbf{Y_T^{DM}}$. Low-order matrices and vectors $\mathbf{F_T} \in \mathbb{R}^{n_T \times n_T}$ (diagonal), $\mathbf{\Gamma_T} \in \mathbb{R}^{n_T \times n_\Pi}$, $\mathbf{G_T} \in \mathbb{R}^{n_T}$, and $\mathbf{H_T} \in \mathbb{R}^{q_T \times n_T}$ will be then identified.

Because of the one-way coupling between temperature and velocity, the identification procedure of the TCRM is performed through two stages:

- One first identifies the FRM defined by Equations 13.87 and 13.88 and characterized by the matrices $\mathbf{\Gamma_V}$, $\mathbf{G_V}$, and $\mathbf{H_V}$. This enables to compute the reduced fluid state vector $\mathbf{Z}$ for a given Reynolds number.
- In the second stage, one identifies the matrices $\mathbf{F_T}$, $\mathbf{\Gamma_T}$, $\mathbf{G_T}$, and $\mathbf{H_T}$ of the TCRM (Equations 13.93 and 13.94). The TCRM also depends on the reduced fluid state vector $\mathbf{Z}$ of size n_V (and not on the velocity field of size $2N_V$).

TABLE 13.2

Matrices to Be Identified, Functional $\mathcal{J}(\theta)$ to Be Minimized, Mean Quadratic Error σ, and Maximum Error ε for Both FRM and TCRM

Model	θ	$\mathcal{J}(\theta)$	σ	ε
FRM	$\mathbf{\Gamma_V, G_V, H_V}$	$\mathcal{J}_V(\theta) = \sum_{i=1}^{2q_V} \sum_{k=1}^{N_R} \left(Y_{Vik} - Y_{Vik}^{DM}\right)^2$ $= \sum_{\ell=1}^{2} \mathcal{J}_{u_\ell}(\theta)$	$\sigma_{u_\ell} = \sqrt{\dfrac{\mathcal{J}_{u_\ell}(\theta)}{q_V \times N_R}}$, $\ell = 1, 2$	$\varepsilon_{u_\ell} = \max_{i,k} \left|Y_{Vik} - Y_{Vik}^{DM}\right|_{u_\ell}$ $\ell = 1, 2$
TCRM	$\mathbf{F_T, \Gamma_T, G_T, H_T}$	$\mathcal{J}_T(\theta) = \sum_{i=1}^{q_T} \sum_{j=1}^{N_t} \sum_{k=1}^{N_R} \left(Y_{Tijk} - Y_{Tijk}^{DM}\right)^2$	$\sigma_T = \sqrt{\dfrac{\mathcal{J}_T(\theta)}{q_T \times N_t \times N_R}}$	$\varepsilon_T = \max_{i,j,k} \left|Y_{Tijk} - Y_{Tijk}^{DM}\right|$

13.10.2.5 RM Identification

In Table 13.2 are given, for each kind of RM, the matrices whose components form the vector θ of parameters to be identified as well as the expressions of the objective functional $\mathcal{J}(\theta)$, the mean quadratic error σ associated with $\mathcal{J}(\theta)$, and the maximum error ε. N_t and N_R are respectively the number of time steps and Reynolds numbers used in the identification. The particle swarm optimization method (Clerc 2005) has been used for solving the minimization problems.

13.10.2.5.1 Fluid Reduced Model

In the present work, as we consider having no prior information on the effect of the Reynolds number on flow patterns, the intuitive choice is to use data velocity fields corresponding to a regular partition of the interval [100;800]. Therefore, eight velocity fields are computed for Reynolds numbers from 100 to 800 by steps of 100 using the DM, involving 144,247 mesh points. All nodes whose x_1 coordinate is included in the range $[-2h; 30h]$ are included in the reduction process. This gives a total of 139,677 nodes. Using these fields as data, a series of seven FRM of orders $n_V = 1$ to 7 can be obtained through the identification procedure. Identification results are summarized in Figures 13.15 and 13.16, showing, respectively, $\mathcal{J}_V(\theta)$ and $\sigma_{u_\ell}, \ell = 1, 2$ with respect to the FRM order n_V. The FRM of order $n_V = 7$ is going to be tested for other values of Re in the following.

The aim is now to validate the FRM of order $n_V = 7$, finding out if it is able to reproduce with accuracy the outputs of the DM, for test Reynolds numbers $150, 250, \ldots, 750$ that have not been used in the identification procedure. Table 13.3 presents the validation results, which assess the FRM ability to reproduce the DM results with accuracy, as shown by the values of mean quadratic errors $\sigma_{u_\ell}, \ell = 1, 2$, maximum errors $\varepsilon_{u_\ell}, \ell = 1, 2$, as well as by the length L_1/h of the main recirculating region and the locations of detachment L_2/h and reattachment L_3/h of the recirculation bubble on the upper wall for $Re \geq 400$.

Figures 13.17 and 13.18 show respectively the fields of velocity components u_1 and u_2, computed by the DM and the FRM of order 7 in the $Re = 550$ case. We note a good agreement on the velocity fields given by both models.

13.10.2.5.2 Thermal Coupled Reduced Model

The aim is now to identify a TCRM able to give the temperatures whatever the value of the Reynolds number in the range [300;800] and whatever the heat flux density $\varphi(t)$ applied at the heater. The reduced fluid state $\mathbf{Z}$ that depends on the Reynolds number is the solution of the state equation of the order 7 FRM previously identified. The TCRM is weakly coupled to the FRM through this reduced fluid state $\mathbf{Z}$. In order to form a set of representative data

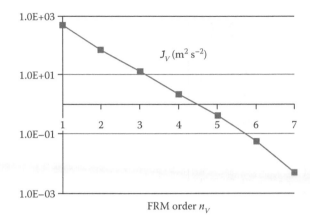

FIGURE 13.15
Objective functional $\mathcal{J}_V(\theta)$ with respect to the FRM order n_V.

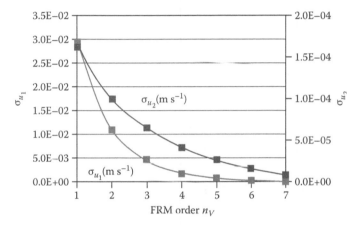

FIGURE 13.16
Mean quadratic errors σ_{u_1} and σ_{u_2} with respect to the FRM order n_V.

of the dynamics of our system, the temperature field is calculated with the DM ($q_T = 144{,}247$) for six Reynolds numbers from 300 to 800 by steps of 100. A step of heat flux density $\varphi(t)$ from 0 at $t = 0$ to 300 W m^{-2} for $t > 0$ is applied during 300 time steps of 0.1 s for each Reynolds number. The obtained temperature fields, from which is subtracted the inlet temperature $T_\infty = 300$ K, are then used as data for the identification. The zone of 213 points indicated by dashed lines in Figure 13.14 ($-2 \leq x_1/h \leq 6$) is considered as the output set of temperatures. Identification results are summarized in Figure 13.19, showing $\mathcal{J}_T(\theta)$ and σ_T, with respect to the TCRM order n_T. The TCRM of order $n_T = 10$ has been retained.

We now want to verify that the TCRM is able to reproduce the behavior of the DM when it is subject to any heat flux density $\varphi(t)$ and for a value of Reynolds number different than those used in the identification process. Here, this test has been performed for $Re = 550$ and for the test heat flux density $\varphi(t)$ shown in Figure 13.20. Figure 13.21 shows the temperature evolution computed with the DM and the TCRM of order 10, for three points C, D, and E defined in Figure 13.14. In Figures 13.22 and 13.23 are respectively plotted the temperature profiles at $x_1/h = \{-2; -1; 0\}$ and $x_1/h = \{1; 2; 3; 4; 5; 6\}$, for time $t = 30$ s. It can be seen that the profiles computed with both the DM and the TCRM are in very good agreement. These results are confirmed by the values of both the mean quadratic error ($\sigma_T = 7.27 \times 10^{-3}$ K) and maximum error ($\varepsilon_T = 3.85 \times 10^{-2}$ K).

TABLE 13.3

Validation of FRM of Order 7: Values of $\sigma_{u_\ell}, \ell = 1, 2, \varepsilon_{u_\ell}, \ell = 1, 2$ and $L_i/h, i = 1, 3$, for Re Test Values

Re	σ_{u_1} (m s^{-1})	σ_{u_2} (m s^{-1})	ε_{u_1} (m s^{-1})	ε_{u_2} (m s^{-1})	L_1/h			L_2/h			L_3/h		
					DM	FRM	Error (%)	DM	FRM	Error (%)	DM	FRM	Error (%)
150	4.0×10^{-4}	2.2×10^{-4}	2.3×10^{-3}	1.4×10^{-3}	3.96	4.04	1.92	—	—	—	—	—	—
250	2.1×10^{-4}	1.2×10^{-4}	1.0×10^{-3}	7.1×10^{-4}	5.87	5.84	0.51	—	—	—	—	—	—
350	1.2×10^{-4}	7.5×10^{-5}	5.3×10^{-4}	3.8×10^{-4}	7.51	7.45	0.75	—	—	—	—	—	—
450	5.8×10^{-5}	3.5×10^{-5}	2.8×10^{-4}	1.8×10^{-4}	8.84	8.77	0.82	7.77	8.21	5.73	11.63	11.29	2.95
550	5.8×10^{-5}	2.7×10^{-5}	2.6×10^{-4}	1.5×10^{-4}	9.88	9.81	0.77	8.19	8.50	3.82	14.48	14.28	1.36
650	1.3×10^{-4}	6.8×10^{-5}	5.0×10^{-4}	3.2×10^{-4}	10.71	10.67	0.44	8.66	8.98	3.68	17.01	16.73	1.67
750	2.6×10^{-4}	1.3×10^{-4}	1.0×10^{-3}	6.1×10^{-4}	11.44	11.39	0.46	9.14	9.42	3.03	19.36	19.07	1.51

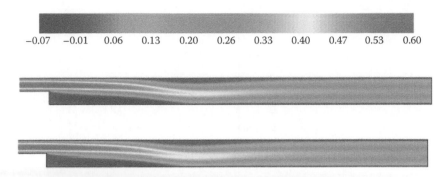

FIGURE 13.17
Horizontal velocity component u_1 field (m s^{-1}) for both the DM (top) and the order 7 RM (bottom) in the $Re = 550$ case.

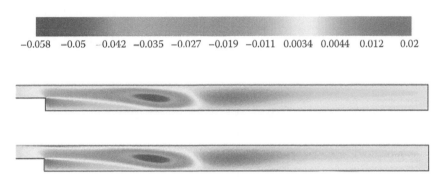

FIGURE 13.18
Vertical velocity component u_2 field (m s^{-1}) for both the DM (top) and the order 7 RM (bottom) in the $Re = 550$ case.

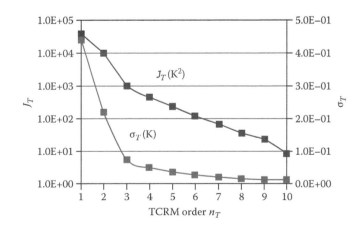

FIGURE 13.19
Objective functional $\mathcal{J}_T(\theta)$ and mean quadratic error σ_T with respect to the TCRM order n_T.

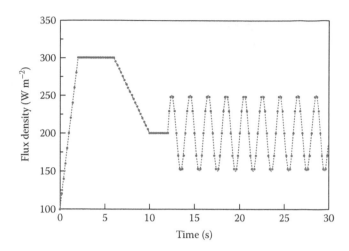

FIGURE 13.20
Heat flux density $\varphi(t)$ for RM validation.

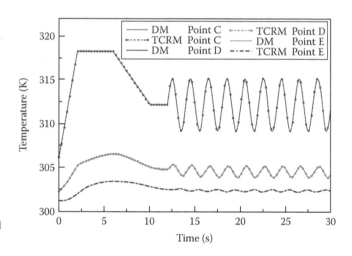

FIGURE 13.21
Temperatures computed by DM and TCRM of order 10 at points C, D, E.

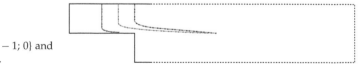

FIGURE 13.22
Temperature profile at $x_1/h = \{-2; -1; 0\}$ and $t = 30$ s for DM and order 10 TCRM.

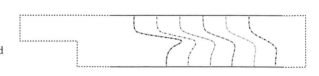

FIGURE 13.23
Temperature profile at $x_1/h = \{1; 2; 3; 4; 5; 6\}$ and $t = 30$ s for DM and order 10 TCRM.

13.11 Transient Nonlinear Inverse Problem Solved with an RM Built by MIM

The RM of order 3 built in Section 13.10.1 is here used for solving a transient nonlinear inverse problem: time-varying strengths of both heat sources are estimated simultaneously from the knowledge of temperature measurements (Girault et al. 2010).

13.11.1 Inverse Problem Resolution

The inverse problem aims at finding an estimation $\hat{\mathbf{U}}$ of the input vector $\mathbf{U}$ from the knowledge of measured temperatures $\mathbf{Y}^{\mathrm{me}}$, corresponding to a part $\mathbf{Y}$ of the output vector $\mathbf{Y_o}$ of the model. $\mathbf{Y}$ contains a selection of $q \leq q_o$ outputs, and an output matrix $\mathbf{H} \in \mathbb{R}^{q \times n}$ is formed with the corresponding lines of matrix $\mathbf{H_o} \in \mathbb{R}^{q_o \times n}$. In the following, we will write $\mathbf{U_k}$ for $\mathbf{U}(t_k)$, $\mathbf{X_k}$ for $\mathbf{X}(t_k)$, $\mathbf{Y_k}$ for $\mathbf{Y}(t_k)$, etc. We adopt a sequential estimation method: From the knowledge of a vector of measured temperatures $\mathbf{Y}_{k+1}^{\mathrm{me}}$ and an estimate $\hat{\mathbf{U}}_k$ of $\mathbf{U_k}$, one looks for an estimate $\hat{\mathbf{U}}_{k+1}$ at time t_{k+1}. The inverse problem aims at finding $\hat{\mathbf{U}}_{k+1}$, minimizing the squared norm of the residual vector $\mathbf{e}_{k+1}$ between the measured data $\mathbf{Y}_{k+1}^{\mathrm{me}}$ and the output vector $\mathbf{Y_{k+1}}$ of the model:

$$\| \mathbf{e}_{k+1} \|_{L_2}^2 = \| \mathbf{Y_{k+1}} - \mathbf{Y}_{k+1}^{\mathrm{me}} \|_{L_2}^2 = \sum_{i=1}^q (e_i(t_{k+1}))^2 = \sum_{i=1}^q \left(Y_i(t_{k+1}) - Y_i^{\mathrm{me}}(t_{k+1}) \right)^2 \tag{13.95}$$

From Equation 13.75, one can write with $\Delta t = t_{k+1} - t_k$ the time step and an Euler implicit scheme:

$$\dot{\mathbf{X}} = \frac{\mathbf{X_{k+1}} - \mathbf{X_k}}{\Delta t} = \mathbf{FX_{k+1}} + \mathbf{C} + \mathbf{GU_{k+1}} + \mathbf{\Omega\Psi}(\mathbf{X_{k+1}})$$

Then,

$$\mathbf{X_{k+1}} = [\mathbf{I} - \mathbf{F}\Delta t]^{-1}[\mathbf{X_k} + \mathbf{C}\Delta t + \mathbf{G}\Delta t\mathbf{U_{k+1}} + \mathbf{\Omega}\Delta t\mathbf{\Psi}(\mathbf{X_{k+1}})] \tag{13.96}$$

Hence, according to Equation 13.76 written for the part $\mathbf{Y}$ of $\mathbf{Y_o}$:

$$\mathbf{Y_{k+1}} = \mathbf{H}[\mathbf{I} - \mathbf{F}\Delta t]^{-1}[\mathbf{X_k} + \mathbf{C}\Delta t + \mathbf{G}\Delta t\mathbf{U_{k+1}} + \mathbf{\Omega}\Delta t\mathbf{\Psi}(\mathbf{X_{k+1}})] \tag{13.97}$$

Inserting Equation 13.97 in Equation 13.95 and letting

$$\mathcal{S} = \mathbf{H}[\mathbf{I} - \mathbf{F}\Delta t]^{-1}\mathbf{G}\Delta t \tag{13.98}$$

and

$$\mathbf{Y_{k+1}} = \mathbf{Y}_{k+1}^{\mathrm{me}} - \mathbf{H}[\mathbf{I} - \mathbf{F}\Delta t]^{-1}[\mathbf{X_k} + \mathbf{C}\Delta t + \mathbf{\Omega}\Delta t\mathbf{\Psi}(\mathbf{X_{k+1}})] \tag{13.99}$$

the squared norm to be minimized becomes

$$\| \mathbf{e}_{k+1} \|_{L_2}^2 = \| \mathcal{S}\mathbf{U_{k+1}} - \mathbf{Y_{k+1}} \|_{L_2}^2 \tag{13.100}$$

meaning that a solution $\hat{\mathbf{U}}_{k+1}$ is sought for the system:

$$\mathcal{S}\mathbf{U}_{k+1} = \mathbf{Y}_{k+1} \tag{13.101}$$

According to Equation 13.99, $\mathbf{Y}_{k+1}$ depends on $\mathbf{X}_{k+1}$, which is unknown at time t_{k+1}, but if an estimate of $\mathbf{X}_{k+1}$ is available, Equation 13.101 is a linear system where $\mathcal{S} \in \mathbb{R}^{q \times p}$ is a dynamic sensitivity matrix, which does not vary with time, but depends on the time step Δt. High values for Δt make matrix $\mathcal{S}$ closer to the static sensitivity matrix $-\mathbf{H}\mathbf{F}^{-1}\mathbf{G}$, making inversion easier.

Assuming that we have at least as many sensors as unknowns, that is, $q \geq p$, the solution is given by linear least squares: $\hat{\mathbf{U}}_{k+1} = [\mathcal{S}^T\mathcal{S}]^{-1}\,\mathcal{S}^T\mathbf{Y}_{k+1}$.

Iterations are thus performed on the following equation:

$$\hat{\mathbf{U}}_{k+1}^{it+1} = [\mathcal{S}^T\mathcal{S}]^{-1}\mathcal{S}^T\mathbf{Y}_{k+1}^{it} \tag{13.102}$$

with

$$\mathbf{Y}_{k+1}^{it} = \mathbf{Y}_{k+1}^{me} - \mathbf{H}[\mathbf{I} - \mathbf{F}\Delta t]^{-1}\left[\hat{\mathbf{X}}_k + \mathbf{C}\Delta t + \mathbf{\Omega}\Delta t\mathbf{\Psi}\left(\hat{\mathbf{X}}_{k+1}^{it}\right)\right] \tag{13.103}$$

where
$\hat{\mathbf{X}}_k$ is the estimate of $\mathbf{X}_k$ obtained with $\hat{\mathbf{U}}_k$ estimated at the previous time step k
$\hat{\mathbf{X}}_{k+1}^{it}$ is the estimate of $\mathbf{X}_{k+1}$ at iteration it

For the first iteration, we set $\hat{\mathbf{X}}_{k+1}^{it=0} = \hat{\mathbf{X}}_k$ as initial guess for the time step. If the system is in steady state at $t = t_0$, then for the first time step, we set $\hat{\mathbf{X}}_0 = \hat{\mathbf{X}}_s^{cv}$ obtained with $\hat{\mathbf{U}}_0 = \hat{\mathbf{U}}_s^{cv}$ estimated in steady state after convergence, and we also set $\hat{\mathbf{X}}_1^{it=0} = \hat{\mathbf{X}}_0$.

Once $\hat{\mathbf{U}}_{k+1}^{it+1}$ obtained, we can compute $\hat{\mathbf{X}}_{k+1}^{it+1}$:

$$\hat{\mathbf{X}}_{k+1}^{it+1} = [\mathbf{I} - \mathbf{F}\Delta t]^{-1}\left[\hat{\mathbf{X}}_k + \mathbf{C}\Delta t + \mathbf{G}\Delta t\hat{\mathbf{U}}_{k+1}^{it+1} + \mathbf{\Omega}\Delta t\mathbf{\Psi}\left(\hat{\mathbf{X}}_{k+1}^{it}\right)\right] \tag{13.104}$$

This sequence is performed until convergence is obtained, that is, $\hat{\mathbf{U}}_{k+1}^{cv} = \hat{\mathbf{U}}_{k+1}^{it+1} \cong \hat{\mathbf{U}}_{k+1}^{it}$. The estimated output vector is then computed using state vector $\hat{\mathbf{X}}_{k+1}^{cv}$ estimated after convergence: $\hat{\mathbf{Y}}_{k+1}^{cv} = \mathbf{H}\hat{\mathbf{X}}_{k+1}^{cv}$.

The introduction of FTS with *Beck's function specification method* is described in Girault et al. (2010).

13.11.2 Test Case and Sensors Choice

First, it should be noted that whereas the estimation of a single heat source is rather easy, the case of two sources is trickier because each measurement point is influenced by both sources. The set of nine temperatures recordings at points 1–9 is considered as possible data set. Figure 13.24 shows the two signals used for testing the identified RM.

13.11.2.1 Sensors Choice

We have to choose which data sets can be used among the whole set of temperatures at points 1–9. Looking at sensitivities will be of great interest for making such a choice. Let us

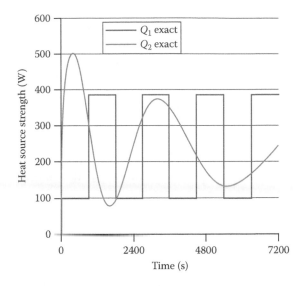

FIGURE 13.24
Heat source strengths to be retrieved.

look at the sensitivity matrix S defined in Section 13.11.1. When no FTS are used ($n_f=0$), S is a $q \times 2$ matrix defined by Equation 13.98, depending on the outputs of RM used for inversion. In the present paper, as q temperatures among those at points 1–9 may be used as data, S will be formed using the q corresponding lines of matrix $\mathbf{H}$. When $n_f>0$, $n_f \times q$ additional lines are included in S.

Note that because the inverse problem is nonlinear, S contains sensitivities of $\mathbf{Y_{k+1}} - \mathbf{H}[\mathbf{I} - \mathbf{F}\Delta t]^{-1}\mathbf{\Omega}\Delta t\mathbf{\Psi}(\mathbf{X_{k+1}})$ with respect to $\mathbf{U_{k+1}}$, according to Equation 13.97, rather than sensitivities of $\mathbf{Y_{k+1}}$. Figure 13.25 shows the components of the sensitivity matrix S for the nine possible data points when $n_f=0$ and those added when using $n_f=1$ and $n_f=2$ FTS. As expected, it can be seen that points 1, 2, and 3 are more sensitive to source 1 than to source 2. Conversely, points 7, 8, and 9 are more sensitive to source 2 than to source 1. Points 4, 5, and 6 have almost the same sensitivities to both sources. Hence, if we use only

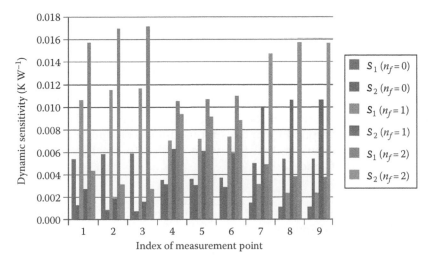

FIGURE 13.25
Dynamic sensitivities S_1 and S_2 of each possible data point to sources 1 and 2, respectively.

points 4, 5, and 6, we should not be able to correctly discriminate the sources. If two points only are to be used, the best choice would be to use points 3 and 9. These deductions are confirmed by the condition number of matrix $\mathbf{S}^{\mathrm{T}}\mathbf{S}$, which is equal to approximately 2 for the couple (3,9) whereas it reaches 6000 for the couple (4,5). Figure 13.25 also shows that adding FTS allows us to include additional higher sensitivities in $\mathbf{S}$.

13.11.2.2 Definition of Mean Quadratic Errors

In a similar way as for the RM identification phase, we define σ_Y as the rms of the residues between the measured temperatures $\mathbf{Y}^{\mathrm{me}}$ and those computed by the RM with the estimated set of heat sources strengths, $\hat{\mathbf{Y}}$:

$$\sigma_Y = \sqrt{\sum_{i=1}^{q} \sum_{j=1}^{N_t - n_f} \left(\hat{Y}_i(t_j) - Y_i^{\mathrm{me}}(t_j)\right)^2 \Big/ (q \times (N_t - n_f))} \qquad (13.105)$$

We recall that q is the number of data points used for inversion ($q \leq q_0$ is here equal to 2), and N_t is the number of time steps of the inverse problem (360 here).

In order to assess the accuracy of the estimations for the present test case, we also define σ_U as the rms of the residues between exact heat sources strengths $\mathbf{U}^{\mathrm{exact}}$ and estimated ones $\hat{\mathbf{U}}$. In the present lab experiment, $\mathbf{U}^{\mathrm{exact}}$ is known (measured), but of course, this quantity is not known in a practical application for which only σ_Y is available.

$$\sigma_U = \sqrt{\sum_{i=1}^{p} \sum_{j=1}^{N_t - n_f} \left(\hat{U}_i(t_j) - U_i^{\mathrm{exact}}(t_j)\right)^2 \Big/ (p \times (N_t - n_f))} \qquad (13.106)$$

Analogous quantities, related to each one of the sources, may be defined. They are called σ_{U_1} and σ_{U_2}.

13.11.2.3 Number of Future Time Steps

As we intend to propose an inversion algorithm able to work in real time, the choice of n_f should be made prior to any inversion using the measured data. For two heat source test signals different than those of the inversion test using real measured data, we have simulated temperature measurements by adding a random Gaussian noise to solutions of the RM of order 4. Then, using these simulated data, we have performed inversions using the RM of order 3 to recover the two test signals. For an added noise of standard deviation $\sigma^m = 0.1$ K, corresponding to the supposed noise on temperature measurements made using the infrared camera, two FTS were needed to obtain values of σ_Y close to $\sigma^m = 0.1$ K, when using T_3 and T_9.

13.11.3 Inversion Results Using RM and Computation of Internal Temperatures with the Estimated Sources

Figure 13.26 shows temperatures at points 3 and 9, measured by the infrared camera, when the test signals of Figure 13.24 are used. The number of time steps is $N_t = 360$ and the time step is $\Delta t = 20$ s.

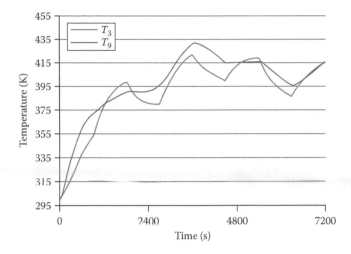

TABLE 13.4

Inversion Results Using RM of Order 3 and T_3 and T_9 as Data

RM Order n	Data for Inversion	n_f	σ_Y (K)	σ_U (W)	σ_{U_1} (W)	σ_{U_2} (W)
3	T_3 and T_9	0	4.5×10^{-14}	43.8	46.3	41.1
		1	0.123	27.5	31.0	23.4
		2	0.191	24.6	28.8	19.5

Inversion results are summarized in Table 13.4. In addition to the $n_f = 2$ case, results for $n_f = 0$ and $n_f = 1$ are also presented in order to assess the improvement in the quality of estimations when using 2 FTS.

Figures 13.27 and 13.28 show respectively estimated heat sources strengths as well as exact signals, for $n_f = 0$ and $n_f = 2$, when the RM of order 3 is used. One can see that both sources are discriminated.

First, let us look at the $n_f = 0$ case. Since there are exactly as many data as unknowns ($q = p = 2$) at each time step, the inverse problem is not over-determined: consequently, temperatures computed with the estimated heat sources strengths fit in quasi perfectly with data ($\sigma_Y \approx 10^{-14}$ K). This is not a good feature because it means that the measurement errors have been fitted, and as a consequence, the estimated signals are pretty bad, as it can be seen by watching at the values of σ_U.

When using one future temperature at each time step of the inversion algorithm ($n_f = 1$), the lagging and damping effects of heat diffusion are taken into account, and inversion results are clearly improved. Moreover, the inverse problem becomes over-determined (more data than unknowns). The value of σ_Y is close to the supposed standard deviation of measurement errors ($\sigma^m = 0.1$ K). With $n_f = 2$, the quality of estimations is improved again. The value of σ_Y is slightly larger than with $n_f = 1$ but it remains of the order of magnitude of σ^m. In fact, for both the $n_f = 1$ and $n_f = 2$ cases, results are very similar.

Thanks to the low-order model, the computing time spent to solve the inverse problem is very low, about 2.5×10^{-2} s for the 360 time steps.

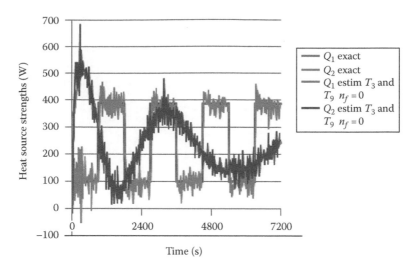

FIGURE 13.27
Exact and estimated heat source strengths, for $n_f = 0$.

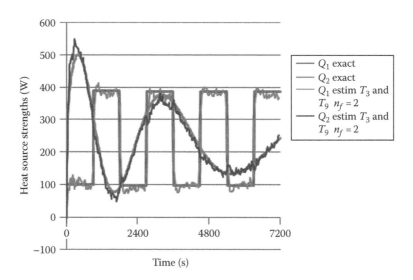

FIGURE 13.28
Exact and estimated heat source strengths, for $n_f = 2$.

Both heat sources intensities estimated from the knowledge of T_3 and T_9 (using $n_f = 2$) can then be used with the RM to compute temperatures at the remaining 10 points of the set $\{T_1 \ldots T_{12}\}$. Of course, in the present study, these temperatures have also been measured in order to assess the quality of the results. Figure 13.29 shows both measured and computed temperatures at point 12.

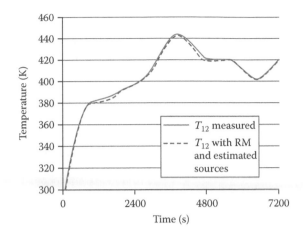

FIGURE 13.29
T_{12} measured and T_{12} computed by RM using identified heat source strengths.

13.12 Conclusion

In this chapter, we have presented the main principles of the model identification method (MIM), which aims at building low-order models through the use of optimization techniques, and we have shown the advantages provided by the use of such RMs to solve transient inverse boundary problems.

The first part of the chapter has dealt with linear systems. The state space representation, and especially its modal form, has been introduced to provide useful background for presenting the MIM. Then, the main stages of the method for linear problems have been developed.

First, the parameter estimation problem leading to the identification of the RM parameters has been described. For each one of the independent thermal inputs (prescribed heat flux densities or temperatures at the boundaries, internal heat sources), it is based on the minimization of a squared residuals functional built on the discrepancy between the low-order model outputs and the ones of the system (simulations of a reference model or in situ measurements) when a known input signal is applied. This phase corresponds to an input/output calibration.

Second, the assembling of the ERMs (one for each single independent input), according to the superposition principle, in order to form a global RM for the whole set of inputs, has been presented.

The use of a low-order model for solving transient inverse boundary problems has been developed. A sequential algorithm has been described, for the estimation of the time-varying input vector gathering thermal loads from the knowledge of output (temperature) data. The use of information at future time steps (FTS) to help solving and regularizing the inverse problem has been introduced. A simple 2D academic example has been proposed to illustrate the approach.

The second part of the chapter has been devoted to the extension of the MIM for nonlinear systems. After a brief presentation of some reduction methods, we have given the general philosophy of the MIM for nonlinear problems. Two examples have then been presented to illustrate the method.

The first was dealing with an experimental setup involving nonlinear heat diffusion with both convective and radiative boundary conditions. After describing the form of the low-order model, we have underlined the benefits of an "experimental modeling," that is, the identification of a low-order model exclusively from in situ measured data, when some crucial parameters such as heat exchange coefficients, thermal contact resistances, and emissivities are not known accurately. Hence, no reference model based on classical spatial discretization of PDE has been needed.

The second example was dealing with a 2D numerical problem: a heated flow on a backward-facing step, involving heat convection and diffusion. The formulations of two types of RM have been described:

- A fluid reduced model (FRM) relative to steady fluid mechanics only, dedicated to the computation of the velocity field, for any constant Reynolds number Re in the range [100;800].

- A thermal coupled reduced model (TCRM) designed to calculate temperature for a time-varying flux $\varphi(t)$ and for different values of $Re \in [300;800]$. The TCRM is therefore weakly coupled to the FRM.

Finally, a nonlinear transient inverse problem has been presented. It was based on the previous experimental setup involving nonlinear heat diffusion with both convective and radiative boundary conditions. A sequential algorithm using the nonlinear RM has been proposed and tested for the simultaneous estimation of two time-varying heat sources from the knowledge of measured temperature data.

Nomenclature

A	state matrix
B	input matrix
C	output matrix
e	residual vector
F	diagonal matrix of eigenvalues
G	input matrix in the modal formulation
H	output matrix in the modal formulation
$\mathcal{J}_{red}$	objective functional
n	order of the reduced model
n_f	number of future time steps
p	number of inputs
q	number of outputs
$\mathcal{S}$	dynamic sensitivity matrix
t	time
T	temperature vector
U	input vector
X	state vector
Y	output vector

References

Alonso, D., Velazquez, A., and Vega, J.M. 2009. Robust reduced order modeling of heat transfer in a back step flow. *International Journal of Heat and Mass Transfer* 52(5–6):1149–1157.

Balima, O., Favennec, Y., Girault, M., and Petit, D. 2006. Comparison between the modal identification method and the POD-Galerkin method for model reduction in nonlinear diffusive systems. *International Journal for Numerical Methods in Engineering* 67(7):895–915.

Beck, J.V., Blackwell, B., and Haji-Sheikh, A. 1996. Comparison of some inverse heat conduction methods using experimental data. *International Journal of Heat and Mass Transfer* 39:3649–3657.

Ben Jaafar, M.T., Pasquetti, R., and Petit, D. 1990. Model reduction for thermal diffusion: Application of the Eitelberg, Marshall and aggregation methods to a heat transmission tube model. *International Journal for Numerical Methods in Engineering* 29:599–617.

Bergmann, M. and Cordier, L. 2008. Optimal control of the cylinder wake in the laminar regime by trust-region methods and pod reduced-order model. *Journal of Computational Physics* 227 (16):7813–7840.

Blanc, G., Raynaud, M., and Chau, T.H. 1998. A guide for the use of the function specification method for 2D inverse heat conduction problems. *Revue Générale de Thermique* 37:17–30.

Bui-Thanh, T., Willcox, K., Ghattas, O., and Van Bloemen Waanders, B. 2007. Goal-oriented, model-constrained optimization for reduction of large-scale systems. *Journal of Computational Physics* 224(2):880–896.

Clerc, M. 2005. *L'optimisation par essaims particulaires*. Hermes-Lavoisier, Paris.

Gill, P.E., Murray, W., and Wright, M.H. 1992. *Practical Optimization*. Academic Press, London.

Girault, M., Maillet, D., Bonthoux, F., Galland, B., Martin, P., Braconnier, R., and Fontaine, J.-R. 2008. Estimation of time-varying pollutant emission rates in a ventilated enclosure: Inversion of a reduced model obtained by experimental application of the modal identification method. *Inverse Problems* 24(1):22 pp.

Girault, M., Maillet, D., Fontaine, J.-R., Braconnier, R., and Bonthoux, F. 2006. Estimation of time-varying gaseous contaminant sources in ventilated enclosures through inversion of a reduced model. *International Journal of Ventilation* 4(4):365–380.

Girault, M. and Petit, D. 2004. Resolution of linear inverse forced convection problems using model reduction by the modal identification method: Application to turbulent flow in parallel-plate duct. *International Journal of Heat and Mass Transfer* 47:3909–3925.

Girault, M., Petit, D., and Videcoq, E. 2003. The use of model reduction and function decomposition for identifying boundary conditions of a linear thermal system. *Inverse Problems in Engineering* 11(5):425–455.

Girault, M., Videcoq, E., and Petit, D. 2010. Estimation of time-varying heat sources through inversion of a low order model built with the modal identification method from in-situ temperature measurements. *International Journal of Heat and Mass Transfer* 53(1–3):206–219.

Han, H. and Yin, D. 2003. A non-overlap domain decomposition method for the forward–backward heat equation. *Journal of Computational and Applied Mathematics* 159(1):35–44.

Holmes, P.J., Lumley, J.L., Berkooz, G., Mattingly, J.C., and Wittenberg, R.W. 1997. Low-dimensional models of coherent structures in turbulence. *Physics Reports* 287(4):337–384.

Joly, F., Quéméner, O., and Neveu, A. 2008. Modal reduction of an advection-diffusion model using a branch basis. *Numerical Heat Transfer, Part B: Fundamentals* 53:1–20.

Langer, U., Discacciati, M., Keyes, D., Widlund, O., and Zulehner, W. 2008. *Domain Decomposition and Engineering XVII*. Springer, New York.

Litz, L. 1981. Order reduction of linear state space models via optimal approximation of the non-dominant modes. *North-Holland Publishing Company Large Scale System* 2:171–184.

Marshall, S.A. 1966. An approximate method for reducing the order of a linear system. *Control* 10:642–643.

Moore, B.C. 1981. Principal component analysis in linear systems: Controllability, observability and model reduction. *IEEE Transactions on Automatic Control* 26(1):17–32.

Osman, A.M., Dowding, K.J., and Beck, J.V. 1997 Numerical solution of the general two-dimensional inverse heat conduction problem (IHCP). *Journal of Heat Transfer* 119:38–45.

Press, W.H., Teukolsky, S.A., Vetterling, W.T., and Flannery, B.P. 2007. *Numerical Recipes: The Art of Scientific Computing*, 3rd edn. Cambridge University Press, Cambridge, U.K.

Rouizi, Y., Girault, M., Favennec, Y., and Petit, D. 2010. Model reduction by the modal identification method in forced convection: Application to a heated flow over a backward-facing step. *International Journal of Thermal Sciences* 49:1354–1368.

Rowell, D. 2002. State-space representation of LTI systems. http://web.mit.edu/2.14/www/Handouts/StateSpace.pdf (Accessed on January 10, 2011).

Rozza, G., Huynh, D.B.P., and Patera, A.T. 2008. Reduced basis approximation and a posteriori error estimation for affinely parametrized elliptic coercive partial differential equations: Application to transport and continuum mechanics. *Archives of Computational Methods in Engineering* 15(3):229–275.

Sirovich, L. 1987. Turbulence and the dynamics of coherent structures parts i, ii, iii. *Quarterly of Applied Mathematics* 45:561–590.

Videcoq, E. and Petit, D. 2001 Model reduction for the resolution of multidimensional inverse heat conduction problems. *International Journal of Heat and Mass Transfer* 44(10):1899–1911.

Videcoq, E., Petit, D., and Piteau, A. 2003. Experimental modeling and estimation of time varying thermal sources. *International Journal of Thermal Sciences* 42(3):255–265.

Videcoq, E., Quéméner, O., Lazard, M., and Neveu, A. 2008. Heat source identification and on-line temperature control by a branch eigenmodes reduced model. *International Journal of Heat and Mass Transfer* 51(19–20):4743–4752.

Willcox, K. and Pereire, J. 2002. Balanced model reduction via the proper orthogonal decomposition. *AIAA Journal* 40(11):2323–2330.

Wrobel, L.C. 2002. *The Boundary Element Method. Volume 1: Applications in Thermo-Fluids and Acoustics.* John Wiley & Sons, New York.

14

Karhunen–Loève Decomposition for Data, Noise, and Model Reduction in Inverse Problems

Elena Palomo del Barrio and Jean-Luc Dauvergne

CONTENTS

14.1 Introduction

Infrared thermography is widely used to measure the thermal diffusivity of materials. A thermographic experiment usually consists in illuminating the front face of a sample by a light beam (point like, line, or other motifs) and detecting the thermal response of the sample on either its front or its rear face. The light beam intensity can be Dirac-like, modulated, pseudorandom, etc. The thermal behavior of the sample is observed using an infrared camera with a focal plane array of infrared detectors. The development of infrared video cameras with fast data acquisition (thousands of images/s) and high lateral resolution (tens of micrometers) provides powerful tools for fast materials testing. The drawback of these cameras is the huge amount of noisy data to be treated. To overcome such a difficulty, experimental conditions are usually chosen so that simple analytical solutions of the heat conduction problem exist. For instance, the diffusivity of homogeneous materials can be easily studied by the so-called slope method when illuminating a thin but large enough sample placed in a vacuum chamber by a point-like or a line-like intensity modulated laser beam (see, e.g., Mendioroz et al. 2009 and references within). Otherwise (i.e. heterogeneous materials, other lighting motifs), the estimation process (heat conduction model inversion) becomes a tricky task: computing time and memory resources required when using standard approaches, i.e., diffusivity estimation by minimization of a chosen residuals norm defined over the whole spatial domain, are generally huge while results are very sensitive to noise when using point-by-point least squares estimation approaches (see Chapter 7). In such a sense, mathematical tools allowing significant reduction of the data set dimension, as well as noise control and parsimonious estimations, could be an interesting alternative for extending characterization of materials based on Infrared thermography to situations with unknown analytical solutions.

Proper orthogonal decomposition techniques are widely used for multivariate data reduction in many areas of application. The reduction starts by choosing an appropriate orthogonal basis allowing identification of some few dominant components (referred to as dominant directions, eigenfunctions, or modes). A low-dimensional approximate description of the whole set of data is thus obtained by projecting the initial high-dimensional set on the dominant eigenfunctions. The choice of the basis makes the main difference among methods.

When dealing with regular signals, those arising out of spectral decomposition of the energy matrix (or covariance matrix) of the multivariate data give the best results. This means that it provides the lowest dimension for a given approximation precision or, alternatively, the best precision for a given dimension. Such a method has been developed about 100 years ago by Pearson (1901) as a tool for graphical data analysis and redeveloped several times since then in different areas of application (Hotteling 1933, Karhunen 1946, Loève 1955), that it has assumed many names such as principal components analysis (PCA), Karhunen–Loève decomposition (KLD), singular value decomposition (SVD), etc. PCA/KLD/SVD is very commonly used today in image processing and signal processing problems for compression and noise reduction (Deprettere 1988). It is also widely used for signals classification, data clustering, and information retrieval problems (Berry et al. 1995, Everitt and Dunn 2001, Everitt et al. 2001). Powerful model reduction techniques based on PCA/KLD/SVD have been also proposed for low-dimensional description of problems described by partial differential equations, mostly in the field of turbulent flows (Berkoz et al. 1993, Holmes et al. 1996). In thermal analysis, SVD-based methods have been developed for efficient reduction of linear and nonlinear heat transfer problems (Ait-Yahia and Palomo del Barrio 1999, 2000; Palomo del Barrio 2000; Dauvergne and Palomo del Barrio

2009, 2010), as well as for solving heat transfer inverse problems dealing with unknown heat sources (Park and Jung 1999, 2001, Palomo del Barrio 2003a,b). For a fairly comprehensive introduction to PCA/KLD/SVD, we recommend the books by Jolliffe (1986) and Deprettere (1988). For more details on the mathematics and computation, good references are Golub and Van Loan (1996), Strang (1998), Berry (1992), and Jessup and Sorensen (1994).

This chapter focuses on the use of KLD techniques in association with infrared thermography for reliable and parsimonious thermal characterization of materials. In Section 14.2, the KLD of infinite-dimensional and finite-dimensional problems is defined. Functions and signal considered are, respectively, space-time-dependent functions and multivariate time series. In the framework of thermal analysis, functions represent the thermal field while multivariate time series are data taken from thermal field sampling. The property of KLD to provide the closest r-dimensional approximation for an infinite-dimensional problem or the closest rank-r approximation for a rank-n ($n > r$) matrix is used. A numerical example illustrates the application of KLD for this problem. In Section 14.4, measurement noise propagation through KLD is analyzed and two KLD-based filters are described. The application of KLD for data filtering is highlighted using a numerical example. A KLD-based method for reliable estimation of the diffusivity of homogeneous materials is described and tested in Section 14.5, while Section 14.6 focuses on thermal characterization of heterogeneous materials. This chapter includes unpublished experiments and results.

14.2 Karhunen–Loève Decomposition

Let $T(x, t)$ be a space-time-dependent function in a bounded region Ω, with $x = (x_1, x_2, x_3)$ and t representing point coordinates and time, respectively. It is assumed that this function satisfies

$$\forall t, \int_{\Omega} T^2(x, t)dx < \infty \tag{14.1}$$

so that it belongs to the infinite-dimensional Hilbert space H associated with Ω. Moreover, we assume that

$$\forall x, \int_{t} T^2(x, t)dt < \infty \tag{14.2}$$

Sampling $T(x, t)$ on Ω leads to the continuous in time n-dimensional vector $\mathbf{T}(t) = \{T_i(t)\}_{i=1\cdots n}(n \times 1)$, whose components represent $T(x, t)$ values at the different n sampling points. In the following, we will assume that sampling is done so that $\mathbf{T}(t)$ provides a good enough approximation of $T(x, t)$.

14.2.1 KLD of Infinite-Dimensional Problems

Let us define the energy function associated with $T(x, t)$ as follows:

$$W(x, x') \equiv \int_{t} T(x, t)T(x', t)dt \tag{14.3}$$

It can be easily proven that $W(x, x') = W(x', x)$ (symmetry property). In fact, $W(x, x')$ is a compact, positive operator on H. We recall that a positive value for $W(x, x')$ means that, for all nonzero functions $\varphi(x) \in$ H, the following inequality is satisfied:

$$\int\int_{\Omega} W(x, x')\phi(x)\phi(x')dxdx' > 0 \qquad (14.4)$$

The well-known spectral theorem states that, if $W(x, x')$ is compact and self-adjoint, then there is a complete orthonormal set in H (Hilbert basis) consisting of the eigenfunctions of $W(x, x')$, noted as $\{V_m(x)\}_{m=1\cdots\infty}$ in the following. Moreover, because $W(x, x')$ is positive, its spectrum consists of 0 (zero) together with a countable infinite set of real and positive eigenvalues: $\sigma_1^2 \geq \sigma_2^2 \geq \cdots > 0$. The problem defining eigenvalues and eigenfunctions of $W(x, x')$ is

$$\int_{\Omega} W(x, x')V_m(x')dx' = \sigma_m^2 V_m(x) \qquad (14.5)$$

with orthogonal condition

$$\langle V_k, V_m \rangle \equiv \int_{\Omega} V_k(x)V_m(x)dx = \delta_{km} \qquad (14.6)$$

Hilbert–Schmidt theorem allows eigenfunction $W(x, x')$ to be written as (see e.g. Intissar 1997)

$$W(x, x') = \sum_{m=1}^{\infty} \sigma_m^2 V_m(x)V_m(x') \qquad (14.7)$$

Besides, all functions belonging to H can be projected in a unique manner on the $W(x, x')$ eigenfunctions set because $\{V_m(x)\}_{m=1\cdots\infty}$ is a complete set in H (Hilbert basis). Since for $\forall t$, $T(x, t) \in$ H, it follows that

$$\forall t, \; T(x, t) = \sum_{m=1}^{\infty} V_m(x)z_m(t) \qquad (14.8)$$

where decomposition coefficients (states in the following) are given by

$$z_m(t) = \langle T(x, t), V_m(x) \rangle \equiv \int_{\Omega} T(x, t)V_m(x)dx \qquad (14.9)$$

Taking into account Equations 14.7 and 14.8, it can be easily proven that states are orthogonal, and they satisfy the following equation:

$$\langle z_m(t), z_k(t) \rangle \equiv \int_t z_m(t)z_k(t)dt = \delta_{mk}\sigma_m^2 \qquad (14.10)$$

Equation 14.8 gives the so-called Karhunen–Loève decomposition of $T(x, t)$, also named singular value decomposition (SVD). As shown in this section, SVD results from $T(x, t)$

expansion on $W(x, x')$ eigenfunctions. The square root of the $W(x, x')$ eigenvalues are the so-called singular values of $T(x, t)$, designated as $\sigma_1, \sigma_2, \dots$

14.2.2 KLD of Finite-Dimensional Problems

Let us now define the energy matrix associated with $\mathbf{T}(t)$ $(n \times 1)$ as follows:

$$\mathbf{W} \equiv \int_t \mathbf{T}(t)\mathbf{T}^t(t)dt \tag{14.11}$$

The meaning of this matrix is close to that of the covariance matrix of $\mathbf{T}(t)$ signals. Diagonal terms represent the energy (close to the variance) of $\mathbf{T}(t)$ components, while non-diagonal terms measure the dynamic likeness among signals. It can be easily demonstrated that $\mathbf{W}(n \times n)$ is a symmetric, definite positive matrix. Accordingly, spectral decomposition of $\mathbf{W}$ leads to a n-dimensional set of orthonormal eigenvectors, $\mathbf{V} = [\mathbf{v}_1 \quad \mathbf{v}_2 \quad \cdots \quad \mathbf{v}_n]$ with $\mathbf{V}^t\mathbf{V} = \mathbf{I}$, and associated eigenvalues satisfying $\sigma_1^2 \geq \sigma_2^2 \geq \cdots \geq \sigma_n^2 > 0$. The energy matrix can hence be written as

$$\mathbf{W} = \mathbf{V}\mathbf{\Sigma}\,\mathbf{V}^t \text{ with } \mathbf{\Sigma} = diag\begin{bmatrix} \sigma_1^2 & \sigma_2^2 & \cdots & \sigma_n^2 \end{bmatrix} \tag{14.12}$$

Moreover, $\mathbf{V} = [\mathbf{v}_1 \quad \mathbf{v}_2 \quad \cdots \quad \mathbf{v}_n]$ is a complete orthonormal set in H. Consequently, elements of $\mathbf{T}(t)$ can be represented as linear combinations of the $\mathbf{W}$ eigenvectors:

$$\forall k, \; T_k(t) = \sum_{m=1}^{n} v_{km} z_m(t) \tag{14.13}$$

Using matrix writing, equations above become

$$\mathbf{T}(t) = \mathbf{V}\mathbf{Z}(t) \tag{14.14}$$

with $\mathbf{Z}(t) = [z_1(t) \quad z_2(t) \quad \cdots \quad z_n(t)]^t$. From Equations 14.11, 14.12, and 14.14, it can be easily proven that states are orthogonal:

$$\int_t \mathbf{Z}(t)\mathbf{Z}^t(t)dt = \mathbf{\Sigma} \tag{14.15}$$

Equation 14.14 is the Karhunen–Loève decomposition (or SVD) of $\mathbf{T}(t)$. It must be noticed that for an appropriate $T(x, t)$ sampling, the elements of $\mathbf{W} = \{w_{ij}\}_{i,j=1\cdots n}$ and the eigenvectors $\mathbf{v}_m (m = 1 \cdots n)$ can be considered as good numerical approximations of $W(x, x')$ and $V_m(x)$, respectively.

14.3 Data Reduction Using KLD

As already mentioned in the introduction, KLD provides the most efficient way of capturing the dominant components of an infinite-dimensional process (or a high-dimensional one) with only a finite number of modes, often surprisingly few. The property of KLD to provide the closest r-dimensional approximation for an infinite-dimensional problem, or the closest rank-r approximation for a rank-n ($n > r$) matrix, is shown in Section 14.3.1. A numerical example for illustrating this property is provided in Section 14.3.2.

14.3.1 Low-Dimensional Approximations

An r-dimensional linear approximation of $T(x,t)$ is achieved when truncating the expansion series given by Equation 14.8 to its first r terms:

$$\forall t, \; T_r(x,t) = \sum_{m=1}^{r} V_m(x)z_m(t) \tag{14.16}$$

This approximation is the orthogonal projection of $T(x,t)$ on the space generated by $\{V_m(x)\}_{m=1\cdots r}$. The approximation error is

$$\forall t, \; e_r(x,t) \equiv T(x,t) - T_r(x,t) = \sum_{m=r+1}^{\infty} V_m(x)z_m(t) \tag{14.17}$$

Taking into account orthogonal properties of eigenfunctions $\{V_m(x)\}_{m=1\cdots\infty}$ (see Equation 14.6) and states $\{z_m(t)\}_{m=1\cdots\infty}$ (see Equation 14.10), it can be easily proven that

$$\|e_r\|^2 \equiv \int_t \int_\Omega e_r(x,t)dxdt = \sum_{m=r+1}^{\infty} \sigma_m^2 \tag{14.18}$$

As $\|T(x,t)\|^2 < +\infty$, the approximation error tends to zero: $\lim_{r\to\infty} \|e_r\|^2 = 0$. Indeed, Allahverdiev theorem (cf. Intissar 1997) for compact operators, say $W(x,x')$, states that all r-dimensional approximations $W_r(x,x')$ of $W(x,x')$ satisfy $\|W - W_r\| \geq \|\sigma_{r+1} \quad \sigma_{r+2} \quad \cdots\|$, where σ_{r+i} is the $(r+i)$th singular value of W and $\|.\|$ represents unitarily invariant norms of the approximation error. The same theorem states that the minimum value $\|W - W_r\| = \|\sigma_{r+1} \quad \sigma_{r+2} \quad \cdots\|$ is achieved by truncation of the Schmidt development of W (Equation 14.7) to its r first terms (those associated with the largest singular values). It follows that all r-dimensional linear approximations of $T(x,t)$ satisfy

$$\|e_r\|^2 \geq \sum_{m=r+1}^{\infty} \sigma_m^2 \tag{14.19}$$

As a result, Equation 14.16 provides the best r-dimensional linear approximation of $T(x,t)$ with regard to the quadratic norm of the error.

In matrix algebra, Fan and Hoffman (1955) and Mirsky (1960) demonstrated that all rank-r approximations for a rank-n ($n > r$) matrix, say $\mathbf{W}$, satisfy

$$\inf_{rank(\mathbf{W}_r)=r} \|\mathbf{W} - \mathbf{W}_r\| = \|diag(\sigma_{r+1} \quad \sigma_{r+2} \quad \cdots \quad \sigma_n)\| \tag{14.20}$$

Indeed, the closest rank-r approximation is provided by $\mathbf{W}_r = \mathbf{V}_r \mathbf{\Sigma}_r \mathbf{V}_r^t$, where $\mathbf{V} = [\mathbf{v}_1 \, \mathbf{v}_2 \quad \cdots \quad \mathbf{v}_r]$ and $\mathbf{\Sigma}_r = diag[\sigma_1^2 \quad \sigma_2^2 \quad \cdots \quad \sigma_r^2]$. It follows that

$$\mathbf{T}_r(t) = \mathbf{V}_r \mathbf{Z}_r(t) \tag{14.21}$$

with $\mathbf{Z}_r(t) = [z_1(t) \quad z_2(t) \quad \cdots \quad z_r(t)]^t$ provides the best r-dimensional approximation of $\mathbf{T}(t)$ in the sense of unitarily invariant norms of the approximation error. As mentioned previously, it can be easily proven that

$$\|\mathbf{e}_r\|^2 \equiv \|\mathbf{T} - \mathbf{T}_r\|^2 = \sum_{m=r+1}^{n} \sigma_m^2 \tag{14.22}$$

14.3.2 Numerical Example

Let us consider the thermal behavior of a squared, homogeneous plate ($L \times L$, $L = 6$ mm) exchanging heat by convection/radiation with an environment at constant temperature, say at 0°C. The temperature field of the plate at time $t = 0$ is represented in Figure 14.1a. Reference axes are chosen so that the energy equation at inner points ($0 < x < L$, $0 < y < L$) can be written as follows:

$$\frac{\partial T(x,y,t)}{\partial t} = \alpha_x \frac{\partial^2 T(x,y,t)}{\partial x^2} + \alpha_y \frac{\partial^2 T(x,y,t)}{\partial y^2} - \beta T(x,y,t) \tag{14.23}$$

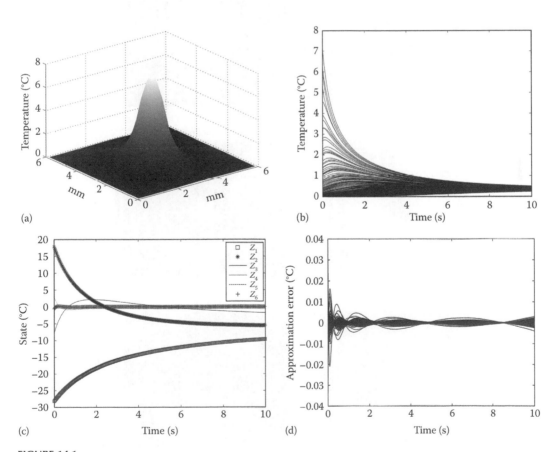

(a)

(b)

(c)

(d)

FIGURE 14.1

Data reduction using KLD: (a) initial temperature field of the plate; (b) temperature behavior in time, one curve by volume of control; (c) KLD states; and (d) error of the 6D KLD approximation.

$\alpha_x = 0.1515 \times 10^{-6} \, \mathrm{m^2 \, s^{-1}}$ and $\alpha_y = 0.3030 \times 10^{-6} \, \mathrm{m^2 \, s^{-1}}$ represent, respectively, the thermal diffusivity in the $\overrightarrow{Ox}$ and $\overrightarrow{Oy}$ directions. $\beta = 0.152 \, \mathrm{s^{-1}}$ is the heat transfer coefficient between the plate and its surroundings. At points on the plate boundaries ($x = 0$, $x = L$, $y = 0$, $y = L$), adiabatic conditions are assumed.

The finite volume method has been applied on an equally spaced $n \times n$ ($n = 30$) grid for discretization of Equation 14.23. This leads to the state-space model:

$$\dot{\mathbf{T}}(t) = \mathbf{A}\mathbf{T}(t) \quad \text{with } \mathbf{A} = \alpha_x \mathbf{L}_x + \alpha_y \mathbf{L}_y - \beta \mathbf{I} \tag{14.24}$$

where $\mathbf{L}_x$ and $\mathbf{L}_y$ represent numerical approximations of $\partial^2/\partial x^2$ and $\partial^2/\partial y^2$, respectively. Results achieved from time integration of Equation 14.24 are reported in Figure 14.1b. It includes the time behavior of the temperature at each control volume, $\mathbf{T}(t) = \{T_i(t)\}_{i=1\cdots n}$, from $t = 0$ to $t = 10 \, \mathrm{s}$ with $\Delta t = 0.5 \times 10^{-2} \, \mathrm{s}$ (2000 sampling times). KLD of this set of signals has been carried out as described in Section 14.2.2: $\mathbf{T}(t) = \mathbf{V}\mathbf{Z}(t)$. Eigenfunctions associated with the six first singular values are depicted in Figure 14.2, while the time evolution of the

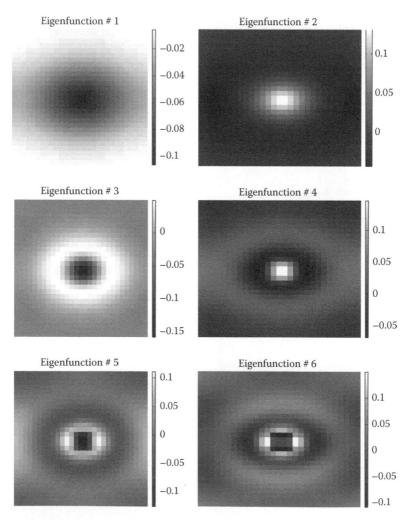

FIGURE 14.2
Eigenfunctions associated to the largest eigenvalues. Maps on the (6 mm × 6 mm) testing plate.

TABLE 14.1

Eigenvalues of the Energy Matrix and Their Contribution to the Total Energy of the Whole Set of Temperature Data

	1	2	3	4	5	6	7
Eigenvalues $\times 10^{-3}$	464.43	59.52	5.07	0.512	0.0587	0.0061	0.001
Contribution to E (%)	87.69	11.24	0.9575	0.0967	0.0111	0.0011	0.0002

corresponding states $\{z_m(t)\}_{m=1,\ldots,6}$ is represented in Figure 14.1c. Table 14.1 includes eigenvalues $\sigma_1^2, \ldots, \sigma_7^2$ as well as the corresponding contribution of the total energy of $\mathbf{T}(t)$ signals: $\sigma_i^2/\left(\Sigma_{m=1,\ldots,n}\sigma_m^2\right)$. It can be noticed that most part of the $\mathbf{T}(t)$ signals energy is captured by the six first KLD components.

Hence, a 6D linear approximation of $\mathbf{T}(t)$ has been considered:

$$\mathbf{T}_{r=6}(t) = \mathbf{V}_{r=6}\mathbf{Z}_{r=6}(t) \Rightarrow \forall k, \quad T_k(t) \approx \sum_{m=1}^{6} v_{km}z_m(t) \tag{14.25}$$

The approximation errors $\mathbf{e}_r(t) = \mathbf{T}(t) - \mathbf{T}_r(t)$ are represented in Figure 14.1d. It can be seen that they are always less than 0.02°C, negligible compared to the values of $\mathbf{T}(t)$ signals. Hence, the whole set of data in $\mathbf{T}(t)$, 900 variables by 2000 sampling times (1.8 millions of data), can be replaced by 6 eigenfunctions (900 × 6 data) and 6 states (6 × 2000 data). Data reduction achieved is greater than 99% accurate. The results are much more spectacular in actual infrared thermography experiences, because the camera focal plane array usually includes 250 × 250 infrared sensors. Following with the example (2,000 sampling times, 6D approximation), this means replacing 125 millions of data by only 17,400 data points.

14.4 Noise Filtering Using KLD

Noise filtering is another striking feature of KLD. Noise propagation through KLD is analyzed in Section 14.4.1. It is proven that spatially uncorrelated noise has no effect on eigenfunctions, the noise being entirely reported on states. Section 14.4.2 explains why KLD truncation acts as a filter and a method for optimal filtering using KLD is described. The last paragraph in this section includes a numerical example that illustrates the powerfulness of KLD for noise reduction purposes.

14.4.1 Noise Propagation through KLD

Let us first consider infinite-dimensional problems with noise-corrupted observations

$$\tilde{T}(x, t) = T(x, t) + \varepsilon(x, t) \tag{14.26}$$

where $x = (x_1, x_2, x_3)$ represents point coordinates. Noise $\varepsilon(x, t)$ is assumed to be independent of $T(x, t)$:

$$\forall x, x' \int_t \varepsilon(x, t)T(x', t)dt = 0 \tag{14.27}$$

as well as spatially uncorrelated

$$\forall x, x' \quad W_\varepsilon(x, x') \equiv \int_t \varepsilon(x, y, t) \, \varepsilon(x', t) dt = \sigma_\varepsilon^2 \delta(x - x') \tag{14.28}$$

Taking into account noise properties, the energy function associated with the observations can be written as

$$\tilde{W}(x, x') \equiv \int_t \tilde{T}(x, t) \tilde{T}(x', t) dt = \int_t T(x, t) T(x', t) dt + \sigma_\varepsilon^2 \delta(x - x') = W(x, x') + W_\varepsilon(x, x') \tag{14.29}$$

Rearranging Equation 14.7 into Equation 14.29 leads to

$$\tilde{W}(x, x') = \sum_{m=1}^{\infty} V_m(x) \tilde{\sigma}_m^2 V_m(x') \quad \text{with } \tilde{\sigma}_m^2 = \sigma_m^2 + \sigma_\varepsilon^2 \delta(x - x') \tag{14.30}$$

which proves that eigenfunctions are not corrupted by noise. From Equation 14.30, it follows that

$$\tilde{T}(x, t) = \sum_{m=1}^{\infty} V_m(x) \tilde{z}_m(t) \tag{14.31}$$

Noise-corrupted states are hence given by

$$\tilde{z}_m(t) = \int_\Omega \tilde{T}(x, t) V_m(x) dx = \int_\Omega T(x, t) V_m(x) dx + \int_\Omega \varepsilon(x, t) V_m(x) dx = z_m(t) + \gamma_m(t) \tag{14.32}$$

where $\gamma_m(t)$ represents the orthogonal projection of the noise on eigenfunction $V_m(x)$. Comparing Equations 14.30 and 14.31, it follows that

$$\forall m, \quad \sigma_{\varepsilon, m}^2 \equiv \int_t \gamma_m^2(t) dt = \sigma_\varepsilon^2 \tag{14.33}$$

For finite-dimensional problems, similar results are achieved starting from noise-corrupted observations given by

$$\tilde{T}(t) = T(t) + \varepsilon(t) \quad \text{with } W_\varepsilon = \sigma_\varepsilon^2 I \tag{14.34}$$

Energy matrix associated with the observations is

$$\tilde{W} = W + W_\varepsilon = W + \sigma_\varepsilon^2 I = V \Sigma V^t + \sigma_\varepsilon^2 I = V \tilde{\Sigma} V^t \tag{14.35}$$

with $\tilde{\Sigma} = diag \begin{bmatrix} \sigma_1^2 + \sigma_\varepsilon^2 & \sigma_2^2 + \sigma_\varepsilon^2 & \cdots & \sigma_n^2 + \sigma_\varepsilon^2 \end{bmatrix}$. Consequently, the KLD of $\tilde{T}(t)$ is

$$\tilde{T}(t) = V \tilde{Z}(t) \tag{14.36}$$

with $\tilde{\mathbf{Z}}(t) = \mathbf{Z}(t) + \boldsymbol{\gamma}(t)$ and $\boldsymbol{\gamma}(t) = \mathbf{V}^t\,\boldsymbol{\varepsilon}(t)$. As before, it can be easily proven that

$$\int_t \boldsymbol{\gamma}(t)\boldsymbol{\gamma}^t(t)dt = \sigma_\varepsilon^2 \mathbf{I} \tag{14.37}$$

14.4.2 Optimal Noise Filtering

Consider the observations given by Equation 14.26, with noise-verifying properties described by Equations 14.27 and 14.28. The energy of the observed thermal response at point x is

$$\tilde{W}(x,x) \equiv \int_t \tilde{T}^2(x,t)dt = \int_t T^2(x,t)dt + \int_t e^2(x,t)dt = \sum_{m=1}^{\infty} V_m^2(x)\left(\sigma_m^2 + \sigma_\varepsilon^2\right) \tag{14.38}$$

The signal/noise ratio is hence

$$I(x) = \frac{\int_t T^2(x,t)dt}{\int_t e^2(x,t)dt} = \frac{\sum_{m=1}^{\infty} V_m^2(x)\sigma_m^2}{\sum_{m=1}^{\infty} V_m^2(x)\sigma_\varepsilon^2} \tag{14.39}$$

Consider now the r-dimensional approximation of $\tilde{T}(x,t)$ given by

$$\tilde{T}_r(x,t) = \sum_{m=1}^{r} V_m(x)\tilde{z}_m(t) \tag{14.40}$$

The energy of $\tilde{T}_r(x,t)$ at point x and the corresponding signal/noise ratio are

$$\tilde{W}_r(x,x) = \sum_{m=1}^{r} V_m^2(x)\left(\sigma_m^2 + \sigma_\varepsilon^2\right) \quad \text{and} \quad I_r(x) = \frac{\sum_{m=1}^{r} V_m^2(x)\sigma_m^2}{\sum_{m=1}^{r} V_m^2(x)\sigma_\varepsilon^2} \tag{14.41}$$

Comparing signal/noise ratios $I(x)$ and $I_r(x)$ leads to

$$\frac{I(x)}{I_r(x)} = \frac{\sum_{m=1}^{\infty} V_m^2(x)\sigma_m^2}{\sum_{m=1}^{r} V_m^2(x)\sigma_m^2}\frac{\sum_{m=1}^{r} V_m^2(x)\sigma_\varepsilon^2}{\sum_{m=1}^{\infty} V_m^2(x)\sigma_\varepsilon^2} = \left(1 + \frac{\sum_{m=r+1}^{\infty} V_m^2(x)\sigma_m^2}{\sum_{m=1}^{r} V_m^2(x)\sigma_m^2}\right)\left(\frac{\sum_{m=1}^{r} V_m^2(x)}{\sum_{m=1}^{\infty} V_m^2(x)}\right) \tag{14.42}$$

The first term into brackets in the equation above tends to 1 because eigenvalues satisfy $\sigma_1^2 \geq \sigma_2^2 \geq \cdots > 0$ and eigenfunctions $V_m(x)$ are $O(1)$. Indeed, first eigenvalue is usually largely dominant ($\sigma_1^2 \gg \sigma_2^2$). Hence we can write

$$\frac{I(x)}{I_r(x)} \approx \left(\frac{\sum_{m=1}^{r} V_m^2(x)}{\sum_{m=1}^{\infty} V_m^2(x)}\right) < 1 \tag{14.43}$$

This equation shows that the signal/noise ratio is improved when truncating the KLD of the observations. This effect is as much important as r is small. However, r must be high enough for avoiding significant bias; that is, for approximation error $e_r(x,t) \equiv T(x,t) - T_r(x,t)$ to be small enough.

Even if simple KLD truncation allows some data filtering, better filtering can be done. Let $\hat{T}(x,t)$ be an estimate of $T(x,t)$ obtained by $\tilde{T}(x,t)$ filtering. The quality of the estimate is usually evaluated by

$$risk \equiv \left\| T(x,t) - \hat{T}(x,t) \right\|^2 \tag{14.44}$$

It has been demonstrated (Wiener theorem, cf. Mallat 2000, p. 434) that

$$\hat{T}(x,t) = \sum_{m=1}^{\infty} V_m(x) \left(\frac{\sigma_m^2}{\sigma_m^2 + \sigma_{\varepsilon,m}^2} \right) \tilde{z}_m(t) \tag{14.45}$$

is a linear filter leading to minimum risk when $W(x,x')$ and $W_\varepsilon(x,x')$ become diagonal in the same Karhunen–Loève basis, as it is the case for $W_\varepsilon(x,x') = \sigma_\varepsilon^2 \delta(x-x')$ (uncorrelated noise).

Equation 14.45 shows that the best linear estimate of $T(x,t)$ is achieved by weighting states $\tilde{z}_m(t)$ with a factor that depends on the signal/noise ratio $\sigma_m^2/\sigma_{\varepsilon,m}^2$ in the $V_m(x)$ direction. Low signal/noise ratio will lead to strong attenuation of $\tilde{z}_m(t)$ values.

14.4.3 Numerical Example

We come back now to the example in Section 14.3.2. The plate thermal behavior simulated by time integration of Equation 14.24 has been corrupted by additive noise: $\tilde{T}(t) = T(t) + \varepsilon(t)$, with $W_\varepsilon = \sigma_\varepsilon^2 I$ (see Figure 14.3a). Noise amplitude is intentionally high ($\pm 0.5°C$) for better illustration of KLD filtering powerfulness. Figure 14.3b shows the statistical correlation of an arbitrary element of $\varepsilon(t)$, say $\varepsilon_i(t)$, with elements $\varepsilon_{j=1,\ldots,n}(t)$. It can be seen that $\forall j \neq i$ correlation is almost zero, but not zero. This means that noise energy matrix W_ε is diagonal dominant, but not strictly diagonal. KLD of $\tilde{T}(t)$ has been carried out as described in Section 14.2.2: $\tilde{T}(t) = \tilde{V}\tilde{Z}(t)$. In Figure 14.3c are represented states $\tilde{z}_1(t), \tilde{z}_2(t), \tilde{z}_3(t)$, while time behavior of $\tilde{z}_4(t), \tilde{z}_5(t), \tilde{z}_6(t)$ is depicted in Figure 14.3d. As expected, signal/noise ratio for states ($\sigma_m^2/\sigma_\varepsilon^2$) reduces as state energy (σ_m^2) decreases. The maps in Figure 14.4 represent the difference between the eigenfunctions (from first to sixth) derived from KLD of $T(t)$ and those from KLD of $\tilde{T}(t)$; they are due to a not strictly diagonal W_ε matrix. As for states, the differences are greater for eigenfunctions associated with lower eigenvalues.

Estimates of $T(t)$ from noise corrupted data in $\tilde{T}(t)$ have been carried out in two different ways, by truncation of $\tilde{T}(t)$-KLD and by truncation and states weighting:

$$\hat{T}_{r=6}(t) = \tilde{V}_{r=6}\tilde{Z}_{r=6}(t) \Rightarrow \forall k, \quad \hat{T}_k(t) \approx \sum_{m=1}^{6} \tilde{v}_{km}\tilde{z}_m(t) \tag{14.46}$$

$$\hat{T}_{r=6}(t) = \tilde{V}_{r=6}\hat{Z}_{r=6}(t) \Rightarrow \forall k, \quad \hat{T}_k(t) \approx \sum_{m=1}^{6} \tilde{v}_{km}\left(\sigma_m^2/(\sigma_m^2 + \sigma_{\varepsilon,m}^2)\right)\tilde{z}_m(t) \tag{14.47}$$

The initial noise $\varepsilon(t) = T(t) - \tilde{T}(t)$ is represented in Figure 14.5a. Noise after filtering by KLD truncation is depicted in Figure 14.5b, while noise after filtering by KLD truncation and states weighting is represented in Figure 14.5c. These three figures illustrate the powerfulness of filters applied, as well as the predominant effect of KLD truncation against states weighting even though strongest noise reduction is achieved by KLD truncation and states weighting.

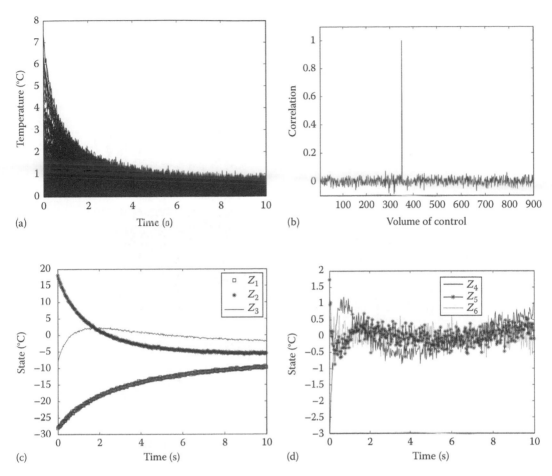

FIGURE 14.3
Highly noised experiment on a homogeneous plate: (a) temperature data (one curve by pixel), (b) one row of the added noise correlation matrix, (c) first to third KLD states associated with the temperature data, and (d) fourth to sixth KLD states associated with the temperature data.

14.5 Thermal Characterization of Homogeneous Materials Using KLD

This section focuses on the use of KLD techniques in association with infrared thermography for reliable and parsimonious thermal characterization of homogeneous materials. The mathematical method proposed for diffusivities estimation takes advantage of the powerfulness of KLD techniques for data and noise reduction. Indeed, orthogonal properties of eigenfunctions and states are intensively used for getting simple diffusivities estimates. As shown later, this leads to an exciting combination of parsimony and robustness to noise. Moreover, no analytical solutions are required for diffusivities estimation (contrary to, i.e., "slope method"), so that few constraints regarding sample illumination (spatial and time patterns) are applicable. The proposed method is, however, restricted to lock-in thermography based on thin, large enough samples; that is, to experimental situations where 2D heat transfer with adiabatic boundary conditions can be assumed.

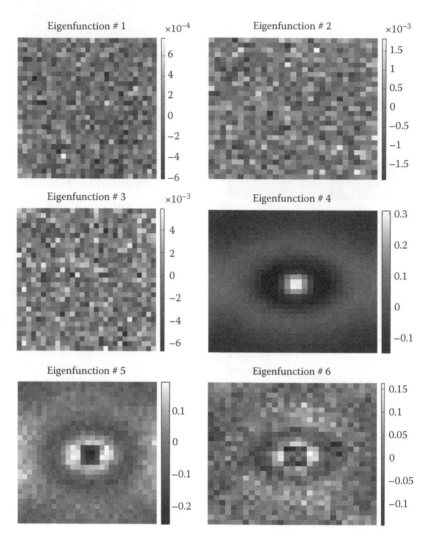

FIGURE 14.4
Noise on the eigenfunctions associated to the largest eigenvalues. Maps on the (6 mm × 6 mm) testing plate.

14.5.1 Problem Statement

Let us first consider the following heat conduction problem in a rectangular plate $L_x \times L_y$. The energy equation at inner points (Ω: $0 < x < L_x$, $0 < y < L_y$) is written as

$$\frac{\partial T(x,y,t)}{\partial t} = \alpha_x \frac{\partial^2 T(x,y,t)}{\partial x^2} + \alpha_y \frac{\partial^2 T(x,y,t)}{\partial y^2} - \beta T(x,y,t) \qquad (14.48)$$

Boundary conditions are assumed to be adiabatic:

$$\left. \frac{\partial T(x,y,t)}{\partial x} \right|_{y=0,L_y} = 0, \quad \left. \frac{\partial T(x,y,t)}{\partial y} \right|_{x=0,L_x} = 0 \qquad (14.49)$$

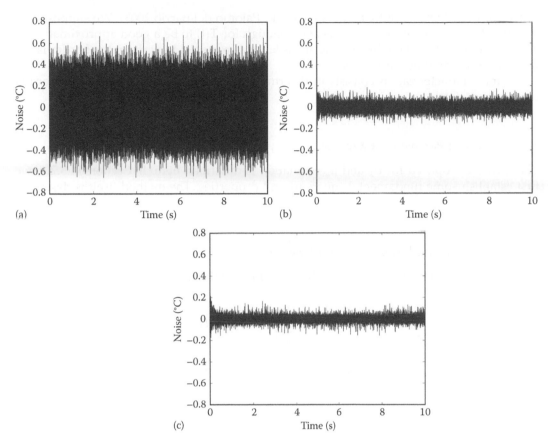

FIGURE 14.5
Noise filtering: (a) noise before filtering, (b) noise after filtering by KLD truncation, and (c) and noise after filtering by KLD truncation and states weighting.

The initial condition is $T(x, y, 0) = T_o(x, y)$. $T(x, y, t)$ represents the excess of temperature with regard to the plate surroundings, which are assumed to remain at constant temperature during the experiment. Rectangular domain and Cartesian coordinates have been intentionally chosen to be in phase with infrared camera images (small square pixels covering a rectangular or a square area). α_x and α_y represent, respectively, the thermal diffusivity in the $\overrightarrow{Ox}$ and $\overrightarrow{Oy}$ directions. β is the heat transfer coefficient between the plate and its surroundings.

To fit to such a problem, experiments must be carried out on thin samples (plates) located in an environment at constant and uniform temperature. Starting from a plate in thermal equilibrium with its environment, the initial condition $T_o(x, y)$ can be established using a light beam with almost arbitrary spatial and time patterns. Thermal relaxation of the plate is thus observed using an infrared camera. At each sampling time, a plate temperature map is recorded and stored: $\tilde{\mathbf{T}}(t) = \mathbf{T}(t) + \boldsymbol{\varepsilon}(t)$. The dimension of vector $\tilde{\mathbf{T}}(t)$ is equal to the number of pixel of the infrared image. Measurements noise is assumed to be spatially uncorrelated: $\mathbf{W}_\varepsilon = \sigma_\varepsilon^2 \mathbf{I}$. This is not true for all infrared cameras, but it has been successfully tested for the CEDIP-ORION type (Wellele and Palomo del Barrio 2008). We suppose that the lateral resolution of the camera is good enough for $\tilde{T}(t)$ to be a good approximation of $T(x, y, t)$. Indeed, the plate dimensions are as large as required to warrant the border is not reached by thermal perturbations.

Thermal characterization consists in determining α_x and α_y values from the recorded temperature data.

14.5.2 Thermal Parameters Estimation

The method proposed for α_x and α_y estimation is described here. In Section 14.5.2.1, we first establish some fundamental equations and properties. The method itself is described in Section 14.5.2.2.

14.5.2.1 Fundamental Equations and Properties

Let us again consider the KLD of the theoretical thermal field:

$$T(x, y, t) = \sum_{m=1}^{\infty} V_m(x, y) z_m(t) \tag{14.50}$$

We remind that eigenfunctions and states verify orthogonal properties:

$$\langle V_k, V_m \rangle \equiv \int_{\Omega} V_k(x) V_m(x) dx = \delta_{km} \quad \langle z_m(t), z_k(t) \rangle \equiv \int_t z_m(t) z_k(t) dt = \delta_{mk} \sigma_m^2 \tag{14.51}$$

Introducing Equation 14.50 into Equation 14.48 leads to

$$\sum_{m=1}^{\infty} V_m(x, y) \frac{dz_m(t)}{dt} = \alpha_x \sum_{m=1}^{\infty} \frac{\partial^2 V_m(x, y)}{\partial x^2} z_m(t) + \alpha_y \sum_{m=1}^{\infty} \frac{\partial^2 V_m(x, y)}{\partial y^2} z_m(t) - \beta \sum_{m=1}^{\infty} V_m(x, y) z_m(t) \tag{14.52}$$

Multiplying this equation by $V_i(x, y)$, integrating over Ω and taking into account eigenfunctions orthogonal property, we obtain

$$\frac{dz_i(t)}{dt} = \alpha_x \sum_{m=1}^{\infty} \left\langle \frac{\partial^2 V_m(x, y)}{\partial x^2}, V_i(x, y) \right\rangle_{\Omega} z_m(t) + \alpha_y \sum_{m=1}^{\infty} \left\langle \frac{\partial^2 V_m(x, y)}{\partial y^2}, V_i(x, y) \right\rangle_{\Omega} z_m(t) - \beta z_i(t) \tag{14.53}$$

The above equation is now multiplied by $z_k(t)$ and integrated over time. Taking into account the orthogonal property of the states, this leads to

$$\frac{1}{\sigma_k^2} \left\langle \frac{dz_i(t)}{dt}, z_k(t) \right\rangle_t = \alpha_x \left\langle \frac{\partial^2 V_k(x, y)}{\partial x^2}, V_i(x, y) \right\rangle_{\Omega} + \alpha_y \left\langle \frac{\partial^2 V_k(x, y)}{\partial y^2}, V_i(x, y) \right\rangle_{\Omega} - \beta \delta_{ik} \tag{14.54}$$

In a similar way, from Equations 14.49 through 14.51, it can be easily proven that

$$\forall m, \left.\frac{\partial V_m(x,y)}{\partial x}\right|_{y=0,\,L_y} = 0, \quad \left.\frac{\partial V_m(x,y)}{\partial y}\right|_{x=0,\,L_x} = 0 \tag{14.55}$$

As demonstrated in Appendix 14.A, the following identity is verified for all (i,k)

$$\frac{1}{\sigma_k^2}\left\langle \frac{dz_i(t)}{dt}, z_k(t)\right\rangle_t = \frac{1}{\sigma_i^2}\left\langle \frac{dz_k(t)}{dt}, z_i(t)\right\rangle_t \tag{14.56}$$

Taking into account Equations 14.54 and 14.56, it follows that

$$\left\langle \frac{\partial^2 V_k(x,y)}{\partial \zeta^2}, V_i(x,y)\right\rangle_\Omega = \left\langle \frac{\partial^2 V_i(x,y)}{\partial \zeta^2}, V_k(x,y)\right\rangle_\Omega, \quad \zeta = x, y \tag{14.57}$$

Accordingly, Equation 14.54 for (i,k) indexes and Equation 14.54 for (k,i) are identical equations. Indeed, we can write

$$\frac{1}{\sigma_k^2}\left\langle \frac{dz_i(t)}{dt}, z_k(t)\right\rangle_t = \alpha_x\left\langle \frac{\partial^2 V_i(x,y)}{\partial x^2}, V_k(x,y)\right\rangle_\Omega + \alpha_y\left\langle \frac{\partial^2 V_i(x,y)}{\partial y^2}, V_k(x,y)\right\rangle_\Omega - \beta\delta_{ik} \tag{14.58}$$

$$\frac{1}{\sigma_i^2}\left\langle \frac{dz_k(t)}{dt}, z_i(t)\right\rangle_t = \alpha_x\left\langle \frac{\partial^2 V_i(x,y)}{\partial x^2}, V_k(x,y)\right\rangle_\Omega + \alpha_y\left\langle \frac{\partial^2 V_i(x,y)}{\partial y^2}, V_k(x,y)\right\rangle_\Omega - \beta\delta_{ik} \tag{14.59}$$

Adding equations above leads to

$$\frac{1}{\sigma_i^2 + \sigma_k^2}\left\langle \frac{dz_i(t)z_k(t)}{dt}\right\rangle_t = \alpha_x\left\langle \frac{\partial^2 V_i(x,y)}{\partial x^2}, V_k(x,y)\right\rangle_\Omega + \alpha_y\left\langle \frac{\partial^2 V_i(x,y)}{\partial y^2}, V_k(x,y)\right\rangle_\Omega - \beta\delta_{ik} \tag{14.60}$$

Consequently, we can write

$$\frac{z_i(t_f)z_k(t_f) - z_i(0)z_k(0)}{\sigma_i^2 + \sigma_k^2} = \alpha_x\left\langle \frac{\partial^2 V_i(x,y)}{\partial x^2}, V_k(x,y)\right\rangle_\Omega + \alpha_y\left\langle \frac{\partial^2 V_i(x,y)}{\partial y^2}, V_k(x,y)\right\rangle_\Omega - \beta\delta_{ik} \tag{14.61}$$

where t_f is the experiment end-time.

Another interesting equation comes from integration of Equation 14.48 over Ω. For adiabatic boundary conditions, this leads to

$$\frac{d\langle T(x,y,t)\rangle_\Omega}{dt} = -\beta\langle T(x,y,t)\rangle_\Omega \tag{14.62}$$

Hence

$$\langle T(x,y,t)\rangle_\Omega = \langle T_o(x,y)\rangle_\Omega \exp(-\beta t) \tag{14.63}$$

14.5.2.2 Estimation Method

For free-noise space-time continuous observations, parameters α_x, α_y, and β can be calculated from any of the three arbitrarily chosen Equations 14.61. On the contrary, for noise-corrupted observations, it is convenient to cast the α_x, α_y, and β estimation problem into a least squares problem.

Taking into account the powerfulness of the spatial-mean operator $\langle ° \rangle_\Omega$ for noise reduction, the best estimate of β is achieved applying the linear least squared method to Equation 14.63. This leads to

$$\hat{\beta} = \langle t^2 \rangle_t^{-1} \left\langle -t \ln\left(\frac{\bar{T}(t)}{\bar{T}_o}\right) \right\rangle_t \tag{14.64}$$

with $\bar{T}(t) = \langle \tilde{T}(x,y,t) \rangle_\Omega$ and $\bar{T}_o = \langle T_o(x,y) \rangle_\Omega$.

The estimation of α_x and α_y could be based on Equation 14.61, but the question is how many of these equations and which ones must to be used. At least two equations are required because the problem involves two unknown parameters. However keeping all of them will be a wrong strategy because there are terms in the KLD of $\tilde{T}(x,y,t)$, which are not significant compared to the noise. As shown in Section 14.4.1, signal/noise rate for states $\tilde{z}_m(t)$ is $\sigma_m^2/\sigma_\varepsilon^2$. On the other hand, eigenvalues σ_m^2 usually decrease quickly, so that $\sigma_1^2 \gg \sigma_2^2 \gg \sigma_3^2 \gg \cdots$ (see Table 14.1). Hence, Equation 14.61 that will be preferred for diffusivities estimation are those involving the states with largest eigenvalues, say $\tilde{z}_1(t)$ and $\tilde{z}_2(t)$. As demonstrated in the previous paragraph, Equation 14.61 with $i=1$ and $k=2$ and Equation 14.61 with $i=2$ and $k=1$ are equivalent. However, for no perfectly uncorrelated noise (see Section 14.4.3), eigenfunctions from KLD of $\tilde{T}(x,y,t)$ will be corrupted by noise, with degradation increasing with decreasing eigenvalues. Consequently, Equation 14.61 with $i=1$ and $k=2$ will be preferred to Equation 14.61 with $i=2$ and $k=1$ in order to limit noise amplification effects due to second-order space derivatives. We can hence write

$$\mathbf{y} = \mathbf{M} \begin{bmatrix} \alpha_x \\ \alpha_y \end{bmatrix} \tag{14.65}$$

with

$$\mathbf{y} = \begin{bmatrix} \Delta\tilde{z}_1\tilde{z}_1/(\tilde{\sigma}_1^2 + \tilde{\sigma}_1^2) + \hat{\beta} \\ \Delta\tilde{z}_1\tilde{z}_2/(\tilde{\sigma}_1^2 + \tilde{\sigma}_2^2) \\ \Delta\tilde{z}_2\tilde{z}_2/(\tilde{\sigma}_2^2 + \tilde{\sigma}_2^2) + \hat{\beta} \end{bmatrix}, \quad \mathbf{M} = \begin{bmatrix} \langle \partial_x^2\tilde{V}_1(x,y),\ \tilde{V}_1(x,y) \rangle_\Omega & \langle \partial_y^2\tilde{V}_1(x,y),\ \tilde{V}_1(x,y) \rangle_\Omega \\ \langle \partial_x^2\tilde{V}_1(x,y),\ \tilde{V}_2(x,y) \rangle_\Omega & \langle \partial_y^2\tilde{V}_1(x,y),\ \tilde{V}_2(x,y) \rangle_\Omega \\ \langle \partial_x^2\tilde{V}_2(x,y),\ \tilde{V}_2(x,y) \rangle_\Omega & \langle \partial_y^2\tilde{V}_2(x,y),\ \tilde{V}_2(x,y) \rangle_\Omega \end{bmatrix} \tag{14.66}$$

and $\Delta\tilde{z}_i\tilde{z}_k = \tilde{z}_i(t_f)\tilde{z}_k(t_f) - \tilde{z}_i(0)\tilde{z}_k(0), \partial_\zeta^2\tilde{V}_i(x,y) = \partial^2\tilde{V}_i(x,y)/\partial\zeta^2$.

The solution of Equation 14.65, in the least squares sense, leads to

$$\begin{bmatrix} \hat{\alpha}_x \\ \hat{\alpha}_y \end{bmatrix} = (\mathbf{M}'\mathbf{M})^{-1}\mathbf{M}'\mathbf{y} \tag{14.67}$$

In the experimental framework of infrared thermography with high lateral resolution, $\tilde{T}(t)$ can be often considered as a good approximation of $\tilde{T}(x,y,t)$. Hence, KLD of $\tilde{T}(t)$ (Equation 14.34) provides appropriate approximations of eigenfunctions $V_m(x,y)$ and states $\tilde{z}_m(t)$. Using previous notation for finite-dimensional problems, with eigenvectors $\tilde{V} = [\tilde{v}_1 \quad \tilde{v}_2 \quad \cdots \quad \tilde{v}_n]$, the elements of the matrix M become

$$\left\langle \partial_\zeta^2 \tilde{V}_i(x,y), \tilde{V}_k(x,y) \right\rangle_\Omega \approx v_k^t L_\zeta v_i \tag{14.68}$$

where L_ζ is the numerical approximation of $\partial^2/\partial\zeta^2$.

For isotropic materials ($\alpha_x = \alpha_y$), the estimate of the diffusivity can be easily obtained by addition of the columns of the matrix M.

14.5.3 Numerical Example

We come back again to the example in Section 14.3.2. The plate thermal behavior simulated by time integration of Equation 14.24 has been corrupted by additive noise: $\tilde{T}(t) = T(t) + \varepsilon(t)$ ($n \times 1$, $n = 900$), with $W_\varepsilon = \sigma_\varepsilon^2 I$. Three different values of noise amplitude have been considered: $\pm 0.5°C$ (bad-quality data), $\pm 0.1°C$ (medium-quality data), and $\pm 0.02°C$ (good-quality data).

The estimation of the parameters α_x, α_y, and β has been carried out by the method described in Section 14.5.2. Step by step this means:

1. Calculation of the KLD of $\tilde{T}(t)$: $\tilde{T}(t) = \tilde{V}\tilde{Z}(t)$ ($n \times 1$), with $\tilde{V} = [\tilde{v}_1 \quad \tilde{v}_2 \quad \cdots \quad \tilde{v}_n]$. This involves calculation and spectral decomposition of the energy matrix: $\tilde{W} = \tilde{V}\tilde{\Sigma}\tilde{V}^t$, with $\tilde{\Sigma} = diag[\tilde{\sigma}_1^2 \quad \tilde{\sigma}_2^2 \quad \cdots \quad \tilde{\sigma}_n^2]$, and thus the calculation of the states: $\tilde{Z}(t) = \tilde{V}^t\tilde{T}(t)$. It must be noticed that the number of the KLD components required for estimations is very low. Hence, algorithms allowing calculation of only the $r \ll n$ largest eigenvalues and associated eigenvectors of W can be used.

2. States filtering as described in Section 14.4.2: $\tilde{Z}(t) \leftarrow F\tilde{Z}(t)$, with $F = diag[\tilde{\sigma}_1^2/(\tilde{\sigma}_1^2 + \tilde{\sigma}_\varepsilon^2) \quad \tilde{\sigma}_2^2/(\tilde{\sigma}_2^2 + \tilde{\sigma}_\varepsilon^2) \quad \cdots \quad \tilde{\sigma}_n^2/(\tilde{\sigma}_n^2 + \tilde{\sigma}_\varepsilon^2)]$.

3. Estimation of the parameter β by applying the linear least squares method to equation $\bar{T}(t) = \bar{T}_o \exp(-\beta t)$, where $\bar{T}(t)$ represents the mean value of $\tilde{T}(t)$ at time t and $\bar{T}_o = \bar{T}(0)$.

4. Estimation of α_x and α_y using Equation 14.67. This involves calculation of vector y and matrix M (Equation 14.66). The elements of M are approached as described by Equation 14.67.

The results achieved are summarized in Table 14.2. It can be seen that even for bad-quality data (noise amplitude equal to $\pm 0.5°C$), estimates of $\hat{\beta}$, $\hat{\alpha}_x$, and $\hat{\alpha}_y$ are very close

TABLE 14.2

Homogeneous Plate: Estimated Values for Thermal Parameters

Noise Amplitude (°C)	$\hat{\beta}$ (s^{-1})	$\left\|\dfrac{\beta - \hat{\beta}}{\beta}\right\| \times 100$	$\hat{\alpha}_x$ ($\times 10^{-6}$ m^2 s^{-1})	$\left\|\dfrac{\alpha_x - \hat{\alpha}_x}{\alpha_x}\right\| \times 100$	$\hat{\alpha}_y$ ($\times 10^{-6}$ m^2 s^{-1})	$\left\|\dfrac{\alpha_y - \hat{\alpha}_y}{\alpha_y}\right\| \times 100$
± 0.50	0.0153	0.91	0.1493	1.44	0.2989	1.37
± 0.10	0.0151	0.028	0.1517	0.15	0.3026	0.15
± 0.02	0.0152	0.017	0.1514	0.097	0.3034	0.124

True values are $\beta = 0.152$ s^{-1}, $\alpha_x = 0.1515 \times 10^{-6}$ m^2 s^{-1}, and $\alpha_y = 0.3030 \times 10^{-6}$ m^2 s^{-1}.

to β, α_x, and α_y values. Maximum bias (1.44%) is observed on $\hat{\alpha}_x$ when using poor-quality data. Such results evidence the robustness to noise of the estimation method proposed.

The method has been already successfully applied on actual experiments for thermal characterization of orthotropic carbon/epoxy materials (Palomo del Barrio and Dauvergne 2009).

14.6 Thermal Characterization of Heterogeneous Materials Using KLD

This section deals with thermal characterization of heterogeneous materials using KLD techniques in association with lock-in thermography experiments based on thin samples testing. As in the previous section, the mathematical method proposed for phase diffusivity estimation takes advantage of the KLD properties for data and noise reduction. Moreover, simple diffusivities estimates are obtained using orthogonal properties of eigenfunctions and states. As for homogeneous materials, this leads to an attractive combination of parsimony and robustness to noise.

14.6.1 Problem Statement

Let us consider a heterogeneous material coming from physical aggregation of n different phases, as well as a thin sample (plate) of this material. Let Ω_i $(i = 1, 2, \ldots, n)$ denote the region of the space occupied by the i-phase so that $\Omega = \Omega_1 \cup \Omega_2 \cup \cdots \cup \Omega_n$ represents the indoor domain of the plate. The boundary separating the medium from its environment is $\partial\Omega$, and the interface between the i-phase and j-phase is referred as $\partial\Gamma_{ij}$. To simplify notation, the phases are supposed to be isotropic, although theoretical analysis and methods in this paper could be applied to anisotropic cases too.

For time $t > 0$ and points belonging to Ω_i $(i = 1, 2, \ldots, n)$, the energy equation can be written as

$$\frac{\partial T(x,t)}{\partial t} = \alpha_i \nabla^2 T(x,t) - \beta_i T(x,t) \qquad (14.69)$$

where
$x = (x_1, x_2, x_3)$ represents point coordinates
$T(x, t)$ is the excess of temperature with regard to the surroundings, which are assumed to remain at uniform and constant temperature during the experiment
α_i and β_i are, respectively, the thermal diffusivity and the heat exchange coefficient between the i-phase and the environment

The initial condition is $T(x, 0) = T_o(x)$. The plate dimensions are as large as required to warrant the border is not reached by thermal perturbations. Hence, adiabatic boundary conditions on $\partial\Omega$ can be assumed to be

$$\nabla T(x,t)\vec{n} = 0 \quad \forall t, \forall x \in \partial\Omega \qquad (14.70)$$

where $\vec{n}$ represents the unit outward-drawn vector normal to $\partial\Omega$ at point x. At the interface $\partial\Gamma_{ij}$, Robin-type conditions are assumed $(\forall t, \forall x \in \partial\Gamma)$:

$$-k_i \nabla T(x,\, t)|^i \vec{n}_{ij} = -h\big(T(x,t)|^i - T(x,t)|^j\big) \tag{14.71}$$

$$-k_j \nabla T(x,t)|^j \vec{n}_{ij} = -h\big(T(x,t)|^j - T(x,t)|^i\big) \tag{14.72}$$

where

k_i and k_j are, respectively, the thermal conductivity of i-phase and j-phase

h represents the inverse of the thermal resistance at the interface

$\vec{n}_{ij}$ is the unit normal vector to the interface directed from Ω_i to Ω_j at point x

As for homogeneous materials, infrared thermography experiments must be carried out on thin samples (plates). Operators ∇ and ∇^2 in the equations above represent 2D gradient and Lapacian, respectively. Starting from a plate in thermal equilibrium with its environment, the initial condition $T_o(x)$ can be established using a light beam with almost arbitrary spatial and time patterns. Thermal relaxation of the plate is thus observed using an infrared camera. At each sampling time, a plate temperature-map is recorded and stored: $\tilde{\mathbf{T}}(t) = \mathbf{T}(t) + \boldsymbol{\varepsilon}(t)$. The dimension of vector $\tilde{\mathbf{T}}(t)$ is equal to the number of pixels of the infrared image. The noise measurement is assumed to be spatially uncorrelated: $\mathbf{W}_\varepsilon = \sigma_\varepsilon^2 \mathbf{I}$. Moreover, we suppose that the lateral resolution of the camera is high enough for $\tilde{\mathbf{T}}(t)$ to be a good approximation of $T(x,t)$.

Thermal characterization consists in determining $\alpha_i(l=1,2,\dots,n)$ values from the recorded temperature data.

14.6.2 Phase Discrimination and Interface Location

The first step toward thermal characterization consists in phase recognition and interface $\partial\Gamma$ location. As shown in Godin et al. (2010), this can be efficiently done applying KLD to temperature data coming from a simple lock-in thermography experiment. The experiment consists in illuminating the sample with a lamp providing uniform and constant intensity over the full area of the sample during a short time, and recording the plate temperature behavior using an infrared camera. Uniform illumination and short times are required to ensure that points belonging to the same phase will show almost identical thermal behaviors. Details on the experimental device, as well as on the expected thermal behaviors, can be found in Godin et al. (2009).

Let $T(x,t)$ be the temperature field of the plate at time t and $\mathbf{T}(t)$ (measurements) a good enough spatial sampling of $T(x,t)$. It has been demonstrated in Godin et al. (2010) that the rank of the energy matrix $\mathbf{W}$ of $\mathbf{T}(t)$ is equal to the number of phases within the plate. Moreover, it has been proven that KLD eigenfunctions do not change of sign within phases: $\forall x, x' \in \Omega_i \Rightarrow \forall m \ V_m(x)V_m(x') > 0$. On the contrary, it has been demonstrated that $\forall x, x'/x \in \Omega_i$ and $\forall x' \in \Omega_j \Rightarrow \exists m / V_m(x)V_m(x') < 0$. This leads to a simple method for phase discrimination based on eigenfunction sign analysis. For instance, in a two-phases plate, it can be shown that points with $sign(V_2(x)) > 0$ belong to one of the phases, say Ω_i, while points with $sign(V_2(x)) < 0$ belong to the other one, say Ω_j.

Let us consider the two-phase (black and white) plate in Figure 14.6. For constant environment temperature, heat transfer equation within the phases can be written as

$$\frac{\partial T(x,y,t)}{\partial t} = \alpha_i \nabla^2 T(x,y,t) - \beta_i T(x,y,t) + \varepsilon_i \varphi, \quad i = \text{black, white} \tag{14.73}$$

FIGURE 14.6
Two-phase heterogeneous plate (6 mm × 6 mm).

with

$$\alpha_i = \frac{k_i}{(\rho c)_i}, \quad \beta_i = \frac{h}{(\rho c)_i e}, \quad \varepsilon_i = \frac{a_i}{(\rho c)_i e},$$

where
φ represents the radiative density flux applied on the plate
a_i is the absorption coefficient of the i-phase
e is the thickness of the plate

Boundary conditions are assumed to be adiabatic. At the interfaces between phases, heat flux continuity and equality of temperatures conditions are applied. Thermal properties are specified in Table 14.3. Two different scenarios have been considered: high contrast and low contrast of physical properties between phases.

Experiments have been simulated using model (14.74). The finite volume method is applied on an equally spaced $n \times n$ ($n = 30$) grid for energy equation discretization. The resulting state-space model is thus integrated in time, from time $t = 0$ to $t = 1$ s. The achieved results, $\mathbf{T}(t)$, are finally corrupted with additive noise: $\tilde{\mathbf{T}}(t) = \mathbf{T}(t) + \boldsymbol{\varepsilon}(t)$, with $\mathbf{W}_\varepsilon = \sigma_\varepsilon^2 \mathbf{I}$. High-quality data (noise amplitude: ±0.02°C) and poor-quality data (noise amplitude: ±0.5°C) have been considered.

The simulated thermal behaviors ($\tilde{\mathbf{T}}(t)$ signals) are depicted in Figure 14.7. Figure 14.7a corresponds to a plate with high contrast in the physical properties of the phases and low-noise amplitude (±0.02°C), while Figure 14.7b represents the thermal behavior of the same plate but with high-noise amplitude (±0.5°C). Figure 14.7c includes $\tilde{\mathbf{T}}(t)$ signals

TABLE 14.3

Heterogeneous Plate: Physical Properties of the White and Black Phases for Phases Discrimination and Interfaces Location within the Plate

	High Contrast Case		Low Contrast Case	
	Black Phase	**White Phase**	**Black Phase**	**White Phase**
Thermal conductivity k (W m^{-1} K^{-1})	20	0.5	0.5	0.5
Thermal capacity ρc (J m^{-3} K^{-1})	0.99×10^6	3.30×10^6	3.15×10^6	3.30×10^6
Absorption coefficient a (−)	0.95	0.80	0.80	0.80
Heat exchange coefficient h (W m^{-2} K^{-1})	10	10	10	10

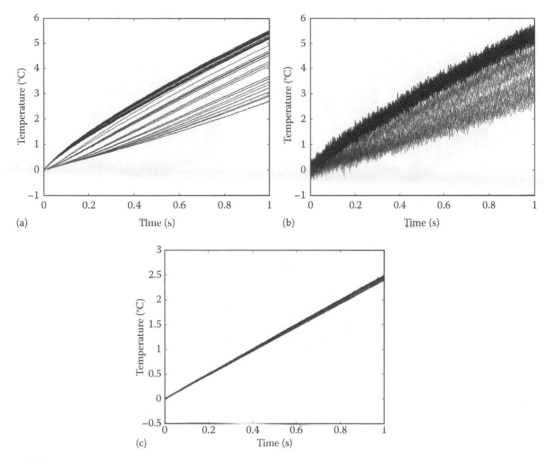

FIGURE 14.7
Simulated temperature behavior (one curve by pixel): (a) plate with a high contrast in the thermal properties of phases and low-noise amplitude, (b) plate with a high contrast in the thermal properties of phases and high-noise amplitude, and (c) plate with a low contrast in the thermal properties of phases and low-noise amplitude.

achieved for a plate with low physical properties contrast and low-noise amplitude ($\pm 0.02°C$). It must be noticed that phases cannot be discriminated to the naked eye in the two last cases.

The second eigenfunctions derived from KLD of $\tilde{T}(t)$ are depicted in Figure 14.8. All of them are characterized by near singularities (sharpness) at points located on the interfaces. In the three cases, the microstructure map reached using the $sign[V_2(x)]$ criterion for phases discrimination match exactly with the plate microstructure in Figure 14.6. However, results achieved for a plate with low contrast in the physical properties of the phases and high-noise amplitude are not satisfactory at all. A deeper analysis on the limits of eigenfunctions sign analysis for phases discrimination can be found in Godin et al. (2010).

14.6.3 Estimation of Thermal Properties

The method proposed for estimation of thermal properties is described in this section. In Section 14.6.3.1, we establish some fundamental equations and properties, while the

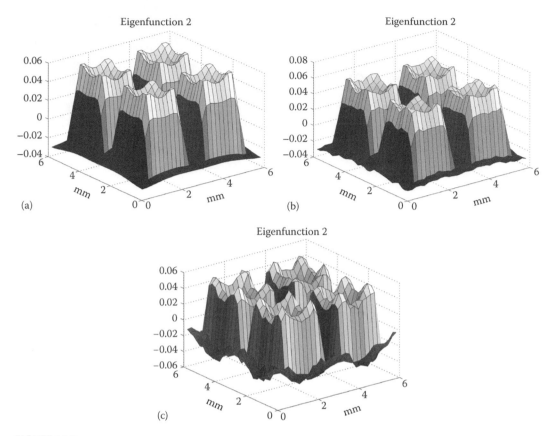

FIGURE 14.8
Second eigenfunction: (a) plate with a high contrast in the thermal properties of phases and low-noise amplitude, (b) plate with a high contrast in the thermal properties of phases and high-noise amplitude, and (c) plate with a low contrast in the thermal properties of phases and low-noise amplitude.

method itself is described in Section 14.6.3.2. Without loss of generality, only two-phase media are considered. We note $\Omega = \Omega_1 \cup \Omega_2$ the plate inner domain, $\partial\Omega = \partial\Omega_1 \cup \partial\Omega_2$ the plate boundary, and $\partial\Gamma$ the interface between the two phases within the plate.

14.6.3.1 Fundamental Equations and Properties

Let us consider the KLD of the theoretical thermal field:

$$T(x,t) = \sum_{m=1}^{\infty} V_m(x)z_m(t) \quad \forall x \in \Omega \tag{14.74}$$

with $\langle z_m(t), z_k(t)\rangle_t = \delta_{mk}\sigma_m^2$ and $\langle V_k(x), V_m(x)\rangle_\Omega = \delta_{km}$. It must be noticed that last inner product refers to the entire spatial domain.

Multiplying Equation 14.69 by $V_k(x)$ and integrating over Ω_i ($i = 1, 2$), leads to

$$\left\langle \frac{\partial T(x,t)}{\partial t}, V_k(x)\right\rangle_{\Omega_i} = \alpha_i \langle \nabla^2 T(x,t), V_k(x)\rangle_{\Omega_i} - \beta_i \langle T(x,t), V_k(x)\rangle_{\Omega_i} \tag{14.75}$$

The second theorem of Green, with adiabatic conditions on $\partial\Omega_i$, allows writing

$$\langle \nabla^2 T(x,t), V_k(x) \rangle_{\Omega_i} = \langle \nabla^2 V_k(x), T(x,t) \rangle_{\Omega_i} + I_k^{(i)}(t) \tag{14.76}$$

with

$$I_k^{(i)}(t) = \int_{\partial\Gamma} [V_k(x)\nabla T(x,t)\vec{n}_{ij} - T(x,t)\nabla V_k(x)\vec{n}_{ij}]\,d\gamma \tag{14.77}$$

Substituting Equation 14.76 into Equation 14.75 and adding resulting equations for $i=1$ and $i=2$, leads to

$$\left\langle \frac{\partial T(x,t)}{\partial t}, V_k(x) \right\rangle_\Omega = \sum_{i=1,2} \alpha_i \langle \nabla^2 T(x,t), V_k(x) \rangle_{\Omega_i} + I_k(t) - \sum_{i=1,2} \beta_i \langle T(x,t), V_k(x) \rangle_{\Omega_i} \tag{14.78}$$

with $I_k(t) = I_k^{(1)}(t) + I_k^{(2)}(t)$. Moreover, it can be demonstrated (Appendix 14.B) that

$$\langle I_k(t), z_k(t) \rangle_t = 0 \tag{14.79}$$

We now replace $T(x,t)$ in Equation 14.78 by its KLD, we multiply by $z_k(t)$ and integrate over time. Taken into account $\langle z_m(t), z_k(t) \rangle_t = \delta_{mk}\sigma_m^2$ and $\langle I_k(t), z_k(t) \rangle_t = 0$, we obtain

$$\frac{1}{\sigma_k^2} \langle \dot{z}_k(t), z_k(t) \rangle_t = \sum_{i=1,2} \alpha_i \langle \nabla^2 V_k, V_k(x) \rangle_{\Omega_i} - \sum_{i=1,2} \beta_i \langle V_k(x), V_k(x) \rangle_{\Omega_i} \tag{14.80}$$

or

$$\frac{z_k^2(t)}{2\sigma_k^2}\bigg|_{t=0}^{t_f} = \sum_{i=1,2} \alpha_i \langle \nabla^2 V_k, V_k(x) \rangle_{\Omega_i} - \sum_{i-1,2} \beta_i \langle V_k(x), V_k(x) \rangle_{\Omega_i} \tag{14.81}$$

The estimation of diffusivities will be based on these equations. It must be noticed that variables and parameters related to heat exchanges at the interface $\partial\Gamma$ does not appear in Equation 14.81. This is a significant advantage for reliable estimation of the diffusivities. First, the number of unknown physical parameters is reduced (i.e., thermal resistances have not to be estimated); second, knowledge or estimation of $V_k(x)$ values on $\partial\Gamma$, which could be a tricky task, is not required.

At last, it can be easily proven that integration over Ω of Equations 14.69 leads to

$$\frac{\gamma_1}{\beta_1} \frac{d\bar{T}_1(t)}{dt} + \frac{(1-\gamma_1)}{\beta_2} \frac{d\bar{T}_2(t)}{dt} = \bar{T}(t) \tag{14.82}$$

$\bar{T}_i(t) = \langle T(x,t) \rangle_{\Omega_i} (i=1,2)$ and γ_1 is the fraction of the plate surface, which is occupied by the phase 1 (surface of Ω_1/surface of Ω). As for homogeneous materials, estimation of β_i ($i=1,2$) parameters will be based on the equation above.

14.6.3.2 Estimation Method

For free-noise observations, parameters α_i and β_i ($i = 1, 2$) can be easily calculated from any four arbitrarily chosen Equations 14.81. On the contrary, for noise-corrupted observations, it is convenient to cast the estimation problem into a least squares problem as done in the case of homogeneous materials.

Let $\tilde{\mathbf{T}}(t)(n \times 1)$ be the vector including temperature measurements (plate temperature map) at time t, where n represents the number of pixels of the image supplied by the infrared camera. As previously, it is assumed $\tilde{\mathbf{T}}(t)$ to be a good approximation of $\tilde{T}(x, y, t)$. Vectors $\tilde{\mathbf{T}}_1(t)$ and $\tilde{\mathbf{T}}_2(t)$ include the elements of $\tilde{\mathbf{T}}(t)$ belonging to Ω_1 and Ω_2, respectively. Mean values of $\tilde{\mathbf{T}}(t)$, $\tilde{\mathbf{T}}_1(t)$, and $\tilde{\mathbf{T}}_2(t)$ at time t are named $\bar{T}(t)$, $\bar{T}_1(t)$, and $\bar{T}_2(t)$ in the following. The distribution of the phases within the plate is assumed to be known.

As mentioned before, the estimation of β_i ($i = 1, 2$) parameters will be based on Equation 14.82. Applying the linear least squares method to Equation 14.82 leads to

$$\begin{bmatrix} 1/\hat{\beta}_1 \\ 1/\hat{\beta}_2 \end{bmatrix} = (\mathbf{M'M})^{-1}(\mathbf{M'y}) \tag{14.83}$$

with

$$(\mathbf{M'M}) = \int_t \left[\gamma_1 \dot{\bar{T}}_1(t) \quad (1 - \gamma_1) \dot{\bar{T}}_2(t) \right]^t \left[\gamma_1 \dot{\bar{T}}_1(t) \quad (1 - \gamma_1) \dot{\bar{T}}_2(t) \right] dt$$

$$(\mathbf{M'y}) = \int_t \left[\gamma_1 \dot{\bar{T}}_1(t) \quad (1 - \gamma_1) \dot{\bar{T}}_2(t) \right]^t \bar{T}(t) dt \tag{14.84}$$

In practice, signals $\tilde{\mathbf{T}}_1(t)$ and $\tilde{\mathbf{T}}_2(t)$ have to be filtered in order to reduce noise amplification due to time derivatives.

The estimation of diffusivities is based on Equations 14.81, assuming parameters β_i ($i = 1, 2$) already known. The first step toward α_i ($i = 1, 2$) estimation is KLD of $\tilde{\mathbf{T}}(t)$. As mentioned previously, this involves calculation and spectral decomposition of the energy matrix: $\tilde{\mathbf{W}} = \tilde{\mathbf{V}} \tilde{\boldsymbol{\Sigma}} \tilde{\mathbf{V}}^t$, with $\tilde{\mathbf{V}} = [\tilde{\mathbf{v}}_1 \quad \tilde{\mathbf{v}}_2 \quad \cdots \quad \tilde{\mathbf{v}}_n]$ and $\tilde{\boldsymbol{\Sigma}} = diag [\tilde{\sigma}_1^2 \quad \tilde{\sigma}_2^2 \quad \cdots \quad \tilde{\sigma}_n^2]$. States can be thus calculated by $\tilde{\mathbf{Z}}(t) = \tilde{\mathbf{V}}^t \tilde{\mathbf{T}}(t)$. Next, they are filtered as described in Section 14.4.2: $\tilde{\mathbf{Z}}(t) \leftarrow \mathbf{F}\tilde{\mathbf{Z}}(t)$, with $\mathbf{F} = diag [\tilde{\sigma}_1^2/(\tilde{\sigma}_1^2 + \tilde{\sigma}_\varepsilon^2) \quad \tilde{\sigma}_2^2/(\tilde{\sigma}_2^2 + \tilde{\sigma}_\varepsilon^2) \quad \cdots \quad \tilde{\sigma}_n^2/(\tilde{\sigma}_n^2 + \tilde{\sigma}_\varepsilon^2)]$.

The second step for diffusivities estimation consists in selecting significant states, i.e., $z_1(t), z_2(t), \ldots, z_r(t)$, those showing high enough signal/noise ratio ($\tilde{\sigma}_k^2/\tilde{\sigma}_\varepsilon^2$). Equations 14.81 for $k = 1, \ldots, r$ are then written in the matrix form:

$$\mathbf{y} = \mathbf{M} \begin{bmatrix} \alpha_1 \\ \alpha_2 \end{bmatrix} \tag{14.85}$$

with

$$\mathbf{y} = \begin{bmatrix} \Delta \tilde{z}_1^2/\tilde{\sigma}_1^2 + \sum_{i=1}^{2} \hat{\beta}_i \langle V_1(x), V_1(x) \rangle_{\Omega_i} \\ \Delta \tilde{z}_2^2/\tilde{\sigma}_2^2 + \sum_{i=1}^{2} \hat{\beta}_i \langle V_2(x), V_2(x) \rangle_{\Omega_i} \\ \cdots \\ \Delta \tilde{z}_r^2/\tilde{\sigma}_r^2 + \sum_{i=1}^{2} \hat{\beta}_i \langle V_r(x), V_r(x) \rangle_{\Omega_i} \end{bmatrix} \quad \Delta \tilde{z}_i^2 = \frac{1}{2} \tilde{z}_i^2 \Big|_{t=0}^{t_f} \tag{14.86}$$

$$\mathbf{M} = \begin{bmatrix} \langle \nabla^2 \tilde{V}_1(x,y),\ \tilde{V}_1(x,y) \rangle_{\Omega_1} & \langle \nabla^2 \tilde{V}_1(x,y),\ \tilde{V}_1(x,y) \rangle_{\Omega_2} \\ \langle \nabla^2 \tilde{V}_2(x,y),\ \tilde{V}_2(x,y) \rangle_{\Omega_1} & \langle \nabla^2 \tilde{V}_2(x,y),\ \tilde{V}_2(x,y) \rangle_{\Omega_2} \\ \cdots & \cdots \\ \langle \nabla^2 \tilde{V}_r(x,y),\ \tilde{V}_r(x,y) \rangle_{\Omega_1} & \langle \nabla^2 \tilde{V}_r(x,y),\ \tilde{V}_r(x,y) \rangle_{\Omega_2} \end{bmatrix} \tag{14.87}$$

For discrete approximations of eigenfunctions, as those coming from KLD of $\tilde{T}(t)$, the inner products in Equations 14.86 and 14.87 become

$$\langle \tilde{V}_k(x,y),\ \tilde{V}_k(x,y) \rangle_{\Omega_i} \approx \mathbf{v}_k^t \mathbf{P}_i \mathbf{v}_k \tag{14.88}$$

$$\langle \nabla^2 \tilde{V}_k(x,y),\ \tilde{V}_k(x,y) \rangle_{\Omega_i} \approx \mathbf{v}_k^t \mathbf{P}_i \mathbf{L} \mathbf{v}_k \tag{14.89}$$

where
 L is the numerical approximation of ∇^2
 $\mathbf{P}_i$ is a 1/0 diagonal matrix, which selects the elements of the eigenvector $\mathbf{v}_k$ associated to the pixels belonging to Ω_i

The solution of Equation 14.85, in the least squares sense, is

$$\begin{bmatrix} \hat{\alpha}_1 \\ \hat{\alpha}_2 \end{bmatrix} = (\mathbf{M}'\mathbf{M})^{-1}\mathbf{M}'\mathbf{y} \tag{14.90}$$

14.6.4 Numerical Example

Let us consider again the two-phase plate in Figure 14.6. Thermal behavior is governed by Equations 14.69 and 14.70. At the interfaces between phases, continuity of temperature and heat flux is assumed. Thermal parameters for the white phase are $\alpha_1 = 1.5152 \times 10^{-7}$ m^2 s^{-1} and $\beta_1 = 0.0061$ s^{-1}, while parameters of the black phase are $\alpha_2 = 5.0505 \times 10^{-7}$ m^2 s^{-1} and $\beta_2 = 0.0202$ s^{-1}. The plate surroundings are assumed to be at uniform and constant temperature. The finite volume method has been applied on an equally spaced $n \times n$ ($n = 30$) grid for discretization of Equations 14.69 and 14.70. The resulting state-space model is thus integrated over time to emulate experiments. As in Section 14.4.3, simulations are corrupted by additive noise: $\tilde{T}(t) = T(t) + \boldsymbol{\varepsilon}(t)$ ($n \times 1$, $n = 900$), with $\mathbf{W}_\varepsilon = \sigma_\varepsilon^2 \mathbf{I}$. Three different values of noise amplitude have been considered: ± 0.5°C (poor-quality data), ± 0.1°C (good-quality data), and ± 0.02°C (validation data). The initial temperature field is depicted in Figure 14.9a (in the case of poor-quality data), while Figure 14.9b includes the time behavior of the temperature at each volume of control, $\tilde{T}(t) = \{\tilde{T}_i(t)\}_{i=1\cdots n}$, from $t = 0$ to $t = 3$ s with $\Delta t = 10^{-3}$ s (3000 sampling times). KLD of this set of signals has been carried out as described in Section 14.2.2: $\tilde{T}(t) = \tilde{V}\tilde{Z}(t)$. The first and second eigenfunctions, those that are being used for parameters estimations, are depicted on the top of Figure 14.10, while time evolution of the states $\{\tilde{z}_m(t)\}_{m=1,\ldots,6}$ is represented in Figure 14.9c and d. It can be seen that compared to noise amplitude there is only five significant states.

The estimation of the parameters α_i and β_i ($i = 1, 2$) has been carried out by the method described in Section 14.6.3.2. The results achieved are reported in Table 14.4. It can be seen that for satisfactory estimation of β_i ($i = 1, 2$) parameters (bias less than 1%), high-quality data are required. On the contrary, whatever may be the noise level, the results achieved for diffusivities are excellent. This can be firstly explained by the low sensitivity of KLD

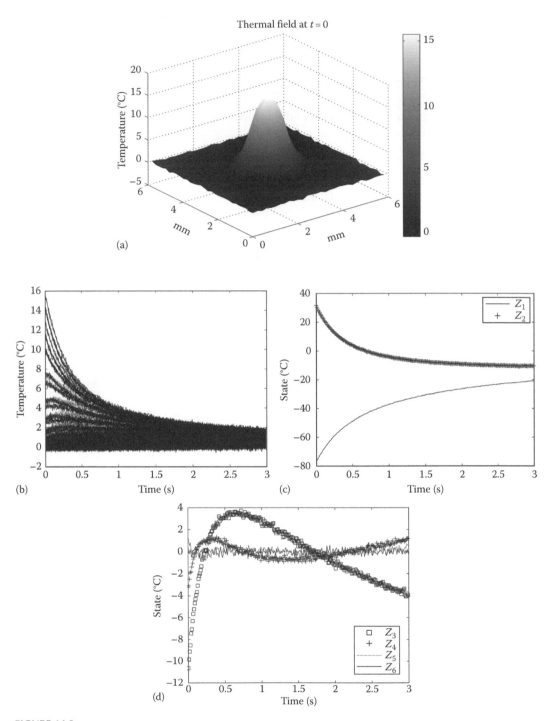

FIGURE 14.9
Highly noisy experiment on a two-phase plate: (a) initial temperature field of the plate; (b) temperature behavior in time, one curve by volume of control; (c) first and second states from KLD of the temperature data; (d) third to sixth states from KLD of the temperature data.

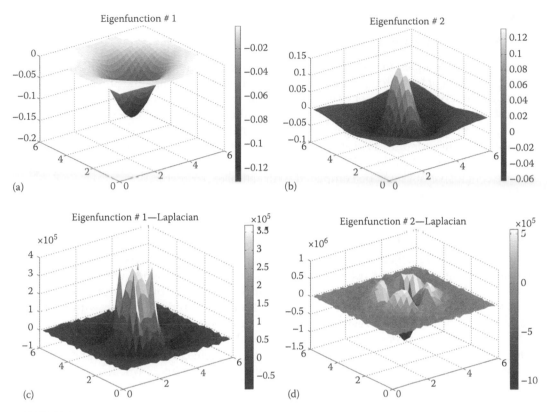

FIGURE 14.10
(*Top*): First (a) and second (b) eigenfunctions obtained by KLD of highly noise-corrupted ($\pm0.5°$C) data. (*Bottom*): Laplacian of first (c) and second (d) eigenfunctions.

eigenfunctions to noise. Figure 14.10 (on the top part) shows that noise effect on first and second eigenfunctions calculated from poor-quality data ($\pm0.5°$C noise amplitude case) is negligible. Secondly, because the sensitivity of Equation 14.81 to β_i values $\left(\langle V_k(x), V_k(x)\rangle_{\Omega_i}\right)$ is very low compared to the sensitivity to α_i values $\left(\langle \nabla^2 V_k(x), V_k(x)\rangle_{\Omega_i}\right)$. As shown in Figure 14.10, $V_k(x)$ ($k=1,2$) are $O(1)$ while $\nabla^2 V_k(x)$ values are $O(10^5)$.

TABLE 14.4

Heterogeneous Plate: Estimated Values for Thermal Parameters

	Noise Amplitude (°)		
	±0.50	±0.10	±0.02
$\hat{\beta}_1$ (s^{-1})	0.0082	0.0076	0.0060
$\left|(\beta_1 - \hat{\beta}_1)/\beta_1\right| \times 100$	34.8	26.1	0.93
$\hat{\beta}_2$ (s^{-1})	0.0266	0.0255	0.0201
$\left|(\beta_2 - \hat{\beta}_2)/\beta_2\right| \times 100$	31.6	26.3	0.67
$\hat{\alpha}_1$ ($\times10^{-6}$ m^2 s^{-1})	0.1524	0.1526	0.1511
$\left|(\alpha_1 - \hat{\alpha}_1)/\alpha_1\right| \times 100$	0.60	0.73	0.23
$\hat{\alpha}_2$ ($\times10^{-6}$ m^2 s^{-1})	0.5001	0.5013	0.5044
$\left|(\alpha_2 - \hat{\alpha}_2)/\alpha_2\right| \times 100$	1.00	0.74	0.12

14.7 Summary and Perspectives

Using KLD techniques in association with infrared thermography experiments for the characterization of homogeneous and heterogeneous materials has some interesting advantages. The first one is the power of KLD for significant reduction of data set dimension. The second advantage is the KLD efficiency for noise filtering. The third one is related with the orthogonal properties of KLD eigenfunctions and states, which allow obtaining simple and efficient estimates for diffusivities. As a result, proposed estimation methods based on KLD are an attractive combination of parsimony and robustness to noise. Numerical tests carried out are conclusive but must be completed with actual experiments.

For the characterization of heterogeneous materials, some further research is still necessary. For instance, phase discrimination and interface location by eigenfunctions sign analysis will be limited by the resolution and the image grid (quadrangular) provided by infrared camera. In such a case, results coming from KLD analysis must be refined using more sophisticated mathematical methods. On the other hand, we think that the estimation of heat exchange coefficients (β_i parameters) has to be improved when working with more than two-phase materials. Moreover, numerical estimation of eigenfunctions spatial derivatives close to the interfaces could be a tricky task when dealing with materials showing significant thermal resistances at the interfaces.

At last, KLD techniques in association with lock-in thermography experiments could offer interesting perspectives for characterization of chemical reactions or phase change phenomena within heterogeneous materials. Some preliminary work has been already done on this problem.

14.8 Appendixes

Appendix 14.A

Let us consider Equation (18.54) for $i \neq k$. Simplifying without loss of generality, we assume $\alpha = \alpha_x = \alpha_y$. Hence, we can write

$$\frac{1}{\sigma_k^2} \left\langle \frac{dz_i(t)}{dt}, z_k(t) \right\rangle_t = \alpha \left\langle \nabla^2 V_k(x,y), V_i(x,y) \right\rangle_\Omega$$

$$\frac{1}{\sigma_i^2} \left\langle \frac{dz_k(t)}{dt}, z_i(t) \right\rangle_t = \alpha \left\langle \nabla^2 V_i(x,y), V_k(x,y) \right\rangle_\Omega$$

Taking into account the second Green theorem, the difference of equations above can be written as

$$\frac{1}{\sigma_k^2} \left\langle \frac{dz_i(t)}{dt}, z_k(t) \right\rangle_t - \frac{1}{\sigma_i^2} \left\langle \frac{dz_k(t)}{dt}, z_i(t) \right\rangle_t = \int_{\partial\Omega} [V_i(x,y)\nabla V_k(x,y) - V_k(x,y)\nabla V_i(x,y)]dxdy$$

As eigenfunctions satisfy Equation 14.55, it yields

$$\frac{1}{\sigma_k^2}\left\langle \frac{dz_i(t)}{dt}, z_k(t)\right\rangle_t - \frac{1}{\sigma_i^2}\left\langle \frac{dz_k(t)}{dt}, z_i(t)\right\rangle_t = 0$$

Appendix 14.B

Introducing KLD of $T(x,t)$ into Equation (18.77) leads to

$$I_k^{(i)}(t) = \int\limits_{\partial\Gamma}\left[V_k(x)\sum_{m=1}^{\infty}z_m(t)\nabla V_m(x)\vec{n}_{ij} - \nabla V_k(x)\vec{n}_{ij}\sum_{m=1}^{\infty}z_m(t)V_m(x)\right]d\gamma$$

Multiplying this equation by $z_k(t)$, integrating over time and taking into account orthogonal property of the KLD states, we obtain

$$\left\langle I_k^{(i)}(t), z_k(t)\right\rangle_t = \int\limits_{\partial\Gamma}\left[\sigma_k^2 V_k(x)\nabla V_k(x)\vec{n}_{ij} - \sigma_k^2 V_k(x)\nabla V_k(x)\vec{n}_{ij}\right]d\gamma = 0$$

Nomenclature

a	absorption coefficient
$e(x,t)$	approximation error
$\mathbf{e}(t)$	vector of approximation errors
h	inverse of the thermal resistance
k	thermal conductivity
t	time
$T(x,t)$	temperature field
$\mathbf{T}(t)$	vector of temperature
$V_m(x)$	eigenfunctions of W
$\mathbf{V}$	matrix of eigenfunctions
$W(x,x')$	energy function
$\mathbf{W}$	energy matrix
x, y	coordinates
$z_m(t)$	states
$\mathbf{Z}(t)$	vector of states

Greek Letters

α	thermal diffusivity
β	thermal loss coefficient
$\varepsilon(x,t)$	noise field
$\boldsymbol{\varepsilon}(t)$	noise vector
ρc	thermal capacity

σ_m^2 eigenvalues of W

σ_ε^2 noise variance

Symbols

$\|\cdot\|$ unitarily invariant norm

$\langle , \rangle$ scalar product

$\cdot$ time derivative

$\sim$ noisy variable

$-$ mean value

$\wedge$ estimated value

Abbreviations

KLD   Karhunen–Loève decomposition

PCA   principal components analysis

SVD   singular values decomposition

References

Ait-Yahia A. and E. Palomo del Barrio. 1999. Thermal systems modelling via singular value decomposition: Direct and modular approach. *Applied Mathematical Modelling* 23: 447–468.

Ait-Yahia A. and E. Palomo del Barrio. 2000. Numerical simplification method for state-space models of thermal systems. *Numerical Heat Transfer, Part B* 37: 201–225.

Berkoz G., P. Holmes, and J.L. Lumley. 1993. The proper orthogonal decomposition in the analysis of turbulent flows. *Annual Reviews of Fluid Mechanics* 25: 539–575.

Berry M.W. 1992. Large-scale sparse singular value computation. *International Journal of Supercomputer Applications* 6: 13–49.

Berry M.W., S.T. Dumais, and G.W. Obrien. 1995. Using linear algebra for intelligent information-retrieval. *SIAM Review* 37: 573–595.

Dauvergne J.L. and E. Palomo del Barrio. 2009. A spectral method for low-dimensional description of melting/solification within shape-stabilized phase change materials. *Numerical Heat Transfer, Part B* 56: 142–166.

Dauvergne J.L. and E. Palomo del Barrio. 2010. Toward a simulation-free P.O.D. approach for low-dimensional description of phase change problems. *International Journal of Thermal Sciences* (doi: 10.1016/j.ijthermalsci.2010.02.006)

Deprettere F. 1988. *SVD and Signal Processing: Algorithms, Analysis and Applications*. Amsterdam: Elsevier Science Publishers.

Everitt B.S. and G. Dunn. 2001. *Applied Multivariate Data Analysis*. London, U.K.: Arnold.

Everitt B.S., S. Landau, and M. Leese. 2001. *Cluster Analysis*. London, U.K.: Arnold.

Fan K. and J. Hoffman. 1955. Some metric inequalities if the space of matrices. *Proceedings of the American Mathematical Society* 6: 111–116.

Godin A., E. Palomo del Barrio, and J.L. Dauvergne. 2009. Thermal analysis at the micro scale: An efficient method for phases and interfaces recognition, 2009. In the *Intermediate Report of the Project: Development of Materials for Thermal Energy Storage at High Temperature*, TREFLE Laboratory, University of Bordeaux, France.

Godin A., E. Palomo del Barrio, E. Ahusborde, and M. Azaiez. 2010. Materials microstructure retrieval using karhunen–Loève decomposition techniques and infrared thermography. *Inverse Problems in Engineering* (Submitted in December 2010).

Golub G. and C. Van Loan. 1996. *Matrix Computation*. Baltimore, MD: Johns Hopkins University Press.

Holmes P., J. Lumley, and G. Berkoz. 1996. *Turbulence, Coherent Structures, Dynamical Systems and Symmetry*. Cambridge Monographs on Mechanics. Cambridge, U.K.: Cambridge University Press.

Hotteling H. 1933. Analysis of complex statistical variables into principal components. *Journal of Educational Psychology* 24: 417–441.

Intissar A. 1997. *Analyse fonctionnelle et théorie spectrale pour les opérateurs compacts non autoadjoints*. Toulouse: Cépaduès-Éditions.

Jessup E.R. and D.C. Sorensen. 1994. A parallel algorithm for computing singular-value decomposition of a matrix. *SIAM Journal on Matrix Analysis and Applications* 15: 530–548.

Jolliffe I.T. 1986. *Principal Components Analysis*. New York: Springer.

Karhunen K. 1946. Uber lineare methoden füer wahrscheiniogkeitsrechnung. *Annales of Academic Science Fennicae, Series A1, Mathematical Physics* 37: 3–79.

Loève M.M. 1955. *Probability Theory*. Princeton, NJ: Van Nostrand.

Mallat S. 2000. *Une exploration des signaux en ondelettes*. Paris: Éditions de l'École Polytecnique.

Mendioroz A., R. Fuente-Dascal, E. Apiñaniz, and A. Salazar. 2009. Thermal diffusivity measurements of tin plates and filaments using lock-in thermography. *Review of Scientific Instruments* 80: 0749004-1/9.

Mirsky L. 1960. Symmetric gauge functions and unitarily invariant norms. *Quarterly Journal of Mathematics Oxford* 11: 50–59.

Palomo del Barrio E. 2003a. An efficient computational method for solving large-scale sensitivity problems. *Numerical Heat Transfer, Part B* 43: 353–372.

Palomo del Barrio E. 2003b. Multidimensional inverse heat conduction problems solution via Lagrange theory and model size reduction techniques. *Inverse Problems in Engineering* 11: 515–539.

Palomo del Barrio E. and J.L. Dauvergne. 2009. Thermal analysis at the micro scale: Validation of the experimental device and estimation method. In the *Intermediate Report of the Project: Development of Materials for Thermal Energy Storage at High Temperature*, TREFLE Laboratory, University of Bordeaux, France.

Park H.M. and W.S. Jung. 1999. On the solution of inverse heat transfer using the Karhunen–Loève Galerkin method. *International Journal of Heat and Mass Transfer* 42: 127–142.

Park H.M. and W.S. Jung. 2001. The Karhunen–Loève Galerkin method for the inverse natural convection problems. *International Journal of Heat and Mass Transfer* 44: 155–167.

Pearson K. 1901. On lines planes of closes fit to system of points in space. *The London, Edinburgh, and Dublin Philosophical Magazine and Journal of Science* 2: 559–572.

Strang G. 1998. *Introduction to Linear Algebra*. Wellesley, MA: Wellesley Cambridge Press.

Wellele O. and E. Palomo del Barrio. 2008. Experimental device for phase change analysis at the pore scale. In the *Intermediate Report of the Project: Development of Materials for Thermal Energy Storage at High Temperature*, TREFLE Laboratory, University of Bordeaux, France.

15

Explicit Formulations for Radiative Transfer Problems

Liliane Basso Barichello

CONTENTS

15.1 Introduction

Many are the applications of interest for inverse radiative transfer problems, in several fields, as cited, for example, in Sarvari and Mansouri (2004), Tito et al. (2004), Ren et al. (2006), Sanchez and McCormick (2008), Gryn (1995), Moura et al. (1998), Chaloub and Campos Velho (2003), Klose and Hielscher (2002), Liu et al. (2008), Özişik and Orlande (2000), and Siewert (2002). Particularly, in Chapter 21, inverse thermal radiation problems are discussed. In this context, a fundamental issue required for several methodologies is a good solution for the direct problem, in this case, a solution for the radiative transfer equation (Chandrasekhar 1960, Özişik 1973, Modest 1993).

The analysis of radiative transport is somewhat more complex than heat transfer by conduction or convection within a medium (Modest 1993). In fact, the temperature is not the main unknown in the balance equation, but the radiation intensity, which is function of space variables and also direction of the particles. Still, radiative properties may depend, along with direction, on the wavelength.

Due to the complexity of the original mathematical model associated with radiative transfer processes (an integro-differential equation where the unknown distribution depends on seven independent variables), several studies have been devoted to the challenge of developing accurate solutions adequate to different geometries. In regard to the integro-differential form of the transport equation, an important result was given by Case (1960), who derived an exact solution for simpler transport models. Although its use is restricted to very simple physical models, it provided important theoretical information, which was very helpful in the development of the preliminar numerical approaches in this field. Nowadays, in general, two ways may be followed: the probabilistic approach and the deterministic approach. In the deterministic approach, one searches for exact solutions of approximated forms of the original equation. In this context, two classical methodologies associated with the solution of radiation problems are very well known and should be mentioned here: the spherical harmonics method and the discrete ordinates method.

The fundamental idea involved in the spherical harmonics method (Davison 1957), also referred to as P_N method, is the approximation (expansion) of the angular dependence in the unknown function, as the radiation intensity, in terms of spherical harmonics functions, or, simply, Legendre polynomials. More recent developments improved important features of the P_N solution, particularly making it more efficient from the computational point of view (Benassi et al. 1983, 1984). The generalization of such approach to the treatment of multidimensional problems and more complex geometries may be a very hard task, if possible. Still, it is always important to emphasize that the spherical harmonics approach provides a solution for the moments of the transport equation instead of the equation itself.

The development of the discrete ordinates method in the solution of the radiative transfer equation may be mostly associated with Chandrasekhar's work (Chandrasekhar 1960), although it seems to be already proposed in Wick's work (Wick 1943). The fundamental idea in the discrete ordinates method is the use of a quadrature scheme to deal with the integral term of the radiative transfer equation, such that the original problem is transformed into a system of differential equations. Under certain restrictions on the quadrature scheme and boundary conditions, it may be shown that the discrete ordinates method is equivalent to the spherical harmonics method (Barichello and Siewert 1998). As extension of the original version of the method, over the years, the discrete ordinates method has been combined with finite-difference techniques (Fiveland 1984, Lewis and Miller 1984), when the spatial dependence of the problem is treated numerically, and multidimensional quadrature schemes have been developed (Lewis and Miller 1984) as well.

In this chapter, we focus our attention on the solution of thermal radiation problems based on a more recent analytical version of the discrete ordinates method: the ADO method (Barichello and Siewert 1999a, Barichello and Siewert 2002). Differently of the Chandrasekhar's approach, the ADO approach (i) does not depend on any special properties of the quadrature scheme, (ii) has the separation constants defined as eigenvalues of a matrix instead of roots of a characteristic equation, and (iii) defines a scaling to avoid positive exponentials that cause *overflows* in numerical calculations. In addition, the ADO formulation leads to eigenvalue systems of reduced order, in comparison with standard discrete ordinates calculations, which results in computational gain. These features have made possible the development of concise and accurate solutions for a wide class of problems, including, with respect to radiative transfer applications, models that consider polarization effects (Barichello and Siewert 1999b) and Fresnel boundary conditions (Garcia et al. 2008), for example.

Here, as an introductory study, simple models will be used for developing the ADO solution, in order to provide a basic scheme for establishing a computational procedure, which may be, however, useful as benchmark case when solving more complex problems with numerical tools. In addition, taking into account that solutions are obtained in a closed form, this formalism may represent important computational gain if used in association with the solution of inverse problems (Barichello and Siewert 1997, Siewert 2002).

In this way, this chapter is organized such that in Section 15.2, we briefly introduce some definitions to formulate the general model of interest. In Section 15.3, we focus our attention in the specific formulation for one-dimensional (plane-parallel) geometry. In Section 15.4, we describe the problem (gray, anisotropic medium) to which we develop the ADO solution in Section 15.5. Continuing, we deal with the simplest model, the isotropic case, in Section 15.6, to show that, in this case, explicit solutions can be found. In Section 15.7, we discuss computational aspects and list numerical results for a test case. Finally, in Section 15.8, we add some general and concluding remarks.

15.2 Formulation

The fundamental balance equation for dealing with radiative transport in participating medium has been derived and presented by many authors (Chandrasekhar 1960, Özişik 1973, Liou 1980, Modest 1993). In this way, here we discuss it briefly, just to introduce the basic mathematical model relevant to the development of the discrete ordinates approach.

We then consider, to describe the radiation field, an amount of radiant energy, dE_λ, in a specified wavelength interval $(\lambda, \lambda + d\lambda)$, which is transported across an element of area dA and in directions confined to an element of solid angle $d\Omega$ (Liou 1980, Modest 1993), which is oriented at an angle θ to the normal of dA, during a time dt (see Figure 15.1). This energy, dE_λ, is expressed in terms of the *specific intensity*, I_λ, or simply, *the intensity*, I_λ, by (Chandrasekhar 1960, Liou 1980)

$$dE_\lambda = I_\lambda \cos \theta \, d\lambda \, dA \, d\Omega \, dt. \tag{15.1}$$

Equation 15.1 defines the (monochromatic) intensity as

$$I_\lambda = \frac{dE_\lambda}{\cos \theta \, d\lambda \, dA \, d\Omega \, dt}, \tag{15.2}$$

which is in units of energy per area per time per frequency and per steradian (units of solid angle). It is clear that the intensity implies a directionality in the radiation stream. In fact, the intensity is said to be confined in a pencil of radiation (Chandrasekhar 1960, Liou 1980).

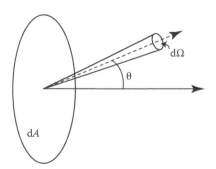

FIGURE 15.1
Characterization of radiation fields.

A pencil of radiation traversing a medium will be weakened by its interaction with matter. If the intensity of radiation I_λ becomes $I_\lambda + dI_\lambda$ after traversing a thickness ds in the direction of its propagation, then (Chandrasekhar 1960, Liou 1980)

$$dI_\lambda = -\beta_\lambda \rho I_\lambda ds, \qquad (15.3)$$

where
  ρ is the *density of the material*
  β_λ denotes the *mass extinction coefficient* for radiation of wavelength λ

This coefficient is the sum of the mass absorption (k_λ) and scattering $(\sigma_{s\lambda})$ coefficients (Chandrasekhar 1960, Liou 1980). In this way, the reduction in intensity is caused by absorption as well as scattering of radiation by the material.

On the other hand, emission of the material $(j_{e\lambda})$ and multiple scattering $(\sigma_{s\lambda})$ from all other directions into the pencil under consideration have to be considered as contribution to strength the intensity

$$dI_\lambda = j_\lambda \rho ds, \qquad (15.4)$$

where j_λ is the *emission coefficient* (Chandrasekhar 1960).

We consider then a small cylindrical element of cross section dA and height ds in the medium. Counting up the gains and losses in the pencil of radiation during its traversal of the cylinder, we have (Chandrasekhar 1960, Liou 1980), upon combining Equations 15.3 and 15.4,

$$dI_\lambda = -\beta_\lambda \rho I_\lambda ds + j_\lambda \rho ds. \qquad (15.5)$$

It is still convenient to define the source function J_λ (Chandrasekhar 1960, Liou 1980),

$$J_\lambda = \frac{j_\lambda}{\beta_\lambda}, \qquad (15.6)$$

such that it has units of radiant intensity, and Equation 15.5 may be rewritten as

$$\frac{1}{\beta_\lambda \rho} \frac{dI_\lambda}{ds} = -I_\lambda + J_\lambda. \qquad (15.7)$$

This is the general form of the equation of transfer processes, without imposing any coordinate system (Chandrasekhar 1960, Liou 1980), which is fundamental in the discussion of thermal radiative transfer process. More details can be found in the books of Özişik (1973) and Modest (1993). Since the source function is dependent on the intensity at a point, the equation of transfer is, in general, an integro-differential equation. In our further discussion, we choose to omit the subscript λ on various radiative quantities. We note also that the time dependence of the radiation intensity may be almost always neglected in heat transfer applications (Modest 1993), and so, for simplicity, we have considered the independence of time in the above relations.

15.3 Plane-Parallel Medium

It follows from the definition of intensity that in a medium that absorbs, emits, and scatters radiation, it may be expected to vary from point to point and also with direction through every point. A case of great interest is a medium stratified in parallel planes in which all the physical properties are invariant over a plane. In this case, we can write (Chandrasekhar 1960)

$$I = I(z, \theta, \phi). \tag{15.8}$$

It is convenient to measure linear distances normal to the plane of stratification. If z is this distance and if we introduce the optical variable τ,

$$d\tau = \beta \rho dz, \tag{15.9}$$

we have, for $\mu = \cos \theta$, the equation of transfer written in the form (Chandrasekhar 1960, Modest 1993)

$$\mu \frac{\partial}{\partial \tau} I(\tau, \mu, \phi) + I(\tau, \mu, \phi) = \frac{\varpi}{4\pi} \int_{-1}^{1} \int_{0}^{2\pi} p(\cos \Theta) I(\tau, \mu', \phi') d\phi' \, d\mu' + (1 - \varpi) B(T). \tag{15.10}$$

Here, we have written explicitly the source term; $\varpi \in [0, 1]$ is the albedo for single scattering (ratio of the scattering coefficient to the extinction coefficient (Özişik 1973)), $\tau \in (0, \tau_0)$ is the optical variable, τ_0 is the optical thickness of the plane-parallel medium, Θ is the scattering angle, $\mu \in [-1, 1]$ is the cosine of the polar angle, as measured from the positive τ-axis, and ϕ is the azimuthal angle. Together, the polar and azimuthal angles define the direction Ω of propagation of the radiation. Still, $B(T)$, which depends on the temperature, is defined in terms of the Planck's function (Chandrasekhar 1960, Özişik 1973). We then seek to establish a solution of Equation 15.10 subject to boundary conditions, which we will define explicitly later on in this chapter.

Still, in regard to Equation 15.10, we may suppose that the phase function can be expanded as a series of Legendre polynomials (Chandrasekhar 1960), that is,

$$p(\cos \Theta) = \sum_{l=0}^{\infty} \beta_l P_l(\cos \Theta), \tag{15.11}$$

where the β_l's are constants.

Two usual forms used to describe the scattering law are the binomial form and the Henyey–Greenstein model. The binomial form

$$p(\cos \Theta) = \frac{L+1}{2^L} (1 + \cos \Theta)^L, \tag{15.12}$$

with $L > 0$, according to Siewert (2002), was introduced by Kaper et al. (1970), where the β coefficients can be computed from a recursion formula (McCormick and Sanchez 1981)

$$\beta_l = \left(\frac{2l+1}{2l-1}\right)\left(\frac{L+1-l}{L+1+l}\right)\beta_{l-1}, \tag{15.13}$$

for $l = 1, 2, \ldots$, with $\beta_0 = 1$. In the Henyey–Greenstein model (Siewert 2002),

$$p(\cos\Theta) = (1-g)^2(1+g^2-2g\cos\Theta)^{-3/2} \tag{15.14}$$

or

$$p(\cos\Theta) = \sum_{l=0}^{\infty} \beta_l P_l(\cos\Theta), \tag{15.15}$$

with $g \in (-1, 1)$ and $\beta_l = (2l+1)g^l$. It is usual to make use of the addition theorem for the Legendre polynomials (Gradshteyn and Ryzhik 1980) and express the phase function, for scattering from $\{\mu', \phi'\}$ to $\{\mu, \phi\}$ in the form

$$p(\cos\Theta) = \sum_{m=0}^{L} (2 - \delta_{0,m}) \sum_{l=m}^{L} \beta_l P_l^m(\mu') P_l^m(\mu) \cos[m(\phi' - \phi)], \tag{15.16}$$

where $P_l^m(\mu')$ are the Legendre functions.

Making use of Equation 15.16, along with a classical Fourier decomposition (Chandrasekhar 1960) of the solution $I(\tau, \mu, \phi)$, in terms of ϕ, we find that each component of the referred decomposition satisfies an equation written in a general form as

$$\mu\frac{\partial}{\partial\tau}I(\tau, \mu) + I(\tau, \mu) = \frac{\varpi}{2}\sum_{l=0}^{L}\beta_l P_l(\mu)\int_{-1}^{1}P_l(\mu')I(\tau, \mu')d\mu' + Q(\tau, \mu), \tag{15.17}$$

where
  ϖ is the albedo for single scattering
  $\tau \in (0, \tau_0)$ is the (dimensionless) optical variable
  τ_0 is the optical thickness of the plane-parallel medium
  $\mu \in [-1, 1]$ is the cosine of the polar angle, as measured from the positive τ-axis, according to the geometry described in Figure 15.2

Still, β_l are the coefficients in the Lth-order expansion of the scattering law, and $Q(\tau, \mu)$ is an inhomogeneous source term. We seek to establish a solution of Equation 15.17 subject to the boundary conditions

$$I(0, \mu) = F_1(\mu) + \rho_1^s I(0, -\mu) + 2\rho_1^d\int_0^1 I(0, -\mu')\mu'\,d\mu' \tag{15.18}$$

and

$$I(\tau_0, -\mu) = F_2(\mu) + \rho_2^s I(\tau_0, -\mu) + 2\rho_2^d\int_0^1 I(\tau_0, -\mu')\mu'\,d\mu' \tag{15.19}$$

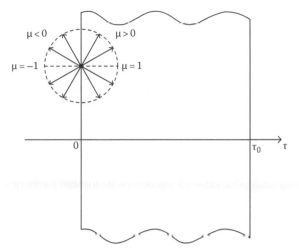

FIGURE 15.2
Slab geometry.

for $\mu \in (0, 1]$. Here, $F_1(\mu)$ and $F_2(\mu)$ refer to known incoming radiation distribution (we assume here that neither $F_1(\mu)$ nor $F_2(\mu)$ contains generalized functions); ρ_α^s, $\alpha = 1, 2$ are the coefficients for specular reflection and ρ_α^d, $\alpha = 1, 2$, for diffuse reflection.

In thermal radiation processes, if one considers local thermodynamic equilibrium, it is usual to have as consequence of the emission, as indicated in Equations 15.10 and 15.17,

$$Q(\tau, \mu) = Q(\tau) = (1 - \varpi)\frac{\sigma}{\pi}T(\tau)^4 \tag{15.20}$$

and for $\alpha = 1, 2$, in the boundary conditions

$$F_\alpha(\mu) = \varepsilon_\alpha \frac{\sigma}{\pi}T_\alpha^4, \tag{15.21}$$

where
  T_1 and T_2 refer to the boundary temperatures
  ε_1 and ε_2 are the emissivities
  σ is the Stefan–Boltzmann constant (Özişik 1973, Modest 1993)

We consider, here that the temperature distribution in the medium $T(\tau)$ is known.

Once we are able to develop a solution for the problem above, we can use the radiation intensity to evaluate, for example, the *radiation density*, given by

$$\rho(\tau) = \int_{-1}^{1} I(\tau, \mu')d\mu' \tag{15.22}$$

and *the partial radiative heat fluxes*

$$q_\pm(\tau) = \pi \int_{0}^{1} I(\tau, \pm\mu')\mu'\,d\mu'. \tag{15.23}$$

15.4 A Basic Problem

For simplicity, in order to have a better understanding of the discrete ordinates method, considering the objective of the proposed text, we choose to work with a simpler mathematical formulation of the problem defined in the previous section. First of all, we consider the homogeneous version of the problem $(Q(\tau, \mu) = 0)$, since the procedure would change basically by adding to the homogeneous solution (developed in this section) a particular solution, which definition can be found in Barichello et al. (2000). We note, however, that the conservative case $(\varpi = 1)$ has to be considered as a special case. Another issue is associated with the boundary conditions. We restrict ourselves to the case where the incoming radiation intensity is known at the boundaries. In this way, we consider the equation

$$\mu \frac{\partial}{\partial \tau} I(\tau, \mu) + I(\tau, \mu) = \frac{\varpi}{2} \sum_{l=0}^{L} \beta_l P_l(\mu) \int_{-1}^{1} P_l(\mu') I(\tau, \mu') d\mu', \tag{15.24}$$

subject to the boundary conditions

$$I(0, \mu) = F_1(\mu) \tag{15.25}$$

and

$$I(\tau_0, -\mu) = F_2(\mu) \tag{15.26}$$

for $\mu \in (0, 1]$.

In the next section, we develop an analytical discrete ordinates solution for the problem formulated above. In particular, the case here treated is a particular case (with azimuthal symmetry) of the solution developed by Siewert (2000).

15.5 A Discrete Ordinates Solution

As a first step to develop a discrete ordinates solution for the problem defined above, we rewrite the integral term in Equation 15.24,

$$\mu \frac{\partial}{\partial \tau} I(\tau, \mu) + I(\tau, \mu) = \frac{\varpi}{2} \sum_{l=0}^{L} \beta_l P_l(\mu) \int_{0}^{1} P_l(\mu')[I(\tau, \mu') + (-1)^l I(\tau, -\mu')] d\mu'. \tag{15.27}$$

We then derive the discrete ordinates version of Equation 15.27 in the form

$$\mu_i \frac{d}{d\tau} I(\tau, \mu_i) + I(\tau, \mu_i) = \frac{\varpi}{2} \sum_{l=0}^{L} \beta_l P_l(\mu_i) \sum_{k=1}^{N} \omega_k P_l(\mu_k) \left[I(\tau, \mu_k) + (-1)^l I(\tau, -\mu_k) \right] \tag{15.28}$$

and

$$-\mu_i \frac{d}{d\tau} I(\tau, -\mu_i) + I(\tau, -\mu_i) = \frac{\varpi}{2} \sum_{l=0}^{L} \beta_l P_l(\mu_i) \sum_{k=1}^{N} \omega_k P_l(\mu_k) \left[(-1)^l I(\tau, \mu_k) + I(\tau, -\mu_k) \right], \quad (15.29)$$

where μ_k and ω_k are the N arbitrary nodes and weights of the quadrature scheme defined in the half-range [0, 1].

Once the integral term in Equation 15.24 was approximated by a quadrature scheme, Equations 15.28 and 15.29 represent a first-order ordinary differential system. We then seek for exponential solutions of the system, in the form

$$I(\tau, \pm\mu_i) = \phi(\nu, \pm\mu_i)e^{-\tau/\nu}. \quad (15.30)$$

In this way, we substitute Equation 15.30 into Equations 15.28 and 15.29, to obtain

$$\mp\mu_i \frac{d}{d\tau} \left[\phi(\nu, \pm\mu_i)e^{-\tau/\nu} \right] + \phi(\nu, \pm\mu_i)e^{-\tau/\nu}$$

$$= \frac{\varpi}{2} \sum_{l=0}^{L} \beta_l P_l(\mu_i) \sum_{k=1}^{N} \omega_k P_l(\mu_k) \left[\phi(\nu, \pm\mu_k) + (-1)^l \phi(\nu, \mp\mu_k) \right] e^{-\tau/\nu}, \quad (15.31)$$

for $i = 1, \ldots, N$. We can simplify the previous expression, to obtain

$$\left(1 \mp \frac{\mu_i}{\nu} \right) \phi(\nu, \pm\mu_i) = \frac{\varpi}{2} \sum_{l=0}^{L} \beta_l P_l(\mu_i) \sum_{k=1}^{N} \omega_k P_l(\mu_k) \left[\phi(\nu, \pm\mu_k) + (-1)^l \phi(\nu, \mp\mu_k) \right], \quad (15.32)$$

for $i = 1, \ldots, N$. At this point, we introduce a matrix notation. We write the $N \times 1$ vectors

$$\mathbf{\Phi}_\pm(\nu) = (\phi(\nu, \pm\mu_1), \ldots, \phi(\nu, \pm\mu_N)), \quad (15.33)$$

$$\mathbf{\Pi}(l) = (Pl(\mu_1), \ldots, P_l(\mu_N)), \quad (15.34)$$

the matrices

$$\mathbf{M} = \text{diag}\{\mu_1, \ldots, \mu_N\} \quad (15.35)$$

and

$$\mathbf{W} = \text{diag}\{\omega_1, \ldots, \omega_N\} \quad (15.36)$$

to rewrite Equation 15.32 in the form

$$\left(\mathbf{I_N} \mp \frac{1}{\nu} \mathbf{M} \right) \mathbf{\Phi}_\pm(\nu) = \frac{\varpi}{2} \sum_{l=0}^{L} \beta_l \mathbf{\Pi}(l) \mathbf{\Pi}^T(l) \mathbf{W} \left[\mathbf{\Phi}_\pm(\nu) + (-1)^l \mathbf{\Phi}_\mp(\nu) \right]. \quad (15.37)$$

Here, $\mathbf{I_N}$ is the $N \times N$ identity matrix and T denotes the transpose operation.

Now, we define the $N \times 1$ vectors

$$\mathbf{U} = \mathbf{\Phi}_+(v) + \mathbf{\Phi}_-(v) \tag{15.38}$$

and

$$\mathbf{V} = \mathbf{\Phi}_+(v) - \mathbf{\Phi}_-(v). \tag{15.39}$$

Continuing, we add the two equations expressed in 15.37

$$\left(\mathbf{I_N} - \frac{1}{v}\mathbf{M}\right)\mathbf{\Phi}_+(v) + \left(\mathbf{I_N} + \frac{1}{v}\mathbf{M}\right)\mathbf{\Phi}_-(v)$$

$$= \frac{\varpi}{2}\sum_{l=0}^{L} \beta_l \mathbf{\Pi}(l)\mathbf{\Pi}^{\mathrm{T}}(l)\mathbf{W}\Big[\mathbf{\Phi}_+(v) + (-1)^l\mathbf{\Phi}_+(v) + \mathbf{\Phi}_-(v) + (-1)^l\mathbf{\Phi}_-(v)\Big], \tag{15.40}$$

which now can be rewritten as

$$\mathbf{U} - \frac{1}{v}\mathbf{MV} = \frac{\varpi}{2}\sum_{l=0}^{L} \beta_l \mathbf{\Pi}(l)\mathbf{\Pi}^{\mathrm{T}}(l)\mathbf{W}[1 + (-1)^l]\mathbf{U}, \tag{15.41}$$

or

$$\left(\mathbf{I_N} - \frac{\varpi}{2}\sum_{l=0}^{L} \beta_l \mathbf{\Pi}(l)\mathbf{\Pi}^{\mathrm{T}}(l)\mathbf{W}[1 + (-1)^l]\right)\mathbf{M}^{-1}\mathbf{MU} = \frac{1}{v}\mathbf{MV}. \tag{15.42}$$

In other words, we get

$$\mathbf{AX} = \frac{1}{v}\mathbf{Z}, \tag{15.43}$$

where

$$\mathbf{A} = \left(\mathbf{I_N} - \frac{\varpi}{2}\sum_{l=0}^{L} \beta_l \mathbf{\Pi}(l)\mathbf{\Pi}^{\mathrm{T}}(l)\mathbf{W}[1 + (-1)^l]\right)\mathbf{M}^{-1}, \tag{15.44}$$

$$\mathbf{X} = \mathbf{MU} \tag{15.45}$$

and

$$\mathbf{Z} = \mathbf{MV}. \tag{15.46}$$

Continuing, we subtract the two equations expressed in 15.37

$$\left(\mathbf{I_N} - \frac{1}{v}\mathbf{M}\right)\mathbf{\Phi}_+(v) - \left(\mathbf{I_N} + \frac{1}{v}\mathbf{M}\right)\mathbf{\Phi}_-(v)$$

$$= \frac{\varpi}{2}\sum_{l=0}^{L} \beta_l \mathbf{\Pi}(l)\mathbf{\Pi}^{\mathrm{T}}(l)\mathbf{W}\Big[\mathbf{\Phi}_+(v) - (-1)^l\mathbf{\Phi}_+(v) - \mathbf{\Phi}_-(v) + (-1)^l\mathbf{\Phi}_-(v)\Big], \tag{15.47}$$

to obtain

$$\mathbf{V} - \frac{1}{\nu}\mathbf{M}\mathbf{U} = \frac{\varpi}{2}\sum_{l=0}^{L}\beta_l\mathbf{\Pi}(l)\mathbf{\Pi}^{\mathrm{T}}(l)\mathbf{W}[1 - (-1)^l]\mathbf{V} \tag{15.48}$$

or

$$\left(\mathbf{I_N} - \frac{\varpi}{2}\sum_{l=0}^{L}\beta_l\mathbf{\Pi}(l)\mathbf{\Pi}^{\mathrm{T}}(l)\mathbf{W}[1 - (-1)^l]\right)\mathbf{M}^{-1}\mathbf{M}\mathbf{V} = \frac{1}{\nu}\mathbf{M}\mathbf{U}, \tag{15.49}$$

such that we can write

$$\mathbf{B}\mathbf{Z} = \frac{1}{\nu}\mathbf{X}, \tag{15.50}$$

where

$$\mathbf{B} = \left(\mathbf{I_N} - \frac{\varpi}{2}\sum_{l=0}^{L}\beta_l\mathbf{\Pi}(l)\mathbf{\Pi}^{\mathrm{T}}(l)\mathbf{W}[1 - (-1)^l]\right)\mathbf{M}^{-1}. \tag{15.51}$$

In summary, from Equations 15.43 and 15.50, we obtain two eigenvalue problems

$$(\mathbf{B}\mathbf{A})\mathbf{X} = \lambda\mathbf{X} \tag{15.52}$$

and

$$(\mathbf{A}\mathbf{B})\mathbf{Z} = \lambda\mathbf{Z}, \tag{15.53}$$

where $\lambda = 1/\nu^2$. We note that the separation constants, ν_j, will appear in $(\pm)$ pairs.

It is important to remark that the eigenvalue problems obtained here are of reduced order (half-order), in general, in comparison with the ones obtained in the standard N-order approximations of the discrete ordinates method, based on full-range quadrature schemes.

Continuing, we choose Equation 15.43 to write

$$\nu\mathbf{A}\mathbf{X} = \mathbf{Z}. \tag{15.54}$$

If now, we add $\mathbf{X}$ to both sides to the above equation, we obtain

$$(\mathbf{I_N} + \nu\mathbf{A})\mathbf{X} = \mathbf{X} + \mathbf{Z}. \tag{15.55}$$

Continuing, we substitute Equations 15.45 and 15.46 into Equation 15.55, from where we see that

$$(\mathbf{I_N} + \nu\mathbf{A})\mathbf{X} = \mathbf{M}(\mathbf{U} + \mathbf{V}), \tag{15.56}$$

and so

$$(\mathbf{I_N} + v_j\mathbf{A})\mathbf{X}(v_j) = 2\mathbf{M}\boldsymbol{\Phi}_+(v_j), \tag{15.57}$$

such that

$$\boldsymbol{\Phi}_+(v_j) = \frac{1}{2}\mathbf{M}^{-1}(\mathbf{I_N} + v_j\mathbf{A})\mathbf{X}(v_j). \tag{15.58}$$

On the other hand, if we subtract $\mathbf{X}$ in Equation 15.54

$$(-\mathbf{I_N} + v\mathbf{A})\mathbf{X} = -\mathbf{X} + \mathbf{Z}, \tag{15.59}$$

and we substitute Equations 15.45 and 15.46 into Equation 15.59

$$(\mathbf{I_N} - v\mathbf{A})\mathbf{X} = \mathbf{M}(\mathbf{U} - \mathbf{V}), \tag{15.60}$$

we can write

$$(\mathbf{I_N} - v_j\mathbf{A})\mathbf{X}(v_j) = 2\mathbf{M}\boldsymbol{\Phi}_-(v_j) \tag{15.61}$$

such that

$$\boldsymbol{\Phi}_-(v_j) = \frac{1}{2}\mathbf{M}^{-1}(\mathbf{I_N} - v_j\mathbf{A})\mathbf{X}(v_j). \tag{15.62}$$

In this way, from the solution of the eigenvalue problem, Equation 15.52 (or Equation 15.53), along with the elementary solutions, Equations 15.58 and 15.62, we can write (in a vector form) the discrete ordinates solution of Equation 15.24,

$$\mathbf{I}_\pm(\tau) = (I(\tau, \pm\mu_1), \ldots, I(\tau, \pm\mu_N)), \tag{15.63}$$

in the form

$$\mathbf{I}_\pm(\tau) = \sum_{j=1}^{N}\left[A_j\boldsymbol{\Phi}_\pm(v_j)e^{-\tau/v_j} + B_j\boldsymbol{\Phi}_\mp(v_j)e^{-(\tau_0-\tau)/v_j}\right]. \tag{15.64}$$

We note that in writing the general solution in this way, we avoid the computational issue of exponential terms, which can lead to *overflow*.

Finally, we substitute Equation 15.64 into the boundary conditions, Equations 15.25 and 15.26, to obtain a $2N$ linear system to determine the arbitrary constants $\{A_j\}$ and $\{B_j\}$:

$$\sum_{j=1}^{N}\left[A_j\boldsymbol{\Phi}_+(v_j) + B_j\boldsymbol{\Phi}_-(v_j)e^{-\tau_0/v_j}\right] = \mathbf{F_1}(\mu_i) \tag{15.65}$$

and

$$\sum_{j=1}^{N} \left[A_j \mathbf{\Phi}_-(v_j) e^{-\tau_0/v_j} + B_j \mathbf{\Phi}_+(v_j) \right] = \mathbf{F}_2(\mu_i), \tag{15.66}$$

for $i = 1, \ldots, N$. Here, the $N \times 1$ vectors, $\mathbf{F}_1(\mu_i)$ and $\mathbf{F}_2(\mu_i)$, have components defined by the known incident radiation, $F_1(\mu)$ and $F_2(\mu)$, given in Equations 15.25 and 15.26.

Having established a discrete ordinates solution, analytical in terms of the spatial variable, for the problem defined in Section 15.4, we note that it would be an easy extension to deal with the boundary conditions defined in Equations 15.18 and 15.19. In fact, the choice of the half-range quadrature scheme is very appropriate for evaluating the integral terms present in those equations.

15.5.1 Radiation Density and Radiative Heat Flux

We can now evaluate some quantities of interest. We want to express in terms of our discrete ordinates solution, the radiation density, Equation 15.22, defined in terms of the intensity. Considering the quadrature scheme we have defined, we write

$$\rho(\tau) = \int_0^1 [I(\tau, \mu') + I(\tau, -\mu')] \, d\mu', \tag{15.67}$$

such that in terms of the discrete ordinates solution, we obtain, to express the radiation density,

$$\rho(\tau) = \sum_{j=1}^{N} \left[A_j e^{-\tau/v_j} + B_j e^{-(\tau_0 - \tau)/v_j} \right] \Phi_0(v_j), \tag{15.68}$$

with

$$\Phi_0(v_j) = \sum_{i=1}^{N} \omega_i [\phi(v_j, \mu_i) + \phi(v_j, -\mu_i)]. \tag{15.69}$$

Still, looking back to Equations 15.23 and 15.33, we write the partial radiative heat fluxes

$$q_\pm(\tau) = \pi \sum_{j=1}^{N} \left[A_j e^{-\tau/v_j} Q_\pm(v_j) + B_j e^{-(\tau_0 - \tau)/v_j} Q_\mp(v_j) \right] \tag{15.70}$$

with

$$Q_\pm(v_j) = \sum_{i=1}^{N} \omega_i \mu_i \phi(v_j, \pm \mu_i). \tag{15.71}$$

15.6 The Isotropic Case

The solution developed in the previous section is valid for a medium where arbitrary anisotropic scattering is considered. Its generalization for the problem without azimuthal symmetry can be found in Siewert (2000).

Now, considering the objective of this chapter, we consider a special (and possibly simplest) case relevant to isotropic scattering (Barichello and Siewert 2002). This case can, in fact, be obtained from using $L = 0$ in the development given in the previous section. However, in addition to dealing with an even simpler formulation, which can lead to special eigenvalue systems, we show here that the ADO method can be used to obtain explicit solutions for the problem—while, in the expression given in Equation 15.64, the elementary solutions are the components of the vectors defined in Equations 15.58 and 15.62.

We then consider the radiative transfer equation written here as (Barichello and Siewert 2002)

$$\mu \frac{\partial}{\partial x} I(x, \mu) + I(x, \mu) = \frac{\varpi}{2} \int_0^1 [I(x, \mu') + I(x, -\mu')] \, d\mu', \tag{15.72}$$

for $x \in (0, x_0)$ and $\mu \in [-1, 1]$. Here, $I(x, \mu)$ is the intensity, and, to make it different from the general solution of the previous section, we use x as the optical (spatial) variable, μ as the direction cosine (as measured from the positive x axis), and ϖ as the albedo for single scattering. In addition, we consider Equation 15.72 with boundary conditions written as

$$I(0, \mu) = F_1(\mu), \quad \mu \in (0, 1], \tag{15.73}$$

and

$$I(x_0, -\mu) = F_2(\mu), \quad \mu \in (0, 1], \tag{15.74}$$

where $F_1(\mu)$ and $F_2(\mu)$ are, again, specified.

We repeat the procedure described in Section 15.4, so we seek exponentials solutions of Equation 15.72; we substitute

$$I(x, \mu) = \phi(\nu, \mu) e^{-x/\nu} \tag{15.75}$$

into the Equation 15.72 to find

$$(\nu - \mu)\phi(\nu, \mu) = \frac{\varpi \nu}{2} \int_0^1 [\phi(\nu, \mu') + \phi(\nu, -\mu')] d\mu'. \tag{15.76}$$

Since it is Equation 15.76 that we wish to solve with the discrete ordinates approximation, we introduce a quadrature scheme (at this point, arbitrary) and rewrite the equation as

$$(\nu - \mu)\phi(\nu, \mu) = \frac{\varpi \nu}{2} \sum_{k=1}^{N} w_k [\phi(\nu, \mu_k) + \phi(\nu, -\mu_k)], \tag{15.77}$$

where the N weights and nodes $\{w_k, \mu_k\}$ are defined for use on the integration interval $[0,1]$. If we now evaluate Equation 15.77 at $\mu = \pm \mu_i$, we can write

$$(\nu \mp \mu_i)\phi(\nu, \pm \mu_i) = \frac{\varpi \nu}{2} \sum_{k=1}^{N} w_k[\phi(\nu, \mu_k) + \phi(\nu, -\mu_k)], \tag{15.78}$$

which can be rewritten as

$$\frac{1}{\nu}M\boldsymbol{\Phi}_+(\nu) = (I_N - \hat{W})\boldsymbol{\Phi}_+(\nu) - \hat{W}\boldsymbol{\Phi}_-(\nu) \tag{15.79}$$

and

$$-\frac{1}{\nu}M\boldsymbol{\Phi}_-(\nu) = (I_N - \hat{W})\boldsymbol{\Phi}_-(\nu) - \hat{W}\boldsymbol{\Phi}_+(\nu), \tag{15.80}$$

where I_N is the $N \times N$ identity matrix,

$$\boldsymbol{\Phi}_\pm(\nu) = (\phi(\nu, \pm \mu_1), \quad \phi(\nu, \pm \mu_2), \dots, \phi(\nu, \pm \mu_N)), \tag{15.81}$$

the elements of the matrix $\hat{W}$ are

$$(\hat{W})_{i,j} = \frac{\varpi}{2}w_j \tag{15.82}$$

and

$$M = \mathrm{diag}\{\mu_1, \mu_2, \dots, \mu_N\}. \tag{15.83}$$

If we now let

$$U = \boldsymbol{\Phi}_+(\nu) + \boldsymbol{\Phi}_-(\nu), \tag{15.84}$$

then we can eliminate between the sum and the difference of Equations 15.79 and 15.80 to find

$$(D - 2M^{-1}\hat{W}M^{-1})MU = \frac{1}{\nu^2}MU, \tag{15.85}$$

where

$$D = \mathrm{diag}\{\mu_1^{-2}, \mu_2^{-2}, \dots, \mu_N^{-2}\}. \tag{15.86}$$

Multiplying Equation 15.85 by a diagonal matrix T, we find

$$(D - 2\hat{V})\hat{X} = \frac{1}{\nu^2}\hat{X}, \tag{15.87}$$

where

$$\hat{V} = M^{-1}T\hat{W}T^{-1}M^{-1} \tag{15.88}$$

and

$$\hat{X} = TMU. \tag{15.89}$$

We can define (Barichello and Siewert 1999a) the elements $t_1, t_2, \ldots, t_N$ of T so as to make $\hat{V}$ symmetric; and therefore, since $\hat{V}$ is a symmetric, rank one matrix, we can write our eigenvalue problem in the form

$$(D - \varpi zz^{\mathrm{T}})\hat{X} = \lambda\hat{X}, \tag{15.90}$$

where $\lambda = 1/v^2$ and

$$z = \left[\left(\frac{1}{\mu_1}\right)w_1^{1/2} \ \left(\frac{1}{\mu_2}\right)w_2^{1/2} \cdots \left(\frac{1}{\mu_N}\right)w_N^{1/2} \right]^{\mathrm{T}}. \tag{15.91}$$

We note that the eigenvalue problem defined by Equation 15.90 is of a form that is encountered when the so-called *divide and conquer* method (Datta 1995) is used to find the eigenvalues of tridiagonal matrices.

Considering that we have found the required eigenvalues from Equation 15.90, we impose the normalization condition

$$\sum_{k=1}^{N} w_k[\phi(v, \mu_k) + \phi(v, -\mu_k)] = 1 \tag{15.92}$$

so that we can write our discrete ordinates solution as

$$I(x, \pm\mu_i) = \sum_{j=1}^{N} \left[A_j\phi(v_j, \pm\mu_i)e^{-x/v_j} + B_j\phi(v_j, \mp\mu_i)e^{-(x_0-x)/v_j} \right], \tag{15.93}$$

where

$$\phi(v_j, \mu_i) = \frac{\varpi v_j}{2} \frac{1}{v_j - \mu_i}. \tag{15.94}$$

Here, the arbitrary constants $\{A_j\}$ and $\{B_j\}$ are to be determined from the boundary conditions, and the separation constants $\{v_j\}$ are the reciprocals of the positive square roots of the eigenvalues defined by Equation 15.90. We note that differently of the previous section, the elementary solutions of the discrete ordinates problem, Equation 15.94, are written here in an explicit form. Of course, the cases where the eigenvalues and the separation constants may be equal have to be avoided.

Now, we can substitute Equation 15.93 into discrete versions of Equations 15.73 and 15.74,

$$I(0, \mu_i) = F_1(\mu_i) \tag{15.95}$$

and

$$I(x_0, -\mu_i) = F_2(\mu_i), \tag{15.96}$$

for $i = 1, 2, \ldots, N$, to define a linear algebraic system we can solve to find the required constants $\{A_j\}$ and $\{B_j\}$. And so our solution is established. Since the intensity is available, we can use, for the isotropic case, Equations 15.92 through 15.94 to express the density radiation

$$\rho(x) = \int_0^1 [I(x, \mu) + I(x, -\mu)] \, d\mu \tag{15.97}$$

as

$$\rho(x) = \sum_{j=1}^{N} \left[A_j e^{-x/\nu_j} + B_j e^{-(x_0-x)/\nu_j} \right] \tag{15.98}$$

and the radiative heat fluxes

$$q_{\pm}(x) = \pi \sum_{j=1}^{N} \left[A_j e^{-x/\nu_j} - B_j e^{-(x_0-x)/\nu_j} \right] \phi_1(\nu_j) \tag{15.99}$$

with

$$\phi_1(\nu_j) = \nu_j \sum_{i=1}^{N} \frac{\omega_i \mu_i}{\nu_j - \mu_i}. \tag{15.100}$$

To conclude this section, we note that while Equation 15.93 is a discrete ordinates expression for the intensity, a better result can be obtained (Barichello and Siewert 1999a). In fact, we can use Equation 15.98 to rewrite Equation 15.72 as

$$\mu \frac{\partial}{\partial x} I(x, \mu) + I(x, \mu) = \frac{\varpi}{2} \sum_{j=1}^{N} \left[A_j e^{-x/\nu_j} + B_j e^{-(x_0-x)/\nu_j} \right] \tag{15.101}$$

which we can solve, after noting Equations 15.73 and 15.74, to find

$$I(x, \mu) = I_0(x, \mu) + \frac{\varpi}{2} \sum_{j=1}^{N} \nu_j \left[A_j C(x : \nu_j, \mu) + B_j e^{-(x_0-x)/\nu_j} S(x : \nu_j, \mu) \right] \tag{15.102}$$

and

$$I(x, -\mu) = I_0(x, -\mu) + \frac{\varpi}{2} \sum_{j=1}^{N} \nu_j \left[A_j e^{-x/\nu_j} S(x_0 - x : \nu_j, \mu) + B_j C(x_0 - x : \nu_j, \mu) \right], \tag{15.103}$$

for $\mu \in (0,1]$. Here, the uncollided components are

$$I_0(x, \mu) = F_1(\mu)e^{-x/\mu} \tag{15.104}$$

and

$$I_0(x, -\mu) = F_2(\mu)e^{-(x_0-x)/\mu}. \tag{15.105}$$

In addition, the S and C functions are given by

$$S(\tau : x, y) = \frac{1 - e^{-\tau/x}e^{-\tau/y}}{x + y} \tag{15.106}$$

and

$$C(\tau : x, y) = \frac{e^{-\tau/x} - e^{-\tau/y}}{x - y}. \tag{15.107}$$

Although our analysis is based on a quadrature approximation, we note that our final results, in this section, for the intensity, radiation density, and radiative heat flux are continuous functions of the independent variables.

15.7 Computational Aspects

The first step to derive the computational procedure, in order to develop the discrete ordinates solution and to obtain the quantities of interest, is to define a quadrature scheme. As in other previous works where the ADO method was used, we choose to map the interval $[0, 1]$ into $[-1, 1]$ to use the well-known Gauss–Legendre quadrature scheme, here denoted $\{y_k, v_k\}$. In this way, the nodes $\mu_k \in [0, 1]$ referred in Equations 15.28, 15.29, and 15.78 are related with the Gauss points, y_k, for $k = 1, \ldots, N$ by

$$2\mu_k = y_k + 1. \tag{15.108}$$

In a consistent form, $\omega_k = (1/2)\, v_k$.

Although some tables are available for Gauss–Legendre quadrature points and weights, here, we mention an efficient way of computing these nodes, which seems to be very useful when higher order quadrature schemes are required. Following previous work (Benassi et al. 1984), where this problem was posed as a tridiagonal eigenvalue problem, we list below an extension of that procedure, which results in simpler symmetric tridiagonal eigenvalue system. The development is based on manipulations of the three-term recursion formula for the generation of Legendre polynomials, such that, we can write

$$x^2 P_n^*(y) = \sqrt{a_{n+1}a_{n+2}}\,P_{n+2}^*(y) + (a_{n+1} + a_n)P_n^*(y) + \sqrt{a_n a_{n-1}}\,P_{n-2}^*(y), \tag{15.109}$$

with $P_n(y) = (2/h_n)^{1/2}P_n^*(y), a_n = n^2/(h_{n-1}h_n)$ and $h_n = 2n + 1$, for $n = 0, 2, \ldots N - 2$ (n even) to generate a tridiagonal symmetric eigenvalue problem for defining the $N/2$ positive roots of $P_N(y)$, N even. In other words, the $N/2$ positive quadrature nodes (we remind they occur in $\pm$ pairs).

For example, for the case $N = 8$, we impose the condition $P_N(y) = 0$ to derive the four positive nodes y, from Equation 15.109,

$$\mathbf{H}\mathbf{v} = y^2\mathbf{v} \tag{15.110}$$

where

$$\mathbf{H} = \begin{pmatrix} a_1 & \sqrt{a_1 a_2} & 0 & 0 \\ \sqrt{a_1 a_2} & a_3 + a_2 & \sqrt{a_3 a_4} & 0 \\ 0 & \sqrt{a_3 a_4} & a_5 + a_4 & \sqrt{a_5 a_6} \\ 0 & 0 & \sqrt{a_5 a_6} & a_7 + a_6 \end{pmatrix} \tag{15.111}$$

and

$$\mathbf{v} = (P_0, P_2, P_4, P_6). \tag{15.112}$$

Having said that, we go back to the point of describing the general computational procedure. The basic steps to follow are:

- To define a quadrature scheme;
- To solve the eigenvalue problem given by Equation 15.52 (or 15.90) using any available linear algebra subroutine (Smith et al. 1976). For the simplest isotropic case, we can use known theoretical (Case and Zweifel 1967) results for checking the expected interval where the eigenvalues should be defined;
- To solve a linear system, Equations 15.65 and 15.66 (Equations 15.95 and 15.96);
- To evaluate quantities of interest, Equations 15.68 and 15.70 (Equations 15.98 and 15.99);

It is important to mention that the ADO solution has shown to be very fast and accurate.

As a simple introductory test case, in order to help the work of evaluating the expressions derived in this chapter, we list in Table 15.1 results obtained for $L = 6$, $\tau_0 = 1$, and $\varpi = 0.99$ and different values of quadrature order N. In regard to the boundary conditions, we consider $F_1(\mu) = 1$ and $F_2(\mu) = 0$. The idea is also to show the behavior of the results as N increases. It was implemented as a Fortran program, which runs in less than one second in a MacBook.

As mentioned before, a good solution of the direct problem, in calculations related to inverse problems in radiative transfer applications, is relevant. In this sense, analytical approaches as described in the previous sections can be very useful, mainly in regard to accuracy and gain in computational time, since, in many cases, the direct problem has to be solved several times.

Analytical approaches have also been used for deriving solutions for multidimensional problems, particularly associated with nodal schemes (Barichello et al. 2009).

TABLE 15.1

Radiation Density ρ

τ	$N = 10$	$N = 20$	$N = 30$
0.0	1.29320	1.29320	1.29320
0.1	1.20038	1.20027	1.20027
0.2	1.13796	1.13798	1.13798
0.3	1.08352	1.08354	1.08354
0.4	1.03279	1.03279	1.03279
0.5	9.83728(−1)	9.83728(−1)	9.83728(−1)
0.6	9.34947(−1)	9.34945(−1)	9.34945(−1)
0.7	8.85085(−1)	8.85070(−1)	8.85070(−1)
0.8	8.32215(−1)	8.32203(−1)	8.32203(−1)
0.9	7.72365(−1)	7.72461(−1)	7.72462(−1)
1.0	6.84710(−1)	6.84711(−1)	6.84711(−1)

15.8 Concluding Remarks

An analytical version of the discrete ordinates method, the ADO method, was used to develop a solution for thermal radiation problems defined in a gray plane-parallel medium. For the isotropic case, an explicit solution is obtained. The choice of using *half-range* arbitrary quadrature schemes leads to special forms of reduced size eigenvalue systems, in comparison with standard discrete ordinates approaches. The solution is easy to implement and fast.

Acknowledgments

The author is grateful to J. T. Reichert, J. F. Prolo Filho, and D. R. Justo for helpful suggestions regarding this text. Still, the author would like to thank CNPq of Brazil for partial financial support to the work presented in this chapter.

Nomenclature

$B(T)$	Planck's function
E_λ	radiant energy
I_λ	specific intensity
$j_{e\lambda}$	emission coefficient
k_λ	absorption coefficient
L	anisotropy degree
$p(\cos \Theta)$	phase function
$q_\pm(\tau)$	partial radiative fluxes
$Q(\tau, \mu)$	source term
T	temperature

x optical variable
z spatial variable

Abbreviation

ADO method   analytical discrete ordinates method

Greek Variables

β_λ extinction coefficient
β_l coefficients of the scattering law
ε_α emissivity at the surface α
ϕ azimuthal angle
μ cosine of the polar angle
ρ_i^s coefficient for specular reflection at the boundary i
ρ_i^d coefficient for diffuse reflection at the boundary i
$\rho(\tau)$ density of radiation
θ polar angle
Θ scattering angle
λ wavelength
ρ density
$\sigma_{s\lambda}$ scattering coefficient
τ optical variable
τ_0 optical thickness
ϖ albedo
Ω unitary vector

References

Barichello, L.B. and Siewert, C.E. 1997. On inverse boundary-condition problems in radiative transfer. *Journal of Quantitative Spectroscopy and Radiative Transfer* 57: 405–410.

Barichello, L.B. and Siewert, C.E. 1998. On the equivalence between the discrete-ordinates and the spherical-harmonics methods in radiative transfer. *Nuclear Science and Engineering* 130: 79–84.

Barichello, L.B. and Siewert. C.E. 1999a. A discrete-ordinates solution for a non-grey model with complete frequency redistribution. *Journal of Quantitative Spectroscopy and Radiative Transfer* 62: 665–675.

Barichello, L.B. and Siewert, C.E. 1999b. A discrete-ordinates solution for a polarization model with complete frequency redistribution. *The Astrophysical Journal* 513: 370–382.

Barichello, L.B. and Siewert, C.E. 2002. A new version of the discrete-ordinates method. *Proceedings of Computational Heat and Mass Transfer 2001 Rio de Janeiro*, I: 340–347.

Barichello, L.B., Cabrera, L.C., and Prolo Filho, J.F. 2009. An analytical discrete ordinates solution for two dimensional problems based on nodal schemes. *International Atlantic Nuclear Conference—INAC 2009*, Rio de Janeiro, Brazil, September 28 to October 02.

Barichello, L.B., Garcia, R.D.M., and Siewert, C.E. 2000. Particular solutions for the discrete-ordinates method. *Journal of Quantitative Spectroscopy and Radiative Transfer* 64: 219–226.

Benassi, M., Cotta, R.M., and Siewert, C.E. 1983. The P_N method for radiative transfer problems with reflective boundary conditions. *Journal of Quantitative Spectroscopy and Radiative Transfer* 30: 547–553.

Benassi, M., Garcia, R.D.M., Karp, A.H., and Siewert, C.E. 1984. A high-order spherical harmonics solution to the standard problem in radiative transfer. *The Astrophysical Journal* 280: 853–864.

Case, K.M. 1960. Elementary solutions of the transport equation and their applications. *Annals of Physics* 9: 1–23.

Case, K.M. and Zweifel, P.F. 1967. *Linear Transport Theory*. Reading, MA: Addison-Wesley Publishing Company.

Chaloub, E.S. and Campos Velho, H.F. 2003. Multispectral reconstruction of bioluminescence term in natural waters. *Applied Numerical Mathematics* 47: 365–376.

Chandrasekhar, S. 1960. *Radiative Transfer*. New York: Dover.

Datta, B.N. 1995. *Numerical Linear Algebra and Applications*. Pacific Grove: Brooks/Cole Publishing Co.

Davison, B. 1957. *Neutron Transport Theory*. London: Oxford University Press.

Fiveland, W.A. 1984. Discrete-ordinates solutions of the radiative transport equation for rectangular enclosures. *Journal of Heat Transfer* 106: 699–706.

Garcia, R.D.M., Siewert, C.E., and Yacout, A.M. 2008. Radiative transfer in a multilayer medium subject to Fresnel boundary and interface conditions and uniform illumination by obliquely incident parallel rays. *Journal of Quantitative Spectroscopy and Radiative Transfer* 109: 2151–2170.

Gradshteyn, I.S. and Ryzhik, I.M. 1980. *Table of Integrals, Series and Products*. New York: Academic Press.

Gryn, V.I. 1995. Inverse problems of steady radiative transfer theory. *Computational Mathematics and Mathematical Physics* 35: 1471–1488.

Kaper, H.G., Shultis, J.K., and Veninga, J.G. 1970. Numerical evaluation of the slab albedo problem solution in one-speed anisotropic transport theory. *Journal of Computational Physics* 6: 288–313.

Klose A.D. and Hielscher, A.H. 2002. Optical tomography using the time-independent equation of radiative transfer—Part 2: Inverse model. *Journal of Quantitative Spectroscopy and Radiative Transfer* 72: 715–732.

Lewis, E.E. and Miller Jr., J.F. 1984. *Computational Methods of Neutron Transport*. New York: John Wiley & Sons.

Liou, K. 1980. *An Introduction to Atmospheric Radiation*. New York: Academic Press.

Liu, D., Wang, F., Yan, J.H., Huang, Q.X., Chi, Y., and Cen, K.F. 2008. Inverse radiation problem of temperature field in three-dimensional rectangular enclosure containing in-homogeneous, anisotropically scattering media. *International Journal of Heat and Mass Transfer* 51: 3434–3441.

McCormick, N.J. and Sanchez, R. 1981. Inverse problem transport calculations for anisotropic scattering coefficients. *Journal of Mathematical Physics* 22: 199–208.

Modest, M.F. 1993. *Radiative Heat Transfer*. New York: McGraw-Hill.

Moura, L.M., Bailis, D., and Sacadura, J.F. 1998. Identification of thermal radiation properties of dispersed media: comparison of different strategies. *Proceedings of 11th International Heat Transfer Conference*, Kyongju, Korea, August 23–28, pp. 409–414.

Özişik, M.N. 1973. *Radiative Transfer*. New York: John Wiley & Sons.

Özişik, M.N. and Orlande, H.R.B. 2000. *Inverse Heat Transfer*. New York: Taylor & Francis.

Ren, K., Bal, G., and Hielscher, A.H. 2006. Frequency domain optical tomography based on the equation of radiative transfer. *SIAM Journal on Scientific Computing* 28: 1463–1489.

Sanchez, R. and McCormick, N.J. 2008. On the uniqueness of the inverse source problem for linear particle transport theory. *Transport Theory and Statistical Physics* 37: 236–263.

Sarvari, S.M.H. and Mansouri, S.H. 2004. Inverse design for radiative heat source in two-dimensional participating media. *Numerical Heat Transfer Part B—Fundamentals* 46: 283–300.

Siewert, C.E. 2000. A concise and accurate solution to Chandrasekhar's basic problem in radiative transfer. *Journal of Quantitative Spectroscopy and Radiative Transfer* 64: 109–130.

Siewert, C.E. 2002. Inverse solutions to radiative-transfer problems based on the binomial or the Henyey–Greenstein scattering law. *Journal of Quantitative Spectroscopy and Radiative Transfer* 72: 827–835.

Smith, B.T., Boyle, J.M., Dongarra, J.J., Garbow, B.S., Ikebe, Y., Klema, V.C., and Moler, C.B. 1976. *Matrix Eigensystem Routines-EISPACK Guide*. Berlin: Springer-Verlag.

Tito, M.J.B., Roberty, N.C., Silva Neto, A.J., and Cabrejos, J.B. 2004. Inverse radiative transfer problems in two-dimensional participating media. *Inverse Problems in Science and Engineering* 12: 103–121.

Wick, G.C. 1943. Über Ebene Diffusionsprobleme. *Zeitschrift für Physik* 120: 702–718.

Part III

Applications

16

Analysis of Errors in Measurements and Inversion

Philippe Le Masson and Morgan Dal

CONTENTS

16.1 Introduction

The difficulty every researcher approaching a physical phenomenon faces is choosing the model that best suits the requirement. For this purpose, knowledge of the parameters and/or the functions of modeling is imperative. These data can, indeed, be obtained not

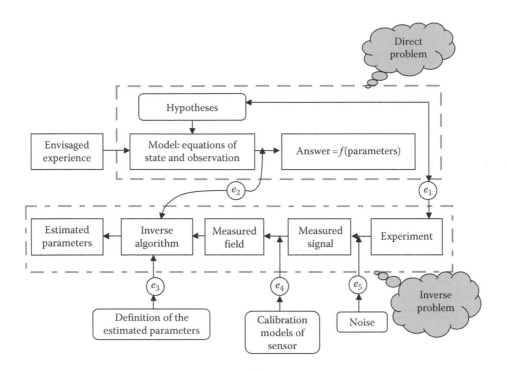

FIGURE 16.1
Summary of the inverse and direct approaches.

only in the literature but also by way of inverse techniques. In an inverse problem, we have information (measurements of temperatures, heat flux, deformations, etc.) which are the effects of the causes we try to estimate. It is, thus, a question of identifying the laws of physics that we study. Although the inverse techniques apply to all the domains of physics, we discuss only the problems of heat diffusion here.

In this chapter, we particularly emphasize the quality of the measurements, which is constantly necessary to try to obtain. Figure 16.1 summarizes the usually followed inverse and direct approaches. We find five types of estimation errors. If we do not take the effects of coupling between these various terms into account, the global error can be defined as the sum of these types of errors:

$$e_g = e_1 + e_2 + e_3 + e_4 + e_5. \tag{16.1}$$

The direct problem is only a mathematical image of our experiment. If the hypotheses made for the definition of this problem are not relevant to the experiment, an error e_1 can lead us to a nonrealistic estimation. Furthermore, the resolution of the direct problem appeals to various methods. The calculated answer is more or less different from the exact answer required by the model. The errors of resolution (error of numerical scheme connected to the steps of time, to the steps of space, and to the precision of the computer) will lead to an error e_2 of estimation.

The first two errors generally occur during the modeling of a phenomenon for which we do not completely know the physics, in particular in the case of coupling between several domains (heat conduction, electricity, phase change, etc.). Other errors for the inverse problems will eventually be added to these errors. First of all, the error e_3 is an error resulting from objectives of the experiment/reverser. The composed questions are as follows:

1. Do we have some knowledge of the shape of the law that we want to estimate?
2. Have we worked on the estimation of a perfectly unknown function or have we estimated the parameters of a law?
3. What is the spatial and/or temporal domain on which we have to estimate our parameters and/or functions?
4. In the case of an estimation on a limited domain, what are the parameter values for the rest of the domain?
5. What errors in the estimation are due to the uncertainties connected to the knowledge of the known data (e.g., spatiotemporal data: position of the sensors and the measures of the forms of the sample)?

Having partly answered these questions by the choice of a parameterization of the function that we want to estimate, by the study of the sensitivity coefficients when faced with the position of the sensors and the uncertainties on the known data, the error is stressed by the fact that the inverse methods themselves cause errors in estimated values. These last errors are engendered by the method.

The fourth component e_4 of the error of the estimation results from the conversion of the signal delivered by the sensor (e.g., temperature). It is thus connected to the characteristics of the sensor (problems of the nonlinearity of the sensor, the reproducibility of a measurement, etc.) and results from the definition of the experimental design of these sensors in the experiment (intrusive character of the sensor). The study of these errors of thermal metrology is an important element to obtain a quality estimation.

The last of the estimation errors e_5 is due to the measuring device of the signal produced by the sensor. This error can be defined under the general term of "noise." Beck and Arnold [1] presented hypotheses on the nature of this noise allowing for the implementation of the methods of estimation.

- The noise is additive:

$$Y_i = T_i + \varepsilon_i, \tag{16.2}$$

  where
  Y_i is the measured temperature
  T_i is the exact temperature
  ε_i is a random noise

- The noise ε_i has a zero mean value:

$$E(\varepsilon_i) = 0. \tag{16.3}$$

- The noises associated with the various measurements are not correlated. Two noises of measures ε_i and ε_j are said to be noncorrelated if their covariance verifies:

$$\text{cov}(\varepsilon_i, \varepsilon_j) \cong E[[\varepsilon_i - E(\varepsilon_i)][\varepsilon_j - E(\varepsilon_j)]] = 0 \quad \text{for } i \neq j. \tag{16.4}$$

In this case, the noise ε_i has no effect on and no relation with the noise ε_j.

- The noise has a constant standard deviation:

$$\sigma_i = \sqrt{E\left[[Y_i - E(Y_i)]^2\right]} = const. \tag{16.5}$$

- The considered noise follows a normal probability distribution.
- We have no information a priori about the unknown.
- Only the measurements used in the estimation procedure constitute the noisy data. Other parameters present in the model are supposed to be exactly known.

These enumerated hypotheses are rarely combined in reality. They are emitted for the validity of the estimation methods.

In this chapter, we thus suggest analyzing the error e_4. We also try to make a link between the errors of measurement and the developed direct model by integrating it into the modeling of the sensor. Working mainly on the measurement of temperature using thermocouples, we will have to take into account several special points such as the drilling of the sample and the quality of the contact between the sensor and the sample.

16.2 Temperature Measurement by Thermocouple and Errors of Measurement

Any temperature measurements by a thermocouple on or in a sample generate errors of measure by their intrusive aspect. First of all, the temperature measure in a point of a material environment in fact involves a measure of a small element surrounding this point. Furthermore, when we have important gradients of temperature in the environment, we suppose that the volume element is small enough so that the temperature is practically uniform. Finally, if the considered element is defined on the surface, it has to take into account the fact that the surface possesses specificities such as its roughness and the state of the surface connected to surface treatments (oxidation and quenching). All these physical specificities underline the difficulties involved, on the one hand, in integrating a sensor whose physical characteristics are often different from the environment where we want to realize the measure and, on the other hand, in estimating the disturbance that this sensor engenders.

In general, following the use of the sensor (strong temperature gradients or not, steady state or transient, and strongly transient phenomenon or not), we have to master the parameters of implementation, namely, its calibration, its sensitivity, and all the involved thermoelectric effects. Numerous works ([1–9], or Chapter 3 by Lanzetta) summarize these various aspects like the size of the sensor, the embedded thermocouple or not in the sample and of the chain of measure.

In this chapter, we will mainly emphasize the presence of the sensor and the disturbance it generates on the phenomenon of heat diffusion. In any theory, the presence of the sensor in an environment supposes a modification of the heat transfers by conduction, convection, or radiation. The local temperature is modified. Concerning more particularly the surface temperature measure, the presence of a sensor modifies the emissivity and the exchanges by radiation. The sensor and the elements transferring the information (e.g., thermocouple wires) are affected by the outside conditions, provoking a measure that can be very different from the one to be measured.

16.2.1 Error during Measurement of a Surface Temperature

As we were able to underline above, during measurement of a surface temperature, the superficial exchanges are modified by the presence of the sensor, and the thermophysical and radiative properties are different from those of the surface material. Furthermore, because of the link between the sensor and the acquisition system, a parasite heat flux is passed on through it and then toward the environment. A generation or an absorption of heat at the level of the sensor can also occur. Furthermore, because of the methodology of the setting up of the sensor, we can have an exothermic or endothermic reaction of chemical origin. All these heat transfers make a disturbance of the surface temperature.

The temperature is not T but T_p. Figure 16.2 presents the heat transfers during a measure of a surface temperature. We can note that, in the presented case, because of heat transfer from the material to the outside environment and thus the pumping of energy engendered by the sensor, the temperature $T_p(t)$ is lower than the true temperature of the surface. Furthermore, generally the temperature sensor is implanted on the surface of the material with a contact which very often remains imperfect. The sensor is thus going to take, in the case of Figure 16.2, an intermediate temperature $T_c(t)$ defined as the temperature between the new temperature of the surface $T_p(t)$ and the temperature of the outside

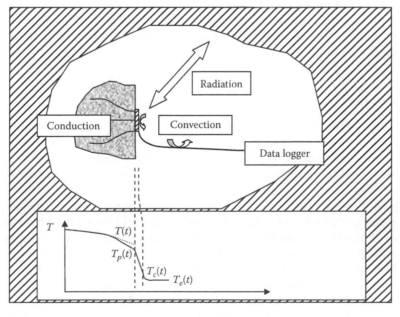

FIGURE 16.2
Heat transfers through the sensor surface.

environment $T_e(t)$. So in the sensor, we have a parasite heat transfer, noted $\varphi(t)$. In reality, during this heat transfer, three effects are conjugated:

- The first effect occurs in the material through the convergence of the lines of heat flux toward the sensor (macro-constriction effect).
- The second effect is connected to the quality of the contact between the sensor and the material. A contact resistance between both elements provokes a temperature drop at the interface.
- The third effect is laid to the exchanges between the sensor, the wires of the sensor, and the ambient. This effect is also mentioned as the fin effect.

So as we were able to highlight above, a measure error appears and is represented by the distance between the true temperature $T(t)$ and the temperature on the sensor $T_c(t)$:

$$\varepsilon(t) = T(t) - T_c(t). \tag{16.6}$$

16.2.2 Error during a Temperature Measure within a Volume

In the case of a measure within a volume, we find the three effects seen during the analysis of the measure of a surface temperature. Indeed, according to the thermophysical characteristics, generally different from those of the material, an effect of macro-constriction between the material and the sensor shall be found. Furthermore, according to the quality of the setting up of the sensor in the middle (sensor simply put in contact, sticking or welding), a contact resistance will provoke a temperature difference between the material and the sensor. Finally, the third effect seen above will globalize transfers among the material, the sensor, and its wires, then between the ambient and the wires of the sensor. Really, a difficulty is added here. The fact of realizing a hole with a diameter always higher than the diameter of the sensor involves having to analyze transfers between both elements (space between the wires and the hole, space between the sensor and the opening, the contact through grease, glue, etc.).

16.2.3 Error Models

We find in the literature a lot of models to describe the errors of measurement by thermocouple. The study of these errors which result from parasite transfers requires the resolution of problems of complex thermal transfers. The variety of configurations and the conditions connected to the environment involves the elaboration of mathematical models adapted to each case.

First of all, we suggest presenting a very simplistic but typical model that will allow us to highlight the respective roles of conduction within the middle, of the imperfection of the contact between the thermocouple and the surface, and finally of the exchanges with the environment. The model described in the study of Cassagne et al. [4] appears through the three effects seen in Section 16.2.1. Most of the conclusions of this study will then be transposed into various configurations usually met in practice.

We shall successively consider the measures in steady-state regime and in transient regime. However, before this, it is necessary to underline that transfers between the wire of the thermocouple and the ambient are defined through a heat transfer coefficient h

and a characteristic temperature T_E, being able to group together the convective and radiative transfers.

$$h = h_r + h_{cv}, \tag{16.7}$$

where
$h_r = 4\varepsilon\sigma T_m^3$
T_m is the medium temperature of the two surfaces

For example, a model in steady-state regime can be developed for a measure of the surface temperature [3,5,6].

Let us suppose a measure of temperature on an isothermal surface (T) of an opaque semi-infinite region. For the needs of the theoretical study, we consider only a single wire (measure by a semi-intrinsic thermocouple). Figure 16.3 presents the theoretic scheme. The surface of the region is supposed to be insulated. We have only heat transfer at the contact with the sensor represented by a bar. The modeling of the contact is defined by the three previously seen effects:

- Effect of the convergence: this effect is defined by a convergence of the heat toward the zone of measure. The temperature T of the insulated region is perturbed by the energy pumping in the wire:

$$T - T_p = r_m\Phi, \tag{16.8}$$

where
r_m is the macro-constriction resistance
Φ is the transferred heat flux

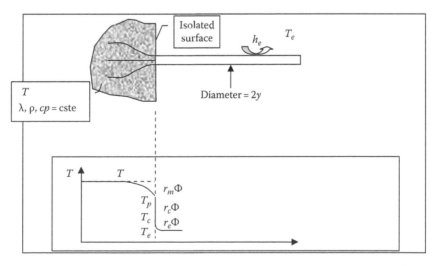

FIGURE 16.3
Model of error in steady-state regime.

We show with the hypothesis of a semi-infinite region, the following two expressions:

$$r_m \cong \frac{1}{4y\lambda} \quad \text{or} \quad r_m \cong \frac{8}{3\pi^2 y\lambda}. \tag{16.9}$$

We suppose the circle of contact to be isothermal or crossed by a uniform heat density (λ is the thermal conductivity of the material). The calculation also shows that the main part of the temperature disturbance is located in the immediate neighborhood of the circle of contact (94% of the temperature drop $T-T_p$ occurs inside the sphere of center O and of radius $10y$).

- Effect of the contact resistance: responsible for the temperature drop between the perturbed temperature and the measured temperature, it is defined by the relation

$$T_p - T_c = r_c \Phi, \tag{16.10}$$

where r_c represents the contact resistance for the contact area s: $r_c = \dfrac{R_c}{s}$. This effect is connected to the imperfection of the contact which results from the roughness of the surface.

- Fin effect: it is defined by heat transfer between the wire and the outside environment. The heat flux is transferred between the face $x=0$ at the temperature T_c and the outside environment at an equivalent temperature T_e:

$$T_c - T_e = r_e \Phi, \tag{16.11}$$

where r_e represents the global thermal resistance of the transfer. It depends on the geometry, on the global heat transfer coefficient h_e (convection + radiation), and on the conductivity λ_e of the wire. For a bar with a radius y, the resistance is defined by the relation:

$$r_e = \frac{1}{\pi y \sqrt{2 h_e \lambda_e y}}. \tag{16.12}$$

With these three equations, we deduct the error of measure:

$$T - T_c = \delta T = (r_c + r_m)\Phi$$

or

$$T - T_e = (r_c + r_m + r_e)\Phi$$

so

$$\frac{T - T_c}{T - T_e} = \frac{\delta T}{T - T_e} = \frac{(r_c + r_m)}{(r_c + r_m + r_e)}, \tag{16.13}$$

$$\delta T = \frac{T - T_e}{\dfrac{(r_c + r_m + r_e)}{(r_c + r_m)}},$$

$$\delta T = \frac{T - T_e}{1 + \dfrac{r_e}{r_c + r_m}}.$$

(16.14)

The committed error is thus proportional to the difference between the temperature to be measured and the outside equivalent temperature. The error will be weak if the sum of the macro-constriction and contact resistances is weak in the presence of the outside resistance.

We show the following:

1. For measures on a metal with a good conductivity $r_m \ll r_c$, it is essentially the contact conditions that set the error.

2. For measures on an insulating material $r_m \gg r_c$, it is the effect of macro-constriction that sets the error.

3. Even in conditions of perfect contact, an error persists depending on the relationship $\dfrac{r_m}{r_e}$.

4. The roles of r_e and of T_e are finally very important. It is necessary to have r_e the biggest possible and T_e the closest of T.

These conclusions, established in the case of a measure on an opaque region and for the configuration defined by a wire having the shape of a perpendicular bar, stay for more complex real configurations. In any case, it would now be necessary to study the case of a measure within the region as well as the cases of the transient regimes.

As the objective of this chapter is to take into account measure errors in the inverse analysis, we will attempt to develop numeric models of error to quantify these and define in transient regime the best configurations of thermocouple design.

16.3 Application

16.3.1 Introduction

Welding is an assembly method in constant evolution. Mainly used by heavy industry, many processes were developed for several years; however, arc welding is the most harnessed.

Moreover, to ensure joint quality, numerical simulations take on a fundamental importance and try by complementing or making it possible to avoid experimental measurements. Actually, mechanical effects such as welding distortions or residual stresses are directly linked to the evolution of the thermal field created by process energy. The difficulties concern, first, the simulation of coupled phenomena such as the arc and the plasma over the liquid metal, and second, the characterization of not well-known parameters. For these reasons, two levels of simulation can be implemented. The first is named

"Multiphysic" and its objective is to model the whole physical phenomena. The second is a simplified thermal simulation with an equivalent heat source.

One way to establish the simplified law of energy distribution is the implementation of an inverse method used in conjunction with thermal measurements. Therefore, the quality of inverse problem results is strongly dependent on the experimental part.

The objective of this work is to investigate the thermal discrepancy between our measured temperatures and temperatures obtained by classical measurement techniques. This study was carried out to propose an optimal configuration for thermocouple installation. Our study was conducted to determine an equivalent heat source for Metal Active Gas (MAG) process with filler material and on "T" configuration [7].

16.3.2 The Goal of the Instrumentation

Before introducing bases of instrumentations, an explanation about the different forms of simulation is required. Two classes of numerical models can be distinguished. The first is designed to simulate the process by modeling the whole phenomena. The second is simplified to a pure conduction model; thus, phenomena occurring in the fused zone are neglected or approximated.

It should be noted that these studies need a large knowledge of physics parameters such as magnetic, hydrodynamic, and thermal properties of gas, liquid, and solid materials. Moreover, the needed numerical resources are also important and not available in the industrial context. For this reason, simplified models are developed using only the thermal diffusion equation and an equivalent heat source term which represent the effects of phenomena in the fused zone.

16.3.2.1 Equivalent Heat Source

The "equivalent source" simulation is based on the assumption that the real heat distribution is closely approximated by a mathematical function. Therefore, the thermal field is assumed to be lowly sensitive to the liquid part, or the equivalent heat source must approximate its effects. Mathematical laws' complexity is dependent on the approximation level. Indeed, the most simplified expression is the "point heat source" in which all the energy is applied at one surface point. If the real energy input is introduced inside the volume, it is possible to simulate it by distributing constant energy along a line. These two kinds of assumptions were mainly used in the past for analytical resolutions. Now numerical simulation and computing evolution allow for the use of more complex laws. When energy has to be applied on the workpiece surface, the Gaussian distribution is commonly used. Indeed, it allows for the set up of the amplitude and the radius. Moreover, for a process that gives energy inside the workpiece volume, the surface equivalent heat source could be used but the theoretical energy had to be distributed along the third dimension. Several cases can be encountered, energy can stay constant along the depth or it can decrease with a mathematical law, for example, linear or Gaussian.

Nevertheless, the shapes of the approximate heat sources depend on the process, which is why several laws are made to be easy to customize. For example, the one named cylindrical involution normal (CIN) is shown in Equation 16.15 in a quasi-steady-state formulation. The source term is assumed to be constant through time, and after the moving of the space referential on it [8,9].

$$S(x, y, z) = \frac{kK_z\eta UI}{\pi(1 - \exp(-K_z e_p))} \exp(-k(x^2 + z^2) - K_z y) * [1 - u(y - e_p)],\qquad(16.15)$$

where
   $u(\cdots)$ is the Heaviside function
   k is the concentration factor
   K_z is the involution factor
   e_p is the depth source application
   U, I, and η are, respectively, the welding voltage (V), current (A), and arc efficiency
   (x, y, z) is the coordinate system in which x and z are tangent and normal to the top
      surface of the filler material

Other kinds of equivalent heat sources try to include the fluid-mechanic effects. An example of these distributions is the "Goldak double ellipsoid" [10], in which the front and the rear ellipsoids (Figure 16.4) are, respectively, written with Equations 16.16 and 16.17:

$$q_f(x, y, z) = \frac{6\sqrt{3}f_f Q}{abc_f \pi\sqrt{\pi}} \exp\left(-\frac{x^2}{a^2}\right) \exp\left(-\frac{y^2}{b^2}\right) \exp\left(-\frac{z^2}{c_f^2}\right),\qquad(16.16)$$

$$q_r(x, y, z) = \frac{6\sqrt{3}f_r Q}{abc_r \pi\sqrt{\pi}} \exp\left(-\frac{x^2}{a^2}\right) \exp\left(-\frac{y^2}{b^2}\right) \exp\left(-\frac{z^2}{c_r^2}\right),\qquad(16.17)$$

where
   f_f and f_r are factors of energy deposited on the front and rear parts of y-axis
   a, b, c_f, and c_r are the Gaussian radius

The difficulty in these approaches is the determination of these previous parameters because they are not directly related to a physical phenomenon. However, they are estimated by an inverse method, but it compels to perform experimentation. By knowing the real temperatures at a few points and having the numerical simulation at these same

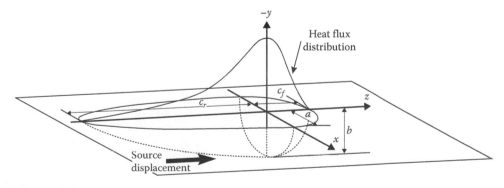

FIGURE 16.4
Goldak's double ellipsoid. (From Goldak, J. et al., *Int. Inst. Weld.*, 603, 1985.)

points, an identification algorithm compares experimental and numerical temperatures to adjust heat source parameters.

In our study, the equivalent heat source is written with the CIN law and the estimation method is based on the *Levenberg–Marquardt* algorithm. It reduces a quadratic criterion defined as the difference between experimental and numerical temperatures.

Actually, a lot of thermal instrumentations are available to solve inverse problems. Measurements must be performed in a high temperature gradient (sensitivity problem), which means close to the liquid/solid interface. In the next part, some of these methods are explained.

16.3.2.2 Welding Measurement Methods

For the characterization of heat diffusion in the fused zone, three main measurement methods can be distinguished: observation with high speed cameras, infrared measurements, and thermoelectric probes.

- High speed cameras are suitable for observations of fluid motion or for measurements of weld pool sizes. However, it gives only surface information. Without temperature measurements, this method is impractical for heat source identification but it is possible to use its observations for constrain estimation.

- Thermography and pyrometry are two methods which allow for measurements of infrared radiations. With appropriate relation, it leads to surface temperatures. But both are limited by the difficulties to determine the emissivity and they are only able to detect surface information directly. The main advantage of thermography is the high number of sensors (for example, 120×160) and its ease of implementation. By measuring radiation at two wavelength values, some pyrometers are not subject to emissivity difficulties.

- The use of thermocouple is an intrusive method. The hot junction (Figure 16.5) has to be in contact with the sample at the exact location where the measurement must be made. So, the way to implement thermocouple inside a solid body is to get it in touch at the bottom of a hole (Figure 16.5). The drawback is the thermal field

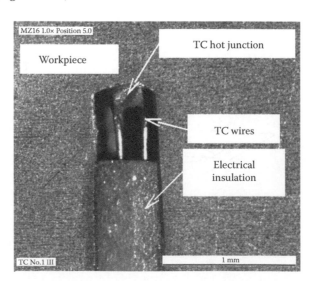

FIGURE 16.5
Example of our thermocouple.

disturbance that can occur all around the probe. Furthermore, different types of thermocouples are available, and their use depends on temperature levels. In our case, type K thermocouples are used (Figure 16.5), and their maximal thermal charge is 1367°C for 0.15 s. Due to this information and because we want measurements that are as sensitive as possible for researched parameters, an implementation plan must be developed.

Figure 16.5 presents a picture showing one of our thermocouples implemented in the sample and the different parts of it. More information is given about this picture in the following sections.

The choice of thermocouple locations is critical for the quality of identification results. The next part explains how we have selected these positions.

16.3.2.3 Thermocouple Locations: Thermal Gradient Measurements

To define a good thermocouple position, several ideas have to be explained. The identification concept is based on observation of the parameter variation effects on the thermal field. This sensitivity decreases when measurements are far off the weld pool. Moreover, a temporal discrepancy is observed. Caused by the thermal capacity of material, those effects can be reduced by introducing thermocouples close to the heat source.

For heat flux identification, one measurement line might permit estimation. Nevertheless, the solution cannot be reliable because several flux shapes can reduce quadratic criterion. For this reason and to stabilize estimation, a second line of thermocouples is installed. So, a three-dimensional (3D) "picture" of thermal gradient is obtained and stability of the estimation procedure is enhanced.

The goal of the identification is to characterize the gradient versus the three directions of diffusion (x, y, z). So, a finite measurement line is defined around the fused zone limit seen on a scheme of a macrograph (Figure 16.6). According to fused zone uncertainty (caused by weld pool instability) and with a previous theoretical simulation, we define a priori this first line near 1100°C–1200°C (Figure 16.6). Then, to define thermal gradients correctly, a second line is fixed to observe isothermal lines at 1000°C–1100°C. The third dimension is along the welding direction. We have considered a quasi-steady-state analysis; therefore, the third dimension is the time multiplying the welding velocity. The z gradient is, in fact, time variations.

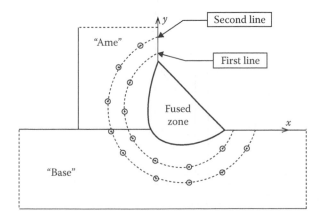

FIGURE 16.6
Measurements locations, scheme of macrograph.

16.3.3 Our Welding Case

As described above, this study takes place in a larger project named MUSICA [7]. This multi-partner work was initiated by the CEA in collaboration with the French "Welding Institute," the CETIM, Esi-Group, and AREVA. The aim of their work is the development of three software tools designed to allow a nonexpert user to realize global simulation of welding processes. Indeed, the first is made for the simulation of welding processes, the second for thermo-mechanical simulation at component sides, and the third for calculation of distortion effects on multicomponent structures. Moreover, the tools coupled to inverse method and with instrumented equipment permit heat source estimation for usual welding processes such as tungsten inert gas (TIG), laser, electron beam welding, or metal inert/active gas (MIG/MAG). These elements need accumulation of knowledge about many welding simulation cases, such as ring-shaped welding or "T" welding.

This welding study deals with the optimal instrumentation for two weld displacements and with the heat source identification in the case of "T" welding with filler material.

It should be noted that the industrial aspect of this work has constrained us to not unveil the process parameters or thermal properties of the used material. Each of the given cases will be specified.

16.3.3.1 Mathematical Model

The best parametric estimation must be performed only with previous experimentations. It allows one to check the "weldability" of the specimen. Two pieces of metal (AISI S355 steel) are used (Figure 16.7): the "Base" ($0.5 \times 0.5 \times 0.01$ m^3) and the "Ame" ($0.5 \times 0.01 \times 0.1$ m^3). After the welding test, the sample is cut out, transversal plans are chemically attacked, and photographs of weld joints are taken. The main objective of this first experimental part is to create a simplified simulation (Figure 16.8), which gives a fused zone as close as possible to the one observed on previous macrographies. The simulation gives a thermal field, which is then used to define the location of an a priori instrumentation for the next experimental part designed for the inverse problem.

In Figure 16.7, we show the two instrumented zones: A for the first transit and B for the second.

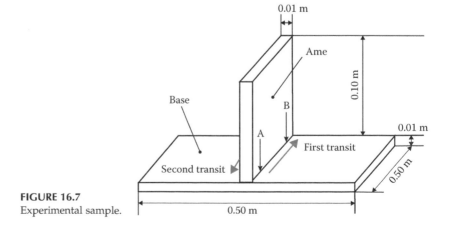

FIGURE 16.7
Experimental sample.

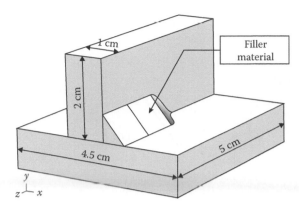

FIGURE 16.8
Simulated sample.

The simulated geometry has to be reduced (Figure 16.8) to decrease calculation time, but dimensions must be sufficient to avoid fused zone disturbances caused by boundary conditions implemented in the model.

It is difficult to simulate the filler material during the welding. So, we shall always suppose it present during the simulation.

A 3D direct problem is solved by the finite-element method with the software Comsol Multiphysic®. This analysis is performed by assuming the heat equation in a quasi-steady-state conduction according to Equation 16.18. To reduce the degrees of freedom solved, the fluid mechanics and other phenomena are neglected. So, the equivalent heat source $S(x, y, z)$, in this heat equation formulation, must represent all these phenomena in the fused zone and approximate the 3D heat distribution. This heat source will be described in the next part. The following is the heat diffusion equation:

$$v\rho(T)cp(T)\frac{\partial T}{\partial z} = \frac{\partial}{\partial x}\left(\lambda(T)\frac{\partial T}{\partial x}\right) + \frac{\partial}{\partial y}\left(\lambda(T)\frac{\partial T}{\partial y}\right) + \frac{\partial}{\partial z}\left(\lambda(T)\frac{\partial T}{\partial z}\right) + S(x, y, z), \qquad (16.18)$$

where
v is the welding speed (m s^{-1}) along z-axis
ρ is the density (kg m^{-3})
cp is the heat capacity (J kg^{-1} K^{-1})
λ is the thermal conductivity (W m^{-1} K^{-1})

Thermal properties are isotropic. Boundary conditions are defined in Figure 16.9. In this quasi-steady-state simulation, the equivalent heat source is centered on the "S" point (Figure 16.9) and its "displacement" is simulated (quasi-steady-state assumption) along the z direction. Consequently, a graph showing temperature along the z-axis is equivalent to a thermogram. This is what we need to compare: real time measurement with this simulation.

- For $z = 0$, before the heat source (Figure 16.9 boundaries A), the temperature is equal to ambient temperature (20°C).

$$T = T_\infty. \qquad (16.19)$$

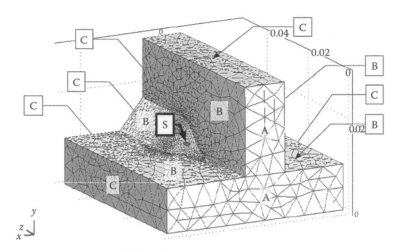

FIGURE 16.9
Mesh and boundary conditions (A: ambient temperature, B: heat losses, and C: thermal symmetry).

- Behind, for $z = 0.05$ (Figure 16.9 boundaries C), boundaries are assumed sufficiently far from heat source to be represented by thermal symmetry. The same assumption is realized for boundaries on $x = -0.02$ and 0.025.

$$\lambda(T)\frac{\partial T}{\partial n} = 0. \tag{16.20}$$

- Lastly, for boundaries that are in contact with air (Figure 16.9 boundaries B), we consider that a correct approximation of temperatures is obtained by classical convection and radiation losses.

$$-\lambda(T)\frac{\partial T}{\partial n} = h(T - T_\infty) - \varepsilon\sigma(T^4 - T_\infty^4), \tag{16.21}$$

where
  h is the coefficient of convection (W m^{-2} K^{-1})
  ε is the emissivity
  σ is the Stefan–Boltzman constant

In the majority of studies, the filler material is ever present in the workpiece geometry, but elements are activated at the rear of the weld pool which can be approximated with the rear of source application. In our case, the software used does not allow us these kinds of numerical tools, but here, the heat source is stationary and we can directly define the front part of the fused zone geometry. This shape is defined thanks to experimental information and created with a *Bézier* surface. The latter is built thanks to observations of the filler material after solidification. During the last part of the welding, when the arc disappears, the fused zone is solidified, but the surface shape of the liquid part stays apparent. The mark left by the maximum of the arc pressure could also be located. So, the surface length is twice the distance between the front of the observed weld pool limit and the hole left by the arc. Consequently, by applying the heat source in the center of the surface, the energy input into the fused zone looks like a good approximation of reality. Indeed, the difficulty is the representation of physical phenomenon which occurs during energy input. The latter is divided into two parts: the first is introduced by the electrical arc (at the surface) and the second by the droplet (in the volume). To simulate these two effects, we located the CIN source at halfway of the filler material in the center of the *Bézier* surface (Figure 16.9).

16.3.3.2 Heat Source Definition

The aim of this chapter is the identification of an equivalent heat source function. The mathematical expression of $S(x, y, z)$ is assumed to be a CIN, as defined in Equation 16.15, but with modified axis, as

$$S(x, y, z) = \frac{kK_z \eta UI}{\pi(1 - \exp(-K_z e_p))} \exp\left(k(v^2 + z^2) - K_\perp w \right) * [1 - u(w - e_p)],$$ (16.22)

(v, w, z) is a modified axis system in which v and w are tangent and normal to the top surface of the filler material. The main advantage of this volume heat source is the small number of unknown parameters: k, K_z, and e_p, the source application depth.

The inverse problem has to find values of CIN parameters; however, the validity of this energy input shape must be checked. The direct problem is solved using different source parameter configurations. Several melted zones are shown in Figure 16.10.

The choice of heat source parameters is not easy, but as shown in Figure 16.10, weld pool observed on a macrograph can be surrounded by the results of two parameter configurations. The *"CIN configuration no. 2"* has good width but the depth is too low; inversely, the *"CIN configuration no. 1"* has an insufficient width but the depth is too great. This means that the mathematical law is able to simulate energy input which validates the shape of the melted zone, but the use of the inverse method is necessary to avoid the manual

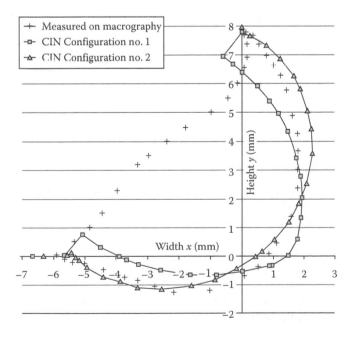

FIGURE 16.10
Melted zones for two sources.

parameterization. Moreover, macrograph allows only for a partial validation (plan 0xy Figure 16.10) of the fused zone limit. Indeed, the shape along the z-axis (or time evolution) cannot be checked this way. So, the good heat source term will be found only after resolution of the inverse problem because estimation is realized with temperature measurements throughout time, which is the third dimension of our quasi-steady-state model.

Model used to obtain Figure 16.10 results is defined with tetrahedral mesh elements. The software Comsol Multiphysics is already fitted with an automatic mesh creator, which allows the definition of the maximal sizes of boundary or volume elements. The numerical sample is too big for a global definition of a thin mesh, but for our case, the interesting zone is near the weld pool. So, only elements on volumes of filler material are refined at 0.0015 mm of maximal size. Other volumes are freely meshed, but with a growing factor equal to 1.1. The number of elements is close to 40,000 elements. The analysis of this mesh is realized by comparing these results to others obtained with the same simulation but discretized more thinly (higher than 60,000 elements). By observing the shape of the phase change temperature, which is very close in both cases, it is possible to conclude that the first mesh is sufficiently thin.

16.3.3.3 Inverse Problem Method and Experimentation Plan

In this study, the inverse problem reduces a quadratic error built on the difference between temperatures which results from the direct problem and experimental measurements, such as Equation 16.23. The used method is the Levenberg–Marquardt algorithm which calculates new parameters with the iterative expression (16.24). It is an association of the steepest and Gauss–Newton methods, which allow for an important estimation speed and good robustness.

$$S(n) = \sum_i (Y(i) - T(i))^2, \tag{16.23}$$

$$p^{n+1} = p^n + [J^T J + \lambda^n \Omega]^{-1} J^T (Y - T(p^n)), \tag{16.24}$$

with

$$J = \frac{T(p^n + \varepsilon p^n) - T(p^n - \varepsilon p^n)}{2\varepsilon_v p^n}, \tag{16.25}$$

where
   n is the iteration step
   P is the estimated parameter
   J is the sensitivity, in our case; it is obtained by the numerical derivation of modeled
      temperature over parameter variation (16.25)
   Y and T are, respectively, measured and simulated temperatures
   λ is a damping factor
   Ω a diagonal matrix defined to offset measurement noises
   ε_v factor is the variation step of parameters

In Equation 16.25, it is possible to observe the link between thermal measurements and parameter estimation. As a consequence, the estimation of the heat source needs a very

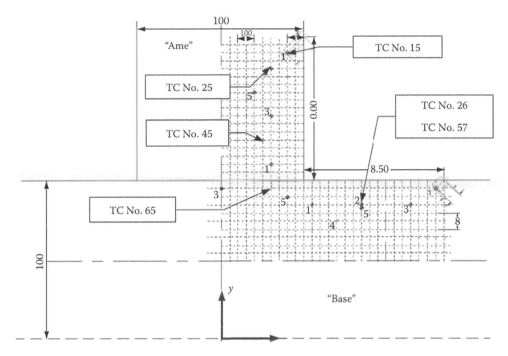

FIGURE 16.11
Theoretical location of thermal measurement.

accurate experimental investigation. Situation and number of temperature measurements have an impact on the parameter quality.

To ensure efficient parameter identification, thermocouple positioning must respect four principles related to the measurement theory and previous observations [3]:

- The thermal field is a picture of energy distribution. Measurements have to be given all around the melted zone.
- For sensitivity reasons, the thermocouple must be as close as possible to the weld pool.
- The thermal gradient measurement in three dimensions can be useful.
- Information observed throughout time gives the third dimension gradient (quasi-steady-state assumption).

With the previous advice, we define an a priori location of thermocouples as shown in Figure 16.11. The validation of this plan is realized by a theoretical inverse problem.

16.3.4 Theoretical Inverse Problem

16.3.4.1 Estimation without Measurement Noise

Before the identifications of the parameters with experimental temperatures, it is necessary to check, theoretically, several points such as, which parameters can possibly be estimated with the previous implementation plan?

For this reason, a sensitivity analysis is made. Two elements are important: amplitudes of sensitivities and linear dependence of parameters. In the first case, no sufficient amplitude signifies that a parameter variation involves too low of a difference on the thermal field; its effect is the difficulty and the increase in the time to estimate. In the second case, if two parameters are linearly dependent, a variation of each of them produces the same effects on the thermal field. Therefore, a small change of the first parameter can influence the second and the algorithm will not be able to estimate parameters.

The previous theoretical model is used as a direct problem and as a "numerical experimentation" with reference parameters. The latter is defined to have a melted zone quite close to the one observed on macrographies.

The parameter choices are realized after few tries. So the melted zone shape is not perfect with regard to macrographies, but sufficient to perform this analysis.

We assume that $k = 57{,}000$ m^{-2}, $K_z = 370$ m^{-1}, and $e_p = 0.012$ m.

With this heat source, the first analysis concerns sensitivity coefficients presented by Equation 16.25. In Figure 16.12, it is possible to observe the sensitivity of the first parameter (k) on the first measurement line at $x = 0.004$ m, $y = 0.018$ m, and along z-axis. It should be noted that the real sensitivity amplitude is obtained by multiplying coefficients by the reference parameter. Obviously, this kind of curve must be realized for each parameter and at each measurement coordinate. Observations are related to theoretical thermocouple locations. First, the k parameter, which is the concentration factor, has more influence over the thermal field near the surface. Consequently, sensitive thermocouples are those that are close to the surface, for example, TC No. 15 in Figure 16.11, and amplitudes of variations observed are sufficient for estimation (approximately 900°C). Second, the K_z parameter is the involution factor and it affects the center part of the heat source by making it more cylindrical or more conical. So, in this case, sensitive thermocouples are those located by the sides of the fused zone, TC No. 45 in Figure 16.11. And third, the depth penetration e_p of heat source produces variation underneath the simulated weld pool, and measurements are sensitive near the lower extremity of heat source, TC No. 65.

The second curve in Figure 16.12 is the ratio between two parameter sensitivities. Here, it is the case of k sensitivity divided by the K_z sensitivity, along the first measurement line. This ratio had to be made for all parameters and for all measurement coordinates. Their aim is the search for linear dependences. If this ratio is equal to a constant value,

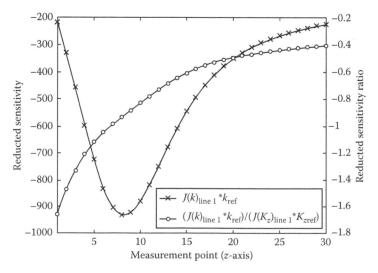

FIGURE 16.12
Check of sensitivity amplitudes and linear dependences.

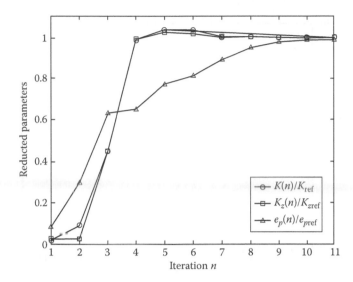

FIGURE 16.13
Parameters' evolution.

parameters are correlated and the estimation will fail. In Figure 16.12, k and K_z are clearly independent on this "thermocouple." Conclusions are the same for other parameters, only several measurements are partially dependent but this is caused by their low sensitivities.

This validation is implemented with theoretical thermocouple locations chosen for the first transit (Figure 16.11). A satisfactory estimation is obtained after 10 iterations with a final criterion equal to 27, which signifies a mean error of 0.24°C. The decrease of the latter can be observed in Figure 16.14.

Figure 16.13 shows evolutions of parameters dividing referential parameters, so as to observe variations in the interval $[0; 1]$ and to be able to compare them. The first two parameters k and K_z have equivalent progress. Their final values, respectively, 57,032 m^{-2} and 370 m^{-1}, are very close to references, and convergences are obtained quite quickly. The third parameter is e_p; its identification takes more time than previous parameters and the final value is not perfect, near 0.01185 m, but tends to evolve correctly.

16.3.4.2 Estimation with Simulated Measurement Noise

The interest of this theoretical method is to test the estimation algorithm. Indeed, the previous test is done with simulated measurements which are not realistic but allow for sensitivity analysis. Nevertheless, before the estimation with experimental temperatures, we have to check if the measurement noise will cause failure of estimation.

For this, inverse and direct problems are the same as before, but an artificial noise is created with a random function. The level of this perturbation is chosen as close as possible to the real measurement noise, which signifies nearly 1% of the maximal signal or 15°C. In this case we only present the criterion decrease. As shown in Figure 16.14, the noise induces estimation after more iterations than before: 15 iterations. Moreover, the criterion is stabilized at a higher value (3800) which is equivalent to a mean error of 5°C in each measurement. Parameters are quite correctly estimated and final values are 56,980 m^{-2}, 370 m^{-1}, and 0.0121 m, respectively, for k, K_z, and e_p. These values are very satisfactory, and with these data, we can assume that the experimental inverse problem will be a success.

Results are summarized in Table 16.1.

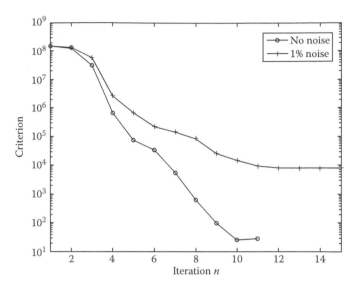

FIGURE 16.14
Criterion reduction during identification.

TABLE 16.1

Results of the Estimations

	k	K_z	e_p	**Criterion**
Reference	57,000	370	0.012	
Without noise	57,032	370	0.01185	27 (0.25°C)
With 1% noise	56,980	370	0.0121	3800 (~5°C)

16.3.5 Experimental Design

16.3.5.1 Experimental Implementation

In our study, we use 32 thermocouples introduced inside the workpiece through the Base. This choice of implementation is imposed by the industrial term of reference; for example, we can only drill the workpiece from its back side. We use type K thermocouples to observe large amplitudes of measurement. TC wires are 50 μm in diameter and the hot junction is 220 μm in diameter. To observe the workpiece temperature, the contact between the thermocouple and workpiece has to be perfect. So TCs are welded (Figure 16.5) to the sample by capacitive discharge.

Following the ideas proposed in Sections 16.3.2.2 and 16.3.2.3, measurements are realized along the 1100°C isotherm, nearly 1 mm after the melted pool limit (Figure 16.11). The location of this isothermal line is realized with a previous experimentation without instrumentation in association with a simplified model. The simulation parameters are approximately chosen to obtain a melted zone close to the one observed on the first macrograph. Finally, the simulation gives us the shape of isothermal lines. This instrumentation method assumes the repeatability of the welding procedure process, which means that two welding procedures with the same process parameters (current and tension) produce a melted zone with the same dimensions.

The high number of thermocouples (15) forces us to divide them into six plans orthogonal to the weld direction and spaced by 4 mm as shown in Figure 16.15.

Experimentations have been realized by the French Welding Institute in Yutz.

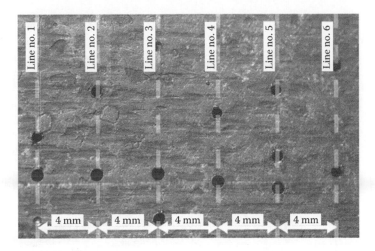

FIGURE 16.15
Hole locations on the back side of the sample "Base."

16.3.5.2 Measurements Results

Results of the inverse problem are directly dependent on thermal measurements. For this reason, experimental temperatures are first analyzed and then the link with estimation difficulties is established.

For the first transit, all thermocouple outputs are presented in Figures 16.16 and 16.17. In the first one, it is possible to observe the failure of TC No. 36 located at $x = 3$ mm and $y = 14$ mm (cf. Figure 16.11 for axis origin). The flat signal shows that the problem is not inherent in the welding conditions.

It should be remarked that the time lag of each curve is a consequence of the space dispersion of thermocouples (Figure 16.15).

On previous curves, another singularity is observed. Two thermocouples, whose coordinates are theoretically the same ($x = 8.5$ mm and $y = 8.5$ mm), have very different maximal temperatures. The discrepancy between TC No. 26 in Figure 16.14 and TC No. 57 in Figure 16.17 is nearly 300°C. The first reason that can explain this error is drilling accuracy. Holes, where thermocouple hot junctions are welded, can be realized with low depth differences, and the very high thermal gradient induces this error. The second reason is the link to the filler material; indeed, droplet configuration is globular and induces variations in melted pool limits along the weld direction. As previously stated, measurements are implemented into six plans spaced by 4 mm. Consequently, thermal field differences appear between plan numbers 2 and 5.

These explanations are also valid for other cases, for example, TC No. 16 in Figure 16.16 and TC No. 45 in Figure 16.17, which are equidistant from the melted pool limit. They observe different temperature amplitudes and have different evolutions throughout time. It should be noted that holes are, respectively, parallel and perpendicular to isothermal surfaces. Moreover, the temperature of perpendicular thermocouple increases more quickly, and maximal thermal level is higher than that of the parallel thermocouple. Nevertheless, the cooling of both is quite equivalent, which leads one to suppose another cause. In fact, when the thermocouple hole is perpendicular to isothermal lines, this means it is along the thermal flow direction. Thus, it blocks thermal diffusion in that direction,

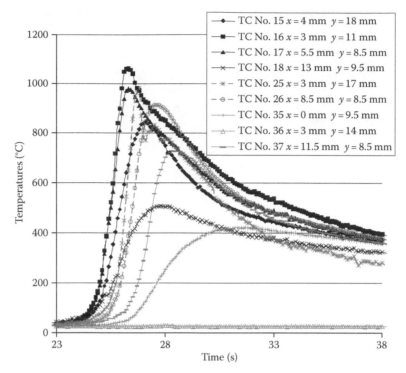

FIGURE 16.16
Thermocouples measurements on three last plans.

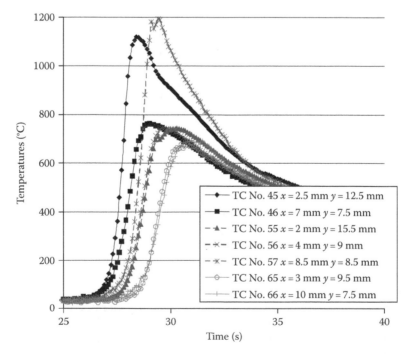

FIGURE 16.17
Measurements on three firsts' plans.

which induces thermal accumulation on the hole head [11]. Conversely, when the hole is parallel to isothermal surfaces, it is perpendicular to the flow, and heat diffusion is less disturbed. Considering this effect, it is possible to understand previous differences, accumulation makes observed temperatures higher and more quickly but there is no effect during the cooling when the flow is reduced.

To check drill accuracy, real positions of hole heads are observed. For this, the workpiece is cut along each measurement plan, polished, and chemically attacked. With these macrographies and knowing the thickness of the Base and the Ame, it is possible to adjust positions of measurements. For example, the two previous thermocouples with same theoretical coordinates are in reality at TC No. 26 $x = 8.2$ mm, $y = 8.4$ mm and TC No. 57 $x = 7.8$ mm, $y = 8.4$ mm. Thus, in reality, the first thermocouple is less close to the melt zone than the second, which explains the thermal difference.

Another phenomenon occurs during the welding and has probably disturbed the thermal measurement. It is the displacement of the vertical part over the horizontal part of the workpiece. Despite previous static spots made to seal samples together, the gap hassled by the first transit is 300 μm in size, which is nearly half of a hole diameter. This displacement implies the use of thermo-mechanical simulation and it becomes a more complex problem. We assume that the thermocouple has moved with the vertical part, and thus relative distances between the melted pool and the thermocouples are constant and can be neglected.

In an inverse problem, it is important to compare equivalent measured and simulated information, which is why simulated temperatures are extracted from the model at the real coordinates.

16.3.5.3 Inverse Problem Results

The Levenberg–Marquardt algorithm has been implemented with the software MATLAB®, and the direct problem is run on Comsol Multiphysic which uses the finite element method. It should be noted that estimated parameters are the concentration factor k, the involution factor K_z, and the depth of source application e_p. Other process parameters, such as weld current and voltage, are known because they were measured during the welding. Moreover, we note the acquisition parameters such as the step time 0.1 s, the total acquisition time 240 s, and the weld speed 5 mm s^{-1}. These values are used to convert the time on third space dimension implemented on the simulation, thanks to the quasi-steady-state assumption. Moreover, before estimation, the time lag observed in Figures 16.16 and 16.17 had to be corrected thanks to the weld velocity and the instrumentation line spacing.

Measurements are realized all along the welding period; it is thus necessary to choose the best interval which allows for an efficient estimation. This time duration is chosen according to sensitivity coefficients.

Figure 16.18 presents the sensitivity of the concentration factor along two measurement lines: the first is TC No. 15 located on $x = 4$ mm and $y = 0.018$ mm and the second is TC No. 65 on $x = 0.003$ mm and $y = 9.5$ mm.

The concentration factor is a parameter that indicates the width of the heat sources (1) and (8) and the two thermocouples—TC No. 15 and TC No. 65—are near and beneath the melted zone, respectively. As previously stated, the melted pool shape is directly linked to heat source function, and variation effects of the concentration factor can be seen near the workpiece surface (Figure 16.18). For this reason, sensitivity is more important on TC No. 15 than on TC No. 65. The same observation could be realized for the involution factor, which also

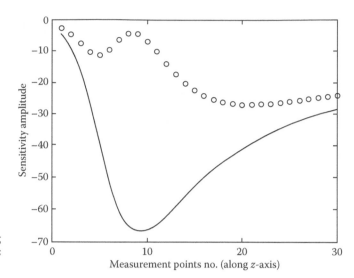

FIGURE 16.18

Sensitivity of concentration factor along two thermocouples (dots, TC No. 65; line, TC No. 15).

defines the shape of the heat source. Indeed, TC No. 65 is more sensitive to depth variations and TC No. 15 does not observe any width variation (Figure 16.18). Abscissa is the number of measurement points used for identification, the step time is 0.2 s with a weld speed of 5 mm s^{-1}; thus the space measurement step used in estimation is 1 mm.

The selection of time (or third space dimension in the model) interval is realized using curves such as that in Figure 16.16. The measurements used are those which have maximal sensitivities and which do not create linear dependences. When measurements are too far from the heat source, all sensitivities are very close to zero and linear dependences appear in each case. As a consequence, it is necessary to keep only the more sensitive measurement points. We have used 15 measurement lines with 30 points per line, which means 450 comparison points.

Results of identification are not satisfactory; the end criterion value is very high. Moreover, the melted zone limit obtained with estimated parameters does not correspond to those observed on macrographs (Figure 16.19). Differences between the two shapes of weld pool are by a majority located on the surface where the heat source is defined by the concentration factor. In the depth of the material, the two shapes are close enough and the algorithm does not seem to be able to enlarge the surface part without increasing depth penetration. This fact leads to reviewing measured temperatures, which gives improper information, and to reassessing effects of thermal accumulation previously cited.

To study thermal disturbances caused by thermocouples, we have to develop a new analysis. Thermocouples were included in simulation, thus we will be able to calculate the level of discrepancies.

16.3.6 Quantification of Thermocouple Disturbance

This work has two objectives: the first concerns quantification of thermal disturbances caused by instrumentation in implementation configuration. Indeed, the theory of thermocouple measurement recommends that holes are drilled parallel to the isothermal surfaces, but in our case the industrial requirements impose a perpendicular direction.

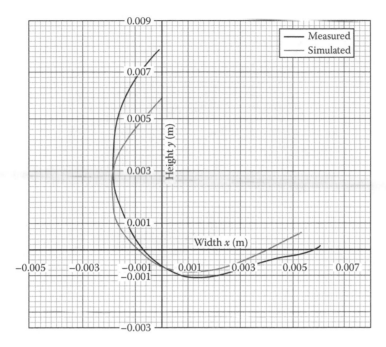

FIGURE 16.19
Measured and estimated melted zone limits.

The second is the comparison between different kinds of thermocouples and the determination of effects caused by their implementation. New simulation needs a new definition of the mathematical problem, which is presented first, and then the two kinds of thermocouples are described; lastly, they are numerically implemented in several configurations.

16.3.6.1 New Mathematical Model

In the present work, thermocouples are simulated with workpiece geometry. As a consequence, the previous assumption concerning quasi-steady state is no longer usable. Indeed, quasi-steady state is obtained when the thermal field is, in each time step, the same for a referential located on heat source axis, which is not realistic when the thermocouple is in the workpiece.

Thus, the solved equation is still the heat conduction equation but now in transient analysis:

$$\rho(T)\,cp(T)\frac{\partial T}{\partial t} = \frac{\partial}{\partial x}\left(\lambda(T)\frac{\partial T}{\partial x}\right) + \frac{\partial}{\partial y}\left(\lambda(T)\frac{\partial T}{\partial y}\right) + \frac{\partial}{\partial z}\left(\lambda(T)\frac{\partial T}{\partial z}\right) + S(x,y,z,t). \qquad (16.26)$$

It is interesting to note that the heat source term is presently a time function and is defined in the following equation:

$$S(x,y,z) = \frac{kK_z\eta UI}{\pi(1 - \exp(-K_z e_p))}\exp\left(-k(v^2 + (z - vt)^2) - K_z w\right) * [1 - u(w - e_p)]. \qquad (16.27)$$

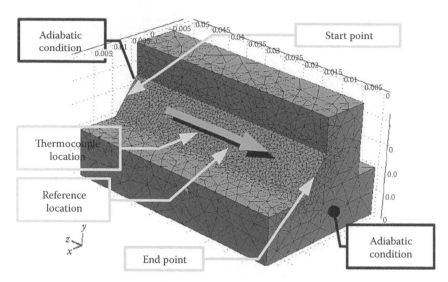

FIGURE 16.20
New boundary conditions and heat source run.

This expression allows a heat flow shape like previously but it travels across the workpiece along the z-axis. Parameters k, K_z, and e_p are not a well-known cause of the failure of the inverse problem. As a consequence, tests have to be realized to set them at coherent values.

Only one boundary condition is modified, the one named A in Figure 16.9. We take in this boundary an adiabatic condition. These adiabatic conditions (Figure 16.20) are also no longer valid because now the heat source is not sufficiently far from them, (at "the beginning" and at "the end"). However, they are kept because thermocouple temperatures are not disturbed by them. On all thermocouple boundaries, convective and radiative heat losses are assumed and defined as Equation 16.21.

Other simulation parameters are kept, only thermocouple geometry has been introduced into the workpiece and will be explained in the next part.

As shown in Figure 16.20, the filler material is now ever present all along the workpiece. The longitudinal shape of the fused zone is thus different than the real case. But, for this analysis, we do not try to simulate the perfect weld pool shape, and an approximation with coherent thermal levels is sufficient.

16.3.6.2 Thermocouple Simulations

The origin of this study is the collaboration with the industrial company. As the time needed to implement laboratory instrumentation is very important, it could be interesting to observe effects caused, on the thermal field, by the use of industrial thermocouples. An experimental investigation could be realized but, as we have previously said, it is very difficult to install two thermocouples at the same coordinates. Therefore, this analysis is performed using numerical simulations.

Laboratory thermocouples are those described in "Instrumentation implementation." Made by ourselves, thermocouples are only composed of the hot junction and the two wires. Materials used are nickel chromium–nickel alloy (type K), mean thermal properties

TABLE 16.2

Thermal Properties

Material	λ (W m^{-1} K^{-1})	cp (J kg^{-1} K^{-1})	ρ (kg m^{-3})
TC	25	445	8600
Sheath (AISI 304L)	15	470	7800

of which are noted in Table 16.2. The hot junction is assumed to be an ellipsoid with 320 μm for high diameter and 240 μm for short diameter, wires are 54 μm in diameter, and the hole is 650 μm in diameter (Figure 16.21-1).

The industrial thermocouple is also constituted of hot junction and wires but it is sheathed with stainless steel AISI 304L (Figure 16.21-2). The sheath is 250 μm in diameter, the hot junction is a 240 μm diameter sphere, and wire holes are the same as before. This thermocouple is selected so as to have an equivalent geometry to a laboratory thermocouple.

Geometry characteristics of laboratory thermocouples are observed on macrographies (Figure 16.21-3) and some features, such as size of contact surface, are measured on pictures realized by scanning electron microscope (SEM). Industrial thermocouple geometry is created using some information found in the manufacturer documentation: *Thermocoax*.

The numerical design of thermal contact between laboratory thermocouples and the workpiece is assumed to be perfect (welded). In the industrial case, the thermocouple is assumed to be put down to the bottom of the hole. This implementation is much faster than that in the laboratory case but induces bad contact quality. This quality is assumed to be symbolized by thermal contact resistance Rc.

It should be noted that the numerical implementation of thermocouples induces an important increase in the number of freedom degrees we have to solve; thus, it is impossible to simulate all of them at once. Moreover, it seems to be obvious that thermal disturbance will increase with thermal gradient; therefore, only TC No. 45 (Figure 16.10),

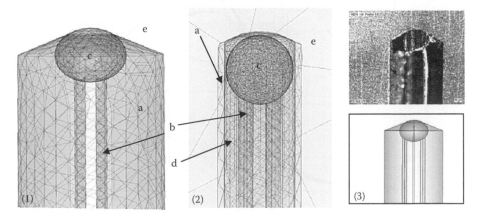

FIGURE 16.21
Geometries for (1) laboratory thermocouple, (2) industrial thermocouple, and (3) comparison between simulation and macrograph (a, hole; b, wires; c, hot junctions; d, sheath material; and e, workpiece).

located on $x = 2.5$ mm and $y = 12.5$ mm is examined. Relevance of the latter is its high temperature level (near 1100°C) and its implementation ease for each configuration.

16.3.6.3 Numerical Tests and Results

An objective of this work is the understanding of inverse problem failure and the observation of effects caused by instrumentation quality. As a consequence, the first test deals with implementation orientation, and the second compares results of the industrial and laboratory thermocouples.

As said in the previous section, the TC No. 45 is interesting because its location allows for different directions of implementation. Moreover, it is the one that poses problem in the "Measurements Results" (Section 16.3.5.2). The maximal temperature observed by this TC seems to be overestimated regarding an equivalent measurement implemented parallel to isothermal surfaces. The method to confirm this point is to simulate two cases: one with real thermocouple implementation, which is perpendicular to isothermal surfaces, and another at the same coordinates but implemented parallel to isothermal surfaces.

For the second analysis, thermocouple evacuation is fixed along the heat flow, which corresponds to the previous second simulation. The bottom of the hole is also positioned at the same coordinates and the contact area is assumed to be maximal. In this case, four values of thermal contact resistances are set: $RC1 = 1 \times 10^{-4}$ K m^2 W^{-1}, $RC2 = 1 \times 10^{-5}$ K m^2 W^{-1}, $RC3 = 1 \times 10^{-6}$ K m^2 W^{-1}, and $RC4 = 1 \times 10^{-7}$ K m^2 W^{-1}, where the first is a bad contact and the last is a good contact.

Results in Figure 16.22 are temperatures observed in hot junction between the two thermocouple wires. These temperature values are then compared with values taken in even coordinates but without influence of thermocouples.

Concerning the first analysis, observation of discrepancy caused by thermocouple implementation, Figure 16.22 clearly shows thermal accumulation on the thermocouple head. The perpendicular TC starts to overestimate temperatures near $t = 5$ s and give again good values near $t = 8$ s, whereas parallel TC observations are merged with referential during

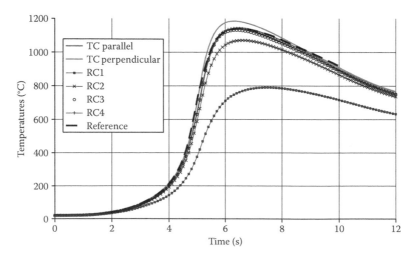

FIGURE 16.22
Thermal measurements.

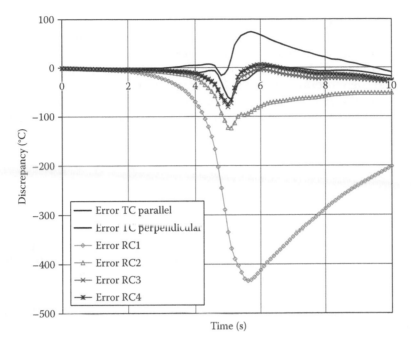

FIGURE 16.23
Thermal discrepancies.

the entire time. This time interval corresponds with the one which is selected for the inverse problem, in other words, when sensitivity coefficients are maximal. Nevertheless, the time delay induced by the diffusion inside the parallel thermocouple induces discrepancy (Figure 16.23) but the time in which it occurs is outside the one used for estimation.

In the experimental case, not all thermocouples are along heat flow; therefore, some temperatures are overestimated and others not. These observations explain why estimation has failed. In fact, algorithms try to reduce the criterion defined as the difference between measured and estimated temperatures by varying parameters. But when temperatures are disturbed and not disturbed, the parameters are different, which leads to previous criterion stagnation with false heat source function.

The second analysis gives information concerning effects of measurement context, industrial or laboratory, and also with regard to implementation quality. In a first time, the worst contact, when $RC = 1 \times 10^{-4}$ K m^2 W^{-1}, induces an important misjudgment of temperatures. The measurement discrepancy of the maximal temperature is higher than 300°C and with nearly 1.1 s delay. These two values lead to understanding temperature discrepancies in "Thermal measurements results," when two measurements at the same coordinates give different thermal evolutions. The time delay (Figures 16.16 and 16.17) between maximal temperature of TC No. 26 and TC No. 57 is about 3.6 s, in which a part results from plan spacing. There is 12 mm between plan nos. 2 and 5 (Figure 16.15) and the speed is 5 mm s^{-1}; thus the delay should be 2.4 s, but it remains 1.2 s. Thus it is possible to conclude that contact quality of TC No. 26 is not perfect. Moreover, its heating and cooling rates are too low according to TC No. 25 which is closer to the melted zone (observed on macrographies); this can be explained by diffusion difficulties to go through contact resistance. The curve with $RC = 1 \times 10^{-5}$ K m^2 W^{-1} shows the same results concerning maximal value and time delay but with less perturbation. When $RC = 1 \times 10^{-6}$ K m^2 W^{-1} and $RC = 1 \times 10^{-7}$ K m^2 W^{-1}, the effects of contact resistance are negligible and information is close to reference. Industrial thermocouples are thus able to measure true temperatures, but the contact quality must be very good by using, for example, heat-sink grease.

The industrial thermocouple is perpendicular to isothermal surfaces, but the ones with best contact do not have overestimated temperatures. Indeed, the important diameter of sheath material (304L) allows for heat evacuation along the thermocouple and helps avoid heat accumulation.

16.4 Conclusions

In this chapter, we have observed effects on the thermal field of disturbances caused by intrusive instrumentation. We have carried out this study because of the failure of the equivalent heat source estimation for the case of "T" welding. Despite several assumptions, our direct problem has been confirmed thanks to macrographies. However, this validation is not significant because it compares a simulated fused zone to a maximal weld pool size observed on the transversal plan. This led us to apply an inverse method which makes 3D temperature comparisons. But theoretical estimations were made before the implementation of the experimental case. This is for two reasons: first, sensitivity amplitudes and independences have been checked, and, second, an artificial measurement noise has been added to theoretical temperature acquisition, in order to check the estimation capacity without a perfect signal. The results in these two cases have been very satisfactory, so we have validated thermocouple location and defined an a priori instrumentation.

Nevertheless, the inverse problem with real information has failed, and thus after experimental result analysis, thermal measurement has been selected as the most important origin of disturbances. Consequently, to observe thermocouple effects, we have chosen to simulate two kinds of them: first, with laboratory shape and second, with industrial characteristics. Moreover, two implementation methods have been studied: the first respects the thermocouple measurement theory by being inserted parallel to isothermal surfaces and the second is our experimental case, imposed by industrial requirement, in which the hole is made perpendicular to isotherms. Results of this analysis are very interesting and help explain our experimental errors. We have shown that a laboratory thermocouple inserted along the thermal flow direction stops the diffusion and induces a thermal accumulation at the end of the hole where the measurement is made. Conversely, when it is perpendicular to the flow, the thermocouple has a temperature that is very close to that of the true value. The link between this cause of error and experimental temperature discrepancies is thus made. Afterward, we have tried to simulate the industrial thermocouple, those with sheath. In this case, perfect contact conditions are difficult to obtain, so a thermal contact resistance has been assumed. Also in this part, simulation has explained some experimental errors; for example, discrepancies between two thermocouples at the same coordinates could result from the bad contact of one over the workpiece.

Finally, this instrumentation simulation has shown why an inverse problem, with a nonoptimal experimental implementation, does not find satisfactory results. So it is possible to define an implementation method which limits temperature disturbances but increases estimation sensitivities.

Measurements had to be realized

- All around the weld pool
- As close as possible to the fused zone
- In such a way as to observe thermal gradient (e.g., along two lines)
- In holes drilled perpendicular to the heat flow
- With the best contact quality

But when this recommendation cannot be applied, two solutions exist. The first is the use of a direct problem which simulates the whole instrumentation inside the workpiece. In this case, the shape of thermocouples induces a very thin mesh and a too important number of meshes. The second is the development of an error model that can be applied to direct problem measurements.

Nomenclature

a, b, c_f, c_r	radii of heat source
cp	heat capacity
e_g	global estimation error
$e_1, \ldots, e_5$	types of estimation error
e_p	depth source application
E	expected value
$f_l f_r$	double ellipsoid front and rear factor
h, h_{cv}, h_e, h_r	heat transfer coefficients
I	process current
J	sensitivity matrix
k	concentration factor
K_z	involution factor
$\vec{n}$	normal vector
n	iteration step
P	estimated parameter
r_c	contact resistance in thermal measurement
r_e	thermal resistance of transfer on wires
r_m	macro-constriction resistance
RC_1, RC_4	contact resistance simulated
$S(x, y, z)$	heat source
$S(n)$	quadratic error
T	temperature
T_∞	ambient temperature used in simulation
T_e	environment temperature
T_p	surface temperature with thermocouple
T_c	temperature measured par thermocouple
TC No. . . .	identification of thermocouple
$u(\cdots)$	Heaviside function
U	process voltage
v	process velocity
Y	measured temperatures

Greek Symbols

δT	thermal measurement error
ε	random noise
ε	measure error
ε	emissivity
ε_v	variation step of parameter
η	process efficiency
λ	thermal conductivity
λ^n	damping factor
λ_e	thermal conductivity of TC wire
ρ	density
σ	statistical deviation
σ	Stefan–Boltzman constant
Φ	heat flux

References

1. Ecole d'Hiver–METTI 99, Métrologie Thermique et Techniques Inverses, 25–30 Janvier, Odeillo Font Romeu, Vols. 1 and 2, Presses Universitaires de Perpignan 1999.
2. Beck J.V., Arnold K.J., *Parameter Estimation in Engineering and Science*, Wiley, New York, 1977.
3. Bardon J.P., Cassagne B. Température de surface: mesures par contact, Techniques de l'Ingénieur, Paris R2732, 1–22, 1981.
4. Cassagne B., Kirsch G., Bardon J.P., Analyse théorique des erreurs liées aux transferts de chaleur, *International Journal of Heat Mass Transfer*, 23, 1207–1217, 1980.
5. Géry A., Thèse de doctorat, Lyon, Avril, 1980.
6. Werling E., Thèse de doctorat, INSA de Lyon, Janvier, 1988.
7. Bouvier A., Thèse de 3ième cycle, Université de Nantes, 1986.
8. Le Masson P., Etude d'une méthode de mesure des transferts convectifs entre un écoulement à grande vitesse et les parois d'un canal de très petites dimensions utilisant des techniques inverses et des microthermocouples semi-intrinsèques, Thèse de doctorat, Université de Nantes, 1991.
9. Cassagne B., Bardon J.P., Beck J.V., Theoretical and experimental analysis of two surface thermocouples, *Proceedings of the 8th International Heat Transfer Conference*, San Francisco, Vol. 2, pp. 483–488, 1986.
10. MUSICA Project: http://www.cea.fr/le_cea/actualites/le_cea_areva_le_cetim_et_la_societe_esi_grou-2699 (Accessed on November 2005).
11. Rouquette S., Guo J., Le Masson P., Estimation of the parameters of a Gaussian heat source by the Levenberg–Marquardt method: Application to the electron beam welding, *International Journal of Thermal Sciences*, 46(2), 128–138, 2007.
12. Guo J., Le Masson P., Artioukhine E., Loulou T., Rogeon P., Carron D., Dumons M., Costa J., Estimation of a source term in a two-dimensional heat transfer problem: Application to an electron beam welding, *Inverse Problems in Engineering*, 14(1), 21–38, 2006.
13. Goldak J., Chakravarti A., Bibby M., A new finite element model for welding heat sources, *International Institute of Welding*, Vol. 603–685, 1985. *Metallurgical and Materials Transactions B*, Vol. 36B, June 2005, pp. 299–305.
14. Li D., Wells M.A., Effect of subsurface thermocouple installation on the discrepancy of the measured thermal history and predicted surface heat flux during a quench operation, *Metallurgical and Materials Transaction B*, 36B, 343–354, 2005.

17

Multisignal Least Squares: Dispersion, Bias, and Regularization

Thomas Metzger and Denis Maillet

CONTENTS

17.1 Introduction

In a thermal characterization experiment, the studied material is stimulated through a steady or transient heat excitation and temperature measurements are compared to a corresponding pertinent model in order to estimate its thermophysical properties. In such an experiment, it is often necessary to set temperature sensors (thermocouples, platinum resistances, etc.) not only at the surface of the sample but also inside the material.

It is usually assumed that measurement noise only affects the temperature signal given by the sensors.

This corresponds to the case where the location of each of these sensors, for example, the hot junction in the case of a thermocouple, is known with a high enough precision that allows neglecting the effect of the location error on the values of the estimated parameters (thermal diffusivity, specific heat, conductivity, contact resistance, heat transfer coefficient, etc.). The interested reader can refer to Chapter 7, where classical estimation methods based on this assumption are presented.

If the previous assumption is not fulfilled, a sensitivity study allows the calculation of the bias, namely, the systematic error on each parameter resulting from a given level of the error on the sensor location. This point is studied in Chapter 9. However, such an assessment of the estimation bias requires knowledge of the location error, which is not known precisely, that is, in a deterministic sense.

Another way to take this location error into account is to measure the exact location of the temperature sensors through a destruction of the sample, once the thermal characterization experiment has been completed. This type of destructive technique can be implemented, for example, for the hot junction of a thermocouple that has been embedded into a resin (polymer sample) or stuck to the bottom of a hole drilled inside the material (metallic sample).

However, situations exist where this type of technique is not possible anymore: friable porous material, textile, nonconsolidated granular medium, etc.

We will illustrate here how this uncertainty on the exact positions of the sensors can be adequately accounted for by the implementation of a modified least squares minimization technique that can be considered either as a *total least square* estimation or as *Bayesian* estimation. This last type of estimation is studied in Chapter 12.

17.2 The Linear Model of a Straight Line

Let us consider a model whose output η, temperature, for example, varies linearly with the considered location τ:

$$\eta = \eta(\tau; \boldsymbol{\beta}) = \beta_1 + \beta_2 \tau \tag{17.1}$$

The notation used in Beck and Arnold (1977) is chosen here: η and τ are the dependent and the independent variables, respectively, and β_1 and β_2 are the two parameters. It is first assumed that experimentally noised data (measurements) y_i are available for locations τ_i ($i = 1$ to n)

$$y_i = \eta(\tau_i; \boldsymbol{\beta}) + \varepsilon_i \tag{17.2}$$

where
$\boldsymbol{\beta} = [\beta_1 \ \beta_2]^t$ is the parameter vector
ε_i is an uncorrelated and unbiased noise of constant standard deviation σ, which can be expressed (using Kronecker's symbol δ_{ij}) as

$$E(\varepsilon_i) = 0 \quad \text{and} \quad E(\varepsilon_i \varepsilon_j) = \sigma^2 \delta_{ij} \tag{17.3}$$

17.2.1 Case of Known Exact Locations

If the exact locations τ_i are known, the best estimation technique consists in minimizing the ordinary least squares sum

$$S_y(\beta_1, \beta_2) = (\boldsymbol{y} - \boldsymbol{X}\boldsymbol{\beta})^t (\boldsymbol{y} - \boldsymbol{X}\boldsymbol{\beta}) = \sum_{i=1}^{m} [y_i - \eta(\tau_i; \boldsymbol{\beta})]^2 \tag{17.4}$$

where y is the measured temperatures vector

$$y = [y_1 \quad y_2 \quad \cdots \quad y_m]^t \tag{17.5}$$

and where $X = [X_1(\tau; \beta) \quad X_2(\tau; \beta)]$ is the sensitivity matrix, the column of which are the sensitivity vectors X_j of model η to the parameters β_j (for $j = 1$–2).

The coefficients of this sensitivity matrix are the sensitivity coefficients X_{ij} that are the values of the sensitivity functions X_j evaluated for the m discrete values τ_i:

$$X_{ij} = X_j(\tau_i; \beta) = [X_j]_i = \frac{\partial \eta(\tau_i; \beta)}{\partial \beta_j} \tag{17.6}$$

Vector $\tau = [\tau_1 \quad \tau_2 \quad \cdots \quad \tau_m]^t$ is the column vector composed of the values τ_i of the independent variable t (the locations of the measurements) for which the temperatures have been measured.

For the straight line model (17.1) corresponding to the least square sum given by (17.4), the sensitivity matrix is

$$X = \begin{bmatrix} 1 & 1 & \cdots & 1 \\ \tau_1 & \tau_2 & \cdots & \tau_m \end{bmatrix}^t \tag{17.7}$$

In this case, the locations are known and the solution is explicit (see Chapter 7):

$$\hat{\beta} = (X^t X)^{-1} X^t y \tag{17.8}$$

The notation "^" (hat) is used here to designate the estimator (or the estimated value), in the statistical sense, of a parameter. This solution can also be written as

$$\hat{\beta}_2 = \frac{s_{\tau y}}{s_\tau^2} \tag{17.9a}$$

$$\hat{\beta}_1 = \bar{y} - \hat{\beta}_2 \bar{\tau} \tag{17.9b}$$

The preceding solution involves the mean values of both measurement locations and measured temperatures

$$\bar{\tau} = \frac{1}{m} \sum_{i=1}^{m} \tau_i \quad \text{and} \quad \bar{y} = \frac{1}{m} \sum_{i=1}^{m} y_i \tag{17.10}$$

as well as the statistical variance of the locations (a measure of their dispersion)

$$s_\tau^2 = \frac{1}{m} \sum_{i=1}^{m} \tau_i^2 - \bar{\tau}^2 \tag{17.11}$$

and the statistical covariance between signal and location of measurement

$$s_{\tau y} = \frac{1}{m} \sum_{i=1}^{m} \tau_i y_i - \bar{\tau}\bar{y} \tag{17.12}$$

The covariance matrix of the estimation error on parameters β_j can be calculated then (Beck and Arnold 1977)

$$\text{cov}(\hat{\boldsymbol{\beta}}) = \sigma^2(\boldsymbol{X}^t\boldsymbol{X})^{-1} = \begin{bmatrix} \sigma_1^2 & \rho_{12}\sigma_1\sigma_2 \\ \rho_{12}\sigma_1\sigma_2 & \sigma_2^2 \end{bmatrix} \qquad (17.13)$$

where σ_1 and σ_2 are the standard deviations of estimators $\hat{\beta}_1$ and $\hat{\beta}_2$, respectively

$$\sigma_1 = \frac{\sigma}{\sqrt{m}}\left(1 + \frac{\bar{t}^2}{s_\tau^2}\right)^{1/2} \quad \text{and} \quad \sigma_2 = \frac{\sigma}{s_\tau\sqrt{m}} \qquad (17.14)$$

and ρ_{12} is their correlation coefficient

$$\rho_{12} = -\frac{1}{(1 + \bar{t}^2/s_\tau^2)^{1/2}} \qquad (17.15)$$

17.2.2 Case of Unknown Exact Locations

If no information is available on the exact locations τ_i, the first idea would be to incorporate these unknowns into a new augmented parameter vector $\boldsymbol{\alpha} = [\boldsymbol{\beta}^t\boldsymbol{\tau}^t]^t = [\beta_1\,\beta_2\,\tau_1\,\tau_2\,\ldots\,\tau_m]^t$, the new matrix for the sensitivity coefficients becoming

$$\boldsymbol{X}_\alpha = \begin{bmatrix} 1 & \tau_1 & \beta_2 & 0 & \ldots & 0 \\ 1 & \tau_2 & 0 & \beta_2 & 0 & 0 \\ \vdots & \vdots & & 0 & \ddots & \vdots \\ 1 & \tau_m & 0 & \ldots & 0 & \beta_2 \end{bmatrix} \qquad (17.16)$$

However, one can easily show that the sensitivity vectors $\boldsymbol{X}_{\alpha j}$ (the column vectors of matrix $\boldsymbol{X}_\alpha$) are linearly dependent:

$$-\beta_2\boldsymbol{X}_{\alpha 1} + 0\boldsymbol{X}_{\alpha 2} + \sum_{j=1}^m \boldsymbol{X}_{\alpha(j+2)} = 0 \qquad (17.17)$$

so that it is impossible to estimate β_1 and β_2 without any information on the exact locations τ_i. In fact, this impossibility of estimating both the n parameters (here $n = 2$) and the m locations holds for any model since the number of data (the m measurements) is lower than the number of unknowns $(n + m)$.

17.2.3 Case of Uncertain Exact Locations

In most practical cases, only approximate values t_i of the exact locations τ_i are available: these values are the *nominal* locations of the sensors that the experimenter tries to respect when the sensors are embedded in the material. It is therefore possible to write

$$t_i = \tau_i + \delta_i \qquad (17.18)$$

where δ_i is an uncorrelated and unbiased noise of standard deviation σ' with

$$E(\delta_i) = 0 \quad \text{and} \quad E(\delta_i\delta_j) = \sigma'^2\delta_{ij} \tag{17.19}$$

Often, one may interpret the nominal locations t_i as design quantities (deterministic quantities) that are used by the experimentalist as setpoints in the control of the true locations τ_i during the embedding operation. Then, the true locations $\tau_i = t_i - \delta_i$ have to be considered as stochastic quantities.

One further assumes that the error on the locations of the sensors δ_i is not correlated with the error on temperature ε_j:

$$E(\delta_i\,\varepsilon_j) = 0 \tag{17.20}$$

In this new light, two kinds of "measurements" are now available, y_i and t_i, and two kinds of parameters are looked for, the initial parameters β_1 and β_2 and, additionally, the unknown locations τ_i.

The location can therefore be considered at the same time as a noiseless deterministic "pseudo-signal" (t_i) and as stochastic unknown parameters (τ_i).

In order to get estimations of all unknown parameters, $\hat{\beta}_j$ and the $\hat{\tau}_i$, two types of residuals (temperature and location) have to be considered within a minimization process. Since such residuals are not expressed in the same units, they have to be normalized by the two standard deviations σ and σ'. The sum that can be minimized with respect to parameter vector α becomes

$$S(\beta,\tau) = \sum_{i=1}^{m}\left(\frac{y_i - \eta(\tau_i;\beta)}{\sigma}\right)^2 + \sum_{i=1}^{m}\left(\frac{(t_i - \tau_i)}{\sigma'}\right)^2 = \frac{1}{\sigma^2}S_y(\beta,\tau) + \frac{1}{\sigma'^2}S_t \tag{17.21}$$

or, in a vector and matrix form

$$S(\beta,\tau) = \frac{1}{\sigma^2}(y - X(\tau)\beta)^t(y - X(\tau)\beta) + \frac{1}{\sigma'^2}(t - \tau)^t(t - \tau) \tag{17.22}$$

Minimization of this sum leads to an estimation problem that is not linear any more. This corresponds to a *total least squares* problem (see Van Huffel and Lemmerling [2002]). The estimators of β and τ are the solutions of the following system of equations

$$\nabla_\beta S_y(\hat{\beta},\hat{\tau}) = 0 \tag{17.23a}$$

$$\nabla_\tau S_y(\hat{\beta},\hat{\tau}) + \frac{1}{Q}\nabla_\tau S_t(\hat{\tau}) = 0 \quad \text{with } Q = \left(\frac{\sigma'}{\sigma}\right)^2 \tag{17.23b}$$

The solution of Equation 17.23b provides the estimator of τ as a function of the estimator of β:

$$\hat{\tau}_i = \frac{t_i + Q(y_i - \hat{\beta}_1)\hat{\beta}_2}{1 + Q\hat{\beta}_2^2} \tag{17.24}$$

It is interesting to notice here that Equation 17.24, for an error-free knowledge of locations t ($Q=0$), leads to its exact value $\hat{\tau} = t$ (whatever the error on y). Conversely, if there is no information on t ($Q \to \infty$) or if measurements y are exact (with the same consequence: $Q \to \infty$), we obtain $y_i = \hat{\beta}_1 + \hat{\beta}_2\hat{\tau}_i$. This means that for each of these two extreme cases, either of the two terms S_y and S_t of the least square sum S—Equation 17.21—is equal to zero. Even if it is not a function that has to be estimated, coefficient Q (or its inverse) behaves as a Tikhonov's regularization coefficient of order zero. Equation 17.22a provides the classical linear estimator of β in terms of $\hat{\tau}$:

$$\hat{\beta} = (X^t(\hat{\tau})X(\hat{\tau}))^{-1}X^t(\hat{\tau})y \tag{17.25}$$

Substituting $\hat{\tau}_i$ given by Equation 17.24 into Equation 17.25 provides the estimation of the two remaining parameters

$$\hat{\beta}_1 = \bar{y} - \hat{\beta}_2\bar{t} \tag{17.26a}$$

$$\hat{\beta}_2^2 - 2Z\hat{\beta}_2 - \frac{1}{Q} = 0 \quad \text{with } Z = \frac{Qs_y^2 - s_t^2}{2Qs_{ty}} \tag{17.26b}$$

and

$$s_y^2 = \frac{1}{m}\sum_{i=1}^{m} y_i^2 - \bar{y}^2 \tag{17.26c}$$

where
The upper bar designates the statistical mean
s_y^2 and s_t^2 are the statistical variances of y and t, respectively
s_{ty} is the statistical covariance of t and y

See Equations 17.10 through 17.12 and 17.26c for the definition of these coefficients. Equation 17.26b has two solutions of opposite signs. The correct solution can be discriminated using the sign of the linear correlation coefficient r between t and y:

$$\hat{\beta}_2 = Z + s\left(Z^2 + \frac{1}{Q}\right)^{1/2} \quad \text{with } r = \frac{s_{ty}}{s_t s_y} \quad \text{and} \quad s = \frac{r}{|r|} \tag{17.27}$$

If σ' tends to zero (i.e., $t \to \tau$ and $Q \to 0$), $\hat{\beta}_2$ and $\hat{\beta}_1$ approach the classical ordinary least squares estimator (17.8). If it is σ that tends to zero (i.e., $y \to \eta$ and $Q \to \infty$), the same holds for interchanged dependent and independent variables and the model can be rewritten as

$$\tau = -\frac{\beta_1}{\beta_2} + \left(\frac{1}{\beta_2}\right)\eta \tag{17.28}$$

This means that the nonlinear estimator given by Equations 17.24, 17.26a, and 17.27 is only a generalization of the ordinary least squares estimator (17.8). It is called by various names: estimator of the *error in variables model*, *orthogonal regression least square estimator*, and *total least squares* estimator. Bibliographic information about this type of estimator can be found

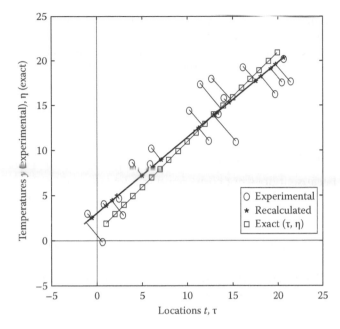

FIGURE 17.1
Estimation example (*S* minimization).

in Van Huffel and Lemmerling (2002), Fuller (1987), Seber and Wild (1988), Cheng and Van Ness (1999), and Emery (2001).

An estimation example is shown in Figures 17.1 and 17.2, for $m = 20$ locations that are uniformly distributed between 1 and 20, in the following case: $\beta_1 = 1\,K$, $\beta_2 = 1\,K\,m^{-1}$, $\sigma = 3\,K$, and $\sigma' = 3\,m$. Noised variables t and y have been generated by use of Equations 17.2 and 17.18—thanks to a stochastic number generator for ε for δ.

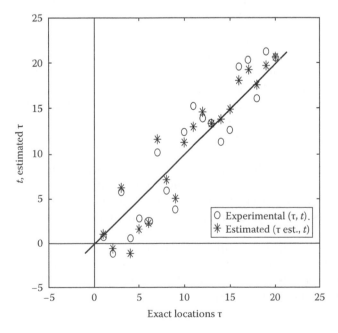

FIGURE 17.2
Estimation example (*S* minimization)—results for the location variable.

In this quite extreme case (strong noise on both t and y, relatively low number of measurement points), the linear estimator (17.8) provides, with $\tau = t$, the estimation (3.79; 0.765) for the (β_1, β_2) pair whereas the nonlinear estimator (17.26) and (17.27) provides the better estimation (3.04; 0.837) that is closer to the true value (1; 1).

17.3 Bias and Variance of the Nonlinear Estimator

The $\hat{\alpha} = [\hat{\beta}^t, \hat{\tau}^t]^t$ estimator is biased because it does not correspond to a linear combination of the data (y_i and t_i). This means that its stochastic mean (its expectation) a priori differs from its exact value:

$$E(\hat{\alpha}) \neq \alpha \tag{17.29}$$

It is possible to calculate (see Maillet et al. [2003]) approximated values of biases $b_j = E(\hat{\beta}_j) - \beta_j$ and standard deviations σ_j of the estimators of the two parameters β_j:

$$b_2 \approx \frac{\beta_2/m}{(s_\tau/\sigma')^2} \tag{17.30a}$$

$$b_1 \approx -\bar{\tau}\frac{\beta_2/m}{(s_\tau/\sigma')^2} \tag{17.30b}$$

$$\sigma_2 \approx \frac{1}{\sqrt{m}}\frac{(\beta_2^2 + 1/Q)^{1/2}}{s_\tau/\sigma'} \tag{17.30c}$$

$$\sigma_1 \approx \frac{\sigma}{\sqrt{m}}(1 + Q\beta_2^2)^{1/2}\left(1 + \frac{\bar{\tau}^2}{s_\tau^2}\right)^{1/2} \tag{17.30d}$$

These approximations have been derived in the case $\sigma/s_\eta = \sigma/(\beta_2 s_\tau) < 1$ and $\sigma'/s_\tau < 1$.

An important factor in these indicators of the quality of the nonlinear estimation is the "signal-over-noise" ratio (s_τ/σ') of the (deterministic) distribution of the location of measurement (the independent variable). In the example that is presented in Figures 17.1 and 17.2, this ratio is close to 2 and the experimental locations t (nominal locations) do not constitute a monotonically increasing function of the exact locations—see Figure 17.2: this explains why the ordering of the estimated locations visible in this figure is nearly blurred.

The preceding equations show that when the number m of measurement points is increased, the biases of the estimators decrease as $1/m$ while their standard deviation decreases as $1/m^{1/2}$. This means that for most situations met in practice, the bias over standard deviation ratio (b_j/σ_j) remains low, if a high enough number of measurements are available.

At last, since in this very particular nonlinear case (the straight line), the theoretical bias is known, it is possible to derive an unbiased corrected estimator: $\hat{\beta}_{jcor} = \hat{\beta}_j - b_j$. The properties of this unbiased estimator can be verified through a great number of Monte Carlo simulations of inversions (see Maillet et al. [2003]).

17.4 Effect of Uncertainty in the Statistical Properties of the Two Sources of Noise: Regularization of the Estimation

In practice, when one tries to implement the nonlinear estimator given by Equations 17.26 and 17.27, which accounts for both errors of measured temperature and of nominal locations, the variance ratio $Q = (\sigma'/\sigma)^2$ or $q = (\sigma/\sigma')^2$ is rarely known.

Usually, only an order of magnitude of the standard deviation σ' on the locations can be evaluated.

In order to test the sensitivity of the estimators to this effect, Monte Carlo simulations have been achieved for the following values: $\beta_1 = 1$ K, $\beta_2 = 100$ K m^{-1}, $\sigma = 3$ K, $\sigma' = 3$ m, and $m = 20$ points (with $Q_{\text{exact}} = (\sigma'/\sigma)^2 = 1$ m^2 K^{-2}).

For each value of the Q ratio, $N = 10,000$ inversions are made, each inversion number k resulting from the 20 (t_i^k, y_i^k) data coming from the exact locations and model output $(\tau_i, \eta_i = \beta_1 + \beta_2 \tau_i)$ that have been noised $(t_i^k = \tau_i + \delta_i^k; y_i^k = \eta_i + \varepsilon_i^k)$ using a random number generator for δ_i and ε_i. Each inversion k provides an estimation $(\hat{\beta}_1^k, \hat{\beta}_2^k, \hat{\tau}_1^k, \hat{\tau}_2^k, \ldots, \hat{\tau}_n^k)$. The statistical mean and standard deviation of the estimates of each parameter are calculated as

$$\bar{\hat{\beta}}_j = \frac{1}{N} \sum_{k=1}^{N} \hat{\beta}_j^k; \quad s_{\hat{\beta}_j} = \left(\frac{1}{N} \sum_{k=1}^{N} \left(\hat{\beta}_j^k \right)^2 - \left(\bar{\hat{\beta}}_j \right)^2 \right)^{1/2} \tag{17.31}$$

They are plotted versus the Q hyperparameter in Figures 17.3 through 17.6. The corresponding means and standard deviations resulting from the linear estimator (see Equation 17.8) are also plotted in the same figures.

It is interesting to note that the mean estimated values depend only very weakly on this parameter as soon as it has reached a high enough level ($Q/Q_{\text{exact}} > 10^{-2}$). The bias of the nonlinear estimator remains low and does not deteriorate too much the dispersion of the estimations when compared to the linear model estimations.

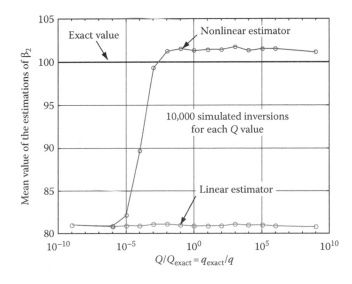

FIGURE 17.3
Monte Carlo inversions—β_2 estimation—mean value.

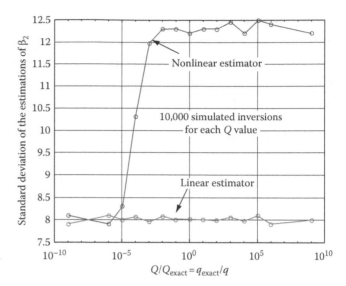

FIGURE 17.4
Monte Carlo inversions—β_2 estimation—
standard deviation.

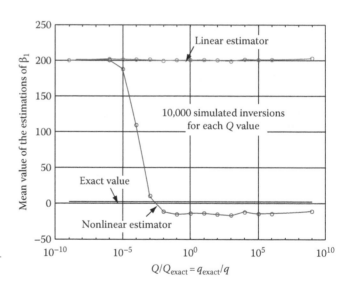

FIGURE 17.5
Monte Carlo inversions—β_1 estimation—
mean value.

17.5 Characterization of Thermal Dispersion in a Granular Medium

17.5.1 Direct Problem and Experimental Setup

Thermal dispersion, that is, heat transfer in a porous medium through which a fluid is flowing, occurs in many natural situations or industrial applications. In the case of process engineering, modeling of this phenomenon is very important for controlling temperature in granular catalyst beds, since chemical conversion and/or catalyst lifetime strongly depend on temperature. Thermal dispersion in a porous model is a complex phenomenon, resulting from conduction in the solid phase and convection and conduction in the moving fluid.

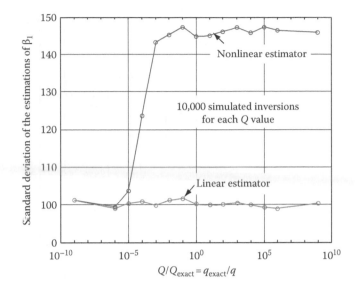

FIGURE 17.6
Monte Carlo inversions—β_1 estimation—standard deviation.

The simplest homogeneous model that can be used in such a situation is based on a local mean temperature that is an average between the local solid and fluid temperatures with a weighting according to their respective heat capacities (see Moyne et al. [2000]). This reduced model requires the definition of a thermal dispersion tensor, the coefficients of which can be considered as pseudo-conductivities that depend on the local Darcy (or filtration/superficial) velocity.

Metzger et al. (2003, 2004) have shown experimentally that this model could be used in the case of water flowing through a bed of glass beads. They estimated the dependence of the longitudinal thermal dispersion coefficient on the reduced Darcy's velocity (the Péclet number). In the water/glass beads case, they could only yield rough estimates of the transverse dispersion coefficient. Testu et al. (2007) used the same setup for air flow through the same bed of glass beads to estimate the corresponding dispersion coefficients.

We will focus here on the estimation technique and the interested reader can refer to Moyne et al. (2000), Metzger et al. (2003, 2004), Testu et al. (2007), and Maillet et al. (2009) for more physical insight.

We consider a fixed granular bed as shown in Figure 17.7 through which a fluid flows downward with a uniform Darcy velocity u. The incoming fluid has constant temperature T_0 and, initially, the whole bed is at the same uniform temperature. An electric heating wire is set along the z-axis (normal to the xy plane of the figure and located at its origin $x = y = 0$). It dissipates heat with a power step of line power intensity Q (W m^{-1}) at time $t = 0$. The medium is treated as infinite and the temperature response to this excitation $\Delta T = T - T_0$ tends to zero at large distances from the source.

The one-temperature model dispersion heat equation (Moyne et al. 2000) can be written as

$$\rho c_t \frac{\partial T}{\partial t} = \lambda_x \frac{\partial^2 T}{\partial x^2} + \lambda_y \frac{\partial^2 T}{\partial y^2} - \rho c_f u \frac{\partial T}{\partial x} + s \qquad (17.32)$$

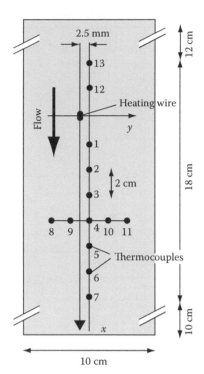

FIGURE 17.7
Dimensions of the granular medium and locations of thermocouples.

Its solution, for a volumetric heat source $s = Q\delta(x)\delta(y)H(t)$, where $\delta(\cdot)$ is the Dirac distribution and $H(\cdot)$ is the Heaviside function, can be calculated using two-dimensional Green's function:

$$\Delta T(x,y,t) = \frac{Q}{4\pi\sqrt{\lambda_x\lambda_y}} \exp\left(\frac{\rho c_f u x}{2\lambda_x}\right) \int_0^{\frac{(\rho c_f u)^2 t}{4\rho c_t \lambda_x}} \exp\left(-\left(\frac{x^2}{\lambda_x}+\frac{y^2}{\lambda_y}\right)\frac{(\rho c_f u)^2}{16\lambda_x}\frac{1}{\theta} - \theta\right)\frac{d\theta}{\theta} \quad (17.33)$$

Here, the total volumetric heat ρc_t of the medium is given by a mixing law based on the volumetric heat of both phases

$$\rho c_t = \varepsilon_f \rho c_f + (1 - \varepsilon_f)\rho c_s \quad (17.34)$$

where ε_f is the void volume fraction of the granular medium (porosity). In Equations 17.32 and 17.33, λ_x and λ_y are the longitudinal and transverse thermal dispersion coefficients, respectively. The integral (with dimensionless integration variable θ) can be calculated by numerical quadrature.

The experimental fixed bed (see Figure 17.7) comprises monodisperse glass beads of diameter $d = 2$ mm and has a porosity $\varepsilon_f = 0.365$. Either water or air can flow downward through it. Table 17.1 gives the thermal properties of the respective phases. The initial bench (Metzger et al. 2004), designed for water flow, has been modified for a gas flow (Testu et al. 2007). Then, a fan is located in a cylindrical duct downstream the setup. It aspires air from a large volume room upstream the lab through a second upstream cylindrical duct. This design ensures a quasi-constant temperature for the inlet air.

TABLE 17.1

Thermal Properties of the Two Phases
of the Granular Medium

	Water	Air	Glass
ρc (KJ m^{-3})	4170	1.2	2080
λ (W m^{-1} K^{-1})	0.607	0.026	1

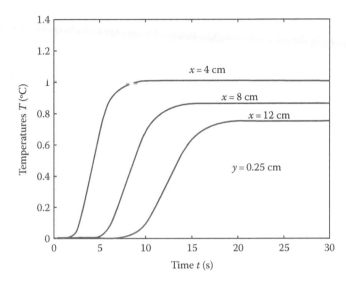

FIGURE 17.8
Thermal responses at three different locations (water flow).

The heating wire is perpendicular to the air flow. Thirteen thermocouples of type E and of 127 μm diameter set parallel to the wire and located mainly downstream the heating wire measure the temperature response of the medium to the power step. Thermocouples 12 and 13 allow to check the constancy of inlet temperature whereas thermocouples 8 and 11 allow verifying that the heated zone does not reach the wall and that the assumption of an infinite medium is valid. The fluid velocity is measured, in the case of water, by a gear flow meter or, in the case of air, by a hot wire anemometer in the downstream cylindrical duct.

The heating level Q is chosen in order not to modify the thermophysical properties of both the fluid and solid (maximum temperature rise on the order of 1 K). Measurements have been made for Péclet numbers ($Pe = \rho c_f u d/\lambda_f$) from 10 to 70 in the case of air flow, which corresponds to maximum filtration velocities close to 0.7 m s^{-1}. For water flow, the Péclet number varies between 10 and 130, with maximum filtration velocities on the order of 7 mm s^{-1}.

Theoretical temperature variations with time—see Equation 17.33—are plotted in Figure 17.8 for three downstream locations (water flow, $u = 6.55$ mm s^{-1}, $\lambda_x = 60$ W m^{-1} K^{-1}, $\lambda_y = 3$ W m^{-1} K^{-1}, and $Q = 300$ W m^{-1}).

17.5.2 Sensitivity Study

As already indicated in Section 17.1, it is impossible to know the exact locations of the 13 thermocouples since they have been installed before filling the experimental volume

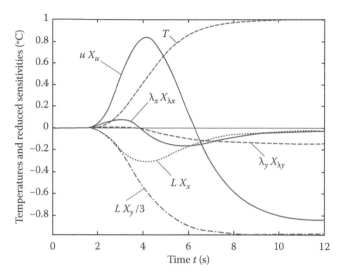

FIGURE 17.9
Variation of temperature and sensitivity coefficients with time for thermocouple located at $x = 4$ cm and $y = 0.25$ cm (water flow).

with glass beads, which caused a displacement of the hot junctions of the thermocouples (in the $z = 0$ plane). This means that their exact locations (x_i, y_i) differ from their nominal locations (x_i^{nom}, y_i^{nom}). As a consequence, the unknown parameters that have to be estimated by inversion of the temperature measurements are not only λ_x, λ_y, and u, which can be put in a parameter vector $\boldsymbol{\beta} = [\lambda_x, \lambda_y, u]^t$ but also the exact locations (x_i, y_i) of the NTc thermocouples. We are therefore trying to estimate the augmented parameter vector $\boldsymbol{\alpha} = \begin{bmatrix} \lambda_x & \lambda_y & u & (x_i, y_i)_{i=1, NTc} \end{bmatrix}^t$.

It is thus necessary to study the behavior of the sensitivity coefficients $X_{\alpha j}$, which are the partial derivative of temperature T with respect to each parameter α_j ($j = 1-5$):

$$X_{\alpha j} = \frac{\partial T}{\partial \alpha_j}.$$

The reduced sensitivity coefficients $\alpha_j X_{\alpha j}$ are plotted in Figure 17.9 for temperature at exact location ($x = 4$ cm; $y = 0.25$ cm), for the same conditions as in Figure 17.8 (water flow). The two sensitivity coefficients to location, X_x and X_y, have been normalized by multiplication by a length $L = 1$ cm, in order to give them the same unit (K) as the other reduced sensitivities. It is very clear that, at this location, the sensitivities of temperature to velocity u and y location (divided by 3 in Figure 17.9) are the highest. But one can notice that sensitivities to λ_y and y are almost proportional. The same is true for sensitivities to u and x (for short times, which bear the majority of the information on the different parameters). This means that, without any information on the exact location of the thermocouples, it will not be possible to estimate the $\boldsymbol{\beta}$ parameters. The same holds if more than one thermocouple is used for inversion because of the addition of new location parameters.

17.5.3 Parameter Estimation Technique

The most classical parameter estimation technique relies on the minimization of the ordinary least squares sum

$$S_T(\boldsymbol{\beta}) = \sum_{i=1}^{NTc} \sum_{k=1}^{Nt} (T_{\exp, ik} - T(x_i, y_i, t_k, \boldsymbol{\beta}))^2 \tag{17.35}$$

where

$T_{\exp, ik}$ is the temperature measured at the location (x_i, y_i) of the ith thermocouple at time t_k

$T(x_i, y_i, t_k, \boldsymbol{\beta})$ is the corresponding theoretical temperature given by solution 17.33 of the direct problem, which depends on the different parameters to be estimated

If the exact locations of the thermocouples are known, it is possible to estimate the two dispersion coefficients as well as the filtration velocity, that is, $\boldsymbol{\beta} = [\lambda_x, \lambda_y, u]^t$.

With the assumption of a normal independent and identically distributed noise (at different times and locations) of standard deviation σ, the opposite of $S_T/(2\sigma^2)$ is equal to the logarithm of the joint probability distribution of these temperature measurements (within an additive constant). Minimizing S_T with respect to parameters β_j (for fixed measurements) corresponds to a maximization of this logarithm ($S_T/(2\sigma^2)$ and hence S_T), which is also called log-likelihood. The resulting estimator $\hat{\boldsymbol{\beta}}$ is called *maximum likelihood* estimator. If the normality hypothesis is relaxed from the above assumptions and, moreover, if one assumes the model to be linear (with respect to the parameters), which can always be verified on a local incremental basis, minimization of S_T yields the *minimum variance* estimator: this is the linear estimator (17.8) that is characterized by the lowest variances of all its components (see Beck and Arnold [1977]).

In practice, the exact locations of the thermocouples are not known and it is necessary to estimate not only λ_x, λ_y, and u but also the unknown locations (x_i, y_i).

In order to estimate the new parameter vector $\boldsymbol{\alpha} = [\lambda_x, \lambda_y, u, (x_i, y_i)_{i=1, NTc}]^t$, the prior knowledge of these nominal locations is incorporated in the new objective function:

$$S(\boldsymbol{\alpha}) = \frac{1}{\sigma_T^2} \sum_{i=1}^{NTc} \sum_{k=1}^{Nt} \left(T_{\exp, ik} - T_{ik}(\boldsymbol{\alpha})\right)^2 + \frac{1}{\sigma_{loc}^2} \sum_{i=1}^{NTc} \left(x_i^{nom} - x_i\right)^2 + \frac{1}{\sigma_{loc}^2} \sum_{i=1}^{NTc} \left(y_i^{nom} - y_i\right)^2 \quad (17.36)$$

where

σ_{loc} is the standard deviation of the locations in x and y of the thermocouple hot junctions in the bed

σ_T is the standard deviation of the temperature measurement

This corresponds to a *total least square* problem (Van Huffel and Lemmerling 2002) already described in Section 17.2.3.

If the probability distributions of the temperature noise, called $\boldsymbol{\varepsilon}$ in Section 17.2, and the distribution of $\boldsymbol{\delta}$, the error on thermocouple location in the same section, are both independent and normal, minimization of S (for fixed measurements) corresponds to a maximization of the conditional joint probability distribution of $\boldsymbol{\varepsilon}$ for a given known distribution $\boldsymbol{\delta}$, if Bayes' theorem is used (see Chapter 12 for more insight into this subject).

The distribution of $\boldsymbol{\delta}$ is called the *prior* distribution, because its knowledge brings information before the minimization of S. So, the resulting estimator is called a *maximum a posteriori* estimator.

Since our problem is nonlinear, an iterative technique has to be used to minimize sum S. A Gauss–Newton algorithm (see Beck and Arnold [1977]) is used:

$$\boldsymbol{\alpha}^{(k)} = \boldsymbol{\alpha}^{(k)} + (X^t(\boldsymbol{\alpha}^{(k-1)})X(\boldsymbol{\alpha}^{(k-1)}) + qR^tR)^{-1}(X^t(\boldsymbol{\alpha}^{(k-1)})(T_{\exp} - T(\boldsymbol{\alpha})) + qR^t(z^{nom} - z))$$

$$(17.37)$$

In the above equation, the nominal and exact position vectors are

$$z^{\text{nom}} = [x_1^{\text{nom}} \cdots x_{NTc}^{\text{nom}} \quad y_1^{\text{nom}} \cdots y_{NTc}^{\text{nom}}]^t, \quad z = [x_1 \cdots x_{NTc} \quad y_1 \cdots y_{NTc}]^t \qquad (17.38)$$

the corresponding theoretical temperatures

$$\boldsymbol{T}(\boldsymbol{\alpha}) = [T(x_1, y_1, t_1, \boldsymbol{\beta}) \cdots T(x_1, y_1, t_{Nt}, \boldsymbol{\beta}) T(x_2, y_2, t_1, \boldsymbol{\beta}) \cdots$$
$$T(x_2, y_2, t_{Nt}, \boldsymbol{\beta}) \cdots T(x_{NTc}, y_{NTc}, t_1, \boldsymbol{\beta}) \cdots T(x_{NTc}, y_{NTc}, t_{Nt}, \boldsymbol{\beta})]^t \qquad (17.39)$$

and the experimental temperatures

$$\boldsymbol{T}_{\text{exp}} = [T_{\text{exp}, 11} \cdots T_{\text{exp}, 1Nt} T_{\text{exp}, 21} \cdots T_{\text{exp}, 2Nt} \cdots T_{\text{exp}, NTc1} \cdots T_{\text{exp}, NTcNt}]^t \qquad (17.40)$$

The sensitivity matrices are defined by

$$X(\boldsymbol{\alpha}) = [X_{\boldsymbol{\beta}} \quad X_z] \quad \text{with} \quad X_{\boldsymbol{\beta}} = \begin{bmatrix} \dfrac{\partial T}{\partial \beta_1} & \dfrac{\partial T}{\partial \beta_2} & \dfrac{\partial T}{\partial \beta_3} \end{bmatrix} \quad \text{and} \quad X_z = \begin{bmatrix} \dfrac{\partial T}{\partial z_1} & \cdots & \dfrac{\partial T}{\partial z_{2NTc}} \end{bmatrix}$$
$$\qquad (17.41)$$

and

$$R = [0 \quad I] \qquad (17.42)$$

The matrix 0 above is a zero matrix of size $(2NTc \times 3)$ and I an identity matrix of dimension $2NTc$. The coeffecint q is equal to $(\sigma_T/\sigma_{\text{loc}})^2$.

The preceding estimator can also be considered as a Gauss Markov estimator since sum (17.36) can also be written as

$$S(\boldsymbol{\alpha}) = (Y - T(\boldsymbol{\alpha}))^t \Omega^{-1} (Y - T(\boldsymbol{\alpha})) \quad \text{with} \quad Y = \begin{bmatrix} T_{\text{exp}} \\ z^{\text{nom}} \end{bmatrix}, \quad \eta(\boldsymbol{\alpha}) = \begin{bmatrix} T(\boldsymbol{\alpha}) \\ z \end{bmatrix} \qquad (17.43)$$

with

$$\Omega = \begin{bmatrix} \sigma_T^2 I_{NTcNt} & 0 \\ 0 & \sigma_{\text{loc}}^2 I_{2NTc} \end{bmatrix} \qquad (17.44)$$

where I_p is the identity matrix of size $(p \times p)$.

An approximation of the covariance matrix of this estimator is (see Beck and Arnold [1977]) as follows:

$$\text{cov}(\hat{\boldsymbol{\alpha}}) = (\tilde{X}^t(\hat{\boldsymbol{\alpha}}) \, \Omega^{-1} \tilde{X}(\hat{\boldsymbol{\alpha}}))^{-1} \quad \text{with} \quad \tilde{X} = \begin{bmatrix} X \\ R \end{bmatrix} \qquad (17.45)$$

The experimental temperature standard deviation ($\sigma_T = 0.02°C$) can be measured in a steady-state situation, that is, without any excitation Q, and it can be assumed that the standard deviation of the location of a hot junction, that is, a measure of its displacement, is on the order of one bead radius ($\sigma_{\text{loc}} = 1$ mm).

Metzger et al. (2003, 2004) have shown that the estimated values depend quite weakly on the choice of the standard deviation σ_{loc}, as soon as this standard deviation becomes larger than a fraction of a millimeter. A low value σ_{loc} (smaller than one micrometer) leads to very poor temperature residuals with estimated locations close to their nominal values ($\hat{x}_i \approx x_i^{nom}$; $\hat{y}_i \approx y_i^{nom}$). In the range between one micrometer and a few tenth of millimeter, a decrease of the residuals and a variation of the estimated values are observed. As soon as σ_{loc} reaches a 1 mm value, both residuals and estimates become good and do not vary any more. At last, for $\sigma_{loc} > 1$ m, one nearly meets the case of ordinary least squares where temperatures only are fitted and the nonlinear inversion algorithm does not converge anymore. We notice that multiplication of sum S by σ_T^2 shows that this minimization can also be considered as some form of Tikhonov zeroth-order regularization where the regularization coefficient is $q = (\sigma_T/\sigma_{loc})^2$.

17.5.4 Monte Carlo Simulations

For Monte Carlo simulations of inversion, nominal thermocouple coordinates (x_i^{nom}, y_i^{nom}) are noised with an uncorrelated additive noise of standard deviation σ_{loc} to produce the exact locations (x_i, y_i). Subsequently, the same technique is applied to the true temperature response of model (17.33) with a noise of standard deviation σ_T to obtain simulated experimental temperatures $T_{exp,ik}$. A Gauss–Newton minimization of S yields an estimation $\hat{\alpha}$ of the parameter vector. If 400 simulations of this type are made with the corresponding inversions, 400 estimates $\hat{\alpha}_j^{(n)}$ are available for the jth parameter of $\hat{\alpha}$, n being the inversion number. It is then possible to assess the statistical distribution of each estimated parameter (via its histogram) and to calculate the dispersion (standard deviation s_j) of each estimate as well as its bias b_j, that is,

$$b_j = \bar{\hat{\alpha}}_j - \alpha_j \quad \text{and} \quad s_j = \frac{1}{400}\sum_{n=1}^{400}\left(\hat{\alpha}_j^{(n)}\right)^2 - \left(\bar{\hat{\alpha}}_j\right)^2 \quad \text{with} \quad \bar{\hat{\alpha}}_j = \frac{1}{400}\sum_{n=1}^{400}\hat{\alpha}_j^{(n)} \tag{17.46}$$

Such estimates are given in Table 17.2 (see Testu et al. [2007]) for air or water flow through the glass beads. They correspond to the use of temperature signal of thermocouples 2–7 in Figure 17.1 and to a time step of 0.15 s, with a final time of 900 s for air, the corresponding values being 0.15 and 45 s for water. One can use here the $(|b_j| + s_j)/\alpha_j$ ratio (relative error) as an index of inversion quality for parameter α_j. The λ_x estimations have the same quality

TABLE 17.2

Monte Carlo Simulations of Inversion for Air or Water Flow through a Bed of Glass Beads

	j	Parameter	Exact Value α_j	Average Estimation $\bar{\hat{\alpha}}_j$	Estimation Bias b_j	Estimation Standard Deviation s_j	Bias/ Dispersion $|b_j|/s_j$ (%)	Relative Error $(|b_j| + s_j)/\alpha_j$ (%)
Air	1	λ_x (W K^{-1} m^{-1})	0.962	0.984	+0.022	0.008	275	3
	2	λ_y (W K^{-1} m^{-1})	0.256	0.246	−0.010	0.003	336	5.2
	3	u (m s^{-1})	0.353	0.355	+0.002	0.004	50	1.7
Water	1	λ_x (W K^{-1} m^{-1})	60	60.321	+0.321	1.009	32	2.2
	2	λ_y (W K^{-1} m^{-1})	3	2.681	−0.329	0.310	106	21
	3	u (mm s^{-1})	6.288	6.306	+0.018	0.033	55	0.8

for air and water with "relative errors" smaller than 3%: bias is larger for air but it is compensated by a lower dispersion.

For λ_y estimations, the relative error is still acceptable for air (5%) but too large for water (21%) to yield precise values. For both fluids, the filtration velocity is the parameter that is estimated with the maximum precision (relative errors lower than 2%). This confirms the possibility of estimating the transverse dispersion coefficient for air, which was not possible for water.

It is interesting to note here that the estimation bias for the different parameters, which is caused by the nonlinear character of the estimator, can be of the same magnitude as or even higher than the standard deviation (see the $|b_j|/s_j$ column in Table 17.2).

17.6 Conclusion

Starting from the linear model of the straight line, it has been shown that the errors in the measured signal (ordinates, dependent variable) and in the locations of the sensors (abscissa, independent variable) can be simultaneously accounted for in the estimation problem if a two-term functional is minimized. The weighting factor in this sum is the ratio of the variances of these two types of variables. It has also been shown that precise knowledge of the weighting factor may not be necessary. This nonlinear estimator is a generalization of the classical ordinary least squares estimator and can be considered as a Bayesian estimator. Bias and standard deviation of the estimated parameters have been derived analytically for this simple case.

The same approach can be used in an experimental estimation problem of thermophysical parameter determination where the location(s) of the sensor(s) is not precisely known. The nonlinear character of this new type of estimator can be taken into account. Monte Carlo simulations of inversion allow an assessment of bias and standard deviation of the estimated parameters.

This type of total least squares estimator can also be very useful for multiphysical estimation problems where the composite signal comes from different types of sensors (temperature, pressure, etc.).

Nomenclature

b	bias
$E(\cdot)$	expectancy
m	number of measurement locations
Nt	number of times of measurements
NTc	number of thermocouples
Q	ratio of variances
s	statistical standard deviation
t	location (Sections 17.2 through 17.4) or time (Section 17.5)
y	noised signal

X sensitivity matrix
X_j sensitivity vector

Symbols

α augmented parameter vector
β parameter vector
δ, ε location and signal noise
δ_{ij} Kronecker's symbol
ε_f porosity
η model
σ, σ' standard deviations of ε and δ
σ_j standard deviation of $\hat{\beta}_j$
τ exact time or location

Superscripts

$\wedge$ estimated value or estimator
-      statistical average
nom   nominal
t transpose

Subscripts

exp   experimental
loc   location

References

Beck, J.V. and K.J. Arnold. 1977. *Parameter Estimation in Engineering and Science*. Chichester: Wiley.

Cheng C.-L. and J.W. Van Ness. 1999. *Statistical Regression with Measurement Errors*. London, U.K.: Arnold.

Emery, A.F. 2001. Parameter estimation in the presence of uncertain parameters and with correlated errors. In *Proceedings of EUROTHERM Seminar No. 68, Inverse Problems and Experimental Design*, March 5–7, 2001, Poitiers, France.

Fuller, W.A. 1987. *Measurement Error Models*. Chichester: Wiley & Sons.

Maillet, D., B. Fiers, and G. Ferschneider. 2009. Characterization of thermal dispersion and wall effects in porous media. *Proceedings (CD-ROM) of the 20th International Congress of Mechanical, Symposium Energy & Thermal Sciences*, Paper LEC0024. Gramado, RS, Brazil: ABCM Publisher.

Maillet, D., T. Metzger, and S. Didierjean. 2003. Integrating the error in the independent variable for optimal parameter estimation. Part I: Different estimation strategies on academic cases. *Inverse Probl. Eng.* 11(3): 175–186.

Metzger, T., S. Didierjean, and D. Maillet. 2003. Integrating the error in the independent variable for optimal parameter estimation. Part II: Implementation to experimental estimation of the thermal dispersion coefficients in porous media with not precisely known thermocouple locations. *Inverse Probl. Eng.* 11(3): 187–200.

Metzger, T., S. Didierjean, and D. Maillet. 2004. Optimal experimental estimation of thermal coefficients in porous media. *Int. J. Heat Mass Transfer* 47: 3341–3353.

Moyne, C., S. Didierjean, H.P. Amaral Souto, and O.T. Da Silveira. 2000. Thermal dispersion in porous media: One-equation model. *Int. J. Heat Mass Transfer* 43: 3853–3867.

Seber, G.A.F. and C.J. Wild. 1988. *Nonlinear Regression*. Chichester: Wiley.

Testu, A., S. Didierjean, D. Maillet, C. Moyne, T. Metzger, and T. Niass. 2007. Thermal dispersion coefficients for water or air flow through a bed of glass beads. *Int. J. Heat Mass Transfer* 50(7–8): 1469–1484.

Van Huffel, S. and P. Lemmerling. 2002. *Total Least Squares and Errors-in-Variables Modeling: Analysis, Algorithms and Applications*. Dordrecht, the Netherlands: Kluwer Academic Publishers.

18

Thermophysical Properties Identification in the Frequency Domain

Valério L. Borges, Priscila F.B. Sousa, Ana P. Fernandes, and Gilmar Guimarães

CONTENTS

18.1 Introduction

Several researchers have been working on the simultaneous determination of thermal diffusivity and thermal conductivity (Guimarães et al., 1995, Huang and Yang, 1995, Dowding et al., 1996, Nicolau et al., 2002, Lima e Silva et al., 2003, Borges et al., 2006). However, the methods proposed in most of these works can only be used to obtain α and k of nonconductor materials. Additional problems occur due to conductive materials: problems such as contact resistance, low sensitivity due to the small temperature gradient, and the heat flux losses are responsible for the difficulty of direct application of these methods.

As in any experimental method, the identification of thermal properties is sensitive to measurement uncertainty. Thus, to guarantee accuracy in the estimation, the design of the experiments should be optimized.

The optimal design is related to the boundary conditions and sensor locations. Beck and Arnold (1977) have shown that the best experiment corresponds to a finite body with a heat flux that produces a temperature change in a surface, keeping the other surfaces insulated. This basic idea for the thermal model is used here. However, to avoid the low sensitivity and the thermal contact resistance problems, the sensors are disposed in different ways using the three-dimensional (3D) model.

Two distinct problems are then established: experimental and thermal model developments. These problems are developed in the next section.

18.2 Fundamentals

18.2.1 Dynamic and Thermal Equivalent System

The technique proposed here is based on the use of an input/output dynamical system (Figure 18.1), given by the convolution integral

$$Y(t) = \int_0^\infty H(t - \tau)X(t)d\tau \tag{18.1}$$

or in transformed frequency–plane

$$Y(f) = H(f) \times X(f) \tag{18.2}$$

where the weighting function, $H(f)$, is equal to 0 for $\tau < 0$, when the system is physically realizable. In frequency domain, $H(f)$ represents the frequency response, which is defined as the Fourier transform of $H(\tau)$:

$$H(f) = \int_0^\infty H(\tau)e^{(-j2\pi)}d\tau \tag{18.3}$$

where $j = \sqrt{-1}$ is the imaginary unit (Bendat and Piersol, 1986).

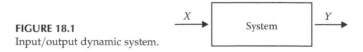

FIGURE 18.1
Input/output dynamic system.

18.2.1.1 Thermal Model

The proposed thermal model to be reproduced experimentally is given by a sample initially at uniform temperature, T_0. The sample is then submitted to a heat flux (W/m^2) while all other surfaces are kept isolated.

The dynamic model can be obtained from a thermal model shown in Figure 18.2, where $q(t)$ represents the heat flux, T represents the temperature, and $i = 1, 2$ is the index used to describe the location of the respective temperature in the sample.

In this thermal model, the input, $X(t)$, and output, $Y(t)$, data are defined as $X(t) = q(t)$ and $Y(t) = T_1(t) - T_2(t)$, respectively.

The 3D thermal model can be obtained by the solution of the diffusion equation

$$\frac{\partial^2 T}{\partial x^2} + \frac{\partial^2 T}{\partial y^2} + \frac{\partial^2 T}{\partial z^2} = \frac{1}{\alpha}\frac{\partial T}{\partial t} \tag{18.4a}$$

in the region $\Re$ $(0 < x < L, 0 < y < W, 0 < z < R)$ and $t > 0$.

Subjected to the boundary conditions:

$$-k\frac{\partial T}{\partial y}\bigg|_{y=W} = q(t); \quad \text{in } A_P \text{ surface} \tag{18.4b}$$

$$-k\frac{\partial T}{\partial y}\bigg|_{y=W} = 0; \quad \text{in } A_0 \text{ surface} \tag{18.4c}$$

$$\frac{\partial T}{\partial x}\bigg|_{x=0} = \frac{\partial T}{\partial y}\bigg|_{x=L} = \frac{\partial T}{\partial y}\bigg|_{y=0} = \frac{\partial T}{\partial z}\bigg|_{z=0} = \frac{\partial T}{\partial z}\bigg|_{z=R} = 0 \tag{18.4d}$$

and the initial

$$T(x, y, z, 0) = T_0 \tag{18.4e}$$

where
A is defined by $(0 < x < L, 0 < z < R)$
x_H and z_H are the boundary of A_P
A_0 surface is defined by $A_0 = A - A_P$

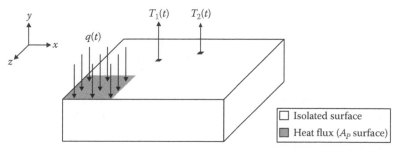

FIGURE 18.2
3D thermal equivalent model.

If $T(\mathbf{r}, t)$ represents $T(x, y, z, t)$, the solution of Equations 18.4 can be given in terms of Green's function as in Beck et al. (1992).

$$T(\mathbf{r},t) = \int_{\Re} G(\mathbf{r}, t|\mathbf{r}', 0) T_0 d\mathbf{r}' + \frac{\alpha}{k} \int_A \int_{\tau=0}^{t} [q(\tau) G(\mathbf{r}|\mathbf{r}', t - \tau)] \, dA \, d\tau \tag{18.5}$$

Since Green's function is available and exists (Beck et al., 1992), the solution of the problem defined by Equations 18.4a through 18.4e can be performed numerically or analytically.

Equation 18.5 reveals that an equivalent thermal model can be associated with the dynamic model given by Equation 18.1. In this case, the equivalent thermal model can be obtained as the convolution product in the frequency domain:

$$H(f) = G_H(f) = \frac{Y(f)}{X(f)} = \frac{T_1(f) - T_2(f)}{q(f)} \tag{18.6}$$

where the variable f indicates that Fourier transform was applied to the variables $T(t)$, $q(t)$, and $G_H(t - \tau)$. A comparison of Equation 18.6 with Equation 18.1 gives

$$G_H(t - \tau) = \frac{\alpha}{k} [G(r_1, t - \tau) - G(r_2, t - \tau)] \tag{18.7}$$

It can be observed that as $T_1(t)$ and $T_2(t)$ are obtained by discrete measurements, Fourier transform can be performed numerically by using the Cooley–Tukey algorithms (Discrete Fast Fourier Transform) for these data (Guimarães et al., 1995).

Observing Equations 18.4 and 18.6, it can be concluded that the frequency response $H(f)$ is strongly dependent on the thermal properties:

$$H(f) = G_H(f) = function(\alpha, k) \tag{18.8}$$

It also should be observed that the transformed impedance in the f plane is a complex variable, which in a polar form can be written by

$$H(f) = G_H(f) = |H(f)| e^{-j\varphi(f)} \tag{18.9}$$

where $|H(f)|$ and $\varphi(f)$ represent, respectively, the amplitude and the phase factor of H. The phase factor can be written by

$$\varphi(f) = \arctan\left(\frac{\Im H(f)}{\Re H(f)}\right) \tag{18.10}$$

where $\Im H(f)$ and $\Re H(f)$ are the imaginary and real parts of $H(f)$, respectively.

The phase of frequency response $H(f)$ and the time evolution of superficial temperatures $T_1(t)$ and $T_2(t)$ are the experimental data used for estimation of thermal diffusivity and thermal conductivity, respectively.

18.2.2 Thermal Diffusivity Estimation: Frequency Domain

The fact that the phase factor is just a function of the thermal diffusivity α is the great convenience of working in the frequency domain. The basic idea here is the observation that the delay between the experimental and theoretical temperature is an exclusive function of α. This condition was first verified by Guimarães et al. (1995), and this effect in the sensitivity coefficient will be shown in Section 18.3. Therefore, the minimization of an objective function, S_φ, based on the difference between experimental and calculated values of the phase is used to determine the thermal diffusivity. This function can be written as

$$S_\varphi = \sum_{i=1}^{Nf} (\varphi_e(i) - \varphi(i))^2 \tag{18.11}$$

where φ_e and φ are the experimental and calculated values of the phase factor of $H(f)$, respectively.

The theoretical values of the phase factor are obtained from the identification of $H(f)$ by Equation 18.10. In this case, the output $Y(f)$ is the Fourier transform of the difference obtained by the numerical solution of Equations 18.4a through 18.4e using the finite volume method (Patankar, 1980). In fact, this procedure avoids the necessity of obtaining an explicit and analytical model of $H(f)$.

The values of α will be supposed to be those that minimize Equation 18.11. In this work, this minimization is done by using the golden section method with polynomial approximation (Vanderplaats, 1984).

18.2.3 Thermal Conductivity Estimation: Frequency Domain

Once the thermal diffusivity value is obtained, an objective function based on least square temperature error can be used to estimate the thermal conductivity. In this case, there is no identification problem as just one variable is being estimated. Therefore, the variable k will be supposed to be the parameter that minimizes the least square function, S_{qH}, based on the difference between the calculated and experimental of the frequency response amplitude defined by

$$S_{qH} = \sum_{j=1}^{s} \sum_{i=1}^{n} (|H_e(i,j)| - |H_t(i,j)|)^2 \tag{18.12}$$

where
 $|H_e(i,j)|$ is the experimental frequency response modulus
 $|H_t(i,j)|$ is the respective calculated values
 n is the total number of frequency measurements
 s represents the number of sensors

The optimization technique used to obtain k is also the golden section method with polynomial approximation (Vanderplaats, 1984).

18.2.4 Thermal Conductivity Estimation: Time Domain

The variable k can also be the parameter that minimizes the least square function, S_{qT}, based on the difference between the calculated and experimental temperature defined in time domain and given by

$$S_{qT} = \sum_{j=1}^{s} \sum_{i=1}^{n} (T_e(i,j) - T_t(i,j))^2 \tag{18.13}$$

where
$T_e(i,j)$ is the experimental temperature
$T_t(i,j)$ is the calculated temperature
n is the total number of time measurements
s represents the number of sensors

18.3 Sensitivity Analysis

Although the thermal contact resistance and the low gradient problems do not represent any difficulties for nonmetallic materials, they must be taken into account in the presence of conductor materials. This section discusses both problems.

Figure 18.3 presents the thermal contact resistance that can appear between sample and sensors in a 1D model (Figure 18.3a) and an alternative 3D model (Figure 18.3b) that avoid this problem. Another advantage in a 3D model is the experimental flexibility, allowing the optimal location of the identification sensors.

In 1D model, a high magnitude of heat flux (input) can be necessary to establish a thermal gradient high enough for the estimation process. Figures 18.4 and 18.5 present a simulation using the same heat flux input. It can be observed that while the temperature

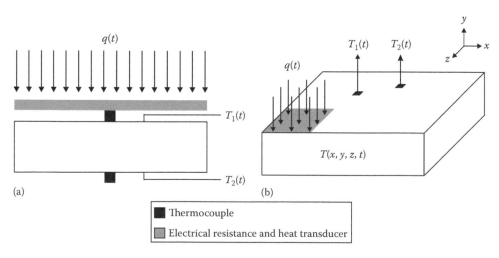

FIGURE 18.3
Experimental scheme for models: (a) 1D model and (b) 3D model.

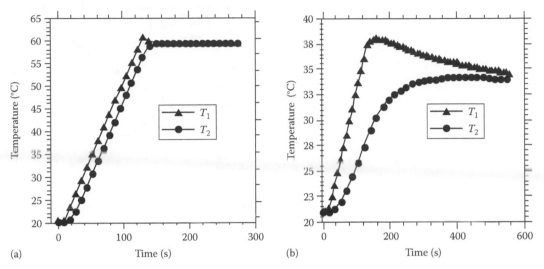

FIGURE 18.4
Temperature evolution for AISI304 sample: (a) 1D model and (b) 3D model.

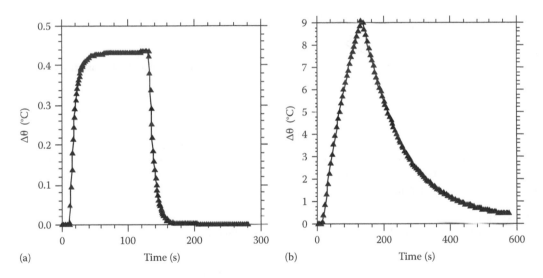

FIGURE 18.5
Temperature difference for AISI304 sample: (a) 1D model and (b) 3D model.

gradient is situated in the region of the uncertainty of thermocouples (0.3 K), for 1D model (Figure 18.3a), the 3D model produces a sufficient gradient to properties estimation (Figure 18.3b). Figure 18.5 presents the difference between the two temperatures involved in each model as shown in Figure 18.3.

This fact can be better analyzed through a sensitivity analysis. Small and/or inaccurate values of temperature difference and heat flux signals produce linear dependence or low values. The linear dependence of two or more coefficients indicates that the parameters cannot simultaneously be estimated. Low values indicate that the estimation is strongly sensitive to the measurements uncertainty (Beck and Arnold, 1977). The sensitivity coefficients involved in this technique are defined as follows and presented in Figures 18.6 and 18.7.

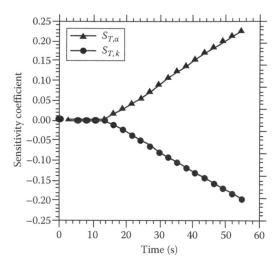

FIGURE 18.6
Sensitivity coefficients for the 3D model to AISI304 sample.

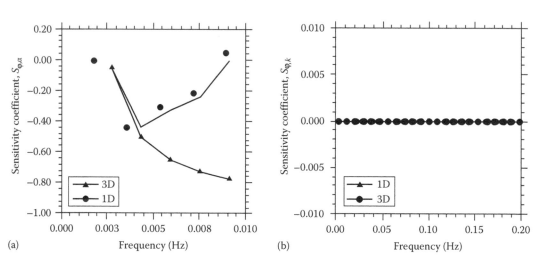

(a)   (b)

FIGURE 18.7
Sensitivity coefficient of phase for 1D and 3D for the AISI304 sample.

$$S_{T,\alpha} = \frac{\alpha}{T} \frac{\partial T}{\partial \alpha}, \quad S_{T,k} = \frac{k}{T} \frac{\partial T}{\partial k}, \quad S_{\varphi,\alpha} = \frac{\alpha}{\varphi} \frac{\partial \varphi}{\partial \alpha}, \quad S_{\varphi,k} = \frac{k}{\varphi} \frac{\partial \varphi}{\partial k} \tag{18.14}$$

$$S_{|H|,\alpha} = \frac{\alpha}{|H|} \frac{\partial |H|}{\partial \alpha}, \quad S_{|H|,k} = \frac{k}{|H|} \frac{\partial |H|}{\partial k} \tag{18.15}$$

Figure 18.6 reveals a linear dependency of $S_{T,\alpha}$ and $S_{T,k}$ as shown by the symmetry. This fact indicates that both thermal properties cannot be estimated simultaneously in time domain, justifying the use of frequency domain for the thermal diffusivity estimation.

It can be observed in Figure 18.7 that the absolute values of the sensitivity of phase related to the thermal diffusivity are higher for the 3D model and there is no possibility to estimate the thermal conductivity in frequency domain due to $S_{\varphi,k} = 0$ for any frequency value. This fact reveals that the phase dependency with thermal diffusivity is unique and exclusive.

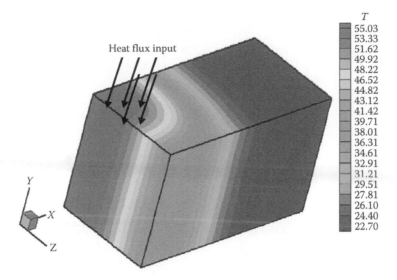

FIGURE 18.8
Spatial temperature in a thin conductor sample.

Another advantage of using a 3D model is the possibility of estimating thermal properties from a thin sample. In the 1D case, for conductor materials, it is very hard to obtain temperature gradients with values high enough to allow a good estimation as in Figure 18.8. For a sample with thin thickness, it can be seen that no temperature variation in the direction y is observed. This fact makes the 1D analysis unpractical.

Another important characteristic of the technique presented here is the very low sensitivity of α related to the amplitude of the signals X and Y. It means that the estimated value of the thermal diffusivity is insensitive to bias error, like uncertainty due to poor calibration of thermocouples or heat flux transducers or both. This fact can be demonstrated by verifying Figures 18.9 and 18.10, which show the behavior of phase factor and modulus

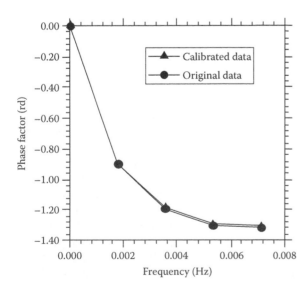

FIGURE 18.9
Phase factor subjected to the original and calibrated pair of input/output data.

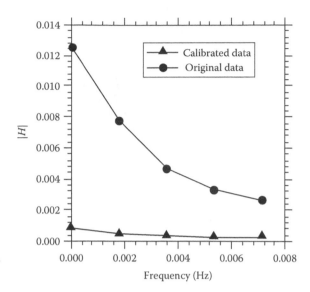

FIGURE 18.10
Absolute value of H for the original and calibrated pair input/output.

of H due to the same input/output signals in both versions: original data, $\mu V/V$, and calibrated data, $°C/W\,m^2$.

It can be observed in Figures 18.9 and 18.10 that there are no changes in the phase factor while the modulus is strongly affected.

18.4 Experimental Determination of Thermal Conductivity and Diffusivity Using Partially Heated Surface Method with Heat Flux Transducer

18.4.1 Experimental Apparatus and Results

18.4.1.1 Conductor Material Application

It should be observed that the boundary conditions present in the theoretical model must be guaranteed in the experimental apparatus. It means that the isolated condition at the reminiscent surface needs to be reached for the success of the estimation techniques. A good way to reach the isolation condition in a vertical direction is the use of a symmetric experiment apparatus. Figure 18.11 presents this scheme.

In this case, the effect of no heat flux lateral loss is reached by placing insulating material such as expanded polystyrene as shown in Figure 18.11. Two AISI304 stainless steel samples were used in a symmetric assembly, both with thickness of 10 mm and lateral dimensions of 139×65 mm. The sample initially in thermal equilibrium at T_0 is then submitted to a unidirectional and uniform heat flux. The heat is supplied by a 318 Ω electrical resistance heater, covered with silicone rubber, with lateral dimensions of 50×50 mm and thickness of 0.3 mm. The heat flux are acquired by a transducer with lateral dimensions of 50×50 mm, thickness of 0.5 mm, and constant time less than 10 ms. The transducer is based on the thermopile conception of multiple thermoelectric junction (made by electrolytic deposition) on a thin conductor sheet (Guimarães et al., 1995). The temperatures are measured using surface thermocouples (type K). The signals of

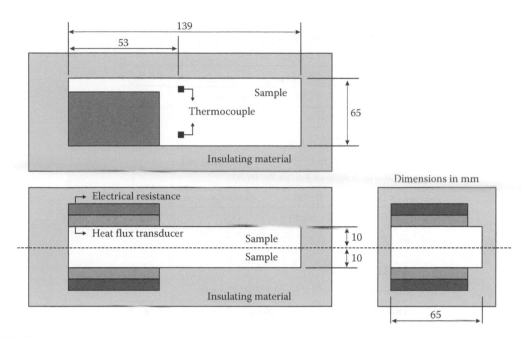

FIGURE 18.11
Schematic of experimental apparatus.

heat flux and temperatures are acquired by a data acquisition system HP Series 75000 with voltmeter E1326B controlled by a personal computer.

Twenty independent runs were performed. In each of the experiments were acquired 1024 points at time intervals of 0.54 s. The time duration of heating, t_h, was approximately 120 s with a heat pulse generated at 90 V(DC).

Figure 18.12 shows the evolution of the input/output normalized signals in function of time for one of the experimental of AISI304 sample.

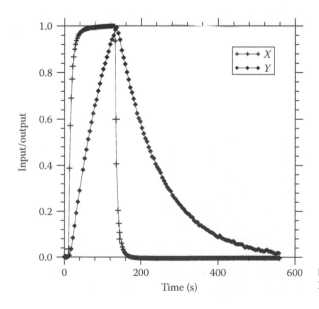

FIGURE 18.12
Input data of a typical run.

TABLE 18.1

Statistical Data of the Average Value of α
(Initial Value, $\alpha = 1.0 \times 10^{-6}$ m^2/s)

$\alpha \times 10^6$ (m^2/s)	Initial S_φ	Final S_φ	$\sigma \times 10^8$ (m^2/s)
3.762	20	0.0016	4.0

TABLE 18.2

Statistical Data of the Average Value of k
(Initial Value of $k = 10$ W/m K)

k (W/m K)	Initial S_{qT}	Final S_{qT}	σ (W/m K)
14.64	14,700	18.3	0.31

TABLE 18.3

Summary of α and k for AISI304 Sample

Thermal Properties	This Work	References	Error (%)
$\alpha \times 10^6$ (m^2/s)	3.762	3.82	1.54
k (W/m K)	14.64	14.90	1.77

Tables 18.1 and 18.2 present respectively the value estimated of α and k for the AISI304 stainless steel sample.

In Table 18.3, a summary of the simultaneous estimation of α and k of the AISI304 sample is presented. In this table, the value obtained for α using the Flash method (Parker et al., 1961) and the value of k from Incropera and DeWitt (1996) are also presented.

It can be observed that there is an excellent agreement between the values of this work and the literature for the thermal diffusivity and the thermal conductivity (error less than 2%).

These results show the potential of the method proposed here. The comparison between the experimental and estimated temperatures for $\alpha = 3.76 \times 10^{-6}$ m^2/s and k is shown in Figure 18.13. In this figure, a good agreement between the data can be observed; the deviated are situated in the range of uncertainty measurement of thermocouples, which in this work is 0.3 K.

18.4.1.2 Nonconductor Material Application

The thermal identification technique can also be applied to nonconductor solid materials. In this case, a 1D model can be used. This section presents some results of and estimation for polyvinyl chloride (PVC) polymers. More details about the 1D model and its respective sensitivity analysis can be found in Borges et al. (2006).

A PVC sample with thickness of 50 mm and lateral dimensions of 305 $\times$ 305 mm is initially at temperature T_0. For times $t > 0$, the sample is submitted to a unidirectional and uniform heat flux on its upper surface. The heat is supplied by a 22 Ω electrical resistance heater, covered with silicone rubber, with lateral dimensions of 305 $\times$ 305 mm and thickness of 1.4 mm. The heat flux and temperature are acquired using sensor and instruments with the same specification of that described here in the conductor material application.

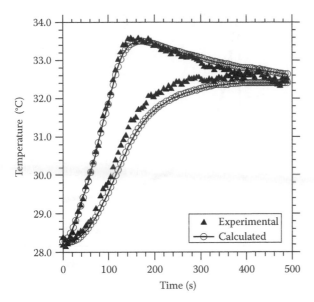

FIGURE 18.13
Comparison of an output of a typical run.

Fifty independent runs for PVC were realized and 1024 points were taken. The time intervals, t, were 7.034 s, and the time duration of heating, t_h, was approximately 150 s with a heat pulse generated at 40 V(DC).

Tables 18.4 and 18.5 present respectively the value estimated of α and k for the 50 runs of PVC, with 99.87% confidence interval. In Table 18.6, a summary of the simultaneous

TABLE 18.4

Statistical Data of the Average Value of α
(Initial Value, $\alpha = 1.0 \times 10^{-8} \ \text{m}^2/\text{s}$)

$\alpha \times 10^7 \ (\text{m}^2/\text{s})$	Initial S_φ	Final S_φ	$\sigma \times 10^{10} \ (\text{m}^2/\text{s})$
1.24 ± 1.88	1.961	0.009243	7.06

TABLE 18.5

Statistical Data of the Average Value of k
(Initial Value of $k = 0.01$ W/m K)

k (W/m K)	Initial $S_{qT} \times 10^{-6}$	Final S_{qT}	$\sigma \ (\text{m}^2/\text{s}) \times 10^5$
0.152	1.351	5.91	4.9

TABLE 18.6

Summary of α and k for PVC Sample with 99.87% Confidence Interval

$\alpha \ (\text{m}^2/\text{s})$	$\alpha \times 10^7 \ (\text{m}^2/\text{s})$ (FM)	k (W/m K)	k (W/m K) (GHP)
$1.24 \pm 1.88\%$	$1.28 \pm 3.1\%$	$0.152 \pm 1.1\%$	0.157

Sources: Parker, W. J. et al., *J. Appl. Phys.*, 32(9), 1679, 1961 (for FM, flash method); NPL, *Certificate of Calibration: Thermal Conductivity of a Pair of Polythene Specimens.* England unpublished: Technical Report No. X2321/90/021. NPL, 1991 (for GHP, guarded hot plate).

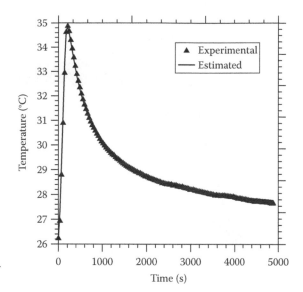

FIGURE 18.14
Temperature evolution: experimental and calculated data.

estimation of α and k of the PVC sample is presented. In this table, the comparison with the values obtained for α by using the Flash method (Parker et al., 1961) and k by using the guarded hot plate method (NPL, 1991) presented discrepancies of 3.22% and 3.30% for α and k, respectively.

The comparison between the experimental and estimated temperatures for $\alpha = 1.24 \times 10^{-7}$ m^2/s and $k = 0.152$ W/m K is shown in Figure 18.14. A good agreement between the data can be observed. It can be noted that the residuals in time domain presented in Figure 18.15 are situated in the range of uncertainty measurement of thermocouples that in this work is 0.3 K.

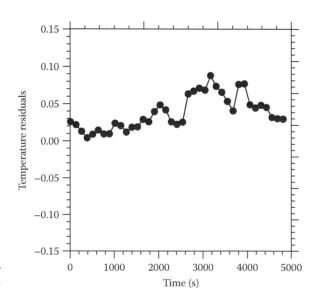

FIGURE 18.15
Temperature evolution: residuals of experimental and calculated data in time domain ($T_{e,1} - T_1$).

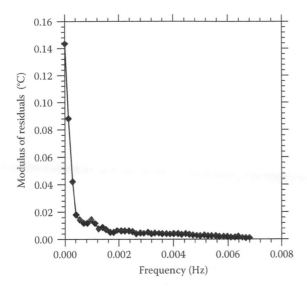

FIGURE 18.16
Temperature evolution: residuals of experimental and calculated data in frequency domain $(T_{e,1} - T_1)$.

Figure 18.16 also shows these residuals in frequency domain. It can be observed in both domains that residuals are correlated, although they are small and do not significantly affect the final estimates.

18.5 Experimental Determination of Thermal Conductivity and Diffusivity Using Partially Heated Surface Method without Heat Flux Transducer

18.5.1 Introduction

This section presents a new experimental technique to obtain the thermal conductivity of conductor and nonconductor materials of small dimensions. As usual, the thermal conductivity estimation involves a thermal model with a known heat flux input. However, here the main contribution of this study is the use of inverse techniques to estimate the heat flux input instead of measuring with heat transducers. It can be observed that the presence of transducers represents an additional experimental limitation for small samples. Besides the experimental difficulties, the smaller the transducer dimensions the more difficult it is to obtain the calibration curves due to the low sensitivity.

The procedure proposed here is based on the following steps: (i) development of experimental apparatus and thermal model considering a heat flux input in part of the sample surface while the remaining surfaces are kept isolated; (ii) estimation of a dimensionless heat flux, $q^+(t)$, proportional to the heat flux input using inverse techniques; (iii) estimation of thermal diffusivity; (iv) comparison between this heat flux, $q^+(t)$, and the total heat flux supplied by the heating element P/S_1 to estimate the thermal conductivity of the sample.

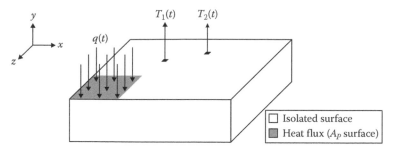

FIGURE 18.17
3D thermal equivalent model.

18.5.2 Direct Thermal Model

The thermal model is the same presented in Figure 18.17.

As shown before, the 3D thermal model can be obtained by the solution of the diffusion equation (18.4).

However, in this new procedure, the following dimensionless quantities are defined as

$$u = \frac{x}{L}; \quad v = \frac{y}{W}; \quad w = \frac{z}{R} \tag{18.16a}$$

$$\mu = \frac{\alpha_{ref}\, t}{W^2}; \quad \Theta(u,v,w,\mu) = \frac{T(x,y,z,t) - T_0}{\dfrac{q_{ref}L}{k_{ref}}}, \quad q^+ = +\frac{q(t)}{k}\frac{k_{ref}}{q_{ref}} \tag{18.16b}$$

where α_{ref}, k_{ref}, and q_{ref} are references quantities for thermal diffusivity, thermal conductivity, and heat flux input, respectively.

It can be observed that these quantities can assume any value, as, for example, unity without lost of generality.

Applying the dimensionless definitions in Equations 18.4a through 18.4e, the dimensionless thermal model can be obtained as

$$\frac{\partial^2 \Theta}{\partial u} + \frac{\partial^2 \Theta}{\partial v} + \frac{\partial^2 \Theta}{\partial w} = \frac{\partial \Theta}{\partial \mu} \tag{18.17a}$$

In region $\Re$, subject to the boundary conditions,

$$-\frac{\partial \Theta}{\partial v}\bigg|_{v=W} = q^+(t) \text{ in } A_P \text{ surface: } \frac{L_1}{L} \le u \le \frac{L_2}{L} \quad \text{and} \quad \frac{R_1}{R} \le w \le \frac{R_2}{R} \tag{18.17b}$$

$$-\frac{\partial \Theta}{\partial v}\bigg|_{v=w} = 0 \quad \text{in} \quad A - A_P \tag{18.17c}$$

$$\frac{\partial \Theta}{\partial u}\bigg|_{u=0} = \frac{\partial \Theta}{\partial u}\bigg|_{u=L} = \frac{\partial \Theta}{\partial v}\bigg|_{v=0} = \frac{\partial \Theta}{\partial w}\bigg|_{w=0} = \frac{\partial \Theta}{\partial w}\bigg|_{w=R} = 0 \tag{18.17d}$$

and the initial condition,

$$\Theta(u, v, w, 0) = 0 \tag{18.17e}$$

where A is defined by $(0 \leq u \leq 1, 0 \leq w \leq 1)$ and A_P is region where the heat flux is applied. Since $q^+(t)$ is unknown, the inverse problem is characterized.

18.5.3 Inverse Problem

Different inverse problem techniques can be used for estimation of the dimensionless heat flux q^+. This work will use the dynamic observer technique based on Green's function, described by Sousa (2006), that will be briefly described as follows.

The thermal model described by Equation 18.17 can be represented by a dynamic system given by a block diagram (Blum and Marquardt, 1997) shown in Figure 18.18.

It can be observed from the block diagram that

1. The unknown dimensionless heat flux $q^+(t)$ is applied to the conductor (reference model), G_H, and results in a measurement signal $v_{e,M}$ corrupted by noise N:

$$\Theta_{e,M} = \Theta_M + N = G_H \times q_M^+ + N \tag{18.18}$$

2. Any solution algorithm determines the estimated dimensionless heat flux, q^+, such that the estimated measured dimensionless temperatures predicted by a reference model (which here is assumed to be known, so that $\hat{G}_H = G_H$) match the real measured dimensionless temperature $\widehat{\Theta_{e,M}}$ (Equation 18.17). Reference model obtaining is described in the next section.

3. The estimate value $\hat{q}^+$ is computed from the output data, $\Theta_{e,M}$. Thus, the estimator can be represented in a closed-loop transfer function of the feedback loop as shown in Figure 18.18 (Sousa, 2006) by

$$\hat{q}^+(s) = \frac{G_C}{1 + G_C G_H} \Theta_{e,M} \tag{18.19}$$

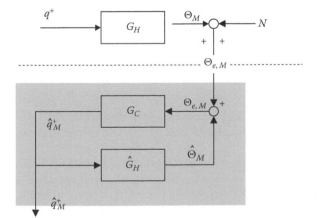

FIGURE 18.18
Frequency–domain block diagram. (From Blum, J. W. and Marquardt, W., *Numer. Heat Transf. B, Fundam.*, 32(4), 453, 1997.)

or

$$\hat{q}^+(s) = G_Q(s)q^+ + G_N(s)N \qquad (18.20)$$

where symbol ($\wedge$) denotes estimates values and G_Q and G_N are, respectively, given by

$$G_Q(s) = \frac{G_C G_H}{1 + G_C G_H} \qquad (18.21)$$

and

$$G_N(s) = G_Q G_H^{-1} \qquad (18.22)$$

$$G_Q(s) = \frac{k_{cheb}}{(s - s_{cheb,1})(s - s_{cheb,2}) \cdots (s - s_{cheb,n_Q})} \qquad (18.23)$$

In Equation 18.20, G_Q is referred to as the signal transfer function and G_N is referred to as the noise transfer function. The variable Θ is the true value of dimensionless heat flux in Laplace domain and N is the random noise due to the temperature measurements.

The transfer function G_Q is chosen to have the behavior of type I *Chebychev* filter, and its frequency response magnitude assumes the form

$$G_Q(s) = \frac{k_{cheb}}{(s - s_{cheb,1})(s - s_{cheb,2}) \cdots (s - s_{cheb,n_Q})} \qquad (18.24)$$

The poles $s_{cheb,i}$ are computed using MATLAB® software package. As mentioned, more details of the observer procedure can be found in Blum and Marquardt (1997) and Sousa (2006).

The optimization procedure can be resumed in the use of the two discrete-time difference equations

$$\hat{q}^+(k) = \sum_{i=0}^{n_n} b_i \Theta_M(k - i) - \sum_{i=0}^{n_n} a_i \hat{q}^+(k - i) \qquad (18.25)$$

and

$$\hat{q}^+(k) = \sum_{i=0}^{n_n} b_i q^+(k - i) - \sum_{i=0}^{n_n} a_i \hat{q}^+(k - i) \qquad (18.26)$$

It is important to observe that in Equations 18.25 and 18.26, $\Theta_{e,M}$ are data related to the measured temperature, and Θ_M is related to the calculated data using the dimensionless definition given by Equation 18.17.

Yet in Equations 18.25 and 18.26, a_i and b_i are coefficients obtained using Equations 18.22 and 18.23. Therefore, in order to complete the inverse procedure, the thermal model G_H must be identified. This identification is carried out here using Green's functions. (Details can be found in Sousa (2006) or Fernandes (2009).)

18.5.4 Thermal Diffusivity Determination

18.5.4.1 Dynamic System and Equivalent Thermal Model

In order to obtain the thermal diffusivity, it is convenient to define a dimensional dynamic model with an input, $X(t)$, and output, $Y(t)$, defined, respectively, as $X(t) = q^+(t)$ and $Y(t) = T_1(t) - T_2(t)$.

In this case, the same procedure described previously can be used to obtain the equivalent thermal model as

$$T_1(f) - T_2(f) = H_{12}(f)q^+(f) \tag{18.27}$$

where the variable f indicates that Fourier transform was applied to the variables T, H_{12}, and Θ. As described previously, H_{12} represents the heat transfer function to the dynamical model and can be also related to the Green's function.

At this point, an observation must be made. The input of the system given by $X(t) = q(t)$ is not yet totally identified. In fact, the inverse procedure has identified only $q^+(t)$, which is proportional to this heat flux.

It means that one can write

$$q(t) = \frac{q^+(t)}{k} \frac{k_{ref}}{q_{ref}} = Kq^+(t) \tag{18.28}$$

where K represents the proportional factor between the quantity $q^+(t)$ and the heat flux $q(t)$.

It means that

$$H^*(f) = KH(f) = \frac{T_1(f) - T_2(f)}{q(f)} \tag{18.29}$$

It can be observed that although the factor K affects directly the modulus of H, no effect can be verified in the phase factor. In this case, it can be demonstrated that the phase factor of H (f) and that of $H^*(f)$ are identical.

This fact indicates that the amplitude has a low sensitivity in relation to the amplitudes of the signals X and Y, which in turn lead to a low sensitivity to deterministic errors such as calibration curve uncertainties of the temperature sensors.

Since the functions $H(f)$ and $H^*(f)$ are complex numbers and provided that Equation 18.27 is valid is easily demonstrated that both functions have the same phase factor as follows:

$$\varphi(f) = \arctan\left[\frac{\Im H(f)}{\Re H(f)}\right] = \arctan\left[\frac{b}{a}\right] = \arctan\left[\frac{\Im H^*(f)}{\Re H^*(f)}\right] = \arctan\left[\frac{Kb}{Ka}\right] = \varphi^*(f) \tag{18.30}$$

18.5.5 Simultaneous Determination of the Heat Flux and the Thermal Conductivity

Once thermal diffusivity and the dimensionless heat flux, $q^+(t)$, have been determinated, it remains to determinate the heat flux, $q(t)$, and the thermal conductivity of the sample.

As mentioned, the basic idea is very simple. It consists of applying a heat flux generated by a heater glued to a surface of the sample. In order that all the heat generated be absorbed by the sample, it is necessary for the surrounding to be a vacuum. Thus, the total heat supplied to the sample can be obtained using the voltage and the current of the heater.

The only inconvenience of this process is due to the thermal capacity of the heater, such that not all the electrical power supplied to the heater is immediately absorbed by the surface of the sample.

Observing the history of the heat flux generated, $P(t)$, at the heater and the history of the admissive heat flux, $q(t)$, at the sample surface as shown, it is possible to conclude that the heat flux will be totally absorbed by the sample only after a certain time interval t_f. Hence, applying the principle of conservation of energy, we have

$$\int_0^{t_f} q(t)dt = \int_0^{t_f} \frac{V(t)I(t)}{A_P} dt \tag{18.31}$$

where $V(t)$ and $I(t)$ represent the voltage and the current supplied, respectively.

But using Equation 18.28 in Equation 18.31 allows the thermal conductivity k to be obtained as

$$k = \frac{\int_0^{t_f} \dfrac{V(t)I(t)}{A_P}}{\left[\dfrac{q_{ref}}{k_{ref}} \int_0^{t_f} q^+(t)dt\right] dt} \tag{18.32}$$

Once k has been determinated, Equation 18.28 can be used to determine $q(t)$.

18.5.6 Experimental Apparatus

As observed before, the boundary conditions of the thermal problem (Equation 18.4 or 18.17) must be guaranteed. It means that the isolated condition (in all the surfaces where the sample is not in contact with the heater) must be obtained for the success of the technique. An efficient form of obtaining experimental isolation is to expose these surfaces to an evacuated atmosphere.

The sample/heater assembly is inserted inside the vacuum chamber as shown schematically in Figure 18.19. Figure 18.20 shows the vacuum chamber.

This test investigates the thermal properties of an AISI304 stainless steel sample with thickness of 10 mm and lateral dimensions of 139×65 mm. The sample initially in thermal

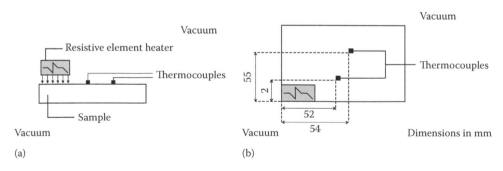

(a)   (b)

FIGURE 18.19
Scheme of the sample/heater assembly: (a) frontal view and (b) superior view.

FIGURE 18.20
Experimental apparatus.

equilibrium at T_0 is then submitted to a unidirectional and uniform heat flux. The heat is supplied by a 318 Ω electrical resistance heater, covered with silicone rubber, with lateral dimensions of 50×50 mm and thickness of 0.3 mm. The temperatures are measured using surface thermocouples (type K).

Two thermocouples, type K, are attached to the frontal surface of the test plate (AISI304) (Figure 18.19) using capacity discharge. Collection and storage of the data from the thermocouples use a microcomputer-based data acquisition system (HP 75000 B E1326B), abbreviated to DAS. The DAS, with a control software, sampled (multiplexed) each thermocouple at intervals of 0.27 s (totaling 2048 points for each thermocouple). The DAS was also used to acquire the voltage and current signals of the heater.

Twenty-one independent runs were realized. The time duration of heating, t_h, was approximately 120 s with a heat pulse generated at 60 V (DC).

Figure 18.21 shows a typical signal of the power generated by the resistive element heater on the conductive sample.

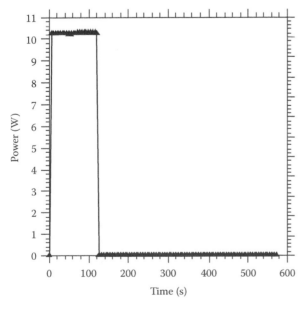

FIGURE 18.21
Typical signal power generated by an on/off resistive heater element.

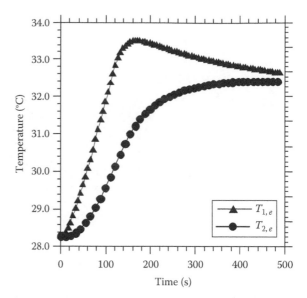

FIGURE 18.22
Temperature evolution for AISI304 sample.

Typical signal values of temperature are shown in Figure 18.22.

Figures 18.22 and 18.23 show the histories of the temperatures measured at the front surface of the AISI304 sample, $T_{e,1}$ and $T_{e,2}$ and the output $Y(t)$. Figure 18.24 shows the dimensionless heat flux $q^+(t)$ (input $X(t)$), estimated using the observer technique.

Using the input signals, $X(t)$, and the output signals, $Y(t)$, the spectral densities S_{xx}, S_{xy} were obtained, and consequently the respond function $H(f)$ can be identified. Figure 18.25 shows the behavior of the phase factor of $H(f)$ in relation to the frequency.

As already described, the thermal conductivity of the sample can be obtained using Equation 18.32.

Figure 18.26 and Table 18.7 show, respectively, the estimated and statistical values of k for AISI304.

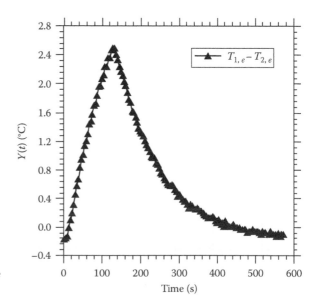

FIGURE 18.23
Output $Y(t)$ given by difference of temperature evolution for AISI304 sample.

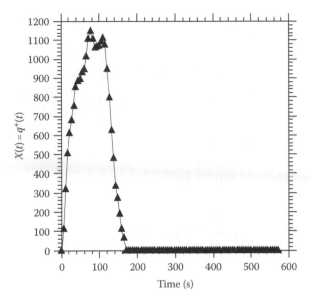

FIGURE 18.24
Estimated dimensionless heat flux, $q^+(t)$, $\alpha_{ref} = 10^{-6}$ m^2/s, using $k = 1.0$ W/m K.

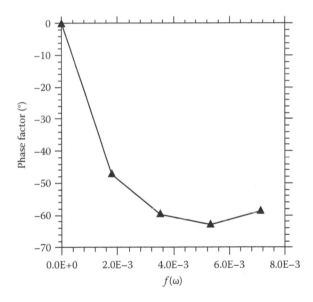

FIGURE 18.25
Relation between phase factor and frequency.

The statistical averages were calculated with 97.5% confidence interval, and the results were compared with literature data (Incropera and DeWitt, 1996).

The value of thermal conductivity obtained then permits the determination of the heat flux, $q(t)$, as shown in Figure 18.27. Thermal diffusivity estimations are shown in Figure 18.28 and statistical data in Table 18.8.

Figure 18.29 shows the experimental temperatures and the estimated temperatures (using $\alpha = 3.77 \pm 0.029$ m^2/s and $k = 14.99 \pm 0.17$ W/m K). Figures 18.30 and 18.31 show the residuals for thermocouple 1 in time and frequency domain, respectively. Again, a small correlation in the data that do not significantly affect the final estimates can be observed.

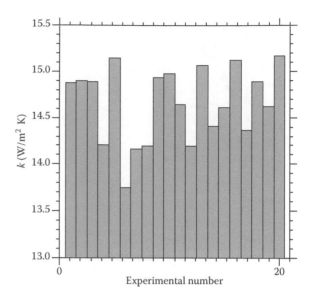

FIGURE 18.26
Estimated values of k for AISI304.

TABLE 18.7

Statistical Data of the Average Value Estimated of k,
21 Experiments, Confidence Interval of 97.5%

k (W/m K) (This Work)	k (W/m K)	σ (W/m K)
14.99 ± 0.17	14.90	0.407

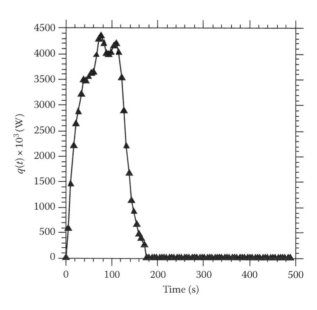

FIGURE 18.27
Estimated heat flux, $q(t)$.

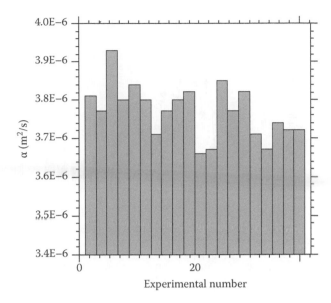

FIGURE 18.28
Estimated values of α for AISI304.

TABLE 18.8

Statistical Data of the Average Value of α, 21 Experiments,
Confidence Interval of 97.5%

$\alpha \times 10^6$ (m^2/s) (This Work)	$\alpha \times 10^6$ (m^2/s)	$\sigma \times 10^6$ (m^2/s)
3.77 ± 0.029	3.82	0.0689

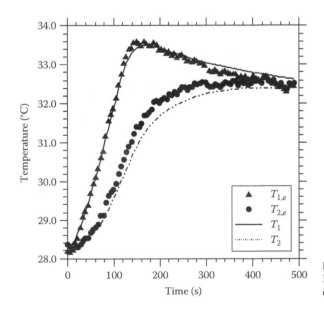

FIGURE 18.29
Experimental and estimated temperature evolution.

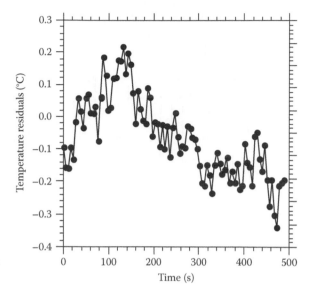

FIGURE 18.30
Residuals of estimated temperature and experimental temperature $(T_{e,1} - T_1)$.

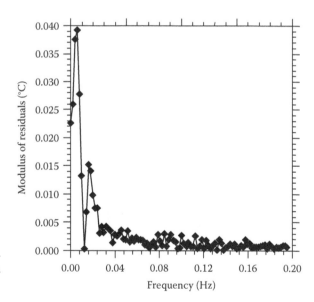

FIGURE 18.31
Temperature evolution: residuals of experimental and calculated data in frequency domain $(T_{e,1} - T_1)$.

18.5.7 Insulation Hypothesis Verification

18.5.7.1 Introduction

An important characteristic of this work is the estimation of thermal properties using a small thermal gradient throughout a sample or between a sample and its neighboring. In this sense, as previously pointed out in this chapter, the technique presented is able to obtain thermal properties from a temperature difference below 3 K.

An experimental apparatus must have enough sensitivity and reproduce the thermal conditions assumed in the thermal model. Therefore, the boundary conditions present in the theoretical model must be guaranteed in the experimental apparatus. This means that the isolated condition at the reminiscent surface needs to be reached for the success of the estimation technique. The aim of this section is to demonstrate the low influence of the losses by convection heat transfer and thermal radiation and, consequently, the validity of assumptions given by Equation 18.4.

18.5.7.2 Experimental Apparatus Design to Estimate the Heat Loss inside the Vacuum Chamber

Figure 18.32 shows schematically the experimental apparatus designed to identify the convection and radiation heat transfer loss from the sample into the chamber. An AISI304 sample with thickness of 10 mm and lateral dimensions of 139×65 mm initially in thermal equilibrium at T_0 is then submitted to a unidirectional and uniform heat flux. A total heat rate, P, is supplied by a 318 Ω electrical resistance heater, covered with silicone rubber, with lateral dimensions of 50×50 mm and thickness of 0.3 mm. The magnitude of P can be obtained just by multiplying the voltage difference versus current value. See Figure 18.21 for details of a typical $P(t)$ evolution.

In order to obtain the heat flux effectively delivered to the sample, $q(t)$, a transducer with lateral dimensions of 50×50 mm, thickness of 0.5 mm, and constant time less than 10 ms is inserted between the sample and the electrical resistance. The transducer is based on the thermopile conception of multiple thermoelectric junctions (made by electrolytic deposition) on a thin conductor sheet (Guimarães et al., 1995). The temperatures are measured using surface thermocouples (type K). The signals of heat flux and temperatures are acquired by a data acquisition system HP Series 75000 with voltmeter E1326B controlled by a personal computer.

The heat transfer loss into the vacuum chamber by convection and/or radiation can then be calculated by the difference between P/S_1 and $q(t)$.

$$\frac{P}{A_P} = q(t) + q_{loss} \tag{18.33}$$

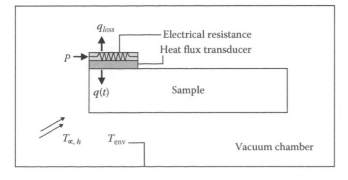

FIGURE 18.32
Scheme of the experimental apparatus designed to identify the convection and radiation heat transfer loss from the sample into the chamber.

As mentioned, due to the existence of thermal inertial of both the electrical resistance and the heat flux transducer, the heat flux will be absorbed by the sample only after a certain time interval t_f. Therefore, the energy loss by convection/radiation will be computed by integrating Equation 18.33 in time interval t_f, that is,

$$\int_0^{t_f} P(t)dt = \int_0^{t_f} V_1(t) \times I(t)dt = \int_0^{t_f} q(t) \times A_P(t)dt + \int_0^{t_f} q_{loss} \times A_P(t)dt \qquad (18.34)$$

or

$$\int_0^{t_f} q_{loss} \times A_P(t)dt = \int_0^{t_f} V_1(t) \times I(t)dt - \int_0^{t_f} q(t) \times A_P(t)dt \qquad (18.35)$$

In this experiment, the total heat rate dissipated is

$$\int_0^{t_f} P(t)dt = 193.5J \qquad (18.36)$$

and the energy equivalent to the heat flux effectively delivered to the sample is calculated as

$$\int_0^{t_f} q(t) \times A_p(t)dt = 190.3J \qquad (18.37)$$

Therefore, the values of the energy loss during all the experiment are situated below 2% of the total energy supplied by the electrical resistance. This means that although heat loss occurs, the insulated thermal boundary is a very good experimental hypothesis and can be used with a minimum influence in the estimation results.

In order to validate the inverse estimation procedure, a comparison between the measured heat flux and the estimated heat flux is also carried out.

Figure 18.33 presents the heat flux measured by the transducer and the heat flux estimated. It must be mentioned that this estimated heat flux is obtained only after estimations of $q(t)$, k, and α. Figure 18.34 shows the residuals between the respective values.

A very good agreement between the data can be observed. It also can be noted that the residuals, presented in Figure 18.34, have values below the uncertainty obtained in the heat flux transducer calibration that is 3 (W/m^2) for the sensor used.

The results observed in Figures 18.33 and 18.34 validate the technique presented here, since they involve both the estimated heat flux $q(t)$ using an inverse technique based on dynamic observer method and the procedure for the determination of the thermal properties.

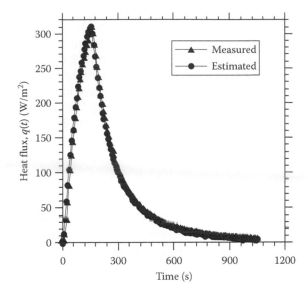

FIGURE 18.33
Comparison between the measured and estimated heat flux.

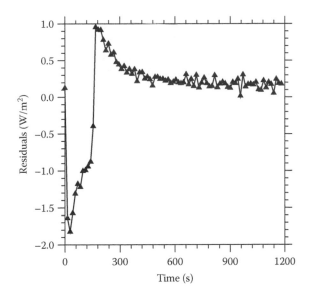

FIGURE 18.34
Measured and estimated heat flux residuals.

18.6 Use of 3D-Transient Analytical Solution Based on Green's Function to Reduce Computational Time in Inverse Heat Conduction Problems

The procedure to obtain the thermal properties involves two optimization problems (inverse problem), which obtain $q^+(t)$ and the respective properties. Both estimations involve the evaluation of temperatures calculated using the thermal model, $T_1(t)$ and $T_2(t)$, which represent the solution of the direct problem. From the viewpoint of application of the experimental technique proposed by Borges et al. (2006), the choice of the method of solution of the direct problem is open. In other words, whichever numerical method such as finite volume or finite elements or if possible analytical solutions can be used. As mentioned, this study proposes the incorporation of analytical solutions in the experimental technique of determination of thermal conductivity and thermal diffusivity using the method of a partially heated surface without heat flux transducer proposed by Borges et al. (2006) reducing the computational cost and increasing the precision of the numerical calculations. In the following, the thermal model represented by Figure 18.2 and the obtaining of its analytical solution using Green's function (Fernandes, 2009) are presented.

18.6.1 Exact Solution of the Thermal Model Using Green's Function

The thermal model proposed to be reproduced experimentally is given by a sample initially at a uniform temperature T_0. The sample is then subjected to a heat flux (W/m^2) while all other surfaces are maintained isolated. This problem is described in Section 18.4.1.

As discussed before, Equation 18.17 represents a direct problem of heat conduction if the dimensionless heat flux $q^+(t)$ is specified. On the other hand, the inverse problem is established when $q^+(t)$ is unknown. The various solution techniques of inverse problems (or optimization problems) have as procedure the evaluation repetitive of direct problems, always used with estimated or calculated values of heat flux density in an iterative process. Thus, the proposed solution of the direct problem given by Equation 18.17 will be presented considering a known transient heat flux. In order to simplify the results, the direct problem solution is presented in its dimensional form considering definitions in Equation 18.16.

The solution of Equation 18.17 can be given in terms of Green's function as in Beck et al. (1992).

$$T(x,y,z,t) = \int_0^L \int_0^W \int_0^R G(x,y,z,t|x',y',z',0)T(x',y',z',0)dx'dy'dz'$$

$$+ \frac{\alpha}{k} \int_0^t \int_{L_1}^{L_2} \int_{R_1}^{R_2} q(\tau)G(x,y,z,t|x',W,z',\tau)dx'dz'd\tau \tag{18.38}$$

where $G(x, y, z, t|x', y', z', \tau) = G_{X22}.G_{Y22}.G_{Z22}$.

$$G_{X22} = \frac{1}{L}\left[1 + 2\sum_{m-1}^{\infty} e^{-\frac{m^2}{L^2}\pi^2\alpha(t-\tau)} \cos\left(\frac{m\pi x}{L}\right)\cos\left(\frac{m\pi x'}{L}\right)\right] \tag{18.39}$$

$$G_{Y22} = \frac{1}{W}\left[1 + 2\sum_{n=1}^{\infty} e^{-\frac{n^2}{W^2}\pi^2\alpha(t-\tau)} \cos\left(\frac{n\pi y}{W}\right)\cos\left(\frac{n\pi y'}{W}\right)\right] \tag{18.40}$$

$$G_{Z22} = \frac{1}{R}\left[1 + 2\sum_{p=1}^{\infty} e^{-\frac{p^2}{R^2}\pi^2\alpha(t-\tau)} \cos\left(\frac{p\pi z}{R}\right)\cos\left(\frac{m\pi z'}{R}\right)\right] \tag{18.41}$$

Therefore, the temperature can be obtained as (Fernandes, 2009)

$$T(x, y, z, t) = T_0 + \frac{\alpha}{k} \frac{1}{LWR} [L_2 - L_1][R_2 - R_1] \int_0^t q(\tau) d\tau + \frac{\alpha}{k} \frac{2}{WR} [R_2 - R_1]$$

$$\times \sum_{m=1}^{\infty} e^{-\frac{m^2}{L^2} \pi^2 \alpha t} \cos\left(\frac{m\pi x}{L}\right) \left[\sin\left(\frac{m\pi L_2}{L}\right) - \sin\left(\frac{m\pi L_1}{L}\right)\right]$$

$$\times \frac{1}{m\pi} \int_0^t \left[q(\tau) e^{\frac{m^2}{L^2} \pi^2 \alpha \tau}\right] d\tau + \frac{\alpha}{k} \frac{2}{LWR} [L_2 - L_1][R_2 - R_1]$$

$$\times \sum_{n=1}^{\infty} e^{-\frac{n^2}{W^2} \pi^2 \alpha t} \cos\left(\frac{n\pi y}{W}\right) \cos(n\pi) \int_0^t \left[q(\tau) e^{\frac{n^2}{W^2} \pi^2 \alpha \tau}\right] d\tau + \frac{\alpha}{k} \frac{2}{LW} [L_2 - L_1]$$

$$\times \sum_{p=1}^{\infty} e^{-\frac{p^2}{R^2} \pi^2 \alpha t} \cos\left(\frac{p\pi z}{R}\right) \left[\sin\left(\frac{p\pi R_2}{R}\right) - \sin\left(\frac{p\pi R_1}{R}\right)\right] \frac{1}{p\pi}$$

$$\times \int_0^t \left[q(\tau) e^{\frac{p^2}{R^2} \pi^2 \alpha \tau}\right] d\tau + \frac{\alpha}{k} \frac{4}{WR} [R_2 - R_1] \sum_{m=1}^{\infty} \sum_{n=1}^{\infty} e^{-\left(\frac{m^2}{L^2} + \frac{n^2}{W^2}\right) \pi^2 \alpha t} \cos\left(\frac{m\pi x}{L}\right)$$

$$\times \left[\sin\left(\frac{m\pi L_2}{L}\right) - \sin\left(\frac{m\pi L_1}{L}\right)\right] \frac{1}{m\pi} \cos\left(\frac{n\pi y}{W}\right) \cos(n\pi)$$

$$\times \int_0^t \left[q(\tau) e^{\left(\frac{m^2}{L^2} + \frac{n^2}{W^2}\right) \pi^2 \alpha \tau}\right] d\tau + \frac{\alpha}{k} \frac{4}{W} \sum_{m=1}^{\infty} \sum_{p=1}^{\infty} e^{-\left(\frac{m^2}{L^2} + \frac{p^2}{R^2}\right) \pi^2 \alpha t} \cos\left(\frac{m\pi x}{L}\right)$$

$$\times \left[\sin\left(\frac{m\pi L_2}{L}\right) - \sin\left(\frac{m\pi L_1}{L}\right)\right] \frac{1}{m\pi} \cos\left(\frac{p\pi z}{R}\right) \left[\sin\left(\frac{p\pi R_2}{R}\right) - \sin\left(\frac{p\pi R_1}{R}\right)\right]$$

$$\times \frac{1}{p\pi} \int_0^t \left[q(\tau) e^{\left(\frac{m^2}{L^2} + \frac{p^2}{R^2}\right) \pi^2 \alpha \tau}\right] d\tau + \frac{\alpha}{k} \frac{4}{LW} [L_2 - L_1] \sum_{n=1}^{\infty} \sum_{p=1}^{\infty} e^{-\left(\frac{n^2}{W^2} + \frac{p^2}{R^2}\right) \pi^2 \alpha t}$$

$$\times \cos\left(\frac{n\pi y}{W}\right) \cos(n\pi) \cos\left(\frac{p\pi z}{R}\right) \left[\sin\left(\frac{p\pi R_2}{R}\right) - \sin\left(\frac{p\pi R_1}{R}\right)\right]$$

$$\times \frac{1}{p\pi} \int_0^t \left[q(\tau) e^{\left(\frac{n^2}{W^2} + \frac{p^2}{R^2}\right) \pi^2 \alpha \tau}\right] d\tau + \frac{\alpha}{k} \frac{8}{W} \sum_{m=1}^{\infty} \sum_{n=1}^{\infty} \sum_{p=1}^{\infty} e^{-\left(\frac{m^2}{L^2} + \frac{n^2}{W^2} + \frac{p^2}{R^2}\right) \pi^2 \alpha t} \cos\left(\frac{m\pi x}{L}\right)$$

$$\times \left[\sin\left(\frac{m\pi L_2}{L}\right) - \sin\left(\frac{m\pi L_1}{L}\right)\right] \frac{1}{m\pi} \cos\left(\frac{n\pi y}{W}\right) \cos(n\pi) \cos\left(\frac{p\pi z}{R}\right)$$

$$\times \left[\sin\left(\frac{p\pi R_2}{R}\right) - \sin\left(\frac{p\pi R_1}{R}\right)\right] \frac{1}{p\pi} \int_0^t \left[q(\tau) e^{\left(\frac{m^2}{L^2} + \frac{n^2}{W^2} + \frac{p^2}{R^2}\right) \pi^2 \alpha \tau}\right] d\tau$$

$$(18.42)$$

where m, n, and p are the number of terms required for the convergence of series.

This equation has been applied in other simpler problem to compare with literature data. These comparisons that can be found in detail in Fernandes (2009) assure the precision of the analytical solution presented here.

18.6.2 DPT Code Using Analytical Solution

In the following will be presented results for the use of analytical solution in the thermal diffusivity and conductivity estimation by using DPT code (Thermal Properties Determination). The comparison with the numerical solution is made using the 3D transient thermal model shown in Figure 18.17. In this case, only part of the superior surface is heated. The cast iron block has dimensions of $L = 0.0805$ m, $W = 0.008$ m, and $Z = 0.06$ m; and the heated region, S_1, is defined by $0 < x < 0.005$ m and $0 < z < 0.005$ m.

As mentioned, two thermocouples are used to measure the temperatures. The thermocouples T_1 and T_2 are located at (0.0378; 0.008; 0.003) and (0.0345; 0.008; 0.0438), respectively. The results obtained numerically and analytically are compared in Figures 18.35 and 18.36.

Figure 18.35a shows the heat fluxes estimated using the analytical and numerical solutions. The good agreement between them is clear. The maximum deviation between the estimated heat fluxes is 3% (Figure 18.35b). Figure 18.36 shows the temperature evolution at locations T_1 and T_2 using the respective estimated heat flux shown in Figure 18.35a.

The maximum deviation observed of 0°C, 12°C (0%, 4%) includes not only numerical dispersion but also the influence of each heat flux on the solution (Figure 18.36b).

In this case, the main contribution is in the computational time as shown in Table 18.9. The difference between the two is of order of 7500%. The analytical solution took 5 min and the numerical solution took 6.4 h. This time corresponds to all procedures to obtain the thermal properties, including the optimization by iteration.

The deviation in temperatures and heat fluxes was responsible for deviation in estimated values of thermal conductivity and diffusivity of 1.7% and 2.6%, respectively.

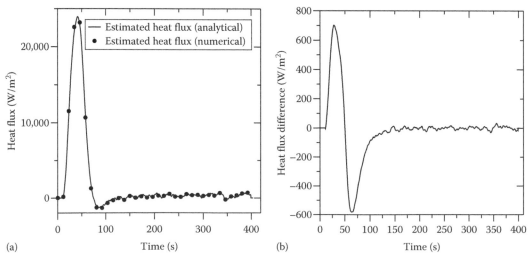

FIGURE 18.35
Heat flux for a cast iron sample: (a) estimated heat flux and (b) residuals.

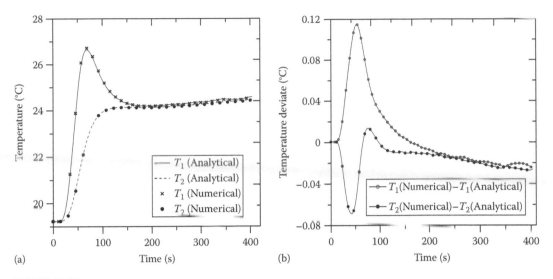

FIGURE 18.36
Estimated temperatures (T_1 and T_2) for a cast iron sample: (a) temperatures and (b) residuals.

TABLE 18.9

Comparison between Estimated Values of Thermal Diffusivity and Conductivity and the Computational Time Using DPT with Numerical and Analytical Solutions

Solution Type	α (m^2/s) $\times 10^5$	k (W/m K)	Computational Time
Analytical	1.13 ± 0.023	42.59 ± 0.51	305.00 (0.085 h)
Numerical	1.10 ± 0.017	43.32 ± 0.39	23,040.00 (6.4 h)

18.7 Conclusions

The experimental techniques proposed were shown to be efficient for the thermal properties identification of AISI304 sample. The procedure gives a great flexibility to the technique, allowing the technique to deal with sample of small dimensions and also to be applied to conductor or nonconductor materials.

The great advantage of the dynamic observers technique is the easy and fast numerical implementation for any 1D, 2D, or 3D model. The robustness and low computational cost and low error sensitivity give this procedure a great potential in inverse techniques application.

This study also shows the great contributions of the use of analytical solutions in inverse problems, once the optimization algorithms usually have to calculate the direct problem several times. The use of analytical solutions not only increases the precision but also reduces "drastically" the computational time.

Acknowledgments

The authors would like to thank CAPES, Fapemig, CNPq, CNPq/PROSUL, French and Brazilian governments.

Nomenclature

A_p	area of heat supply, m^2
f	frequency, Hz
G	Green's function, m^2 K/W
G_H	heat conductor transfer function, m^2 K/W
G_N	noise transfer function, m^2 K/W
G_Q	signal transfer function, m^2 K/W
h	heat transfer coefficient, W/m^2 K
$H(f)$	frequency response function, m^2 K/W
$Im(S_{xy})$	imaginary component of the cross-spectral density function, K^2
k	thermal conductivity, W/m K
L, R, W	plate dimensions, m
S_H	objective function based on difference between estimated and measured absolute value of H
S_T	objective function based on difference between estimated and measured phase temperature, K^2
S_φ	objective function based on difference between estimated and measured phase factor values, rad^2
t	time, s
T_0	initial temperature, °C
q	heat flux, W/m^2
q^+	dimensionless heat flux
x, y, z	Cartesian coordinates, m
$X(f)$	input signal in frequency domain, °C
$X(t)$	input signal in time domain, °C
$Y(f)$	output signal in frequency domain, °C
$Y(t)$	output signal in time domain, °C

Greek Symbols

α	thermal diffusivity, m^2/s
θ	temperature difference, °C
μ, ν, ω	dimensionless Cartesian coordinates
ρ	density, kg/m^3
ϕ	heat flux, W/m^2
φ	phase angle, rad
Θ	dimensionless temperature difference, °C

Subscripts

1	relative to the thermocouple 1
2	relative to the thermocouple 2
E	relative to experimental data
m	relative to integer variables
M	relative to discrete time
T	relative to calculated data
ref	relative to reference values
s_{cheb}	poles of Chebyshev filter

References

Beck, J.V. and K.J. Arnold. 1977. *Parameter Estimation in Engineering and Science*. New York, NY: Wiley.

Beck, J.V., K.D. Cole, A. Haji-Sheik, and B. Litkouhi. 1992. *Heat Conduction Using Green's Function*. Washington, DC: Hemisphere Publishing.

Bendat, J.S. and A.G. Piersol. 1986. *Analysis and Measurement Procedures*. New York, NY: Wiley-Intersience.

Blum, J.W. and W. Marquardt. 1997. An optimal solution to inverse heat conduction problems based on frequency–domain interpretation and observers. *Numerical Heat Transfer Part B, Fundamentals* 32(4):453–478.

Borges, V.L., S.M.M. Lima e Silva, and G. Guimarães. 2006. A dynamic thermal identification method applied to conductor and nonconductor materials. *Inverse Problems in Science and Engineering* 14(5):511–527.

Dowding, K.J., J.V. Beck, and B.F. Blackwell. 1996. Estimation of directional-dependent thermal properties in a carbon–carbon composite. *International Journal of Heat and Mass Transfer* 39 (15):3157–3164.

Fernandes, A.P. 2009. Green's function: IHCP analytical solution applications (in Portuguese). Master's thesis, Universidade Federal de Uberlandia, Uberlandia, Brazil.

Guimarães, G., P.C. Philippi, and P. Thery. 1995. Use of parameters estimation method in the frequency domain for the simultaneous estimation of thermal diffusivity and conductivity. *Review of Scientific Instruments* 66(3):2582–2588.

Huang, C.H. and J.Y. Yang. 1995. An inverse problem in simultaneously measuring temperature dependent thermal conductivity and heat capacity. *International Journal of Heat and Mass Transfer* 38(18):3433–3441.

Incropera, F.P. and D.P. DeWitt. 1996. *Fundamentals of Heat and Mass Transfer*. New York, NY: John Wiley & Sons Inc.

Lima e Silva, S.M.M., T.H. Ong, and G. Guimarães. 2003. Thermal properties estimation of polymers using only one active surface. *Journal of the Brazilian Society of Mechanical Sciences and Engineering* 25(1):9–14.

Nicolau, V.P., S. Güths, and M.G. Silva. 2002. Medição da condutividade térmica e do calor específico de materiais isolantes. In *Congresso Brasileiro de Engenharia e Ciências Térmicas*, ed. 9, vol. CIT02-0529. Caxambu-MG: ABCM—Associação Brasileira de Ciências Mecânicas.

NPL. 1991. *Certificate of Calibration: Thermal Conductivity of a Pair of Polythene Specimens*. England unpublished: Technical Report No. X2321/90/021.

Parker, W.J., R.J. Jenkins, C.P. Butler, and G.L. Abbott. 1961. Flash method of determining thermal diffusivity, heat capacity and thermal conductivity. *Journal of Applied Physics* 32(9):1679–1684.

Patankar, S.V. 1980. *Numerical Heat Transfer and Fluid Flow*, 2nd edn. Washington, DC: Hemisphere Publishing Corporation.

Sousa, P.F.B. 2006. Development of dynamic observers based on Green's functions to be applied in inverse heat conduction problem (in Portuguese). Master's thesis, Universidade Federal de Uberlândia, Uberlandia, Brazil.

Vanderplaats, G.N. 1984. *Numerical Optimization Techniques for Engineering Design*. New York, NY: McGraw-Hill College.

19

Front Face Thermal Characterization of Materials by a Photothermal Pulse Technique

Fabrice Rigollet and Christophe Le Niliot

CONTENTS

19.1 Introduction

The rear flash method (or "rear face pulsed method") is widely used to measure the thermophysical properties of materials (Beck 1998, Maldague 2001). When only the front face of the material is accessible for excitation and measurement, the method is called "front face pulsed method." The objective of this chapter is to design such a method for the characterization of the front face of a bilayer material. The two layers are conducting (steel) and insulating (PVC) and they will be alternatively front and rear face. The measurements will be realized with an infrared camera.

After a description of all the components of the experiment and the resulting measurements, the corresponding model is described, with its associated parameters that contain the searched thermophysical properties (diffusivity and effusivity). The minimization of the ordinary least squares (OLS) objective function between measurements and model response is presented (see Chapter 9). The optimal design of experiment-identification is then detailed to define which parameters may be identified or must be fixed, and what is the optimal duration of the experiment. Results are presented with simulated noisy data.

19.2 Principles of Parameter Estimation from Experimental Data

19.2.1 Real System, Measurements, and Measurement Errors

The sample studied here is composed of two layers (steel and PVC) whose nominal thermophysical properties are presented in Table 19.1. The experiment used in this chapter to illustrate the different tools of parameter estimation is a front face pulsed photothermal experiment. A radiative source provides a uniform energy density Q on the front face of the bilayer sample for a short duration (compared to the heat diffusion time of the sample's first layer). The sample is initially isotherm at T_0 (see Figure 19.1). The decrease of the front face temperature, consecutive to its sharp increase, is analyzed to deduce some thermophysical data of the sample. An infrared camera measures this front face temperature decreasing with time $T(t)$.

Two positions of excitation are possible: one on the steel side ("steel/PVC" case) and the other on the PVC side ("PVC/steel" case). The two corresponding simulated perfect signals (without measurement errors) are plotted in Figure 19.2 on log–log scales. After a common $t^{-1/2}$ decreasing regime, representative of heat diffusion in the semi-infinite-like front layer, the two signals have different transitions, representative of the rear layer encountered by heat. For the PVC/steel case, the rear layer is more effusive than the front one while the cooling of the front surface is faster when compared to the steel/PVC case. For this second case, after the $t^{-1/2}$ regime, a quasi-constant temperature is obtained, as if the sample was isolated, before convection acts to cool the sample.

The vector $(n \times 1)$ of experimental measurements $\boldsymbol{y} = [y_1 \cdots y_i \cdots y_n]^t$ is built by removing the initial temperature to all temperature measurements $(y_i = y(t_i) = T(t_i) - T_0)$. Measurements are thus described with respect to n various discrete values of time (the independent variable time) $\boldsymbol{t} = [t_1 \cdots t_i \cdots t_n]^t$ (vector $(n \times 1)$) with $t_i = idt$, $i = 1, \ldots, n$. The shortest time step dt allowed by the infrared camera's acquisition system is $dt = 1/25$ s $= 40$ ms. Let ε_i be the (unknown) error associated with the measurement y_i $(i = 1, \ldots, n)$, then the

TABLE 19.1

Thermophysical Properties of Each Layer

	k (W m^{-1} K^{-1})	ρC_p (J m^{-3} K^{-1})	a (m^2 s^{-1})	b (W s$^{1/2}$ m^{-2} K^{-1})	e (m)	e/k (m^2 K W^{-1})	e^2/a (s)	$\rho C_p e$ (J m^{-2} K^{-1})
Steel	30.0	3.601×10^6	8.33×10^{-6}	1.0394×10^4	0.005	1.67×10^{-4}	3.00	18.0×10^3
PVC	0.19	1.570×10^6	0.121×10^{-6}	0.0546×10^4	0.001	52.6×10^{-4}	8.26	1.57×10^3

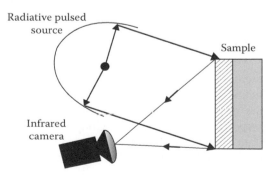

FIGURE 19.1

Principle of the front face pulsed photothermal experiment.

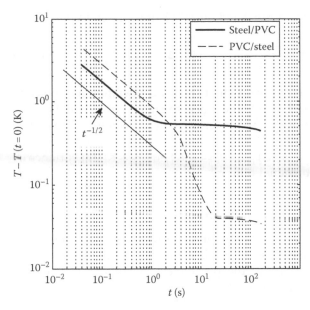

FIGURE 19.2
Example of simulated signals (Q is adjusted as explained in (19.2.1)).

measurement errors vector ($n \times 1$) is $\boldsymbol{\varepsilon} = [\varepsilon_1 \cdots \varepsilon_i \cdots \varepsilon_n]^t$. The first assumption on measurement errors is that they are purely additive:

$$y = y_{perfect} + \boldsymbol{\varepsilon} \tag{19.1}$$

Here, $y_{perfect}$ represents the vector ($n \times 1$) of (unknown) errorless measurements.* Moreover, measurement errors are assumed to be the realizations of a random variable with a Gaussian distribution with zero mean, that is, $E[\boldsymbol{\varepsilon}] = 0$ (errors unbiased), $E[\cdot]$ being the expected value operator (representing the mean of a large number of realizations of the random variable). The covariance matrix ($n \times n$) $\boldsymbol{\psi} = E\big[(\boldsymbol{\varepsilon} - E[\boldsymbol{\varepsilon}])(\boldsymbol{\varepsilon} - E[\boldsymbol{\varepsilon}])^t\big] = E[\boldsymbol{\varepsilon}\boldsymbol{\varepsilon}^t]$ of error measurements contains on its main diagonal, the variance σ_ε^2 of each measurement that is supposed constant for each time t_i, $i = 1, \ldots, n$. This variance may or may not be known. Finally, measurement errors are assumed uncorrelated (error at time t_i independent of error at time t_j ($E[\varepsilon_i \varepsilon_j] = 0$ for $i \neq j$), then $\boldsymbol{\psi}$ is a diagonal matrix:

$$\boldsymbol{\psi} = diag(\sigma_\varepsilon^2, \ldots, \sigma_\varepsilon^2, \ldots, \sigma_\varepsilon^2) = I\sigma_\varepsilon^2 \tag{19.2}$$

In order to have a measure of the relative magnitude of measurement errors, its standard deviation σ_ε is compared to the particular value of the "perfect" measurement at an early time t_s (i.e., a short time compared to the heat diffusion time of the first layer, typically $t_s = 0.1 e_1^2 / a_1$). For such a short time, the first layer may be considered to be a semi-infinite material for which the perfect measurement is given (for a 1D heat transfer) by

$$y_{perfect}(t_s) = \frac{Q}{b_1\sqrt{\pi t_s}} \quad \text{with } b_1 = \sqrt{k_1 \rho C_1} = \frac{k_1}{\sqrt{a_1}} \text{ the effusivity of first layer} \tag{19.3}$$

---

* The objective of "direct" modelization is to give the best approximation of $y_{perfect}$.

A measure of the noise-to-signal ratio (or "normalized noise") is thus given by

$$\sigma_\varepsilon^* = \frac{\sigma_\varepsilon}{y_{perfect}(t_s)} = \frac{\sigma_\varepsilon b_1 \sqrt{\pi t_s}}{Q} = \sqrt{0.1\pi}\, \frac{\sigma_\varepsilon}{\Delta T_{1,adiab}} \quad \text{with } \Delta T_{1,adiab} = \frac{Q}{\rho C_{p1} e_1} \qquad (19.4)$$

For a given energy density Q, and a given measurement standard deviation σ_ε (typically $\sigma_\varepsilon = 0.05°C$ for the infrared camera used here), the noise-to-signal ratio will be thus proportional to the thermal capacitance of the first layer $\rho C_{p1} e_1$ (according to Table 19.1, signal will be about 10 times noisier for steel/PVC than for PVC/steel, for a given value of Q). Here, Q will be adjusted in order to have $y_{perfect}(t_c = 0.1 e_1^2/a_1) = 1$ K; the variance σ_ε will thus be directly equal to the noise-to-signal ratio σ_ε^*.

19.2.2 The Signal Modelization and the Associated Parameters

The objective of such a model is to give a mathematical expression, noted $\eta(t, \boldsymbol{\beta})$ of the perfect measurements $y_{perfect}(t)$ mentioned above. This model is a function of the independent variable (time) and of q parameters composing the parameters vector $(q \times 1)$ noted as $\boldsymbol{\beta} = [\beta_1 \cdots \beta_q]^t$. The model vector $(n \times 1)$ is then given by $\boldsymbol{\eta}(\boldsymbol{\beta}) = [\eta_1(t_1, \boldsymbol{\beta}) \cdots \eta_i(t_i, \boldsymbol{\beta}) \cdots \eta_n(t_n, \boldsymbol{\beta})]^t$. The thermal quadrupoles formalism (Maillet et al. 2000) is used to give this mathematical expression. It helps to solve the heat equation in the Laplace domain using a time Laplace transformation of temperatures and fluxes. Then, the Laplace inversion is realized by Stehfest's numerical method (Maillet et al. 2000) to provide the solution in the time domain. This formalism is well adapted to the treatment of multilayer materials in imperfect contact or otherwise. Linear boundary conditions of the convector-radiative type (through the use of a unique convector-radiative coefficient h) are also easily taken into account. The more general model that may be needed here is the front face temperature of a finite length bilayer with an interface contact resistance R_c and a convector-radiative transfer through a coefficient h supposed to be the same on each face (see the schematic representation of the physical model in Figure 19.3 and its quadrupolar representation in Figure 19.4).

The mathematical expression obtained in the Laplace domain (p is the Laplace variable in s^{-1}) is developed below. It shows that six "natural" parameters are present. Their expression and their nominal values are presented in Table 19.2 for the two cases PVC/steel and steel/PVC.

$$\eta(p, \boldsymbol{\beta}) = \bar{\theta}_{front}(p, \boldsymbol{\beta}) = \frac{\beta_2}{\beta_3 \sqrt{\beta_1}} \frac{U}{U + V} \qquad (19.5)$$

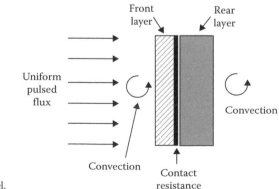

FIGURE 19.3
Physical model.

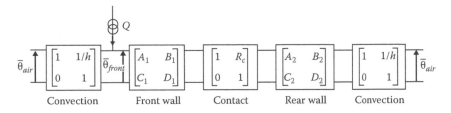

FIGURE 19.4

Quadrupolar representation of the physical model ($\bar{\theta}_{air} = 0$).

TABLE 19.2

Nominal Values of the Natural Parameters in Each Configuration

	β_1 (s^{-1})	β_2 (K $s^{1/2}$)	β_3 (−)	β_4 (−)	β_5 (s^{-1})	β_6 (−)
Expression (1 for front layer, 2 for rear layer)	$\dfrac{a_1}{e_1^2}$	$\dfrac{Q}{b_1}$	$\dfrac{he_1}{k_1}$	$\dfrac{b_1}{b_2}$	$\dfrac{a_2}{e_2^2}$	$\dfrac{R_c k_1}{e_1}$
Nominal value PVC/steel	0.121	1.611	0.053	0.053	0.333	0.0019
Nominal value steel/PVC	0.333	0.971	0.0017	19.03	0.121	0.06

Values of the convecto-radiative coefficient and the contact resistance are respectively $h = 10$ W m^{-2} K^{-1} and $R_c = 10^{-5}$ m^2 K W^{-1}.

with

$$U = (1 + \beta_3\beta_6)ch_1ch_2 + \left(\frac{\beta_6}{\beta_4}\sigma_1 + \frac{\beta_3\beta_4}{\sigma_1}\right)ch_1sh_2 + \frac{\beta_3}{\sigma_1}sh_1ch_2 + \frac{1}{\beta_4}sh_1sh_2 \qquad (19.6)$$

$$V = ch_1ch_2 + \frac{1}{\beta_3\beta_4}\sigma_1ch_1sh_2 + \left(\beta_6 + \frac{1}{\beta_3}\right)\sigma_1sh_1ch_2 + \left(\beta_4 + \frac{\beta_6}{\beta_3\beta_4}\sigma_1^2\right)sh_1sh_2 \qquad (19.7)$$

$$ch_i = \cosh(\sigma_i), \quad sh_i = \sinh(\sigma_i) \quad \text{for } i = 1, 2 \quad \text{and} \quad \sigma_1 = \sqrt{p/\beta_1}, \quad \sigma_2 = \sqrt{p/\beta_5} \qquad (19.8)$$

Parameters β_1 and β_5 are the inverse of the characteristic heat diffusion times of the front and rear layers. Parameter β_3 is the Biot number associated with the front layer. Parameter β_2 is the "semi-infinite" parameter present in (19.3). Parameter β_4 is the front/rear effusivities ratio. Parameter β_6 is the thermal contact resistance normalized by the thermal resistance of the front layer. In cases where the aim is to characterize the front layer with a rear layer supposed known, the parameters of interest will be β_1 and β_4. Parameter β_5 will then always be supposed known. Special cases can be obtained from this general model: perfect contact ($\beta_6 = 0$, 5 parameters left), same layers in imperfect contact ($\beta_5 = \beta_1$ and $\beta_4 = 1$, 4 parameters left), single layer ($\beta_5 = \beta_1$, $\beta_4 = 1$, and $\beta_6 = 0$, 3 parameters left), semi-infinite rear layer ($\beta_5 = 0$, 5 parameters left), and isolated sample ($\beta_3 = 0$, 5 parameters left). It can be seen that the model is nonlinear with respect to all its parameters except β_2. In order to take into account some fixed parameters in the parameter vector $\boldsymbol{\beta}$, a division into two subvectors, $\boldsymbol{\beta}_r$ ($r \times 1$) (containing the parameters to estimate) and $\boldsymbol{\beta}_c$ (($q - r) \times 1$) (containing the fixed parameters) are made

$$\boldsymbol{\beta} = [\boldsymbol{\beta}_r \vdots \boldsymbol{\beta}_c]^t \qquad (19.9)$$

19.2.3 The Objective Function

Assuming the model has the right form (or "right structure," given per the resolution of the "right" partial differential equations describing the "right" physical phenomena) and is calculated with the right values of parameters, then $\eta(\beta) = y_{perfect}$ and (19.1) becomes

$$y = \eta(\beta) + \varepsilon \qquad (19.10)$$

Since the n measurement errors composing ε are not known, the problem of finding the values of the r components of β_r of $\beta = [\beta_r \vdots \beta_c]^t$, given n measurements verifying (19.10), is underdetermined (n equations, $n + r$ unknowns). The new problem consisting in minimizing the difference between measurements y and model $\eta(\beta)$ (n equations, r unknowns, $n > r$) is overdetermined, and its solution is needed to solve the minimization problem. Without any a priori information on the parameters, and given the above assumptions for measurements errors, the OLS estimator can be used.[*] It means that the objective function to minimize in order to have an estimation of β_r, noted $\hat{\beta}_r$, is

$$S_{OLS}(\hat{\beta}_r, \beta_c) = (y - \eta(\hat{\beta}))^t (y - \eta(\hat{\beta})) = \sum_{i=1}^{n} (y_i - \eta_i(\hat{\beta}))^2 \quad \text{with } \hat{\beta} = [\hat{\beta}_r \vdots \beta_c]^t \qquad (19.11)$$

The parameters β_c are supposed to be fixed to their exact value. The study of the influence of a systematic error (a bias) on these fixed parameters will be presented later. Thus, the optimal value of $\hat{\beta}_r$ will be the one that minimizes the scalar function $S_{OLS}(\hat{\beta}_r, \beta_c)$; it will have to verify

$$\frac{\partial S_{OLS}(\hat{\beta}_r, \beta_c)}{\partial \hat{\beta}_r} = 0 \qquad (19.12)$$

19.2.4 Inverse Problem Resolution: Minimization of the Objective Function

For a nonlinear model, with respect to its parameters, the resolution of (19.12) is not explicit in $\hat{\beta}_r$ and needs to use an iterative method (Aster et al. 2005; Beck and Arnold 1977). The Gauss–Newton method is used here. It is a gradient type method with the optimization of the parameters step based on the hypothesis of model linearity with respect to $\hat{\beta}_r$ near the solution. From an initial estimation $\hat{\beta}_r^{(0)}$, a suit of estimations $\hat{\beta}_r^{(1)}, \ldots, \hat{\beta}_r^{(k)}, \ldots, \hat{\beta}_r^{(k+1)}, \ldots, \hat{\beta}_r^{(K)}$ is obtained, corresponding to the continuously decreasing values of $S_{OLS}(\hat{\beta}_r^{(k)}, \beta_c)$ ($k = 1, \ldots, K$) until a stop criterion, based on the relative variation of $\hat{\beta}_r^{(K)}$ compared to $\hat{\beta}_r^{(K-1)}$, is verified. The relation that gives the parameters step from iteration k to $k + 1$ is

$$\hat{\beta}_r^{(k+1)} = \hat{\beta}_r^{(k)} + \Delta\hat{\beta}_r^{(k)} \qquad (19.13)$$

---

[*] It is here the more *efficient*, i.e. with the minimal variance.

with

$$\Delta\hat{\boldsymbol{\beta}}_r^{(k)} = P_r^{(k)}\big[X_r^{(k)t}(\boldsymbol{y} - \boldsymbol{\eta}^{(k)})\big] \quad \text{with } \boldsymbol{\eta}^{(k)} = \boldsymbol{\eta}\Big(\hat{\boldsymbol{\beta}}_r^{(k)}\Big) \tag{19.14}$$

$$P_r^{(k)} = \big[H_r^{(k)}\big]^{-1} \tag{19.15}$$

$$H_r^{(k)} = X_r^{(k)t}X_r^{(k)} \quad \text{with } X_r^{(k)} = X_r\Big(\hat{\boldsymbol{\beta}}_r^{(k)}\Big) \tag{19.16}$$

The final value of the estimated parameters $\hat{\boldsymbol{\beta}}_r^{(K)}$, obtained for the fixed value $\boldsymbol{\beta}_c$, is noted as $\hat{\boldsymbol{\beta}}_{r,opt}\,(\boldsymbol{\beta}_c)$. The main difficulty in the above algorithm, if ill-conditioned, lies in the inversion of the matrix $H_r^{(k)}$ present in Equation 19.15. The matrix X_r ($n \times r$) is the sensitivity matrix to estimated parameters. It is a part of the "complete" sensitivity matrix X, relative to all the parameters (estimated $\boldsymbol{\beta}_r$ ($r \times 1$) and the fixed $\boldsymbol{\beta}_c$ (($q-r) \times 1$))

$$X = [X_r \vdots X_c] = \begin{bmatrix} \begin{bmatrix} X_1(t_1) & \dots & X_r(t_1) \\ \vdots & \dots & \vdots \\ X_1(t_n) & \dots & X_r(t_n) \end{bmatrix} \vdots \begin{bmatrix} X_{r+1}(t_1) & \dots & X_q(t_1) \\ \vdots & \dots & \vdots \\ X_{r+1}(t_n) & \dots & X_q(t_n) \end{bmatrix} \end{bmatrix} \tag{19.17}$$

Column k contains the n values of the sensitivity coefficient of the model with respect to the parameter β_k, given by

$$X_k(t_i) = \frac{\partial \eta_i(t_i, \boldsymbol{\beta})}{\partial \beta_k}, \quad k = 1, \dots, q \quad \text{and} \quad i = 1, \dots, n \tag{19.18}$$

This coefficient is a measure of the "influence" of parameter β_k on the response of the model $\eta(t, \boldsymbol{\beta})$. If the r sensitivity coefficients relative to the estimated parameters are of "high" magnitude and "independent" from each other, the conditioning of P_r will be correct and the simultaneous estimation of the r parameters composing $\boldsymbol{\beta}_r$ will be possible. The meaning of "high" and "independent" will be developed later.

Remark 19.1

If the model is linear with respect to its parameters, it can be written as follows:

$$\boldsymbol{\eta}(\boldsymbol{\beta}) = X\boldsymbol{\beta} = X_r\boldsymbol{\beta}_r + X_c\boldsymbol{\beta}_c \tag{19.19}$$

with X_r and X_c independent of $\boldsymbol{\beta}$. In this case, the OLS estimator is found directly without iteration, and is written as

$$\hat{\boldsymbol{\beta}}_r = \big[X_r^t X_r\big]^{-1} X_r^t (\boldsymbol{y} - X_c\boldsymbol{\beta}_c) \tag{19.20}$$

Note that OLS is more deeply presented in Chapter 7 and nonlinear estimation in Chapters 9 and 10.

Remark 19.2

When convergence is attained, a last iteration is realized in a sequential way. In this iteration, a series of n estimations is built (one per time step), $\boldsymbol{\beta}_r^{(i)}$ ($i=1,\dots,n$), which enables the appreciation of the stability of estimations with respect to each new measurement considered at each new time step. This suite converges toward $\hat{\beta}_{r,opt}$ (β_c). An estimation is said to be acceptable if such a suite is stable along the last third of the total duration of the experiment.

19.2.5 Confidence in Estimations (Variance and Bias of Estimator)

Let $\hat{\beta}_{r,opt}$ ($\tilde{\beta}_c$) be the estimated parameters for a value of fixed parameters $\tilde{\beta}_c$ different from their exact value β_c. Let e_r be the vector ($r \times 1$) of the error estimation (the difference between estimated and exact values of β_r), and let e_c be the deterministic error (the bias) on the values of the fixed parameters:

$$e_r = \hat{\beta}_{r,opt}(\beta_c) - \beta_r \qquad (19.21)$$

$$e_c = \tilde{\beta} - \beta_c \qquad (19.22)$$

Using the hypothesis of quasi-linearity of the model near $\hat{\beta}_{r,opt}$ ($\tilde{\beta}_c$), one can write

$$\hat{\beta}_{r,opt}(\tilde{\beta}_c) \approx A_{r,opt}(y - X_{c,opt}\tilde{\beta}_c) \quad \text{with } A_{r,opt} = \left[P_{r,opt}\right]^{-1}X_{r,opt}^t \qquad (19.23)$$

and

$$X_{r,opt}^t = X_r^t(\hat{\beta}_{r,opt}(\tilde{\beta}_c), \tilde{\beta}_c) \quad \text{and} \quad X_{c,opt}^t = X_c^t(\hat{\beta}_{r,opt}(\tilde{\beta}_c), \tilde{\beta}_c)$$

Assuming β_r is not too far from $\hat{\beta}_{r,opt}$ ($\tilde{\beta}_c$) (little estimation error e_r), one may develop (19.10):

$$y = \eta(\beta) + \varepsilon \approx X_{r,opt}\beta_r + X_{c,opt}\beta_c + \varepsilon \qquad (19.24)$$

Combining (19.24) and (19.23), the error of estimation (19.21) may then be approached by

$$e_r = \hat{\beta}_{r,opt}(\tilde{\beta}_c) - \beta_r \approx A_{r,opt}\varepsilon - A_{r,opt}X_{c,opt}e_c = e_{r1} + e_{r2} \qquad (19.25)$$

The first term $e_{r1} = A_{r,opt}\varepsilon$ is the random contribution to the total error; it represents the error due to measurement errors ε whose covariance matrix ψ is given by (19.2). The second term $e_{r2} = -A_{r,opt} X_{c,opt} e_c$ is the nonrandom (deterministic) contribution to the total error vector due to the deterministic error on the fixed parameters e_c. The expected value of e_{r1} is

$$E[e_{r1}] = A_{r,opt}E[\varepsilon] = 0 \qquad (19.26)$$

meaning that no systematic bias is introduced by the random measurement errors. The covariance matrix of e_{r1} is approached by

$$C_1 = \text{cov}(e_{r1}) = E[e_{r1}e_{r1}^t] \approx A_{r,opt}E[\varepsilon\varepsilon^t]A_{r,opt}^t = A_{r,opt}\psi A_{r,opt}^t = P_{r,opt}\sigma_\varepsilon^2 \qquad (19.27)$$

The matrix $P_{r,opt}$ given by (19.27) may thus be seen as the "amplification" of the measurement errors matrix. The expected value of e_{r2} is

$$E[e_{r2}] = -A_{r,opt}X_{c,opt}e_c \neq 0 \qquad (19.28)$$

if $e_c \neq 0$, meaning that the estimation of β_r is biased because of the wrong values of the fixed parameters β_c. This bias is computed using the corresponding sensitivity coefficients matrix $X_{c,opt}$. The covariance matrix (($q - r$) × ($q - r$)) of e_{r2} error is $C_2 = \text{cov}(e_{r2}) = 0$ because

e_c is not a random error. Finally, the total bias associated with the estimation $\hat{\beta}_{r,opt}(\tilde{\beta}_c)$ is due to the biased value of $\tilde{\beta}_c$, and its value is approached by

$$E[e_r] = E[e_{r2}] = -A_{r,opt}X_{c,opt}e_c = P_{c,opt}e_c \quad \text{with } P_{c,opt} = -P_{r,opt}X_{r,opt}^t X_{c,opt} \tag{19.29}$$

The matrix $P_{c,opt}$ ($r \times (q - r)$) given by (19.29) may thus be seen as the "amplification" of the bias on the fixed parameter e_c. For a fixed value of $\tilde{\beta}_c$, the covariance matrix C_r of the estimations errors is

$$C_r = \text{cov}(e_r) = E\left[(e_r - E[e_r])(e_r - E[e_r])^t\right] = E\left[e_{r1}e_{r1}^t\right] = \text{cov}(e_{r1}) = C_1 \tag{19.30}$$

The covariance matrix components are

$$C_r = \begin{bmatrix} \sigma_1^2 & \text{cov}(e_{r1}, e_{r2}) & \cdots & \text{cov}(e_{r1}, e_{rr}) \\ & \sigma_2^2 & & \text{cov}(e_{r2}, e_{rr}) \\ & & \ddots & \vdots \\ sym & & & \sigma_r^2 \end{bmatrix} \tag{19.31}$$

Its main diagonal elements contain the individual variance of error associated with each component of the estimated vector $\hat{\beta}_r$ and the other elements are the covariance of crossed errors. Expression (19.27) shows that a knowledge of the variance of measurement errors σ_ε^2 is needed in order to compute the covariance matrix. If σ_ε^2 is not measured before the experiment, an estimation of it may be obtained at the end of the estimation thanks to the final value of the objective function $S_{OLS}(\hat{\beta}_{r,opt}, \tilde{\beta}_c)$. Indeed, this estimation is based on the fact that, at the end of the estimation, the only difference that subsists between measurements and model (if its structure and its parameters are correct) must be the measurement errors. This difference is called the residuals of estimation:

$$e(\hat{\beta}_{r,opt}, \tilde{\beta}_c) = y - \eta(\hat{\beta}_{r,opt}, \tilde{\beta}_c) \tag{19.32}$$

and have statistical properties close to those of measurement errors. A nonbiased estimation of σ_ε^2 for the estimation of the p parameter from the use of n measurements is thus given by

$$\hat{\sigma}_\varepsilon^2 = \frac{e(\hat{\beta}_{r,opt}, \tilde{\beta}_c)^t e(\hat{\beta}_{r,opt}, \tilde{\beta}_c)}{n - r} = \frac{S_{OLS}(\hat{\beta}_{r,opt}, \tilde{\beta}_c)}{n - r} \tag{19.33}$$

19.2.6 Confidence Intervals of Estimations

Since measurement errors are assumed to have a normal probability density, the quantity $(\hat{\beta}_{r,opt}(\tilde{\beta}_c) - \beta_{r,k})/\sigma_k$ ($k = 1, \ldots, r$) has the $t_{1-\alpha/2}$ ($n - r$) distribution. The quantity $t_{1-\alpha/2}$ ($n - r$) is the t-statistic for $n - r$ degrees of freedom at the confidence level of $100(1 - \alpha)\%$ and σ_k ($k = 1, \ldots, r$), that is, the standard deviation of the estimation $\hat{\beta}_{r,opt}(\tilde{\beta}_c)$ is the square root of the kth diagonal element of $C = C_r$ given by (19.31) and (19.27). Thus, for a fixed value of $\tilde{\beta}_c$, the 95% ($\alpha = 0.05$) confidence interval (due to measurement errors) associated with the estimation $\hat{\beta}_{r,opt}(\tilde{\beta}_c)$ of the exact value $\beta_{r,k}$ is

$$\hat{\beta}_{r,opt,k}(\tilde{\beta}_c) \pm CI_k t_{1-0.975}(n - r)\sigma_k, \quad \text{with } CI_k = t_{1-0.975}(n - r)\sigma_k, \quad k = 1, \ldots, r \tag{19.34}$$

19.2.7 Correlation of Estimation Errors

The variance σ_i^2 of the estimation error associated to $\hat{\beta}_{r,i}$ may not be arbitrarily low independently of σ_j^2 ($j \neq i$) if $\text{cov}(e_{ri}, e_{rj}) \neq 0$ because $\sigma_i \sigma_j \geq \text{cov}(e_{ri}, e_{rj})$ (estimations $\hat{\beta}_{r,i}$ and $\hat{\beta}_{r,j}$ are said to be correlated). The correlation level between estimations $\hat{\beta}_{r,i}$ and $\hat{\beta}_{r,j}$ is thus measured by the quantity

$$\rho_{ij} = \frac{\text{cov}(e_{ri}, e_{rj})}{\sigma_i \sigma_j} = \frac{C_{r,ij}}{\sqrt{C_{r,ii} C_{r,jj}}} = \frac{P_{r,ij}}{\sqrt{P_{r,ii} P_{r,jj}}}, \quad i, j = 1, \ldots, r \quad (19.35)$$

that lies between -1 and 1. One considers that estimations are highly correlated when $|\rho_{ij}| \geq 0.9$ (Beck and Arnold 1977). This quantity is independent of the magnitude of measurement errors.

19.3 Identifiability of Parameters and Design of Experiment

19.3.1 The Qualitative Sensitivity Coefficients Analysis

The *reduced* sensitivities Z_k ($k = 1, \ldots, q$) of the model $\boldsymbol{\eta}(\boldsymbol{\beta})$ with respect to its parameters are built as

$$Z_k(t_i) = \beta_k X_k(t_i) = \beta_k \frac{\partial \eta_i(t_i, \boldsymbol{\beta})}{\partial \beta_k} = \frac{\partial \eta_i(t_i, \boldsymbol{\beta})}{\frac{\partial \beta_k}{\beta_k}}, \quad k = 1, \ldots, q \quad \text{and} \quad i = 1, \ldots, n \quad (19.36)$$

where X_k is the sensitivity coefficient given by (19.18). Equation 19.36 shows that these reduced sensitivity coefficients Z_k have the same unit as the model. They may be seen as the absolute variation of the model η induced by a relative variation of β_k. If that model variation is less than the magnitude of the measurement errors, it means that the influence of the considered parameter on the model response will not be accurately measurable. Consequently, the estimation of this parameter through the use of experimental measurements, if it is possible, will be highly inaccurate. Rapid information may then be given by comparing the magnitude of each reduced sensitivity coefficient to the magnitude of noise measurement, with respect to the independent variable (time).

The reduced sensitivity coefficients compose the reduced sensitivity matrix as

$$\mathbf{Z} = [\mathbf{Z}_r \vdots \mathbf{Z}_c] \begin{bmatrix} \begin{bmatrix} X_1(t_1) & \cdots & X_r(t_1) \\ \vdots & \cdots & \vdots \\ X_1(t_n) & \cdots & X_r(t_n) \end{bmatrix} \vdots \begin{bmatrix} X_{r+1}(t_1) & \cdots & X_q(t_1) \\ \vdots & \cdots & \vdots \\ X_{r+1}(t_n) & \cdots & X_q(t_n) \end{bmatrix} \end{bmatrix} = \mathbf{XB} \quad (19.37)$$

where

$$\mathbf{B} = diag(\boldsymbol{\beta}) = \begin{bmatrix} \beta_1 & 0 & 0 \\ 0 & \ddots & 0 \\ 0 & 0 & \beta_q \end{bmatrix} = diag(\mathbf{B}_r \vdots \mathbf{B}_c) = diag(\beta_1, \ldots, \beta_r, \beta_{r+1}, \ldots, \beta_q)$$

This reduced sensitivity matrices are also divided into two submatrices $\mathbf{Z}_r$ ($n \times r$) and $\mathbf{Z}_c$ ($n \times (q - r)$), containing respectively the reduced sensitivity coefficients of the model

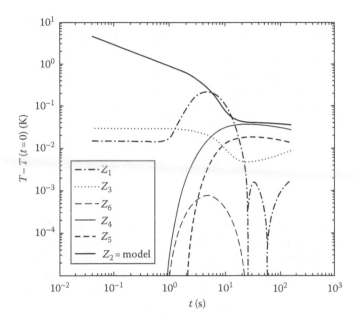

FIGURE 19.5

Absolute value of reduced sensitivity coefficients for PVC/steel case. See Table 19.2 for parameter numbers.

with respect to the parameters $\boldsymbol{\beta}_r$ and $\boldsymbol{\beta}_c$. The reduced sensitivity coefficients are plotted for the PVC/steel case in Figure 19.5 and for steel/PVC in Figure 19.6.

PVC/steel case (Figure 19.5): Sensitivity to β_6 (contact resistance) is less than the other. Its estimation will be difficult if the noise-to signal-ratio σ^* is less than 10^{-3}. Parameters β_4 (effusivities ratio) and β_5 (heat diffusion frequency of rear layer) seem to have the same influence on the model along time (their correlation coefficient presented later is over 0.9). It would be difficult to simultaneously identify the properties of the rear and front layers if

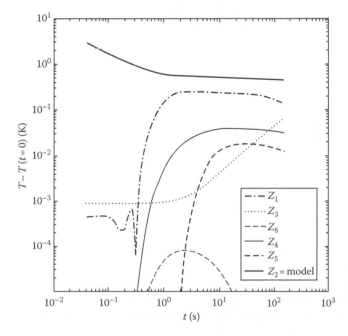

FIGURE 19.6

Absolute value of reduced sensitivity coefficients for steel/PVC case. See Table 19.2 for parameter numbers.

the rear layer was unknown. Sensitivity to the Biot number is relatively low. In this case, β_5 and β_6 will be fixed to their nominal value, while β_1, β_2, β_3, and β_4 will be estimated.

Steel/PVC case (Figure 19.6): Sensitivity to β_6 (contact resistance) is again less than the other. Its estimation will be difficult if the noise-to-signal ratio σ^* is less than 10^{-4}. Sensitivity to β_4 (convection losses) is very low for short times and increases up to the other sensitivities for longer times. This parameter will be fixed to its nominal value. Correlations are more difficult to apprehend here except for long times when many coefficients seem to have the same transient regime. The study of the components of the correlation matrix defined later will help to detect the problems. The parameter β_5 is still fixed.

19.3.2 Quantitative Criteria

The objective here is to present some quantitative criteria that may be used in order to apprehend the feasibility of the simultaneous identification of several parameters, for their nominal value. All these criteria are based on the analysis of the two amplification matrices (of variance of measurement errors (P_r (19.27)) and of the bias on the fixed parameters (P_c (19.29)) and the components of the correlation matrix (ρ_{ij} (19.35)), all defined from the Hessian matrix $H_r = X_r^t X_r$. In fact, we will work with analogue matrices, but built from the reduced sensitivity matrix Z. The *reduced* Hessian matrix J_r relative to the parameter to be estimated is then

$$J_r = Z_r^t Z_r = B_r H_r B_r \tag{19.38}$$

The reduced amplification of the measurement errors matrix R_r is written as

$$R_r = J_r^{-1} = B_r^{-1} H_r^{-1} B_r^{-1} = B_r^{-1} P_r B_r^{-1} \tag{19.39}$$

The square root of the diagonal elements of R_r is the term of amplification of noise-to-signal ratio that will give the *relative* standard deviation of each estimation composing $\hat{\beta}_r$:

$$\sqrt{R_{r,ii}} = \frac{\sigma_{r,ii}/\beta_{r,i}}{\sigma_\varepsilon^*}, \quad i = 1, \ldots, r \tag{19.40}$$

One optimal experiment criterion may thus consist in searching to reduce such amplification terms corresponding to the parameter of interest in $\hat{\beta}_r$ ($\hat{\beta}_{r,1}$ and $\hat{\beta}_{r,4}$ containing the diffusivity and effusivity of the front layer for instance). The elements of the correlation matrix are written as

$$\rho_{ij} = \frac{\mathrm{cov}(e_{ri}, e_{rj})}{\sigma_i \sigma_j} = \frac{C_{r,ij}}{\sqrt{C_{r,ii} C_{r,jj}}} = \frac{P_{r,ij}}{\sqrt{P_{r,ii} P_{r,jj}}} = \frac{R_{r,ij}}{\sqrt{R_{r,ii} R_{r,jj}}}, \quad i, j = 1, \ldots, r \tag{19.41}$$

They are the same when they are calculated with reduced sensitivities or "normal" sensitivities. Another possible criterion to observe in order to search for the best condition number of the Hessian matrix is to try to limit the possible correlations between parameters by trying to reduce some ρ_{ij}. Some parameters may then have to be fixed because of their very high correlation with others. If some parameters have to be fixed to an arbitrary value, one should try to limit the amplification of their possible bias. The relative bias on estimations $B_r^{-1}e_r$, caused by a relative bias on fixed parameters $B_c^{-1}e_c$, is given by

$$B_r^{-1}e_r = R_c(B_c^{-1}e_c) \tag{19.42}$$

with

$$R_c = -\left[Z_r^t Z_r\right]^{-1} Z_r^t Z_c = -R_r Z_r^t Z_c \tag{19.43}$$

R_c is thus the amplification matrix $(p \times (q - r))$ of a relative bias on fixed parameters. Element $R_{c,ij}$ is the amplification of the relative bias on a fixed parameter number j for the evaluation of the relative bias on the estimated parameter number i.

All the preceding criteria will now be observed with respect to the independent variable (time) in order to search for a more interesting duration of experiment for estimation, in the two cases PVC/steel and steel/PVC. It is shown that under the hypothesis of a high number of measurements n, and for a constant time step dt, the quantity

$$g_{t,i}(t_n) = \sqrt{\frac{R_{r,ii}}{dt}} = \frac{\sigma_{r,ii}/\beta_{r,i}}{\sigma_\varepsilon^*} \frac{1}{\sqrt{dt}}, \quad i = 1, \ldots, r \tag{19.44}$$

is only dependent on the duration of the experiment t_n.* This quantity is plotted in Figures 19.7 and 19.8. It enables, for a given time step and a given noise-to-signal ratio, the

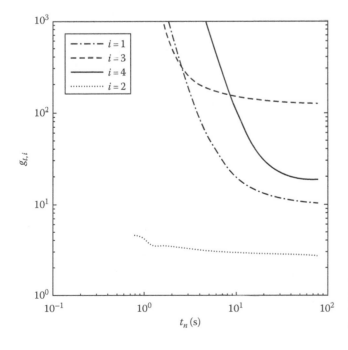

FIGURE 19.7
PVC/steel case: Amplification of measurement errors with respect to the duration of experiment t_n (Equation 19.44).

---

* Discrete summations that compose the elements of J_r and R_r can be approached by integrals whose evaluation is independent of time sampling.

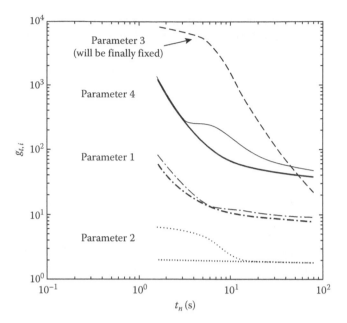

FIGURE 19.8
Steel/PVC: Amplification of measurement errors with respect to the duration of the experiment (Equation 19.44). Bold lines: estimation of β_1, β_2, β_4. Thin lines: estimation of β_1, β_2, β_3, β_4.

discovery of the value of the relative standard deviation of the estimation $\beta_{r,ii}$ and the duration of the experiment for which it will be minimum.

PVC/steel case (Figures 19.7, 19.9, and 19.10): Figure 19.7 shows that these coefficients always decrease with the duration of the experiment until a quasi-constant value, indicating that "long experiments" will minimize the amplification of measurement errors. For the PVC/steel, measurement errors will cause greater inaccuracy in the estimation of effusivity (β_4) than in the estimation of diffusivity (β_1). The estimation of the Biot number (β_3) will be highly inaccurate, this is due to its low sensitivity (see Figure 19.5). Figure 19.9 shows that correlation coefficients (given by (19.41)) are far from 0.9 for long enough

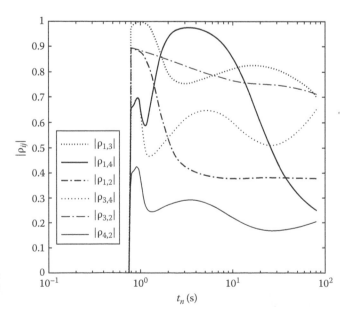

FIGURE 19.9
PVC/steel case: Correlation coefficients with respect to the duration of the experiment t_n.

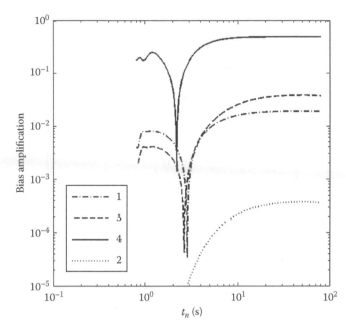

FIGURE 19.10
PVC/steel case: Amplification of 1% bias on each fixed parameter with respect to the duration of the experiment t_n.

experiments. A duration of 40 s seems to be a good compromise. Figure 19.10 shows the relative biases on estimations $B_r^{-1}e_r$ ((19.42)) due to a relative bias of 1% on each fixed parameter β_5 and β_6 (inverse of diffusion time of the rear layer and normalized contact resistance). They all have attained a fixed value for long times. These values have to be multiplied by the real value of bias e_c if it is not equal to 1%.

Steel/PVC case (Figures 19.8, 19.11, and 19.12): The coefficients of the amplification of measurement errors (Figure 19.8) also decrease with time, which makes long experiments seem better. The problem here comes from the correlation between the parameters plotted in Figure 19.11. They are, for many of them, most of the time over 0.9, especially the

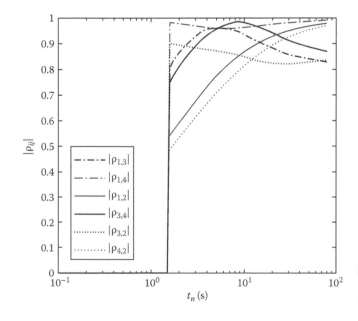

FIGURE 19.11
Steel/PVC case: Correlation coefficients with respect to the duration of the experiment t_n.

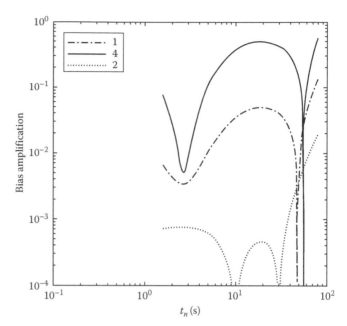

FIGURE 19.12
Steel/PVC: Amplification of 1% bias on each fixed parameter with respect to the duration of the experiment t_n.

coefficient ρ_{14} between the two parameters of interest, that is, even sometimes very near 1. The only durations of experiment that enable a slight decrease of this correlation are around 7–8 s, and not at all for long periods. Unfortunately, for these durations, very high correlations exist between β_3 and β_1, and between β_3 and β_4.

We then finally choose to fix the parameter β_3 (Biot number) to its nominal value in order to try to estimate β_1, β_4, and β_2. We will choose a duration of experiment of 8 s. Figure 19.12 shows the relative biases on estimations $B_r^{-1} e_r$ (19.42) due to a relative bias of 1% on each fixed parameter β_2, β_5, and β_6.

19.4 Identifications with Simulated Noisy Measurements

Simulated measurements have been used here to test the optimal experiment designed above.

PVC/steel case (Table 19.3): The first identification tested is for the following conditions: noise-to-signal ratio $\sigma_\varepsilon^* = 5\%$, $dt = 1/25$ s, $n = 1000$ ($t_n = 40$ s). It gives correct estimations of β_1 and β_4 with confidence intervals of respectively 7.5% and 13.6% (diffusivity is more accurate than effusivity). Relative bias induced by a bias of 10% on fixed parameters β_5 and β_6 are respectively −0.2% and −5% (still better for diffusivity than for effusivity).

But β_3 is very badly estimated, with a huge relative confidence interval (174%). This was suspected during the design of the experiment. Two solutions are then tested.

The first is to work with a lower noise-to-signal ratio ($\sigma_\varepsilon^* = 1\%$); this solution implies some experiment modifications (energy density Q larger or the repetition of the same experiment to provide a mean signal) that are not always possible. In this case, estimations are better, and confidence intervals are reduced (1.5% and 2.7%), even for the Biot number (18%).

TABLE 19.3

Results of Estimation for the PVC/Steel Case

		β_1 (s^{-1})	β_2 (K s$^{1/2}$)	β_3 (–)	β_4 (–)	β_5 (s^{-1})	β_6 (–)
Expression (1 for front layer, 2 for rear layer)		$\dfrac{a_1}{e_1^2}$	$\dfrac{Q}{b_1}$	$\dfrac{he_1}{k_1}$	$\dfrac{b_1}{b_2}$	$\dfrac{a_2}{e_2^2}$	$\dfrac{R_c k_1}{e_1}$
Nominal value PVC/steel		0.121	1.611	0.053	0.053	0.333	0.0019
Initial value	$\sigma_\varepsilon^* = 5\%$	0.150 (e)	1.800 (e)	0.060 (e)	0.060 (e)	0.366 (f)	0.0021 (f)
	$\sigma_\varepsilon^* = 5\%$	0.150 (e)	1.800 (e)	0.060 (f)	0.060 (e)	0.366 (f)	0.0021 (f)
	$\sigma_\varepsilon^* = 1\%$	0.150 (e)	1.800 (e)	0.060 (e)	0.060 (e)	0.366 (f)	0.0021 (f)
Identified	$\sigma_\varepsilon^* = 5\%$	0.128	1.598	0.027	0.045	—	—
	$\sigma_\varepsilon^* = 5\%$	0.123	1.612	—	0.0057	—	—
	$\sigma_\tilde{\varepsilon}^* = 1\%$	0.124	1.612	0.053	0.051	—	—
Relative confidence interval	$\sigma_\varepsilon^* = 5\%$	7.5%	2%	174%	13.6%	—	—
	$\sigma_\varepsilon^* = 5\%$	4.4%	1%	—	9.5%	—	—
	$\sigma_\varepsilon^* = 1\%$	1.5%	0.4%	18%	2.7%	—	—
Relative bias due to fixed parameters	$\sigma_\varepsilon^* = 5\%$	−0.2%	0.003%	0.6%	−5%	10% (f)	10% (f)
	$\sigma_\varepsilon^* = 5\%$	−1%	0.2%	10% (f)	−3.8%	10% (f)	10% (f)
	$\sigma_\varepsilon^* = 1\%$	−0.2%	0.003%	0.3%	−5%	10% (f)	10% (f)

(e) Estimated. (f) Fixed.

The second solution is to fix the parameter whose estimation is inaccurate to its nominal value, and to evaluate the influence of a bias on this fixed value on the estimations. The results are that β_1 and β_4 are much better estimated with lower confidence intervals (1.5% and 2.7%) and with an acceptable bias (−1% and −3.8%) despite a 10% biased value of each fixed parameter.

Steel/PVC case (f)—Fixed (Table 19.4): Remembering that for a given energy density Q of heat source, the noise-to-signal ratio is better in this case than in the PVS/steel case, we use here a value of 1% for $\sigma_\varepsilon^* = 1\%$.

Table 19.4 shows that in these conditions, an estimation of β_1 and β_4 is possible in spite of their relatively high correlation, if β_3 is fixed. The confidence interval associated to β_4 is relatively large (17.5%) and can only be reduced with a lower noise-to-signal ratio. Fixed parameters do not induce biases that are too large on estimations, the larger being again for β_4 (3%).

TABLE 19.4

Results of Estimation for the Steel/PVC Case

	β_1 (s^{-1})	β_2 (K s$^{1/2}$)	β_3 (–)	β_4 (–)	β_5 (s^{-1})	β_6 (–)
Expression (1 for front layer, 2 for rear layer)	$\dfrac{a_1}{e_1^2}$	$\dfrac{Q}{b_1}$	$\dfrac{he_1}{k_1}$	$\dfrac{b_1}{b_2}$	$\dfrac{a_2}{e_2^2}$	$\dfrac{R_c k_1}{e_1}$
Nominal value steel/PVC	0.333	0.971	1.7×10^{-3}	19.03	0.121	0.060
Initial value	0.360 (e)	1.100 (e)	2×10^{-3} (f)	21.00 (e)	0.13 (f)	0.068 (f)
Identified	0.339	0.969	—	18.2	—	—
Confidence interval	2.5%	0.4%	—	17.5%	—	—
Bias (fixed parameters)	0.3%	0.003%	10% (f)	−3%	10% (f)	10% (f)

(e) Estimated. (f) Fixed.

19.5 Conclusion

As a complement to Chapter 9, some useful tools applied in parameter estimation have been presented and illustrated in the case of the thermophysical characterization of a bilayer material by a front face pulsed experiment. The design of this experiment is greatly helped by the study of the reduced sensitivity coefficients of the model with respect to all its parameters. These coefficients are the base of all optimization criteria used here: the minimization of noise errors amplification during estimation, the minimization of the correlation between parameters and the minimization of bias induced by fixed parameters. These tools have helped to choose the parameter to be fixed and to evaluate the optimal duration of the experiment. In our test case, the contact resistance was too low to be identified, and the Biot number has always been fixed. Results have been verified with simulated measurements.

Nomenclature

a thermal diffusivity, $m^2\,s^{-1}$
b effusivity, $W\,s^{1/2}\,m^{-2}\,K^{-1}$
C_p thermal capacity, $J\,kg^{-1}\,K^{-1}$
dt time step, s
e thickness, m
$E[\cdot]$ expected value operator
$g_{t,i}$ amplification of measurement standard deviation for a given dt,
           Equation 19.38, $s^{-1/2}$
h convector-radiative coefficient, $W\,m^{-2}\,K^{-1}$
n number of measurements
p Laplace variable, s^{-1}
q total number of parameters
Q energy density, $J\,m^{-2}$
r number of parameters to estimate
R_c contact thermal resistance, $m^2\,K\,W^{-1}$
S_{OLS} ordinary least squares objective function, $°C^2$
t time, s
t_c short time for normalization of measurement errors
t_n duration of experiment, s
T temperature, $°C$
T_0 initial temperature, $°C$
$X_i(t)$ sensitivity coefficient of model with respect to parameter β_i, $°C\,[\beta_i]^{-1}$
$y(t)$ measurements, $°C$
$Z_i(t)$ reduced sensitivity coefficient of model with respect to parameter β_i, $°C$

Symbols

β exact parameter
$\hat{\beta}$ estimated parameter
$\tilde{\beta}$ fixed parameter

ε_i error measurement at ith time step, °C
$\eta(t)$ model, °C
ρ mass density, kg m^{-3}
ρ_{ij} correlation coefficient between errors on parameters β_i and β_j
σ_ε standard deviation of measurement errors, °C
σ_ε^* normalized standard deviation of measurement errors
$\overline{\theta}_{front}$ Laplace transform of heating $T{-}T_0$

Matrix and Vectors

B $= diag(\boldsymbol{\beta})$ diagonal matrix of parameters $\boldsymbol{\beta}$ $(q \times q)$
C covariance matrix of estimations $(r \times r)$
e_r error vector on estimations (r)
e_c error vector on fixed parameters $(q-r)$
H $X^t\, X$ Hessian matrix $(q \times q)$
J $Z^t\, Z$ reduced Hessian matrix $(q \times q)$
P H^{-1} $(q \times q)$ amplification matrix of measurement errors
R J^{-1} $(q \times q)$ reduced amplification matrix of measurement errors
X sensitivity matrix $(n \times q)$
y measurements vector (q)
Z reduced sensitivity matrix $(n \times q)$
β parameters vector (q)
η model vector (n)
ψ covariance matrix of error measurements $(n \times n)$

Subscripts

r relative to estimated parameters
opt relative to optimal parameters
c relative to fixed parameters

Abbreviation

OLS ordinary least squares

References

Aster R. C., Borchers B., Thurber C. H., 2005, *Parameter Estimation and Inverse Problems*, London, U.K.: Elsevier Academic Press.

Beck J. V., 1998, Parameter estimation concepts and modeling: Flash diffusivity application, *Inverse Problem in Engineering: Theory and Practice*, FL: ASME.

Beck J. V. and Arnold K. J., 1977, *Parameter Estimation in Engineering and Science*, New York: John Wiley & Sons.

Maillet D., André S., Batsale J. C., Degiovanni A., Moyne C., 2000, *Thermal Quadrupoles: Solving the Heat Equation through Integral Transforms*, Chichester: Wiley & Sons.

Maldague X., 2001, *Theory and Practice of Infrared Technology for Nondestructive Testing*, New York: John Wiley & Sons.

20

Estimation of Space Variable Thermophysical Properties

Carolina P. Naveira-Cotta, Renato M. Cotta, and Helcio R. Barreto Orlande

CONTENTS

20.1 Introduction

The analysis of diffusion problems in heterogeneous media involves formulations with spatially dependent thermophysical properties in different ways, such as large-scale variations in functionally graded materials (FGM), abrupt variations in layered composites, and random variations due to local concentration fluctuations in dispersed phase systems (Lin 1992, Divo and Kassab 1998, Fudym et al. 2002, Chen et al. 2004, Kumlutas and Tavman 2006, Fang et al. 2009). For instance, composite materials consisting of a dispersed reinforcement phase embedded in a bulk matrix phase have been providing engineers with increased opportunities for tailoring structures to meet a variety of property and performance requirements. As the composite material morphology in the realm of applications presents endless possibilities of design tailoring, manufacturing processes, and even self-structuring, the characterization of their physical properties is to be made almost case to case (Progelhof et al. 1976, Tavman 1996, Tavman and Akinci 2000, Danes et al. 2003, Kumlutas et al. 2003, Weidenfeller et al. 2004, Serkan Tekce et al. 2007).

The usefulness of such materials in heat transfer applications is nevertheless limited by the precise knowledge of the corresponding thermophysical properties and boundary condition coefficients that are fed into the corresponding models, and quite often need to be determined by the appropriate inverse problem analysis (Flach and Ozisik 1989, Huang and Ozisik 1990, Lesnic et al. 1999, Divo et al. 2000, Huang and Chin 2000, Rodrigues et al. 2004, Huttunen et al. 2006, Huang and Huang 2007). Among the various available solution techniques of inverse problems (Beck and Arnold 1977, Alifanov 1994, Ozisik and Orlande 2000, Kaipio and Somersalo 2004, Zabaras 2006), a fairly common approach is related to the minimization of an objective function that usually involves the quadratic difference between measured and estimated values, such as the least squares norm, or some variations with the inclusion of regularization terms. Although very popular and useful in many situations, the minimization of the least squares norm is a non-Bayesian estimator. A Bayesian estimator is basically concerned with the analysis of the posterior probability density, which is the conditional probability of the parameters given the measurements, while the likelihood is the conditional probability of the measurements given the parameters (Kaipio and Somersalo 2004). This work illustrates the use of Bayesian inference, already discussed in detail in Chapter 12, in the estimation of spatially variable equation and boundary condition coefficients in diffusion problems, by employing the Markov chain Monte Carlo (MCMC) method (Migon and Gamerman 1999, Kaipio and Somersalo 2004, Gamerman and Lopes 2006, Fudym et al. 2008, Orlande et al. 2008, Paez 2010). The Metropolis–Hastings algorithm is applied for the sampling procedure (Metropolis et al. 1953, Hastings 1970), implemented in the *Mathematica* platform (Wolfram 2005). This sampling procedure used to recover the posterior distribution is in general the most expensive computational task in solving an inverse problem by Bayesian inference, since the direct problem is calculated for each state of the Markov chain.

In the context of variable properties identification, the use of a fast, accurate, and robust computational implementation of the direct solution is extremely important. The accurate representation of the heat conduction phenomena requires a detailed local solution of the temperature distribution, generally with the aid of discrete numerical solutions with sufficient mesh refinement and computational effort and/or semi-analytical approaches for specific or simplified functional forms. Analytical solutions of linear diffusion problems have been analyzed and compiled in Mikhailov and Ozisik (1984), where seven different classes of heat and mass diffusion formulations were systematically solved by the classical integral transform method. The obtained formal solutions are applicable to a very broad range of problems in heat and mass transfer. Later on, the classical integral transform approach gained a hybrid numerical–analytical implementation, referred to as the generalized integral transform technique (GITT) (Cotta and Ozisik 1986, Cotta 1990, 1993, 1998, Cotta and Mikhailov 1997, Cotta and Mikhailov 2006), offering more flexibility in handling nontransformable problems, including among others, the analysis of nonlinear diffusion and convection–diffusion problems. The methodology to be employed here forms the basis of the mixed symbolic–numerical computational code called "*UN*ified *I*ntegral *T*ransforms" (UNIT) (Sphaier et al. 2009), which was intended to bridge the gap between simple problems that allow for a straightforward analytical solution, and those more complex and involved situations that would otherwise require expensive commercial software systems. The open source UNIT code is then an implementation and development platform for researchers and engineers interested in the hybrid integral transform solutions of diffusion and convection–diffusion problems.

Thus, the integral transformation approach discussed above becomes very attractive for combined use with the Bayesian estimation procedure, since all steps in the method are determined analytically at once by symbolic computation, and the single numerical repetitive task is the solution of an algebraic matrix eigenvalue problem (Naveira-Cotta et al. 2009). Also, instead of seeking the function estimation in the form of a sequence of local values for the variable coefficients, an alternative path is followed based on the eigenfunction expansion of the coefficients to be estimated themselves (Naveira-Cotta et al. 2009, 2011a,b), and then seeking the estimation of the corresponding series coefficients.

Another novel aspect in the present work is the alternative analysis of the inverse problem in the transformed temperature field instead of employing the directly measured temperature data along the domain (Naveira-Cotta et al. 2011b). Thus, the experimental temperature values at each time value are first integrally transformed to yield transformed temperature values of increasing order. This procedure is particularly advantageous when a substantial amount of experimental measurements are available, such as in thermographic sensors, permitting a remarkable data compression after the integral transformation process.

Typical applications were selected to illustrate the robustness of the proposed combination of methodologies related to the estimation of thermal properties in two-phase dispersed media, complementing in scope what has been discussed in Chapters 1 and 2. Simulated experimental results were used in inverse analysis allowing for the inspection of the identification problem behavior in terms of the parameters to be estimated.

20.2 Direct Problem: Integral Transforms

We consider a one-dimensional form of the general formulation on transient heat conduction presented in Naveira-Cotta et al. (2009), for the temperature field $T_m(x, t)$, in a region $x \in [0, L_x]$. The formulation includes space variable thermal conductivity and heat capacity, as shown in problem (20.1). The volumetric heat capacity $w(x) = \rho(x)c_p(x)$ and the conductivity $k(x)$ are thus responsible for the information related to the heterogeneity of the medium. The heat conduction equation with initial and boundary conditions are given by

$$\rho(x)c_p(x)\frac{\partial T_m(x,t)}{\partial t} = \frac{\partial}{\partial x}\left(k(x)\frac{\partial T_m(x,t)}{\partial x}\right) - \frac{h_{\text{eff}}(x)}{L_z}(T_m(x,t) - T_\infty) + \frac{q(x,t)}{L_z}, \quad 0 < x < L_x; t > 0$$

(20.1a)

$$T_m(x,0) = T_\infty \tag{20.1b}$$

$$\left.\frac{\partial T_m(x,t)}{\partial x}\right|_{x=0} = 0, \quad t > 0 \tag{20.1c}$$

$$\left.\frac{\partial T_m(x,t)}{\partial x}\right|_{x=L_x} = 0, \quad t > 0 \tag{20.1d}$$

Problem (20.1) covers a typical one-dimensional transient thermophysical properties experimental setup for a thermally thin plate, including prescribed heat flux at one surface and convective heat losses at the opposite surface, as illustrated in Figure 20.1, and based on a

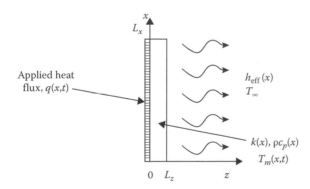

FIGURE 20.1
Schematic representation of the experimental setup for the determination of thermophysical properties, employed in direct/inverse problem solutions.

lumped formulation across the sample thickness. The space variation of the temperature distribution is then promoted by the variation of the applied heat flux, and eventually of the effective heat transfer coefficient. The exposed plate surface is required so as to allow for temperature measurements via infrared thermography (Fudym et al. 2008). Before providing the integral transform solution of problem (20.1), a simple filtering solution is employed for the improved convergence behavior of the eigenfunction expansions, in the form

$$T(x,t) = T_\infty + T^*(x,t) \tag{20.2}$$

Here, the term "filtering" stands for the extraction of analytical information from the original problem formulation, aimed at minimizing the effects of boundary and equation source terms in the convergence behavior of the filtered potentials. A more complete analytical filter may be preferred (Cotta 1993, Cotta and Mikhailov 1997) that fully homogenizes the original equation (20.1a), eliminating the source terms, but the above choice was already quite effective in the present situation.

The filtered temperature problem formulation is then given by

$$w(x)\frac{\partial T^*(x,t)}{\partial t} = \frac{\partial}{\partial x}\left(k(x)\frac{\partial T^*(x,t)}{\partial x}\right) - d(x)T^*(x,t) + P(x,t), \quad 0 < x < L_x;\ t > 0 \tag{20.3a}$$

$$T^*(x,0) = 0 \tag{20.3b}$$

$$\left.\frac{\partial T^*(x,t)}{\partial x}\right|_{x=0} = 0, \quad t > 0 \tag{20.3c}$$

$$\left.\frac{\partial T^*(x,t)}{\partial x}\right|_{x=L_x} = 0, \quad t > 0 \tag{20.3d}$$

where

$$w(x) = \rho(x)c_p(x) \tag{20.3e}$$

$$d(x) = \frac{h_{\text{eff}}(x)}{L_z} \tag{20.3f}$$

$$P(x,t) = \frac{q(x,t)}{L_z} \tag{20.3g}$$

The formal exact solution of problem (20.3) is then obtained with the classical integral transform method (Mikhailov and Ozisik 1984), and is written as

$$T(x,t) = T_\infty + \sum_{i=1}^{\infty}\tilde{\psi}_i(x)\int_0^t \overline{g}_i(t')e^{-\mu_i^2(t-t')}dt' \tag{20.4}$$

where the eigenvalues μ_i and eigenfunctions $\psi_i(x)$, are obtained from the eigenvalue problem that contains the information about the heterogeneous medium, in the form

$$\frac{d}{dx}\left[k(x)\frac{d\psi_i(x)}{dx}\right] + \left(\mu_i^2 w(x) - d(x)\right)\psi_i(x) = 0, \quad x \in [0, L_x] \tag{20.5a}$$

with boundary conditions

$$\frac{d\psi_i(x)}{dx} = 0, \quad x = 0 \tag{20.5b}$$

$$\frac{d\psi_i(x)}{dx} = 0, \quad x = L_x \tag{20.5c}$$

Also, the other quantities that appear in the exact solution (20.4) are computed after solving problem (20.5), such as

$$\tilde{\psi}_i(x) = \frac{\psi_i(x)}{\sqrt{N_i}}, \quad \text{normalized eigenfunctions} \tag{20.6a}$$

$$N_i = \int_0^{L_x} w(x)\psi_i^2(x)dx, \quad \text{normalization integrals} \tag{20.6b}$$

$$\bar{g}_i(t) = \int_0^{L_x} P(x,t)\tilde{\psi}_i(x)dx, \quad \text{transformed source terms} \tag{20.6c}$$

It is quite desirable to employ a flexible computational approach to handle eigenvalue problems with arbitrarily variable coefficients, such as problem (20.5). Thus, the GITT is here employed in the solution of the Sturm–Liouville problem (20.5) via the proposition of a simpler auxiliary eigenvalue problem, and by expanding the unknown eigenfunctions in terms of the chosen basis (Naveira-Cotta et al. 2009). Also, the variable equation coefficients are themselves expanded in terms of known eigenfunctions (Naveira-Cotta et al. 2009), so as to allow for a fully analytical implementation of the coefficient matrices in the transformed system. The solution of problem (20.5) is thus proposed as an eigenfunction expansion, in terms of a simpler auxiliary eigenvalue problem, given by

$$\frac{d^2\Omega_n(x)}{dx^2} + \lambda_n^2\Omega_n(x) = 0, \quad x \in [0, L_x] \tag{20.7a}$$

with boundary conditions

$$\frac{d\Omega_n(x)}{dx} = 0, \quad x = 0 \tag{20.7b}$$

$$\frac{d\Omega_n(x)}{dx} = 0, \quad x = L_x \tag{20.7c}$$

that is chosen to allow for an analytical solution of the auxiliary problem, with a normalized eigenfunction, eigenvalues, and norms given by

$$\tilde{\Omega}_n(x) = \frac{\cos(\lambda_n x)}{\sqrt{M_n}}, \quad \lambda_n = \frac{n\pi}{L_x}, \quad \text{with } n = 0, 1, 2, \ldots,$$

$$M_0 = L_x \quad \text{and} \quad M_n = \frac{L_x}{2}, \quad \text{with } n = 1, 2, \ldots, \tag{20.8a-d}$$

The proposed expansion of the original eigenfunction is then given by

$$\psi_i(x) = \sum_{n=1}^{\infty} \tilde{\Omega}_n(x)\overline{\psi}_{i,n}, \quad \text{inverse} \tag{20.9a}$$

$$\overline{\psi}_{i,n} = \int_0^{L_x} \psi_i(x)\tilde{\Omega}_n(x)dx, \quad \text{transform} \tag{20.9b}$$

The integral transformation is thus performed by operating Equation 20.5a with $\int_0^{L_x} \tilde{\Omega}_n(x) - dx$ to yield, after some manipulation (Naveira-Cotta et al. 2009), the following algebraic problem in matrix form:

$$(\mathbf{A} - \nu\mathbf{B})\overline{\psi} = 0, \quad \text{with } \nu = \mu^2 \tag{20.10a}$$

$$\overline{\psi} = \{\overline{\psi}_{n,m}\} \tag{20.10b}$$

$$\mathbf{B} = \{B_{n,m}\} \tag{20.10c}$$

$$B_{n,m} = \int_0^{L_x} w(x)\tilde{\Omega}_n(x)\tilde{\Omega}_m(x)dx \tag{20.10d}$$

$$\mathbf{A} = \{A_{n,m}\} \tag{20.10e}$$

$$A_{n,m} = \int_0^{L_x} \tilde{\Omega}_m(x)\frac{d}{dx}\left[k(x)\frac{d\tilde{\Omega}_n(x)}{dx}\right]dx - \int_0^{L_x} d(x)\tilde{\Omega}_n(x)\tilde{\Omega}_m(x)dx \tag{20.10f}$$

The algebraic problem (20.10a) can be numerically solved to provide results for the eigenvalues and eigenvectors upon truncation to a sufficiently large finite order M, which will be combined by the inverse formula (20.9a), to provide the desired original eigenfunctions (Mikhailov and Cotta 1994, Sphaier and Cotta 2000, Naveira-Cotta et al. 2009).

It is also of interest to express the variable coefficients themselves as eigenfunction expansions (Naveira-Cotta et al. 2009, 2011a,b). This is particularly advantageous in the evaluation of the algebraic system coefficients, $A_{n,m}$ and $B_{n,m}$. All the related integrals can then be expressed in terms of eigenfunctions, allowing for straightforward analytical evaluations. For instance, the coefficient $w(x)$ can be expanded in terms of eigenfunctions, together with a filtering solution to enhance convergence, in the following form:

$$w(x) = w_f(x) + \sum_{k=1}^{\infty} \tilde{\Gamma}_k(x)\overline{w}_k, \quad \text{inverse} \tag{20.11a}$$

$$\overline{w}_k = \int_0^{L_x} \hat{w}(x)[w(x) - w_f(x)]\tilde{\Gamma}_k(x)dx, \quad \text{transform} \tag{20.11b}$$

where $\hat{w}(x)$ is the weighting function for the chosen normalized eigenfunction $\tilde{\Gamma}_k(x)$. For instance, the eigenfunction basis may be chosen by employing the same auxiliary problem equation, but with first-order boundary conditions throughout, while the filtering function $w_f(x)$ would be a simple analytic function that satisfies the boundary values for the original coefficients. Again, the sense of filtering in this context would be the extraction of analytical information from the original functional behavior, here in terms of the property values at the boundaries. Then, once the transformed coefficients have been obtained through the transform formula, Equation 20.11b, computations may be carried on with the inverse expression for the variable coefficient (Equation 20.11a). The two remaining coefficients are equally expanded, in terms of eigenfunctions, to yield

$$k(x) = k_f(x) + \sum_{k=1}^{\infty} \tilde{\Gamma}_k(x)\bar{k}_k, \quad \text{inverse} \tag{20.11c}$$

$$\bar{k}_k = \int_0^{L_x} \hat{w}(x)\lfloor k(x) - k_f(x)\rfloor \tilde{\Gamma}_k(x)dx, \quad \text{transform} \tag{20.11d}$$

$$d(x) = d_f(x) + \sum_{k=1}^{\infty} \tilde{\Gamma}_k(x)\bar{d}_k, \quad \text{inverse} \tag{20.11e}$$

$$\bar{d}_k = \int_0^{L_x} \hat{w}(x)[d(x) - d_f(x)]\tilde{\Gamma}_k(x)dx, \quad \text{transform} \tag{20.11f}$$

The coefficient matrices may then be rewritten in terms of the expanded functions, such as for the elements of matrix **B**:

$$B_{n,m} = \int_0^{L_x} w_f(x)\tilde{\Omega}_n(x)\tilde{\Omega}_m(x)dx + \sum_{k=1}^{\infty} \overline{w}_k \int_0^{L_x} \tilde{\Gamma}_k(x)\tilde{\Omega}_n(x)\tilde{\Omega}_m(x)dx \tag{20.12a}$$

and for matrix **A**:

$$
\begin{aligned}
A_{n,m} = & \int_0^{L_x} \tilde{\Omega}_m(x)\frac{d}{dx}\left[k_f(x)\frac{d\tilde{\Omega}_n(x)}{dx}\right]dx + \sum_{k=1}^{\infty}\left[\int_0^{L_x} \tilde{\Omega}_m(x)\frac{d}{dx}\left[\tilde{\Gamma}_k(x)\frac{d\tilde{\Omega}_n(x)}{dx}\right]dx\right]\bar{k}_k \\
& - \int_0^{L_x} d_f(x)\tilde{\Omega}_n(x)\tilde{\Omega}_m(x)dx - \sum_{k=1}^{\infty}\left[\int_0^{L_x} \tilde{\Gamma}_k(x)\tilde{\Omega}_n(x)\tilde{\Omega}_m(x)dx\right]\bar{d}_k
\end{aligned} \tag{20.12b}
$$

Also, the normalization integrals are then computed from

$$N_i = \sum_{n=1}^{\infty}\sum_{m=1}^{\infty} \overline{\Psi}_{i,n}\overline{\Psi}_{i,m}\left\{\int_0^{L_x} w_f(x)\tilde{\Omega}_n(x)\tilde{\Omega}_m(x)dx + \sum_{k=1}^{\infty}\left[\int_0^{L_x} \tilde{\Gamma}_k(x)\tilde{\Omega}_n(x)\tilde{\Omega}_m(x)dx\right]\overline{w}_k\right\} \tag{20.12c}$$

This procedure shall also be handy in the function estimation task, when the transformed coefficients of the series in Equations 20.11a,c,e will be the parameters to be estimated.

The present study is also aimed at advancing the solution of the inverse problem in the transformed temperature field, from the integral transformation of the experimental temperature data, thus compressing the experimental measurements in the spatial domain into few transformed modes (Naveira-Cotta et al. 2011b). Once the experimental temperature readings have been obtained, one proceeds to the integral transformation of the temperature field at each measured time. For this purpose, the temperature measurements can be interpolated in the spatial domain, generating the continuous functions $T_{exp}(x, t)$, which are then integrally transformed according to the integral transform pair as follows:

$$\text{Transform } \overline{T_{\exp,i}}(t) = \int_0^{L_x} w(x)\tilde{\psi}_i(x)\left[T_{\exp}(x,t) - T_\infty\right]dx \qquad (20.13a)$$

$$\text{Inverse } T_{\exp}(x,t) = T_\infty + \sum_{i=0}^{NT} \tilde{\psi}_i(x)\overline{T_{\exp,i}}(t) \qquad (20.13b)$$

20.3 Inverse Problem: Bayesian Inference

A variety of techniques is nowadays available for the solution of inverse problems (Beck and Arnold 1977, Alifanov 1994, Ozisik and Orlande 2000, Kaipio and Somersalo 2004, Zabaras 2006). However, one common approach relies on the minimization of an objective function that generally involves the squared difference between measured and estimated variables, like the least-squares norm, as well as some kind of regularization term. Despite the fact that the minimization of the least-squares norm is indiscriminately used, it only yields *maximum likelihood* estimates if the following statistical hypotheses are valid: the errors in the measured variables are additive, uncorrelated, normally distributed, with zero mean and known constant standard-deviation; only the measured variables appearing in the objective function contain errors; and there is no prior information regarding the values and uncertainties of the unknown parameters.

Although very popular and useful in many situations, the minimization of the least-squares norm is a non-Bayesian estimator. A Bayesian estimator (Kaipio and Somersalo 2004) is basically concerned with the analysis of the *posterior probability density*, which is the conditional probability of the parameters given the measurements, while the likelihood is the conditional probability of the measurements given the parameters. If we assume the parameters and the measurement errors to be independent Gaussian random variables, with known means and covariance matrices, and that the measurement errors are additive, a closed form expression can be derived for the posterior probability density. In this case, the estimator that maximizes the posterior probability density can be recast in the form of a minimization problem involving the *maximum a posteriori objective function*. On the other hand, if different *prior* probability densities are assumed for the parameters, the posterior probability distribution may not allow an analytical treatment. In this case, MCMC methods are used to draw samples of all possible parameters, so that inference on the posterior probability becomes inference on the samples. In this work, we illustrate the use of Bayesian techniques, as discussed in Chapter 12, for the estimation of parameters in heat conduction within heterogenous media, via MCMC methods (Migon and Gamerman 1999,

Kaipio and Somersalo 2004, Gamerman and Lopes 2006, Fudym et al. 2008, Orlande et al. 2008, Paez 2010), as applied to the simultaneous identification of thermophysical properties and boundary condition coefficients. The Metropolis–Hastings algorithm (Metropolis et al. 1953, Hastings 1970) is employed for the sampling procedure, implemented in the *Mathematica* platform (Wolfram 2005).

Consider the vector of parameters appearing in the physical model formulation as

$$\mathbf{P}^T \equiv \begin{bmatrix} P_1, P_2, \ \ldots, \ P_{N_p} \end{bmatrix} \tag{20.14a}$$

where N_p is the number of parameters. For the solution of the inverse problem of estimating $\mathbf{P}$, we assume available the measured temperature data given as

$$(\mathbf{Y} - \mathbf{T})^T = \left(\vec{Y}_1 - \vec{T}_1, \ \vec{Y}_2 - \vec{T}_2, \ldots, \vec{Y}_{N_t} - \vec{T}_{N_t} \right) \tag{20.14b}$$

where $\vec{Y}_i$ contains the measured temperatures for each of the N_x sensors at time t_i, $i = 1, \ldots, N_t$, that is,

$$(\vec{Y}_i - \vec{T}_i) = (Y_{i1} - T_{i1}, \ Y_{i2} - T_{i2}, \ldots, Y_{iN_x} - T_{iN_x}) \quad \text{for } i = 1, \ldots, N_t \tag{20.14c}$$

so that we have $N_m = N_x N_t$ measurements in total. In the present transient state estimation procedure, the sensors are assumed to be distributed along the plate length and the measurements are taken at various time values within the measurement period.

Bayes' theorem can then be stated as (Paez 2010)

$$p_{\text{posterior}}(\mathbf{P}) = p(\mathbf{P}|\mathbf{Y}) = \frac{p(\mathbf{P})p(\mathbf{Y}|\mathbf{P})}{p(\mathbf{Y})} \tag{20.15}$$

where

$p_{\text{posterior}}(\mathbf{P})$ is the posterior probability density, that is, the conditional probability of the parameters $\mathbf{P}$ given the measurements $\mathbf{Y}$

$p(\mathbf{P})$ is the prior density, that is, a statistical model for the information about the unknown parameters prior to the measurements

$p(\mathbf{Y}|\mathbf{P})$ is the likelihood function that gives the relative probability density (loosely speaking, relative probability) of different measurement outcomes $\mathbf{Y}$ with a fixed $\mathbf{P}$

$p(\mathbf{Y})$ is the marginal probability density of the measurements, which plays the role of a normalizing constant

In this work, we assume that the measurement errors are Gaussian random variables, with known (modeled) means and covariances, and that the measurement errors are additive and independent of the unknowns. With these hypotheses, the likelihood function can be expressed as

$$p(\mathbf{Y}|\mathbf{P}) = (2\pi)^{-M/2} |\mathbf{W}|^{-1/2} \exp\left\{ -\frac{1}{2} [\mathbf{Y} - \mathbf{T}(\mathbf{P})]^T \mathbf{W}^{-1} [\mathbf{Y} - \mathbf{T}(\mathbf{P})] \right\} \tag{20.16}$$

where $\mathbf{W}$ is the covariance matrix of the measurement errors.

When it is not possible to analytically obtain the corresponding marginal distributions, one needs to use a method based on simulation (Migon and Gamerman 1999, Gamerman and Lopes 2006). The inference based on simulation techniques uses samples to extract information about the posterior distribution $p(\mathbf{P}|\mathbf{Y})$. The numerical method most used to explore the space of states of the posteriori is the Monte Carlo approach. The Monte Carlo simulation is based on a large number of samples of the probability density function (in this case, the function of the posterior probability density $p(\mathbf{P}|\mathbf{Y})$). Several sampling strategies are proposed in the literature, including the Monte Carlo method with the Markov chain (MCMC), adopted in this work, where the basic idea is to simulate a random walk in the space of $p(\mathbf{P}|\mathbf{Y})$ that converges to a stationary distribution, which is the distribution of interest in the problem.

A Markov chain is a stochastic process $\{\mathbf{P}_0, \mathbf{P}_1, \dots\}$, such that the distribution of $\mathbf{P}_i$, given all previous values $\mathbf{P}_0, \dots, \mathbf{P}_{i-1}$, depends only on $\mathbf{P}_{i-1}$. That is, it interprets the fact that for a process satisfying the Markov property of Equation 20.17, given the present, the past is irrelevant to predict its position in a future instant (Gamerman and Lopes 2006):

$$p(\mathbf{P}_i \in A|\mathbf{P}_0, \dots, \mathbf{P}_{i-1}) = p(\mathbf{P}_i \in A|\mathbf{P}_{i-1}) \qquad (20.17)$$

The most commonly used MCMC method algorithms are the Metropolis–Hastings, here employed, and the Gibbs sampler (Migon and Gamerman 1999, Gamerman and Lopes 2006). The Metropolis–Hastings algorithm uses the same idea of the rejection methods, that is, a value is generated from an auxiliary distribution and accepted with a given probability. The Metropolis–Hastings algorithm uses an auxiliary probability density function, $q(\mathbf{P}^*|\mathbf{P})$, from which it is easy to obtain sample values. Assuming that the chain is in a state $\mathbf{P}$, a new candidate value, $\mathbf{P}^*$, is generated from the auxiliary distribution $q(\mathbf{P}^*|\mathbf{P})$. The new value $\mathbf{P}^*$ is accepted with probability given by Equation 20.18, where the ratio that appears in this equation was called by Hastings as the ratio test, and today it is called ratio of Hastings "*RH*":

$$RH(\mathbf{P}, \mathbf{P}^*) = \min\left[1, \frac{p(\mathbf{P}^*|\mathbf{Y})q(\mathbf{P}^*|\mathbf{P})}{p(\mathbf{P}|\mathbf{Y})q(\mathbf{P}|\mathbf{P}^*)}\right] \qquad (20.18)$$

where $p(\mathbf{P}|\mathbf{Y})$ is the a posteriori distribution of interest. An important observation in Equation 20.18 is that we only need to know $p(\mathbf{P}|\mathbf{Y})$ up to a constant, since we are working with ratios between densities.

In practical terms, the simulation of samples of $p(\mathbf{P}|\mathbf{Y})$ by using the Metropolis–Hastings algorithm can be outlined as follows (Gamerman and Lopes 2006):

1. Boot up the iterations counter of the chain $i = 0$ and assign an initial value $\mathbf{P}^{(0)}$.
2. Generate a candidate value $\mathbf{P}^*$ of the distribution $q(\mathbf{P}^*|\mathbf{P})$.
3. Calculate the probability of acceptance of the candidate value $RH(\mathbf{P}, \mathbf{P}^*)$ by Equation 20.18.
4. Generate a random number u with uniform distribution, that is, $u \sim \mathrm{U}(0, 1)$.
5. If $u \leq RH$, then the new value is accepted and we let $\mathbf{P}^{(i+1)} = \mathbf{P}^*$. Otherwise, the new value is rejected and we let $\mathbf{P}^{(i+1)} = \mathbf{P}^{(i)}$.
6. Increase the counter of the number of states from i to $i+1$ and return to step 2.

The transition core $q(\mathbf{P}^*|\mathbf{P})$ defines only a proposal for a movement that can be confirmed by $RH(\mathbf{P}, \mathbf{P}^*)$. For this reason, it is usually called the proposal or density distribution. The success of the method depends on the not-so-low acceptance rates and proposals that are easy to simulate. The method replaces a difficult to generate $p(\mathbf{P}|\mathbf{Y})$ by several generations of the proposal $q(\mathbf{P}^*|\mathbf{P})$. In this study, we have chosen to adopt a symmetrical proposal density, that is, $q(\mathbf{P}^*|\mathbf{P}) = q(\mathbf{P}|\mathbf{P}^*)$ for all $(\mathbf{P}^*, \mathbf{P})$. In this case, Equation 20.18 reduces to the ratio of the posterior densities calculated at the previous and proposed chain positions, and does not depend on $q(\mathbf{P}^*|\mathbf{P})$.

If the inverse problem is solved with the integral transformed measured temperatures as given by Equation 20.13a, we have to reformulate the likelihood function given by Equation 20.16. Thus, we compare in Equations 20.19, the likelihood expressions as traditionally obtained directly from the temperature measurements, Equation 20.19a, and as calculated from the transformed temperature fields, Equation 20.19b, both weighted by the adequate experimental standard deviations in each field, which were assumed constant.

Likelihood in the temperature field is

$$\propto \exp\left[-\frac{1}{2} \sum_{s}^{\text{No. Sensors}} \sum_{m}^{\text{No. Measur.}} \frac{1}{\sigma_s^2} \left(T_{\exp}(x_s, t_m) - T_{\text{calc}}(x_s, t_m)\right)^2\right] \tag{20.19a}$$

Likelihood in the transformed temperature field is

$$\propto \exp\left[-\frac{1}{2} \sum_{i}^{NT} \sum_{m}^{\text{No. Measur.}} \frac{1}{\overline{\sigma}_i^2} \left(\overline{T_{\exp, i}}(t_m) - \overline{T_{\text{calc}, i}}(t_m)\right)^2\right] \tag{20.19b}$$

where NT is the number of modes used in representing the transformed temperature.

20.4 Results and Discussions

Two applications are considered below to illustrate the alternative approaches of estimating the space variable properties, either directly with measured temperature data or with integrally transformed experimental temperatures. The first inverse problem solution illustrated here involves the analysis of an abrupt variation of particles concentration in a two-phase dispersed system (see also Chapter 1). In order to examine the accuracy and robustness of the proposed inverse analysis, we have made use of simulated measured transient temperature data along the length of the domain. Such measurements were obtained from the solution of the direct (forward) problem by specifying the functions and values for the filler concentration distribution and thermophysical properties. The simulated data were disturbed by an error with a zero mean and a constant and known variance. For the results of the inverse analysis to be presented below, we have employed the parameter values shown in Table 20.1 for the generation of the simulated measured data, as extracted from Kumlutas and Tavman (2006) and Tavman (1996) for a polyethylene matrix filled with alumina particles.

TABLE 20.1

Parameter Values Used to Generate Simulated Measurement Data for Cases
1–3 Extracted from Kumlutas and Tavman (2006) and Tavman (1996)

Length	$L_x = 0.04$ m
Percentual filler concentration at $x = 0$	$\phi_0 = 0$
Percentual filler concentration at $x = L_x$	$\phi_L = 45$
Matrix properties (HDPE)	$\rho_m = 968$ kg/m^3
	$c_{pm} = 2300$ J/kg °C
	$k_m = 0.545$ W/m °C
Filler properties (Alumina)	$\rho_d = 3970$ kg/m^3
	$c_{pd} = 760$ J/kg °C
	$k_d = 36$ W/m °C
Effective thermal conductivity model	Lewis and Nielsen ($A = 1.5$; $\phi_m = 0.637$)
Parameters in filler concentration function	$\gamma = 25$ m^{-1}
	$x_c/L_x = 0.2$
Effective heat transfer coefficient	$h_{eff} = 16.7$ W/m^2 °C
Parameters in applied heat flux function	$\gamma_q = 100$ m^{-1}
	$x_{c,q}/L_x = 0.5$
	$q_0 = 0$
	$q_L = 598$ W/m^2
Ambient and initial temperature	$T_\infty = 23$°C
Plate thickness	$L_z = 0.003$ m

Sources: Kumlutas, D. et al., *J. Thermoplast. Compos. Mater.*, 19, 441, 2006; Tavman, I.H.,
        J. Appl. Polym. Sci., 62, 2161, 1996.

The abrupt variation of the concentration of the filler into the matrix is governed by the
parameter γ in the following function:

$$\phi(x) = \phi_{x=0} + (\phi_{x=L_x} - \phi_{x=0})\delta(x) \tag{20.20a}$$

$$\delta(x) = \frac{1}{1 + e^{-\gamma(x-x_c)/L_x}} \tag{20.20b}$$

with x_c being the transition position between the regions of low and high concentrations of
the filler. For γ sufficiently large, the two property values at the boundaries are approximately recovered. Equation 20.20b thus provides a continuous transition, more or less
abrupt depending on the value of the parameter γ, between two limiting values.

From the availability of the filler concentration distribution along the domain, Equation
20.20a, the heat capacity along the space coordinate is deterministically made from the
theory of mixtures. Thus, the coefficient $w(x)$ is considered as known in the direct problem
analysis to produce simulated experimental data, given as

$$w(x) = 1 + \left(\frac{\rho_d c_{pd}}{\rho_m c_{pm}} - 1\right)\phi(x) \tag{20.21}$$

For thermal conductivity determination, the volumetric content of the filler is not sufficiently informative to yield a good prediction of this physical property (Tavman 1996,

Kumlutas and Tavman 2006), especially for the higher concentration values. Many theoretical and empirical models have been proposed to predict the effective thermal conductivity of two-phase dispersed systems, and comprehensive review articles have discussed the applicability of many of these models.

In Lewis and Nielsen (1970), a model is proposed that attempts to include the effect of the shape of the particles and the orientation or type of packing for a two-phase system. The resulting expression for effective thermal conductivity is given as

$$
k_c = k_m \left[\frac{1 + AB\phi}{1 - B\phi\psi} \right], \quad \text{where } B = \frac{(k_d/k_m) - 1}{(k_d/k_m) + A} \quad \text{and} \quad \psi = 1 + \left(\frac{1 - \phi_m}{\phi_m^2} \right) \phi \quad (20.22a\text{--}c)
$$

The values of A and ϕ_m are suggested in Lewis and Nielsen (1970) for a number of different geometric shapes and orientations, such as $A = 1.50$ for spheres and $\phi_m = 0.637$ for random packing.

Figure 20.2a through c illustrate the behavior of the filler concentration distribution employed in the simulations that follow, besides the corresponding behavior of the dimensionless heat capacity and thermal conductivity according to Equations 20.21 and 20.22, by using $\gamma = 25 \text{ m}^{-1}$ and $x_c/L_x = 0.2$.

The estimations in the present work involved the coefficients of the eigenfunction expansion of $w(x)$ and $k(x)$ and the two values of each coefficient at the boundaries used in the linear filter function of the expansion process. The effective heat transfer coefficient (to be estimated in the form of the parameter d, see Equation 20.3f) was considered to be

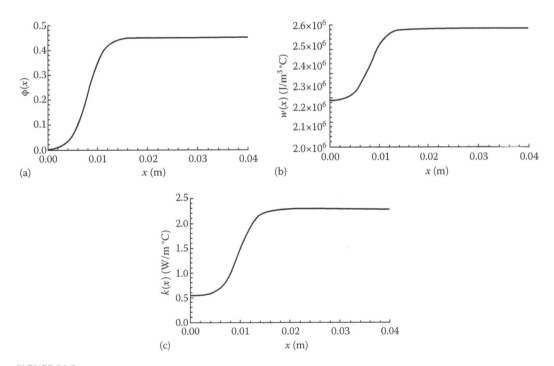

FIGURE 20.2
Sample behavior of the (a) filler concentration distribution, (b) thermal capacity, and (c) thermal conductivity, according to Table 20.1 (cases 1–3).

constant in the present inverse analysis. Thus, the parameters and the number of parameters to be estimated are given by

$$\mathbf{P} = \left[w_{x=0}, w_{x=L}, \overline{w}_1, \overline{w}_2, \overline{w}_3, \dots, \overline{w}_{N_w}, k_{x=0}, k_{x=L}, \overline{k}_1, \overline{k}_2, \overline{k}_3, \dots, \overline{k}_{N_k}, d \right],$$
$$\text{with } N_P = N_w + N_k + 5 \tag{20.23a}$$

where

$$w(x) = w_f(x) + \sum_{k=1}^{N_w} \tilde{\Gamma}_k(x)\overline{w}_k \tag{20.23b}$$

$$k(x) = k_f(x) + \sum_{k=1}^{N_k} \tilde{\Gamma}_k(x)\overline{k}_k \tag{20.23c}$$

The prescribed heat flux is also considered to be governed by an abrupt behavior in the space coordinate, such as in Equations 20.20, practically reproducing a step function with $\gamma_q = 100 \text{ m}^{-1}$ and $x_{c,q}/L_x = 0.5$, for heating in half of the plate length. The two extreme values for the heat flux were taken as $q_0 = 0$, as in Table 20.1, and $q_L = q_w$. The heat flux q_w was not estimated in the present analysis due to linear dependency with the remaining parameters. Therefore, the other parameters were all divided by the assumed value of q_w, in light of the linearity of the problem formulation. This does not impose restrictions on the estimation of the other parameters, which can be recovered by multiplying the estimated values by the available value of q_w.

In the proposed inverse approach, the truncation orders of the thermal capacity and conductivity expansions, N_w and N_k, respectively, control the number of parameters to be estimated. The convergence analysis of the $w(x)$ and $k(x)$ expansions, Equations 20.23b and c, is shown in Figure 20.3a through c, for three different truncation orders, $N_w = N_k = 4, 7$, and 10. It can be observed that the expansions with these three increasing truncation orders are able to recover the behavior of the chosen thermal properties, following the abrupt change in filler concentration. On the other hand, the results for the lowest truncation order, $N_w = N_k = 4$, still show some oscillation around the exact functions, while for $N_w = N_k = 7$, a much closer agreement between the expanded and the exact functions is observed and practically full convergence is achieved with the largest truncation order ($N_w = N_k = 10$).

Before addressing the estimation of the unknown parameters, the behavior of the determinant of the information matrix $\mathbf{J}^T\mathbf{J}$ (Ozisik and Orlande 2000) needs to be analyzed in order to inspect the influence of the number of parameters to be estimated in the solution of the inverse problem. The sensitivity matrix $\mathbf{J}$ is defined as

$$\mathbf{J(P)} = \left[\frac{\partial \mathbf{T}^T(\mathbf{P})}{\partial \mathbf{P}} \right]^T = \begin{bmatrix} \dfrac{\partial T_1}{\partial P_1} & \dfrac{\partial T_1}{\partial P_2} & \dfrac{\partial T_1}{\partial P_3} & \cdots & \dfrac{\partial T_1}{\partial P_{N_P}} \\[2mm] \dfrac{\partial T_2}{\partial P_1} & \dfrac{\partial T_2}{\partial P_2} & \dfrac{\partial T_2}{\partial P_3} & \cdots & \dfrac{\partial T_2}{\partial P_{N_P}} \\[2mm] \vdots & \vdots & \vdots & & \vdots \\[2mm] \dfrac{\partial T_I}{\partial P_1} & \dfrac{\partial T_I}{\partial P_2} & \dfrac{\partial T_I}{\partial P_3} & \cdots & \dfrac{\partial T_I}{\partial P_{N_P}} \end{bmatrix} \tag{20.24}$$

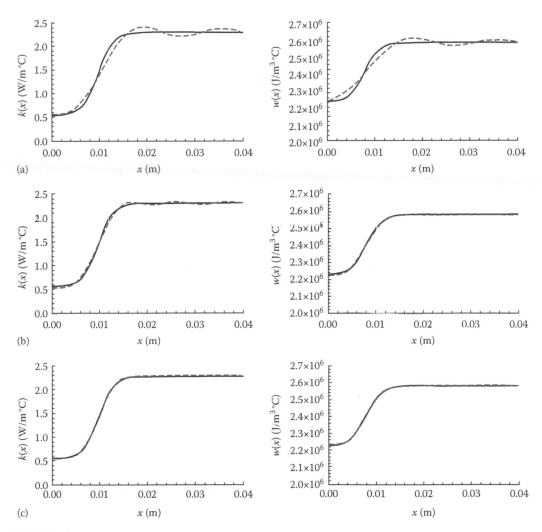

FIGURE 20.3
Convergence behavior of the thermal conductivity and heat capacity expansions; cases 1–3 (original function-solid line, expansion-dashed line): (a) N_w and $N_k = 4$, (b) N_w and $N_k = 7$, and (c) N_w and $N_k = 10$.

The sensitivity coefficients $J_{ij} = \dfrac{\partial T_i}{\partial P_j}$ give the sensitivity of T_i with respect to changes in the parameter P_j. A small value of the magnitude of J_{ij} indicates that large changes in P_j yield small changes in T_i. It can be easily noticed that the estimation of the parameter P_j is extremely difficult in such cases, because basically the same value for T_i would be obtained for a wide range of values of P_j. In fact, when the sensitivity coefficients are small, $|\mathbf{J}^T \mathbf{J}| \approx 0$, and the inverse problem is ill-conditioned (see Chapter 7). It can also be shown that $|\mathbf{J}^T \mathbf{J}|$ is null if any column of $\mathbf{J}$ can be expressed as a linear combination of other columns. Therefore, the maximization of the determinant of the information matrix can be used for the design of the experiment, in a procedure called the D-optimum experimental design (Beck et al. 1977, Ozisik and Orlande 2000). The reduced sensitivity coefficients obtained from the multiplication of the original sensitivity coefficients by the parameters that they are referred to, are

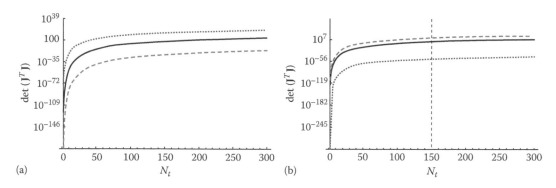

FIGURE 20.4
Evolution of the sensitivity matrix determinant with the number of measurements in time: (a) $N_x = 40$ sensors, $\Delta t = 10$ s ($N_P = 13$, 19, and 25 parameters); (b) $N_P = 19$ parameters, fixed $\Delta t = 10$ s, and $N_x = 4$, 40, and 160 sensors.

preferred for the analysis of linear dependence and small magnitudes. Such is the case because the reduced sensitivity coefficients can be directly compared in magnitude to the measured variables (Beck et al. 1977, Ozisik and Orlande 2000). The analysis of the determinant of the information matrix presented below was performed with reduced sensitivity coefficients.

Based on possible experimental procedures, we will consider the following two cases for the analysis of the determinant of the information matrix: (i) variation of the number of parameters to be estimated with a fixed number of spatial measurements (sensors) and a fixed frequency of measurements (Figure 20.4a) and (ii) variation of the number of spatial measurements (sensors) with a fixed frequency of measurements and a fixed number of parameters (Figure 20.4b).

Figure 20.4a shows the evolution in time of the information matrix determinant for a total of 12,000 measurements ($N_x = 40$ along the domain, and $N_t = 300$ in time). The three curves stand for an increasing number of parameters, $N_P = 13$, 19, and 25 (from top to bottom), which correspond respectively to $N_w = N_k = 4$, 7, and 10, plus the two end values of thermal capacity and conductivity that are filtered from the expansion and the parameter d. Clearly, the gradual increase on the number of parameters decreases the value of the determinant. Therefore, it has been observed, as expected, that increasing the number of parameters significantly affects the conditioning of the estimation procedure.

Figure 20.4b presents the information matrix determinant for the case of $N_P = 19$ parameters, but with a variable number of equally spaced measurements along the domain ($N_x = 160$, 40, and 4, from top to bottom). The lowest value of N_x has been considered to inspect the possibility of employing traditional temperature measurement techniques, such as thermocouples, while the higher values represent a thermographic type of temperature measurement. This figure shows that the determinant of the sensitivity matrix significantly decreases by reducing the number of measurements along the domain.

A relevant aspect in the use of the eigenfunction expansion coefficients as a parameter estimation procedure is the definition of maximum and minimum values for the coefficients to be estimated, from the corresponding maximum and minimum values of the thermal capacity and conductivity, w_{max}, k_{max} and w_{min}, k_{min}. For instance, the parameterized form of the thermal conductivity used in this application is given by

$$k(x) = \left(\frac{k_{x=L} - k_{x=0}}{L} \right) x + k_{x=0} + \sum_{k=1}^{N_k} \bar{k}_k \tilde{\Gamma}_k(x) \qquad (20.25a)$$

which can be rewritten as

$$\sum_{k=1}^{N_k} \bar{k}_k \tilde{\Gamma}_k(x) = k(x) - \left(\frac{k_{x=L} - k_{x=0}}{L}\right)x + k_{x=0} \tag{20.25b}$$

Operating with $\displaystyle\int_0^L \tilde{\Gamma}_i(x)\,dx$ on both sides of the above equation, we have

$$\bar{k}_i = \int_0^L \tilde{\Gamma}_i(x)k(x)dx - \left(\frac{k_{x=L} - k_{x=0}}{L}\right)\bar{g}_i - k_{x=0}\bar{f}_i \tag{20.26a}$$

where

$$\bar{g}_i = \int_0^L x\tilde{\Gamma}_i(x)dx \tag{20.26b}$$

$$\bar{f}_i = \int_0^L \tilde{\Gamma}_i(x)dx \tag{20.26c}$$

Thus, for a bounding maximum or minimum $k(x)$, $k_b = k_{\min}$ or $k_b = k_{\max}$, respectively, we have

$$\bar{k}_{i,b} = (k_b - k_{x=0})\bar{f}_i - \left(\frac{k_{x=L} - k_{x=0}}{L}\right)\bar{g}_i \tag{20.27}$$

Since the values of the thermal capacity or conductivity at the boundaries are not known a priori, to either maximize or minimize the values of the transformed coefficients in Equation 20.27, we need to take into consideration the signs of the coefficients $\bar{g}_i$ and $\bar{f}_i$. Thus, from the analysis of the expression above, and the specific forms of the transformed quantities, $\bar{g}_i$ and $\bar{f}_i$ for odd or even indices, one may get conservative upper and lower limits for the expansion coefficients, $\bar{k}_{i,\max}$ and $\bar{k}_{i,\min}$.

The parameters were estimated by using the Metropolis–Hastings algorithm as described above. To estimate the maximum and minimum ranges for each parameter, we have conservatively adopted as the upper limit, the filler thermal capacity and conductivity, $w_{\max}(x) = w_d$ and $k_{\max}(x) = k_d$, and as the lower limit the matrix capacity and conductivity $w_{\min}(x) = w_m$ and $k_{\min}(x) = k_m$. Alternatively, the theoretical models previously discussed could have been used to narrow the intervals $[w_{\min}, w_{\max}]$ and $[k_{\min}, k_{\max}]$, but at the present stage of tools demonstration, we have preferred to employ the wider range. As initial states in the Markov chain for the coefficients $w(x)$ and $k(x)$, we have considered a constant function given by the average value of the coefficients in the range defined by their upper and lower bounds.

Gaussian distributions were provided as priors for each of the coefficients to be estimated. In building the priors for the thermophysical properties, we have assumed a Gaussian uncertainty for the filler concentration distribution, with a standard deviation of 20% of the exact value, which was then propagated to the thermal capacity and conductivity prior distributions.

TABLE 20.2

Exact Values, Initial States in the Markov Chain, and the Lower
and Upper Bounds for Each Parameter (Cases 1–3)

Parameter	Exact	Initial State	P_{min}	P_{max}
h_{eff}, W/m^2°C	16.694	18.364	10	20
$k_{x=0}$, W/m°C	0.54897	0.60386	0.545	5.7856
$k_{x=L}$, W/m°C	2.2929	2.5221	0.545	5.7856
$\bar{k}_1$, W/m°C	0.10972	0.12069	−0.9436	0.9436
$\bar{k}_2$, W/m°C	0.00204	0.00225	−0.2359	0.2359
$\bar{k}_3$, W/m°C	−0.02825	−0.03108	−0.3145	0.3145
$\bar{k}_4$, W/m°C	−0.02661	−0.02927	−0.1180	0.1180
$\bar{k}_5$, W/m°C	−0.01328	−0.01461	−0.1887	0.1887
$\bar{k}_6$, W/m°C	−0.00107	−0.00118	−0.07864	0.07864
$\bar{k}_7$, W/m°C	0.00485	0.00534	−0.1348	0.1348
$w_{x=0}$, J/m^3°C	2.2288×10^6	2.4517×10^6	2.226×10^6	2.938×10^6
$w_{x=L}$, J/m^3°C	2.5823×10^6	2.8405×10^6	2.226×10^6	2.938×10^6
$\bar{w}_1$, J/m^3°C	25,047.5	27,552.2	−128,155	128,155
$\bar{w}_2$, J/m^3°C	4,370.18	4,807.2	−32,038.7	32,038.7
$\bar{w}_3$, J/m^3°C	−2,701.11	−2,971.23	−42,718.2	42,718.2
$\bar{w}_4$, J/m^3°C	−4,449.02	−4,893.93	−16,019.3	16,019.3
$\bar{w}_5$, J/m^3°C	−3,613.83	−3,975.21	−25,630.9	25,630.9
$\bar{w}_6$, J/m^3°C	−1,955.27	−2,150.79	−10,679.6	10,679.6
$\bar{w}_7$, J/m^3°C	−512.218	−563.44	−18,307.8	18,307.8

The proposal densities used for the generation of candidates for the parameters at each state of the Markov chain were also Gaussian with a standard deviation of 1% of the standard deviation of the Gaussian priors. With the lower and upper bounds computed as described above, the prior densities were null outside these limits for each parameter. Such bounds are presented in Table 20.2, together with the exact and initial states of the Markov chain for each parameter.

Three illustrative cases were analyzed below, aiming at the validation and demonstration of the proposed methodology, which are summarized in Table 20.3. Note that case 1 involves idealized conditions with $NT = 15$ and $N_w = N_k = 4$ ($N_P = 13$ parameters), in both the simulated results and the inverse analysis. In addition, the simulated temperature data were generated with 0.1°C of uncertainty, and uncertainties in the filler concentration prior were not considered. The priors for the thermal capacity and conductivity were then allowed

TABLE 20.3

Definition of Input Data for Test Cases 1–3

Input Data	Case 1	Case 2	Case 3
NT, N_w, N_k (*simul. data*)	15, 4, 4	50, 14, 14	50, 14, 14
NT, N_w, N_k (*inverse sol.*)	15, 4, 4	15, 7, 7	15, 7, 7
N_P (*inverse sol.*)	13	19	19
Std. dev. priors (w, k, h_{eff})	40%, 40%, 20%	40%, 40%, 20%	40%, 20%, 20%
Exp. uncertainty	0.1°C	0.5°C	0.5°C

to have a standard deviation of 40%, while the effective heat transfer coefficient prior included a standard deviation of 20%. Cases 2 and 3 are closer to the actual thermophysical properties identification task in the present application, where the number of parameters to be estimated was increased to $N_P = 19$, with $N_w = N_k = 7$, and the uncertainty on the temperature measurements was of 0.5°C. For cases 2 and 3, the simulated measurements were computed with $NT = 50$ terms in the temperature expansion, and $N_k = 14$ terms in both coefficient expansions for $k(x)$ and $w(x)$. The difference between the two cases is that the standard deviation of the Gaussian prior for the thermal capacity was lowered to 20% in case 3, which lies on the fact that the uncertainty on such property is less affected by changes on the filler concentration, as well as less sensitive to the thermal conductivity models.

By using a burn in period of 10,000 states and a total of 50,000 states in the Markov chains, estimates for the parameters were obtained given by the sample average of the remaining 40,000 states. Table 20.4 summarizes the estimates obtained for each parameter for the three cases examined. As expected, case 1 provided the best estimates because it was aimed at validating the constructed computational procedure, where the inverse crime was allowed for and the uncertainty of the simulated temperature data was very low (0.1°C). Cases 2 and 3 both involved the same uncertainty of 0.5°C in the simulated data, but now the number of parameters has been increased. The prior for the heat capacity was allowed to have a smaller standard deviation in the last case, being reduced from 40% to 20%. Some improvement was then observed from cases 2–3, in particular for the estimated coefficients in the thermal capacity expansion. It should be recalled, that since the inverse crime has been avoided for the two last cases, the solution approach is not expected to exactly recover the same parameters employed to generate the simulated data.

TABLE 20.4

Estimated Parameter Values with 50,000 States in Markov Chains Obtained by Neglecting the First 10,000 States (Cases 1–3)

P	Exact	Case 1	Case 2	Case 3
h_{eff}, W/m²°C	16.694	16.690	16.692	16.692
$k_{x=0}$, W/m°C	0.54897	0.55742	0.56523	0.57677
$k_{x=L}$, W/m°C	2.2929	2.3041	2.3023	2.3359
$\bar{k}_1$, W/m°C	0.10972	0.10801	0.10723	0.10327
$\bar{k}_2$, W/m°C	0.00204	0.00225	0.00205	0.00232
$\bar{k}_3$, W/m°C	−0.02825	−0.02912	−0.02969	−0.03080
$\bar{k}_4$, W/m°C	−0.02661	−0.02636	−0.02728	−0.02658
$\bar{k}_5$, W/m°C	−0.01328	—	−0.01275	−0.01351
$\bar{k}_6$, W/m°C	−0.00107	—	−0.00111	−0.00105
$\bar{k}_7$, W/m°C	0.00485	—	0.00580	0.00589
$w_{x=0}$, J/m³°C	2.2288×10^6	2.2341×10^6	2.2814×10^6	2.2471×10^6
$w_{x=L}$, J/m³°C	2.5823×10^6	2.5872×10^6	2.6184×10^6	2.5947×10^6
$\bar{w}_1$, J/m³°C	25,047.5	24,264.8	15,923.9	22,196.6
$\bar{w}_2$, J/m³°C	4,370.18	4,928.48	4,892.01	5,009.25
$\bar{w}_3$, J/m³°C	−2,701.11	−3,156.08	−2,405.1	−2,622.2
$\bar{w}_4$, J/m³°C	−4,449.02	−5,132.59	−4,654.52	−4,857.93
$\bar{w}_5$, J/m³°C	−3,613.83	—	−3,912.02	−4,337.8
$\bar{w}_6$, J/m³°C	−1,955.27	—	−2,367.64	−2,283.89
$\bar{w}_7$, J/m³°C	−512.22	—	−610.09	−529.54

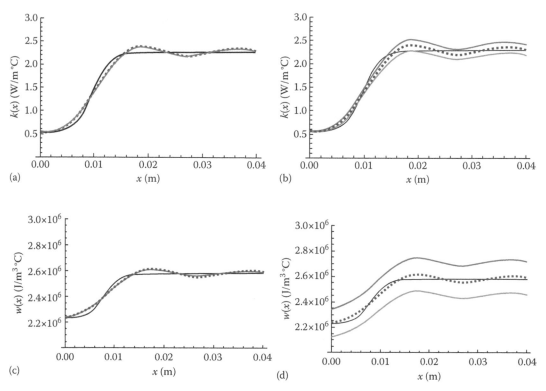

FIGURE 20.5
Estimated functions in case 1: (a) $k(x)$—exact (solid thin), exact expanded (solid thick), and estimated (dashed); (b) $k(x)$—exact (solid thin), confidence bounds (solid thick), and estimated (dashed); (c) $w(x)$—exact (solid thin), exact expanded (solid thick), and estimated (dashed); (d) $w(x)$—exact (solid thin), confidence bounds (solid thick), and estimated (dashed).

Figures 20.5 through 20.7 summarize the estimated functions for the two properties, their exact original and expanded variations, together with the curves of the confidence intervals at a degree of 99% of confidence, for the three cases examined, respectively. Clearly, case 1 (Figure 20.5) provides the best set of results, offering further evidence of the algorithm verification. It is noticeable from Figure 20.5a and c that the exact functions, $k(x)$ and $w(x)$, expanded with four terms, are fully recovered by the inverse analysis, also with $N_w = N_k = 4$. The thin solid line represents the exact original function for the coefficients, which is not to be recovered, but shown just for reference purposes.

In Figure 20.6a through d, related to case 2, we illustrate the behavior of the estimation of the properties with expansions of $N_w = N_k = 7$, which clearly provide a much better reproduction of the original functions (thin solid line), practically overwritten by the exact expansions to these same orders. Fairly large standard deviations of 40% were provided for the priors on both $k(x)$ and $w(x)$, so as to challenge the proposed approach, and still quite reasonable estimates were achieved. Only for the thermal capacity estimation some deviations are noticeable at the two boundary positions. The exact value for $w_{x=0}$ is 2.22878×10^6 J/m^3 °C and for $w_{x=L}$, it is 2.58226×10^6 J/m^3 °C, while the estimation provides the values 2.28139×10^6 J/m^3 °C and 2.61837×10^6 J/m^3 °C, with 99% confidence intervals $[2.22093 \times 10^6, 2.34185 \times 10^6]$ J/m^3 °C and $[2.58306 \times 10^6, 2.65367 \times 10^6]$ J/m^3 °C, at $x = 0$ and $x = L$, respectively, which clearly include the exact boundary values. Case 3

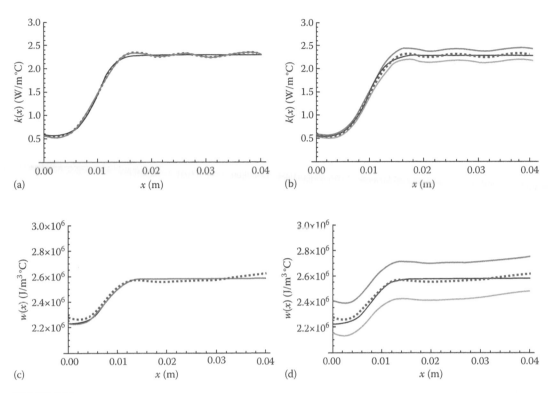

FIGURE 20.6
Estimated functions in case 2: (a) $k(x)$—exact (solid thin), exact expanded (solid thick), and estimated (dashed); (b) $k(x)$—exact (solid thin), confidence bounds (solid thick), and estimated (dashed); (c) $w(x)$—exact (solid thin), exact expanded (solid thick), and estimated (dashed); (d) $w(x)$—exact (solid thin), confidence bounds (solid thick), and estimated (dashed).

(Figure 20.7) results in a more accurate estimation, also obtained with $N_w = N_k = 7$, but with a reduction on the standard deviation to 20% in the thermal capacity prior. Such a reduction promotes a noticeable improvement on the estimated thermal capacity function (Figure 20.7c), with respect to the previous case (Figure 20.6c). The estimation of the thermal capacity boundary values now provides 2.24709×10^6 J/m^3°C and 2.59468×10^6 J/m^3°C, with the 99% confidence intervals $[2.21344 \times 10^6, 2.28073 \times 10^6]$ J/m^3°C and $[2.57038 \times 10^6, 2.61898 \times 10^6]$ J/m^3°C, at $x = 0$ and $x = L_x$, respectively. The estimation of the thermal conductivity function was fairly accurate in both cases 2 and 3, while the estimates for the heat transfer coefficient were very accurate throughout the cases analyzed.

The second selected test configuration, aimed at demonstrating inverse analysis with the transformed temperatures, deals with a thermally thin plate of thickness $L_z = 1$ mm heated by an electrical resistance on one of its surfaces, up to a fraction $x_c = L_x/3$ of its total length, $L_x = 12$ cm. The opposite surface of the plate experiences heat losses by both natural convection and radiation, here taken into account in a linearized form, while all lateral bounderies are considered to be insulated. This problem was thus modeled as a one-dimensional transient heat conduction formulation, given by Equations 20.1, and represented in Figure 20.1, after lumping in the transversal direction. The power dissipated in the resistance per unit area was considered to be known, q_{inf}, and the spatial distribution of the applied heat flux was also available, being uniform up to $x = x_c$ and zero for $x > x_c$,

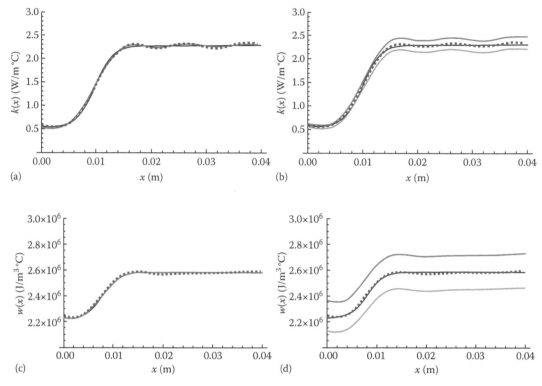

FIGURE 20.7
Estimated functions in case 3: (a) $k(x)$—exact (solid thin), exact expanded (solid thick), and estimated (dashed); (b) $k(x)$—exact (solid thin), confidence bounds (solid thick), and estimated (dashed); (c) $w(x)$—exact (solid thin), exact expanded (solid thick), and estimated (dashed); (d) $w(x)$—exact (solid thin), confidence bounds (solid thick), and estimated (dashed).

while an exponential function in time, $f(t)$, modeled the delay due to the thermal capacitance of the resistance assembly, that is,

$$q_w[x,t] = q[x]f[t] \qquad (20.28a)$$

$$q[x] = \begin{cases} q_{\text{inf}} & 0 < x < L_x/3 \\ 0 & L_x/3 < x < L_x \end{cases} \qquad (20.28b)$$

$$f[t] = 1 - ae^{-bt} \qquad (20.28c)$$

We then seek the simultaneous estimation of the thermal capacity, thermal conductivity, effective heat transfer coefficient, and coefficients of the time delay of the applied heat flux, respectively, $w(x)$, $k(x)$, $h_{\text{eff}}(x)$, a and b.

In the present inverse problem analysis, the test case was chosen in the form of a polymeric matrix (HDPE) with alumina nanoparticles (Al_2O_3) dispersed in the matrix, with the concentration distribution governed by an exponential function, and the sample has essentially just the polymer at $x = 0\%$ and 60% of nanoparticles at $x = L_x$. The polymer has thermal capacity and conductivity, respectively, of $w_m = 2.2264 \times 10^6$ J/m^3°C and $k_m = 0.545$ W/m°C, while the alumina nanoparticles have thermophysical properties given by $w_p = 3.0172 \times 10^6$ J/m^3°C and $k_p = 36$ W/m°C. By employing the theory of mixtures and the Lewis and Nielsen correlation (Kumlutas et al. 2003, Kumlutas and Tavman 2006) to compute the effective thermal capacity and conductivity throughout the domain, we obtain at $x = L_x$ the effective values $w_{x=L_x} = 2.7008 \times 10^6$ J/m³ °C and $k_{x=L_x} = 9.078$ W/m/°C. In the present test case, the thermophysical properties were chosen to vary in the following exponential forms:

$$k(x) = k_0 \exp\left[2\beta_k\left(1 - \frac{x}{L_x}\right)\right], \quad \beta_k = 1.4064 \tag{20.29a}$$

$$w(x) = w_0 \exp\left[2\beta_w\left(1 - \frac{x}{L_x}\right)\right], \quad \beta_w = 0.0966 \tag{20.29b}$$

which are illustrated in Figure 20.8a and b for the thermal conductivity and capacity, respectively.

The effective heat transfer coefficient was also estimated, accounting for natural convection and linearized radiation at the horizontal plate, yielding the behavior shown in Figure 20.9a. Figure 20.9b illustrates possible behaviors for the time lag in the applied heat flux, by

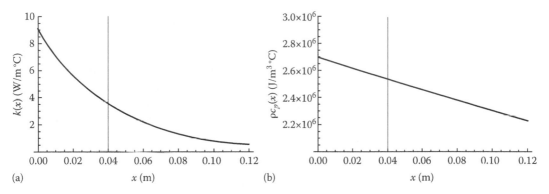

FIGURE 20.8
Spatial behavior of thermophysical properties (cases 4–6): (a) thermal conductivity and (b) thermal capacity.

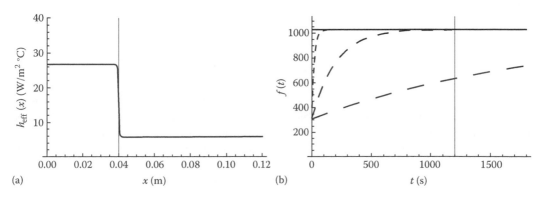

FIGURE 20.9
Behavior of additional coefficients (cases 4–6): (a) effective heat transfer coefficient and (b) time lag function in applied heat flux (Equation 20.28c).

TABLE 20.5

Data Employed to Generate
Simulated Experimental Data
for Cases 4–6

t_{final}	3600 s
L_x	0.12 m
L_y	0.04 m
L_z	0.001 m
x_c	0.04 m
E	0.97
A	0.7
B	0.005 s^{-1}
q_{inf}	1030.9 W/m^2
T_{∞}	23.4°C

varying the parameter a and fixing $b = 0.005$. The remaining data that defines the next test cases are provided in Table 20.5, with which the simulated experimental data were generated.

The filter functions were chosen in such a way to incorporate the values of the coefficients at the two boundaries, $x = 0$ and L_x, (k_{x0}, k_{xL}, w_{x0}, w_{xL} and d_{x0}, d_{xL}), which are unknown and should be estimated together with the eigenfunction expansion coefficients, so as to make the boundary conditions of the chosen eigenfunction homogeneous. For the thermophysical properties $k(x)$ and $w(x)$, we have again employed a simple linear filter without any sort of prior information on the coefficients variation. For the heat loss coefficient, $d(x)$, a more informative filter was adopted in the form of a steep variation approaching a step function, since this behavior is physically expected, in light of the functional form of the applied heat flux.

For the generation of the simulated data, we have employed 50 terms in the eigenfunction expansions ($NT = 50$) for temperature and 10 terms ($N_k = 10$, $N_w = 10$, and $N_d = 10$) in the expansions of $k(x)$, $w(x)$, and $d(x)$. Two different levels of experimental error were examined, namely, with an uncertainty of 0.01°C for validation purposes, and an uncertainty of 0.5°C as a more realistic error level (corresponding to a standard deviation of almost 0.2°C).

The integral transformation process on the experimental data was performed according to Equation 20.13a, after interpolating the experimental points with cubic splines in the spatial domain. In the inverse analysis that follows, 241 sensors were employed in the integral transformation of the simulated experimental data, such as in a thermographic type measurement. A considerable compression on the experimental data set is then achieved, as illustrated in Table 20.6, where the total number of experimental points is shown for a fixed number of sensors (241), available for the estimation procedure when performed directly in the temperature field or for the estimation on the transformed temperature field, by varying the number of terms in the temperature eigenfunction expansion ($NT = 10$, 20, and 40). It can be noticed that a reduction of more than 10 times is achieved when the plain temperature data is replaced by the transformed temperature field with the truncation order of $NT = 20$.

Figure 20.10 and Table 20.7 present the determinant of the information matrix by using the integral transformed data in the inverse analysis. The three curves shown in

TABLE 20.6

Comparison of the Number of Experimental Points for the Estimation
in the Temperature Field and in the Transformed Temperature Field (Cases 4–6)

	Number of Time Measurements, N_t	Total Number of Experimental Points, N_m
Number of spatial measurements, N_x		
241	120	25,680
	200	48,200
	300	72,300
Number of modes in temperature expansion		
$NT = 10$	120	1,200
	200	2,000
	300	3,000
$NT = 20$	120	2,400
	200	4,000
	300	8,000
$NT = 40$	120	4,800
	200	8,000
	300	12,000

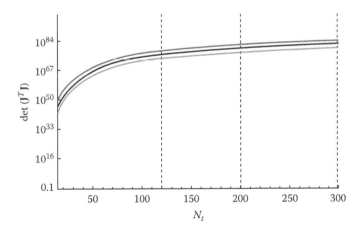

FIGURE 20.10
Determinant of the information matrix for estimation in the transformed temperature field (from top to bottom, $NT = 40$, 20, and 10 terms).

TABLE 20.7

Analysis of the Determinant of the Information Matrix with the Estimation
in the Transformed Temperature Field with Different Temperature
Transformation Modes and Different Number of Measurements in Time

	Number of Sensors $N_x = 241$, $N_P = 15$		
Number of Time Measurements N_t	Determinant		
	$NT = 10$	$NT = 20$	$NT = 40$
120	9.34×10^{73}	3.50×10^{76}	2.17×10^{78}
200	4.19×10^{77}	1.69×10^{80}	1.09×10^{82}
300	8.35×10^{79}	3.63×10^{82}	2.80×10^{84}

Figure 20.10 correspond from top to bottom to $NT = 40$, 20, and 10. We can notice that an increase in the number of time measurements leads to a large increase in the values of the determinant, as a result of the increase on the number of experimental points in the transformed domain. On the other hand, by doubling the number of modes available for the transformed data, a relatively small increase is observed in the determinant values for a fixed number of measurements in time. Therefore, we preferred to perform the inverse problem solution by keeping $NT = 10$ modes.

Figure 20.11a shows simulated experimental temperature data with the 0.5°C uncertainty level, for selected times along the plate. For the sake of comparison, Figure 20.11b illustrates the time evolution of the first 10 transformed temperature fields, with the abscissa represented by the number of time intervals, again for the 0.5°C uncertainty case. This figure shows the more significant role of the first five transformed potentials.

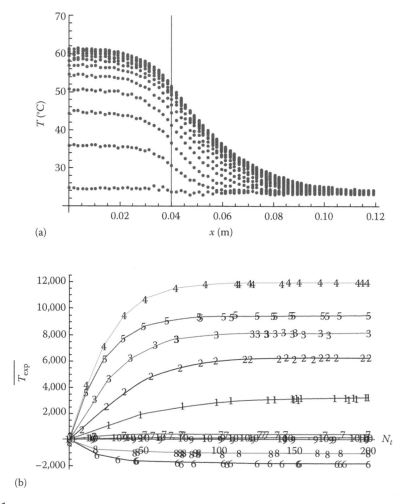

FIGURE 20.11
(a) Simulated temperature data for selected times along the plate, for the 0.5°C uncertainty and (b) simulated transient variation of the transformed temperature data up to $NT = 10$ and for the 0.5°C uncertainty (numeric symbols indicate the transformed temperature mode).

TABLE 20.8

Test Cases Examined (Second Example, Cases 4–6)

Case	Field	NT	N_t	N_x	Prior k_{x0}, k_{xL}, $\bar{k}_j$, w_{x0}, w_{xL}, $\bar{w}_j$, d_{x0}, d_{xL}, $\bar{d}_j$, a, b
Test case with temperature uncertainty of 0.01°C					
4	Temperature	10	120	61	N, N, U–N, N, U–N, N, U–U, U
Test cases with temperature uncertainty of 0.5°C					
5	Transformed	10	200	241	N, N, U–N, N, U–N, N, U–U, U
6	Transformed	10	200	241	U, U, U–U, U, U–N, N, U–U, U

N, normal (Gaussian) prior; U, uniform prior.

The three test cases examined in this second example are summarized in Table 20.8. Test case 4 was chosen for verification purposes, since the temperature uncertainty was kept very low, 0.01°C, and the experimental data were generated with the same number of terms in the expansions that were employed in the estimations, that is, $NT = 10$, $N_k = 3$, $N_w = 3$, and $N_d = 1$, respectively for $T(x, t)$, $k(x)$, $w(x)$, and $d(x)$. Case 4 involves the estimation in the original temperature field, for the sake of comparison against the other estimations with the transformed temperature measurements. For the more realistic test cases 5 and 6, the experimental data were generated with $NT = 50$, $N_k = 10$, $N_w = 10$, and $N_d = 10$, while the estimation was performed with the same reduced number of terms as in test case 4, in order to avoid the inverse crime. The difference between these two cases lies on the type of prior information provided for the boundary values of the thermophysical properties, that is, uniform distributions for case 6 or normal (Gaussian) distributions for case 5. In all three cases, a noninformative prior (uniform) was utilized for the transformed coefficients of the two thermophysical properties and of the heat loss coefficient, besides the two heat flux parameters a and b. The uniform prior was defined with the minimum and maximum allowable limits in the search procedure, as detailed in Table 20.9. In case 5,

TABLE 20.9

Input Data for Estimation (Cases 5 and 6)

P	Prior	Exact	Min	Max	Initial
k_{x0}, W/m°C	Uniform or normal ($\sigma = 5\%$)	9.0780	0.463	10.440	8.6157
k_{xL}, W/m°C	Uniform or normal ($\sigma = 5\%$)	0.545	0.463	10.440	0.5028
$\bar{k}_1$, W/m°C	Uniform	−0.6677	−3.111	3.111	−0.7256
$\bar{k}_2$, W/m°C	Uniform	−0.1111	−0.778	0.778	−0.1082
$\bar{k}_3$, W/m°C	Uniform	−0.04091	−1.037	1.037	−0.04433
w_{x0}, J/m³°C	Uniform or normal ($\sigma = 5\%$)	2.701×10^6	1.892×10^6	3.106×10^6	2.686×10^6
w_{xL}, J/m³°C	Uniform or normal ($\sigma = 5\%$)	2.226×10^6	1.892×10^6	3.106×10^6	2.282×10^6
$\bar{w}_1$, J/m³°C	Uniform	−2,894.68	−378,487.0	378,487.0	−2,810.39
$\bar{w}_2$, J/m³°C	Uniform	−34.942	−94,621.8	94,621.8	−33.045
$\bar{w}_3$, J/m³°C	Uniform	−107.57	−126,162.0	126,162.0	−104.67
h_{x0}, W/m²°C	Normal ($\sigma = 20\%$)	26.620	13.310	53.241	26.601
h_{xL}, W/m²°C	Normal ($\sigma = 20\%$)	5.7286	2.8643	11.457	6.2323
$\bar{h}_1$, W/m²°C	Uniform	0	-3×10^{-12}	3×10^{-12}	0
$a^* = aq_{inf}$ W/m²	Uniform	721.65	0	1,237.1	700.89
b, s^{-1}	Uniform	0.005	0	0.1	0.00521

for the values of thermophysical properties and of the heat loss coefficient at the boundaries, Gaussian prior distributions were adopted, centered at the expected values for each parameter, with a standard deviation of 5% of the respective exact value for the $k_f(x)$ and $w_f(x)$ boundary values and of 20% for the $d_f(x)$ boundary values. Table 20.9 also presents the expected exact values, the upper and lower allowable bounds, and the initial states in the Markov chain used in the inverse analysis for each parameter in cases 5 and 6.

Table 20.10 presents the results obtained for test cases 4–6. The exact values of the parameters are well recovered in these test cases. However, one may observe a better estimation for the parameters in test cases 5 and 6, which made use of integral transformed measured data. Therefore, the data compression through the integral transformation of the measured data, which was introduced in this work, does not affect the spatial information conveyed by the local temperature measurements, and is still capable of resulting in very accurate estimations for the parameters. The analysis of the results for the more challenging test cases 5 and 6, with uncertainty of the temperature measurements of 0.5°C, reveals that better estimates are achieved with the use of the Gaussian priors for the thermal properties (test case 5). This in fact does not represent practical limitations, since in most real situations, some sort of information is in general available to be somehow accounted for in the Gaussian prior, either from previously obtained direct measurements of the thermophysical properties or from theoretical considerations in terms of the constituents. Nevertheless, the results obtained for test case 6, with essentially noninformative uniform prior distributions for the thermal properties, demonstrate that the present approach, even in such cases, can still provide reasonable estimates in terms of the unknown coefficients. This is also clear from the excellent agreement achieved between the experimental and estimated temperatures, for both test cases 5 and 6, as demonstrated in Figure 20.12a and b, respectively, for times $t = 120$, 600, and 1200 s.

TABLE 20.10

Estimated Parameters

P	Exact	Test Case 4	Test Case 5	Test Case 6
k_{x0}, W/m°C	9.078	10.281	9.3645	10.404
k_{xL}, W/m°C	0.545	0.592	0.5186	0.7424
$\bar{k}_1$, W/m°C	−0.668	−0.804	−0.6742	−0.8135
$\bar{k}_2$, W/m°C	−0.111	−0.147	−0.1015	−0.1197
$\bar{k}_3$, W/m°C	−0.0409	−0.0494	−0.02804	−0.03674
w_{x0}, J/m³°C	2.7009×10^6	2.872×10^6	2.791×10^6	3.093×10^6
w_{xL}, J/m³°C	2.2264×10^6	2.308×10^6	2.290×10^6	2.258×10^6
$\bar{w}_1$, J/m³°C	−2894.7	−3025.4	−2789.49	−2823.6
$\bar{w}_2$, J/m³°C	−34.942	−38.779	−31.272	−32.303
$\bar{w}_3$, J/m³°C	−107.57	−124.94	−107.78	−110.12
h_{x0}, W/m²°C	26.620	26.503	26.551	26.434
h_{xL}, W/m²°C	5.7286	6.023	5.9186	6.2039
$\bar{h}_1$, W/m²°C	0	-1.31×10^{-13}	1.316×10^{-15}	-4.700×10^{-14}
$a^* = aq_{\text{inf}}$ W/m²	721.65	701.08	710.44	677.37
b, s⁻¹	0.005	0.00510	0.00505	0.00519

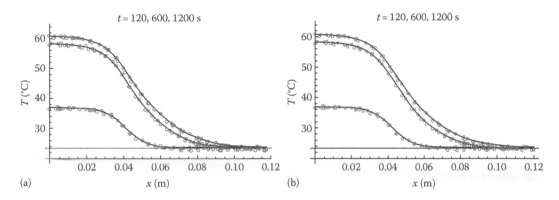

FIGURE 20.12
(a) Comparison between experimental (circles) and estimated temperatures (solid lines) at different times for test case 5 and (b) comparison between experimental (circles) and estimated temperatures (solid lines) at different times for test case 6.

20.5 Conclusions

The combined use of integral transforms and Bayesian inference was illustrated for the inverse problem of simultaneously estimating the space variable thermal capacity and conductivity in two-phase dispersed systems, undergoing a transient one-dimensional heat conduction process. The direct problem solution was analytically obtained with the classical integral transform technique, while the related eigenvalue problem, that carries the information on the medium heterogeneities, was solved with the GITT.

The inverse problem solution was based on the MCMC method. The Metropolis–Hastings algorithm was employed for the sampling procedure, all implemented in the *Mathematica* symbolic computation platform. Instead of seeking the function estimation in the form of a set of local values, an alternative approach was employed by using eigenfunction expansions of the thermal properties themselves. This approach significantly reduces the number of parameters to be estimated in comparison to the strategy of employing local values. Gaussian and uniform distributions were used as priors with fairly large standard deviations, together with simulated experimental data with added uncertainty, in order to demonstrate the robustness of the inverse analysis.

The results obtained with simulated temperature data reveal that the proposed inverse analysis approach can provide accurate estimation of the thermophysical properties variation and is robust with respect to measurement errors, even for noninformative prior distributions.

Also introduced here is the use of experimental information in the transformed domain, that is, the discrete temperature measurements along the space coordinate are integrally transformed into a small set of modes that represent the experimental transformed temperatures. A significant data compression is achieved through this transformation procedure, thus accelerating the inverse problem algorithm, without loss of information in the reconstruction of the local information on the spatial behavior of the thermophysical properties. The approach implemented here, should in principle be directly extendable to multidimensional situations, following the well-established application of the direct problem solution in two- and three-dimensional problems, thus taking further advantage of the abundant spatially distributed data set available through thermographic temperature measurements.

Acknowledgments

The authors would like to acknowledge the financial support provided by CNPq, CAPES, and FAPERJ, Brazilian agencies for the fostering of sciences.

Nomenclature

a	coefficient in the time lag function of the applied heat flux (Equation 20.28c)
b	coefficient in the time lag function of the applied heat flux (Equation 20.28c)
c_p	specific heat (Equation 20.1a)
$d(x)$	heat loss operator coefficient (Equation 20.3a)
$f(t)$	time lag function in applied heat flux (Equation 20.28a)
$h_{\text{eff}}(x)$	effective heat transfer coefficient (Equation 20.1a)
$k(x)$	space variable thermal conductivity (Equation 20.1a)
L_x	plate length
L_z	plate thickness
M	truncation order in the eigenvalue problem expansion
M_n	normalization integrals in the auxiliary eigenvalue problem
NT	truncation order in temperature expansion
N_w, N_k	truncation orders in coefficients expansions, $w(x)$ and $k(x)$, respectively
N_P	number of parameters to be estimated
N_x	number of measurements along the spatial domain (sensors)
N_t	number of measurements in time
N_m	total number of measurements
N_i	normalization integrals in the original eigenvalue problem
$\mathbf{P}$	vector of unknown parameters
$P(x,t)$	source term (Equation 20.3a,g)
$q_w(x,t)$	applied heat flux (Equation 20.1a)
q_{inf}	heat flux dissipated from electrical resistance (Equation 20.28b)
t	time variable
$\mathbf{T}$	vector of estimated temperatures
$T_m(x,t)$	temperature distribution
$w(x)$	thermal capacity (Equation 20.3a)
$w_f(x)$	filter for thermal capacity expansion
$\mathbf{W}$	covariance matrix of the measurement errors
x	space coordinate
$\mathbf{Y}$	vector of measurements

Greek Letters

γ	parameter in the heat flux or the linear dissipation coefficient spatial variation
ε	emissivity
λ	eigenvalues of the auxiliary problem
μ	eigenvalues of the original problem
ρ	density

ψ	eigenfunctions of the original problem
Ω	eigenfunctions of the auxiliary problem

Subscripts and Superscripts

d	dispersed phase (filler) properties
f	filtering function in coefficients expansion
i, n, m	order of eigenquantities
$-$	integral transform
$\sim$	normalized eigenfunction
m	matrix phase properties

References

Alifanov, O.M. 1994. *Inverse Heat Transfer Problems*, New York: Springer-Verlag.

Beck, J.V. and K. Arnold. 1977. *Parameter Estimation in Engineering and Science*, New York: Wiley Interscience.

Chen, B., L. Tong, Y. Gu, H. Zhang, and O. Ochoa. 2004. Transient heat transfer analysis of functionally graded materials using adaptive precise time integration and graded finite elements. *Numer. Heat Transf., B: Fundam.* **45**:181–200.

Cotta, R.M. 1990. Hybrid numerical–analytical approach to nonlinear diffusion problems. *Numer. Heat Transf., B: Fundam.* **127**:217–226.

Cotta, R.M. 1993. *Integral Transforms in Computational Heat and Fluid Flow*, Boca Raton: CRC Press.

Cotta, R.M. 1998. *The Integral Transform Method in Thermal and Fluids Sciences and Engineering*, New York: Begell House.

Cotta, R.M. and M.D. Mikhailov. 1997. *Heat Conduction: Lumped Analysis, Integral Transforms, Symbolic Computation*, New York: Wiley-Interscience.

Cotta, R.M. and M.D. Mikhailov. 2006. Hybrid methods and symbolic computations. In *Handbook of Numerical Heat Transfer*, W.J. Minkowycz, E.M. Sparrow, and J.Y. Murthy, Eds. 2nd edn., pp. 493–522, New York: Wiley.

Cotta, R.M. and M.N. Ozisik. 1986. Laminar forced convection in ducts with periodic variation of inlet temperature. *Int. J. Heat Mass Transf.* **29**:1495–1501.

Danes, F., B. Garnier, and T. Dupuis. 2003. Predicting, measuring and tailoring the transverse thermal conductivity of composites from polymer matrix and metal filler. *Int. J. Thermophys.* **24**: 771–784.

Divo, E. and A. Kassab. 1998. Generalized boundary integral equation for transient heat conduction in heterogeneous media. *J. Thermophys. Heat Transf.* **12**:364–372.

Divo, E., A. Kassab, and F. Rodriguez. 2000. Characterization of space dependent thermal conductivity with a BEM-based genetic algorithm. *Numer. Heat Transf., A: Appl.* **37**: 845–875.

Fang, J., G.F. Zhao, J. Zhao, and A. Parriaux. 2009. On the truly meshless solution of heat conduction problems in heterogeneous media. *Numer. Heat Transf., B: Fundam.* **55**:1–13.

Flach, G.P. and M.N. Ozisik. 1989. Inverse heat conduction problem of simultaneously estimating spatially varying thermal conduction and heat capacity per unit volume. *Numer. Heat Transf., A: Appl.* **16**:249–266.

Fudym, O., B. Ladevie, and J.C. Batsale. 2002. A seminumerical approach for heat diffusion in heterogeneous media: One extension of the analytical quadrupole method. *Numer. Heat Transf., B: Fundam.* **42**:325–348.

Fudym, O., H.R.B. Orlande, M. Bamford, and J.C. Batsale. 2008. Bayesian approach for thermal diffusivity mapping from infrared images processing with spatially random heat pulse heating. *J. Phys. Conf. Ser.* (Online) **135**:12–42.

Gamerman, D. and H.F. Lopes. 2006. *Markov Chain Monte Carlo: Stochastic Simulation for Bayesian Inference*, 2nd edn., Boca Raton: Chapman & Hall/CRC.

Hastings, W.K. 1970. Monte Carlo sampling methods using Markov chains and their applications. *Biometrika* **57**:97–109.

Huang, C.H. and S.C. Chin. 2000. A two-dimensional inverse problem in imaging the thermal conductivity of a non-homogeneous medium. *Int. J. Heat Mass Transf.* **43**:4061–4071.

Huang, C.H. and C.Y. Huang. 2007. An inverse problem in estimating simultaneously the effective thermal conductivity and volumetric heat capacity of biological tissue. *Appl. Math. Model.* **31**:1785–1797.

Huang, C.H. and M.N. Ozisik. 1990. A direct integration approach for simultaneously estimating spatially varying thermal conductivity and heat capacity. *Int. J. Heat Fluid Flow* **11**:262–268.

Huttunen, J.M.J., T. Huttunen, M. Malinen, and J. Kaipio. 2006. Determination of heterogeneous thermal parameters using ultrasound induced heating and MR thermal mapping. *Inst. Phys. Publish.: Phys. Med. Biol.* **51**:1011–1032.

Kaipio, J. and E. Somersalo. 2004. *Statistical and Computational Inverse Problems*, New York: Springer-Verlag.

Kumlutas, D. and I.H. Tavman. 2006. A numerical and experimental study on thermal conductivity of particle filled polymer composites. *J. Thermoplast. Compos. Mater.* **19**:441–455.

Kumlutas, D., I.H. Tavman, and M.T. Çoban. 2003. Thermal conductivity of particle filled polyethylene composite materials. *Compos. Sci. Technol.* **63**:113–117.

Lesnic, D., L. Elliot, D.B. Ingham, B. Clennell, and R.J. Knioe. 1999. The identification of the piecewise homogeneous thermal conductivity of conductors subjected to a heat flow test. *Int. J. Heat Mass Transf.* **42**:143–152.

Lewis, T. and L. Nielsen. 1970. Dynamic mechanical properties of particulate-filled polymers. *J. Appl. Polym. Sci.* **14**:1449–1471.

Lin, S.H. 1992. Transient conduction in heterogeneous media. *Int. Commun. Heat Mass Transf.* **10**:165–174.

Metropolis, N., A.W. Rosenbluth, M.N. Rosenbluth, A.H. Teller, and E. Teller. 1953. Equations of state calculations by fast computating machines. *J. Chem. Phys.* **21**:1087–92.

Migon, H.S. and D. Gamerman. 1999. *Statistical Inference: An Integrated Approach*. London/New York: Arnold/Oxford.

Mikhailov, M.D. and R.M. Cotta. 1994. Integral transform method for eigenvalue problems. *Commun. Numer. Methods Eng.* **10**:827–835.

Mikhailov, M.D. and M.N. Ozisik. 1984. *Unified Analysis and Solutions of Heat and Mass Diffusion*, New York: John Wiley; also, Dover Publications, 1994.

Naveira-Cotta, C.P., R.M. Cotta, H.R.B. Orlande, and O. Fudym. 2009. Eigenfunction expansions for transient diffusion in heterogeneous media. *Int. J. Heat Mass Transf.* **52**:5029–5039.

Naveira-Cotta, C.P., R.M. Cotta, and H.R.B. Orlande, 2011a. Combining integral transforms and Bayesian inference in the simultaneous identification of variable thermal conductivity and thermal capacity in heterogeneous media. *ASME J. Heat Transf.*, in press.

Naveira-Cotta, C.P., R.M. Cotta, and H.R.B. Orlande, 2011b. Inverse analysis with integral transformed temperature fields: Identification of thermophysical properties in heterogeneous media. *Int. J. Heat Mass Transf.*, in press

Orlande, H.R.B., M.J. Colaço, and G.S. Dulikravich. 2008. Approximation of the likelihood function in the Bayesian technique for the solution of inverse problems. *Inverse Probl. Sci. Eng.* **16**:677–692.

Ozisik, M.N. and H.R.B. Orlande. 2000. *Inverse Heat Transfer: Fundamentals and Applications*, New York: Taylor & Francis.

Paez, M. 2010. Bayesian approaches for the solution of inverse problems, chap. 12, in *Thermal Measurements and Inverse Techniques*, H.R.B. Orlande, O. Fudym, D. Maillet, and R.M. Cotta, Eds., Boca Raton, FL: CRC Press.

Progelhof, R.C., J.L. Throne, and R.R. Ruetsch. 1976. Methods for predicting the thermal conductivity of composite systems: A review. *Polym. Eng. Sci.* **16**:615–625.

Rodrigues, F.A., H.R.B. Orlange, and G.S. Dulikravich. 2004. Simultaneous estimation of spatially-dependent diffusion coefficient and source-term in a nonlinear ID diffusion problem. *Math. Comput. Simul.* **66**:409–424.

Serkan Tekce, H., D. Kumlutas, and I.H. Tavman. 2007. Effect of particle shape on thermal conductivity of copper reinforced polymer composites. *J. Reinf. Plast. Compos.* **26**:113–121.

Sphaier, L.A. and R.M. Cotta. 2000. Integral transform analysis of multidimensional eigenvalue problems within irregular domains. *Numer. Heat Transf., B: Fundam.* **38**:157–175.

Sphaier, L.A., R.M. Cotta, C.P. Naveira-Cotta, and J.N.N. Quaresma. 2009. The UNIT (Unified Integral Transforms) symbolic-numerical computational platform for benchmarks in convection–diffusion problems. *Proceedings of the 30th CILAMCE, 30th Iberian–Latin–American Congress on Computational Methods in Engineering*, Búzios, Brazil, November.

Tavman, I.H. 1996. Thermal and mechanical properties of aluminum powder-filled high-density polyethylene composites. *J. Appl. Polym. Sci.* **62**:2161–2167.

Tavman, I.H. and H. Akinci. 2000. Transverse thermal conductivity of fiber reinforced polymer composites. *Int. Commun. Heat Mass Transf.* **27**:253–261.

Weidenfeller, B., M. Hofer, and F.R. Schilling. 2004. Thermal conductivity, thermal diffusivity, and specific heat capacity of particle filled polypropylene. *Compos., A* **35**:423–429.

Wolfram, S. 2005. *The Mathematica Book, Version 5.2*. Cambridge, U.K.: Cambridge-Wolfram Media.

Zabaras, N. 2006. Inverse problems in heat transfer. In *Handbook of Numerical Heat Transfer*, W.J. Minkowycz, E.M. Sparrow, and J.Y. Murthy, Eds., 2nd edn., pp. 525–557. New York: Wiley.

21

Inverse Thermal Radiation Problems: Estimation of Radiative Properties of Dispersed Media

Luís Mauro Moura

CONTENTS

21.1 Introduction

Radiative transfer in participating media is important in many industrial applications. In many processes, radiative heat transfer is the main form of transfer. This situation is well illustrated either by porous materials or media containing particles, which play a key role in radiative transfer process. Some examples are fluidized and packed beds, combustors, surface pigmented coatings, soot and fly ash, sprayed fluids, porous sintered materials, microspheres, ceramic foam, and fibrous insulations (Moura et al., 1998a).

In recent years, many works on semitransparent media (STM) have been carried out considering the medium without interface, that is, the distance among the particles is important so that the porosity is extremely high. Consequently, the effects of reflection and refraction at interfaces cannot be considered. However, many materials, such as ceramic, thin films, coatings, and metals with low porosity, are far from being considered with unit refractive index. Besides the direction change due to refraction, in some problems it is necessary to take into account the increase of local blackbody emission by a factor of the refractive index squared.

In addition, some works can be found referring the study of radiative transfer within materials without unitary refractive index. Special conditions at interfaces must then be taken into account due to refraction (Hottel et al., 1968, Wu et al., 1994, Liou and Wu, 1996).

Baillis and Sacadura (2000) present a review on the determination of radiative properties in STM. Two different techniques can be used to evaluate the radiative transfer in STM. The first one considers radiative transfer as a term that must be added into the heat conduction equation (Tong and Tien, 1980). Although this method is simpler, it always demands the determination of experimental parameters and these parameters are restricted to the range of experimental analysis. The second technique consists of using the radiative transfer equation (RTE) coupled to the heat conduction equation and/or Navier Stokes equations. The solution of RTE requires the knowledge of the radiative properties of the medium. These properties can be determined using two different techniques: (i) using the Maxwell equations (electromagnetic field), when it is necessary to know the morphologic parameters and the spectral optical properties of the medium; (ii) measuring the radiative intensities field emitted, transmitted, and/or reflected by a sample with an experimental device and identifying the radiative properties by inversion techniques. The first one is normally used when morphologic parameters and the spectral optical properties are easier to define such as spherical or cylindrical geometries and glass materials. The second one is more adequate to the complex geometries and/or heterogeneous materials. The second approach is the aim of this work.

Even if analytical or numerical techniques are available to solve the coupled heat transfer in participating media, some difficulty remains in determining the radiative properties of STM.

Tong and Tien (1980) employ a radiative conductivity model considering radiation as a conductive process. Although the expressions are simple, most of them contain a parameter that has to be experimentally determined.

The works of Cunnington and Lee (1996), Boulet et al. (1996), and Doermann and Sacadura (1996) calculate spectral radiative properties using the electromagnetic theory and knowing the porosity, particle size, particle shape, spectral optical properties, and particle orientation distribution. Hendricks and Howell (1996) have calculated the radiative properties of reticulated porous ceramics with inverse analysis techniques, based on discrete ordinates radiative models and measurements of hemispherical reflectance and transmittance. Hahn et al. (1997) have also measured the hemispherical reflectance and transmittance to determine radiative properties of ceramic materials by using a three-flux approximation. Silva Neto and Özisik (1992) have analyzed the estimation of optical thickness, albedo, and the coefficients of Legendre phase function for anisotropic scattering from simulated measurements. The direct problem is solved by P_N approximation and one of the boundaries is subjected to an isotropic incident radiation. An analysis with different number of terms in the phase function show strong deviations from exact phase function shape for a number of coefficients greater than four. Nicolau et al. (1994), using measurements of the spectral bidirectional transmittance and reflectance, have determined fiber radiative properties. They solved the RTE using the discrete ordinates method (DOM) with a fine angular quadrature to take into account the highly forward and backward peaked scattering, which is commonly observed for fiber or foam materials. The phase function has been approximated by a combination of two Henyey–Greenstein (HG) functions coupled with an isotropic component. The choice of this phase function is made because the identification of the high orders of classical Legendre polynomial values is very difficult.

21.2 Radiative Transfer Equation

The RTE, which describes the variation of the spectral radiation intensity I_λ (in a solid angle Ω, function of optical depth τ) in an absorbing-emitting-scattering medium, can be written as follows:

$$\frac{1}{\beta_\lambda}\Omega\nabla I_\lambda(\tau,\Omega) + I_\lambda(\tau,\Omega) = (1-\omega_\lambda)I_{b_\lambda}(\tau) + \frac{\omega_\lambda}{4\pi}\int\limits_{\Omega'=4\pi} I_\lambda(\tau,\Omega)p_\lambda(\Omega',\Omega)d\Omega' \qquad (21.1)$$

where
   β_λ is the spectral extinction coefficient
   ω_λ is the spectral albedo
   p_λ is the spectral phase function
   I_{b_λ} is Planck's blackbody function (in order to simplify the notations, the subscript λ can be omitted in the text)

These properties are those of a pseudo-continuum medium equivalent, in terms of radiative transport, to the real dispersed material.

The RTE in a plane parallel geometry with azimuthal symmetry can be written as follows (Özisik, 1973):

$$\mu\frac{\partial I_\lambda(\tau,\mu)}{\partial\tau} + I_\lambda(\tau,\mu) = S(\tau,\mu) \quad \text{in} \begin{cases} 0 \leq \tau \leq \tau_0 \\ -1 \leq \mu \leq 1 \end{cases} \qquad (21.2)$$

where

$$S(\tau,\mu) = (1-\omega_\lambda)I_{\lambda b}[T_p(\tau)] + \frac{\omega_\lambda}{2}\int\limits_{-1}^{1} p(\mu,\mu')_\lambda I_\lambda(\tau,\mu')d\mu' \qquad (21.3)$$

and the phase function can be expressed by a Legendre polynome

$$p(\mu,\mu')_\lambda = \sum_{n=0}^{N} a_{n,\lambda}P_{n,\lambda}(\mu)P_{n,\lambda}(\mu'); \quad a_0 = 1 \qquad (21.4)$$

An explicit formulation for inverse radiative problems in participating medium has been derived and discussed in Chapter 15.

21.2.1 Boundary Conditions

Different strategies can be considered to identify radiative properties (Figure 21.1). We will analyze five experimental devices, consequently five boundary condition cases (Moura et al., 1998a):

   Case a: normally incident collimated beam onto the sample with bidirectional transmittance and reflectance measurements
   Case b: collimated beam with different angles of incidence onto the sample with hemispherical transmittance and reflectance measurements
   Case c: diffuse irradiation onto the sample with bidirectional transmittance and reflectance measurements

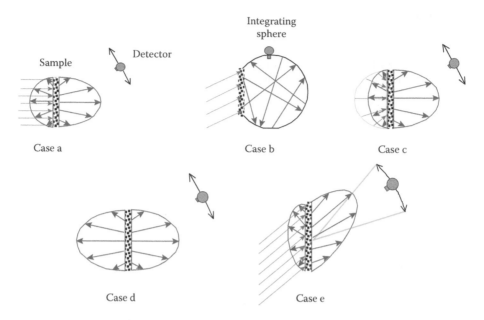

FIGURE 21.1
Different experimental measurement conditions. (From Moura, L.M. et al., Identification of thermal radiation properties of dispersed media: Comparison of different strategies, *Proceedings of the 11th IHTC*, August 23–28, Korea, 1998a.)

Case d: measurements of bidirectional radiation flux emitted by a hot sample (self-emission)

Case e: inclined incident collimated beam onto the sample with bidirectional transmittance and reflectance measurements

21.2.2 Phase Function

The phase function due to particles randomly oriented in space depends only on the scattering angle, θ_p, which is the angle between the incident and the scattered radiation. A classical approach to represent the phase function consists of developing this function in a limited series of Legendre polynomials. Unfortunately, the phase function expansion by Legendre polynomials for fibers or foams needs a large number of expansion coefficients.

The phase functions for fibrous or foam media always exhibit a strong peak in the direction of incident radiation and a fair backscattering, showing that the scattering is highly anisotropic. An alternative solution requires the use of simpler phase functions, such as HG, or a combination of different phase functions (Nicolau et al., 1994, Hendricks and Howell, 1996). In the current work, we employ a combined phase function proposed by Nicolau et al. (1994):

$$p(\theta_d) = f_1 f_2 p_{\mathrm{HG},g_1}(\theta_d) + (1 - f_1) f_2 p_{\mathrm{HG},g_2}(\theta_d) + (1 - f_2) \tag{21.5}$$

where the parameters g_1 and g_2 govern the shape of HG functions (p_{HG,g_1} and p_{HG,g_2}) in the forward and backward directions (Figures 21.2 and 21.3). f_1 is the weighting factor between forward and backward anisotropy in the phase function, f_2 is the weighting factor between anisotropic and isotropic scattering.

Forward

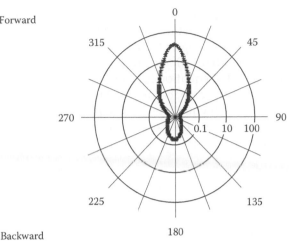

FIGURE 21.2
Phase function ($f_1 = 0.9$, $g_1 = 0.84$, $f_2 = 0.95$, $g_2 = -0.6$). (From Moura, L.M. et al., Identification of thermal radiation properties of dispersed media: Comparison of different strategies, *Proceedings of the 11th IHTC*, August 23–28, Korea, 1998a.)

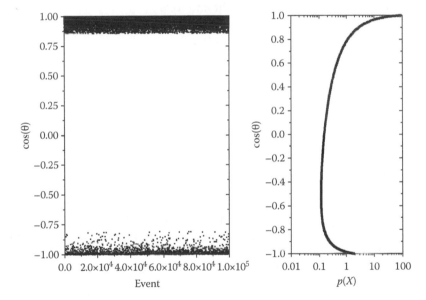

FIGURE 21.3
Anisotropic phase function to 100,000 statistical realizations. $g_1 = 0.86$, $g_2 = 0.8$, $f_1 = 0.96$, $f_2 = 0.96$. (From Xavier Filho, O. and Moura, L.M., A comparative analysis between the Monte Carlo method and the discrete ordinate method applied to solve the radiative transfer equation, *Proceedings of the 19th International Congress of Mechanical Engineering*, 2007.)

21.3 Discrete Ordinate Method

The discrete ordinate method (DOM) was initially used by Schuster (1905) and Schwarzschild (1906) for studying radiative transfer in stellar atmospheres (apud Siegel and Howell, 2002), and later, Chandrasekar (1960) extended the formulation to astrophysics problems. Carlson and Lathrop in 1968 have developed a solution to the neutron transport equation.

The majority of works uses the RTE formulation presented by Chandrasekhar (1960) and Özisik (1973). These techniques of solution of the RTE can be found in Moura et al. (1997, 1998c). The RTE solution by DOM is constituted of two stages: (i) an angular discretization, where the integral term of RTE is substituted by a radiative intensities weighted sum in the angular directions; in this way, the integro-differential equation is transformed into a set of first-order ordinary partial differential equations; (ii) a space discretization, considering control volumes, for solution of the partial differential equations. In Chapter 15, the DOM method is used to solve the set of differential equation by the explicit method. Considering a "cold media," Equation 21.2 can be rewritten as follows:

$$I_{i+1/2,j} = \frac{1}{(1+f\alpha_j)} \left[f\alpha_j \frac{\omega}{2\beta} \left[\sum_{n=1}^{N} w_n (p_{nj} I_{i+1/2,n}) \right] + I_{i,j} \right] \tag{21.6}$$

where $i+1/2$ represents the control volume center coordinate, f is the interpolation function that can be: upwind ($f=1$), linear ($f=1/2$), integral (f is function of the α_j calculated from integration of RTE) or exponential (f is function of the α_j calculated from the solution of RTE) (Moura et al., 1998c), w is the weight and α_j is

$$\alpha_j = \frac{\Delta\tau_{i+1/2}}{\mu_j} \tag{21.7}$$

where $\Delta\tau$ is the optical thickness of the control volume.

21.4 Monte Carlo Method

Radiative heat transfer by Monte Carlo method (MCM) is based on probability concepts applied to the physical phenomena, such as emission, reflection, and absorption of the photon. When solving RTE by MCM, radiative energy is not treated as a continuous energy flux but is considered a pack of photons, each with a fixed amount of energy. To quantify the radiative energy attenuation in a monochromatic source, Beer's law is considered, which expresses the attenuation of radiant energy inside a volume. In the MCM, Beer's law can be modified to express the radiative energy extinction probability emitted from a point in the media (or surface) and to travel over the distance, s (Yang et al., 1995):

$$Rs = 1 - e^{\beta s} \tag{21.8}$$

where
  Rs is the random number
  β is the extinction coefficient
  s is the distance traveled by the photon up to when it is absorbed or scattered

If the value of s is bigger than the distance taken from the photon point emission then the photon was absorbed or scattered. To determine if the photon is absorbed or scattered, a uniform random number for scattering albedo, R_ω, is used. If R_ω is bigger than albedo the photon is absorbed, otherwise the photon is scattered (Brewster and Tien, 1992). The Monte

Carlo algorithm for an anisotropic media with a collimated beam incident onto a slab surface has the following steps (Xavier Filho and Moura, 2007):

1. Determine the direction of propagation by the photon inside the solid angle of the incident beam onto the face of the slab.
2. Determine by Equation 21.8 if it was absorbed, scattered, or remains in its trajectory.
3. If the photon is absorbed or has reached the boundaries, it will initiate a new photon analysis.
4. Verify the absorption or scattering criteria. If the photon was scattered, to choose a new direction (random choice by the phase function and to repeat step 2, considering the current position of the photon).

Repeat these steps for a number of photons to assure accuracy.

An isotropic and conservative (unitary albedo, $\omega = 1$) case is used to compare the MCM and DOM. The solution of MCM was implemented in Xavier Filho and Moura (2007). Figure 21.4 presents the results to this case. The P_1 solution presented by Modest (2003) is used as a reference.

Modest (2003) shows that the analytical solution is close to the P_1 solution. In such a way, the P_1 solution is compared with MOD and MCM solutions as a function of the optical thickness. Two different divergence angles of the incident beam, θ_o, two different spatial discretizations of the control volumes, dx, and two different numbers of packages, np, are used to investigate a probable influence of these parameters on the MCM solution. It can be observed that they don't present a significant influence in the nondimensional flux.

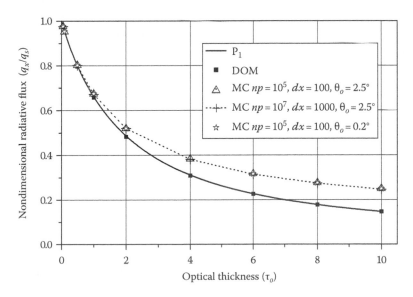

FIGURE 21.4
Comparison between DOM and MCM for a nondimensional radiative heat flux in a purely scattering layer with a normal collimated irradiation and isotropic scatter. (From Xavier Filho, O. and Moura, L.M., A comparative analysis between the Monte Carlo method and the discrete ordinate method applied to solve the radiative transfer equation, *Proceedings of the 19th International Congress of Mechanical Engineering*, 2007.)

As presented by Modest (2003), Lataillade et al. (2002), and Eymet et al. (2005), the MCM errors (difference between MCM and reference solution) increase strongly with the optical thickness. They presented a boundary-based net-exchange MCM applied to solve a scattering media of optically thick absorption (and/or for quasi-isothermal configurations). They presented to the MCM the number of statistical realizations needed in order to get a 1% standard deviation over the slab emission value as a function of slab total optical thickness. They showed that the number of statistical realizations increase strongly with the optical thickness. In case of very short photon mean free paths, most bundles are absorbed in the vicinity of their emission positions, which means that only very few bundles effectively participate in distant radiative transfers. The consequence is that MCM based on bundle transport formulations requires very large numbers of statistical realizations for sufficiently accurate radiative exchange estimations. The computational costs impose constraints to geometrical grid sizes that cannot be reduced sufficiently for the optically thin assumption to be valid. They presented the exchange formulation in order to reduce the number of statistical realizations to the acceptable values. It can also be observed that the DOM presents a close agreement with the reference solution.

21.5 An Experimental Device (Nicolau et al., 1994, Moura, 1998)

An experimental device used to measure infrared bidirectional transmittance and reflectance is schematically shown in Figure 21.5. The objective is to determine the fraction of the infrared incident collimated beam reflected by the sample (the range $2.0-16.0$ μm of wavelength region is considered). The spectrometer is a FTS 60 A (Bio-Rad Inc.) type, based on Fourier transform spectroscopy. The source of radiation, characterized by a blackbody emission spectrum at 1300°C, is a tungsten filament inside a silica tube. An entrance slit with four movable holes (1.2, 2.7, 4, 7 mm diameter) determines the solid angle

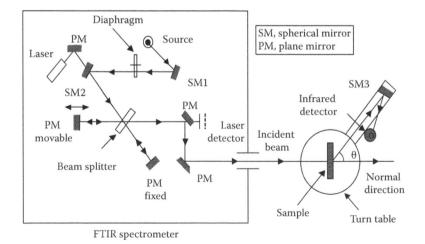

FIGURE 21.5

Experimental device using an FTIR spectrometer to measure reflectance. (From Nicolau, V.P. et al., *Int. J. Heat Mass Transfer*, 37, 311, 1994; Moura, L.M., Identification des proprietes radiatives des materiaux semi-transparent en situation de non-symetrie azimutale du champ radiatif, PhD diss., INSA, Lyon, 1998. With permission.)

of the infrared collimated beam. The Michelson interferometer principle is used so that the exit infrared beam can be measured by a detector as a function of path difference between the fixed and movable mirrors (Figure 21.5). The detection system, composed of a spherical mirror collecting the beams and concentrating them on a quantic detector, HgCdTe, is mounted on a goniometric arm to allow bidirectional reflection measurements and detection of reflected radiation. Both the spectrometer and the detection system are purged with dry air and connected to a data acquisition system.

The experimental arrangement presented allows the measurement of spectral reflectance. Two measurements are carried out: one without the sample (to know the incident energy) and the other with the sample (reflected energy). The ratio between the reflected and incident energies provides the reflectance.

21.6 Parameter Identification

The five different strategies described in Section 21.2.1 (cases a, b, c, d, and e) can be used in order to determine the radiative properties of a semitransparent sample. For each strategies and sample thickness, the model described above in which the material radiative properties ($\tau_o = \beta_\ell$, ω, g_1, f_1, g_2, f_2) should be given, is used to calculate the theoretical transmittances and reflectances, $T_t(\theta, \phi)$. The experimental bidirectional transmittances and reflectances, $T_{ed}(\theta, \phi)$ for an incident radiation, the experimental hemispherical transmittance T_{eh}, and the experimental directional emittance, $\varepsilon_{ed}(\theta, \phi)$, are defined by the following expressions:

$$T_{ed}(\theta, \varphi) = \frac{I(\theta, \varphi)}{I_o \, d\omega_o}$$

$$T_{eh} = \frac{\int_0^{2\pi} I(\theta, \varphi) \cos \theta \, d\Omega}{I_o \, d\omega_o} \tag{21.9}$$

$$\varepsilon_{ed}(\theta, \varphi) = \frac{I(\theta, \varphi)}{I_b}$$

where
I is the transmitted, reflected, or emitted intensity
I_o is the intensity of the beam incident onto the sample within a solid angle, $d\omega_o$
I_b is the blackbody emission intensity

The irradiation is presented only inside the solid angle, $d\omega_o$.

For cases a, b, and e, $d\omega_o$ depends on the experimental device, for case c, $d\omega_o$ is the half hemisphere (2π).

The identification of the radiative parameter, $\hat{\chi}_k$, is based on the minimization of the quadratic error between the measured, T_{en}, and calculated, T_{tn}, transmittances and reflectances over the N measurements for the K parameters:

$$F(\hat{\chi}_k) = \sum_{n=1}^{N} [T_{tn}(\hat{\chi}_k) - T_{en}]^2, \quad k = 1, \ldots, K \tag{21.10}$$

where $\hat{\chi}_k$ represents the radiative parameter vector that must be identified.

In cases a, c, d, and e the summation, Equation 21.10, is performed on different bidirectional measurements. In case b, the summation is on the hemispherical transmittance and reflectance measurements for the different angles of incidence onto the sample.

The method adopted to achieve this minimization is the Gauss linearization method that minimizes $F(\hat{\chi}_k)$ by setting the derivatives to zero with respect to each of the unknown parameters, $\hat{\chi}_k$. As the system is nonlinear, an iterative process is performed over m iterations (Nicolau et al., 1994).

$$
\left[
\begin{array}{cccc}
\sum_{n=1}^{N}\left(\dfrac{\partial T_{tn}}{\partial \chi_1}\right)^2 & \sum_{n=1}^{N}\dfrac{\partial T_{tn}}{\partial \chi_1}\dfrac{\partial T_{tn}}{\partial \chi_2} & \cdots & \sum_{n=1}^{N}\dfrac{\partial T_{tn}}{\partial \chi_1}\dfrac{\partial T_{tn}}{\partial \chi_K} \\[4mm]
\sum_{n=1}^{N}\dfrac{\partial T_{tn}}{\partial \chi_1}\dfrac{\partial T_{tn}}{\partial \chi_2} & \sum_{n=1}^{N}\left(\dfrac{\partial T_{tn}}{\partial \chi_2}\right)^2 & \cdots & \sum_{n=1}^{N}\dfrac{\partial T_{tn}}{\partial \chi_2}\dfrac{\partial T_{tn}}{\partial \chi_K} \\[4mm]
\hline
\sum_{n=1}^{N}\dfrac{\partial T_{tn}}{\partial \chi_K}\dfrac{\partial T_{tn}}{\partial \chi_1} & \sum_{n=1}^{N}\dfrac{\partial T_{tn}}{\partial \chi_K}\dfrac{\partial T_{tn}}{\partial \chi_2} & \cdots & \sum_{n=1}^{N}\left(\dfrac{\partial T_{tn}}{\partial \chi_K}\right)^2
\end{array}
\right]^m
\cdot
\left[
\begin{array}{c}
\Delta\chi_1 \\ \Delta\chi_2 \\ -- \\ \Delta\chi_K
\end{array}
\right]^{m+1}
= -
\left[
\begin{array}{c}
\sum_{n=1}^{N}(T_{tn}-T_{en})\dfrac{\partial T_{tn}}{\partial \chi_1} \\[4mm]
\sum_{n=1}^{N}(T_{tn}-T_{en})\dfrac{\partial T_{tn}}{\partial \chi_2} \\[4mm]
\hline
\sum_{n=1}^{N}(T_{tn}-T_{en})\dfrac{\partial T_{tn}}{\partial \chi_K}
\end{array}
\right]^m
$$

$$(21.11)$$

The different methods and tools for nonlinear estimation are discussed in depth in Chapters 8 through 11.

The solution of this system gives the increments $\Delta\chi_{k=1,\ldots,K}$ to be added to each parameter, χ_k, at each step of the iterative process, as follows:

$$
\chi_k^{m+1} = \chi_k^m + \alpha_k \Delta\chi_k^m, \quad k = 1,\ldots,K \tag{21.12}
$$

where α_k is an under-relaxation coefficient for the parameter k to assure convergence. Convergence is obtained when $\Delta\chi_k^m / \chi_k^m$ is less than a convergence tolerance.

The matrix on the left-hand side is composed of the sensitivity coefficient products calculated from the theoretical model, and it does not directly depend on the experimental values. This matrix, M, can be used in the sensitivity analysis to verify possible linear dependences between the sensitivity coefficients calculated for each parameter. The calculation of a condition number, CN, of this matrix can be used to determine the degree of ill-posedness of the identification problem (McCormick, 1992):

$$
\mathrm{CN}(M) = \left\| M^{-1} \right\| \cdot \left\| M \right\| \tag{21.13}
$$

where the norm $\|M\|$ is calculated from the elements $M_{k',k}$, as

$$
\|M\| = \max_{k'=1,K} \sum_{k=1}^{K} M_{k',k} \tag{21.14}
$$

The CN is greater than one. The larger the CN is, the worse ill-conditioned the system is. Small changes in the right-hand side of Equation 21.5, i.e., in the measurements, result in very large change in the solution vector, i.e., the increments $\Delta\hat{\chi}_k$. It is then almost impossible to simultaneously determine all of the unknown parameters. Poor conditioning occurs when at least two of the sensitivity coefficients are quasi-linearly dependent or when at least one is very small or very large compared to the others.

The performances of the five cases are shown in Figures 21.6 through 21.18. The CN as a function of optical thickness (τ_o) and the signal intensity measured for different directions and/or optical thickness for the four cases are analyzed. It should be noted that the diffuse intensities are very small compared to the incident intensity; this poses a real practical difficulty in measuring the intensity. The six parameters case corresponds to the determination of $((\hat{\chi}_k)_{k=1,\ldots,6} = \omega, g_1, f_1, f_2, g_2, \tau_o)$. The five parameters case correspond to the determination of $((\hat{\chi}_k)_{k=1,\ldots,5} = \omega, g_1, f_1, f_2, g_2)$ and so on. Obviously, the CN depends on the choice of the parameters to identify. For this reason, the optical thickness is left as the

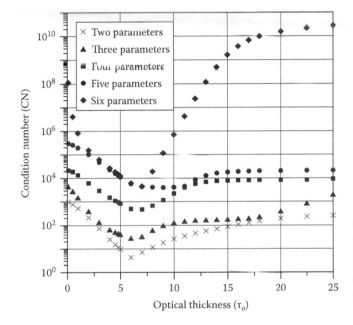

FIGURE 21.6
Condition number for case a. (From Moura, L.M. Identification des proprietes radiatives des materiaux semi-transparent en situation de non-symetrie azimutale du champ radiatif, PhD diss., INSA, Lyon, 1998.)

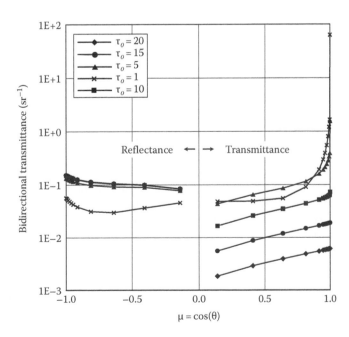

FIGURE 21.7
Bidirectional transmittance and reflectance for case a. (From Moura, L.M. et al., Identification of thermal radiation properties of dispersed media: Comparison of different strategies, *Proceedings of the 11th IHTC*, August 23–28, Korea, 1998a.)

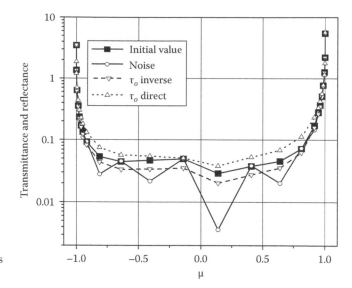

FIGURE 21.8
Transmittance and reflectance functions for $\tau_o = 5$.

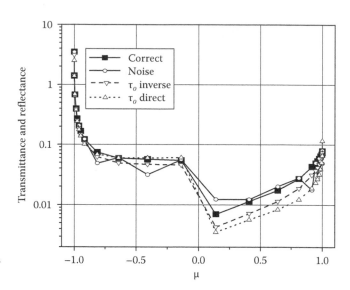

FIGURE 21.9
Transmittance and reflectance functions for $\tau_o = 15$.

last parameter. Indeed, the optical thickness can be determined directly by Beer's law or with corrections to subtract the scattered intensity from the total intensity in the incident direction (Nicolau et al., 1994).

The fifth parameter is g_2. Nicolau et al. (1994) showed that this parameter induces a smaller variation of radiation intensity than the variations of the other parameters. One suitable possibility is to set $g_2 = g_1$. This analysis depends on the values of the radiative properties. Here, we consider the values given by Nicolau et al. (1994) for fiber insulation (Table 21.1).

As expected, the increase in the number of parameters simultaneously identified results in an increase of the CN (Figures 21.6, 21.11, 21.14, 21.15, and 21.17).

Figures 21.6 through 21.9 correspond to case a. In Figure 21.6 the CN is represented. The estimated value of τ_o increases suddenly for values greater than 6. It is due to the extinction

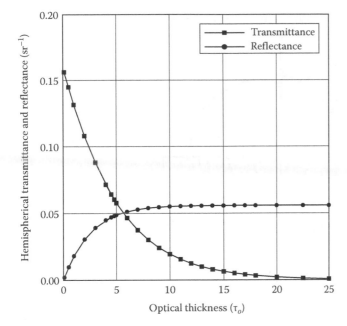

FIGURE 21.10

Hemispherical transmittance and reflectance under normal incidence (case b). (From Moura, L.M. et al., Identification of thermal radiation properties of dispersed media: Comparison of different strategies, *Proceedings of the 11th IHTC*, August 23–28, Korea, 1998a.)

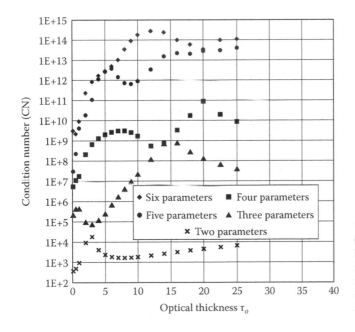

FIGURE 21.11

Condition number for case b. (From Moura, L.M. et al., Identification of thermal radiation properties of dispersed media: Comparison of different strategies, *Proceedings of the 11th IHTC*, August 23–28, Korea, 1998a.)

of collimated intensity for the optically thick samples. This figure also indicates that the optimal interval of τ_o is between 5 and 8.

Figure 21.7 shows the bidirectional transmittances versus the azimuthal angle for different optical thicknesses. It shows that the collimated intensity is important for optically thin samples. A change in the transmittance and reflectance shape is also noted for optically thin samples. This can be used if an estimation is performed to take into account several different thicknesses. The reflectance increases suddenly and presents a semi-infinite

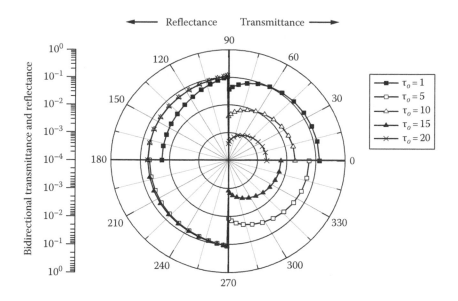

FIGURE 21.12
Bidirectional transmittance and reflectance for case c.

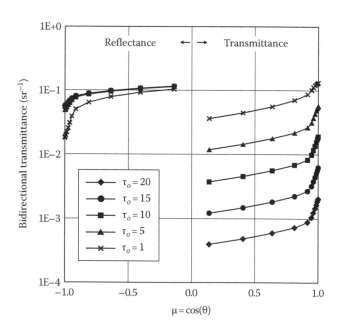

FIGURE 21.13
Bidirectional transmittance and reflectance for case c. (From Moura, L.M. et al., Identification of thermal radiation properties of dispersed media: Comparison of different strategies, *Proceedings of the 11th IHTC*, August 23–28, Korea, 1998a.)

medium behavior for τ_o greater than 15. It should be noted that the solid angle of incidence is on the order of 10^{-3}. This results in a signal intensity 10^5 smaller than the signal intensity of the incident beam for a bidirectional transmittance of 10^{-2}.

Tables 21.2 and 21.3 present the numerical identification analysis (simulated) and Figures 21.8 and 21.9 the respective graphical representation.

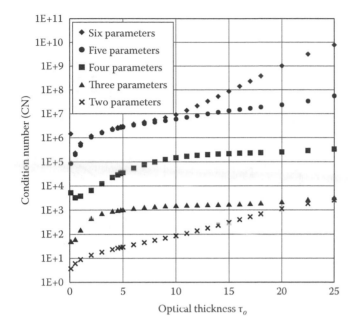

FIGURE 21.14
Condition number for case c. (From Moura, L.M. et al., Identification of thermal radiation properties of dispersed media: Comparison of different strategies, *Proceedings of the 11th IHTC*, August 23–28, Korea, 1998a.)

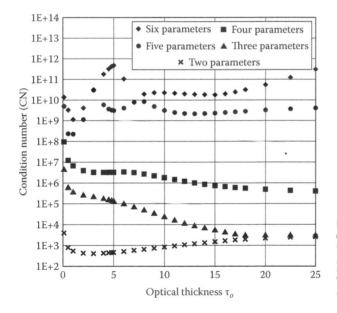

FIGURE 21.15
Condition number for case d. (From Moura, L.M. et al., Identification of thermal radiation properties of dispersed media: Comparison of different strategies, *Proceedings of the 11th IHTC*, August 23–28, Korea, 1998a.)

In Figures 21.10 and 21.11, case b is presented. The measurements over 12 directions are used in the summation of Equation 21.10. The hemispherical transmittance and reflectance are considered with a normal incident beam. Moreover, 10 hemispherical transmittances are considered with an oblique incident beam onto the sample. The directions considered are between 5° and 50° with a step of 5°. The error is calculated by the expression

$$\text{Error } (\%) = \frac{\chi_k(\text{correct}) - \chi_k(\text{estimated})}{\chi_k(\text{correct})} \times 100 \qquad (21.15)$$

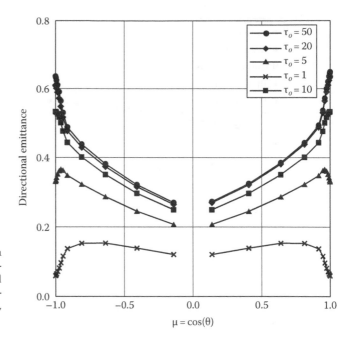

FIGURE 21.16
Directional emittance for case d. (From Moura, L.M. et al., Identification of thermal radiation properties of dispersed media: Comparison of different strategies, *Proceedings of the 11th IHTC*, August 23–28, Korea, 1998a.)

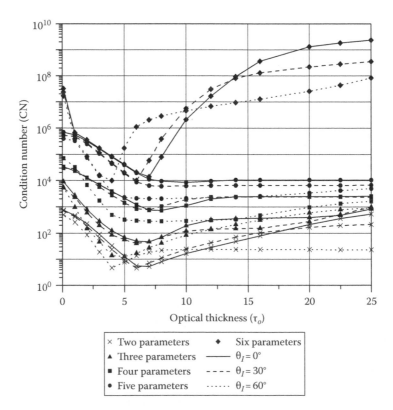

FIGURE 21.17
Condition number for case e.

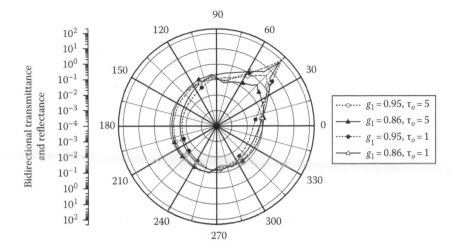

FIGURE 21.18
Bidirectional transmittance and reflectance for case e and $\theta_I = 45°$, $\omega = 0.95$, $g_2 = -0.6$, $f_1 = 0.9$, $f_2 = 0.95$. (From Moura, L.M., Identification des proprietes radiatives des materiaux semi-transparent en situation de non-symetrie azimutale du champ radiatif, PhD diss., INSA, Lyon, 1998.)

TABLE 21.1

Radiative Properties Used in This Analysis

Radiative Properties	
ω	0.95
g_1	0.84
f_1	0.9
g_2	-0.6
f_2	0.95

TABLE 21.2

Identification Results to $\tau_o = 5$

	Correct Value	Initial Value	τ_o Inverse	Error (%)	τ_o Direct	Error (%)
ω	0.95	0.85	0.931	2.0	0.973	-2.4
g	0.95	0.8	0.951	-0.1	0.914	3.8
f_1	0.9	0.98	0.900	0.0	0.861	4.3
f_2	0.95	0.98	0.966	-1.7	0.990	-4.2
ω_o	5.0	—	5.019	-0.4	3.965	20.7
Number of iterations	—	—	13	—	10	—
CN	—	—	100	—	85	—

TABLE 21.3

Identification Results to $\tau_o = 15$

	Correct Value	Initial Value	τ_o Inverse	Error (%)	τ_o Direct	Error (%)
ω	0.95	0.85	0.920	3.2	0.881	7.3
g	0.95	0.8	0.942	0.8	0.923	2.8
f_1	0.9	0.98	0.871	3.2	0.765	15.0
f_2	0.95	0.98	0.939	1.2	0.819	13.8
ω_o	15.0	—	12.344	17.7	8.524	43.2
Number of iterations	—	—	78	—	15	—
CN	—	—	2.8×10^6	—	2100	—

Figures 21.12 through 21.14 correspond to case c. In this case, the bidirectional transmittances and reflectances are considered in the estimation procedure. Table 21.4 presents the numerical identification analysis (simulated) for a sample test. Measurements of reflectance under a diffuse incidence should be difficult to accomplish experimentally. Unfortunately, identification with only transmittance measurements showed a poor CN. CN for case c, Figure 21.14, presents better results for optically thin samples. Generally, case c is less well conditioned than case a.

Case d presents an inverse behavior of case c. Generally, the CN is greater for the optically thick samples, except for two parameters (Figure 21.15). The directional emittance increases with the sample thickness and becomes nearly constant for values of thickness greater than about $\tau_o = 20$. The signal also increases with the absolute temperature of the sample. The radiative properties can change with the temperature. An advantage of case d is that radiative properties can be obtained for different sample temperatures. But the sample temperature must be sufficiently high in order to have enough energy to measure the directional emission (Figure 21.16).

Case e is shown in Figures 21.17 and 21.18 and it presents a better result when compared with case a. The anisotropic conditions improved the identification method.

The optical thickness could be identified using a direct model, like a Beer's law model or a second-order model (Nicolau et al., 1994). The second-order model uses a second-order polynomial to determine the diffuse transmittance in the directions near the incident beam and extrapolate these values to the direction of incidence. For a divergence angle of the incident beam $\theta_o = 0.38°$, it was shown that the optical thickness was estimated correctly for values going up to 20. However, the accuracy of the estimation depends on the divergence angle of the incident beam and the anisotropy of the phase function. Figures 21.19 and 21.20 show the errors using direct models to determine the optical thickness for two different coefficients g_1. Figure 21.19 shows that the second-order model gives better identification than Beer's law to anisotropy factor, $g_1 = 0.86$. However, when the anisotropy factor increases, $g_1 = 0.95$, the optical thickness identification can present uncertainties higher than 20% (especially when the divergence angle of the incident beam is close to unity) (Figure 21.20). For this reason, the optical thickness should be identified by inverse method. Increasing one parameter in the identification procedure it will increase the ill-conditioned of the identification procedure. To assure convergence, an under-relaxation coefficient, α_k, is used. This parameter should increase the number of iterations.

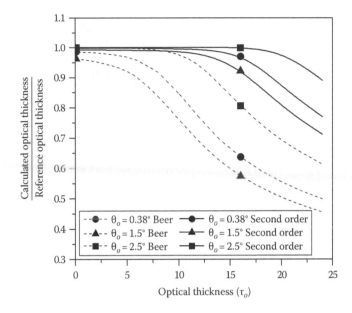

FIGURE 21.19
Optical thickness ratio for $\omega = 0.95$, $g_1 = 0.86$, $f_1 = 0.9$, $f_2 = 0.95$, $g_2 = -0.6$. (Moura, L.M., Identification des proprietes radiatives des materiaux semi-transparent en situation de non-symetrie azimutale du champ radiatif, PhD diss., INSA, Lyon, 1998.)

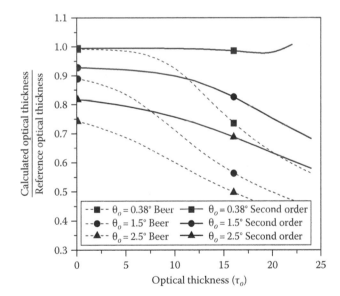

FIGURE 21.20
Optical thickness ratio for $\omega = 0.95$, $g_1 = 0.95$, $f_1 = 0.9$, $f_2 = 0.95$, $g_2 = -0.6$. (Moura, L.M., Identification des proprietes radiatives des materiaux semi-transparent en situation de non-symetrie azimutale du champ radiatif, PhD diss., INSA, Lyon, 1998.)

21.7 Radiative Transfer in a Medium with Refractive Index and Specular Reflection at the Boundaries

In recent years, many works on RTE solutions have been carried out considering the medium without interface, consequently, to the high porosity medium with the radiative index close to unity. However, many usual materials like coatings, glasses, and thin films present refractive index higher than unity, $n > 1$. Conventional paints transmit or absorb most of the incident infrared radiation. A coating can be optimized to produce interesting effects in the thermal performances. Respecting the color of the coating on the visible range, the scattering on the infrared radiation can be changed in order to improve the reflection. This effect can be applied to automobiles, oil and gas tanks, and building coatings. In some countries, minimum building separation distances must be respected by consideration of the fire radiant flux (Berdahl, 1995).

The radiative behavior of the medium, $n > 1$, needs special interface conditions that must be used to take into account the refractive index (Hottel et al., 1968, Wu et al., 1994, Liou and Wu, 1996). Furthermore, when using the DOM, a special discretization must be employed to consider the changes of the radiation direction, even when the radiation beam has a normal incidence.

Hottel et al. (1968) analyzed the effects of Fresnel reflection at the interface of a slab containing anisotropic scattering medium using the DOM. In 1970, Hottel et al. compared the measurements of monodisperse polystyrene spheres confined between two parallel glass slides with the values predicted from Mie theory. Orel et al. (1990) measured the solar absorptance by Fourier transform infrared (FTIR) spectroscopy of paint coatings for solar collectors. Wu et al. (1994) analyzed a radiative transfer problem in an isotropic media with no unit refractive index and Fresnel boundaries. They used a set of DOM quadratures to treat the strongly angular dependence of the radiative intensity around the critical angles.

Oppenheim and Feiner (1995) investigated the polarized infrared (IR) reflectivity of painted and rough surfaces. Values of the bidirectional reflectance function are measured for sandblasted aluminum, concrete, painted metal, and asphalt surfaces. In the same year, Berdahl presented a study on the pigments that reflect the infrared radiation from fire. He used the Mie theory to analyze the radiative transfer by the particles of titanium dioxide, iron oxide, chromium oxide, and silicon, with particle diameters of 1–2 μm. Shah and Adroja (1996) measured the diffuse reflectance of titanium dioxide pigment dispersions in the visible region. They used the Mie theory to predict the reflectance. Theoretical and experimental values of the reflectance are compared with reasonable agreement.

Liou and Wu (1996) analyzed the radiative transfer in a two-layer scattering medium with Fresnel conditions. They used the DOM with composite quadratures to take into account the refractive index differences. Their results presented the effects of the albedo and the refractive index.

More recently, Abdallah and Le Dez (2000) analyzed by the ray-tracing method the intensity of radiation emitted from a nonscattering semitransparent plate with a refractive index varying with position. Lemonnier and Le Dez (2002) analyzed the same problem now using the DOM and achieved good agreement. In this case, the directions of radiation propagation in the medium vary greatly due to the variable refractive index. Lacroix et al. (2002) analyzed a similar problem, but now considering the conduction heat transfer coupling. Garcia et al. (2008) studied the Fresnel boundary and interface conditions for multilayered media. They use an "analytical" discrete ordinates method, ADOM, presented by Barichello and Siewert (1999).

Traditionally, in literature there are two different major materials involving refractive index solutions: metallic and dielectric materials (Özisik, 1973, Modest, 2003, McCluney, 1994, Siegel and Howell, 2002). All these solutions are obtained from Maxwell equations, Planck (1914), and Hulin et al. (1998). In this case, the coating can be considered like a dielectric material. This method involves the use of Snell equation to obtain the angular quadratures of DOM due to the refractive index considerations. Even when there is a normal incidence onto the sample, the solid angle of the incident beam changes. This must be taken into account in the quadrature setup, and the use of classic quadratures, like Gauss and Radau, is not possible. Moura et al. (1999) showed that in the higher anisotropical materials, the small variations in the solid angle of the incident beam can result in significant errors on phase function identification. In this way, it is recommendable to use adaptive quadratures, and consequently the quadrature weights, with respect to solid angle of the incident beam.

Wu et al. (1994) and Liou and Wu (1996) have used a linear transformation to correct the quadratures to consider the refractive index changes in the directions μ ($\mu = \cos\theta$) and the quadrature weights, w, are written as

$$\mu'_j = \frac{(\mu_j + 1)}{2}, \quad w'_j = \frac{w_j}{2} \tag{21.16}$$

$$\int_0^1 f(\mu') \, d\mu' = \int_{-1}^1 f\left(\frac{\mu + 1}{2}\right) d\mu \tag{21.17}$$

The divergence angle, θ'_2, inside the medium is obtained by the incident divergence angle, θ_2, by the following equation:

$$\theta'_2 = \text{arcsen}\left(\frac{\text{sen } \theta_2}{n_2} n_1\right) \tag{21.18}$$

Figure 21.21 presents a sketch of the physical conditions analyzed. A collimated incident beam irradiates the coating surface. The coating substrate can have specular or diffuse reflection. Figure 21.22 shows the hemispherical bidirectional function of the optical thickness to albedo, $\omega = 0.95$, and the phase function $g_1 = 0.84$, $g_2 = -0.6$, $f_1 = 0.9$, and $f_2 = 0.95$. The hemispherical reflectance decreases with the refractive index and with optical thickness. For the optical thickness up to 10, the hemispherical reflectance remains constant. This limit will be defined as optical thickness limit (OPL point).

Figure 21.23 shows the surface specular reflectivity to a media with refraction index, n_2, bigger than n_1. Between 30° and 53.13°, the Brewster angle can be observed and for angles greater than 53.13°, the reflection is 100% (critical angle). With the increase of the refraction index, the increase of the critical angle and the reduction of the divergence angle of the incident beam can be observed.

Figure 21.24 presents the CN considering the identification of the refraction index with the values listed in Table 21.5. Figure 21.25 shows the comparative results for this case.

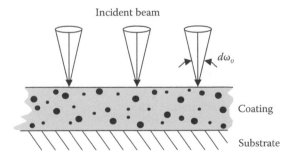

FIGURE 21.21
Physical conditions.

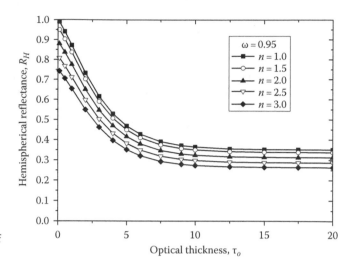

FIGURE 21.22
Hemispherical reflectance function of optical thickness for $\omega = 0.95$.

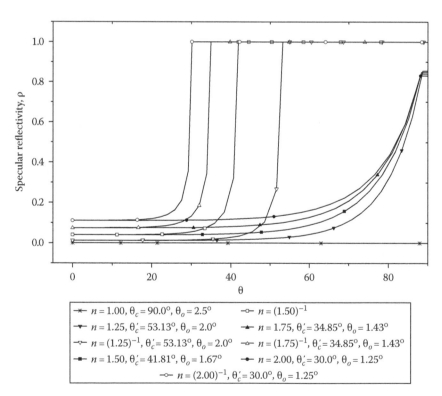

FIGURE 21.23
The interface specular reflectivity function of the incidence for different refractive indexes.

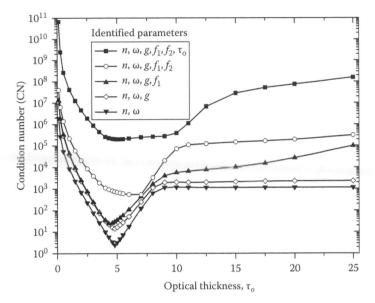

FIGURE 21.24
Condition number (CN) function of the optical thickness for $n = 1.5$, $\omega = 0.95$, $g = 0.84$, $f_1 = 0.9$, and $f_2 = 0.95$.

TABLE 21.4

Identification Results of Five Parameters for Case c

Parameter	True Value	Initial Value	Estimated Value	$|\varepsilon_a|$ (%)	$|\varepsilon_S|$ (%)
ω	0.95	0.85	0.948	0.08	0.18
g	0.84	0.8	0.862	0.64	2.66
f_1	0.90	0.98	0.869	0.09	3.49
f_2	0.95	0.98	0.927	0.24	2.46

TABLE 21.5

Identification Results to a Medium with Refractive Index
for $\tau_o = 5$ and $n = 1.5$

Parameter	Correct Value	Initial Value	Identified Value	Standard Deviation (95%)	Bias (%)
n	1.5	1.3	1.520	0.16	1.3
ω	0.95	0.85	0.950	0.009	−0.03
g	0.84	0.8	0.842	0.019	0.20
f_1	0.9	0.8	0.902	0.016	0.19
f_2	0.95	0.8	0.943	0.053	−0.67
τ_o	5.0	2° order	5.023	0.028	0.46

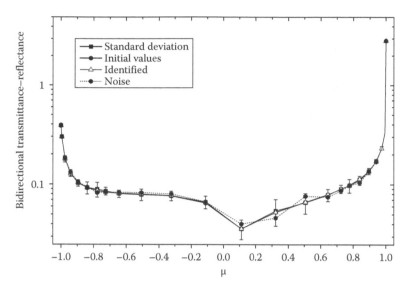

FIGURE 21.25
Bidirectional transmittance and reflectance for $n = 1.5$.

21.8 Parameter Identification for Packed Spheres Systems

Radiative heat transfer through randomly packed spheres system has received much attention due to its many industrial applications. Baillis and Sacadura (2000) have reviewed in a recent work a number of theoretical and experimental studies reported in the literature. These studies have shown that radiative heat transfer involves complex radiative interactions between the individual spheres due to the close packed system (Tien, 1988). Results of previous studies have led to a better understanding of the radiative heat transfer mechanism in a packed spheres system.

To this end, Brewster and Tien (1982) applied a two-flux model to calculate the hemispherical transmittances of a packed bed of spheres with the radiative properties predicted from uncorrelated scattering theory. Yang et al. (1983) utilized a MCM to simulate the energy bundle traveling through the voids of bed, showing a strong dependence of the packing structure and the size and emissivity of constituent spheres on the thermal radiative properties.

Kamiuto (1990) has proposed an heuristic correlated scattering theory for packed bed consisting of relatively large spheres. Their results showed that the transfer calculations based on correlated scattering theory provide better agreement with the experimental results than their previous transfer calculations based on uncorrelated scattering theory.

Singh and Kaviany (1992) have presented an approach for modeling-dependent scattering radiative heat transfer in beds of large spherical particles, having shown that the dependence properties for a bed of opaque spheres can be obtained from their independent properties by scaling the optical thickness, while leaving the albedo and the phase function unchanged. The scaling factor was found to depend mainly on the porosity and was almost independent of the reflectivity. They also concluded from this study that the results obtained from the Kamiuto correlated scattering theory do not generally show good agreement with the results obtained from the MCM.

Jones et al. (1996) measured the spectral directional distribution of radiation intensity, using a direct radiometric technique, at the exposed boundary of a packed bed of opaque spheres. Intensity exiting the bed was numerically simulated using a discrete ordinate solution to the RTE, with radiation-conduction combined mode solution of the coupled energy conservation equation. Radiative properties were computed using the correlated scattering theory for large size parameter from Kamiuto (1990). The measured intensity results showed good agreement with computed results in near-normal directions, though agreement in near-grazing directions is poor. With these results, they concluded that either radiative transfer near the boundaries of this medium might not be adequately represented by a continuous form of the RTE, or that the properties derived from correlated scattering theory were insufficient.

Taking into account these conclusions from previous works, this study aims to compare the measured reflectance to theoretical predictions for a dispersion of oxidized bronze spherical particles. Theoretical prediction of directional spectral reflectance of absorbing and scattering sample in the infrared range is calculated by a radiative model that uses the discrete ordinates method to solve the RTE.

Radiative properties for the packed spheres systems are computed using the correlated scaling factor theories for large size parameter and for large porosity proposed by Singh and Kaviany (1992) and Kamiuto (1990). The extinction and scattering coefficients are calculated for opaque particle diameters lying from 100 to 400 μm. These models require the knowledge of morphological characteristics (particle dimensions and porosity) and hemispherical spectral reflectivity. The first can be obtained from morphological analysis, but the solid hemispherical spectral reflectivity is very difficult to obtain directly. An identification method (Gauss linearization) is used to identify this parameter.

This method uses experimental and theoretical results of directional spectral reflectance of randomly packed bed of spheres (Figure 21.26, Moura et al., 1998b). The discrete ordinates method associated to the control volume is used to solve the RTE.

The radiative properties of packed bed that are required for solving the RTE are the spectral volumetric scattering and absorption coefficients and the spectral phase function. They can be obtained from the radiative properties of the packed bed particles by adding up the effects of all the particles of different sizes (Brewster, 1992). Consequently, the radiative properties can be determined from the following parameters: particle diameter (d), the volume fraction (f_v), and the particle spectral hemispherical reflectivity (ρ_λ). Indeed, when the particle size x is much larger than unity and when the refractive index is not too small ($x|\tilde{n} - 1| \gg 1$), then the series expansions used to evaluate the expression in Mie theory converge very slowly. For these cases, it is preferable to use geometric optics theory

FIGURE 21.26
Microscopic analysis obtained from packed spheres sample (Poral30), with a magnification of 45. (Moura, L.M. et al., Parameter identification for packed spheres systems: Solid hemispherical spectral reflectivity, *Proceedings of the LATCYM 98*, Argentina, 1998b.)

to predict the radiative properties. The details of the independent radiative properties prediction model are given by Brewster (1992). If a packed bed consists of large diffuse spheres, then the uncorrelated radiative properties become

$$\beta_u = \frac{1.5 f_v}{d} \qquad (21.19)$$

$$\omega_u = \rho_\lambda \qquad (21.20)$$

$$p(\mu) = \left(\frac{8}{3\pi}\right)\left(\sqrt{1-\mu^2} - \mu \cos^{-1}\mu\right) \qquad (21.21)$$

Sing and Kaviany (1992) showed scaling factors so that the independent radiative properties can be scaled to give the dependent properties of the particulate media. The scaling factor γ is assumed to be scalar and scales the optical thickness leaving the phase function and the albedo unchanged. The scaling factor is defined by the following expression:

$$\beta_c = \gamma \beta_u \qquad (21.22)$$

The values of γ for $\rho_\lambda = 0.9$ and $f_v < 0.7$ can be fitted as

$$\gamma = 1 + 1.84 f_v - 3.15 f_v^2 + 7.2 f_v^3 \qquad (21.23)$$

Since the effect of reflectivity on γ is small (Sing and Kaviany, 1992), Equation 21.22 can be used to obtain the value for other reflectivities. The parameters d and f_v can be easily determined from microscopic analysis and photographs (Figure 21.26). The main difficulty remains to determine ρ_λ. The spectral hemispherical reflectivity cannot be obtained from

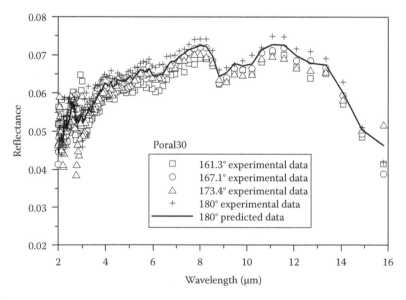

FIGURE 21.27
Comparison of measured and predicted spectral reflectance of the Poral30. (Moura, L.M. et al., Parameter identification for packed spheres systems: Solid hemispherical spectral reflectivity, *Proceedings of the LATCYM 98*, Argentina, 1998b.)

direct measurement. Moreover, it cannot be accurately obtained from literature because of the great dispersion of the reported data. So it is preferable to determine ρ_λ by using an identification method. Figure 21.27 shows spectral directional comparison between measurement and computation prediction for packed spheres sample (named Poral30).

21.9 Conclusion

The problem of identification of radiative properties for dispersed materials such as foams, fibers, coatings, and packed beds has been investigated. The main difficulty is the necessity to employ few parameters, capable of representing the real (physical) medium. An analysis of the CN for different strategies has shown the advantages and limitations of each of them.

The optimal optical thickness can be determined for the different strategies. If the sample thickness is imposed by experimental constraints, we can choose the best strategy to identify the radiative parameters. However, the CN cannot be used only to choose the better strategies. Some high condition number for a strategy can show better results than another strategy and an analysis of the identification must be performed to reach general conclusions.

Nomenclature

a	Legendre polynomial coefficient
d	particle diameter
$d\omega_o$	solid angle of collimated incident beam
dx	thickness of the control volumes
f	interpolation function
f_v	volume fraction
f_1	weighting factor between forward and backward anisotropy in the phase function
f_2	weighting factor between anisotropic and isotropic scattering
g	anisotropic parameter for Henyey–Greenstein phase function
g_1	anisotropic parameter in the forward directions
g_2	anisotropic parameter in the backward directions
I	radiative intensity
I_b	Planck's blackbody function
I_o	intensity of the beam incident onto the sample
I_1^C	radiative intensity incidence
K	number of parameters to identify
l_x	thickness
m	iterations
M	sensitivity matrix
N	number of measurements
n	real refractive index
$\tilde{n}$	complex refractive index
np	number of packages
p	phase function
Rs	random number

s	distance traveled by a beam
S	source function
T	transmittance and reflectance
T_p	temperature
w	weight
x	particle size

Greek Variables

β	extinction coefficient
γ	scaling factor
ε	emissivity
θ	polar angle
θ_o	divergence angle of the incident beam
θ_I	incidence angle
θ_p	angle between incidence and scatter direction
μ	$\cos\theta$
ρ_λ	particle spectral hemispherical reflectivity
τ	optical coordinate
τ_o	optical thickness
ϕ	azimuth angle
χ	parameters to identify
φ, ξ, η	direction cosines
ω	albedo
Ω	solid angle

Subscripts

λ	wavelength
I	incidence
i	position
j	direction
HG	Henyey–Greenstein function
t	theoretical
e	experimental
k	parameter

Superscripts

$'$	new angular position

References

Abdallah, P.B. and Le Dez, V., 2000. Radiative flux inside an absorbing-emitting semi-transparent slab with variable spatial refractive index at radiative conductive coupling. *Journal of Quantitative Spectroscopy and Radiative Transfer* 67:125–137.

Baillis, D. and Sacadura, J.-F., 2000. Thermal radiation properties of dispersed media: Theoretical prediction and experimental characterization. *Journal of Quantitative Spectroscopy and Radiative Transfer* 67:327–363.

Barichello, L.B. and Siewert, C.E., 1999. A discrete-ordinates solution for a non-grey model with complete frequency redistribution. *Journal of Quantitative Spectroscopy and Radiative Transfer* 62:665–675.

Berdahl, P., 1995. Pigments to reflect the infrared radiation from fire. *Journal of Heat Transfer* 117: 355–358.

Boulet, P., Jeandel, G., and Morlot, G., 1996. Etude théorique de l'influence des infibrés sur le comportement radiatif des isolants fibreux. *International Journal of Heat and Mass Transfer* 39(15):3221–3231.

Brewster, M.Q., 1992. *Thermal Radiative Transfer and Properties.* New York: John Wiley and Sons.

Brewster, M.Q. and Tien, C.L., 1982. Radiative transfer in packed fluidized beds: Dependent versus independent scattering. *Journal of Heat Transfer* 104:570–579.

Carlson, B.G. and Lathrop, K.D., 1968. Transport theory—The method of discrete ordinates. In *Computing Methods in Reactor Physics*, H. Greespan, C.N. Kleber, and D., Okrent, Eds. New York: Gordon and Breach.

Chandrasekar, S., 1960. *Radiative Transfer.* New York: Dover Publications.

Cunnington, G.R. and Lee, S.C., 1996. Radiative properties of fibrous insulations: Theory versus experiment. *Journal of Thermophysics and Heat Transfer* 10(3):460–465.

Doermann, D. and Sacadura, J.F., 1996. Heat transfer in open cell foam insulation. *Journal of Heat Transfer* 118:88–93.

Eymet, V., Fourniera, R., Blancoa, S., and Dufresne, J.L., 2005. A boundary-based net-exchange Monte Carlo method for absorbing and scattering thick media. *Journal of Quantitative Spectroscopy and Radiative Transfer* 91:27–46.

Garcia, R.D.M., Siewert, C.E., and Yacout, A., 2008. On the use of Fresnel boundary and interface conditions in radiative-transfer calculations for multilayered media. *Journal of Quantitative Spectroscopy and Radiative Transfer* 109:752–769.

Hahn, O., Raether, F., Arduini-Schuster, M.C., and Fricke, J., 1997. Transient coupled conductive/radiative heat transfer in absorbing, emitting and scattering media: Application to laser-flash measurements on ceramic materials. *International Journal of Heat and Mass Transfer* 40(3): 698–698.

Hendricks, T.J. and Howell, J.R., 1996. Absorption/scattering coefficients and scattering phase functions in reticulated porous ceramics. *Journal of Heat Transfer* 118:79–87.

Hottel, H.C., Sarofim, A.F., Evans, L.B., and Vasalos, I.A., 1968. Radiative transfer in anisotropically scattering media: Allowance for Fresnel reflection at the boundaries. *Journal of Heat Transfer* 56–62.

Hottel, H.C., Sarofim, A.F., Vasalos, I.A., and Dalzell, W.H., 1970. Multiple scatter: Comparison of theory with experiment. *Journal of Heat Transfer* 285–291.

Hulin, M., Hulin, N., and Perrin, D., 1998. *Équations de Maxwell: Ondes Électromagnetiques*, 3rd edn. Paris: Dunod.

Jones, P.D., McLeod, D.G., and Dorai-Raj, D.E., 1996. Correlation of measured and computed radiation intensity exiting a packed bed. *Journal of Heat Transfer* 118:94–102.

Kamiuto, K., 1990. Correlated radiative transfer in packed-sphere systems. *Journal of Quantitative Spectroscopy and Radiative Transfer* 43(1):39–43.

Lacroix, D., Parent, G., Asllanaj, F., and Jeandel, G., 2002. Coupled radiative and conductive heat transfer in a non-grey absorbing and emitting semitransparent media under collimated radiation. *Journal of Quantitative Spectroscopy and Radiative Transfer* 75:589–609.

Lataillade, A., Dufresne, J.L., El Hafi, M., Eymet, V., and Fournier, R., 2002. A net-exchange Monte Carlo approach to radiation in optically thick systems. *Journal of Quantitative Spectroscopy and Radiative Transfer* 74:563–584.

Lemonnier, D. and Le Dez, V., 2002. Discrete ordinates solution of radiative transfer across a slab with variable refractive index. *Journal of Quantitative Spectroscopy and Radiative Transfer* 73:195–204.

Liou, B.-T. and Wu, V.-Y., 1996. Composite discrete-ordinate solutions for radiative transfer in a two-layer medium with Fresnel interfaces. *Numerical Heat Transfer*, Part A 30:739–751.

McCluney, R., 1994. *Introduction to Radiometry and Photometry*. Boston: Artech House.

McCormick, N.J., 1992. Inverse radiative transfer problem: A review. *Nuclear Science and Engineering* 112:185–198.

Modest, M.F., 2003. *Radiative Heat Transfer*. 2nd edn. New York: McGraw-Hill.

Moura, L.M., 1998. Identification des proprietes radiatives des materiaux semi-transparent en situation de non-symetrie azimutale du champ radiatif. PhD diss., INSA, Lyon.

Moura, L.M., Bailis, D., and Sacadura, J.F., 1997. Analysis of the discrete ordinate method: Angular discretization (in Portuguese). *Proceedings of the 14th COBEM*, December 8–12.

Moura, L.M., Bailis, D., and Sacadura, J.-F., 1998a. Identification of thermal radiation properties of dispersed media: Comparison of different strategies. *Proceedings of the 11th IHTC*, August 23–28, Korea.

Moura, L.M., Lopes, R., Bailis, D., and Sacadura, J.F., 1998b. Parameter identification for packed spheres systems: Solid hemispherical spectral reflectivity. *Proceedings of the LATCYM 98*, Argentina, October 5–8.

Moura, L.M., Bailis, D., and Sacadura, J.F., 1998c. Analysis of the discrete ordinate method: Spacial discretization (in Portuguese). *Proceedings of the VII ENCIT*, RJ, Brazil, November 3–6.

Moura, L.M., Bailis, D., and Sacadura, J.F., 1999. Analysis of the sensitive parameters in the thermal radiative properties identification. *Proceedings of the XV COBEM*, Brazil, 22–26.

Nicolau, V. de P., 1994. Identification des propriétes radiatives des materiaux semi-transparents diffusants. PhD diss., INSA, Lyon.

Nicolau, V.P., Raynaud, M., and Sacadura, J.F., 1994. Spectral radiative properties identification of fiber insulating materials. *International Journal of Heat and Mass Transfer* 37:311–324.

Oppenheim, U.P. and Feiner, Y., 1995. Polarization of the reflectivity of paints and other rough surfaces in the infrared. *Applied Optics* 34(10):1664–1671.

Orel, Z.C., Jerman, R., Hodoscek, M., and Orel, B., 1990. Characterization of TSSS paint coatings for solar collectors by FTIR spectroscopy. *Solar Energy Materials* 20:435–454.

Özisik, M.N., 1973. *Radiative Transfer and Interaction with Conduction and Convection*. New York: John Wiley & Sons.

Planck, M., 1914. *The Theory of Heat Radiation*. Philadephia, PA: Blakiston Son and Co.

Shah, H.S. and Adroja, D.G., 1996. Estimation of reflectance spectra of particulate matter from size and refractive index of particles. *Indian Journal of Pure and Applied Physics* 31:560–563.

Siegel, R. and Howell, J.R., 2002. *Thermal Radiation Heat Transfer*, 4th edn. New York: Taylor & Francis.

Silva Neto, A.J. and Özisik, M.N., 1992. An inverse analysis of simultaneously estimating phase function, albedo and optical thickness. *ASME-HTD—Developments in Radiative Heat Transfer*, Vol. 203, pp. 23–30.

Singh, B.P. and Kaviany, M., 1992. Independent theory versus direct simulation of radiation heat transfer in packed beds. *International Journal of Heat and Mass Transfer* 35(6):1397–1405.

Tien, C.L., 1988. Thermal radiation in packed and fluidized beds. *Journal of Heat Transfer* 110: 1230–1242.

Tong, T.W. and Tien, C.L., 1980. Analytical models for thermal radiation in fibrous insulations. *Journal of Thermal Insulation* 4:28–44.

Wu, C.-Y., Liou, B.-T., and Liou, J.-H., 1994. Discrete-ordinate solutions for radiative transfer in a scattering medium with Fresnel boundaries. *Proceedings of the 10th International Heat Transfer Conference*, Brighton, U.K., Vol. 2, pp. 159–164.

Xavier Filho, O. and Moura, L.M., 2007. A comparative analysis between the Monte Carlo method and the discrete ordinate method applied to solve the radiative transfer equation. *Proceedings of the 19th International Congress of Mechanical Engineering*.

Yang, Y.S., Howell, J.R., and Klein, D.E., 1983. Radiative heat transfer through a randomly packed bed of spheres by the Monte Carlo method. *Journal of Heat Transfer* 105:325–332.

Yang, W.-J., Hiroshi T., and Kazuhiko K., 1995. *Radiative Heat Transfer by the Monte Carlo Method*. San Diego: Academic Press.

Index